# The Physics of Electronic and Atomic Collisions

*Invited Lectures, Review Papers,*
*and Progress Reports of the IX*
*International Conference on the Physics*
*of Electronic and Atomic Collisions*

Edited by
JOHN S. RISLEY AND R. GEBALLE

IX ICPEAC

SEATTLE
24–30 JULY 1975

UNIVERSITY OF WASHINGTON PRESS
SEATTLE AND LONDON

Copyright © 1976 by the University of Washington Press
Library of Congress Catalog Card Number 75-39962
ISBN 0-295-95455-8
Printed in the United States of America

# THE PHYSICS OF ELECTRONIC
# AND ATOMIC COLLISIONS

## PREFACE

This volume contains the texts of 58 invited lectures presented at the IXth International Conference on the Physics of Electronic and Atomic Collisions, held at the University of Washington, Seattle, from 24 July through 30 July, 1975. The invited program for this ICPEAC was considerably larger than those of earlier conferences in this series because the Program Committee anticipated a smaller number of contributed papers. Despite the universal decline in research support, however, the contributed program with 566 papers also turned out to be larger than any earlier one. The magnitude, scope and quality of both the invited and contributed programs testify to the continuing vitality of the field of electronic and atomic collisions.

The Editors are indebted to the contributors for making it possible to publish their papers; we anticipate that the volume will be a valuable resource for many years. We also express our gratitude to Bridget Lynch for assisting with the preparation of the manuscript and to the staff of the University of Washington Press for skillful and timely production of the volume.

The IX ICPEAC is indebted to those agencies, both public and private, which have made financial contributions to its support. Those include the International Union of Pure and Applied Physics, the U.S. National Science Foundation, the U.S. Energy Research and Development Administration, the U.S. Army Research Office, the U.S. Office of Naval Research, the Battelle Memorial Foundation, the International Business Machines Corporation, the Boeing Company and the Pacific Northwest Bell Telephone Company.

Special thanks are owed President John R. Hogness of the University of Washington for making the facilities of the University available to the conference and for direct support of many other kinds.

November, 1975
                              Ronald Geballe
                              Local Chairman

                              John S. Risley
                              Local Secretary-Treasurer
                              IX ICPEAC

INTERNATIONAL CONFERENCE ON THE PHYSICS
OF ELECTRONIC AND ATOMIC COLLISIONS

## ORGANIZATION 1973–1975

OFFICERS:

Chairman: P. G. Burke, The Queen's University of Belfast, Belfast, Northern Ireland

Vice Chairman: F. T. Smith, Stanford Research Institute, Menlo Park, California, USA

Secretary: R. Geballe, Department of Physics, University of Washington, Seattle, Washington, USA

Treasurer: F. J. de Heer, FOM Institute for Atomic and Molecular Physics, Amsterdam, The Netherlands

EXECUTIVE COMMITTEE:

Officers and

V. V. Afrosimov, A. F. Ioffe Physical-Technical Institute, USSR Academy of Sciences, Leningrad, USSR

M. Barat, Institut d'Electronique, Bat. 220 Faculte des Sciences, University of Paris, 91 Orsay, France

B. Cobic, (Past Local Chairman) Ion Physics Laboratory, Boris Kidric Institute of Nuclear Sciences—Vinca Belgrade, Yugoslavia

H. Ehrhardt (Past Chairman) Universität Trier|Kaiserlautern Naturwissenschaftlich-Technische Fakultät, 675 Kaiserlautern, West Germany

E. Gerjuoy, Department of Physics, University of Pittsburgh, Pittsburgh, Pennsylvania, USA

M. E. Rudd, Department of Physics, University of Nebraska, Lincoln, Nebraska, USA

I. C. Percival (1975), Theoretical Physics Department, University of Stirling, Stirling, Scotland

A. C. H. Smith (1975), Department of Physics, University College, London WC1 GBT

U.S.A.

R. Geballe (1975) (Secretary and Local Chairman), Department of Physics, University of Washington, Seattle, Washington, 98195

M. Inokuti (1977), Radiological and Environmental Research, Argonne National Laboratory, Argonne, Illinois 60439

N. F. Lane (1975), Department of Physics, Rice University, Houston, Texas 77001

W. C. Lineberger (1977), JILA, University of Colorado, Boulder, Colorado 80304

M. E. Rudd (1975), Department of Physics, University of Nebraska, Lincoln, Nebraska 68508

F. T. Smith (1975) (Vice Chairman), Stanford Research Institute, Menlo Park, California 94025

W. W. Smith (1977), Department of Physics, University of Connecticut, Storrs, Connecticut 06268

R. N. Zare (1977), Department of Chemistry, Columbia University, 308 Havemeyer, New York, New York 10027

U.S.S.R.

V. V. Afrosimov (1975) (Executive Committee), A. F. Ioffe Physical-Technical Institute, USSR Academy of Sciences, Leningrad

Yu. N. Demkov (1975), Department of Physics, Leningrad University, Leningrad

N. Penkin (1975), Department of Physics, Leningrad State University, Leningrad

E. E. Nikitin (1977), Institute of Chemical Physics, Vorob Yevskaye Shosse, 2-A, Moscow, V-334

YUGOSLAVIA

B. Cobic (1975) (Past Local Chairman), Ion Physics Laboratory, Boris Kidric Institute of Nuclear Sciences - Vinca, Belgrade

M. V. Kurepa (1977), Institute of Physics, Belgrade

PROGRAM COMMITTEE
FOR THE IX ICPEAC

The Executive Committee and the Local
Committee plus

Yu. N. Demkov
Leningrad State University
Leningrad, USSR

G. H. Dunn
JILA
University of Colorado
Boulder, Colorado, USA

A. C. Gallagher
JILA
University of Colorado
Boulder, Colorado, USA

T. M. Miller
Stanford Research Institute
Menlo Park, California, USA

J. R. Peterson
Stanford Research Institute
Menlo Park, California, USA


LOCAL COMMITTEE
FOR THE IX ICPEAC

Chairman
Ronald Geballe
Department of Phyyics
University of Washington
Seattle, Washington

Secretary-Treasurer
John S. Risley
Department of Physics
University of Washington
Seattle, Washington

Members
David F. Burch
Department of Physics
University of Washington
Seattle, Washington

Kenneth C. Clark
Department of Physics
University of Washington
Seattle, Washington

Bernd Crasemann
Department of Physics
University of Oregon
Eugene, Oregon

E. Norval Fortson
Department of Physics
University of Washington
Seattle, Washington

Larry H. Toburen
Battelle Northwest Laboratory
Richland, Washington

Lawrence Wilets
Department of Physics
University of Washington
Seattle, Washington


SYMPOSIA ORGANIZERS
FOR THE IX ICPEAC

Electron-Atom Scattering

E. Gerjuoy
Department of Physics
University of Pittsburgh
Pittsburgh, Pennsylvania, USA


Inner Shell Ionization Phenomena

E. Merzbacher
Department of Physics
University of North Carolina
Chapel Hill, North Carolina, USA


Secondary Electron Yields from Electron
and Ion Impact

M. Inokuti
Radiological and Environmental
Research
Argonne National Laboratory
Argonne, Illinois, USA


Atomic Collision Aspects of Energy
Related Research

B. Bederson
Department of Physics
New York University
New York, New York, USA

# CONTENTS

## *Special Topics in Atomic Collisions*

Negative ions, positive electrons    3
     PROFESSOR SIR HARRIE MASSEY

Analysis of electron correlations    27
     U. FANO

Decaying states    40
     PROFESSOR SIR RUDOLF PEIERLS

Parity violation in atoms due to neutral weak currents    51
     ERNEST M. HENLEY

Interstellar molecule formation    62
     ERIC HERBST AND WILLIAM KLEMPERER

## *Electron-atom Scattering*

Elastic and total scattering of electrons by noble gases    79
     F. J. DE HEER

Measurement of absolute collision cross sections of electrons elastically
scattered by gases    98
     J. PHILIP BROMBERG

Recent advances in polarized-electron experiments    112
     JOACHIM KESSLER

Electron impact excitation of light atoms    126
     M. R. C. MCDOWELL

Inelastic electron-hydrogen atom experiments    139
     J. F. WILLIAMS

Theory of low-energy electron-atom collisions and related processes    151
     J. N. BARDSLEY

Electron scattering by laser-excited atoms    158
     I. V. HERTEL, H. W. HERMANN, W. REILAND, A. STAMATOVIĆ, AND
     W. STOLL

Energy exchanges between two escaping electrons    176
     FRANK H. READ

Angular correlation of electrons coming from ionization by electron impact    194
     DEREK PAUL, K. JUNG, E. SCHUBERT, AND H. EHRHARDT

## *Electron-molecule Scattering*

The theory of low energy electron-molecule scattering    219
     KAZUO TAKAYANAGI

Multiple scattering approach to the vibrational excitation of molecules by
slow electrons    231
     G. DRUKAREV

Classification of Feshbach resonances in electron-molecule scattering    241
     DAVID SPENCE

## Molecular Collisions and Reactive Scattering

Progress in the quantum dynamics of reactive molecular collisions   259
ARON KUPPERMANN

Non-adiabatic effects in collisional vibrational relaxation of diatomic
molecules   275
E. E. NIKITIN

## Heavy Particle Collisions at Intermediate Energies

Some aspects of the molecular approach to atomic collisions   295
V. SIDIS

Asymptotically exact theory of electron exchange in distant collisions   313
YU. N. DEMKOV

Optical emission in slow atomic collisions   327
V. KEMPTER

Electron ejection in slow heavy particle collisions   345
R. MORGENSTERN

## Inner Shell Ionization Phenomena

Experimental studies of inner-shell excitation in slow ion-atom collisions   361
BENT FASTRUP

The theory of inner-shell excitation in slow ion-atom collisions   384
J. S. BRIGGS

Experimental studies of target and projectile x radiation in high velocity
atomic collisions   408
JAMES R. MACDONALD

Study of K, L and M inner-shell ionization by proton impact   419
V. S. NIKOLAEV, V. P. PETUKHOV, E. A. ROMANOVSKY, V. A. SERGEEV,
I. M. KRUGLOVA, AND V. V. BELOSHITSKY

The impact parameter dependence of inner shell excitation   432
H. O. LUTZ

High-resolution x-ray and Auger electron measurements in ion-atom collisions   447
C. FRED MOORE

Quasimolecular K x rays   470
W. E. MEYERHOF

Radiative processes in transient quasi-molecules   481
B. MÜLLER

Experimental evidence for anisotropy of non-characteristic x rays   501
P. H. MOKLER, P. ARMBRUSTER, F. FOLKMANN, S. HAGMANN, G. KRAFT,
AND H. J. STEIN

Radiative and nonradiative electron capture from and into outer and inner
shells in heavy ion-atom collisions   520
H.-D. BETZ, M. KLEBER, E. SPINDLER, F. BELL, H. PANKE,
AND W. STEHLING

Important problems in future heavy ion atomic physics   531
W. BETZ, G. HEILIGENTHAL, J. REINHARDT, R. K. SMITH, AND
WALTER GREINER

## Photon Interactions

Atomic physics with synchrotron radiation     563
R. P. MADDEN

Photodetachment threshold processes     584
W. C. LINEBERGER

Multiphoton processes with polarized light     593
P. LAMBROPOULOS

Laser photodissociation of hydrogen molecule ions with fragment kinetic
energy analysis     609
J. DURUP

## Time-Dependence and Anisotropy of Collision Products

Alignment and orientation in atomic collisions     627
JOSEPH MACEK

Analysis of inelastic electron-photon coincidence experiments     641
H. KLEINPOPPEN, K. BLUM, AND M. C. STANDAGE

Electronic state alignment, orientation, and coherence produced by beam-foil
collisions     660
D. A. CHURCH

## Recent Theoretical Advances

On Glauber and Glauber-related methods in atomic physics     675
F. W. BYRON, JR.

Variational principles, subsidiary extremum principles, and variational bounds     685
LARRY SPRUCH

## Highly Excited Rydberg States

Ionization of highly excited atoms by atomic particle impact     701
B. M. SMIRNOV

Hyperfine structure of the highly excited states of alkali atoms     712
R. GUPTA

The production and detection of highly excited states     726
JAMES E. BAYFIELD

## Secondary Electrons from Electron and Ion Impact

Basic aspects of secondary-electron distributions     741
YONG-KI KIM

Resonances seen in secondary electron spectra     756
W. MEHLHORN

The (e,2e) reaction     766
I. E. MCCARTHY

Decay of autoionization states in collisions of heavy atomic particles     779
G. N. OGURTSOV

Collisionally produced autoionizing and autodetaching states of neutral atoms
and their negative ions     790
A. K. EDWARDS

## Collision Processes in Gas Lasers and Plasmas

Determination of effective cross-sections of various elementary processes from
   low temperature plasma data                                                  803
     N. P. PENKIN

Study of momentum transfer distributions in rotationally inelastic collisions    820
     WILLIAM K. BISCHEL AND CHARLES K. RHODES

## Atomic Collision Aspects of Energy Related Research

Atomic collisions and fission technology                                         833
     SHELDON DATZ

Atomic physics in the controlled thermonuclear research program                  846
     C. F. BARNETT

## Methods in Low Energy Collisions

The ion-storage collision technique                                              857
     HANS G. DEHMELT

Application of ion cyclotron resonance spectroscopy to studies of collision
   processes                                                                     871
     J. L. BEAUCHAMP

Application of the variable-temperature flowing afterglow and flow-drift tube
   techniques to studies of the energy dependence of ion-molecule reactions      889
     F. C. FEHSENFELD AND D. L. ALBRITTON

Author Index                                                                     901

# SPECIAL TOPICS IN ATOMIC COLLISIONS

# NEGATIVE IONS, POSITIVE ELECTRONS

Professor Sir Harrie Massey

University College London
Department of Physics and Astronomy
Gower Street, London WC1E 6BT  England.

## Negative Ions

While the title of this review is equally shared between negative ions and positive electrons by far the major part of it will be concerned with atomic collisions involving the latter.  The association of the two is purely personal in that one of my earliest research interests concerned negative ions while my most recent have been concerned, both theoretically and experimentally, with slow positrons in gases.  Moreover as I have recently completed a new, greatly enlarged edition of my book on Negative Ions which is now in course of publication it seemed worthwhile to begin with a few remarks on the development of that subject which has been especially striking in the 25 years since the last (2nd) edition was published.

At that time no experimental measurements had been made of radiative processes involving negative ions.  Since then the experimental study of photodetachment, stimulated in the first instance by the realisation of the importance of absorption by $H^-$ in the solar atmosphere, has become a most important source of information about negative ions.  In recent years the availability of laser sources of light of appropriate frequency, including tunable dye lasers, has extended the scope of the work so that it is one of the liveliest branches of the subject (1).  Values of electron affinities are being obtained which are of reliability and precision unrivalled by any other technique.  This is true not only for atomic ions but also for molecular ions for which the high resolution makes possible, in many cases, the clear distinction between the true electron affinity and the vertical detachment energy.

Perhaps the biggest development during the period has been in the study of autodetaching states of negative ions (2).   This has not only become a very fertile and still very active branch of atomic physics, both experimental and theoretical, but the importance of autodetachment in determining the rates of dissociative attachment reactions of electrons with gas molecules, first realized by Herzenberg and Mandl (3), has been properly appreciated and exploited.

The experimental study of dissociative attachment reactions, although one of the earliest aspects of the physics of negative ions to be studied experimentally, has enjoyed a new and very fruitful lease of life thanks to the introduction of new techniques.  Many of the difficulties of interpretation have been removed and the influence of temperature on observed reaction rates demonstrated.   Three body attachment, as to molecular oxygen, has also been thoroughly studied experimentally both by swarm and recently also by beam methods.  Even associative detachment is being studied effectively in the laboratory.

Collision processes involving the production and destruction of negative ions have been investigated using charged and neutral beams of several keV while, at much lower energies, the determination of threshold energies for charge transfer reactions and for ion-pair production in neutral-neutral collisions has yielded valuable information about electron affinities.  Measurements have also been made of the rates of mutual neutralization and other ionic reactions and of detachment by electron impact and by electrostatic fields.   In all, this constitutes a formidable list of experimental achievement.

On the theoretical side there has been substantial progress, assisted greatly by the availability of high speed computers, but depending also on the introduction of more penetrating choices of trial functions in the variational calculations involved.   Only a few years ago it was considered quite an achievement to obtain a positive value for an electron affinity of an atom such as oxygen, but much better results are now being obtained by methods which take into account the vitally important contributions from correlation effects.

This expansion in basic knowledge of negative ions has made it possible to understand, in much greater detail, the variety of phenomena in which they play a significant role. Thus the conditions under which negative ions are of importance in stellar atmospheres are quite well known.  It is now clear that, in the terrestrial ionosphere, negative

ions are unimportant at altitudes above 80 km. At lower
altitudes they are important but analysis of the situation
is complicated by ion clustering. Direct in situ analysis
of negative ion composition at these altitudes by rocket-
borne equipment is difficult, partly because of the
relatively high pressure and partly because a rocket is
usually at a negative potential to its surroundings, inimical
to the collection of the negative ions. Although a great
deal of information is already available from laboratory
experiment on reaction rates of complex ions there are still
outstanding problems to be solved before the lower
ionosphere is thoroughly understood.

Probably because of the complexity of the problems,
relatively little advance has been made in understanding how
the presence of negative ions affects the behaviour of
electric discharges. In 1961 Thompson (4) published the
results of an interesting experimental study of discharges
in oxygen using a variety of diagnostic techniques, which
have made it possible to distinguish and interpret a number
of features due to negative ions. However, in this work the
discharges were controlled by wall losses. Very recently,
stimulated by the need to understand the behaviour of laser-
generated discharges in $CO_2$, a considerable step forward
has been taken by Nighan and Wiegand (5) who applied known
data on the relevant reaction rates to analyse the effect of
negative ions on volume-controlled discharges in gas mixtures
involving $CO_2$, CO and $O_2$. In particular, the effect of
negative ions on the stability of such discharges was
recognized and analysed. Further developments in this
aspect of the subject can be expected.

Negative ions are employed to advantage in tandem and
circular particle accelerators and may even prove to be
useful in magnetically-confined controlled thermonuclear
fusion devices such as Tokomak. However, one type of
application is perhaps more surprising and that is the use
of the high rate of attachment of slow electrons to certain
molecules containing three or more halogen atoms, to detect,
with high sensitivity, halogen-containing pesticides such as
DDT which are so important in environmental studies. In
fact much of the material which first became available was
obtained (6) using a very simple attachment detector
designed by Lovelock and Lipsky (7). The very high rate
attachment of very slow electrons to $SF_6$ is being used to
study air motion by following that of a sample of injected
$SF_6$ (8).

## Positive Electrons

### Introduction

The study of collision processes for slow positrons in gases is of interest and importance because it provides new material which may be used to check the current procedures for calculating cross sections for collisions of particles of electronic mass with atoms and molecules. The opposite sign of the charge means that, in many cases, the approximation used must be more sophisticated than for electron collisions. Thus, contributions to the scattered amplitude from different physical processes tend to interfere due to sign differences whereas for electrons they are additive. Furthermore, in addition to elastic and inelastic collisions involving excitation and ionization, for positrons we must consider also collisions in which positron annihilation occurs as well as those which lead, through electron capture, to formation of positronium. The former are of especial interest because, while being rather easily measurable, they are more difficult to predict theoretically than elastic cross sections, depending strongly as they do on the magnitude of correlation effects between the incident positron and the atomic electrons. Because of this the adequacy of approximate theory is more readily tested.

In addition to the collisions of free positrons with atoms and molecules it is possible to study collisions of positronium (Ps) atoms with other atoms and molecules. Because the centres of mass and charge coincide for Ps there is no mean static interaction between a ground state Ps atom and another neutral atom so that the significant interactions arise through electron exchange and distortion of the colliding systems during the impact. A wealth of possibilities arise in these collisions in which positron annihilation, electron exchange or an essentially chemical reaction may occur. In all of these the conditions are novel in slow atomic collision phenomena so considerable interest attaches to their study.

Because positrons are produced by $\beta^+$ decay, with energies of many keV a considerable loss in flux is necessarily involved in moderating their energies to a few eV. Up to the present it has therefore only been possible to study collision processes involving slow positrons or Ps atoms using particle-counting techniques. Many of the new results which will be presented in this review have come from the introduction of fast counting methods and sophisticated techniques for removing random background signals. We shall begin by discussing the major experimental developments before considering the new results obtained by their use.

Electron collisions with atoms are carried out either
using electron beams of well-defined energy or by observing
the behaviour of electron swarms, the latter technique being
especially valuable for studying impact at low (near thermal)
energies. Until quite recently, positron collisions could
only be studied by swarm methods, especially by the observation of the variation of the rate of annihilation of positrons
diffusing in a gas with time from emission from a radioactive
source (annihilation spectra). However, beam experiments
are now possible and have already yielded a number of
important results. For details of the various experiments
which have been carried out up to 1970 reference can be
made to Electronic and Ionic Impact Phenomena by Massey,
Burhop and Gilbody, Vol. 5, 1974, Chap 26 pp 3123-3229.

In my laboratory at University College London we have
concentrated particularly on improving the accuracy of
observation by introducing fast counting techniques which
must be coupled with a much more sophisticated procedure
for removing background counts than the simple one of
subtracting a constant background. Such a procedure has
been described by Coleman, Griffith and Heyland (9) and is
employed in the analysis of all data obtained in our work.
In this way it has not only been possible to increase very
markedly the accuracy of annihilation rate measurements but
also to carry out time-of-flight measurements of total
positron scattering cross sections in a number of gases over
a wide energy range (4-400 eV).

## Total cross sections—measurements and results

The first preliminary observations of a total cross
section for slow positrons of fairly well defined energy
were made by Costello, Groce, Herring and McGowan (10) for
helium. Their success depended on the use of a source of
moderated positrons which, in modified form, has proved to
be suitable for providing positrons of quite well defined
energy for time-of-flight collisions measurements.

A typical arrangement (11) used in these experiments is
as follows. Positrons from an $^{22}$Na source are detected in
a plastic scintillator about 0.17 mm thick which gives a
'start' pulse to the timing sequence. They then pass through
a Melinex window of 0.075 mm thickness, aluminized on the
surface, and a thin aluminium foil, to fall on a system of
gold vanes, as in a photomultiplier, coated with a thin layer
of magnesium oxide. For reasons which are still far from
clear, positrons of energy 1.0 $\pm$ 0.5 eV emerge from this

system and are accelerated to the desired energy by application of a D.C. potential. The emerging positrons enter the time of flight tube consisting of a straight section 70 cm long followed by a sector 15 cm long curved in an arc of 25 cm radius. They are confined to paths close to the axis throughout the full flight path by application of a suitable magnetic field and are detected by the annihilation radiation produced in an aluminium foil by a sodium iodide counter. About 1 in $10^5$ of the positrons emitted from the source enter the flight tube as slow positrons.

A time-of-flight spectrum of the positrons is obtained by using a time to amplitude converter and multichannel

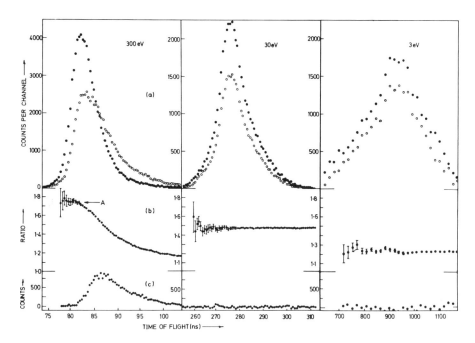

Fig. 1. (a) The time spectrum for 3, 30 and 300 eV positrons in helium after random background correction. Experimental points ● for vacuum runs, o for gas runs (b) Attenuation ratio obtained as discussed in the text. (c) Time distribution of the scattered positrons derived from the difference between the gas and scaled-down vacuum runs.

analyser, the start and stop signals in the sequence being
as described above. By observing the spectrum with and
without gas present at a suitable pressure the total collision
cross section can be obtained to a good accuracy. One
advantage of the method is that the path length pursued
by the positrons through the system may be obtained from the
flight time and collision energy, independent of geometry.

It is important in this work to investigate in some
detail the shape of the 'lines' in the time of flight
spectrum in order to ensure that the measurements refer to
the total scattering cross section, including small angle
scattering. With the longitudinal magnetic field used, all
positrons which are moving in the forward direction eventually
reach the target at the end of the flight path and it is
only positrons scattered through angles greater than 90°
which will fail to do so. We therefore might expect a long
tail of delayed positrons when gas is present. Fig 1 shows
typical observations (12) in helium for three positron
energies. At 30 eV the line shape remains almost the same
with and without gas present, suggesting that most of the
scattering is backward. This being so the nearly constant
ratio (see Fig 1) of the signals for different flight times
is a reliable measure of the attenuation due to gas scattering.
The situation is different at 300 eV. The line shape with
gas present now exhibits a marked 'tail' at long flight times
due to forward scattered positrons. In this case the
attenuation may be measured from the ratio of signals, with
and without gas, on the short time side of the peak, a ratio
which does not vary appreciably, within experimental error,
with time. At the very low energy of 3 eV the shape of the
'line' is likely to be seriously affected by the energy
spread of the positrons as well as by the form of the
differential scattering cross section. Because of this,
results at these energies are less reliable. As we shall
see, theory predicts that for energies between 4 eV and
20-30 eV most of the elastic scattering is in the backward
direction which is consistent with the absence of a marked
tail in the line shape at 30 eV.

To investigate further the relation of the differential
cross section to the line shape, a modified form of time of
flight apparatus has been introduced in which the major part
(about 80%) of the scattering comes from a short section
only (about 8 cm) of the total flight path. The general
arrangement is illustrated in Fig. 2 showing the differential
pumping designed to achieve this. It is hoped that with this
system more definite information can be derived about the
angular distribution of the scattering of positrons with
energies near or below 5 eV. Up to the present, however,

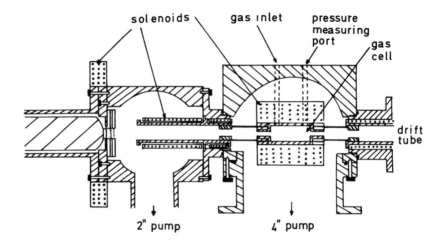

Fig. 2    Schematic diagram of modified time-of-flight
          apparatus using differential pumping

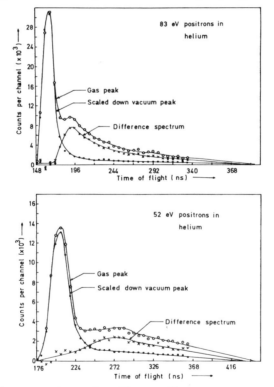

Fig 3.    Time spectrum for 52 and 83 eV positrons in helium
          observed with the equipment shown in Fig 2.

most attention has been paid to the study of positrons with
energies above 40 eV. Under these conditions a second peak
is observed in the time of flight spectrum as seen for
helium in Fig 3. This peak is predominantly due to inelastic
scattering in the forward direction and is displaced between
25 and 30 eV from the peak in vacuo. New opportunities for
investigating the relative contribution of elastic and
inelastic collisions are opened up by these observations
which are still at a preliminary stage. So far there is
no evidence at 52 and 83 eV of any distortion in shape of
the main peak which would be expected from the occurrence
of forward elastic scattering. Particular attention is being
given to the elucidation of this result.

Using these techniques total cross sections have been
measured for positrons in the rare gases (12) (13) as well
as in $N_2$, CO, $CO_2$, $O_2$, $H_2$ and $D_2$ (12, 14). Special attention
has been paid to helium because it is possible for this case
to carry out accurate theoretical calculations by variational
procedures and also to apply a dispersion relation as a
check which is more definite than for electrons. Thus if
$Q_t (k)$ is the total scattering cross section for collisions
of positrons of wave number $k$ with helium atoms then (15)

$$a + f_B = -\frac{1}{2\pi} \int_0^\infty Q_t(k)\, dk . \qquad (1)$$

$a$ is the zero energy scattering length and $f_B$ the Born
approximation to the forward scattered amplitude. For
electrons a further term must be added on the left hand side
which is the exchange amplitude calculated in the Born-
Oppenheimer approximation. This is much more difficult to
calculate accurately than $f_B$ which, for positrons, is equal
in magnitude and opposite in sign to that for electrons.
It has been calculated by Bransden and McDowell (16) as
$-0.791\ a_o$. Humberston (17) has calculated $a$ using an
elaborate variational wave function which allows for
correlation between the electron and positron motions and
obtains $-0.472 a_o$. This is consistent with extrapolation
of the results of Coleman et al (12) from their lowest
observed energy (2 eV) to the zero energy limit using the
theoretical expansion (18)

$$Q_t = 4\pi \left\{ a^2 + (2\pi/3a_o)\alpha a k + (8/3a_o)\alpha a^2 k^2 \ln(ka_o) + b k^2 + c k^4 + \cdots \right\} \quad (2)$$

$\alpha$ is the polarizability of the atom. Bransden and Hutt
(19) found that a good fit is obtained with

$$Q_t = \left\{ 0.883 - 7.5 k a_o - 8.97 k^2 a_o^2 \ln(ka_o) + 5.45 k^2 a_o^2 + 1.44 k^4 a_o^4 \right\} \pi a_o^2, \quad (3)$$

consistent with the calculated value of $a$ and the observed

value of $\alpha$ .  The left hand side of ( l ) therefore has
the value  -1.263 $a_o$.

To calculate the right hand side the observed results
of Coleman et al (12) may be used up to 400 eV at which
energy they are so close to the predictions of Born's
approximation that smooth extrapolation via this approximation
to high energies may be made.  In this way Bransden and Hutt
(19) find a value -1.24 $\pm$ 0.005 $a_o$ for the right hand side
which is so close to that for the left hand side ( -1.263 $a_o$ )
as to provide very good confirmation of the observed results.
Measurements of total cross sections for helium for positrons
with energies between 4 and 270 eV have also been made by
Jaduszliwer and Paul (20) using a less direct technique.
These results give values between -1.23 and -1.32 $a_o$ for the
right hand side of ( l ).   Measurements between 50 and
400 eV have also been made by Dutton, Harris and Jones (21)
with a method rather similar to the classical Ramsauer method
for electrons.

Elaborate theoretical calculations of the  s  and  p
wave shifts for elastic scattering of positrons by helium
atoms have been carried out recently by Humberston (17),
(22), over an energy range extending from zero to the
threshold for positronium formation.   These calculations
use the Kohn variational method with a trial wave function
depending on the distances $r_1$, $r_2$, $r_3$, $r_{12}$ and $r_{13}$, the
electrons being distinguished by the suffixes 2 and 3 and
the positron by the suffix 1.   Fig 4 shows the calculated
total cross section derived from these phase shifts combined
with the somewhat less accurate  d  phase shifts calculated
by Drachman (23).  Comparison with the cross sections
measured by Canter et al (11) and Coleman et al (12) shows
remarkably good agreement, while the empirical formula
obtained by Bransden and Hutt (19) gives almost indistinguish-
able results.  It seems on this evidence that the variational
phase shifts must be very close to the correct ones, a
conclusion which is reinforced when the variational wave
functions are applied to calculate the annihilation cross
section as discussed below.  Again the calculated
distributions of scattering per unit angle show, for energies
between 2 eV up to the positronium formation threshold, a
predominance of backward scattering consistent with the line
shape observations (see Fig 1).   For positron energies
below the positronium threshold the cross sections for
elastic scattering of positrons by helium atoms appear to be
very well known and understood.

The observed total cross section rises with a somewhat
steeper slope as the positron energy increases above the

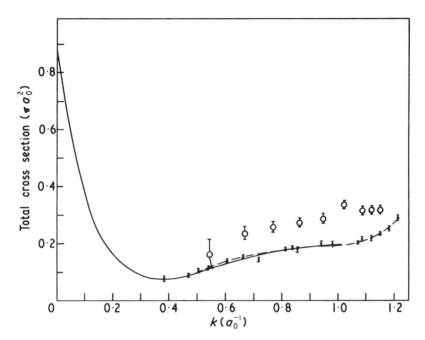

Fig 4.    Variation with wave number k of the total cross
section for elastic scattering of positrons in helium
●    observed by Canter et al (13)    o    observed by
Jadusliwer and Paul (20)    ——    calculated as
described in text  --- least squares fit of Bransden
et al (19) to the data of Canter et al (13)


threshold and it is likely that this is mainly due to the
onset of positronium formation.    A further increase in
slope is observed above the threshold for excitation and
a very tentative analysis (24) into separate contributions
from these two inelastic processes and from elastic scattering
is presented in Fig 5.    This is consistent with evidence
from observations of the probability of positronium formation
by positrons in helium which are discussed below.

For neon, a similar dispersion relation applies as for
helium and Bransden and Hutt (19) have carried out a similar
analysis.    In this case the observed total cross section at
the highest energy of observation (400 eV) is still well
below that predicted by Born's approximation.    Nevertheless,
a satisfactory extrapolation procedure is possible.    The
right hand side of (1) is found,using the latest results of
Coleman et al (12) to be $-3.71 \pm 0.08$ a$_o$.    $f_B$    has been

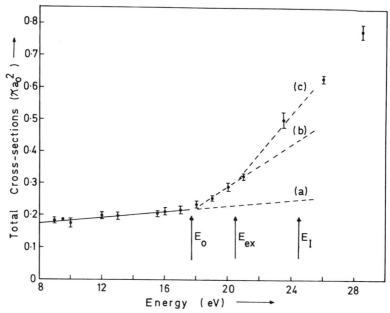

Fig 5.    Observed total cross sections for positrons in helium
          (Canter et al (13) ).    Arrows indicate thresholds,
          $E_o$ for positronium formation, $E_{ex}$ for helium excit-
          ation and $E_I$ for helium ionization.    Broken lines
          are postulated extrapolations to give (a) as elastic,
          (b) as Ps formation and (c) as excitation cross
          sections.

calculated by Inokuti and McDowell (25), from elaborate
neon wave functions, as $-3.32 \ a_o$ so that for satisfaction of
(1), a should be $-0.39 \ a_o$.    The value calculated by
Montgomery and Labahn (26) using the polarized orbital method
is $-0.6 \ a_o$ which is perhaps not unsatisfactory.

     For all the rare gases the cross section falls with
positron energy at low energies down to the lowest observed
energy (3 eV).    For helium there is evidence (see Fig 4) that
at still lower energies the cross section rises again.    In
other words, a Ramsauer-Townsend effect exists.    It is likely
that this is so also for the heavier rare gases.    Evidence in
support of this for argon has been derived by Canter and
Heyland (27) from analysis of the annihilation spectra for
positrons in argon at room temperature and at 100°K and is
described below.

Application of Annihilation Spectra to the Study of Collisions
involving Positrons and Positronium.

We recall briefly the form of a typical annihilation
spectrum which represents the variation of the rate of
annihilation in a gas of positrons from a suitable source,
with time t since emission of the source. At very short
times t there is a strong peak known as the prompt peak
after which there is a gradual change referred to as the
'shoulder' before the spectrum settles down to a combination
of two exponential decay curves i.e. is of the form

$$a e^{-\lambda_1 t} + b e^{-\lambda_2 t} \qquad (4)$$

with $\lambda_1 > \lambda_2$.

The shoulder occurs over the time interval during which
the positrons are slowing down due to collisions with gas
atoms and terminates when the positrons come to thermal
equilibrium. With molecular gases the shoulder is much
shorter because the rate of energy loss of the positrons
is enhanced by excitation of molecular vibration and rotation.

Once thermal equilibrium is reached the free positrons
decay at a rate proportional to the gas density through
annihilating collisions with the gas atoms. If n is
the concentration of gas atoms the rate may be written

$$\pi n \, \overline{Z}_a(T) \, (e^2/mc^2)^2 c \qquad (5)$$

where $\overline{Z}_a(T)$ is the mean number of annihilation electrons per
atom at the temperature T. The constant $\lambda_1$ in (4) is
then given by (5).

In addition to the free positrons there will also be
ortho-positronium atoms which have sufficiently long lives
to provide a significant proportion of annihilation events,
particularly at large t. Thus $\lambda_2$ in (4) is a measure
of the rate at which orthopositronium atoms, in thermal
equilibrium with the gas, decay through one process or
another. In general we have

$$\lambda_2 = \frac{1}{\tau_0} + n \, \overline{Q}_q \, \overline{v} \qquad (6)$$

where $\overline{Q}_q$ is a mean cross section for quenching of
orthopositronium by one process or another in gas atom
collisions. $\overline{v}$ is the r.m.s. velocity and $\tau_0$ is the
lifetime of free orthopositronium (o-Ps).

It follows that, if $\lambda_1$ and $\lambda_2$ may be measured,
information may be obtained about $\overline{Z}_a$, $\tau_0$ and $\overline{Q}_q$. Again,

from measurement of the ratio b/a  the fraction of positrons
which form positronium may be derived, provided proper
allowance is made for the different detecting efficiencies
of the system for annihilation involving 2 and 3 gamma ray
production (see (30) below).  Additional information is
provided if the annihilation spectra are measured in an
electric field because the mean value $\overline{Z}_a$ at temperature T
will depend on the field.  This will occur because the field
will change the equilibrium velocity distribution of the
positrons by an amount depending on the momentum transfer
cross section $Q_d$.   The new value of $\overline{Z}_a$ will therefore depend
on both $Q_d$ and $Z_a$ as functions of the positron velocity v.
Still further information may be obtained from observations
of the variation of $\lambda_1$ and $\lambda_2$ with gas temperature  T
while the study of annihilation in gas mixtures can also
provide useful material concerning $Z_a$ and $Q_d$.

For further details of the form of annihilation spectra
and of the processes leading to quenching of orthopositronium
and of the results obtained in earlier experiments, reference
may be made to Electronic and Ionic Impact Phenomena 2nd
edition, 1974, Vol. V. pp. 3129-3145 and 3191-3213.   We
wish here to draw attention to the improved accuracy which
has been attained in the last few years by application of
fast counting techniques.   The need for high statistical
accuracy is apparent from the fact that it is necessary to
analyse the spectrum in the region past the shoulder into
two exponential decay terms, particularly if the slower decay
rate is to be well determined.  Until recently a weak source
has been usually employed to maintain a large signal to noise
ratio and to make justifiable the assumption of a constant
background term.  However, the slow rate of data accumulation
then requires runs of several days to achieve a useful
statistical accuracy.  Under these conditions ageing and
thermal drift of the electronics may become serious.  Coleman,
Griffith, Heyland and Killeen (28) have designed and are
operating a simple system with a relatively fast rate of
data accumulation with which they have considerably improved
the accuracy attainable.   This has depended also on the use
of the more sophisticated procedure for elimination of random
background signals referred to earlier.

Fig 6 shows the general arrangement of their apparatus.
Gas, at pressures up to 60 atm, is contained in a small
pressure vessel (30 mm internal diameter, 70 mm long)
machined from a driven copper rod.   The inner surface is
electroplated with gold.   Positrons are emitted from a
$5\mu Ci$ $^{22}Na$ source deposited on a thin gold spatula mounted
in contact with the wall of the vessel.  Immediately behind
the source is the start pulse detector, consisting of a large

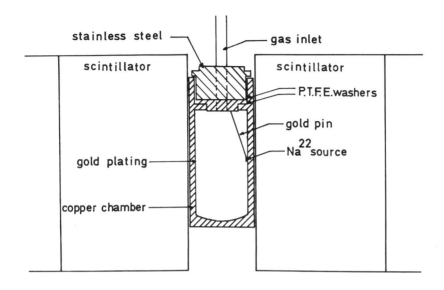

**Pressure  chamber  and  counters**

Fig 6.    Schematic diagram of apparatus for observing
          annihilation times spectra (Coleman et al (28) ).

diameter plastic scintillator which is set to detect the
1.28 MeV gamma ray  emitted with negligible delay after the
positron.  With this arrangement the efficiency of the start
counter is increased above that for the source in the centre
of the chamber.  A similar scintillator mounted opposite
the first provides the stop pulses.  It is set to respond to
pulses above 75 keV and hence to the annihilation gamma rays.
Spectra with a resolution of 1.5 ns are obtained.  Start and
stop pulse rates are $10^4$ and $2 \times 10^4$ $s^{-1}$ giving a coincidence
rate of 1200 events $s^{-1}$.

Since 1 MeV positrons have a 50% chance of backscattering
with reduced energy from a gold surface, the gold plating of
the inner surface and deposition of the source on a gold
backing greatly increases the number of positrons annihilating
in the gas.

Fig 7 shows annihilation spectra observed (29) with
this equipment in helium at a density of 43.6 amagat and
room temperature.  The raw data and that obtained after
application of the procedure for removal of random
coincidences are shown separately as well as the results of
analysing the latter into the two exponentials.  Fig 8 shows

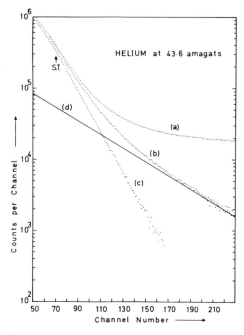

Fig 7.    Annihilation spectrum observed, with the apparatus
          of Fig. 6 for pure helium at a density of 43.6
          amagat (Coleman et al (29).   Curve (a), unprocessed
          data ,    curve (b),data minus background-
          curve (c),free positron component, curve (d),
          fitted o-Ps component.   Each channel corresponds
          to 1.94 ns.    ST denotes the terminus of the shoulder

that the rate constants $\lambda_1$ and $\lambda_2$ behave quite accurately
according to (5) and (6) as functions of the gas pressure.
From these plots it is found, by least squares fitting, with
infinite weight given to the origin,

$$\lambda_1 = 0.793 \pm 0.001 \, \rho \text{ per } \mu s \qquad (7)$$

with $\rho$ the gas density in amagats.

        Similarly, with infinite weight given to the vacuum
intercept taken as the theoretical value, 7.24 per $\mu s$,

$$\lambda_2 = 7.24 + (0.1006 \pm 0.003) \, \rho \text{ per } \mu s. \qquad (8)$$

For $\rho = 10.5$ amagat, $\lambda_1 = \lambda_2$ and it was confirmed by
observations at this density that the annihilation spectrum
beyond the shoulder was a single exponential with the

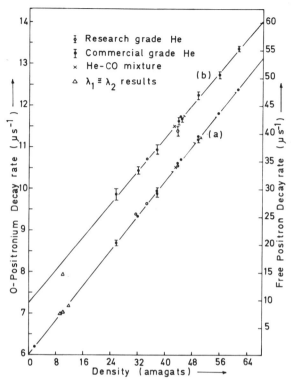

Fig 8.    Variation with density of (line a) the free positron
and (line b) the o-Ps decay rate.    Observed points
(Coleman et al (29),     research grade He;
        commercial grade He;     x He-Co mixture
        results obtained when     $\lambda_1 = \lambda_2$    (see text)


appropriate value of $\lambda$ .


      The value of $\bar{Z}_a$ at room temperature obtained from (7) is
3.94 $\pm$ 0.02.    This is to be compared with $\bar{Z}_a$ = 3.9 calculated
by Campeanu and Humberston (22).


      Fig 9 shows the observed variation of $Z_a$ with time over
the shoulder region both with research grade helium and with
helium containing 1% CO.    The marked reduction in the
importance of the shoulder is due to the rapid moderation of
positrons through rotational excitation of CO.


      Results of similar precision have been obtained for the
other rare gases.    Coleman, Griffith, Heyland and Killeen
(24) have used the data to extract information about the

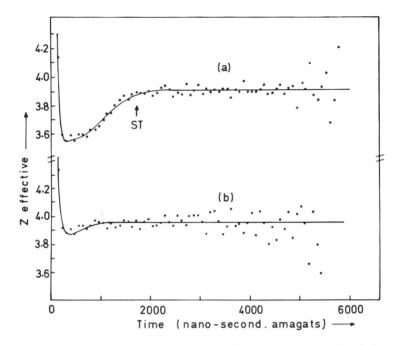

Fig 9.    Variation of $Z_a$ with time (Coleman et al (29) ) for,
         (a) research grade helium  (b)  helium containing
         1% CO.    The curves are hand fitted to the data
         points.    ST denotes the terminus of the shoulder.

fraction of positrons which form positronium in the gas.    They
are careful to take account of the different efficiencies of
the counter system for the detection of 2 and 3 gamma ray
events.   Let $N(oPs), N(e^+)$ be the total numbers of o-Ps atoms
and free positrons in the gas and $N_m(oPs)$, $N_m(e^+)$ the measured
numbers.   If $\alpha(e)$ and $\alpha(e)\eta$ are the respective efficiencies of
detection of 2 and 3 $\gamma$ events respectively, $\eta$ being independ-
end of gas pressure,   then

$$\frac{N_m(oPs)}{N_m(e^+)} = W = \frac{N(oPs)}{N(e^+)} \left\{ (1-x)\eta + x \right\} \tag{9}$$

x  being the fraction of oPs atoms decaying by emission of
2 $\gamma$ rays.   Hence

$$\phi = \frac{N(oPs)}{N(e^+)} = W / \left\{ (1-x)\eta + x \right\}, \tag{10}$$

so

$$W = \phi\eta + x\phi(1-\eta). \tag{11}$$

A plot of W against x should then be linear and give both $\phi$ and $\eta$, provided $\eta$ is independent of gas density. Such linear plots were found for all the rare gases and values of 0.24, 0.26, 0.33, 0.19 and 0.07 obtained for the full positronium fraction F, including para positronium, which is equal to $4\phi/(4\phi+3)$ for helium, neon, argon, krypton and xenon respectively. These results are considerably more accurate than earlier observations but, allowing for this, the earlier data for these lighter gases are not in disagreement. For the first three gases the new values are consistent with what would be expected from Ö re gap theory assuming the initial distribution of positrons before entering the gap to be uniform in energy, but for krypton and xenon considerably less positronium is produced than would be expected. The reason for this is not yet clear.

With the improved accuracy now available it is possible to measure with some precision the natural lifetime $\tau_o$ of o-Ps which appears in (6). For this purpose the determination of the density-independent part of $\lambda_2$ in (6) may be assisted by working in freon for which $\lambda_1$ is so large that most of the annihilation spectrum has the form of a single exponential. In the first experiments of this kind Coleman and Griffith (30) found

$$1/\tau_o = 7.262 \pm 0.015 \text{ per } \mu\text{s},$$

in good agreement with earlier measurement by Beers and Hughes (31) who found

$$1/\tau_o = 7.275 \pm 0.015 \text{ per } \mu\text{s}.$$

The best calculated value (32, 33) including corrections up to $e^8$, is

$$1/\tau_o = 7.241 \pm 0.010 \text{ per } \mu\text{s}.$$

Further experimental work is being carried out in my laboratory to improve the accuracy and already substantial progress has been made.

With data available from both beam and annihilation spectrum measurements of all possible kinds it is possible by seeking for self consistency in the analysis to go far towards the determination of $Z_a$ and the momentum transfer cross section $Q_d$ as functions of positron energy up to the threshold for positronium formation. Although data of comparable accuracy are not yet available from all kinds of experiment Hara and Fraser (34) have carried out an analysis of this type for argon.

As referred to earlier, Canter and Heyland (27) have obtained evidence for the existence of a minimum in the momentum transfer cross section for argon at a positron energy below 2 eV.  If a deep minimum exists at an energy $E_m$ say, then the time taken for positrons to slow down to an energy $E_r < E_m$ will be much greater for positrons with initial energy $> E_m$ than for those with initial energy $< E_m$. Canter and Heyland took advantage of the fact that the annihilation spectra for argon at low temperatures (34) shows a delayed peak at an energy close to 0.1 eV after which the free positron function decays exponentially at a rate which is not proportional to the gas density.   At room temperature the peak is much less prominent and the subsequent exponential decay is less rapid and proportional to the gas density. Assuming that the time-dependent annihilation of a positron is the same at the two temperatures until its energy reaches 0.1 eV, the difference between the two spectra, normalized to the same number of free positron events, should depend on the distribution of slowing-down times for positrons which have energies below the threshold of positronium formation. Canter and Heyland made a careful series of measurements of the difference spectra between room temperature and 100K and found this to consist of two peaks near 100 and 200 ns amagat.  These can be interpreted as arising respectively from positrons with initial energies above and below $E_m$, the separation between  the peaks being determined by the width of the minimum in  the cross section.   Determination of $E_m$ depends on what assumption is made about the initial distribution of the positrons in passing out of the Öre gap. If it is uniform in energy space $E_m = 1.3 \pm 0.1$ eV but is considerably smaller, $0.18 \pm 0.02$ eV, if uniform in momentum space.

While the nature of the processes leading to the delayed peak in the low temperature argon annihilation spectrum was immaterial for the above analysis it is nevertheless of considerable interest to understand what is involved.  Similar behaviour has been observed for other rare gases and poly-atomic moledules and seems likely to be due to complex formation around the positron.  A detailed study has recently been made for methane and methane-argon mixtures (36) in which it is shown that only single stage formation of a complex is consistent with the data.   There is scope for much further work in this aspect of the subject.

Apart from measurements of cross sections for annihila-tion of the positron in 0-Ps in collision with a gas atom (pick-off quenching) some information is also available from experiment about the scattering length in the rare gases and in $N_2$.  Spektor and Paul (37) have introduced a new method

for this purpose in which a positron source is placed in the
centre of a chamber divided into cells by a series of parallel
plates in the form of thin aluminium foils. Measurements
are made of the annihilation decay rate as a function of the
partial pressure of the gas under study, mixed with a isobutane
at a pressure of a few tens of torr to eliminate the contribution from free positron decay. The decay coefficient, when
the decay is exponential can be written in the form

$$\lambda = \frac{1}{\tau_0} + \lambda_g + \lambda_w$$

where $\tau_0$ is as in ( 6 ), $\lambda_g$ is the known contribution from
pick-off quenching and $\lambda_w$ that due to annihilation by the
walls of a cell. $\lambda_w$ is determined by the diffusion
coefficient D of o-Ps in the gas and by the probability P
of annihilation on impact with the walls. D and P are
determined so as to provide the best fit to the observed
data in terms of diffusion theory. For helium the scattering
length obtained in this way is 0.06 $a_0$ which is much smaller
than that (1.45 $a_0$) estimated by Canter (38) from analysis
of low temperature annihilation data in helium in terms of
bubble formation. The most elaborate theoretical calculation
by Drachman and Houston (39), which takes into account
correlation effects in a semi-empirical procedure, gives
1.39 $a_0$. This same theory gives for the effective number
$\overline{Z}_a$ of annihilation electrons in pick-off quenching, 0.098
compared with 0.133 observed (see (8) ).

Further Recent Developments. The first definite observation
of radiation emitted from Ps atoms in the first excited state
have been reported by Canter, Mills and Berko (40) and
already (41) measurements have been made of the $2^3S_1 - 2^3P_2$
energy splitting by magnetic resonance techniques. However,
at the moment this beautiful work does not really fall within
the ambit of the present conference and will not be enlarged
upon here.

Golden and Epstein (42) have investigated the binding
of positrons to a number of atoms and molecules by relating
the problem to that in which the positrons are replaced by
protons. They conclude that normal helium, neon and nitrogen
cannot bind positrons while for krypton such binding seems
likely. These results are of importance in considering the
role of complex formation in positron annihilation within
the gas.

The possibilities opened up by Stein, Kauppila and
Roellig (43) for the production of monochromatic low energy
positron beams by using the $^{11}B(p,n)$ $^{11}C$ reaction. The
bombarding proton beam is produced at an energy between
4.5 and 5.0 MeV by a van der Graaf machine. Positrons are
produced by the $\beta^+$ decay of the $^{11}C$ nuclei which have a half

life of 20.3 min.  More than $10^{-7}$ of the positrons emitted
from a boron target at room temperature emerge with an energy
less than 1 eV and an energy width of 0.10 eV (44).  With
this source relatively very intense positron beams may be
obtained with low associated background radiation.

Throughout the preparation of this review I have had the
benefit of many valuable discussions with Drs. Griffith,
Heyland and Coleman.  I am also indebted to Mrs. Harding for
the preparation of the typescript in an unduly short time and
to the Photographic section of the Department of Physics and
Astronomy at University College London for the preparation of
figures and slides.

## References

1.   See W.C. Lineberger, Chem. and Biol. Applicns,of Lasers,
     Ed. C.B.Moore, Academic Press, 1974— H-P Popp,
     Physics Reports, 16C, 170 1975

2.   G.J.Schulz, Rev. Mod. Phys. 45, 378 and 425, 1973.

3.   See J.N. Bardsley and F. Mandl, Repts on Prog. in
     Physics, 31, 471, 1968.

4.   J.B.Thompson, Proc. Roy. Soc. A 262, 503, 1961.

5.   W.L. Nighan and W.J. Wiegand, Phys. Rev. A., 10, 922,
     1974.

6.   E.S.Goodwin, R. Coulden and J.G.Reynolds, Analyst,
     86, 697, 1971.

7.   J.E.Lovelock and S.R.Lipsky, J. Amer. Chem. Soc. 82,
     431, 1960;  J.E.Lovelock Anal. Chem. 33, 162, 1961.

8.   J.E.Lovelock., R.J. Maggs and E.R.Adlard, Anal. Chem.
     43, 1962, 1971.

9.   P.G.Coleman, T.C.Griffith and G.R.Heyland, Appl. Phys.
     5, 223, 1974.

10.  D.G.Costello, D.E. Grace, D.F.Herring and J.W.McGowan.,
     Can. J. Phys. 50, 23, 1972.

11.  P.G.Coleman, T.C.Griffith and G.R.Heyland, Proc. Roy.
     Soc. Lond A. 331, 561-69, 1973 and K.F.Canter,
     P.G.Coleman, T.C.Griffith and G.R.Heyland, J.Phys. B.5
   . L167, 1972.

12.  P.G.Coleman, T.C.Griffith, G.R.Heyland and T.L. Killeen, Atomic Physics 4 (Plenum, New York 1975).

13.  K.F.Canter, P.G.Coleman, T.C.Griffith and G.R.Heyland, J.Phys. B.6 L201, 1973.

14.  P.G.Coleman, T.C.Griffith and G.R.Heyland, Appl. Phys. 4, 89, 1974.

15.  B.H. Bransden, P.K. Hutt and K.H. Winters, J.Phys. B.7 L129, 1974.

16.  B.H.Bransden and M.R.C.McDowell, J.Phys. B.3, 29, 1970.

17.  J.W.Humberston, J. Phys. B.6, L305, 1973.

18.  T.F.O'Malley, L.Spruch and L.Rosenberg. J.Math. Phys. 2, 491, 1961.

19.  B.H.Bransden and P.K. Hutt,  J.Phys. B.8, 603, 1975.

20.  B.Jadusliwer, W.C.Keever and D.A.L. Paul, Can. J. Phys. 50, 1414, 1972; B.Jadusliwer and D.A.L. Paul, Can. J. Phys. 51, 1565, 1974., 52, 272., 1047, 1974;  and 53, 962 (1975) Appl. Phys. 3, 281, 1974.

21.  J. Dutton, F.M. Harris and R.A.Jones. J. Phys. B.8, L65, 1975.

22 . R.L. Campeanu and J.W.Humberston, J.Phys. B. in course of publication.

23.  R.J. Drachman, Phys. Rev. 144, 25, 1966.

24.  P.G. Coleman, T.C.Griffith, G.R.Heyland and T.L. Killeen, J.Phys. B. Lett 8, 1975.

25.  M. Inokuti and M.R.C. McDowell, J.Phys. B.7, 2382, 1974.

26.  R.E.Montgomery and R.W. La Bahn, Can. J. Phys. 48, 1288, 1970.

27.  K.F.Canter and G.R.Heyland, Appl. Phys. 5, 231, 1974.

28.  P.G.Coleman, T.C.Griffith, G.R.Heyland and T.L. Killeen, Appl. Phys. 5, 271, 1974.

29.  P.G.Coleman, T.C.Griffith, G.R.Heyland and T.L. Killeen, J.Phys. B.8, 1734, 1975 .

30. P.G.Coleman and T.C.Griffith. J.Phys. B. 6, 2155, 1973.

31. R.H.Beers and V.W.Hughes, Bull. Am. Phys. Soc. 13, 633, 1968.

32. M.A. Stroscio and J.M. Holt. Phys. Rev. A., 10, 749, 1974.

33. M.A.Stroscio, Phys. Lett 50A, 81, 1974.

34. S.Hara and P.A.Fraser, J. Phys. B.8, 219, 1974.

35. K.F. Canter and L.O. Roellig. Phys. Rev. Lett. 25, 328, 1970.

36. J.O. McNutt, V.B. Summerour, A.D.Ray and P.H. Huang, J. Chem. Phys. 62, 1777, 1975.

37. D.M. Spektor and D.A.L. Paul, Appl. Phys. 5, 383, 1975. Can. J. Phys. 53, 13, 1975.

38. K.F.Canter, Thesis, Wayne State University, Detroit, 1970.

39. F.J. Drachman and S.K. Houston, J.Phys. B.3., 1657, 1970.

40. K.F.Canter, A.P. Mills and S. Berko, Phys. Rev. Lett. 34 , 177, 1975.

41. A.P. Mills, S. Berko and K.F. Canter, Phys. Rev. Lett. 34, 1541, 1975.

42. S.Golden and I.R.Epstein, Phys. Rev. A 10, 761, 1974.

43. J.S.Stein, W.E. Kauppila and L.O. Roellig, Rev. Sci. Instrum. 45, 951, 1974.

44. T.S.Stein, W.E. Kauppila and L.O. Roellig, Physics Lett. 51A, 327, 1975.

# ANALYSIS OF ELECTRON CORRELATIONS*

U. Fano

The University of Chicago

Chicago, Illinois 60637

## I. Introduction

This report aims at illustrating some aspects of a program whose development was outlined at the Heidelberg Conference last year.[1] The program deals with the inelastic collisions of low energy electrons and with the related phenomenon in which an electron excited by photo-absorption transfers part of its energy to a second electron. Both processes involve a close correlation of two electrons; unraveling this correlation is, I feel, a prerequisite for a successful general treatment of either process. We have been dealing explicitly with the two-electron complexes formed in the simplest examples:

$$h\nu + H^- \leftrightarrows H^{-**} \rightleftarrows H^* + e \quad , \quad h\nu + He \rightleftarrows He^{**} \rightleftarrows He^{+*} + e \, ,$$

with the side processes $h\nu + He^*$ above, $e + H$ below the first, and $e + He^+$ below the second. (1)

but we have in mind the relevance of results to atoms or molecules with many electrons. When the two electrons in the doubly excited complex, $H^{-**}$ or $He^{**}$, have comparable energies, they remain correlated until the complex dissociates into any of the channels radiating from it.

A hint that hidden simplicities might permit unraveling of the correlations came from the early studies of $He^{**}$, which detected Rydberg series of levels with unknown classification but with striking, novel selection rules; this point was stressed in a 1968 review.[2] At that time, Macek had just traced the origin of the various series to the approximate separability of a particular coordinate in the Schroedinger equation, namely a hyperspherical radius of the complex, $R = (r_1{}^2 + r_2{}^2)^{\frac{1}{2}}$ (where $\vec{r_1}$ and $\vec{r_2}$ are the electron positions with respect to the nucleus).[3] He treated this radius R as an adiabatic parameters, just as one treats the internuclear distance of a diatomic molecule. He first diagonalized the Hamiltonian numerically at fixed R, obtaining eigenvalues $U_\mu(R)$ and eigenfunctions $\Phi_\mu(R;\Omega)$, where $\Omega$ indicates any set of five coordinates that identify $(\vec{r_1}, \vec{r_2})$ at fixed R, and then solved the residual Schroedinger equation in R, using $U_\mu(R)$ as a potential and obtaining energy eigenvalues $E_{\mu n}$ and radial eigenfunctions $F_{\mu n}(R)$. For a few values of $\mu$ that were tested, the calculated $E_{\mu n}$ agreed with those of levels that had been previously grouped together on purely empirical grounds. Thus it appeared that the products $\Phi_\mu(R;\Omega) F_{\mu n}(R)$ should represent fair approximations to actual wave functions and that the terms of the Hamiltonian disregarded in this adiabatic treatment should be responsible for the slow — or even very slow — autoionization transitions. Interpretation of the quantum number $\mu$, and unraveling of the correlations in the $\Phi_\mu(R;\Omega)$, proved unaccessible to Macek owing to numerical complexities.

It may cause surprise that encouraging results have emerged from this adiabatic treatment, where no small parameter replaces the electron/nucleus mass ratio of the molecular analog. However, one may see from current treatments of this analog[4] that the coupling of adiabatic states does not actually hinge on any such simple parameter but on more delicate combinations of reaction velocities, such as appear in the Landau-Zener formula.

Unraveling of the correlations of an electron pair has now emerged from Lin's remark[1,5,6] that radial and angular correlations are nearly independent in the following sense. A useful approximation to Macek's eigenfunctions $\Phi_\mu(R;\Omega)$ is constructed from products of spherical harmonics and of an eigenfunction which depends only on the ratio of radial distances, $r_2/r_1$, at constant R. Macek's empirical index $\mu$ is then largely replaced by a pair of orbital quantum numbers $(\ell_1, \ell_2)$ of the separate electrons, complemented by a new quantum number m, the number of nodes of the eigenfunction of $r_2/r_1$. The quantum numbers $(\ell_1 \ell_2 m)$ play thus the role of "correlation eigenvariables" which we had discussed at Belgrade.[5] Unfamiliarity of the quantum number m and of the eigenfunction itself hinder the interpretation of this otherwise rather simple analysis; we aim here at reducing this obstacle.

## 2. Angular and Radial Correlations

As an introductory example, consider the evaluation of the mean value of the product $\vec{r_1} \cdot \vec{r_2}$ in a stationary state. This mean value is the essential ingredient of the correlation coefficient of the two electron positions as defined in statistics; it also determines the dipole-dipole component of the electrons' interaction. Insofar as angular and radial correlations are independent, one may write

$$\langle \vec{r_1} \cdot \vec{r_2} \rangle = \langle r_1 r_2 \cos \theta_{12} \rangle \sim \langle r_1 r_2 \rangle \langle \cos \theta_{12} \rangle . \qquad (2)$$

Further, insofar as R can be treated as an adiabatic parameter, we see that $\langle r_1 r_2 \rangle$ is proportional to $R^2$. The proportionality constant depends on the intrinsic radial correlation, that is, on the statistical distribution of the ratio $r_2/r_1$. We find it more convenient, however, to replace this ratio by a variable with a finite range, i.e., by a polar coordinate angle $\alpha$, setting

$$r_1 = R \cos \alpha , \quad r_2 = R \sin \alpha , \quad 0 \leqslant \alpha = \operatorname{arctg} \left( \frac{r_2}{r_1} \right) \leqslant \frac{\pi}{2} . \qquad (3)$$

Equation (2) thus takes the form

$$\langle \vec{r_1} \cdot \vec{r_2} \rangle \sim R^2 \langle \sin \alpha \cos \alpha \rangle \langle \cos \theta_{12} \rangle . \qquad (4)$$

Nonzero values of $\langle \sin \alpha \cos \alpha \rangle$ and of $\langle \cos \theta_{12} \rangle$ are typical manifestations of nonvanishing correlations, radial and angular, respectively. A full description of the correlations is provided by probability distributions of the variables $\alpha$ and $\theta_{12}$. The mean values of complete sets of operators, such as the set of polynomials $P_k(\cos \theta_{12})$ for the angular correlations, provide however equivalent information, more closely related to observables.

In the simple case of two-electron S states, the angular correlation is fully represented by the dependence of the wave function on the single variable $\theta_{12}$. Insofar as the orbital momentum $\ell$ of each electron is well defined, this dependence takes the form of the Legendre polynomial $P_\ell(\cos \theta_{12})$. More generally, we shall refer loosely to the number of nodes of the distribution in $\theta_{12}$ as an index of the excitation — and of the sharpness — of angular correlations. For states with total orbital momentum $L > 0$, the dependence of the wave function on $\theta_{12}$ is not readily disentangled from its dependence on the orientation of the electron pair in a laboratory frame. Considering then the set of parameters $\langle P_k(\cos \theta_{12}) \rangle$, we note that large (small) values of the high-k parameters imply sharp (smooth) correlations. Positive (negative) values of the odd-k parameters imply that the two electrons lie prevalently on the same (opposite) side of the atom. Positive values of the even-k parameters imply prevalence of extreme values of $\theta_{12}$ ($\sim 0$ or $\sim \pi$). Any dependence of $\langle P_k(\cos \theta_{12}) \rangle$ on the other variables R and $\alpha$ manifests variations of the angular correlation pattern.

Analytical expressions of the $\langle P_k(\cos \theta_{12}) \rangle$ are provided by spectroscopy insofar as a pair $(\ell_1 \ell_2)$, with $\ell_1 \geqslant \ell_2$, serves as a quasi-good quantum number. The orbital wave function of two electrons with these orbital momenta coupled to total $(L, M)$ is indicated by $Y_{\ell_1 \ell_2 LM}(\hat{r}_1, \hat{r}_2)$.

Symmetry under permutation of $\vec{r}_1$ and $\vec{r}_2$ introduces, however, an exchange correlation between this orbital wave function and the radial correlation wavefunction which we have called[1] $g_m^{\ell_1 \ell_2}(R;\alpha)$ and which will be discussed in the next section. Thereby Macek's eigenfunction[6] $\Phi_\mu(R;\Omega)$ is now replaced by the approximation

$$\Phi_{\ell_1 \ell_2 m}(R;\Omega) = g_m^{\ell_1 \ell_2}(R;\alpha) Y_{\ell_1 \ell_2 LM}(\hat{r}_1,\hat{r}_2)$$

$$+ (-1)^S g_m^{\ell_1 \ell_2}(\tfrac{1}{2}\pi - \alpha) Y_{\ell_1 \ell_2 LM}(\hat{r}_2,\hat{r}_1), \quad \text{for } \ell_1 \neq \ell_2 , \qquad (5)$$

$$\Phi_{\ell \ell m}(R;\Omega) = [1 + (-1)^{S+L+m}] g_m^{\ell \ell}(R;\alpha) Y_{\ell \ell LM}(\hat{r}_1,\hat{r}_2), \quad \text{for } \ell_1 = \ell_2 = \ell ,$$

where S is the spin quantum number of the electron pair. The expression (5) yields, in the language of spectroscopy,[7]

$$\langle P_k(\cos\theta_{12})\rangle = (-1)^L (\ell_1 \| C^{[k]} \| \ell_1)(\ell_2 \| C^{[k]} \| \ell_2) \left\{ \begin{array}{ccc} \ell_1 & \ell_2 & L \\ \ell_2 & \ell_1 & k \end{array} \right\}$$

$$\times \left\{ [g_m^{\ell_1 \ell_2}(R;\alpha)]^2 + [g_m^{\ell_1 \ell_2}(R;\tfrac{1}{2}\pi - \alpha)]^2 \right\}, \quad \text{for even } k , \qquad (6)$$

$$\langle P_k(\cos\theta_{12})\rangle = (-1)^{S+L}(\ell_1 \| C^{[k]} \| \ell_2)^2 \left\{ \begin{array}{ccc} \ell_1 & \ell_2 & L \\ \ell_1 & \ell_2 & k \end{array} \right\}$$

$$\times 2 g_m^{\ell_1 \ell_2}(R;\alpha) g_m^{\ell_1 \ell_2}(R;\tfrac{1}{2}\pi - \alpha) , \quad \text{for odd } k . \qquad (7)$$

Note that the odd-k parameters result from exchange terms and vanish, owing to selection rules, unless $\ell_1$ and $\ell_2$ have opposite parity.

The study of radial correlations is less familiar, nor has any convenient set of radial parameters analogous to the $\langle P_k(\cos\theta_{12})\rangle$ been identified yet. Radial correlations have, however, an important bearing on the binding of valence electrons. In the notable example of H⁻, no binding is obtained from a variational wavefunction with the uncorrelated form $\exp[-\beta(r_1 + r_2)]$, or with any other uncorrelated form, but substantial binding results from the simple correlated Eckart wave function

$$y(r_1,r_2) = \exp(-\beta r_1 - \gamma r_2) + \exp(-\gamma r_1 - \beta r_2) . \qquad (8)$$

Figure 1 shows a contour plot of this function. The curvature of the contour lines depends on the ratio $(\beta-\gamma)^2/(\beta+\gamma)^2$. The value 0.16

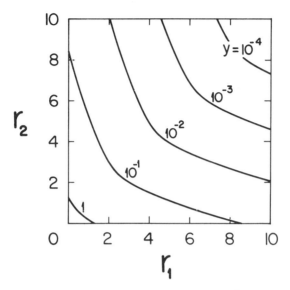

Fig. 1. Contour line plot of the approximate Eckart wavefunction for the He ground state, $y(r_1, r_2) = \exp(-\beta r_1 - \gamma r_2) + \exp(-\gamma r_1 - \beta r_2)$ with $\beta = 2$ and $\gamma = 1.34$. (Courtesy C. E. Theodosiou)

of this ratio, corresponding approximately to the He ground state, was used in Fig. 1; a zero value, implying lack of correlation, would yield straight contour lines at 45° obliquity. On the other hand, the H⁻ value, 0.38, would yield contour lines much more sharply curved; the long tails of nonzero probability amplitude extending along each coordinate axis represent the fact that either one of the two electrons of H⁻ may range much farther away from the nucleus than the other electron. This effect is present, to a lesser but significant extent, in all atoms with two or more electrons in their valence shell; its evaluation would improve Hartree–Fock values of the ionization potentials.

Upon replacement of $(r_1, r_2)$ with $(R; \alpha)$ according to (3), the Eckart function (8) constitutes a variational approximation to the eigenfunction $g_m^{\ell_1 \ell_2}(R; \alpha)$ with $\ell_1 = \ell_2 = m = 0$, to within an unessential factor. By following the variation of $y(r_1, r_2)$ in Fig. 1 along circles with radius R centered at the origin, we note that $g_0^{00}(R; \alpha)$ dips to a minimum at $\alpha = 45°$; we shall see later that this minimum is a general feature of all $g_m^{\ell_1 \ell_2}(R; \alpha)$ at large R.

More elaborate examples of radial correlation plots are shown in Fig. 3 of Ref. 1. One notices there nodal lines running either at constant R, corresponding to nodes of Macek's $F_{\mu n}(R)$, or in a generally radial direction, corresponding to the two nodes of $g_2^{00}$ (R;α) as a function of α. Oscillatory variations of $F_{\mu n}(R)$ represent a radial excitation of the two-electron system <u>as a whole</u>; oscillatory variations of $g_m^{\ell_1 \ell_2}(R;α)$ as a function of α represent an excitation of the <u>relative</u> radial motions of the two electrons, and sharpen their radial correlation.

### 3. The Radial Correlation Functions $g_m^{\ell_1 \ell_2}(R;α)$

The eigenfunction $g_m^{\ell_1 \ell_2}(R;α)$ is defined by the requirement that the function $\Phi_{\ell_1 \ell_2 m}(R;\Omega)$, with the form (5) and with g = 0 at α = $(0, \frac{1}{2}\pi)$, constitutes a variational approximation to an eigenfunction of the operator

$$U(R) = \frac{1}{R^2}\left[-\frac{d^2}{dR^2} + \frac{\vec{\ell}_1^{\,2}}{\cos^2\alpha} + \frac{\vec{\ell}_2^{\,2}}{\sin^2\alpha}\right] - \frac{C(\alpha, \theta_{12})}{R}. \qquad (9)$$

Here $\vec{\ell}_1^{\,2}$ and $\vec{\ell}_2^{\,2}$ are the squared orbital momentum operators of the two electrons, and

$$- C(\alpha, \theta_{12}) = R\left(-\frac{2Z}{r_1} - \frac{2Z}{r_2} + \frac{2}{r_{12}}\right)$$

$$= -\frac{2Z}{\cos\alpha} - \frac{2Z}{\sin\alpha} + \frac{2}{(1 - \sin 2\alpha \cos\theta_{12})^{\frac{1}{2}}} \qquad (10)$$

represents the Coulomb interaction of two electrons and one nucleus in Rydberg units, multiplied by R; Fig. 4 of Ref. 1 shows a relief map of $- C(\alpha, \theta_{12})$. The term of (9) in brackets represents a generalized centrifugal energy, i.e., the kinetic energy of the electrons' motion at constant R, which is positive definite. The potential function $-C$ ranges from $-\infty$ at α = $(0, \frac{1}{2}\pi)$, where one electron lies at the nucleus, to $+\infty$ at (α = π/4, $\theta_{12}$ = 0) where the electron positions coincide; its two negative valleys are separated by a most important flat saddle point at (α = π/4, $\theta_{12}$ = π) where $-C = -\sqrt{2}(4Z - 1)$.

These properties of the operator (9), and especially the fact that it consists of one term with a factor $1/R^2$ and one with $1/R$, have a determining influence on the R-dependence of the approximate eigenvalues $U_{\ell_1 \ell_2 m}(R)$ and eigenfunctions $g_m^{\ell_1 \ell_2}(R;α)$, from the small-R ("condensed atom") limit to the large-R ("dissociated atom") limit. This dependence characterizes, in turn, the evolution of correlations in the course of a collision or following photoabsorption. This evolution, and the variations themselves of $U_{\ell_1 \ell_2 m}(R)$ and of $g_m^{\ell_1 \ell_2}(R;α)$ as functions of R, also resemble in many ways the variations of the corresponding quantities in the molecular analog as the internuclear distance ranges from the united-atom to the separate-atoms limit.

In the dissociated-atom limit, we consider our two-electron system as resolved into one electron bound to the nucleus in a hydrogenic state with energy $-Z^2/n^2$ and one electron at a large distance away. We are thus in a situation where either: a) $r_1 >> r_2$, $\alpha << 1$, $r_1 \sim R$, $r_2 \sim R\alpha$, or b) $r_1 << r_2$, $\frac{1}{2}\pi - \alpha << 1$, $r_1 \sim R(\frac{1}{2}\pi - \alpha)$, $r_2 \sim R$. The main dependence on $\alpha$ of the total wave function, and hence of the radial correlation function $g_m^{\ell_1 \ell_2}(R;\alpha)$, is then concentrated in two narrow ranges of $\alpha$, near 0 and near $\frac{1}{2}\pi$, where $g(\alpha)$ must coincide with a hydrogenic radial eigenfunction $P_{n\ell}$. Inspection of Eqs. (9) and (10) confirms that we must have

$$\lim_{R \to \infty} g_m^{\ell_1 \ell_2}(R;\alpha) = \begin{cases} c\, P_{n\ell_2}(R\alpha), & \text{for } \alpha << 1, \\ c'\, P_{n\ell_1}(R[\frac{1}{2}\pi - \alpha]), & \text{for } \frac{1}{2}\pi - \alpha << 1, \end{cases} \tag{11}$$

where c and c' are normalization coefficients which are still undetermined; the relation between n and m is also undetermined at this point. The wave functions $P_{n\ell_2}$ and $P_{n\ell_1}$ vanish exponentially as $\alpha$ draws away from 0 and $\frac{1}{2}\pi$ respectively.

This separation of $g_m^{\ell_1 \ell_2}$ in two parts confined within narrow, widely separated ranges of $\alpha$, in the large-R limit, is analogous to the separation of a molecular orbital into two atomic orbitals in the separate atoms limit. In both cases the exponential tails of the separate parts actually join in the intermediate range to form a single wave function; the joint may or may not involve a node, depending on the relative sign of the coefficients c and c' in (11). In the molecular analog the orbital is bonding (antibonding) in the absence (presence) of a node; in our problem m equals $2n - \ell_1 - \ell_2 - 2$ or $2n - \ell_1 - \ell_2 - 1$, respectively, and the state characterized by the g function is designated + or −. For $\ell_1 > \ell_2$, both of these alternatives occur normally, except for the lowest value of n, namely, $\ell_2 + 1$, when only a single, nodeless function exists, with a + character. For $\ell_1 = \ell_2 = \ell$, on the other hand, the parity of m must equal that of L+S, according to (5), whereby m equals $2(n - \ell - 1)$ or $2(n - \ell - 1) + 1$, depending on its parity.

Figure 2 shows plots of the functions $g_1^{10}(\alpha)$ and $g_2^{10}(\alpha)$ for $^1P$ states and for a few values of R; the plots for R=15 illustrate the large-R limit. Here both functions approach a 2s radial orbital for $\alpha < \pi/4$ and a 2p orbital for $\alpha > \pi/4$, but the sign of the 2p orbital differs for the + and the − function. Both wave functions are small in the range of $\alpha$ around $\pi/4$ which is "classically inaccessible" because the eigenvalue $U_{10m} = -Z^2/2^2$ lies below the saddle point of the potential function $-C(\alpha, \theta_{12})/R$.

As the system develops away from the large-R limit, the dipole component of the electrons' interaction becomes appreciable, whereby the bound electron's orbit is polarized by the other electron. In the case of $\ell_1 \neq \ell_2$, the eigenvalues $U_{\ell_1 \ell_2 m}$ of a + and a − state having a

34   U. FANO

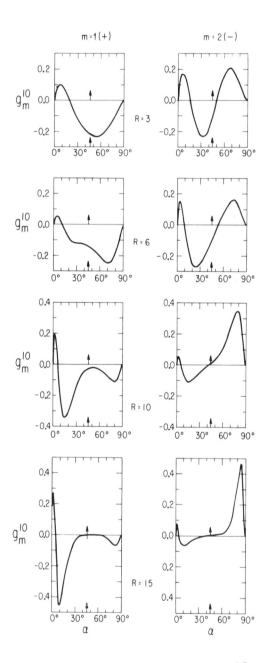

Fig. 2. Approximate radial correlation functions $g_1^{10}(R;\alpha)$ and $g_2^{10}(R;\alpha)$ of He $^1$P states, for different values of R.  (Courtesy C. D. Lin)

common limit $-Z^2/n^2$ depart from each other and from this limit as R recedes to finite values. The separation of the two eigenvalues is propor-
tional to the mean dipole product $\langle \vec{r}_1 \cdot \vec{r}_2 \rangle$ and hence to $\langle P_1(\cos\theta_{12}) \rangle$
according to (4). Moreover the expression (7) of $\langle P_1(\cos\theta_{12}) \rangle$ clearly
has opposite sign for the + and the − state, meaning that the electrons
are on the same or on the opposite side of the atom in the two cases,
whereby the dipole interaction raises one of the eigenvalues and lowers
the other. A curious result may occur here, and does occur in the important example of sp $^1$P states shown in Fig. 2, namely, that the dipole
interaction raises the eigenvalue of the + state, which has a lower nodal
index m, <u>above</u> that of the − state quasi-degenerate with it. This inversion of the normal order of eigenvalues is, however, counteracted by the
contribution of the mean value of the kinetic energy (i.e., of the first)
term of (9), which is an increasing function of the number of nodes m.
The contribution of this term to the eigenvalue $U_{\ell_1 \ell_2 m}(R)$ becomes increasingly important as R decreases further, so that the energy of the −
state rises again above that of the + state, thus restoring the normal order.
An important consequence of this phenomenon for the sp $^1$P autodetaching
states of H$^-$ has been pointed out by Lin.[8]

The plots of $g_1^{10}(\alpha)$ and of $g_2^{10}(\alpha)$ for successively lower values of
R, shown in Fig. 2, illustrate the increasing influence of the kinetic
energy term of Eq. (9) in smoothing out the oscillations of the eigenfunctions. In the R=0 limit, where the kinetic energy contribution predom-
inates altogether, the eigenfunctions have an analytic representation in
terms of Jacobi polynomials, i.e., of generalized Legendre functions.
References 1 and 6 have emphasized the role of the additional node of
the − functions, which remains close to $\alpha = \pi/4$; the presence of this
node determines the selection rules which make the formation and decay
of − states unlikely in the usual processes of electron collision or photoabsorption.

## 4. Mixing of Angular and Radial Correlations

We consider here the influence of that portion of the electron interaction which induces changes of the orbital quantum numbers $(\ell_1, \ell_2)$
of an electron pair but is disregarded in Lin's approximation (5) to
Macek's eigenfunctions $\Phi_\mu(R;\alpha)$. In general this portion of the interaction only causes an actual eigenfunction $\Phi_\mu$ to consist primarily of a
single function $\Phi_{\ell_1 \ell_2 m}$ with a small admixture of other functions
$\Phi_{\ell_1' \ell_2' m'}$. However, a large admixture of two (or more) $\Phi_{\ell_1 \ell_2 m}$ occurs
when two (or more) eigenvalues $U_{\ell_1 \ell_2 m}(R)$ with different $(\ell_1 \ell_2)$ coincide, even only approximately.

This quasi-degeneracy of eigenvalues occurs at large R where different eigenvalues generally converge to the same limit $-Z^2/n^2$; the resulting eigenfunctions have been studied repeatedly, particularly in the

context of close-coupling calculations,[9] and will not be discussed here. Degeneracy occurs also in the opposite, condensed-atom limit of $R=0$, but it is generally removed as soon as the potential energy becomes non-negligible at nonzero $R$, because this interaction separates out states with different $(\ell_1, \ell_2)$, especially if they have also different values of $\ell_1 + \ell_2$ (see Sec. VI of Ref. 6).

Here we direct our attention to a class of eigenvalue crossings that occur at rather small values of $R$, in the general proximity of the classical inner turning points of the radial wavefunctions $F_{\mu n}(R)$. These crossings occur because the (negative) potential energy contribution to each eigenvalue $U_{\ell_1 \ell_2 m}(R)$ increases in magnitude with increasing $m$. Thus it happens for a pair of eigenvalues, $U_{\ell_1 \ell_2 m}$ and $U_{\ell'_1 \ell'_2 m'}$, with $m > m'$ and $\ell_1 + \ell_2 < \ell'_1 + \ell'_2$, that $U_{\ell_1 \ell_2 m}$ lies higher at very small $R$ but falls rapidly below $U_{\ell'_1 \ell'_2 m'}$ as $R$ increases. Under these circumstances the full electron interaction — including the terms that couple the quantum number pairs $(\ell_1 \ell_2)$ and $(\ell'_1 \ell'_2)$ — causes the crossing to be avoided by a substantial amount. This effect replaces the pair of approximate eigenvalues $U_{\ell_1 \ell_2 m}$ and $U_{\ell'_1 \ell'_2 m'}$ by a pair of exact (adiabatic) eigenvalues, $U_{\mu}(R)$ and $U_{\mu'}(R)$, which depart appreciably from the approximate ones over a substantial range of $R$ values (see Figs. 7 and 8 of Ref. 6). Over the same range of $R$ each of the exact eigenfunctions, $\Phi_{\mu}(R;\Omega)$ and $\Phi_{\mu'}(R;\Omega)$, is represented by a superposition of two expressions (5) with comparable coefficients.

Consequently the pattern of angular and radial correlations of either of the exact $\Phi_{\mu}$ and $\Phi_{\mu'}$ changes progressively from that which pertains to the expression (5) with one set of quantum numbers to that which pertains to the other set. Figure 3 illustrates this transition for $^1S$ pairs corresponding to $s^2$ configurations ($\ell_1 = \ell_2 = 0$, $m = 2$) and to $p^2$ configurations ($\ell_1 = \ell_2 = 1$, $m = 0$). At very small $R$ the eigenstate $\Phi_1(R;\Omega)$ has $p^2$, $m=0$, character — with a node at $\theta_{12} \sim \frac{1}{2}\pi$ and no node at $\alpha \sim$ const. — while $\Phi_2(R;\Omega)$ has $s^2$, $m=2$, character — with two nodes at $\alpha \sim 30°$ and $60°$ and no node with $\theta_{12} \sim$ const.. The patterns of both wave functions evolve progressively so as to interchange eventually at large $R$, while remaining orthogonal throughout. Analogous displays could be constructed for other $S$ states, though not for states with $L > 0$ in which case the dependence of the wavefunction on $\theta_{12}$ and on other variables cannot be readily disentangled, as noted in Sec. 2.

Some of the observable, and observed, consequences of this class of avoided crossings have been pointed out in Sec. VI of Ref. 6, but we surmise that many other known effects are related to the same phenomenon. These probably include the admixture of $(\ell+1)^2$ excited configurations in the ground states of full $\ell$ subshells, which has a major effect on photoabsorption spectra;[10] we have stressed elsewhere[11] that the admixture is confined to short radial distances, where the energy of $(\ell+1)^2$ pairs with lower radial excitation should fall below that of $\ell^2$ pairs with

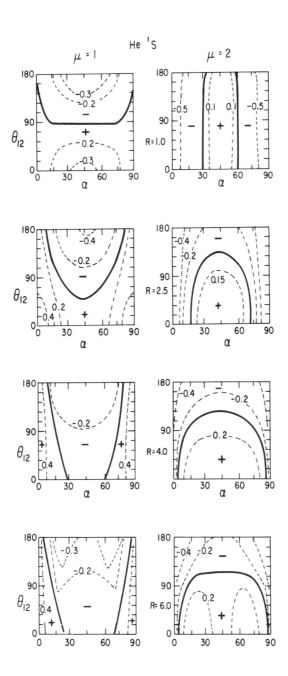

Fig. 3. Contour line plots of the joint radial and angular correlation functions $\Phi_1$ and $\Phi_2$ of $^1$S states of He for different values of R. (Courtesy C. D. Lin)

higher radial excitation, according to the above discussion. The exceptional frequency of double photoexcitations of the $6s^2$ ground state of Ba, observed at 21.2 eV,[12] might conceivably be traced to doubly excited sp $^1P$ states receiving a strong admixture of singly excited $5p^5d$ states at short radial distances. Such surmises can, however, be checked only by future extensions of our analysis to atoms with many electrons.

## 5. Concluding Remarks

The main purpose of this paper was to introduce and discuss the plots in Figs. 1, 2, and 3. We have omitted any reference to the better known correlation pattern which has been studied in the Wannier–Rau–Peterkop theory of threshold ionization and which pertains to the restricted range of variables $R \gtrsim 25$, $\alpha \sim \pi/4$, $\theta_{12} \sim \pi$. This pattern is quite different from those discussed in the present paper and we perceive only dimly the connection between the two types of correlation pattern. Filling this gap is one of the unsolved problems raised in Ref. 1. The material covered in the present paper needs itself further elaboration and sharpening.

A further task is to adapt the results obtained for two-electron atoms to the two-electron excitations of atoms with many electrons. One may anticipate that these phenomena are rather localized in a limited portion of the many-electron configuration space, in which the prevailing conditions resemble those of the space considered here. Even in this region, however, the wavefunctions should differ from those of the two-electron problem, because they must extend continuously into the small-R region where they depend on the spin and orbital momenta of the ionic core. Analysis of the numerous resonances observed in the three-electron system $e + He \to He^{-**} \to He + e + h\nu$ [13] may afford an introduction to this problem. Three-electron excitations, and especially their remarkable properties discovered recently by Read's group,[14] pose a still more distant challenge.

## References

*Work supported by the U. S. Energy Research and Development Administration, Contract No. COO-1674-111.

1.   U. Fano and C. D. Lin, "Correlations of Excited Electrons", in Atomic Physics 4 (Plenum Press, New York, 1975) p. 47.
2.   U. Fano, "Doubly Excited States of Atoms" in Atomic Physics 1 (Plenum Press, New York, 1969) p. 209.
3.   J. Macek, J. Phys. B 2, 831 (1968).
4.   J. B. Delos, W. R. Thorson, and S. K. Knudson, Phys. Rev. A 6, 709 (1972).
5.   U. Fano and C. D. Lin, "Correlations in He**, e + He and in Related Systems", in Physics of Electronic and Atomic Collisions (Belgrade, 1973) p. 229.

6.  C. D. Lin, Phys. Rev. A 10, 1986 (1974).
7.  A. R. Edmonds, Angular Momenta in Quantum Mechanics
    (Princeton University Press, 1957) p. 114.
8.  C. D. Lin, submitted to Phys. Rev. Letters.
9.  P. G. Burke and D. D. McVicar, Proc. Phys. Soc. (London) 86,
    989 (1965); V. L. Jacobs and P. G. Burke, J. Phys. B 5, 2272 (1972).
10. M. Ya. Amusia, N. A. Cherepkov, and L. V. Chernisheva,
    Sov. Phys.—JETP 33, 90 (1971).
11. T. N. Chang and U. Fano, submitted to Phys. Rev. A.
12. B. Brehm and K. Hoefler, Int. J. Mass. Spectry. and Ion Phys. 18,
    xxx (1975); H. Hotop and D. Mahr, submitted to J. Phys. B.
13. D. W. O. Heddle, "Resonance Series in Helium" in Electron and
    Photon Interactions with Atoms , ed. by H. Kleinpoppen and M. R.
    C. McDowell (Plenum Publ. Corp., 1975).
14. G. C. King, F. H. Read, and R. C. Bradford, J. Phys. B, in press.

# DECAYING STATES

Professor Sir Rudolf Peierls

University of Washington

Seattle, Washington  98195

I speak with some diffidence to this gathering of distinguished atomic physicists, since I am a stranger to your subject.  I have some familiarity with problems of nuclear physics, and there have been occasions when concepts developed in the study of nuclei have proved of use in atomic problems.  I have in mind particularly the concept of compound resonances in collisions.  This encourages me to present here a summary of results and ideas relating to decaying states, which occur in both fields, but which have been studied in particular detail by nuclear theorists.

I should, however, preface this review with the caution that many of the exact results I shall quote are derived on the assumption that the forces in the system studied have a finite range, and vanish beyond it.  This is a satisfactory approximation for many nuclear problems, although the corrections for the small, usually exponential, "tails" of the forces can sometimes cause annoying complications.  In atomic problems, on the other hand, one is usually concerned with long-range forces, such as Coulomb or dipole terms, and this can profoundly change the analytic character of the functions we are dealing with, so that few of the mathematical results I am about to mention will survive in the atomic applications. The analogy between the two fields may nevertheless be suggestive from the physical point of view.

The first appearance of decaying states in quantum mechanics was in Gamow's work on alpha decay, and they are often referred to as "Gamow states" or resonance states.

Consider, as the simplest example, the radial Schrödinger equation for zero angular momentum,

$$\frac{d^2 u}{dr^2} + (k^2 - v(r))u = 0 \tag{1}$$

where $u(r)$ is the radial wave function, which must vanish at the origin, $k^2$ is a measure of the energy ($k$ being the wave vector of a free particle of that energy) and $v(r)$ a central potential (in suitable units) assumed to vanish beyond a distance R. It follows that at greater distances, the most general solution is of the form

$$u(r) = Ae^{ikr} + Be^{-ikr} , \quad r > R \tag{2}$$

Here the A term represents a radially outgoing, the B term a radially incoming wave. The decaying state represents a situation in which there are no particles incident, and therefore $B = 0$, which corresponds to the condition

$$\frac{du}{dr} - iku = 0, \quad r = R \tag{3}$$

Evidently we cannot satisfy the wave equation (1) with the boundary condition (3) for any real k. This would correspond to a steady state with a steady stream of outgoing particles. One easily verifies that for any solution of (1) and (3) other than a bound state solution

$$\text{Im } k = -|u(R)|^2 / \int_0^R |u(r)|^2 dr \tag{4}$$

If we choose the real part of k positive, which corresponds with our interpretation of the two terms of (2) as outgoing and incoming, respectively, the imaginary part of the energy, i.e. of $k^2$, is negative, corresponding to an exponential decrease of the modulus of the wave function with time, as expected.

However, a negative imaginary part of k means also that the radial dependence of the wave function, given at large r by the factor $e^{ikr}$, increases exponentially in amplitude. This property of the eigenfunction for decaying states has often led people to reject them as physically unreasonable. Yet this increase is entirely reasonable, because it reflects the fact that we are assuming an exponentially decaying state, and thus we see at distance r the particles emitted by the system a time r/v earlier, where v is their velocity, and these are more numerous by a factor $e^{r/v\tau}$; $\tau$ being the mean life. One easily verifies that this agrees with the increase in intensity due to the imaginary part of k.

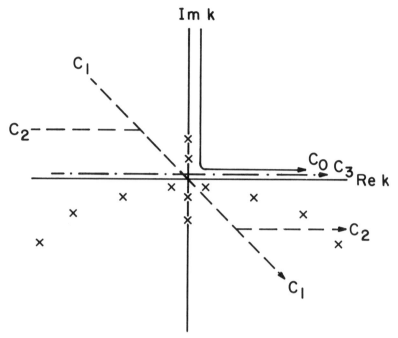

Fig. 1. Location of poles and integration contours.

This increase means that the function u cannot be normalized in the usual way because the integral of its square modulus diverges at large r.  The function therefore cannot represent a physical state of the system.  In any real physical situation the decaying state cannot have existed forever, but must have been created by some other physical process in the past.  Nevertheless the pure decaying state is a convenient idealization if we are not interested in how it was obtained, just as it is convenient to use a plane wave, which also cannot be normalized, when we are not interested in the practical limitations to our knowledge of the momentum.

One can show that for the problem I have specified, the values of k for which (1) and (3) have solutions are distributed in the complex plane as in fig. 1.  There are usually an infinite number of eigenvalues in the lower half plane, placed in pairs symmetrically about the imaginary axis, except that there may (but need not) be unpaired values on the imaginary axis.  Bound states are also included since positive imaginary k corresponds to a negative energy, and to a radially decaying wave function by the first term of (2).

The properties of decaying states which I have sketched

so far have been known for a long time, and, apart from the
determination of decay constants, are of rather academic
interest.  In recent years, however, one has learnt to use
them as tools for calculation in cases in which decaying
states may be intermediate states of some reaction.  Since
the eigenfunctions of decaying states do not possess the
usual orthonormality properties of eigenfunctions for real
energy, one cannot use them in expansions of the conventional
kind.  A convenient way to connect them with more general
questions is via the resolvent, or Green function, of the
Schrödinger equation.

The Green function of equation (1) with outgoing boundary
condition (which is the Fourier transform of the retarded
resolvent for the time-dependent wave equation) satisfies the
equation

$$\frac{\partial^2 G(r,r',k)}{\partial r^2} + (k^2 - v(r))G(r,r',k) = \delta(r-r') \qquad (5)$$

and the boundary condition

$$\frac{\partial G}{\partial r} - ikG = 0, \quad r = R. \qquad (6)$$

In its dependence on k, G becomes singular at any point for
which an eigenfunction of (1) and (3) exists, because an
inhomogeneous equation generally has no solution when there
exists a solution of the corresponding homogeneous problem.
In fact, the Green function has poles, and its behavior near
the pole $k_n$ can be shown to be of the form

$$G(r,r',k) = \lambda_n \frac{u_n(r)u_n(r')}{k^2 - k_n^2} + G_1 \qquad (7)$$

where $G_1$ is regular at $k_n$.  $\lambda_n$ is a constant, which evidently
depends on the, so far arbitrary, choice of the multiplicative
factor in $u_n$.

For a bound state, the factor $\lambda$ is unity, if $u_n$ is
normalized.  Indeed for a system which has only discrete
eigenstates, the Green function is given exactly by a sum of
pole terms of the type (7), with $\lambda = 1$.  It therefore appears
convenient to choose the factor in the definition of all $u_n$
in such a way as to make the factor $\lambda$ in the pole residue
equal to unity.  This can be shown to amount to the condition

$$\int_0^R u_n^2 \, dr - \frac{1}{2ik_n} u_n(R)^2 = 1 \qquad (8)$$

This condition can be imposed at any R exceeding the range of the potential $v(r)$, and it is easy to verify that, by virtue of the boundary condition (3) the left-hand side of (8) does not depend on R.

A normalization convention equivalent to (8) was first suggested by Zel'dovich[1] using an argument in terms of analytic continuation.   It was also used by Berggren.[2]

Note that (8) contains $u^2$, not $|u|^2$.  In exceptional cases the expression may vanish, so that no choice of factor will bring it to unity, and this very exceptional situation needs special treatment.   (See ref. 3 for a very similar case.) For bound states, for which $u_n$ can be chosen real, (8) reduces to conventional normalization, because the second term then equals the integral of $u_n^2$ over its exponential tail beyond R.

The choice for the normalization convention (8) was based entirely on mathematical convenience.   However, it has also a useful physical consequence, noticed already by Hokkyo[4]:   If one adds to the potential $v(r)$ a small perturbation $v_1(r)$, which is assumed also to vanish beyond R, then the change in the energy of a bound state is, to first order in $v_1$, by standard perturbation theory

$$\delta k_n^2 = \int v_1(r) u_n^2 dr$$

provided $u_n$ is correctly normalized.   With the convention (8) the first-order shift of a resonance level is given by the same expression.

Let us now turn to the retarded Green function of the time-dependent wave equation, which serves to obtain the time evolution of the wave function.   This is given by

$$g(r,r',t) = \frac{i}{2\pi} \int_{-\infty}^{\infty} e^{-iEt} G(r,r',k) dE = \int \frac{i}{\pi} e^{-ik^2 t} G(r,r',k) k dk$$

(9)

where the integration in the E plane goes along the real axis, but passing above any bound-state poles.   In the complex k plane this corresponds to the full line marked $C_o$ in fig. 1.

Since we are interested in the physics associated with the known poles, we are tempted to try to close the integration contour in the k plane.   However, the factor $\exp(-ik^2 t)$ diverges at large distances in the upper right and lower left quadrants, while it decreases rapidly (for reasonable t) in the other two quadrants.   One can verify that the variation of G with k is at most exponential, so that the behavior of the integrand in (9) at large k is dominated by the first factor. It is therefore not possible to remove the whole contour to infinity, but we can deform it in the second and fourth quadrants. One useful choice is to change it from $C_0$ to $C_1$, shown by a broken line in fig. 1. This is the line of steepest descent of the first factor, and should therefore lead to a moderately rapidly converging integral for not too short times. By deforming the contour we have passed over some of the poles of G, including all those belonging to bound states, and those Gamow states for which the energy has a positive real, and negative imaginary part, as is appropriate for resonance states.

We may now write the time-dependent Green function in the form

$$g(r,r',t) = \sum_n u_n(r)u_n(r') \, e^{-iE_n t} + \frac{1}{\pi}\int_{-\infty}^{\infty} e^{-k^2 t} G(r,r',\gamma k)k\,dk$$

$$(10)$$

Here the summation includes all bound states and all resonance states with Re $E_n > 0$, Im $E_n < 0$. $\gamma$ is an abbreviation for

$$\gamma = \sqrt{-i} = \frac{1}{\sqrt{2}}(1 - i)$$

This simple representation of the time-dependent Green function was derived in a thesis by Scheffler.[5]

This function yields the solution of the initial-value problem in the form

$$\psi(r,t) = \int g(r,r',t)\psi(r',0)dr' \qquad (11)$$

Inserting for g from (10), the resonance terms yield exponentially decaying contributions, as expected, and their coefficients can be evaluated in terms of the resonance eigenfunctions, almost as if we were dealing with bound states.

The contribution from the integral term is of interest. After a long time t, the exponential is a very steeply decreasing function of k and we may expect G to vary slowly in comparison. If we treat G as constant, the integrand becomes an odd function of k and the integral vanishes. We must therefore include the variation of G by taking the next term in its Taylor series, and the integral becomes

$$\frac{\gamma}{\pi} \left(\frac{dG(r,r',k)}{dk}\right)_{k=0} \int e^{-k^2 t} k^2 dk = \frac{1}{2\sqrt{\pi}} \left(\frac{dG}{dk}\right)_{k=0} t^{-3/2} \tag{12}$$

This term in (10) therefore contributes to $\psi$ a long-lived contribution which must ultimately dominate over all terms from decaying states.

This behavior is well known. For example, Newton[6] demonstrates the existence of such terms in resonance scattering by an argument which is in principle identical with the above, but less transparent because it uses relativistic kinematics.

As Newton points out, this paradox has its origin in the uncertainty principle. If the initial state of the particle is well localized, so that we know it is inside the "nucleus" at time 0, it is impossible to exclude the possibility that its energy might happen to be close to 0, so that it would emerge with very low velocity, and take a very long time to reach any detector.

It is in line with this interpretation that the result (12) depends on the behavior of G near k = 0, i.e. on the dynamics of a particle emerging with zero velocity. We note that it is not only the wave function at large r which shows this long-lived term, but also the wave function inside the nucleus, which is related to the probability that the particle is still inside. The intuitive interpretation of the paradox therefore requires that a particle which attempts to emerge from a potential barrier with very low kinetic energy will not only take very long to move after it has emerged, but will also take long to cross the barrier.

Returning to our expression (10), it is convenient to combine, in the integral term, the contributions from k and -k, so that the integral becomes

$$\frac{1}{\pi} \int_0^\infty e^{-k^2 t} [G(r,r',\gamma k) - G(r,r',-\gamma k)] k dk \tag{13}$$

Now the difference between the two Green functions in the bracket is as a function of r a solution of the homogeneous Schrödinger equation, as can be seen by subtracting (5) for opposite values of k. Thus the difference is proportional to $u(r,E)$, the solution of the Schrödinger equation for energy $E = -ik^2$, which vanishes at the origin. Furthermore, because of the symmetry of the Green functions, it is also proportional to $u(r',E)$. The solution of the wave equation

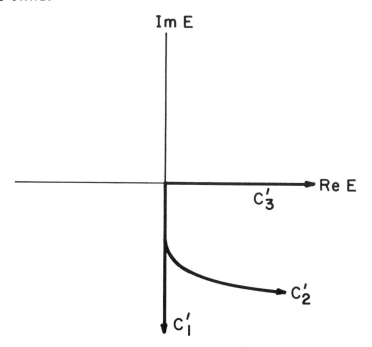

Fig. 2. "Continuous spectrum" contours in the energy plane.

contains an arbitrary factor, and we shall choose this so as
to make the difference of Green functions come exactly to
$-2\pi i u(r,E)u(r',E)$. With this convention, (10) finally reduces
to

$$g(r,r',t) = \sum_n u_n(r)u_n(r')e^{-iE_n t} + \int_{C_1'} dE u(r,E)u(r',E)e^{-iEt}$$

(14)

where the integration goes over the contour $C_1'$ (vertical line
in fig. 2)

This allows us to write the solution of the initial-value
problem, in which the wave function is given at time $t = 0$,
in the form

$$\psi(r,t) = \int g(r,r',t)\ \psi(r',0)dr'$$

$$= \sum_n a_n u_n(r)e^{-iE_n t} + \int_{C_1} b(E)u(r,E)e^{-iEt} dE \qquad (15)$$

where

$$a_n = \int u_n(r)\psi(r,o)dr \; ; \quad b(E) = \int u(r,E)\psi(r,0)dr \qquad (16)$$

If the initial wave function $\psi(r,0)$ has a finite range and vanishes beyond a certain distance, then the integrals (16) involve only finite integrations and raise no problems. If, however, the initial wave function decreases exponentially, say as $e^{-\alpha r}$, for large $r$, then the integrals diverge if the imaginary part of $k_n$, or $k$, respectively, is less than $-\alpha$. In that case, we can replace the contour $C_1$ by $C_3$, which runs parallel to the real axis from some suitable point with an imaginary part between $-\alpha$ and $\alpha$. The summation in (15) then includes only poles above $C_2$, and the integral runs over energy along the corresponding contour labelled $C_2'$ in fig. 2.

The normalization convention which we have specified, can be expressed in the form

$$\left(\frac{du}{dr}\right)^2 + k^2 u^2 = \frac{k}{\pi} \; , \quad r \gtreqless R \qquad (17)$$

It depends only on E, but not on the shape of the contour.

A special case is the contour $C_3$, for which the sum includes only bound states, and the integral extends over real positive energies. One easily verifies that the rule (17) for the normalization of $u(r,E)$ then reduces to the usual rule for normalization in the continuous spectrum.

The relations (15) and (16) can, of course, be applied at time 0, and then amount to expanding a general function in terms of a sum over resonance terms, and an integral over a continuum, so that we come close to treating the resonance states as part of a complete set of eigenfunctions.

In this we should note some limitations: The completeness in terms of a specific contour applies only if we restrict the space of functions to be expanded to those decreasing fast enough at infinity. The most convenient contour $C_1$ (which presumably gives the least oscillatory integral term) goes only with functions decreasing faster than any exponential. If we want to be able to expand any function decreasing exponentially however slowly, we are forced back on the conventional contour $C_3$.

The other limitation is that the coefficients $a_n$ and $b(E)$ cannot be interpreted as probability amplitudes, since the sum of their square moduli does not add up to the norm of the expanded function. This is related to the occurrence in all these equations of the square of an eigenfunction, not its square modulus.

Nevertheless resonance eigenstates can be useful in dealing with practical problems, such as inelastic scattering of particles, resulting in a resonance state of the target. This was shown in the thesis of García.[7]

The idea of this treatment is best explained in terms of the simplest possible model. We assume a "nucleus" consisting of one particle, coordinate $\underline{r}$, in a potential $v(r)$ in the same units as above, initially in its ground state $\chi_0(r)$. The nucleus is hit by a projectile, radius vector $\underline{\xi}$, which interacts only with the $\underline{r}$ particle, by an interaction $w(\underline{r} - \underline{\xi})$. If w is regarded as small, and its square neglected, this amounts to first-order Born approximation. To first order we then have to solve the equation,

$$\{i\frac{\partial}{\partial t} - T_\xi - H_r\}\psi = \omega(\underline{r} - \underline{\xi})e^{i\underline{k}\cdot\underline{\xi}}\chi_0(\underline{r})e^{-i(E_0+k^2)t} \qquad (18)$$

where $T_\xi$ is the kinetic energy of the projectile, and $\underline{k}$ its initial wave vector. We want to know the transition cross section for a collision in which the projectile is emitted with wave vector $\underline{k}' = \underline{k} + \underline{q}$. Then we may project from (18) the Fourier component of wave vector $\underline{k}'$ in $\underline{r}$, and are left with a function $\phi(\underline{\xi},t)$ which satisfies the equation

$$\left(i\frac{\partial}{\partial t} - H_r\right)\phi(\underline{r},t) = W(\underline{q})\, e^{i\underline{q}\cdot\underline{r}}\chi_0(r)e^{-iE't} \equiv F(\underline{r},t) \qquad (19)$$

where $E' = E_0 + k^2 - k'^2$ is the final energy of the "nucleus".

We shall denote the right-hand side of (19), which acts as the "source" in our inhomogeneous equation, by $F(\underline{r},t)$. The left-hand side of (19) is the time-dependent equation for a Hamiltonian of the same type as that discussed above. The most obvious suggestion would be to use the time-dependent Green function, whose properties we know to solve (19) for $\phi$. However, the transition probability to all states in which the projectile ends up in the state $\underline{k}'$, which is the quantity of interest, is given by the norm of $\phi$, and this would mean computing the space integral of the square modulus of an expression of the form (15). This would be very impractical.

Instead we notice that, by the usual continuity argument, one can derive from (19) the relation

$$\frac{\partial}{\partial t} \int \phi^*\phi d^3\underline{r} = 2\mathrm{Im} \int \phi F^* d^3\underline{r} \qquad (20)$$

so that the norm of $\phi$, at time t after the beginning of the collision, is

$$2\mathrm{Im} \int_0^+ dt' \int d^3\underline{r} F^*(\underline{r},t')\phi(\underline{r},t') \qquad (21)$$

This is now a linear relation in $\phi$, in which it is convenient to substitute the expansion (15), based on a suitable contour. The resonance states now contribute clearly defined terms, each peaked at an energy appropriate to the nucleus being excited to an energy given by the real part of $E_n$, the width of the peak coming from the imaginary part. The integral term, as well as the distant terms in the sum contribute a background, which is, in principle, well defined, and which will vary slowly with energy. The contribution of each resonance state to the differential cross section is real, as it should be, but not necessarily positive. However, if a level is narrow, compared to its distance from adjacent resonances, its wave function is predominantly real, and its contribution manifestly positive. If the levels overlap, there is no reason why some contributions should not be negative, as long as the total cross section remains positive.

The approach which I have reviewed can easily be generalized to systems containing more than one particle. Apart from a more cumbersome notation, because of the greater number of variables, the main new problem is the presence of further branch points in the Green function, due to the existence of further thresholds for emission. These thresholds lie at real positive energies, if the zero of energy is still taken at the threshold for the emission of the most easily removed particle. The integration contour should not pass over these branch points unless we are willing to extend the definition of our function over the whole Riemann surface. One convenient choice is to add to the contour $C_1'$ of fig. 2, other lines running vertically down in the E plane, and if necessary link them by lines of constant Im $k$, as in $C_2'$.

In the preparation of the written version of this talk, I had the benefit of help from Dr. G. García-Calderon.

1. Ya.B. Zel'dovich, Zh.E.T.F. <u>39</u>, 776 (1960) Sov. Phys. – JETP <u>12</u>, 542 (1961).
2. T. Berggren, Nucl.Phys. <u>A 109</u>, 265 (1968); Phys.Lett. <u>33B</u>, 547 (1970).
3. R. Peierls, Proc.Camb.Phil.Soc. <u>44</u>, 242 (1948).
4. N.H. Hokkyo, Prog.Theo.Phys. <u>33</u>, 1116 (1965).
5. T.B. Scheffler, D.Phil. thesis, Univ. of Oxford, 1970.
6. R.G. Newton, Scattering of Waves and Particles, New York 1966, section 19.2.
7. G. García-Calderon, D.Phil thesis, Univ. of Oxford, 1973.

# PARITY VIOLATION IN ATOMS DUE TO NEUTRAL WEAK CURRENTS

Ernest M. Henley

Department of Physics, University of Washington

Seattle, Washington 98195 U.S.A.

## I.  INTRODUCTION

Mirror symmetry is one of several beautiful aspects of nature.  Its beauty is perhaps even enhanced by small, but noticeable imperfections.  Among the basic physical interactions, it is the weak interactions that are responsible for the imperfections which break mirror symmetry.

It is well known that atoms are dominated by the electromagnetic interactions.  But what if the weak and electromagnetic forces are but two different manifestations of the same basic force?  Although this suggestion has been made many times, it has been given a solid mathematical basis and received considerable impetus by the work of Weinberg[1,2] and of Salam,[3] and by some experimental confirmation of their ideas.  It follows from the Weinberg-Salam theories that observable atomic parity-violating (PV) effects should occur.  Although the older currents responsible for weak forces were always charged, the fast growing younger set include neutral currents as well.  It is the realization that neutral currents can give rise to coherent parity-violating effects in atoms which has caused a flurry of experimental activity.  The hunt for PV effects has been undertaken in laboratories around the world.

There are several features which make atomic tests particularly attractive.  To-date the evidence for neutral weak currents, comes mainly from high energy experiments in which neutrinos are present in both the initial and final states,[4]

$$\nu_\mu + e \rightarrow \nu_\mu + e$$

$$\nu_\mu + p \rightarrow \nu_\mu + p$$

$$\nu_\mu + p \rightarrow \nu + n + \pi^+ \quad .$$

Although these experiments demonstrate the existence, and give
a measure of the strength of the weak interactions due to neutral currents, they leave many questions unanswered:  Are neutral current effects also induced by electron neutrinos?  Are
these neutral currents the same as those induced by muons?  Is
there a neutral current weak interaction between charged leptons $(\mu^\pm, e^\pm)$ or between charged leptons and hadrons (e.g. nu-
cleons)?  Nor do the experiments carried out to-date give us
clear indications of the space-time properties of the weak currents.  Are they true vectors (V), like the electromagnetic
current, are they axial vectors (A), or hopefully for atomic
tests are they of mixed parity, e.g. mixed V and A.  If so, is
the coupling V-A as for the charged currents, or is it V+A?  Or
is it of a totally different form?  We are quite in the dark
as to the correct answers to most of these questions.

Thus, although the basic framework of the Weinberg theory
appears to be correct, considerable freedom remains in covering this frame.  Atomic physics experiments may reveal the ar-
chitectural details of the weak interactions, and it is this
aspect which gives these tests their importance.  To cite  but
two examples:  Experiments will certainly reveal whether the
weak force due to neutral currents violate mirror symmetry, and
if so tests on various atoms can also give us the isospin (N &
Z) dependence of the weak neutral currents.

## II.  BASIC FRAMEWORK

In analogy to the ordinary electromagnetic interaction
between charged particles, the weak interactions are assumed
to be of a current-current type, $[j^\mu = (j^0, \vec{j})\; j_\mu = (j^0, -\vec{j}); $ a
sum over double indices is implied]

$$H_{int} = \frac{g^2}{c^2} \int d^3x\, d^3x'\; j^\mu(x)\; f(x-x')\; j_\mu(x') \tag{1}$$

with

$$f_{em}(r) = 1/r \tag{1a}$$

and

$$f_{wk}(r) = \frac{1}{r} e^{-m_W cr/\hbar} \underset{\sim}{} 4\pi \left(\frac{\hbar}{m_W c}\right)^2 \delta^3(r) \quad , \tag{1b}$$

where $m_W$ is the mass of the "intermediate" boson and g is a coupling constant; $g_{em} = e$ for the electromagnetic force. The weak interaction can be written as

$$H_{wk} \underset{\sim}{} \frac{G}{\sqrt{2}} \int d^3x \; j^\mu(x) \; j_\mu(x) \tag{2}$$

with

$$G \underset{\sim}{} \sqrt{2} \; 4\pi \left(\frac{\hbar}{m_W c}\right)^2 g_{wk}^2 \quad .$$

The weak coupling constant $G = 1.43 \times 10^{-49}$ erg-cm$^3$ = 89.6 eV – fm$^3$. If $g_{wk} = e/\sqrt{2}$, one obtains $m_W c^2 \underset{\sim}{} 60$ GeV. It is difficult to compare the weak and electromagnetic forces directly because their form (e.g. range) is totally different. Nevertheless, in atomic physics the mass and lengths scales are determined by the mass of the electron $m_e$, and by $\alpha$. One thus finds that the ratio of the weak force to electromagnetic force effects is roughly

$$\frac{\langle H_{wk}\rangle}{\langle H_{em}\rangle} \underset{\sim}{} \frac{G\, m_e^2}{\alpha} \frac{c}{\hbar^3} \underset{\sim}{} 3 \times 10^{-10} \quad . \tag{3}$$

This is indeed a small, but, as we shall see meaningful number.

The weak current has both charged (V-A) and neutral parts. Neutral currents or neutral (heavy) boson exchange, shown in Fig. 1, leads to e-p scattering. That is, in addition to the

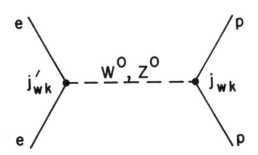

Fig. 1   Electron proton weak force

Coulomb and other electromagnetic forces between an electron
and proton, there is now a short range weak force.  The range,
of order $\hbar/m_W c < 10^{-15}$cm, is essentially zero for any experiment accessible to us today.  In scattering experiments of
electrons on protons or atoms, such a very short range force
contributes only at very high energies and high momentum transfers.  A high energy accelerator like that at Stanford (SLAC)
is required, and experiments which seek a small mirror asymmetry in electron-proton scattering have been proposed at this
installation.[5,6]  It is exactly as for the Rutherford alpha
scattering experiments on atoms, except for a scale factor.
On the other hand, electrons in stationary s-states in atoms
do get very close to the nucleus and one is able to use this
overlap to detect the presence of a weak electron-nucleon force.
The search is greatly facilitated by a lack of mirror symmetry
of the weak interactions.  The parity-conserving part of the
weak force cannot be detected readily.  In the Weinberg theory,[2]
the PV part is

$$H_{PV} = \frac{G}{2\sqrt{2}} \left[ \int \bar{\psi}_e \gamma^5 \gamma^\mu \psi_e \ \bar{\psi}_N \gamma_\mu (2 \sin^2\theta_W + \tau_3(2 \sin^2\theta_W - 1)) \psi_N d^3x \right.$$

$$(4)$$

$$\left. + \int \bar{\psi}_e \gamma^\mu \psi_e \ \bar{\psi}_N \gamma^5 \gamma_\mu (2 \sin^2\theta_W + \tau_3(2 \sin^2\theta_W - 1)) \psi_N d^3x \right] \quad ,$$

where $\sin^2\theta_W \approx 0.35$, and $\bar{\psi}$ and $\psi$ are Dirac creation and destruction operators.  If we neglect nuclear finite size and nu-
clear spin-dependent effects,[7] then one obtains an effective
single particle electron-nucleus PV potential[8]

$$V_{PV} = \frac{G}{4\sqrt{2}} \frac{Q}{m_e c} [\vec{\sigma}\cdot\vec{p} \ \delta^3(r) + \delta^3(r)\vec{\sigma}\cdot\vec{p}] \qquad (5)$$

where $\vec{\sigma}$ and $\vec{p}$ are electron spin and momentum operators, and
where for the Weinberg theory $Q = Q_W$,[2]

$$Q_W = -N + (1-4 \sin^2\theta_Q)Z$$

with N= neutron number and Z= atomic number.  There is also a
PV electron-electron force, but it is neglibile,[8] in great part
because the Coulomb repulsion keeps electrons apart.  In addition, because of the short-range of the weak electron-nucleon
force, it is effective primarily for electrons in s and p
states.  Such single electron states lose their unique parity
assignments for j = 1/2 only, because $V_{PV}$ is a scalar operator;

they become

$$|\bar{s}_{1/2}> = |s_{1/2}> + F|p'_{1/2}>$$

$$|\bar{p}_{1/2}> = |p_{1/2}> + F'|s'_{1/2}> \quad .$$

(6)

The impurity of opposite parity can be calculated by perturbation theory, since it is caused by a very weak force,

$$F|p'_{1/2}> = \sum_{n} |np_{1/2}> \frac{<np_{1/2}|V_{PV}|s_{1/2}>}{E_s - E_{np}} \quad .$$

(7)

Indeed, it is simple to make a rough estimate of $F$ or $F'$ for a medium to heavy atom . Such estimates have been made by the Bouchiats[8] in the quantum defect approximation,

$$<np_{1/2}|V_{PV}|s_{1/2}> \approx \frac{\mathcal{R}^*_{np}(0)}{r} \mathcal{R}_s(0) \frac{3iQ_W GK}{16\pi m_e c\sqrt{2}}$$

$$\approx \frac{i\hbar G\ Z^2 Q_W K}{\sqrt{2}\ 4\pi m_e c\ a_o^4 (\nu_{np}\nu_s)^{3/2}}$$

(8)

where $\mathcal{R}(0)$ is a radial wavefunction of the electron evaluated at the origin, K is a relativistic correction[8] (K $\approx$ 9 for Z = 82), and $\nu$ is an effective radial quantum number, given in terms of the quantum defect, $\mu$, by $n-\mu(\varepsilon_n)$; $a_o$ is the Bohr radius. There are clearly experimental advantages to the use of heavy atoms since $F \propto Z^3K$ . (We assume $Q_W \propto Z$. Experiments should be able to reveal the Z and N dependence of Q.)

$$|F| \sim 10^{-10} \text{ for } Z \approx 83 \quad ,$$

$$|F| \sim 10^{-18} \text{ for } Z \approx 1 \quad ,$$

unless nearly degenerate, e.g. 2s-2p, levels are used in which case it is

$$|F| \sim 10^{-11} \text{ for } Z = 1.$$

### III.  DETECTION OF ATOMIC PARITY VIOLATION

Since the PV mixing in atomic states is tiny, it is necessary to find experimental consequences in which the effects are enhanced and appear to lowest order.  Several general types of experiments are feasible:

1)  In electron elastic or inelastic scattering one searches for non-vanishing helicity effects.  Polarized incident electrons or targets are required and high momentum transfers are a great asset, e.g.

$$\sigma = \sigma_o + a <\vec{s}\cdot\hat{p}>$$

where $<\vec{\sigma}>$ is a measure of the electron polarization and $\hat{p}$ is the direction of the incident beam.

2)  In the capture of polarized electrons by ions, the search is for a correlation of the direction of photons, $\hat{k}$, emitted in resonant capture

$$R = R_o + a <\vec{s}\cdot\hat{k}> \qquad . \qquad\qquad (9)$$

3)  Measurement of the circular polarization $P_\gamma$ of radiation emitted from unpolarized excited atomic states

$$P_\gamma = a<\vec{S}_\gamma\cdot\hat{k}> \qquad\qquad (10)$$

is also possible.

4)  With the advent of tunable lasers it is feasible to resonantly excite atomic states.  In this case one can search for the dependence of the absorption on the polarization state (R or L) of the incident photons.  Moreover, one can seek an optical rotation   in non-optically active substances,[5] i.e. rotation of the plane of linear polarization.  The first of these is just the reverse of the emission experiment and the second one is a variation of it.  One advantage of the optical rotation  is that one can use coherence from a gaseous target to measure the effect over a distance of roughly one mean free path.  As is well known from classical electromagnetic theory, the dichroism is a measure of the difference of the real part of the index of refraction for right (R) and left (L) circularly polarized light of wave-number k,

$$\zeta = \frac{k}{2} Re(n_R - n_L) \qquad . \qquad\qquad (11)$$

In all of the above experiments, except for the first one, it is feasible to enhance the PV effect by seeking transitions in which the normal rate is strongly retarded.  For instance, the normal transition may be a retarded M1 matrix, so that the parity forbidden transition is of the normal E1 type.  The effect sought is then of order

$$a \approx 2 F \frac{<E1>}{<M1>} >> F \quad . \tag{12}$$

Specific laser-induced transitions in a variety of atoms have been suggested;[5,8,9,10] they are the $6s_{1/2} \to 7s_{1/2}$ transition in Cs, the $6p_{1/2} \to 6p_{3/2}$ transition in Tl, the $(6p^2)^3P_0 \to (6p^2)^1S_0$ transition in lead, various transitions in Bi; even hydrogen has been suggested as a possible target. For these atoms, the relative strength of the PV signal, a in Eq. (12), has been estimated;[5,8,9,10] it is $10^{-4}$ in Ce, and in the $2s_{1/2}-1s_{1/2}$ transition in hydrogen and about $10^{-6}$ in Tl, Pb, and Bi. The increased value of a in Cs and H is due to the high degree of forbiddeness for the normal (parity-allowed) M1 transition. This same feature therefore makes it more diffi- cult to observe these transitions. Before I discuss some of the experimental difficulties, let me describe how one computes the strength of the PV signal. I will use Bi as an example, because experiments and detailed calculations are being carried out here by N. Fortson[10] and L. Wilets and myself, respectively. It is a somewhat complicated example to use as illustration be- cause Bi has three valence electrons rather than one, but I will pass over these complications. The relevant level structure of Bi is shown in Fig. 2. The first excited state, a $^2D_{3/2}$ is res- onantly excited by a tunable laser. The transition can be M1 or E2, but is known to be primarily M1, and has been observed in Fortson's laboratory, who has measured the lifetime of the state, which is about 1/30 sec. The parity-violating potential mixes in primarily $(6p^2ns)_{3/2}$ electrons into both initial and final states, with mixing coefficients indicated by F and F' in Fig. 2. This small even parity component of the wavefunc- tions allows E1 matrix elements to contribute to the excitation of the $^2D_{3/2}$ states, as shown in the figure. The interference of the E1 and M1 matrix elements gives rise to a different rate of absorption for left and right circularly polarized light, to a circular polarization of the decay photon, and to optical rotation. It is this rotation of the plane of polarization which is being used by Forton in the laboratory.

If one considers only the mixing of the two bound J = 3/2 states, $6p^27s$, also shown in Fig. 2, then the expected optical rotation can readily be found. The mixing coefficient is found from Eq. (8) adapted to Bi to be $\sim 5 \times 10^{-11}$. To estimate the E1 matrix element between the $^2D_{3/2}$ and $^4S_{3/2}$ states, one can employ the measured decay of the $(6p^27s)^4P_{1/2}$ state at 4.04 eV to the ground ($^4S_{3/2}$) state. This rate is $1.7 \times 10^8$ sec$^{-1}$. When corrected for the difference in energy ($\sim k_\gamma^3$), one obtains

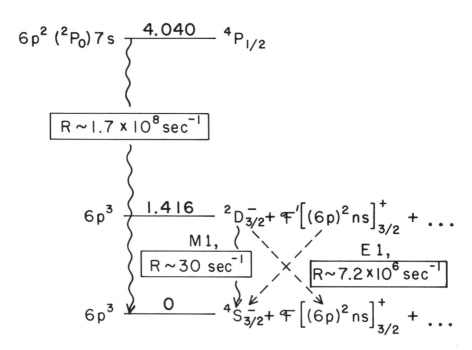

eV

$6p^2 (^3P_1) 8s$  //7.539//  $^4P_{3/2}$
                   //7.288//  Ionization

$6p^2 (^3P_2) 7s$  ___6.131___  $^2P_{3/2}$

$6p^2 (^3P_1) 7s$  ___5.562___  $^4P_{3/2}$

$6p^2 (^2P_0) 7s$  ___4.040___  $^4P_{1/2}$

$R \sim 1.7 \times 10^8 \, sec^{-1}$

$6p^3$  ___1.416___  $^2D_{3/2}^- + \mathcal{F}'[(6p)^2 ns]_{3/2}^+ + \ldots$

M1,
$R \sim 30 \, sec^{-1}$

E 1,
$R \sim 7.2 \times 10^6 \, sec^{-1}$

$6p^3$  ___0___  $^4S_{3/2}^- + \mathcal{F}[(6p)^2 ns]_{3/2}^+ + \ldots$

Fig. 2.  Level diagram for Bi.  Only pertinent levels approximately up to the ionization energy are shown.  Decay rates are
indicated by R; that for the PV transition is found by scaling
the $^4P_{1/2} \rightarrow {}^4S_{3/2}$ rate.

$R(^2D_{3/2} \rightarrow {}^4S_{1/2}) \sim 7.2 \times 10^6 sec^{-1}$. The circular polarization is thus found to be

$$P_\gamma \sim 2 \frac{<E1>}{<M1>} \quad F \sim 5 \times 10^{-8} \quad . \tag{13}$$

The optical rotation   in one mean free path is found to be the same, since the angle of rotation of the plane of polarization $\zeta$ and absorption coefficients $\alpha$ are given by

$$\zeta \stackrel{\sim}{\sim} 2N\ell \; F \; \frac{<M1> \; <E1>}{\hbar^2 c} \quad \frac{\omega - \omega_o}{(\omega - \omega_o)^2 + (\Gamma/2)^2} \tag{14a}$$

$$\alpha \stackrel{\sim}{\sim} 2N \; \frac{|<M1>|^2}{\hbar^2 c} \quad \frac{\Gamma/2}{(\omega - \omega_o)^2 + (\Gamma/2)^2} \tag{14b}$$

$$\zeta/\alpha\ell = \; F \; \frac{<E1>}{<M1>} \frac{\omega - \omega_o}{(\Gamma/2)} \tag{14c}$$

where N is the number density of atoms, $\omega_o$ and $\Gamma$ are the resonance frequency and width, respectively, and $\ell$ is the length of the gas traversed.

For the computation of  F, it is quite feasible to include higher excited states in the continuum, by means of Greens' function techniques,[8] and L. Wilets and I have set up such a calculation but do not yet have any numerical results to report.

The appartus used by Fortson and his co-workers is sketched in Fig. 3.  He uses a tunable laser and hand-chosen calcite prisms as polarizer and analyzer.  An a.c. driven Fraday cell is interposed between the Bi vapor cell and the polarizer in order to remove slow drift effects.  The polarized light passes through about 1 m of Bi vapor maintained at a pressure of a few Torr.  The beam is then split; it is the ratio of the beam which passes through the analyzer to that shunted off prior to it which is analyzed for a rotation of the plane of polarization.

To-date no rotation has been observed to a few parts in a million.  This achievement already constitutes an improvement of about two orders of magnitude over previously published results.[8]  The sensitivity of the apparatus should allow Fortson to detect PV effects down to $\sim 10^{-8}$.  At these levels instrumental uncertainties are crucial.  For instance, stray magnetic fields may induce an optical rotation.  However, it is possible to reverse the beam direction, to check the frequency response,

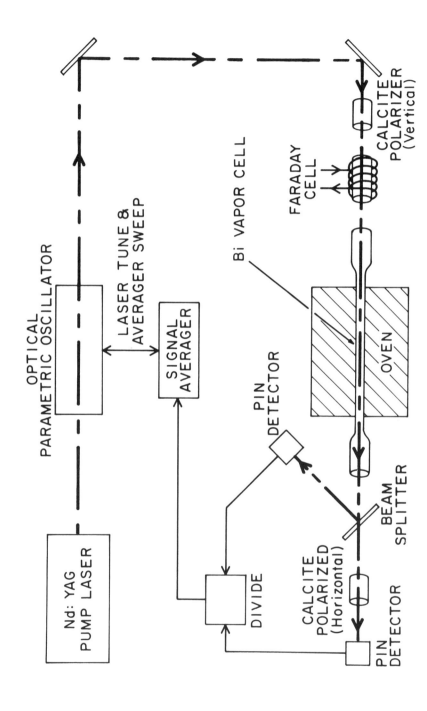

Fig. 3. Sketch of appartus used by N. Fortson to detect parity violation in Bi (courtesy N. Fortson).

and to carry out other tests to make certain that any observed effect is a genuine rather than a systematic one.

Although atomic experiments sensitive to a small parity violation are clearly difficult ones, the knowledge which can be gleaned from them is both important and basic. It is the fact that the results may help unravel details of the weak interactions which makes the hunt for such tiny effects exciting and worthwhile.

The author is grateful to Professors N. Fortson and L. Wilets for numerous helpful discussions and to Professor Fortson for making available details and results of his experiment prior to publication.

## REFERENCES

1) S. Weinberg, Phys. Rev. Letters 19, 1264 (1967); Phys. Rev. Letters 27, 1688 (1971).

2) S. Weinberg, Phys. Rev. D5, 1412 (1972).

3) A. Salam, Elementary Particle Physics, N. Svartholm, ed. (Almqvist and Wiksell, Stockholm, 1968) p. 367.

4) See e.g. D.C. Cundy, XVII[th] Intern. Confer. on High Energy Physics, London, 1974 (Sci. Res. Council, Rutherford Lab., 1974) p. IV-131.

5) Ya. B. Zel'dovich, Zh. Eksperim. i Teor. Fiz. 36, 964 (1959) [Transl. Soviet Phys. JETP 36, 682 (1959)]; Ya. B. Zel'dovich and A.M. Perelomov, Zh. Eksperim i Teor. Fiz 39, 1115 (1960) [Transl. Soviet Phys. JETP 12, 777 (1961)]; F.C. Michel, Phys. Rev. 138, B408 (1965).

6) E.M. Henley, A.H. Huffman, and D.U.L. Yu, Phys. Rev. D7, 943 (1973); L. Wolfenstein, Nucl. Phys. B72, 111 (1974).

7) For other terms, see e.g. G. Feinberg and M.Y. Chen, Phys. Rev. D10, 190 (1974).

8) See e.g. M.A. Bouchiat and C. Bouchiat, Phys. Lett. 48B, 111 (1974); J. de Phys. 35, 899 (1974) and (to be published).

9) A.N. Moskalev, Zh. Eksperim. i Teor. Fiz.Pis. Red. 20, 229 (1974) [Transl. JETP Lett. 19 141 (1974)]; I.B. Khriplovich, Zh. Eksperim. i Teor. Fiz. Pis. Red. 20, 686 (1974) [Transl. Sov. Phys. JETP Lett. 20, 315 (1974).

10) E.N. Fortson (private communication); D.C. Soreide and E.N. Fortson, Bull. Am. Phys. Soc. 20, 491 (1975).

# INTERSTELLAR MOLECULE FORMATION

Eric Herbst
Department of Chemistry, William & Mary College
Williamsburg, Virginia
William Klemperer
Dept. of Chemistry, Harvard Univ., Cambridge, MA

The universe is $10^{10}$ years old; thus has a radius of $10^{10}$ light years or $10^{28}$ cm. The matter of the universe is very non uniformly distributed. It appears that much of the matter in the universe exists in galaxies which occupy $10^{-7}$ of the volume in the universe. We shall be concerned here with the diffuse low density matter which exists presumably in most, if not all, galaxies.[1] This is most easily studied in our own galaxy, although most of the discussion is likely to be relevant to most galaxies.

It is generally assumed that the distribution of the

TABLE I

| | |
|---|---|
| H | 1 |
| He | .15 |
| C | $3 \times 10^{-4}$ |
| N | $9 \times 10^{-5}$ |
| O | $7 \times 10^{-4}$ |
| Si | $3 \times 10^{-5}$ |
| S | $1.6 \times 10^{-5}$ |

Cosmic Abundance of the Major Elements, Relative Number Abundance

## TABLE II

| | | | | |
|---|---|---|---|---|
| $H_2$ | $H_2O$ | $HCO^+$ | $NH_3$ | $CH_3OH$ |
| CH | HCN | $HN_2^+$ | $H_2CO$ | $CH_3CN$ |
| $CH^+$ | $H_2S$ | HNC | HNCO | $HCONH_2$ |
| OH | $SO_2$ | CCH | $H_2CS$ | $CH_3OCOH$ |
| CN | OCS | | $H_2CNH$ | $CH_3CH_2$ |
| CO | | | HCOOH | $CH_3COH$ |
| CS | | | HCCCN | $H_2CCHCN$ |
| SiO | | | | $(CH_3)_2O$ |
| SiS | | | | $CH_3CH_2OH$ |
| SO | | | | |

Observed Interstellar Molecules

elements is relatively constant throughout the galaxy and also throughout the universe. A short table of cosmic abundances are given in Table I.[2] It is clear that hydrogen will occupy a special place in discussions of the chemistry of the interstellar matter in view of its abundance.

The average density of matter in the galaxy is $5 \times 10^{-24}/cm^3$. Approximately 50% of the matter is interstellar. The interstellar medium is highly heterogeneous. If it simply reflected the average density, the chemistry would be exceedingly simple, namely, only atomic species would exist. Table II lists the molecular species observed in the interstellar medium[3] at this writing. The list shows a richness unexpected a decade ago.

The vast majority of the molecular species listed in Table II have been observed by radio astronomy. The primary reason for this is that a large column number (greater than $10^{14}$) of a molecule (other than the most abundant species) only occurs in the dense interstellar clouds. These dense clouds are opaque to visible light; thus stellar radiation cannot be used as a source for absorption spectroscopy. The usual requirement for a molecule to show a radio or microwave frequency spectrum of observable intensity is the existence of an electric dipole moment. The list of Table II is therefore

biased.  Species such as $N_2$, $CO2$, $CH4$ (methane) HCCH (acetylene)
are non polar; the question of their abundance in the interstellar medium cannot be <u>directly</u> answered by radio observation.

Secondly, a number of species are listed for which laboratory spectra do not exist.  Thus, the assignment of microwave
transitions is not direct and must be discussed individually.
A number of these species are, we believe, of critical
importance in deducing adequate models of the interstellar
medium.  That these species are not readily observed under
laboratory conditions, reflects the rather unusual conditions
that obtain in the interstellar medium.

Our concern here is in providing a model for the formation
of molecules in a low density, $10-10^6$ atoms/$cm^3$, low temperature,
$30°K$, medium.[4]  The necessity for forming molecules efficiently
will become clear when their relatively short ($10^2 - 10^5$ years)
lifetime is discussed.

That molecules must be formed relatively efficiently is
required since many observed species have lifetimes of $10^1$ to
$10^5$ years under a wide variety of intersteller conditions.

Typical lifetime estimates are as follows:  CH, in the
cloud in front of $\xi$ OPH, $5 \times 10^2$ years.[5]  The lifetime of CH
is determined both by photodissociation and photoionization by
the general galactic radiation field and by chemical reaction
of $C^+$, i.e., $C^+ + CH = C_2^+ + H$.  The rate of these processes
are comparable.

For a species which is destroyed by electrons such as
$HCO^+$, the lifetime is of the order of $[10^{-6} \times 10^{-3}]^{-1} = 10^9$
sec = 30 years.[6]  A species such as $HN_2^+$ which is destroyed by
CO, the lifetime is $[10^{-9} \times 5 \times 10^1]^{-1} = 2 \times 10^7$ sec.  A long
lived species destroyed by $C^+$ attach will have a lifetime of
$[10^{-9} \times 10^{-3}]^{-1} = 10^{12}$ sec.

Although different species can be destroyed by a variety
of processes, it is difficult to imagine that species will not
be destroyed continually in all types of interstellar regions.
Possible exceptions are CO and $H_2$ whose lifetime can approach
$10^{16}$ sec = $3 \times 10^7$ years, a time long compared to cloud lifetimes.[6]

In discussing the chemistry of the interstellar medium,
we first discuss hydrogen.  Hydrogen is by far the most
abundant reactive element; thus to a considerable extent the
nature of chemical processes will depend upon the condition of
hydrogen $H^+$ ($H^{II}$), H ($H^I$) and $H_2$.  $H^+$, ionized hydrogen, can
occur only as an appreciable fraction in a region near a hot

(O type) star. The general galactic radiation field is too
dilute (the sum of all stellar radiation) to support the
ionization of H; therefore radiation of wavelength less than
912 Å (13.595 ev) is virtually absent.[7]  The chemistry of an
$H^{II}$ region appears to be limited to atomic species.  Collisional
processes of interest appear to be those affecting ionization
equilibrium of the atomic species, such as charge exchange.

The boundary between an ionized hydrogen region ($H^{II}$
region) and an $H^{I}$ region is sharp - being of the order of
terrestrial dimensions.  Thus, it will not be necessary in
discussing the chemistry of an $H^{I}$ region to be concerned with
appreciable amounts of $H^{+}$.

The boundary between an $H^{I}$ and an $H_2$ region is also sharp.
The reaction forming $H_2$ from H atoms is recombination on grain
surfaces.[8-10]

$$2H = H_2$$

This reaction is known to occur facilely on many surfaces.[11]
The low polarizability and obvious closed shell character of
$H_2$ together with the high mobility of H at low temperatures
causes surface recombination at the low interstellar temperatures to be orders of magnitude more efficient in forming $H_2$
than any other process.

The dissociation of $H_2$ by optical radiation is uniquely
interesting.  The first allowed optical transition of $H_2$, the
$X\,'\Sigma_g^+ \rightarrow B\,'\Sigma_u^+$ system has a dissociation limit of $D_o(H_2) + E(H2s)$
= 14.67 ev.  Thus $H_2$ cannot be directly photodissociated by
the radiation field present in the intersteller medium.  The
photodissociation process instead consists of a two step
process:  1) discrete $X \rightarrow B$ bound-bound absorption followed by
2) bound-free emission $B \rightarrow X$.  In this two-step process $H_2$ is
thus photodissociated by line radiation.  The consequence is
that a molecular $H_2$ region is self shielding.  The boundary
between H and $H_2$ is relatively sharp.

We discuss chemical processes in three types of interstellar clouds.
   1. An $H^{I}$ region
   2. A molecular $H_2$ region in which starlight penetrates
   3. A dense region $H_2 \sim 10^4 - 10^6/cm^3$, which is opaque
      to starlight.

For the formation of all species, other than $H_2$, we
consider only gas phase two-body processes which can occur
spontaneously in a very low temperature medium.  Our reason
for doing so is the following:  1) the rate constant for all

of these processes are obtainable either from direct laboratory observation or feasible theoretical calculations; 2) the rates of the reactions are fast compared to surface processes. It appears likely that while material may both accrete on grains and be evaporated from grains, it will be processed by gas phase reactions. Thus, it is unlikely that any species formed on grains (other than $H_2$) can persist without change in the gas; 3) the composition and surface character of the grains is unknown. Surface reactions at the low grain temperatures on an unknown surface material appear to be totally unknown; 4) since gas phase reactions occur rapidly, a model of the interstellar chemistry involving surface reactions must also include virtually all of the gas phase reactions. It appears to us most profitable at this time, therefore, to develop a complete purely gas phase model which may then be compared with the astronomical observations. Discrepancies between model predictions and observations may then serve to indicate <u>specific</u> non-gas phase processes.[12]

The temperatures of the clouds vary from $10°$ to $100°$K. We choose a canonical temperature of $30°$K. The reactions that are possible must be exothermic and also require no activation energy. It is not anticipated that metastable species can play a significant role since in general the time between collisions permits most ordinarily forbidden radiative processes to occur.

The first two regions are generally observed simultaneously in stellar spectra. The column density of molecular hydrogen and atomic hydrogen in a number of interstellar clouds has been determined by the far ultraviolet observations of the Princeton copernicus satellite. The results are in good agreement with previously discussed model of surface recombination of H atom and photodissociation of $H_2$.[15]

The molecules observed in these clouds are $CH^+$, CH, CN and CO. OH has tentatively been observed. We now consider the gas phase synthesis of these species. The major form of carbon, nitrogen and oxygen is $C^+$, N and O. The species CN and CO are formed by the highly exothermic reactions

$$CH^+ + N = CN + H^+ \qquad k = 10^{-9} \ sec^{-1}cm^{-3}$$

$$and \ CH + N = CN + H \qquad k = 4 \times 10^{-11}$$

$$CH + O = CO + H$$

(The reaction scheme can be more complex; for example,

$$CH + O = HCO^+ + e$$

$$HCO^+ + e = CO + H \; ).$$

The species $CH^+$ and $CH$ are the fundamental species. Since the dissociation energy of both of these species is less than that of $H_2$ they cannot be found by elementary exchange reaction with $H_2$.

$CH^+$ is formed by radiative association of $C^+ + H$. Solomon and Klemperer have calculated the rate constant of this reaction to be $7 \times 10^{-17}$. Since the reactions,

$$CH^+ + H_2 = CH_2^+ + H$$

$$CH_2^+ + H_2 = CH_3^+ + H$$

occur at the Langevin rate ($k=2 \times 10^{-9}$), it is apparent that $CH^+$ cannot coexist with $H_2$. Thus in the diffuse clouds containing both H and $H_2$ regions, $CH^+$ occurs in the atomic hydrogen region. Physically this is because the outer region exposed to galactic radiation field.[16]

The formation of CH is most likely to occur by the mechanism suggested by Black and Dalgarno,[17] namely,

$$C^+ + H_2 = CH_2^+ + h\nu$$

(Radiative association)

$$CH_2^+ + H_2 = CH_3^+ + H$$

$$CH_3^+ + e = CH + H_2 \; .$$

Since the electron density is equal to or slightly greater than the total carbon density in the diffuse clouds, the rate of formation of CH is determined entirely by the radiative association of $C^+$ and $H_2$. The rate constant of this most important reaction is not known, but is likely to be $10^{-16 \pm 1}$.

The destruction of the species is both by chemical attack and photodissociation and photoionization. In particular, $CH^+$ is primarily destroyed by reaction with O and N while CH is both photodissociated photoionized and destroyed by $C^+$ (and slightly by O and N).

The diffuse clouds have shown no polyatomic molecules thus far. Thus, if molecular complexity is taken to be the criterion of interest, it is the dark clouds which must be considered.

The condition of the dark clouds is no optical radiation and, of course, $H_2$ as the only appreciable form of hydrogen. The temperature is low.  The observed existence of the species CH, $HCO^+$, $C_2H$, $HN_2^+$, HNC, clearly indicates that these regions are not at chemical equilibrium.  Thus, a source of energy is required to maintain disequilibrium.

The flux of high energy cosmic rays (with energy above 100 Mev) is assumed to be roughly constant throughout the galaxy.  The ionization produced by the flux is about $10^{-17}$ $n_H$. Since even the dense clouds, with perhaps the exception of the galactic center, have insufficient column density of matter to stop these particles, ionization occurs throughout the cloud.

The chemistry of the dense clouds is then primarily coupled ion molecule reactions together with electron ion dissociative attachment.[18]

The primary ionization processes are,

$$H_2 + \gamma = \begin{array}{l} H_2^+ + e \qquad .95 \\ H^+ + H + e \quad .05 \end{array} \qquad \xi = 10^{-17}$$

$$He + \gamma = He^+ + e \qquad \xi = 10^{-17} \ .$$

Only these are considered as primary events since all other species are far less abundant than $H_2$ and He.  The rate constant for ionization $10^{-17}$ appears to be relatively well established for high energy cosmic rays.  The flux of cosmic rays with lower energy is not well known.  For dense clouds this is not vitally important since low energy cosmic rays will not penetrate the dense clouds.

The immediate secondary reactions are,

$$H_2^+ + H_2 = H_3^+ + H \qquad k = 2.1 \times 10^{-9}$$

$$He^+ + H_2 = \begin{array}{l} H_2^+ + He \\ H + H^+ + He \end{array} \qquad k = 1.1 \times 10^{-13} \ (300°K)$$

The rate constant for $H_3^+$ formation is Langevin.  It is well studied.  The reaction of $He^+ + H_2$ has been studied.[19]  The value of $k = 1.1 \times 10^{-13}$ is at 300°K.  In view of the observations that this reaction has increasing rate at higher $He^+$ energies, the value of $1.1 \times 10^{-13}$ must be an upper limit at the interstellar cloud temperatures of 30°K.

$He^+$ is a quite energetic, reactive ion.  The reactions,

$$He^+ + CO = C^+ + O + He \qquad k = 2.0 \times 10^{-9}$$

$$He^+ + N_2 = N^+ + N + He \qquad k = 1.2 \times 10^{-9}$$

$$N_2^+ + He$$

have been well studied. The large abundance of He together
with the lack of reactivity of $He^+$ with $H_2$ then provides an
efficient mechanism for producing $C^+$, $N^+$ and perhaps $O^+$. It
is especially the reaction,

$$He^+ + CO = C^+ + O + He$$

which in our view is responsible for the large number of
organic (carbon containing) molecules listed in Table II. We
note at this point that carbon has an ionization potential of
11.25 volts, and that $C^+$ does not react (rapidly) with $H_2$ or
CO. Thus it is relatively stable in all interstellar regions.
We shall discuss reactions of $C^+$ below.

The immediate reactions of $H_3^+$ with components of the
interstellar clouds are of extreme interest. In the dense
clouds the principal form of carbon is CO. It is likely that
the principal form of nitrogen is $N_2$. The reactions of $H_3^+$
with CO and $N_2$ are well studied and proceed at the Langevin
rate.

$$CO + H_3^+ = HCO^+ + H_2$$

$$N_2 + H_3^+ = HN_2^+ + H_2$$

The ion $HCO^+$ is quite stable and has high proton affinity.
The reaction,

$$HN_2^+ + CO = HCO^+ + N_2 \qquad k = 1.5 \times 10^{-9}$$

is responsible for the destruction of $HN_2^+$. Both species are
destroyed by dissociative electron attachment with rate
constants near $10^{-6}$. The electron density, nearly equal to
the $C^+$ density, is about $10^{-3}/cm^3$. The observation of both of
these species is in good agreement with calculated abundance
and is perhaps one of the most direct demonstrations of the
importance of gas phase ion-molecule reactions in the chemistry
of the interstellar medium.

In 1970 Snyder & Buhl observed a line at 89.1 GHz.[20] The
unknown carrier of the line was designated xogen. The line
is a single narrow line. It was suggested that the carrier of
89.1 GHz was $HCO^+$. As such, no hyperfine structure should
exist. Proof of this assignment was obtained only this year.

Snyder, Hollis, Ulich, Lovas and Buhl observed a line at 86754
MHz, close to the expected $H^{13}CO^+$ transition frequency in a
number of xogen sources.[22]  Even more recently, C. Woods
observed in a laboratory discharge through $H_2$ + CO an absorption
line at 89188.5 MHz.  This latter observation is of extreme
importance since it is, to the best of our knowledge, the
first laboratory observation of the microwave spectrum of a
polyatomic ion.

The ion $HN_2^+$, isoelectronic with $HCO^+$ (and also HCN) was
observed in the interstellar medium by the J=1 → 0 transition
at 93.174 GHz.[24]  The line shows hyperfine structure.  The
assignment of this line to $HN_2^+$ is discussed fully by Green,
Montgomery and Thaddeus,[25] who made an accurate self-consistent
field electron structure calculation for the geometry
(rotational constant) and the two nitrogen quadrupole coupling
constants.  By observing this line in an extremely non turbulent
interstellar source, OMC-2, Thaddeus and Turner were able to
partially resolve the smaller nitrogen quadrupole hyperfine
structure.

We take these two discoveries of polyatomic ions as the
most direct demonstration of the importance of ion molecule
reactions for the chemical behavior of the interstellar medium.
The observation of $HCO^+$ is expected from the large abundance
of CO.  The abundance of $HN_2^+$ is in good agreement with the
existence of a substantial fraction of nitrogen in the form of
$N_2$.  Since $N_2$ has not yet been directly observed, the observation of $HN_2^+$ is of importance in providing indirect evidence
for the existence of $N_2$ in the dark interstellar clouds.

The vast majority of the observed interstellar species
are neutral molecules, frequently stable, well studied species.
Their synthesis via a series of ion molecule reactions together
with one or more ion-electron dissociative recombination
reactions is now considered.

It is apparent that a very large number of ion molecule
reactions must be considered.  Of particular importance are
the following general types.

1) $A^+ + H_2 = AH^+ + H$

where $A^+$ is an arbitrary ion.  If this reaction proceeds at a
rate near $10^{-9}$, then $A^+$ is efficiently converted to $AH^+$ (which
may add further H atoms by continuation of this process.)
Known laboratory examples are,

$$CH^+ + H_2 = CH_2^+ + H$$

$$CH_2^+ + H_2 = CH_3^+ + H$$

$$N^+ + H_2 = NH^+ + H$$

$$NH^+ + H_2 = NH_2^+ + H$$

$$NH_2^+ + H_2 = NH_3^+ + H$$

and many more. The large abundance of $H_2$ makes it essential to know the rate constants of all exothermic ion - $H_2$ hydrogen abstraction reactions.

2) $CO + AH^+ = ACO^+ + H$

This type of reaction increases the heavy atoms in a molecular ion. Since CO is abundant it is an efficient method of producing larger molecules.

3) $C^+ + AH = AC^+ + H$

$C^+$ is expected to be the principal ion of the interstellar medium since it is stable with respect to $H_2$, CO and $N_2$. The recombination with electrons is a slow, radiative process.

4) Radiative association

In general, radiative association of an ion with $H_2$ is the most important reaction of this type. We have mentioned previously the importance of the reaction

$$C^+ + H_2 = CH_2 + h\nu \ .$$

It is known from scattering of $C^+$ and $H_2$ that a long lived complex is formed by these species. The rate constant for this radiative association reaction is not known. The electronic spectrum of $CH_2^+$ has not been observed. Thus questions such as the oscillator strength of free-bound transitions which directly determine the radiative association rate is unknown. A satisfactory completely theoretical estimation of the rate constant for radiative associative formation of a polyatomic molecule is extremely difficult presently. The reason for this is the low rate constant or equivalently small cross section. Basically, the radiative association cross section is of the order of magnitude of the ion-molecule collision frequency times the ratio of collision complex lifetime to the radiative lifetime of the emitting state of the complex. The latter ratio is of the order of $10^{-6}$ for an allowed electronic emission in a triatomic complex. Thus, all entrances into the emitting state with entrance probability of the order of $10^{-6}$ must be considered.

The direct laboratory measurement of the radiative association reaction rate constant or cross section, applicable for well-aged $C^+$ (i.e., $^2P_{1/2}C^+$) at $30°K$ appears to be a formidable task. It appears likely that a reliable estimate for the rate constant of this process will be obtained by combining a detailed study of the optical spectrum of $CH_2^+$ together with semi-empirical calculations. Radiative association of any ion $A^+$ with $H_2$ is generally of importance if the rate constant for $A^+ + H_2 = AH_2^+ + h\nu$ is near $10^{-16}$. The exceptions occur if the reaction

$$A^+ + CO = BCO^+ + H \quad (\text{or } H_2)$$

$$\text{or } A^+ + CO = HCO^+ + B$$

occurs (with Langevin rate). While radiative association cannot in general deplete an ion, if it occurs it leads to formation of larger species. For example, Herbst and Klemperer estimated that for

$$HCO^+ + H_2 = H_3CO^+ \quad k = 10^{-17}.$$

This step is followed by

$$H_3CO^+ + e = H_2CO + H.$$

This sequence of reactions would be typical for systems in which radiative association could occur.

5) Electron-Ion dissociative recombination

In general the rate constant for this type of process is $10^{-6} - 10^{-7}$. Of most interest is the branching ratio of neutral product species. For small polyatomic ions, $AH_2^+$, it is likely that

$$e + AH_2^+ = AH + H$$

$$A + H_2 .$$

In general <u>photodissociation</u> of a small neutral species

$$AH_2 \xrightarrow{h\nu} AH + H$$

$$\text{or } A + H_2 .$$

It is likely therefore that this will occur in dissociative

recombination. For larger species, i.e., those with several heavy atoms, it is not clear how large a polyatomic neutral species may occur in dissociative recombination.

Likely examples of this process are,

$$HCN + C^+ = CCN^+$$

$$CCN^+ + H_2 = HCCN^+ + H$$

$$HCCN^+ + H_2 = H_2CCN^+ + H$$

$$H_2CCN^+ + H_2 = H_3CCN^+ + H$$

$$H_3CCN^+ + H_2 = H_4CCN^+ + H$$

$$H_4CCN^+ + e = H_3CCN + H$$

and

$$H_3CCN + C^+ = H_2CCCN^+$$

$$H_2CCCN^+ + e = HCCCN + H$$

We note that the species $H_2CNH$, $HCCCN$, $CH_3CN$, $CH_3NH_2$, $H_2CCH$ $CN$ have been observed in dense clouds.

We have listed only one likely route to their synthesis. Other routes utilizing $CH_3^+$ in one case and $NH_3^+$ in another are also possible.

Summary

Our present model for the specific synthesis of interstellar molecules may be summarized as cosmic ray ionization of $H_2$ and He followed by a variety of ion-molecule reactions. Neutral molecular species are in general formed from ions by dissociative recombination of ions and electrons. These reactions are entirely gas-phase two body processes. Several specific exceptions do exist. Hydrogen atom recombination to form molecular hydrogen must occur on grain surfaces. It is the one instance where a specific surface reaction occurs. There are probably some neutral-neutral reactions of interstellar consequence. These are reactions with no activation energy. An example is $N + NO = N_2 + O$.

The reason for the preponderance of carbon compounds is the stability of $C^+$ due to the low ionization potential of C. The interesting question of how large organic molecules can be produced by $C^+$ or $CH_3^+$ reaction with organic molecules or CO reaction with organic molecular ions is presently unanswered. It certainly merits further laboratory study. It is likely that dissociative electron ion recombination will also place limits to the size of molecule that can be formed.

References

1.  An excellent concise survey of the interstellar medium is L. Spitzer, Jr., Diffuse Matter in Space, Interscience (John Wiley, 1968).

2.  This list is an abridgement from C.W. Allen, Astrophys. Quantities, 2nd ed., Univ. London, 1962.

3.  P. Solomon, Phys. Today 26, 32 (1973). The list given in the text is from L. Snyder.

4.  A detailed survey and discussion of interstellar processes relevant to and including molecule formation is given by W.D. Watson, Lectures at the 1974 Les Houches Summer School for Theoretical Physics, Les Houches, France.

5.  W. Klemperer in Highlights of Astronomy, DeJager (Ed.), P. 421 (1971). P. Solomon and W. Klemperer, Ap. J. 178, 389 (1972).

6.  E. Herbst and W. Klemperer, Ap. J. 185, 505 (1973).

7.  H.J. Habing, B.A.N. 19, 421 (1968).

8.  T.P. Stecher and D.A. Williams, Ap. J. (Lett.) 149, L29 (1967).

9.  D.J. Hollenbach, M.W. Werner, E.E. Salpeter, Ap. J. 163, 165 (1971).

10.  M. Jura, Ap. J. 191, 375 (1974).

11.  T.J. Lee, Nature 237, 99 (1972).

12.  The reader is cautioned that this view is by no means universal. Other models of molecule formation are given in Refs. 13 and 14.

13.  P. Aannestad, Ap. J. (Suppl.) 25, 205 (1973).

14.  W.D. Watson and E.E. Salpeter, Ap. J. 174, 321 (1972); 175, 659 (1972).

15.  D.C. Morton, Ap. J. 197, 85 (1975).

16. This is discussed by W.D. Watson, Ref. 4.

17. J.H. Black and A. Dalgarno, Ap. Lett. 15, 79 (1973).

18. This discussion is taken from E. Herbst and W. Klemperer, Ref. 6.

19. R. Johnson and M.A. Biondi, J. Chem. Phys. 61, 2112 (1974).

20. L.E. Snyder and D. Buhl, Nature 227, 862 (1970).

21. W. Klemperer, Nature 227, 1230 (1970).

22. L.E. Snyder, J.M. Hollis, B.L. Ulich, F.J. Lovas, D. Buhl, Bull. A.A.S. (1975) (To be published, Ap. J.).

23. R.C. Woods, T.A. Dixon, R.J. Saykally, P.G. Szanto, (To be published).

24. B.E. Turner, Ap. J. (Lett.) 193, L83 (1974).

25. S. Green, J.A. Montgomery, Jr., P. Thaddeus, Ap. J. (Lett.) L89 (1974).

26. P. Thaddeus and B.E. Turner, Ap. J. (To be published).

# ELECTRON-ATOM SCATTERING

# ELASTIC AND TOTAL SCATTERING OF ELECTRONS BY NOBLE GASES

F.J. de Heer

F.O.M. Instituut voor Atoom- en Molecuulfysica,

Kruislaan 407, Amsterdam, The Netherlands

## 1. INTRODUCTION AND DISPERSION RELATIONS

Recently there has been a kind of renaissance in the field of elastic and total scattering of electrons, both experimentally and theoretically. Experimentally, because data were only available in a limited energy and angular range, theoretically, because new methods have been introduced, which for some part were already known in nuclear physics. We shall in this connection refer to the Glauber theory, the eikonal-Born-series, the optical model and the use of second order potentials. From an experimental and fundamental point of view the problem has become even more interesting because of the study of the forward dispersion relation. For noble gases, where there is no bound state for the negative ion below the elastic threshold, the forward dispersion relation reads (see e.g. Bransden and McDowell [1]) and Byron and Joachain [2]):

$$\mathrm{Re} f(E,o) = f_B(E,o) - g_B(E,o) + \frac{1}{\pi} P \int_o^\infty \frac{\mathrm{Im} f(E',o)}{E'-E} \, dE', \qquad (1)$$

where $f(E,o)$ is the forward elastic scattering amplitude at electron impact energy $E$ and angle $o$, and where $f_B(E,o)$ and $g_B(E,o)$ are the first Born contributions to the direct and exchange amplitudes respectively. $f_B(E,o)$ is independent of $E$. Although in the non-relativistic case a rigorous proof has not been given, it is generally assumed that the forward dispersion relations are valid for elastic scattering from a composite system with Coulomb forces (see Gerjuoy [3]) and Gerjuoy and Krall [4])). The imaginary part from the forward amplitude follows from the optical theorem:

$$\text{Imf}(E,o) = \frac{k}{4\pi} \sigma_t(E), \tag{2}$$

where k is the wave number of the incident electrons and $\sigma_t(E)$ is the cross section for total scattering of electrons. Different direct measurements on total scattering of electrons, using a kind of beam attenuation method, have been carried out for several noble gases. At low energies, where the inlastic processes do not yet occur, some results have been summarized by Andrick [5]). Normand [6]) has done measurements between 0.5 and 400 eV and his data have been used in the dispersion relation analysis of Bransden and McDowell [7]). However, indications were present that Normand's data could contain serious errors (see also Inokuti and McDowell [8])). Recently de Heer and Jansen [9,10,11]) succeeded in getting fairly accurate ( $\lesssim$ 10 per cent) cross sections for total scattering by adding experimental total cross sections for ionization, excitation and elastic scattering. These calculations have been carried out between zero and 3000 eV and will be discussed in section 3. The data can be used for evaluation of Imf(E,o) by means of (2). The differential elastic cross section in the forward direction is given by

$$\sigma_{el}(E,o) = \{\text{Ref}(E,o)\}^2 + \{\text{Imf}(E,o)\}^2. \tag{3}$$

Measurements of $\sigma_{el}(E,o)$ do not exist, but it is possible to extrapolate experimental cross sections of $\sigma_{el}(E,\theta)$ at small angles to angle zero. It has been found empirically, see Bromberg[12,13]), Jansen et al.[11,14]) and Huo [15]), that in many cases near zero degree the differential elastic cross section is an exponential function of scattering angle:

$$\sigma_{el}(E,\theta) = \sigma_{el}(E,o) \exp(-\beta\theta), \tag{4}$$

where $\sigma_{el}(E,o)$ and $\beta$ are connected with the polarizability of the target. Applying (3) and using the evaluated $\sigma_{el}(E,o)$ and Imf(E,o) values one can determine also the real part of the forward amplitude. The quantities thus obtained can be compared with the corresponding theoretical ones and at the same time the dispersion relation (1) can be numerically tested.

    At such low energies, $\lesssim$ 15 eV, where no inelastic channel is open, one can apply a phase shift analysis to electron-atom scattering. Such an analysis has been made by Bransden and McDowell [1]) and by Andrick and Bitsch [16]) for e-He. In the theory of partial waves the scattering amplitude can be presented by

$$f(E,\theta) = -\frac{i}{2k} \sum_{\ell=o}^{\infty} (2\ell + 1)\{\exp(2i\delta_\ell) - 1\} P_\ell(\cos\theta), \tag{5}$$

where $P_\ell$ are the Legendre polynomials and $\delta_\ell$ are the scattering phase shifts, which are real for pure elastic scattering.

On the basis of a minimum difference criterion, Bransden and
McDowell parameterized their phase shifts, comparing calculated and experimental values of some observables, namely total
elastic cross sections, diffusion cross sections and elastic
differential cross sections. At low energy scattering the
first few partial waves will dominate. The higher partial
waves can be considered in a simple approximation [16,17] and
are predominated by the long range part of the electron-atom
interaction, presented by the dipole polarization potential.
If the phase shifts are small they can be computed with the
Born approximation [16,17]. There are different differential
elastic scattering experiments in this low energy region which
cover the angular range between about 20⁰ and 150⁰ [5,16].
Having fitted the phase shifts to the experimental angular
distribution, it is possible to calculate $f(E,\Theta)$ at any angle,
according to (5), and so $\sigma_{el}(E,o)$ can also be obtained. At the
same time $\sigma_t(E)$ can be evaluated by means of the equation

$$\sigma_t(E) = \sigma_{el}(E) = \frac{4\pi}{k^2} \sum_{\ell=o}^{\infty} (2\ell + 1) \sin^2 \delta_\ell, \qquad (6)$$

where $\sigma_{el}(E)$ stands for the total elastic scattering cross
section and is identical with $\sigma_t(E)$ because the inelastic
channels are closed. Just as before $Imf(E,o)$ and $Ref(E,o)$ can
be evaluated. More recently Naccache and McDowell [18] have
extended the phase shift analysis for electron scattering on
He and Ne at energies between 20 and 100 eV, where the inelastic channels are open, using an optical potential with
variable parameters.

At zero impact energy the dispersion relation in (1)
reduces to a sum rule [1]:

$$Ref(o,o) - f_B(o,o) + g_B(o,o) = \frac{1}{\pi} \int_{o}^{\infty} \frac{Imf(E',o)}{E'} \, dE'. \qquad (7)$$

According to O'Malley et al. [19]

$$\lim_{E \to o} Ref(E,o) = -A, \qquad (8)$$

where A is the scattering length.
One of the practical problems in applying the dispersion
relation is the calculation of the exchange amplitudes
dependent on the accuracy of the wavefunctions used. Therefore
positron scattering experiments are useful for testing the
dispersion relations, the exchange amplitudes being zero in
that case. However, the experiments are much more difficult
than with electrons and up to now only total scattering cross
sections have been measured (see for instance Dutton et al. [20])
and references therein).

In the following sections we shall discuss experimental results, which are related to the dispersion relations. This implies that we shall confine ourselves mainly to differential elastic scattering at relatively small angles and to total scattering of electrons. It is not our meaning to make a review of the large number of experiments and theoretical calculations which have been performed in this field. For the differential scattering of electrons a good review above 100 eV impact energy is given in the thesis of Jansen [11]). For low energies ($\lesssim$ 15 eV) we have already referred to the review article of Andrick [5]). Many theoretical calculations and experiments are described in the Journals of Physics B of 1974 and 1975, by Bromberg in the Journal of Chemical Physics (see for instance references 12 and 13) and in the latter volumes of the Physical Review. For total cross sections much of the older work has been referred to in the articles of Golden and Bandel [21,22]).

## 2. DIFFERENTIAL ELASTIC SCATTERING OF ELECTRONS

### 2.1. *Classical and Quantummechanical Criteria*

In this section we discuss some results on differential elastic scattering of Jansen et al.[11,14]) between 100 and 3000 eV and angles between 5° and 55° for He, Ne, Ar, Kr and Xe. The usual way of analyzing experimental results is plotting the differential cross section $\sigma(\Theta)$ versus momentum transfer K. When the Born approximation is valid this method should yield a universal curve for each gas, i.e. all experimental points should lie on the same curve independent of impact energy. Jansen et al. did indeed find such a universal curve for He, except for the low energy and small angle results where exchange of electrons and polarization of the target are important. This is illustrated in fig. 1. Because these latter effects become more important at smaller energy, this explains the increase of $\sigma(\Theta)$ at fixed K with decreasing impact energy. The plots for Ne, Ar, Kr and Xe (only illustrated for Ar in fig. 1) all show a K dependence completely different from that of He (see also Bromberg [13])). Here the curves lie just in reversed order compared with He: the curves for the higher energies lie above those for the lower energies. The Born limit is not yet reached at 3 keV. To understand this behaviour van Wingerden et al. [23]) applied the criterion for Born scattering. Using a screened Coulomb potential, the criterion for the validity of the Born approximation is given by Merzbacher [24]) and it was found that in the considered energy region the Born approximation could only be valid for He but never for heavier noble atoms. Applying the criteria for classical scattering as given by Mott and Massey [25]) van Wingerden et al. showed that over a large energy and angular

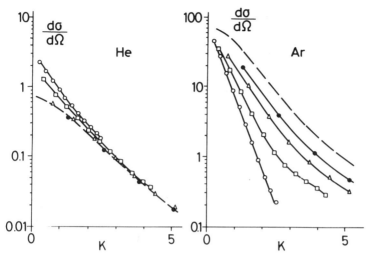

Fig. 1. Absolute differential cross sections for electrons
elastically scattered by He and Ar vs. momentum transfer at
impact energies of 100 (○), 300 (□), 1000 (△) and 3000 (●)
eV; dashed curve (- - -): first Born approximation. Atomic
units are used (from refs. 11, 14).

range the scattering from the heavier noble gases could be
described classically. Smith et al. [26]) showed that in such an
approach, using a forward impact expansion, there exists a
universal relation between $\sigma(\theta) \sin\theta/E$ and $E\theta$. In this way
the results of Jansen et al. have been given in double loga-

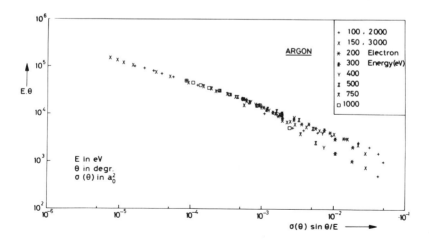

Fig. 2. Absolute differential cross sections for electrons
elastically scattered by Ar presented in a $E\theta$ vs. $\sigma(\theta) \sin \theta/E$
plot (from refs. 11,23).

rithmic plots, as illustrated for Ar (see fig. 2). From these
plots one can see that this universal relation holds for the
heavier noble gases over a large angular and energy range,
consistent with the classical criteria. It can be simply derived for classical scattering that in the case of an interaction potential $V = C/r^n$ a scaling of $\sigma(\Theta)$ vs. K leads to
higher cross sections at higher impact energies as found experimentally (see fig. 1 for Ar). For more details the reader
is referred to the work of van Wingerden et al. [23,11]).

The semilogarithmic plots of $\sigma(\Theta)$ versus K in fig. 1 show
that at small angles there is a linear relation as remarked
in the introduction (see (4)).

### 2.2. High Energy Differential Scattering from He

It has been discussed before that in the keV region the
Born approximation holds over a large angular range for differential elastic scattering by He. The Born cross section is
given by

$$\sigma(\Theta) = \frac{4}{(K)^4} \left| Z - F(K) \right|^2 , \qquad (9)$$

where Z is the nuclear charge number. F(K) is the atomic form
factor which accounts for the screening of the nucleus by the
atomic charge cloud. When $K \to \infty$, the formfactor becomes
negligible with respect to Z and we have pure electron–nucleus
scattering (Rutherford scattering). When the exchange of
incident and atomic electrons is considered, the cross section
in the Born-exchange approximation, according to Vriens et al.
[27]), is found by replacing $Z - F(K)$ in (9) by $Z - F(K) + RF(K)/2E$,
where E is the impact energy and R the Rydberg energy. Jansen
et al. [11]) have calculated $\sigma(\Theta)$ in the Born and Born-exchange
approximation, using F(K) values from the International Tables
for X-ray Christallography [28]). In fig. 3 a comparison is made
between the experimental data of Jansen et al. and the theoretical calculations at 3 keV impact energy; $K^4\sigma(\Theta)$ is plotted
versus K. According to the Rutherford formula the limit for
He(Z = 2) corresponds to $K^4\sigma(\Theta) = 16$ atomic units. The Born
exchange values are a few percent larger than the Born values
and both calculations approach asymptotically the Rutherford
limit. Included are also data evaluated with the eikonal–Born
series [29]). The close agreement of the experimental data of
Jansen et al. with the Rutherford theory at relatively large
K values is a good proof for the consistency of the absolute
scale in their differential elastic cross sections.

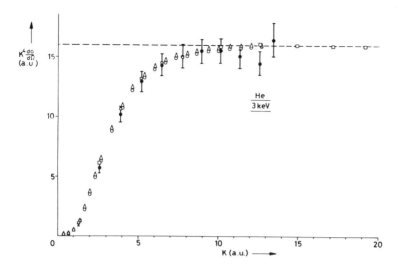

Fig. 3. The differential cross section for elastic electron-helium scattering at 3 keV, plotted as $K^4 d\sigma/d\Omega$ vs. K;
● experiment; - - - Rutherford; ○ Born; △ Born-exchange; □ eikonal-Born-series. Atomic units are used (from ref. 11).

### 2.3. *Comparison of Experimental and Theoretical Results*

When we compare experiment and theory it is useful to explain the physical meaning of the different theoretical approaches (see for instance Jansen [11]), Joachain [30]) and Schneider et al. [31]). For a noble gas one can present the differential cross section by

$$\sigma(\Theta) = \left| f(\Theta) - g(\Theta) \right|^2 , \qquad (10)$$

where $f(\Theta)$ is the direct and $g(\Theta)$ the exchange amplitude. In many theories the direct scattering amplitude is solved in different orders of perturbation theory, taking into account the effects of target polarization and absorption (removing from the elastic channel) of the incident electrons, and the exchange amplitude is treated in a first order approximation. We shall first start with theoretical approximations used above about 50 eV impact energy. Byron and Joachain [2]) have introduced the eikonal-Born-series (EBS) to calculate cross sections in e-H and e-He scattering. In this approach the direct amplitude is written as

$$f = \overline{f}_{B1} + (\mathrm{Re}\overline{f}_{B2} + i\mathrm{Im}\overline{f}_{B2}) + \overline{f}_{G3} + O(k^{-3}), \qquad (11)$$

which is derived from the Born and Glauber series. The ampli-

tude in the first Born approximation, $\overline{f}_{B1}$, is connected with
the static interaction between projectile and target only.
It is real valued and only a function of momentum transfer K.
The second order Born terms $\mathrm{Re}\overline{f}_{B2}$ and $\mathrm{Im}\overline{f}_{B2}$ are related to
polarization and absorption effects and of order $1/k$ and
$\ln k/k$ respectively at small momentum transfers. The third
order Born term (of order $k^{-2}$) is maintained and approximated
by the third order Glauber term $\overline{f}_{G3}$, also of order $1/k^2$. This
EBS series is more complete than the second Born approximation,
which misses the term $\mathrm{Re}\overline{f}_{B3}$, and the Glauber series (in which
the phase is integrated along an axis perpendicular to $\vec{K}$)
where the terms $\overline{f}_{Gn}$ are alternatively pure real and pure
imaginary. It can be proved that $\overline{f}_{B1} = \overline{f}_{G1}$ and that for large
k $\overline{f}_{G2} = i\mathrm{Im}\overline{f}_{B2}$ and $\overline{f}_{G3} = \mathrm{Re}\overline{f}_{B2}$. Since $\mathrm{Re}\overline{f}_{B2}$ has no analogue
in the Glauber series and $\overline{f}_{G2}$ diverges at K = 0, it is clear
that the Glauber series suffers from serious deficiencies.
Byron and Joachain[2]) have made a second approximation in the
calculation of $\overline{f}_{B2}$ by means of the introduction of an average
excitation energy and subsequent use of closure properties.
In fig. 4 a comparison is made for e-He scattering at 500 eV
between experiment and Born, Glauber and EBS theory.

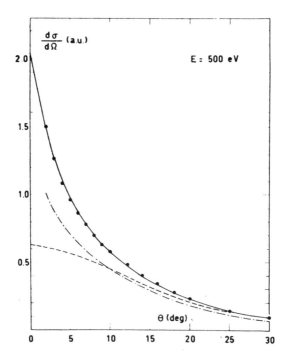

Fig. 4. Differential cross section (in a.u.) for elastic scattering from He at 500 eV. ——— EBS, - - - first Born,
-.-.- Glauber, • experiment[13]) (from ref. 2).

It is clear that the Born approximation underestimates the cross sections at small angles due to the neglect of polarization and exchange effects. The Glauber theory with its deficiencies still differs from experiment, while good agreement is obtained for the EBS series.

For more complex atoms than He, the EBS method has not been used because of the complication of numerical problems. For that purpose Byron and Joachain [32]) developed an ab-initio optical model, which will be discussed in the frame work of second order potentials. In this approach the full Hamiltonian of the system is replaced by the equivalent one-body Schrödinger equation for elastic scattering, namely

$$[-\tfrac{1}{2} \nabla^2 + V_{opt} - \tfrac{1}{2} k^2] \, \psi_i^{(+)} = 0, \tag{12}$$

where $\psi_i^{(+)}$ is the elastic scattering wave function describing the motion of the projectile in the optical potential $V_{opt}$. Apart from the exchange contribution the optical potential $V_{opt}^{direct}$ is written to second order in a multiple scattering expression, in terms of the full interaction between project-ile and target, as

$$V_{opt}^{direct} = V^{(1)} + V^{(2)} \tag{13}$$

where $V^{(1)}$ is the static potential and $V^{(2)}$ accounts for pola-rization and absorption effects. $V^{(2)}$ has a non-local charac-ter, which makes it difficult to solve (12) to second order. Winters et al. [33]) have used the second order potential approximation of Bransden and Coleman [34]) in a partial wave formalism to calculate the elastic scattering of electrons from hydrogen and helium, but not for heavier atoms because of complications. For simplification Byron and Joachain [32]) and later Vanderpoorten [35]) have used an approximation for $V^{(2)}$ in the form of a local, central, complex pseudo-potential, which is derived by using the properties of the Born and Glauber series:

$$V^{(2)} = V_p + i \, V_{abs} \, . \tag{14}$$

In this eikonal optical model (EOM), at large distances,

$$V_p(r) = - \frac{\bar{\alpha}}{r^4} \, (1 + \frac{6a^2}{r^2} + \dots \dots) \tag{15}$$

with $a = k/2 \, \Delta$, where $\Delta$ is the average excitation energy. $\bar{\alpha}$ is the dipole polarizability of the target atom in the closure approximation, $\bar{\alpha} = 2 < o|Z^2|o > / \Delta$ . The parameter $\Delta$ is fixed by requiring that $\bar{\alpha}$ is equal to the experimental dipole pola-rizability of the target atom. With $V_{opt}$ a full wave calculation

of the scattering amplitude is performed by using the partial
wave method to solve (12). When we consider e-He scattering,
the agreement between experiment (see for instance Bromberg
[13]) and Jansen et al. [11,14]) and the optical model (OM) calculations is the best at relatively large impact energies
(this holds also for EBS). In figure 5, comparison is made
for He at 100 eV. We have only included the experimental data
of Jansen et al. [11,14]) and of Crooks [36]), considering the
first as the more accurate ones. We see that all OM and the
EBS calculations fall above Jansen's data, except the one in
which only the real part of $V^{(2)}$ is included. It has been
noted by Byron and Joachain [29,38]) that $V^{(1)}$ (static) and the
complete $V_{opt.}$ lead to almost similar results at larger angles.
This is explained by the fact that absorption, polarization and
exchange effects must cancel out each other approximately.
Besides the static potential will dominate at larger angles.
This has been confirmed by Jost et al. [39]) showing that the
angular distribution for K and Ar is almost similar at larger
angles, although the polarizability of the two atoms is quite
different. In fig. 6 EOM results of Byron and Joachain [38])
for e-Ne at 500 eV agree very well with experimental results.
According to Byron and Joachain the agreement between experiment and EOM extends down to impact energies of 200 eV, where

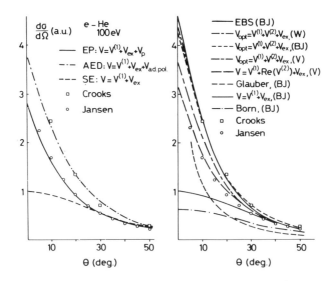

Fig. 5. Differential cross section (in a.u.) for elastic
electron-helium scattering at 100 eV. For explanation see text.
Theoretical curves in left drawing → LaBahn and Callaway [44]);
theoretical curves in right drawing: BJ → Byron and Joachain
[37]), W → Winters et al. [33]), and V → Vanderpoorten [35]).
(Partly from refs. 44 and 35).

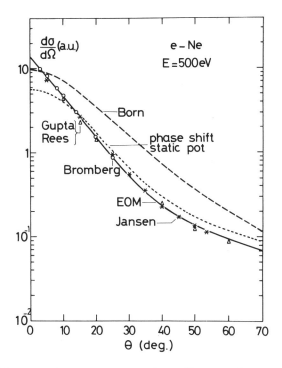

Fig. 6. Differential cross section (in a.u.) for elastic
electron-neon scattering at 500 eV. Experiments from Gupta
and Rees [40]), Bromberg [13]) and Jansen et al. [11,14]). Theory
from Byron and Joachain [38]). (From ref. 38.)

higher order terms are required in the construction of the
optical potential. For Ar the agreement between EOM calculations and experiment is equally well [11,14,41]). Lewis et al.
[42]) made calculations based on the phenomenological OM theory
of Furness and McCarthy [43]), which contains two adjustable
parameters and also accounts for exchange, polarization and
absorption (see further below).

At lower energies, 20 - 100 eV, mostly phase shift calculations have been applied with a real potential (see LaBahn
and Callaway [44])), taking into account effects of both adiabatic and non-adiabatic distortion effects (polarization
effects) of the target atom and exchange interaction to first
order. Because of the use of a real potential, these calculations satisfy elastic unitarity only and do not make allowance
for inelastic transitions. In fig. 5 we show the calculations
of LaBahn and Callaway for e-He scattering at 100 eV compared
with experiment. Here, SE means static exchange and corresponds
to $V = V^{(1)} + V_{ex}$. AED stands for adiabatic exchange dipole and

in this case $V = V^{(1)} + V_{ex} + V_{ad.pol.}$ · EP means extended polarization and $V = V^{(1)} + V_{ex} + V_p$ includes a non-adiabatic correction to $V_{ad.pol.}$ related to the finit speed of the perturbing charge (the symbols in V differ from ref. 44). When the polarization is not taken into account (SE), we note that the cross section is underestimated. Including $V_{ad.pol.}$ (AED) only, the cross section is overestimated, but accounting for non-adiabatic effects the agreement between experiment (Jansen et al. [11,14]) and theory is very good. Remark the similarity between EP and EOM calculations with $V = V^{(1)} + ReV^{(2)} + V_{ex}$ in fig. 5. Although the EP phase shift method does not make allowance for inelastic transitions, it is remarkable that the method gives good agreement with differential cross section measurements in e-He collisions over a large energy range (0 - 200 eV), see also the data of McConkey and Preston[45]), Sethuraman et al. [46]) and those of Andrick and Bitsch [16]) discussed further on. In fig. 7, for Ar, experimental data of Dubois and Rudd [47]) at 49.1 eV are compared to calculations of Walker [48]) (SE + relativistic effects), of Thompson [49]) (SE) and of Furness and McCarthy [43]) (phenomenological OM). The shape of the curves

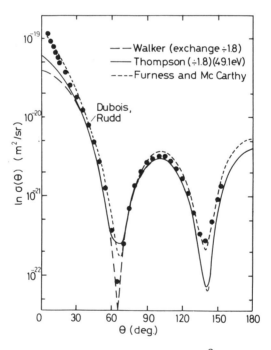

Fig. 7. Differential cross section in (m$^2$/sr) for elastic electron-argon scattering at 50 eV; experimental data from Dubois and Rudd [47]), theoretical calculations from Walker [48]), Thompson [49]) and Furness and McCarthy [43]), see text. (From ref. 47.)

is similar in theory and experiment, but only the absolute
cross sections of Furness and McCarthy are close to the experimental ones.

At low energies ($\lesssim$ 20 eV) the AED and EP approximation
have found their origin in the polarized orbital method of
Temkin [50]). Perhaps the best formal justification of the
polarized orbital method is given by the recent work of
Schneider et al. [31]) on the hierarchy of coupled Green's
functions in the generalized random phase approximation (GRPA),
considering polarization, absorption and configuration interaction effects. We compare theoretical calculations with the
accurate experimental cross sections ($\lesssim$ 5 per cent) of
Andrick and Bitsch [16]) for He. The experimental results have
been put on an absolute scale by means of a phase shift
analysis (see also section 1). In this low energy region, in
stead of comparing cross sections, phase shifts are often
considered. Because of the low energy, only a few partial
waves contribute to the cross section. In fig. 8 experiment
and theory are compared for s- and p-wave phase shifts. The
best agreement is obtained with the GRPA method [51]), the variational procedure of Sinfailam and Nesbet [52]) using continuum
Bethe-Goldstone equations and the EP calculations of
Callaway et al. [53]) and Duxler et al. [54]). The first two
methods account for polarization and electron correlation,
while the latter only considers polarization in a single
channel approach. The close coupling method of Burke et al. [55])
and the R-matrix method of Burke and Robb [56]) lead  to poorer
results, for the p-wave, because the polarization
has been neglected. Because of the near coincidence of values,

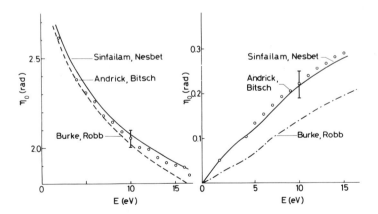

Fig. 8. Experimental and theoretical s- and p-wave phase
shifts. Experiment, Andrick and Bitsch [16]); theory,
Sinfailam and Nesbet [52]) and Burke and Robb [56]).

we have not plotted all the theoretical data in fig. 8. For
the s-waves most of the theoretical values obtained by the
methods mentioned, taking into account polarization, agree
with each other within about one per cent. For the p-waves the
corresponding values differ slightly more. For instance at
13.6 eV $\eta_1$ is 0.2646 for Sinfailam and Nesbet [52]), 0.2520 for
Yarlagadda et al. [51]), 0.2508 for Callaway et al. [53]),
0.2749 for Duxler et al. [54]) and 0.2690 for the experiment [16]).
  Studies on Ar, related to the GRPA method have been
carried out by Pindzola and Kelly [57]).


### 3. TOTAL CROSS SECTIONS FOR ELECTRON SCATTERING

  In the introduction it has been mentioned that de Heer
and Jansen [9,10]) recently succeeded in getting fairly accurate
($\lesssim$ 10 per cent) cross sections for total scattering by adding
experimental total cross sections for ionization, $\sigma_{ion}$, excitation, $\sigma_{exc}$, and elastic scattering, $\sigma_{el}$. The total elastic
cross section has been calculated from the integral

$$\sigma_{el} = 2\pi \int_o^\pi \sigma_{el}(\Theta) \sin \Theta \, d\Theta \qquad (16)$$

It is clear that below the threshold for inelastic processes
$\sigma_{el} = \sigma_{tot}$. Semi-empirical, direct experimental and theoretical results for He are compared in table I. Andrick and
Bitsch [16]) used their experimental phase shifts to get $\sigma_{el}$.
From all these data it is possible to use a consistent set of
total cross sections for application in the dispersion relations and to determine Imf(E,o). At low energies the accurate
calculations of Sinfailam and Nesbet [51]) and of LaBahn and
Callaway [44]) (see also section 2.3) are consistent with the
data of Ramsauer and Kollath [58]), above 5 eV with Andrick
and Bitsch [16]) and below 5 eV with Golden and Bandel [21]). The
data of de Heer and Jansen [9]) agree reasonably with those of
the EBS [37]) and with Brode [59]) at higher energies, going over
smoothly in the Born cross sections above 1000 eV. The recommended value (in a.u.) for the scattering length A is (1.15 +
0.03) or $\sigma(0,0)=4\pi A^2=16.61$ [60]). De Heer and Jansen [10]) have
also calculated $\sigma_{tot}$ for Ne, Ar, Kr and Xe between 100 and
1000 eV; the data will be extended below 100 eV in the near
future. At 100 and 200 eV the agreement between their semi-
empirical values and OM calculations for Ne [38]) and Ar [41]) is
not so good, with deviations of about 15 to 50% and smaller
deviations at larger energies. For Ne, Ar, Kr and Xe the Born
approximation is not valid in the energy region (100 - 3000 eV)
considered, as follows from the calculations for Ne by Inokuti
and McDowell [8]) (see also ref. 23).

TABEL I. Total Scattering Cross Sections for He in units of $a_0^2$.

| E(eV) | Semi emp. A,B [16] | Experimental R,K [58] | Experimental G,B [21] | Theory S,N [51] | Theory L,C [44] |
|---|---|---|---|---|---|
| 1 |  | 19.7 | 19.9 | 20.8 | 20.8 |
| 5 | 20.7 | 20.2 | 17.4 | 18.4 | 18.5 |
| 10 | 15.9 | 15.3 | 13.8 | 15.6 | 15.3 |
| 15 | 13.2 | 12.8 | 11.3 | 12.4 | 12.7 |
| 20 | 11.2 |  | 9.34 |  | 10.6 |
|  | dH,J [9] | Br [59] | Nor [6] | B,J [32] | I,MD [8] |
|  |  |  |  | E.B.S. | Born |
| 30 | 8.61 | 7.36 | 6.74 |  |  |
| 50 | 6.35 | 5.50 | 5.18 |  |  |
| 100 | 4.05 | 3.94 | 3.43 | 4.57 | 5.00 |
| 200 | 2.68 | 2.63 | 2.14 | 2.90 | 2.98 |
| 300 | 2.03 | 1.90 | 1.37 | 2.14 | 2.20 |
| 400 | 1.66 |  | 0.933 | 1.72 | 1.73 |
| 500 | 1.39 |  |  | 1.44 | 1.45 |
| 1000 | 0.786 |  |  | 0.82 | 0.817 |

## 4. NUMERICAL TEST OF THE DISPERSION RELATION FOR FORWARD SCATTERING

In several papers McDowell and Bransden and coworkers [1,7,18,60,61] have done numerical tests on the dispersion relation for He, Ne and Ar for electrons and positrons at variable impact energy. The general trend is that the dispersion relations (see section 1) might hold. However in an accurate test problems are often caused by the fact that the experimental and theoretical data used may contain large errors and then a definite conclusion about the validity of dispersion relations from a numerical point of view is not possible. We shall confine our attention to some recent work on He. In (7) of section 1 the dispersion relation is given in the form of a sum rule (see also (8))

$$-A- f_B(o,o) + g_B(o,o) = \frac{1}{\pi} \int_o^\infty \frac{\text{Im} f(E',o)}{E'} dE' \qquad (7)$$

Rabheru et al. [60] have used this equation for a test. We shall follow their line of operation. According to them one can adopt the values (all in a.u.) A = (1.15 $\pm$ 0.03) and $f_B(o,o) = 0.791$. They indicate the problems in calculating $g_B(o,o)$, adopting 3.92 $\pm$ 0.05 as a result of the application of different wave functions. These values lead to 1.98 $\pm$ 0.08 at the left hand

side of (7). For the right hand side we make use of the $\sigma_t$
values as recommended by de Heer and Jansen [9]) (see also
section 3) and the optical theorem ((2) of section 1) and
above 1000 eV of analytical expressions as given by Bransden
and Hutt [61]) (see also Inokuti and McDowell [8])). The accuracy
estimated in the total cross sections is better than five per
cent [9]) and the right hand side of (7) becomes $2.57 \pm 0.13$.
The difference between the left and right hand side is larger
than the combined uncertainties. A similar deviation was found
by Rabheru et al. [60]), using total cross sections of Golden
and Bandel [21]), Normand [6]) and de Heer et al. [62]). This shows
that the sum rule is not valid for e-He scattering.

In the same way, the sum rule (7) has been considered for
positron scattering on He by Branden and Hutt [61]). In that
case the exchange term $g_B(o,o)$ is zero. Bransden and Hutt [61]),
using total scattering cross sections from recent positron-
helium experiments, conclude that (7) is satisfied to an
accuracy better than 10%. The numbers used in (7) for this
case are $A = -0.472$ $f_B(o,o) = -0.791$ and for the integral
on the right hand side 1.24 or 1.39, dependent on the set
of experimental data used. The fact that (7) hold for positrons
and not for electrons has given rise to further study of the
desired analyticity properties of the exchange amplitude
throughout the entire energy plane. It has been found in
model problems by Byron et al. [63]) and by Burke [64]) that
singularities in $g_B$ are present at negative energies so that
the relevant analyticity is unlikely to be present for the
model exchange amplitude. This may be the explanation for the
failure of (7) for electron atom collisions.

McDowell et al. [65]) have also found by analysis of e-H
and e-Li collisions, that the sum rule does not hold in these
cases.

It is important to consider the dispersion relation (1) in
section 1 at intermediate and higher impact energies for
electron-atom scattering, where the contribution of the
exchange term becomes relatively small. Such a study has been
performed by Bransden and Hutt [61]) for He. However, in stead
of (1), they used the so called subtracted disperion relation
(see also (29) in ref. 1), following from (1) and (7):

$$\text{Re}f(E,0) + A = g_B(E,0) - g_B(0,0) + \frac{E}{4\pi} P \int \frac{\text{Im}f(E',0)}{E'(E'-E)} dE'$$

In view of the invalidity of the sum rule, as indicated $\qquad$ (17)
above, the dispersion relation at higher energies should also
be applied in form (1) and the related results of Bransden
and Hutt [61]) have to be reconsidered.

## ACKNOWLEDGEMENTS

The author is grateful for help and valuable comments by drs. D. Andrick, F.W. Byron, C.J. Joachain, M.R.C. McDowell, R.K. Nesbet, K. Nygaard, E. Weigold and K.H. Winters; further to dr. B. Crompton who indicates the importance of consideration of his electron diffusion measurements in relation with phase shifts and total scattering cross sections. The author is thankful to dr. R.H.J. Jansen, drs. H.R. Blaauw and Mr. R.W. Wagenaar for assistance in some numerical calculations.

This work is part of the research program of the Stichting voor Fundamenteel Onderzoek der Materie (Foundation for Fundamental Research on Matter) and was made possible by financial support from the Nederlandse Organisatie voor Zuiver-Wetenschappelijk Onderzoek (Netherlands Organisation for the Advancement of Pure Research).

## REFERENCES

[1]) B.H. Bransden and M.R.C. McDowell, J.Phys.B (At.Mol.Phys.) 2 (1969) 1187.
[2]) F.W. Byron and C.J. Joachain, Phys.Rev. A 8 (1973) 1267.
[3]) E. Gerjuoy, Anal.Phys. 5 (1958) 58.
[4]) E. Gerjuoy and N.A. Krall, Phys.Rev. 119 (1960) 705.
[5]) D. Andrick, 1973, in *Advances in Atomic and Molecular Physics*, Eds. D.R. Bates and I. Esterman, Vol. 9 (Academic Press, New York, U.S.A.), p. 207.
[6]) C.E. Normand, Phys.Rev. 35 (1930) 1217.
[7]) B.H. Bransden and M.R.C. McDowell, J.Phys.B (At.Mol.Phys.) 3 (1970) 29.
[8]) M. Inokuti and M.R.C. McDowell, J.Phys.B (At.Mol.Phys.) 7 (1974) 2382.
[9]) F.J. de Heer and R.H.J. Jansen, FOM report 37173, 1975.
[10]) F.J. de Heer and R.H.J. Jansen, FOM report 37174, 1975.
[11]) R.H.J. Jansen, Ph.D. Thesis, University of Amsterdam, Amsterdam 1975.
[12]) J.P. Bromberg, J.Chem.Phys. 51 (1969) 4117.
[13]) J.P. Bromberg, J.Chem.Phys. 61 (1974) 963.
[14]) R.H.J. Jansen, F.J. de Heer, H.J. Luyken, B. van Wingerden and H.J. Blaauw, J.Phys.B (At.Mol.Phys.), to be published.
[15]) W.M. Huo, J.Chem.Phys. 56 (1972) 3468.
[16]) D. Andrick and A. Bitsch, J.Phys.B (At.Mol.Phys.) 8 (1975) 393.
[17]) D.G. Thompson, Proc.R.Soc. A 294 (1966) 160.
[18]) P.F. Naccache and M.R.C. McDowell, J.Phys.B (At.Mol.Phys.) 7 (1974) 2203.
[19]) T.F. O'Malley, L. Rosenberg and L. Spruch, Phys.Rev. 125 (1962) 1300.

[20]) J. Dutton, F.M. Harris and R.A. Jones, J.Phys. B $\underline{8}$ (1975) L 65.

[21]) D.E. Golden and H.W. Bandel, Phys.Rev. A $\underline{138}$ (1965) 14.

[22]) D.E. Golden and H.W. Bandel, Phys.Rev. $\underline{149}$ (1966) 58.

[23]) B. van Wingerden, F.J. de Heer, R.H.J. Jansen and J. Los in *Proceedings of the Stirling Symposium in honour of Ugo Fano, 1975*, Ed. H. Kleinpoppen and M.R.C. McDowell, Plenum Press.

[24]) E. Merzbacher, 1970, *Quantum Mechanics*, (John Wiley and Sons, New York, USA) second edition, p. 229.

[25]) N.F. Mott and H.S.W. Massey, 1965, *The Theory of Atomic Collisions* (Clarendon Press, Oxford, England, third edition), p. 110.

[26]) F.T. Smith, R.P. Marchi and K.G. Dedrick, Phys.Rev. $\underline{150}$ (1966) 79.

[27]) L. Vriens, C.E. Kuyat and S.R. Mielczarek, Phys.Rev. $\underline{170}$ (1968) 163.

[28]) *International Tables for X-ray Crystallography*, 1962, Vol. III, § 3.3 (Keynoch Press, Birmingham, England).

[29]) F.W. Byron and C.J. Joachain, Phys.Rev. A8 (1973) 3266 and private communication.

[30]) C.J. Joachain, 1975, *Quantum Collision Theory* (North-Holland Publ.Comp., Amsterdam, The Netherlands).

[31]) B. Schneider, H.S. Taylor and R. Yaris, Phys.Rev. A $\underline{1}$ (1970) 855.

[32]) F.W. Byron and C.J. Joachain, Phys.Rev. A $\underline{9}$ (1974) 2559.

[33]) K.H. Winters, C.D. Clark, B.H. Bransden and J.P. Coleman, J.Phys. B (At.Mol.Phys.) $\underline{7}$ (1974) 788.

[34]) B.H. Bransden and J.P. Coleman, J.Phys.B (At.Mol.Phys.) $\underline{5}$ (1972) 537.

[35]) R. Vanderpoorten, J.Phys.B (At.Mol.Phys.) $\underline{8}$ (1975) 926.

[36]) G.B. Crooks, Ph.D. Thesis, University of Nebraska, Lincoln 1972.

[37]) F.W. Byron and C.J. Joachain, F.O.M.-report 37521, 1975.

[38]) F.W. Byron and C.J. Joachain in *Proceedings of the Stirling Symposium in honour of Ugo Fano*, 1975, Editor H. Kleinpoppen, and to be published.

[39]) K. Jost, private communication.

[40]) S.C. Gupta and J.A. Rees, J.Phys.B (At.Mol.Phys.) $\underline{8}$ (1975) 417.

[41]) C.J. Joachain, K.H. Winters and F.W. Byron, Phys.Rev., to be published, and J.Phys.B(At.Mol.Phys.)8(1975)L289.

[42]) B.R. Lewis, J.B. Furness, P.J.O. Teubner and E. Weigold, J.Phys.B (At.Mol.Phys.) $\underline{7}$ (1974) 1083.

[43]) J.B. Furness and I.E. McCarthy, J.Phys.B (At.Mol.Phys.) $\underline{6}$ (1973) 2280.

[44]) R.W. LaBahn and J. Callaway, Phys.Rev. $\underline{180}$ (1969) 91 and ibid. A $\underline{2}$ (1970) 366.

45) J.W. McConkey and J.A. Preston, J.Phys.B (At.Mol.Phys.)
    8 (1975) 63.
46) S.K. Sethuraman, J.A. Rees and J.R. Gibson, J.Phys. B
    (At.Mol.Phys.) 7 (1974) 1741.
47) R.D. Dubois and M.E. Rudd, J.Phys.B.(At.Mol.Phys.)8(1975)1474.
48) D.W. Walker, Adv.in Physics 20 (1971) 257.
49) D.G. Thompson, J.Phys.B (At.Mol.Phys.) 4 (1971) 468.
50) A. Temkin, Phys.Rev. 107 (1957) 1004.
51) B.S. Yarlagadda, G. Csanak, H.S. Taylor, B. Schneider and
    R. Yaris, Phys.Rev. 7 (1973) 146.
52) A.L. Sinfailam and R.K. Nesbet, Phys.Rev. A6 (1972) 2118.
53) J. Callaway, R.W. LaBahn, R.T. Pu and W.M. Duxler, Phys.
    Rev. 168 (1968) 12.
54) W.M. Duxler, R.T. Poe and R.W. LaBahn, Phys.Rev. A4
    (1971) 1935.
55) P.G. Burke, J.W. Cooper and S. Ormonde, Phys.Rev. 183
    (1969) 245.
56) P.G. Burke and W.D. Robb, J.Phys.B (At.Mol.Phys.) 5 (1972)
    44.
57) M.S. Pindzola and H.P. Kelly, Phys.Rev. A 9 (1974) 323.
58) C. Ramsauer and R. Kollath, Ann.Phys. (Leipzig) 12 (1932)
    529.
59) R.B. Brode, Phys.Rev. 25 (1925) 636.
60) A.J. Rabheru, M.M. Islam and M.R.C. McDowell (1975) report.
61) B.H. Bransden and P.K. Hutt, J.Phys.B (At.Mol.Phys.)
    8 (1975) 603.
62) F.J. de Heer, H.J. Blaauw and R.H.J. Jansen (1974),
    Memorandum on total cross sections.
63) F.W. Byron, F.J. de Heer and C.J. Joachain, F.O.M. report
    38017 , 1975 and submitted for publication.
64) P.G. Burke, private communication.
65) M.R.C. McDowell, P.K. Hutt, A. Rabheru and M. Islam, 1975,
    to be submitted to J.Phys.B (At.Mol.Phys.)

MEASUREMENT OF ABSOLUTE COLLISION CROSS SECTIONS OF ELECTRONS
ELASTICALLY SCATTERED BY GASES

J. Philip Bromberg

Carnegie Mellon Institute of Research
4400 Fifth Avenue
Pittsburgh, Pennsylvania    15213

## I.    INTRODUCTION

The results of absolute differential elastic cross section
measurements of electrons scattered by gases are important from
two points of view.  Firstly, they are of interest in themselves
and serve to elucidate various aspects of theoretical studies
of electron scattering, and secondly, they serve as a calibration standard for placing inelastic spectra on an absolute
basis.  For the latter purpose, the elastic peak is included
in an energy loss spectrum, and from the known absolute cross
section of the elastic peak and the relative intensities of the
elastic and inelastic peaks, the absolute collision cross
sections may be conveniently determined.  Much of the progress
in recent theoretical studies, both in elastic scattering and
in spectroscopy as studied by electron impact loss spectra may
be attributed to the present availability of absolute cross
section data.

The earliest work in this field was carried out in the
early 1930's by a number of workers using a variety of target
gases.  All of these experimental investigations were on a
*relative* rather than an *absolute* basis;  where these early data
were reported on an absolute basis, they were made absolute by
comparison with calculated cross sections.  These older relative
cross sections usually agree with modern measurements, but only
at larger angles.  The available scientific talent soon turned
its attention to other areas such as nuclear physics and high
energy physics.  Interest in electron scattering remained
relatively dormant for the next 30 years, and electron scattering concerned itself mainly with structural studies using
electron diffraction techniques.

Interest in this area has revived, and once again electron scattering is an active branch of science.  Some 15 years ago E.N.Lassettre and his coworkers, first at Ohio State, and later at the Mellon Institute began a series of beautiful experiments in electron impact spectroscopy which clearly demonstrated the power and utility of this technique.  It was to establish these relative inelastic spectra on an absolute basis in order to provide a clear comparison between the theory and experiment which served as the original impetus for the development of an electron spectrometer capable of accurately measuring absolute differential cross sections of electrons scattered by gases.

In the remainder of this brief review we shall first examine the factors which must be given consideration in order to successfully measure absolute cross sections, and then discuss some of the results.  Space limitations preclude a complete bibliography, and more complete listings may be had in the references cited.  Much of the data presented are from our own work;  this reflects the ready availability of suitably drawn figures for our own work.  All of our unreferenced work has not previously appeared in the scientific literature.

## II. EXPERIMENTAL

### General Considerations

The measurement of *absolute* cross sections is much more difficult than that of *relative* cross sections.  A number of

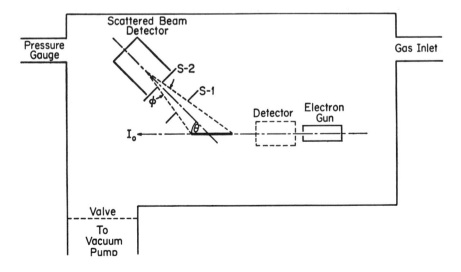

Fig. 1. General schematic spectrometer.

conditions are imposed upon the apparatus which must be met if the measured cross sections are to be reliable.  Figure 1 indicates schematically the relevant and crucial components.[1] The absolute differential cross section at angle $\theta$ is given by

$$\sigma(\theta) = I/I_oPn\Omega\ell \qquad\qquad (1)$$

where $I$ is the electron current scattered into the solid angle $\Omega$ at scattering angle $\theta$;  P is the pressure and n is the number of molecules per unit length scattering volume per unit pressure; $\Omega$ is the solid angle subtended by S-2 about the rotation axis; $\ell$ is the length of the scattering volume and varies as $1/\sin\theta$; $I_o$ is the incident beam current density.  Each of these variables must be measured accurately and absolutely, and we shall examine each one in succession.

## Particular Considerations

The _incident_ _beam_ current density, $I_o$, takes the form of a narrow pencil of electrons of diameter equal to that of the exit pinhole of the electron gun chamber.  The exact diameter of this pencil is not of great concern, as the cross sectional term in $I_o$ is exactly cancelled by that term in n.  The detector must detect _all_ the electrons entering the scattering chamber, and must reject extraneous charges such as secondary electrons. Incident beam currents are normally about 10(E-6)A, and a Faraday cylinder can be used to collect the charge;  by leading the current to ground through a precision resistor, the current can be determined by measuring the voltage drop across the resistor. The cylinder must be properly aligned, and it should be floated below ground to reject secondaries.  The opening should be sufficiently large to accept all electrons in the incident beam, and the length sufficiently long to minimize losses by back-scattering.  Placing the cylinder immediately in front of the electron gun exit pinhole will minimize losses due to scattering from the target gas.  We have noted that $I_o$ may not remain constant during a run, and may be a slowly varying function of the pressure, the angle, and the time, hence means must be pro-vided to periodically measure $I_o$ during an actual run.  Whereas $I_o$ does change, we note that the ratio $I/I_o$ is constant.

The _scattered_ _current_, $I$, is typically less than 10(E-11)A. Scattered currents can be conveniently and absolutely measured with vibrating reed electrometers.  The scattered currents are also collected by a Faraday cylinder, and to minimize errors in the ratio $I/I_o$ this cylinder should have the same geometry as the primary beam cylinder.  Since $I$ and $I_o$ enter as the ratio of the two, highest precision requires that the two detection systems be intercalibrated.  While in principle it should be possible to calibrate an electron counting system as the

scattered beam detector, the practical difficulties may introduce large errors in absolute cross sections.  It is not
necessary to measure all angular points absolutely, but only a
sufficient number so that a complete set of relative cross
sections may be made absolute by comparing the ratios.

The velocity selector must transmit *all* the elastically
scattered electrons and reject *all* others such as inelastically
scattered electrons.  Figure 2 shows an incident beam impinging
on S-2, the entrance to the velocity selector, in this case a
hemispherical analyzer.  We assume the pinhole to be infinit-
esimally small.  Due to the energy spread of the incident beam
a finite image of diameter 2Y is formed at the exit pinhole, and
the detector will be right in front of this exit pinhole, S-3.
In order to insure that all electrons entering at S-2 exit at
S-3, the transmission of the analyzer must be unity.  This will
be the case if the diameter of S-3 exceeds that of S-2 by at
least 2Y.  The resolution of the analyzer should be at least
sufficient to separate the elastic from the inelastic peaks,
but yet sufficiently poor that the entire energy spread of the
elastic peak is transmitted.  Since the resolution of an
analyzer depends upon the velocity with which the electrons
pass through, this condition can be checked by noting whether
the scattered signal intensity is constant for a given incident
beam as the velocity with which they pass through is varied.

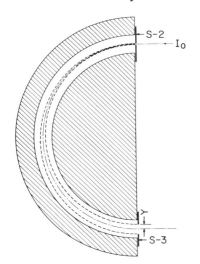

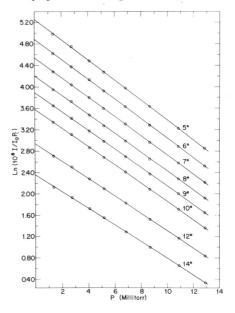

Fig. 2. Electron path through
the analyzer showing the
divergence caused by the
energy spread in the incident
beam.

Fig. 3. Plots of $ln(I/I_0 P)$ vs P
for 500 eV electrons scattered
by CO. The straight lines are
the least squares fitted lines.

   Pressure, typically in the range 0.1-20 mTorr, must be
determined absolutely to provide a value for the density of
target molecules in the scattering gas.  The most suitable
instrument for this range is the capacitance manometer, which
permits continuous and accurate monitoring of the pressure.
The liquid nitrogen trapped McLeod gauge suffers from the
serious streaming error, and is inconvenient and discontinuous
in use;   further it is useless for gases such as $H_2O$, $NH_3$, and
Hg.  We have carried out careful intercalibrations of a
commercial capacitance manometer, and have found them to be far
superior to the McLeod gauge[2]

   While we have verified that at constant pressure the ratio
$I/I_o$ is independent of $I_o$, we noted that at constant $I_o$ the
ratio $I/I_oP$ is a strong function of pressure.  This is due to
the effect of the electron scattering analogue of the Lambert-
Beer law, and the scattered currents are given by

$$I = I_oAPexp(-BP\lambda) \qquad (2)$$

where $\lambda$ is the path length of the electron through the gas,
B is a constant related to the total cross section, and A is the
ratio $I/I_oP$ in the limit as P approaches zero.  An experimental
procedure to determine A is obtained by rearranging (2) to get

$$ln(I/I_oP) = lnA - BP\lambda \qquad (3)$$

Plots of $ln(I/I_oP)$ vs P yield straight lines with intercepts
given by $lnA$.  Eq. (1) for the cross sections becomes

$$\sigma(\theta) = A/n\Omega\ell \qquad (4)$$

and least squares fits of Eq. (3) to the data provide values
for A and the standard deviation.  The standard deviation is a
measure of the uncertainty in *relative* cross sections.  We have
found that Eq. (3) is valid out to 60 mTorr for helium, and is
presumably valid at even higher pressures.  In our work a
sufficient number of points are taken at each angle over a range
of pressure such that the standard deviation in A is less than
1%.  A number of such plots for CO is indicated in Figure 3.
Particularly noteworthy is the fact that the slopes of the
straight lines may vary with angle, and this can introduce large
errors if measurements are made at only one pressure.  For
example, in our work on Kr, if cross sections at 500 eV had been
measured at the *single* pressure of 1 mTorr, then the error in
*relative* cross section between the 3° and 25° points would have
been 6%.  This error is proportional to the pressure and the
magnitude of the cross section, and would be different for
different spectrometers.

The slit system determines a wedge of angle $\phi$ which inter-
sects $I_0$ and determines the length of the scattering volume, $\ell$.
The solid angle subtended by S-2 about the rotation axis is
determined from the area of the pinhole and the distance from
the rotation axis. The distance is easily and conveniently
measured with a scale and is of the order of 14 cm. The
dimensions of the pinholes may be measured from photomicrographs,
by optical comparators, and microscopes with calibrated stages.
It is to be noted that an error of 1% in the diameter corresponds to an error twice as large in the area. Our procedure
was to use the combined results of all three methods, the photo-
micrograph result being obtained by measuring the area of the
pinholes with a planimeter. An error in slit measurement
contributes a systematic error to the final results. To minimize this error we originally used three different sets of slits
and pinholes to obtain a weighted average dimension; we found
the results from the three slit systems to agree within the
experimental error. To provide a critical check on whether the
transmission of the analyzer was, indeed, unity, we made two
sets of runs using the same entrance pinhole, but different
sized exit pinholes (S-3), and found the results to be the same.

The angle $\theta$ presents no difficulty as the rotating detector
system (in some cases the electron gun is rotated) is arranged
through a series of precision gears through vacuum feedthroughs.
Angular errors may be minimized by measuring scattered currents
at $+\theta$ and $-\theta$, and taking the average result. This will cancel
any errors in setting the zero angle, provided the error is
small (of the order of 0.1°).

The total fractional error will be the composite error due
to errors in pressure, beam currents, slit dimensions, temperature, background corrections, and random errors in determining
the constant A, and will be given by the expression

$$\delta^2_{fract} = \delta^2_P + \delta^2_I + \delta^2_{slit} + \delta^2_T + \delta^2_{bg} + \delta^2_{rand} \qquad (5)$$

The total estimated uncertainty for data reported by our
laboratory at the Mellon Institute is of the order of 3%. The
shapes of our cross section curves are of necessity checked by
E. N. Lassettre and his coworkers in their normalization of
their inelastic cross sections, and close agreement has always
been found[3]. Comparison between our data and recently determined
absolute data of other laboratories discussed later produces
agreement to within the combined uncertainties of the results.


III. ABSOLUTE CROSS SECTION DATA

Figure 4 reproduces the results of the first extensive set
of absolute measurements of elastic cross sections, in this case

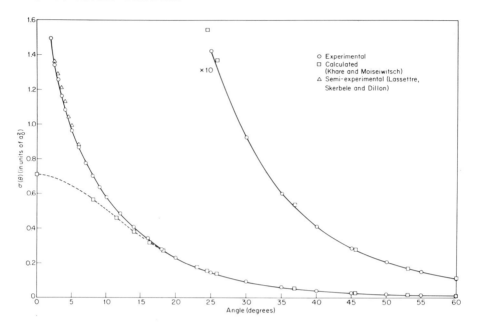

Fig. 4. Absolute elastic cross sections for 500-eV electrons
        scattered by He vs angle.

for helium at 500 eV.[1]  There are several points worth noting.
Above 20° there is no significant difference between these
results and the older data of Hughes McMillan and Webb;[4] nor is
there any significant difference between these and the calculated cross sections of Khare and Moiseiwitsch[5] based on a
first Born approximation with first-order exchange.  At smaller
angles the experimental curve rises much more steeply than the
theoretical curve.  Interestingly, the smaller angle data of
Hughes (at 9.5° and 12°) were too low by an amount just
sufficient to place them on the theoretical curve.  The points
shown by triangles at small angles indicates the close agreement
(within 3%) with cross sections determined by extrapolating
oscillator strengths determined from inelastic spectra to the
optical value.[6]  The use of our absolute elastic cross sections
as a calibration standard for inelastic spectra to determine
generalized oscillator strengths produces substantial agreement
with optical oscillator strengths.

     Our results for mercury clearly indicate the advantage of
plotting the data in terms of the natural variable of momentum
change rather than the artificial variable of angle.  The
momentum change in atomic units is given by

$$\Delta P = [0.292576 E \sin^2 (\tfrac{1}{2}\theta)]^{\frac{1}{2}} \qquad (6)$$

where E is the electron energy in electron volts, and $\theta$ is the scattering angle. Our small angle results for Hg are shown in Figure 5 where we have plotted the natural logarithm of the cross section vs $\Delta P$.[7] Mittleman and Watson[8] have shown that an effective potential (the optical potential) exists from which elastic scattering can be calculated. In general this potential depends upon incident energy, but this dependence is negligible at high kinetic energy. That the small angle cross sections

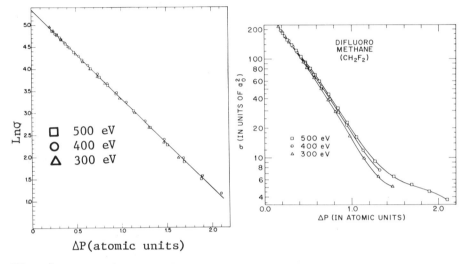

**Fig. 5.** Natural log. of cross sections of electrons scattered by Hg vs $\Delta P$.

Fig. 6. Absolute elastic cross sections of electrons scattered by $CH_2F_2$ vs $\Delta P$.

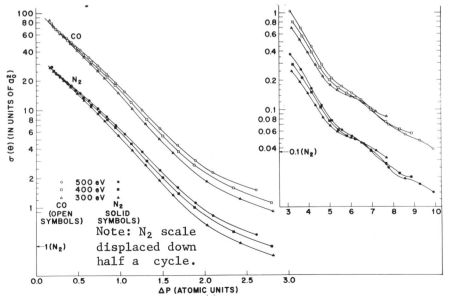

**Fig. 7.** Cross sections of electrons scattered by $N_2$ and CO.

are independent of momentum change immediately suggests a Born approximation. As shown in Figure 5 the cross sections all fall on a straight line, and can be expressed by

$$\sigma(\Delta P) = \sigma_o exp(-\beta\Delta P) \qquad (7)$$

By inverting the sine transform as expressed in the simple Born approximation using this expression for $\Delta P$ we can obtain a simple expression for the potential in the form

$$V(r) = 4Cb/\pi(r^2 + b^2)^2 \qquad (8)$$

where $C = (\sigma_o)^{\frac{1}{2}}$ and $b=\frac{1}{2}\beta$. Remarkably, this purely phenomeno-logical approach produces the polarization potential in the limit as $r\to\infty$. A polarization potential is a long range potential, and small angle scattering is due to electrons which pass the target atoms at large distances. The polarizability should be given by $4Cb/\pi$; for Hg the polarizability so obtained from electron scattering data is one half the value obtained from other experiments[9]. For other gases studied by us and by others this procedure produces polarizability values quite close to the static polarizability determined by other techniques.

It appears that for most gases thus far studied, the cross sections become independent of incident energy, and depend on the momentum change in the limit of small momentum change, indicating the adequacy of a Born approximation description. The merging of the curves at small $\Delta P$ is shown for $CH_2F_2$ in Figure 6, and for CO and $N_2$ in Figure 7[10]. CO and $N_2$ are iso-electronic, and their cross sections are the same to within the

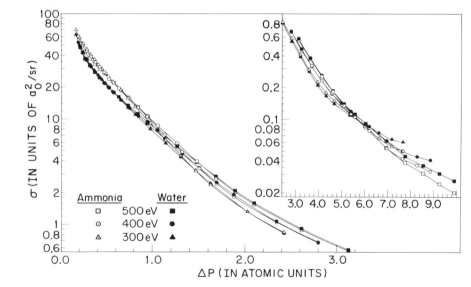

Fig. 8. Cross sections of electrons scattered by $H_2O$ and $NH_3$.

experimental errors at large angles.  At small angles, the cross
sections for CO rise at a slightly faster rate, reflecting the
small dipole moment of CO.

In attempting to note the effect of large dipole moments
on elastic scattering cross sections we measured the cross
sections for the polar (and also isoelectronic) pair of
molecules, $H_2O$ and $NH_3$, and these results are shown in Figure 8.
Again, the cross sections become approximately independent of
incident energy in the limit of small $\Delta P$, but here we note the
rapid increase in cross sections at small $\Delta P$ caused by the high
dipole moments of the molecules.

Of particular importance in theoretical studies are the
inert gases, particularly helium.  As can be seen from the small
angle data of Figure 9, helium is a somewhat anomalous case.[11]
In contradistinction to the other gases, the cross sections for
helium diverge at small $\Delta P$, and the lower energy curves lie
higher, though the curves appear to become independent of
incident energy at large $\Delta P$ as shown in Figure 10 for 200-700 eV
at angles out to 110°.  This behavior has also been noted by
Jansen[12] who has studied the cross sections out to energies
of 3000 eV at angles out to 50°.  It is for this reason that
the first Born approximation has been found valid for helium
only at large angles as previously noted, and as indicated by
the dashed curve of Figure 11.

The advent of absolute data for the elastic cross sections
has had a stimulating effect on theoretical calculations.  The
observed peaking of the cross sections in the forward direction

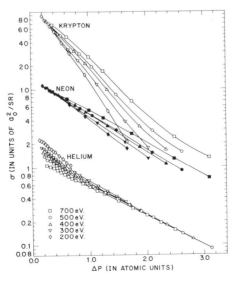

Fig. 9. Cross sections of electrons scattered by He, Ne, and Kr.

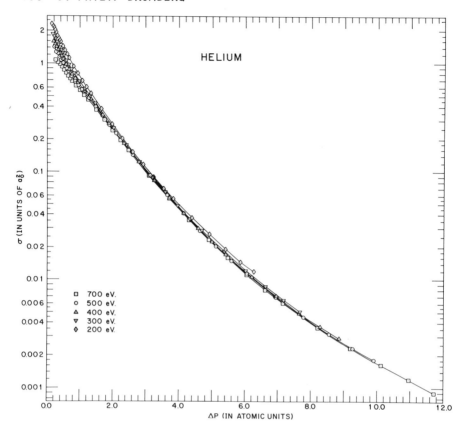

Fig. 10. Elastic cross sections of electrons scattered by He.
at 200–700 eV at angles from 2°–110° (Bromberg).

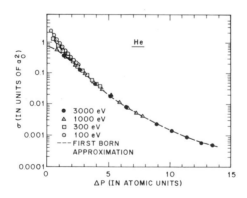

Fig. 11. Elastic cross sections of electrons scattered by He
at 100–3000 eV at angles from 5°–50° (Jansen, Ref. 12).

has engendered improved and more sophisticated theoretical treatments of scattering phenomena. This is dramatically illustrated in Figure 12 where it can be seen that the theoretical results of Byron and Joachain[13] based on the eikonal optical model of elastic scattering provides excellent agreement with the experimental data over the entire angular range at 500 eV

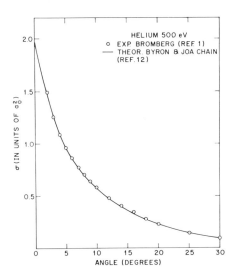

Fig. 12. Comparison of calculated and experimental cross sections for He at 500 eV.

for helium. Close agreement between theory and experiment is also being obtained by a number of other laboratories,[14] though the agreement at lower energies is still problematical. Theoretical calculations for the higher inert gases[15] have not yet been as successful as those for helium.

Approximately one year ago, in the spring of 1974, we wrote, *It is apparent, then, that except for the results of Jansen and deHeer[16] at 500 eV the helium data presented here are in disagreement with the data of other laboratories, not only with respect to absolute values, but also with regard to the shapes of the cross section curves.*[11] This situation has by now changed for the better. We have compared the results of Bromberg and of Jansen for helium between 200-700 eV at angles from 5-50°. We took the ratios of Bromberg's data to that of Jansen's; for the 30 points (6 at each energy) the average ratio was 1.013 with a standard deviation of 0.032, well within the uncertainties of measurements. For neon this ratio was 1.000 with a standard deviation of 0.028. In comparing three sets of absolute data for $N_2$ we find very good agreement between the data of Bromberg,[19] of Jansen,[12] and of DuBois.[17]

The first set of cross section data determined on an experimentally absolute basis first appeared in print some six years ago in 1969.[1] The heightened interest in this area is evidenced by the increasingly large number of laboratories devoting their efforts to experimental[18] and theoretical[14] investigations. While the measurement of an absolute cross section is certainly not a routine procedure, the increasingly large body of reliable data can presently serve as an adequate basis for comparison with theoretical studies of elastic cross sections, and for the calibration of electron loss spectra to obtain absolute values for inelastic collision cross sections.

*Supported in part by the National Science Foundation under grant GP-23517X2

[1] J. P. Bromberg, J. Chem. Phys. 50, 3906 (1969).

[2] J. P. Bromberg, J. Vac. Sci. Technol. 6, 801 (1969).

[3] E. N. Lassettre and A. Skerbele, J. Chem. Phys. 54, 1597 (1971). See also Ref. 11 below.

[4] A.L.Hughes,J.H.McMillan, and G.M.Webb, Phys. Rev. 41, 154 (1932).

[5] S.P.Khare and B.L.Moiseiwitsch, Proc. Phys. Soc. (London) 85, 821 (1965).

[6] E.N.Lassettre,A.Skerbele, and M.A.Dillon, J. Chem. Phys. 49, 2382 (1968)

[7] J.P.Bromberg, J. Chem. Phys. 51, 4117 (1969).

[8] M.H.Mittleman and K.M.Watson, Phys. Rev. 113, 198 (1959)

[9] This discrepancy is discussed in W. Huo, J. Chem. Phys. 56, 3468 (1972). Note that Ref. 7 contains an error of a factor of two yielding incorrectly a polarizability in agreement with the static values.

[10] J. P. Bromberg, J. Chem. Phys. 52, 1243 (1970).

[11] J. P. Bromberg, J. Chem. Phys. 61, 963 (1974).

[12] R.H.J. Hansen, Ph.D. Thesis, University of Amsterdam, Amsterdam, (1975).

[13] F.W.Byron and C.J.Joachain, Phys. Rev. A8, 1267 (1973); also Phys. Rev. A (1975) to be published.

[14] See for example: R.W.LaBahn and J. Callaway, Phys. Rev. A2, 366 (1970); K.H.Winters, C.D.Clark, B.H.Bransden and J.P. Coleman, J. Phys. B (At. Mol. Phys.) 7, 788 (1974); B.D.Buckley, and H.R.J.Walters, J. Phys. B (At. Mol. Phys.) 7, 1380 (1974); S.P.Khare and P. Shobha, J. Phys. B (At. Mol. Phys.) 4, 208 (1971). See also Refs. 5 and 13.

[15] See for example: S.P.Khare and P.Shobha, J. Phys. B (At. Mol. Phys.) 7, 420 (1974); F.W. Byron and C.J.Joachain, Phys. Lett.

49A, 306 (1974);   M. Fink and A.C. Yates, Atomic Data 1, 385
(1970); D.W.Walker, Adv. Phys. 20, 257 (1971); B.R.Lewis, J.B.
Furness, P.J.O.Teubner, and E. Weigold, J. Phys. B (At. Mol.
Phys.) 7, 1083 (1974).

[16]R.H.J. Jansen and F.J. DeHeer, Abstracts of Eighth ICPEAC
Conference (Institute of Physics, Belgrade, Yugoslavia, 1973)
p 269.

[17]R.D. DuBois, Ph.D. Thesis, University of Nebraska, Nebraska
(May, 1975).

[18]See for example: S.C.Gupta and J.A.Rees, J. Phys. B (At. Mol.
Phys. 8, 417 (1975); J.F.Williams and B.P.Willis, J. Phys. B
(At. Mol. Phys.) 8 (1975); M.V.Kurepa, L.Dj. Vuskovic and S.D.
Kalezic, Ref. 16, p267.  Also Refs. 11, 12, and 17.

# RECENT ADVANCES IN POLARIZED-ELECTRON EXPERIMENTS

Joachim Kessler

Physikalisches Institut der Universität Münster*

Münster, West Germany

The field of spin-polarized electrons has rapidly expanded in the past few years.   Let me first discuss this with
the use of a little table (Fig. 1):   One of the processes

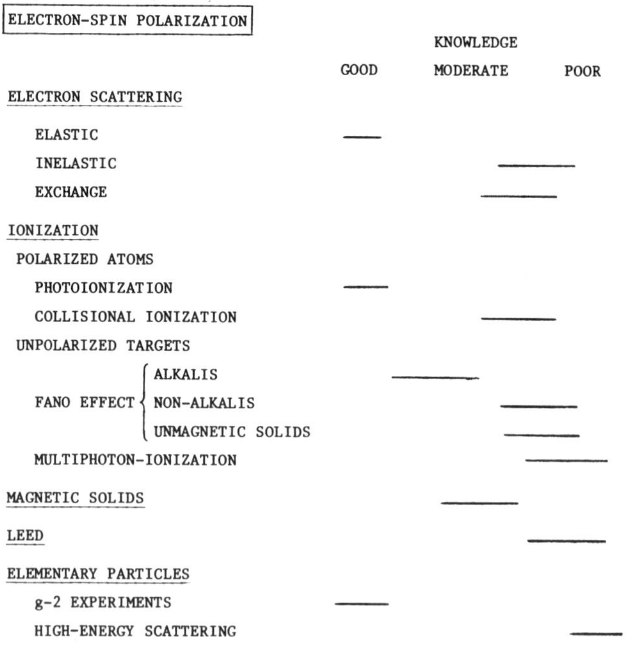

| ELECTRON-SPIN POLARIZATION | KNOWLEDGE | | |
| --- | --- | --- | --- |
| | GOOD | MODERATE | POOR |
| ELECTRON SCATTERING | | | |
|   ELASTIC | — | | |
|   INELASTIC | | — | |
|   EXCHANGE | | — | |
| IONIZATION | | | |
| POLARIZED ATOMS | | | |
|   PHOTOIONIZATION | — | | |
|   COLLISIONAL IONIZATION | | — | |
| UNPOLARIZED TARGETS | | | |
| FANO EFFECT { ALKALIS | — | | |
|   NON-ALKALIS | | — | |
|   UNMAGNETIC SOLIDS | | — | |
|   MULTIPHOTON-IONIZATION | | — | |
| MAGNETIC SOLIDS | | — | |
| LEED | | — | |
| ELEMENTARY PARTICLES | | | |
|   g-2 EXPERIMENTS | — | | |
|   HIGH-ENERGY SCATTERING | | | — |

Fig. 1.   Electron polarization in various fields of physics.

*Visiting Fellow, 1974-75, Joint Institute for Laboratory
Astrophysics, Boulder, Colorado.

from which polarized electrons arise is elastic scattering of
an unpolarized electron beam from an unpolarized target.  The
origin of the polarization is easy to see:  The unpolarized
primary beam may be considered as a mixture of equal numbers
of spin-up and spin-down electrons (e↑, e↓); the polarization
$P = (N_\uparrow - N_\downarrow)/(N_\uparrow + N_\downarrow)$ of such a mixture is zero.  Owing to the
spin-orbit interaction of the incident electrons in the field
of the atom one has a scattering potential that is spin dependent.  One therefore obtains a spin-dependent scattering
cross section.  In other words, different numbers of e↑ and
e↓ are scattered into the direction of observation.  This
means the scattered beam is polarized.  Examples may be found
in Refs. 1-3.  The good agreement between theory and experiment shown there is typical:  Owing to extensive experimental
and theoretical studies in the past decade, the polarization
in elastic scattering is now quite well understood, at least
at energies above 100 eV.[2,3]

    On the other hand, polarization studies on inelastic
scattering are only at the beginning.  Figure 2 shows, for

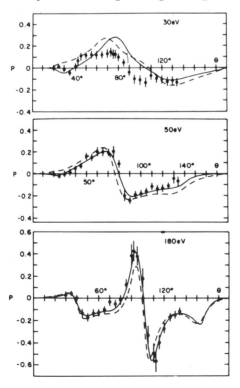

Fig. 2.   Polarization of electrons scattered inelastically
          from Hg atoms (excitation of the $6^1P_1$ state; energy
          loss 6.7 eV).  Experimental[4] and theoretical[5] values.

various incident energies, the polarization vs. scattering angle of electrons that have excited the $6^1P_1$ level of Hg (energy loss 6.7 eV). The experiments have been done by W. Eitel[4] in my group, the theoretical results are those of Madison and Shelton.[5] Bonham[6] published similar theoretical results last year. Strong polarization effects have also been observed in resonance scattering of slow electrons.[7]

Polarized electrons are an ideal tool for studying exchange collisions, as I will discuss later. I will also talk about a few of the many aspects of spin polarization in ionization processes. You can produce polarized atoms by Stern-Gerlach-type magnets or by optical pumping with circularly polarized light and then eject the oriented atomic electrons by photoionization or by collisional ionization. Polarized electrons produced by collisional ionization of polarized metastables in gas discharges have been used as a diagnostic tool for analyzing collision processes by Walters and his group at Rice University.[8,9] Another collision process will be discussed by Obyedkov later. One can also obtain polarized electrons from unpolarized atoms by using circularly polarized light for photoionization. I will discuss this Fano effect later on. I will also briefly describe how polarized electrons arise from multiphoton ionization.

There are other aspects of polarized-electron physics which seem not to belong in a conference of the kind we have here: One can extract polarized electrons from magnetized solids. The field of electron-spin spectroscopy, which has been developed by a group in Zürich, yields interesting and even surprising results on the band structure of magnetic solids.[10] Several groups have started to work with polarized electrons in low-energy electron diffraction (LEED). I will later present the first experimental result.

Figure 1 would not be complete if it did not contain the impressive measurements of the electron g factor by means of polarized electrons.[11] This is one of the fields for which the Seattle Physics Department also is well known. Last but not least, experiments with polarized electrons are being prepared at several high-energy accelerators.

From this broad spectrum I am going to pick out a few topics where significant experimental progress has been made during the two years since the last ICPEAC. Let me first describe studies of exchange scattering.

In a conventional electron-atom scattering experiment it is impossible to decide whether the observed electron has suffered a direct or an exchange collision. For this reason

direct and exchange cross sections cannot be measured separately.  This is not so if the colliding electrons are distinguishable:  If one works with spin-polarized electrons or
atoms and if the colliding electrons retain their spin direction during the scattering event, then it is possible to distinguish between direct and exchange scattering.  This has
been done by Bederson's group[12] in a well-known series of
experiments and by a group at the Joint Institute for
Laboratory Astrophysics.[13]  Both groups scattered unpolarized
electrons by polarized atoms.  This can, of course, only be
done with atoms that have unsaturated spins, since only these
atoms can be polarized (alkalis, for example).  If one, instead, scatters polarized electrons from unpolarized atoms,
one does not have this restriction and can use any atoms one
wants.

   Such an experiment has been performed by Hanne in Münster
with Hg atoms.[14]  The idea was, very roughly, as follows:
Take a polarized electron beam and fire it at Hg atoms in
their ground state.  Observe the electrons scattered in the
forward direction after excitation of the $6^3P$ state of the
mercury.  We chose the forward direction for two reasons.
First, there is the maximum of the scattered intensity.
Second, the spins of the electrons scattered in the forward
direction are not affected by spin-orbit interaction; this
has been established in earlier experiments.

   The excitation of the triplet state by exchange scattering is illustrated in Fig. 3 which shows that the outgoing

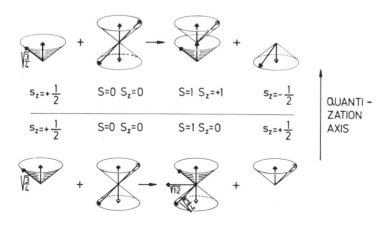

Fig. 3.   Excitation of triplet states by exchange collisions.

electron has the opposite spin direction of the incoming electron, if the sublevel $M_S = 1$ is excited. If $M_S = 0$ is excited, the spin directions of incident and scattered electrons are the same. A simple calculation that considers these two processes shows that the ratio $P'/P$ of the initial to the final polarization should be $-1/3$, if exchange scattering were the only way to excite the triplet state. In mercury, however, the spin-dependent forces are not negligible. The excitation of a triplet state can in this case not only occur by an exchange of electrons but also by a direct process, in which the spin of one of the atomic electrons flips during excitation. This affects, of course, the value of $P'/P$ so that a measurement of this ratio yields the extent to which the exchange processes discussed above still contribute to the excitation. This has been studied in a triple-scattering experiment.

Figure 4 is a schematic diagram of the apparatus. Scattering from a mercury-vapor beam, as described at the beginning of my talk, is used to produce a polarized electron beam. The polarized electrons are decelerated to energies between 5 and 15 eV and focused on a second mercury target. From the electrons scattered here an energy analyzer selects those that have been scattered in the forward direction after excitation of the $6^3P$ states of the mercury atoms (energy loss $\sim$ 5 eV). The polarization of these electrons is analyzed by a Mott detector.

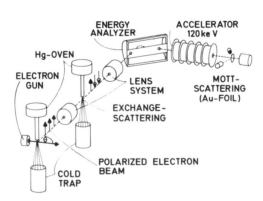

Fig. 4. Triple scattering experiment for direct observation of exchange scattering by spin flip of polarized electrons in excitation of Hg.[14]

Figure 5 illustrates an experimental result, which shows
that at incident energies below 8 eV there are a great number
of spin-flip processes.  Near 6 eV the aforementioned value
of P'/P = -1/3 is observed within the experimental error
limits.  That means that at this energy nearly all the excitation processes of the $6^3P$ levels occur by exchange scattering.
On the other hand, the exchange excitation discussed above no
longer plays an appreciable role at energies above 10 eV.
Figure 6 gives these facts directly for the $6^3P_1$ state.  It
is the evaluation of a measurement in which the fine structure
of $6^3P$ has been resolved.

Needless to say, experiments of this kind are rather
delicate and need careful checks in order to ensure that the
depolarization observed is not spurious.  An essential check
of the experiment discussed is shown in Fig. 7.  Here the excitation of the $6^1P_1$ level (energy loss 6.7 eV) has been studied in the same apparatus.  In the excitation of a singlet
state from a singlet ground state no change of spin directions
can occur, no matter whether the excitation takes place by a
direct or an exchange process.  This is observed in the experiment, which shows that no spurious depolarization effects
occur.

This exchange experiment contains other interesting
points which we are now studying but which I have to skip.
All I can do here is to point out that polarized electrons
provide a means for direct observation of exchange scattering,
a field about which quantitative information is meager.

So far we have been discussing electron scattering by
free atoms.  Even though this is an atomic collisions conference, I think we should not be so specialized as to ignore
an interesting experiment which is connected to surface
physics:  The group of Walters[15] at Rice University obtained
the very first results on polarization effects in low-energy
electron diffraction (LEED).  I mentioned at the beginning
the high polarization arising in electron-atom scattering.
There is no reason why this should occur only in scattering
from free atoms.  There have been theoretical papers[16,17]
predicting high spin polarization of the Bragg reflections
obtained in LEED from solid targets.  Whereas in scattering
from free atoms the polarization is determined solely by the
atomic field, in LEED the scattering process is influenced by
several additional factors:  The periodicity of the crystal
lattice, the surface potential barrier, multiple scattering
and inelastic processes.  The electron polarization is therefore a sensitive probe of these factors.

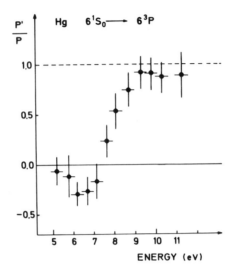

Fig. 5.   Measured values of depolarization vs. incident energy
for $6^1S_0 \to 6^3P$.

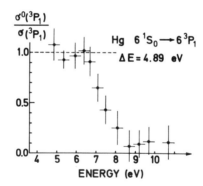

Fig. 6.   Contribution of the exchange processes illustrated
by Fig. 3 to the excitation of the $6^3P_1$ state.   $\sigma^0$
is the differential cross section for excitation
by these exchange processes, $\sigma$ is the complete differential cross section for excitation of $6^3P_1$.

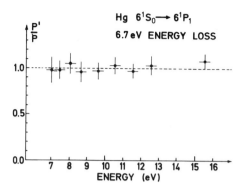

Fig. 7.    Measured values of depolarization vs. incident energy
for $6^1S_0 \rightarrow 6^1P_1$.

Figure 8 gives an example of an experimental result.    It
has been obtained with a clean tungsten (001) surface in
ultrahigh vacuum.    The incident angle $\theta$ of the primary electron beam was varied and the polarization of the specularly
reflected beam was studied with a Mott detector.    The solid

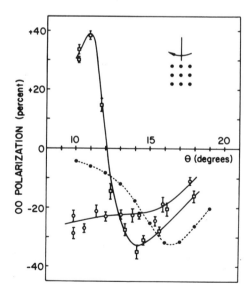

Fig. 8.    Polarization of the 00 beam as a function of angle
of incidence $\theta$ for two values of the incident electron energy:    o, 69 eV; and □ , 82 eV.[15]    Dotted
line:    Calculated polarization for scattering of
100-eV electrons from free tungsten atoms.

lines are experimental polarizations for two different primary energies, 69 eV and 82 eV.  The comparison of the 82-eV curve with the corresponding theoretical curve (corrected for the inner potential) for scattering from free tungsten atoms shows little agreement.  This is not surprising, since the polarization in scattering from a solid target is influenced by all those factors I mentioned before.

The experimental results have not yet been quantitatively explained.  The quantitative theory is certainly much more difficult than in the case of free atoms.  But there is no question that polarization experiments in LEED, in conjunction with theories as they are being developed now, are a new promising technique in surface studies.  This explains why several labs, among them the National Bureau of Standards, have gone into this field.

Let me now talk about polarized electrons arising from photoionization.  This field received a great impact from Fano's discovery that one does not have to photoionize polarized atoms in order to obtain polarized photoelectrons.  One can also start from ordinary unpolarized atoms, if circularly polarized light is used for photoionization.[18]  This Fano effect is caused by the spin-orbit interaction of the photoelectron in the continuum.  Spin-orbit interaction has two effects:

1) It can cause a spin flip of the ejected electron during the photoionization process.

2) It leads to a dependence of the photoionization cross section on the direction of the atomic spins relative to the photon spins.  In other words:  The ionization probability of atoms with spins parallel to the direction of the incident photon spins differs from that of atoms with spin antiparallel to this direction.

Accordingly, if we photoionize an unpolarized atomic beam (i.e. a mixture of equal numbers of spin-up and spin-down atoms) with circularly polarized light, we produce different numbers of e↑ and e↓, that means, we obtain a spin polarization of the photoelectrons.  An example of an experimental result may be found in Ref. 19.  It shows the polarization of the photoelectrons from cesium atoms as a function of the wavelength.  Near 2900 Å a polarization of 100% was obtained within the experimental error limits.  These results together with similar results by Baum, Lubell and Raith[20] allow us to say that the Fano effect is fairly well understood for alkali atoms.

In the subsequent research, there were two basic aspects:  First, the Fano effect was utilized for building

sources of polarized electrons.  Several groups have been
successful recently.  The group at Bonn has published its
result.[21]  By photoionization with a frequency doubled pulsed
laser they obtained $3 \cdot 10^9$ electrons/pulse of 90% polarization.
The second aspect of the further studies is:  How about elements other than alkalis, do they yield polarized photoelectrons, too?  Again the theoreticians were faster than the
experimentalists.  The question was answered with "yes" in
Leningrad,[22] Chicago[23,24] and Belfast.[25]

   In discussing the photoelectron polarization in these
cases we must take into account the fact that for many atoms
autoionizing transitions play a part in the photoionization
process.  They may cause a resonance structure of the polarization curve, as is shown in Fig. 9 for the case of thallium, for which the first experimental result has now been
obtained.[26]  Dr. Heinzmann will explain this later in more
detail.  It is an interesting feature that owing to this
resonance behavior, one frequently has polarization peaks
at those wavelengths where one has maxima of the photoionization cross section.  The magic rule for most other polarization phenomena, that polarization maxima are associated
with intensity minima, is thus broken.  As will be explained
in the following talk, polarization measurements in autoionizing transitions yield information on autoionization that
is not obtainable from cross-section measurements.

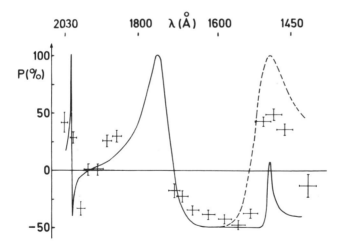

Fig. 9.  Polarization of electrons produced by photoionization
         of Tℓ atoms with circularly polarized light.  Experi-
         mental results and values predicted on the basis of
         experimental cross sections.

Another possibility for producing polarized photoelectrons is multiphoton ionization.  This process, in which several photons are absorbed simultaneously, occurs if one works with high-intensity light sources like lasers.  Figure 10 shows the simplest example I can think of.  It has been suggested by Farago and Walker.[27]  Take an atom like thallium in a $P_{1/2}$ ground state and consider a two-photon process with, say, circularly polarized $\sigma^+$ light.  Then you have the selection rule $\Delta m_j = +1$.  This means that in Fig. 10 the arrows go upwards to the right.  By the first absorption process with a suitable wavelength, you reach the sublevel $m_j = 1/2$ of an S state; no other transitions are allowed.  Since for an S state $m_j$ equals $m_s$, the excited state has the spin-orientation quantum number $m_s = +1/2$; in other words, the excited atoms are totally polarized.  These totally polarized atoms are ionized by absorption of the second photon so that totally polarized electrons are to be expected.

This is by no means an exceptional case.  By playing around with selection rules, energy levels, and the number of photons you can easily find innumerable other transitions which yield polarized electrons.  It seems, however, that quite a few problems come up when one tries to verify these predictions experimentally:  P. Lambropoulos has shown that the polarization may be strongly affected by high intensity effects of the radiation field.[28]  Preliminary experimental results have been obtained with sodium in JILA[28] and with cesium in Münster.[29]  They verify that electron polarization is obtained by multiphoton ionization.  For a quantitative test of the just-mentioned intensity effects of the laser light these measurements must be improved.

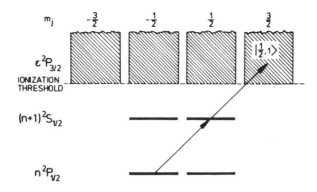

Fig. 10.  Polarized photoelectrons by two-photon transition in trivalent atoms.

Let me conclude with another digression into solid-state
physics.  After we had made the Fano-effect experiment with
cesium atoms we were curious enough to try the same with a
cesium surface:  We evaporated our cesium beam on a substrate,
under the dirty conditions of normal high vacuum, and measured
the polarization of the photoelectrons produced by circularly
polarized light.  It turned out that even under these conditions we obtained a polarization.[30]  (See Fig. 11.)  It was
only 5%, but the polarization maximum was now in the experimentally convenient visible region -- not in the uv as with
free atoms -- and the intensity from this dense solid target
was much higher than that obtained with an atomic beam.  Other
alkalis gave similar results.  The subsequent work at Münster
by Koyama and Merz showed that the polarization is due to
spin-orbit splitting of the energy bands.[31]  This occurs in
many solids so that many materials should yield polarized
electrons in photoemission.

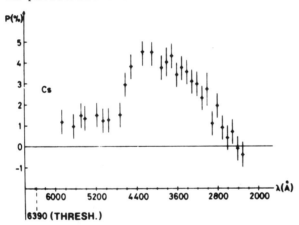

Fig. 11.   Polarization of photoelectrons from solid cesium
produced by circularly polarized light.[30]

This has been demonstrated for the case of GaAs by the
aforementioned group in Zürich.[32]  Figure 12 shows the result
they obtained in ultrahigh vacuum.  The shape of the polarization curve can be easily explained by the well-known band
structure of GaAs.  For the many materials whose band structure is less well known, electron-polarization measurements
provide information on the acute problem of energy-band splitting in solids.

I did not emphasize in this report the aspect of polarized-electron sources since I have reviewed this area at an
earlier conference.[33]  Rather I have tried to show that from

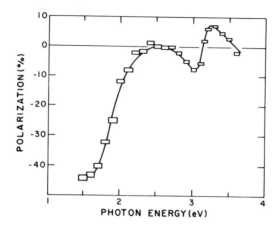

Fig. 12.  Polarization of photoelectrons from GaAs + CsOCs at
T ≲ 10 K produced by circularly polarized light.[32]

experiments with polarized electrons interesting things can
be learned in quite different fields of physics.  A less condensed survey on the physics of polarized electrons will appear early in 1976.[34]


REFERENCES

1.  K. Jost and J. Kessler, Z. Physik 195, 1 (1966).

2.  D. W. Walker, Advances in Phys. 20, 257 (1971).

3.  J. Kessler, Rev. Mod. Phys. 41, 1 (1969).

4.  W. Eitel and J. Kessler, Z. Physik 221, 305 (1969).

5.  D. H. Madison and W. N. Shelton, Phys. Rev. A 7, 514
    (1973).

6.  R. A. Bonham, J. Electron  Spectrosc. 3, 85 (1974).

7.  T. Heindorff, J. Höfft and E. Reichert, J. Phys. B: Atom.
    Molec. Phys. 6, 477 (1973).

8.  P. J. Keliher, F. B. Dunning, M. R. O'Neill, R. D. Rundel
    and G. K. Walters, Phys. Rev. A 11, 1271 (1975).

9.  P. J. Keliher, R. E. Gleason and G. K. Walters, Phys.
    Rev. A 11, 1280 (1975).

10.  H. C. Siegmann, Phys. Reports 17, 37  (1975).

11.  A. Rich and J. C. Wesley, Rev. Mod. Phys. 44, 250 (1972).

12.  B. Bederson, in Atomic Physics 3, edited by S. J. Smith
     and G. K. Walters, Plenum Press, New York 1973, p. 401.

13. D. Hils, M. V. McCusker, H. Kleinpoppen and S. J. Smith, Phys. Rev. Letters 29, 398 (1972).

14. G. F. Hanne and J. Kessler, Phys. Rev. Letters 33, 341 (1974).

15. M. R. O'Neill, M. Kalisvaart, F. B. Dunning and G. K. Walters, Phys. Rev. Letters 34, 1167 (1975).

16. P. J. Jennings and B. K. Sim, Surface Science 33, 1 (1972) and references therein.

17. R. Feder, Phys. Stat. Sol. 62, 135 (1974) and references therein.

18. U. Fano, Phys. Rev. 178, 131 (1969).

19. U. Heinzmann, J. Kessler and J. Lorenz, Z. Physik 240, 42 (1970).

20. G. Baum, M. S. Lubell and W. Raith, Phys. Rev. A 5, 1073 (1972).

21. W. v. Drachenfels, U. T. Koch, R. D. Lepper, T. M. Müller and W. Paul, Z. Physik 269, 387 (1974).

22. N. A. Cherepkov, Sov. Phys. JETP 38, 463 (1974).

23. H. A. Stewart, Phys. Rev. A 2, 2260 (1970).

24. C. M. Lee, Phys. Rev. A 10, 1598 (1974).

25. V. L. Jacobs, J. Phys. B: Atom. Molec. Phys. 5, 2257 (1972).

26. U. Heinzmann, H. Heuer and J. Kessler, Phys. Rev. Letters 34, 441 (1975).

27. P. S. Farago and D. W. Walker, J. Phys. B: Atom. Molec. Phys. 6, L280 (1973).

28. P. Lambropoulos and M. Lambropoulos, Proc. Internat. Symposium on Electron and Photon Interactions with Atoms, Stirling, 1974.

29. H. D. Zeman, Proc. Internat. Symposium on Electron and Photon Interactions with Atoms, Stirling, 1974.

30. U. Heinzmann, K. Jost, B. Ohnemus and J. Kessler, Z. Physik 251, 354 (1972).

31. K. Koyama and H. Merz, Z. Physik B 20, 131 (1975).

32. D. T. Pierce, F. Meier and P. Zürcher, Physics Letters 51A, 465 (1975).

33. J. Kessler, in Atomic Physics 3, edited by S. J. Smith and G. K. Walters, Plenum Press, New York 1973, p. 523.

34. J. Kessler, Polarized Electrons, Springer-Verlag, Berlin, 1976.

# ELECTRON IMPACT EXCITATION OF LIGHT ATOMS

M.R.C. McDowell

Mathematics Department

Royal Holloway College, Englefield Green,
Surrey, U.K.

Since the VIII[th] ICPEAC conferenee in Belgrade, 1973,
considerable progress has been made in some aspects of our
understanding of electron interaction with atomic systems, by
a combination of theoretical and experimental advances, but
many difficult problems remain.

To begin with a success story, and with what is in
principle the simplest theoretical problem, we can consider the
excitation of atomic hydrogen by electron impact. Very recently
Callaway et al. ([1]) have reported hybrid calculations in which
an eleven state basis (1s, 2s, 2p, 3d, three s-pseudostates,
three p-pseudostates, and a d-pseudostate) has been employed
in algebraic variational calculations for total orbital angular
momentum $L \leq 3$ and a Distorted Wave polarized orbital (DWPO)
approach is used for $L > 3$. The total cross sections obtained
are in essentially complete agreement with experiment both for
$1s \rightarrow 2p$ and $1s \rightarrow 2s$ (fig. 1), as are the computed values for
the polarization of Ly$\alpha$ (fig. 2). At energies above 15 eV
these results agree well with those of the 5-pseudostate
calculation of Burke and Webb ([2]) indicating the rapid
convergence of the pseudo-state approach. The agreement with
experiment ([3,4]) persists to high energy. The absolute
differential cross sections for $n = 2$ excitation reported by
Williams and Willis ([5]) are in good agreement with the results
of a DWPO calculation at angles out to $90°$. At larger angles
both the unitarized hybrid method, the unitarized 3-state close
coupling calculations ([6]) and the second order optical potential
calculations of Bransden and Winters ([7]) in which distortion is
included in both initial and final channels, predict results
close to the measured values. At higher energies (> 100 eV)

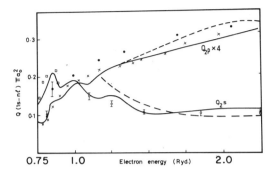

Fig. 1.  Electron excitation of the  n = 2  states of  H.  Full curves hybrid method (1), dashed curves Burke and Webb (2). Experiment  2p. ● Long et al. (4). x McGowan et al. (4), William and Willis (5) 2s Kaupilla et al. ⚡(3), D. Kochsmeider et al. ☐ (3).  The  2s  results include cascade contributions.

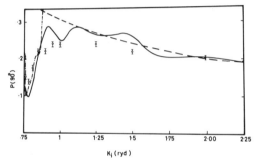

Fig. 2.  Polarization of  Lyα  at 90° to incident beam. Experimental points, Ott et al.(21).  Solid curve, hybrid (1), --- DWPO (20), -.-. Burke, Taylor and Ormonde, J. Phys. B. 1, 325, 1968.

the Eikonal-Born-Series approach of Byron and Joachain has been applied to this problem by Byron and Latour (8) and also gives good agreement with experiment.  It is clear from this work that the dominant effect in large angle inelastic differential scattering is elastic scattering from an intermediate state, which is predominantly either the ground or the excited state.

It may be possible to distinguish between these several sophisticated and successful theoretical models by comparing their predictions for the orientation and alignment parameters which could be measured in coincidence experiments with both Lyα  and  Hα  photons and scattered electrons.

For higher excited states, reasonable agreement exists, and is reported elsewhere at this conference for the total  Hα

cross section, though severe discrepancies still exist for
e + H(1s) → e + H(3d)  and for the polarization of  Hα (57,58).

One might suppose that in view of the good agreement that
now exists for the  n = 2  states of atomic hydrogen, that a
similar success story could be told for  He⁺. This is, however,
not the case.  The position has recently been examined by Seaton
(10).  Measurements are available (11,12) only for the total
cross section for  1s → 2s  excitation (corrected for cascade).
The low energy theoretical close-coupling calculations (13,14)
some of which include short-range correlation effects, agree
well among themselves, both in absolute value, and in their
predictions for the resonance positions.  When folded with the
experimental electron beam energy distribution (Gaussian, FWHM =
1.4 eV) the predicted energy dependence is in agreement with
experiment below the  n = 4  threshold.  The absolute values
are however a factor of two higher than the experimental ones,
which were normalized to scaled Born, corrected for cascade,
at energies from 450 to 1000 eV.  However the high energy
theoretical calculations (Three state close coupling (10),
DWPO (15,16)) are also in close agreement, lying some 10% below
the scaled Born results at 100 Ryd.  Their shape closely fits
the seven measured points at  E > 450° eV.  Either there are
important effects omitted in the low energy models, or in the
high energy models, or both, or there are in addition (or
separately) experimental difficulties in maintaining
normalization over a wide energy range.

Similar but less severe discrepancies (of 34% at threshold)
exist for the excitation of the resonance transition (17) in
Ca⁺.  Here it is likely that the explanation lies in the failure
of the present close coupling calculations (18,19) to correctly
account for the position and widths of the resonances in the
¹P  continuum of  Ca.

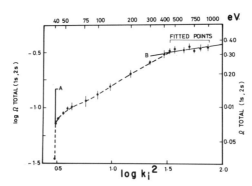

Fig. 3.  Collision strength  Ω  for  He⁺(1s → 2s), from Seaton
(5).  A  is low energy fit,  B  high energy fit.  Data those
of Dolder and Peart (12).

We noted above that the high energy models for $He^+(1s \to 2s)$ tended to give results lying appreciably below the scaled Born values even at very high energies.  This effect has also been noted by McDowell et al. ([20]) in $H(1s \to 2p)$ who showed that inclusion of adiabatic polarization distortion of the target was required to account for the observed values ([21]) of the polarization of $Ly\alpha$.  The resultant cross section behaves asymptotically as

$$EQ \underset{E \to \infty}{\to} 4M^2_{2p} \left[ \ell nk_i^2 + 0.1688 + 0.375k_i^{-2} + .0(k_i^{-3}) \right]$$

whereas ([12]) the Born result is

$$EQ \to 4M^2_{2p} \left[ \ell nk_i^2 + 0.4893 + O(k_i^{-4}) \right]$$

the discrepancy, $\delta Q = (0.72/k^2) \pi a_o^2$, being greater than 5% at energies up to 8 keV.  The predicted decrease in the cross section at high energies is as much as 50% for the $1s \to 3d$ transition ([9]) and for highly polarizable systems such as $K(4s \to 7p)$ and $Cs(6s - 8p)$ ([23]).  Non-adiabatic effects may modify these results.

Turning now to electron excitation of helium we are now able to calculate total and differential cross sections for singlet-singlet cross sections which are in good agreement with the recent absolute measurements ([24-27]).  At low energies the DWPO calculations of Scott and McDowell ([28]) give the most accurate total cross sections, but at energies above 100 eV the first order many-body theory of Taylor and his colleagues in which the HF transition density is replaced by the RPA transition density ([29]) the second-order optical potential approach ([7]) and the eikonal Born series calculations of Byron and Joachain ([30]) are equally successful in this respect, and give superior large angle differential cross sections.  The leading order correction in all these models is dipole distortion of the target.  The Glauber results, while a considerable improvement on the FBA, are of poor quality at large angles at energies below 100 eV.
The many-body theory approach of Taylor leads naturally to a formulation in which both the incident and scattered electron see the field of the ground state, whereas in the other distorted wave models, the scattered electron is either an undistorted wave (DWPO) or sees the field of the exit channel.  Taylor's model is appealing for high energy collisions, the transit time being much shorter than the response time of the atom.  Again, intermediate state elastic scattering dominates the large angle inelastic differential cross sections.

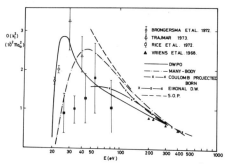

Fig. 4. Total cross section for  e + He(1$^1$S) → e + He(2$^1$S).

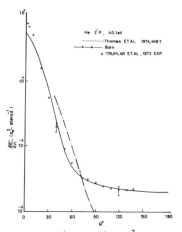

Fig. 5. Electron excitation of  He(2$^1$P).  Differential cross sections.  Solid curve, many-body theory (29), -.-., Born. The experimental points are those of Trajmar (24).

However when one attempts to extend these models to 1$^1$S → n$^3$S, n$^3$P  transitions, disaster strikes.  This is clearly because in singlet-singlet transitions, the non-exchange part of the scattering amplitude, which dominates, is treated to second order, but only a first order treatment of the exchange term has been possible so far, and in the singlet-triplet transittions no direct term occurs.  Second-order effects in the many-body theory models, and exchange-polarization effects in the DWPO model are under investigation, and may provide the answer.  At low energies close-coupling calculations by Burke and his colleagues look promising. (56).

A further feature of interest in intermediate energy inelastic collisions of electrons with both  H  and  He  atoms is the marked forward enhancement of the differential cross

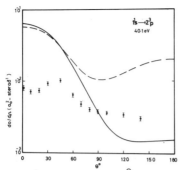

Fig. 6.   As for Fig. 5, but   He($2^3$P) :   --- Many Body Theory (29)
                                               ___ DWPO II., (32).


section predicted by second order theories, compared to the Born
approximation.   Theoretical results for   H(1s → 2s) at 100 eV
are shown in Fig. 7   in a variety of theoretical models.   The
Born, Coulomb-projected-Born and DWPO (without target distortion)
all predict a relatively small forward cross section
($\simeq 8 \times 10^{-1}$ $a_o^2$ -sterad$^{-1}$), but models which take into account
virtual  s - p  transitions (Glauber, DWPO II, Eikonal-Born
Series) predict a much higher value, the last two theories
agreeing on a value close to   $2.6a_o^2$ sterad$^{-1}$.

     There are no experimental measurements in atomic hydrogen
for   θ < 20°, but results on similar transitions in  He  have
been reported by Pochat et al. (31).   Scott and McDowell (32)
have shown that target distortion effects completely account
for the observed enhancement for  $1^1S$ → $4,5^1S$,   at energies
E ≥ 200 eV, though not at lower energies.   Further work is
required.

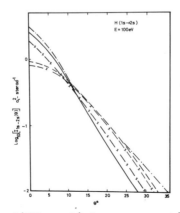

Fig. 7. Small angle differential cross sections for  H(1s → 2s)
-.-. EBS Byron, (private communication), ——— DWPO II (20),
-+-+ Glauber, --- Coulomb projected Born, --x-- DWPO I.

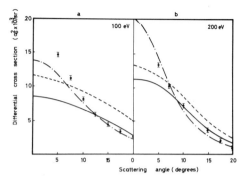

Fig. 8.  Small angle differential cross sections for He($4^1$S).
-.-. DWPO II (32), _____ DWPO I (32), --- FBA (Bell, Kennedy and
Kingston, J. Phys. B. $\underline{2}$, 26, (1969, Experimental points those
of Pochat (31).


Sophisticated tests of our theoretical models are provided
by the electron-photon coincidence experiments (33) of Eminyan
et al. in which measurements yield the ratio of the differential
cross sections for the magnetic sub-levels of the excited
state, e.g.

$$\lambda(2^1P) = \sigma_0/(\sigma_0 + 2\sigma_1)$$

and the relative phase of the components of the excited state,
$\chi$.  Both the first Born and Glauber approximations give  $\chi \equiv 0$.
The calculations to date, in different second order models,
show similar trends but are in no means perfect agreement
with experiment; the most satisfactory being the ten-state
eikonal calculations of Flannery and McCann. (59)

Experimental measurements of inelastic differential cross
sections of the  n = 3  states (for both  H  and  He) are now
required, over a wide range of energies and angles, to further
test the theories.

Considerable progress has been made since the last
conference in making accurate measurements of electron
scattering by the alkali metals.  Results are now available for
ionization cross sections (34) total cross sections (36) and
in some cases for the resonance transition (37,38).  While
good agreement exists for the ionization cross sections (39)
of  Li  and  Na; for the resonance transition in  Li (40,41) at
energies above 15 eV; for the polarization of the resonance
(6708 Å) line and for the resonance cross section and optical
polarization of the resonance line in  Na (42,43) there is a
major discrepancy regarding the total cross sections at
intermediate energies.

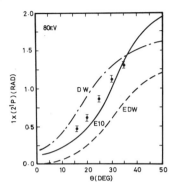

Fig. 9. Modulus, $|\chi|$ of the relative phase of components of He($2^1$P) excited by 80 eV electrons. Experimental points are those of Eminyan et al. (33). Solid curve 10-state eikonal (Flannery and McCann 1975), --- Eikonal distorted wave (Joachin and Vanderpoorten 1974) — · — distorted wave Born (Madison and Shelton, Phys. Rev. A7, 499, (1973).

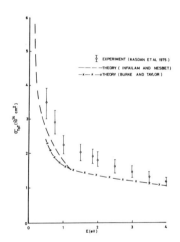

Fig. 10. Total cross section for e - Li scattering. Experimental Points of Kasdan et al. (35), --- Sinfailam and Nesbet (44), -x-x Burke and Taylor (40).

The measured (35) total cross sections for e + Li agree well with the algebraic variational calculations (44) and close-coupling calculations (40) below the inelastic threshold.

By 50 eV the measured total cross section is 60 x $10^{-16}$ cm$^2$. The sum of the measured ionization and resonance cross sections is 25 x $10^{-16}$ cm$^2$, in close agreement with theory, and with the Bethe-Born sum rule value (45) of 28 x $10^{-16}$ cm$^2$. It follows that the elastic contribution to measured total must be closely $35 \times 10^{-16}$ cm$^2$. However Burke and Taylor's close coupling calculation gives $5 \times 10^{-16}$ cm$^2$, and a recent optical potential calculation (46) by Walters yields an even lower value. Inokuti and McDowell (47) have calculated the elastic cross section due to a simple polarization potential, $V(r) = \alpha/r^4$, $r \geqq r_o$, and a random phase approach at small $L$, and obtain

$$Q_{el}(k) = 3.223(\frac{\alpha}{2k})^{2/3} \pi a_o^2$$

which yields a value of $33 \times 10^{-16}$ cm$^2$ at 50 eV, in close agreement with experiment. Similar accuracy is achieved (48) for Na and K, but the model breaks down at high energies. Walters (46) shows that in a model two-state calculation non-adiabatic effects greatly reduce the elastic cross section, and give results in an extension of Inokuti and McDowell's model close to the close-coupling values. If these non-adiabatic effects produce a corresponding reduction in a full optical potential calculation the experimental results at 20 eV and above would be difficult to understand. Many-body and optical potential approaches are urgently required to resolve the discrepancy.

Table I Cross sections for e - Li scattering $10^{-16}$ cm$^2$

|  |  | 5 | 10 | 20 | 50 |
|---|---|---|---|---|---|
|  | Energy (eV) | 5 | 10 | 20 | 50 |
| Burke and Taylor | Elastic | 49 | 21 | 9 | 5 |
|  | Inelastic | 44 | 47 | 35 | 18 |
|  | Total | 93 | 68 | 44 | 23 |
| Inokuti and | Elastic | 73 | 59 | 47 | 33 |
| McDowell | Inelastic | 41 | 38 | 31 | 22 |
|  | Total | 114 | 97 | 78 | 55 |
| Kasdan et al. | Exp. (total) | 112 | 89 | 66 | 60 |

Very sophisticated calculations are now being undertaken for complex atoms. Matrix variational calculations on C, N, and O have been reported by Nesbet and his colleagues (49) and in an R-matrix approach by the Belfast and Meudon group. (50).

Nesbet (49) uses a basis set involving, for O, the $2s^2 2p^4$, $2s 2p^5$ and $2p^6$ target configurations and allows for all possible $2p \rightarrow n'\ell'$ excitations. Le Dourneuf et al. (50) however include both the target configurations and those built out of 3s, 3p, 3d pseudo-states. Both sets

of workers agree on the importance of arranging accurate
polarizabilities for each term of the ground configuration,
for short range correlation effects, and for exchange. Very
many (> 100) configurations may need to be included, if
accurate total cross sections, and positions and widths
of resonances are to be obtained.

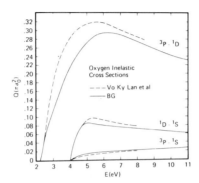

Fig. 11.  Integral inelastic cross sections for  e - O
collisions.  (From Thomas and Nesbet (49).

Thomas et al. (48) find a slightly bound $^3P^e$ state of  N$^-$,
but in the latest calculations of the Belfast group (50) this
has moved up to become a resonance within  0.1 eV  of
threshold.  Its existence depends on accurately locating the
$^3P^e$ state of  N$^-$  relative to the ground  $^4S^o$  state of  N.
Above 1 eV the two sets of calculations are in close agreement,
but the computed total cross section is a factor of 1.5
higher than the available experiments.  Nesbet and Thomas (49)
have also published results for atomic oxygen and show that
by including short-range correlations together with all
configurations with a virtual  2p → n'ℓ'  excitation the
calculated elastic cross section at low energies is reduced
by a factor of three from the obtained using  C.I. target
eigenstates alone.  The elastic cross sections are in good
agreement with Henry's (51) polarized orbital results, and
with the inelastic results of Vo ky Lan et al. (52), except
for  $^3P → ^1D$.

Calculations  of comparable complexity have been
carried out by several groups interested in astrophysical
applications.  Of particular interest are the  2 → 3

transitions in the lithium-like ion, NV. Close coupling
results in 2-state and 5-state models have been published
by Burke et al. (53), Hayes (54) and Henry (55)). Henry (55)
has extended the model to include eight states. The
calculations differ in the representation of the approximate
target states, and in the numerical methods employed.
Selected results at  $k^2$ = 6.0  are shown in table II.

There is no clear pattern of convergence, though the
8-state results without exchange are in reasonable agreement
with the 5-state non-exchange values, except for  2s - 3s.
Further work is required.

Table II.

| No of States | | Collision strengths for  NV  at  $k^2$ = 6.0 | | |
| --- | --- | --- | --- | --- |
| | | 2s - 3s | 2s - 3p | 2s - 3d |
| 2(NE) | (a) | .352 | 109  (b) | .532 |
| 2(E) | (a) | | .170 | |
| | (b) | | .110 | |
| | (c) | | .124 | |
| 5(NE) | (a) | .305 | .196 | .546 |
| 5(E) | (b) | .309 | .192 | .468 |
| | (c) | .253 | .179 | .440 |
| 8(NE) | (c) | .275 | .208 | .556* |

(a)  Burke et al.    (b)  Hayes.    (c)  Henry.

*  Revised results due to van Wyngaarden and Henry, this
   conference.

REFERENCES

1.   J. Callaway, M.R.C. McDowell and L.A. Morgan, J. Phys. B.
       in press, 1975.
2.   P.G. Burke and T.G. Webb, J. Phys. B. 3, L131, (1970).
3.   W.E. Kaupilla, W.R. Ott and W.L. Fite, Phys. Rev. A3,
       1099 (1970). H. Kochsmeider, V. Raible and
       H. Kleinpoppen, Phys. Rev. A8, 1365 (1973).
4.   P.L. Long, D.M. Cox, and S.J. Smith, J. Res. N.B.S. 72,
       521 (1968), J.W. McGowan, J.F. Williams, and E.K.
       Curley, Phys. Rev. 180, 132, (1969).
5.   J.F. Williams and B.A. Willis, J. Phys. B.7 L61, (1974).
6.   P.G. Burke, W.C. Fon and A. Kingston, private comm.
7.   B.H. Bransden and K.H. Winters, J. Phys. B. 8, 1236, (1975).
8.   F.W. Byron Jr. and L. Latour, this conference.
9.   R.F. Syms, M.R.C. McDowell, L.A. Morgan and V.P. Myerscough,
       this conference.

10. M.J. Seaton, to be published in Advances in Atomic and Molecular Processes, (1975).
11. D.R. Dance, M.F.A. Harrison and A.C.H. Smith, Proc. Roy. Soc. A290, 74, (1966).
12. K.T. Dolder and B. Peart, J. Phys. B. 6, 2415, (1973).
13. S. Ormonde, W. Whitaker and L. Lipsky, Phys. Rev. Lett. 19, 1161, (1967).
14. P.G. Burke and A.J. Taylor, J. Phys. B. 2, 44 (1969).
15. M.R.C. McDowell, L.A. Morgan and V.P. Myerscough, J. Phys. B. 6, 1435, (1973).
16. M.R.C. McDowell, L.A. Morgan and U. Narain, J. Phys. B. 7, L195, (1974).
17. P.O. Taylor and G.H. Dunn. Phys. Rev. A8, 2304, (1973).
18. P.G. Burke and D.L. Moores, J. Phys. B. 1 575, (1968).
19. D. Wells, Thesis, London (1973).
20. M.R.C. McDowell, L.A. Morgan, and V.P. Myerscough, J. Phys. B. in press (1975).
21. W.R. Ott, W.E. Kaupilla and W.C. Fite, Phys. Rev. A1, 1089, (1970).
22. M. Inokuti, Rev. Mod. Phys. 43, 297, (1971).
23. I.L. Beigmand and V.P. Shevelko, Optikili Spek. 37, 621 (1974).
24. S. Trajmar, Phys. Rev. A8, 191, (1973).
25. R.I. Hall, G. Joyez, J. Mazeau, J. Reinhardt and C. Seheman, J. de Phys. 34, 827, (1973).
26. C.B. Opal and E.C. Beaty, J. Phys. B. 5, 627, (1972).
27. H. Suzuki and T. Tagyanaki, Abstracts VIII[th] ICPEAC, I, 286 (1973).
28. T.A. Scott and M.R.C. McDowell, J. Phys. B, in press (1975a).
29. L.D. Thomas, Cy Csanak, H.S. Taylor and B.S. Yarlaggada, J. Phys. B. 7, 1719 (1974).
30. F.J. Byron Jr. and C.J. Joachain, J. Phys. B. 7, L212 (1974) and this conference.
31. A. Pochat, D. Rozuel and J. Peresse, J. de Phys. 34, 701, (1973).
32. T. Scott and M.R.C. McDowell, J. Phys. B. in press (1975b).
33. M. Eminyan, K.B. McAdam, J. Slevin and H. Kelinpoppen, J. Phys. B. 7, 1519, (1974).
34. R. Jalin, Thesis, Paris (1972).
35. A. Kasdan, T.M. Miller, S.Z. Zon and B. Bederson, Phys. Rev. A ., in press (1975).
36. A. Kasdan, T.M. Miller and B. Bederson, Phys. Rev. A8, 1562, (1973).
37. E.A. Enemark and A. Gallaher, Phys. Rev. A6, 192 (1972)
38. D.A. Leap and A. Gallaher, submitted to Phys. Rev.,(1975).
39. M.R.C. McDowell, Case Studies in Atomic Collisions Physics, I, (1969).
40. P.G. Burke and A.J. Taylor, J. Phys. B. 2 869 (1969).

41.  N. Feautrier, Thesis, Univ. of Paris (1970), J. Phys.
      B. 3 L152, (1970).
42.  D.L. Moores and D.W. Norcross, J. Phys. B. 5 1482, (1972).
43.  J. Kennedy, V.P. Myerscough and M.R.C. McDowell to be
      submitted to J. Phys. B. (1975).
44.  A.L. Sinfailam and R.K. Nesbet, Phys. Rev. 1987, (1973).
45.  M. Inokuti, R.P. Saxon and J.L. Dehmer, Int. J. Rad.
      Phys. Chem. (1974).
46.  H.R.J. Walters, J. Phys. B. in press, (1975).
47.  M. Inokuti and M.R.C. McDowell, J. Phys. B. 7 2382 (1974)
48.  M.R.C. McDowell, unpublished.
49.  L.D. Thomas and R.K. Nesbet , Phys. Rev. A11, 170 (1975).
      L.D. Thomas, R.S. Oberoi and R.K. Nesbet and P.G. Burke,
      Phys. Rev. A10, 1605 (1974).
50.  M. Le Dourneuf, Vo Ky Lan, K.A. Berrington and P.G. Burke
      this conference.  K.A. Berrington, P.G. Burke and
      W.D. Robb, private communication, (1975).
51.  R.J.W. Henry, Phys. Rev. 162, 56, (1967).
52.  Vo Ky Lan, N. Feautrier, M. Le Dorneuf and H. van Regemorter
      J. Phys. B. 5 1506, (1972).
53.  P.G. Burke, J.H. Tait and B.A. Lewis, Proc. Phys. Soc.
      87, 209, (1966).
54.  M. Hayes, J. Phys. B., 8, L8, (1975).
55.  R.J.W. Henry, J. Phys. B. 7 L439, (1974).
56.  K.A. Berrington, P.G. Burke and A.L. Sinfailam, J. Phys.
      B. 8, K159, (1975).
57.  R. Syms, M.R.C. McDowell, V.P. Myerscough and L.A. Morgan,
      this conference, and J. Phys. B, in press.
58.  M.R. Flannery and K.J. McCann, J. Phys. B. 7, L522 (1974).
59.  M.R. Flannery and K.J. McCann, to be published in Phys. Rev.
60.  C.J. Joachain and R. Vanderpoorten, J. Phys. B. L.528
      (1974) and private communication to H. Kleinpoppen.

# INELASTIC ELECTRON-HYDROGEN ATOM EXPERIMENTS

J.F. Williams

Dept of Pure & Applied Physics, Queen's University,

Belfast BT7 1NN, Northern Ireland.

Experimental studies of inelastic scattering of electrons from hydrogen atoms can be considered from the two viewpoints of (i) the determination of absolute cross section values[1,2,3] and of radiation polarization[3] and (ii) the investigation of scattering phenomena such as resonances[4], coherent excitation[5] and angular correlations of scattered electrons and photons[6]. Such experiments have so far been confined to the n = 2 and 3 states. The bases of the experimental methods can be deduced from the following representation of the scattering processes:

$$e(E_1) + H(1s) \rightarrow H^{-*}(n\ell, n^1\ell^1) \rightarrow H^*(n = 3) + e(E_1 - 12.08 \text{ eV})$$

$$H(1s) \underset{h\upsilon}{\overset{\downarrow}{\leftarrow}} H^*(n = 2) + e(E_1 - 10.20 \text{ eV})$$

which indicate that one observes either the scattered electrons[1], photons[2] or metastable atoms[7]. The considerable experimental difficulty arises from the need to modulate the hydrogen atom beam whose number density is generally of the same order of magnitude as the background gas number density at the interaction region. Signal-to-noise ratios as low as $10^{-3}$ can become limiting factors on the measurements.

## 1. The Measurement of Absolute Cross Section Values

Since the review by Smith[8] in 1969, a new method[9,10,11] is available for the determination of absolute cross section values. The difficulty of such determination for atomic hydrogen beams arises because a detector of the absolute

number density of hydrogen atoms in a beam geometry is not
available.  It is then not possible to make an absolute
calibration of the differential cross section, Q, by the
separate measurement of each of the terms in the simplified
relationship $Q = N_s/N_i n \ell$ where $N_s$ and $N_i$ are the numbers of
scattered particles and incident electrons, respectively, n is
the atomic hydrogen beam density in the interaction region of
length $\ell$.  This difficulty can be evaded by normalizing
relative angular or energy distributions in one of the
following procedures.

(i)   The measured relative energy distributions for 2p
photon production (for example) can be normalized[2] to
the first Born approximation value at some high energy.
Similarly the measured relative angular distributions
for production of electrons which have lost 10.2 eV in
exciting the n = 2 states have been normalized[12] to the
first Born differential cross section value at $20^0$ and
200 eV.  Such procedures are unsatisfactory because the
theory does not predict its own accuracy which must be
established by comparison with experiment.

(ii)   The relative inelastic angular distribution is
measured as a ratio with respect to the elastic scattering
angular distribution at a lower energy below the n = 2
threshold where normalization may be made to the 'exact'
variationally calculated cross section for which theory
allows bounds to be placed upon the cross section value.

(iii)   For a partially dissociated hydrogen beam, a
knowledge of the dissociation fraction and of the ratio
of the atomic to molecular hydrogen scattering
distribution permits the atomic cross section to be
normalized[13] against the molecular cross section which,
in turn, has usually been determined by other workers
using the method of separate determination of all the
scattering parameters.

However, a new method is available which avoids the
measurement of absolute beam number density.  The ratio of
the inelastic cross section at energy $E_1$ to the elastic cross
section at energy $E_2$ is determined.  In our most recent work[14]
the elastic cross section is then obtained directly from the
phase shifts which are fitted to elastic angular distributions
for well resolved $^1S$, $^3P$ and $^1D$ resonances below the n = 2
level.  In earlier work[15] an indirect calibration method
required the measurement of (a) the ratio of the elastic
cross section in atomic H to that in He at 19.3 eV for which
the phase shift analysis method[9,10] allows an absolute value
to be established and (b) the relative number densities in the
H and He beams.  References 14 and 15 give a detailed
discussion of the experimental method and error estimation.

When photons are detected an absolute calibration of the photon detector is required. For Balmer-alpha at $6562^O$A an optical standard can be used, while for Lyman alpha at $1216^O$A the method of Samson and Cairns[14] or Christofori et al[15] (see later section) has been used.

1.2                    Excitation of n = 2 states

Experimental and theoretical studies of electron impact excitation of the n = 2 states have been discussed by Moiseiwitsch and Smith[18], Smith[8], Geltman and Burke[19] and Williams[11].

The total 2p state and the total 2s state excitation cross sections, Q(2p) and Q(2s) have generally been studied by observing either the prompt (1.6 x $10^{-9}$ sec) decay photons of the 2p state or the delayed photons obtained from the electrostatic field mixing of the 2s state with the 2p state. For electric dipole radiation the percentage polarization P and the angular distribution Q($\theta$) are related through the expression

$$Q(\theta) \quad = \quad 3 \ Q_T \ \frac{(100 - P \cos^2\theta)}{(300 - P)}$$

which indicates the observations at angles of $54.7^O$ or $125.3^O$ to the axis of the incident electron beam will have the same energy dependence as the total cross section $Q_T$ (integrated over angle) whereas observations at other angles must take account of the angular anisotropy which is most readily accomplished using the polarization. These photon detection experiments generally determine excitation functions, which, for electron energies greater than the excitation energy of the n = 3 states, must be corrected for cascade contributions from the higher states before comparison can be made with theoretical values.

Observations of the 2p photons at $54.7^O$ (with relative errors of 3%) by McGowan and Williams[16] and Long, Cox and Smith[2] have obtained an energy dependence of Q(2p) from near threshold to 200 eV which are in good qualitative agreement with one another and with the shape of the Born approximation values above 200 eV. These measurements have then relied upon the Born approximation value calculated at some high energy (for example, 0.486 $\pi a_0^2$ at 200 eV for Q(2p) for normalization and for corrections for cascade contributions (about 2% of Q(2p) at 200 eV, see Moiseiwitsch and Smith[18]). These experimental values are then found to be about 20% lower than the best theoretical values[19] at low energies between the n = 2 and n = 3 levels. See figure 1.

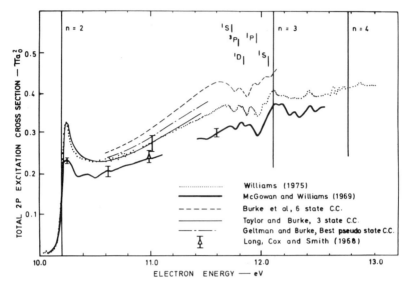

Fig.1   Total 2P state excitation cross section.

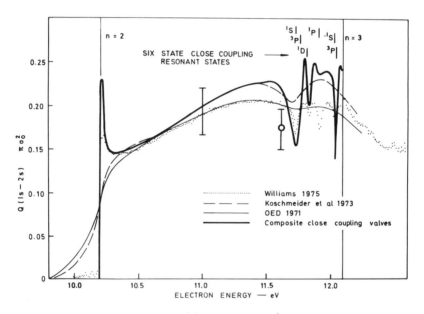

Fig.2 Total 2S state excitation cross section
The data point ⌀ is from Kauppila et al[3].

The close coupling calculations[19,20,21] have successively incorporated the low energy physically significant aspects (polarization and coupling to open and closed channels) of the n = 2 excitation process. Thus Burke et al[21] suggested that the experimental 2s and 2p excitation cross sections could be normalized to those theoretical values just above threshold. Support for this suggestion was obtained by Kauppila et al[3] whose values of the ratio of Q(2s)/Q(2p) agreed, within the experimental accuracy of $\pm$ 10%, with the theoretical values over the range 10.4 to 11.4 eV. Recently direct confirmation of the theoretical values has been provided by the absolute cross section values measured by Williams and Willis[17]. The major advances of the phase shift and resonance analysis calibration methods (see section 1) permit an absolute inelastic cross section measurement. In their experimental arrangement the elastic cross section, Q(1s) and the inelastic Q(2p) could be measured simultaneously which ensured that parameters in the beam overlap integral were identical for the determination of both Q(1s) and Q(2p). Figure 1 shows that at 11.02 eV, the calibrated value of Q(2p) is $0.276 \pm ^{0.026}_{0.016} \pi a_0^2$ which is about 20% above the previous experimental value of McGowan Williams and Curley[16] and about 15% above that of Long, Cox and Smith[2] but is in excellent agreement with the 3 state plus 20 correlation term close coupling value and the pseudo state close coupling value[19]. All of the above experimental error of +9% and -6% arose from the calibration of the photon detector. A further $\pm$6% error arises from the absolute cross section calibration. The implications of this work are similar to those stated by Geltman and Burke. Further study should be made of the cascade contributions to the Lyman alpha signal over the entire energy range, of sources of systematic experimental error from threshold to 500 eV and of the accuracy of the Born approximation at high energies.


1.3                    The 2s State

The state of understanding of the 2s excitation measurements was clarified by the work of Kauppila, Ott and Fite[3]. Figure 2 shows their total Q(2s) cross section maximum value at 11.6 eV which was normalized via their measured polarization fractions and the ratio of Q(2s)/Q(2p) at 90° to the value of Q(2p) (total) of $0.486 \pi a_0^2$ at 200 eV. If the absolute value of Williams and Willis is used for normalization, the data of Kauppila et al would lie about 20% higher and agree with the theoretical values in figure 2. Kauppila et al suggest that the experimental effect, which Stebbings et al[7] identified but perhaps did not fully eliminate, is perhaps still present in some of the measurements. This effect of spurious photon production may operate at all incident energies when secondary electrons are produced at an energy for which

the H(2s) excitation cross section is large. The search for
and the study of similar systematic errors appears to be the
direction for further study. The present absolute values are
in good agreement with theory near threshold.

1.4                Differential Cross Sections

Differential cross sections for electron impact
excitation of the n = 2 (and higher) states, in which the
orbital angular momentum substates are not resolved, may be
obtained from an integration of the area within the n = 2 (and
higher) energy loss peaks[11]. The continuum background arises
mainly from the molecular hydrogen background which has been
shown to be relatively flat through the n = 2 atomic hydrogen
peak. The signal to noise ratio is strongly dependent on the
degree of dissociation of the molecular hydrogen source, the
incident electron energy and scattering angle.

Figure 3 shows the n = 2 excitation differential cross
section for an incident electron energy of 100 eV. The
relative angular distributions of K.G. Williams (1969) have
been normalized at 25° to the present values which were
measured absolutely by the method (ii) of section 1. Similar
measurements from 54 to 680 eV show[11]
    (i)  The first Born approximation values do not describe
the measured cross sections.
    (ii)  At large angles the angular dependence of the
inelastic scattering is similar to that of elastic
scattering. For both processes, the nuclear interaction
is then probably the controlling interaction for which a
$cosec^4 \theta/2$ dependence is observed.
    (iii) The 'Coulomb projected' Born (CPB) approximation[28]
which includes the nuclear interaction shows a markedly
better large angle behaviour than the first Born
approximation. Both theory and experiment show an $E^{-3}$
energy dependence. The distorted wave polarized orbital
values calculated by McDowell et al[31] are slightly lower
than the CPB values.
    (iv)  The effect of open and closed channels upon the
scattering process, as included in the three state
close coupling calculation by Burke et al gives
reasonable agreement with experiment at 54 eV. Improved
calculations yielding values in better agreement with
experiment, will be reported at this conference by
Fon, Kingston and Burke.
    (v)  The effect of a large number of open channels
upon the excitation process in the energy region below
200 eV has been studied by Bransden et al[29], who have
accounted for several strongly coupled states by a
truncated eigenfunction expansion while the remaining

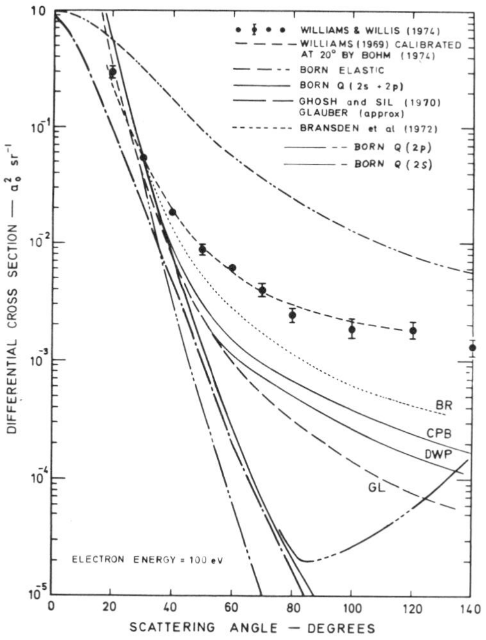

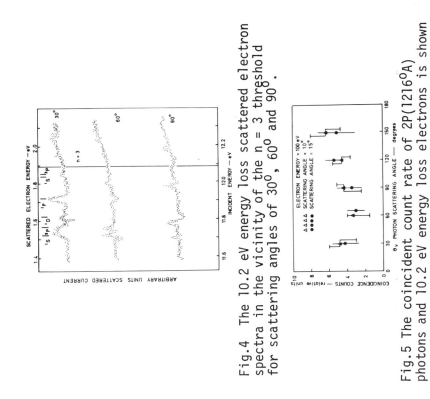

Fig.4 The 10.2 eV energy loss scattered electron spectra in the vicinity of the n = 3 threshold for scattering angles of 30°, 60° and 90°.

Fig.5 The coincident count rate of 2P(1216°A) photons and 10.2 eV energy loss electrons is shown as a function of photon scattering angle.

Fig.3. Differential n = 2 states excitation cross section for 100 eV incident electrons.

(infinite) states are allowed for by a second order
potential.  At small angles their predictions tend
towards the Born approximation while at large angles
their values underestimate (like all the other theories)
the measured cross section but, at 54 eV, the agreement
is better than other theories.
(vi)  At 680 eV, the distorted wave approximation,so
named because of its allowance for distortion of plane
waves by static and dynamic polarization effects  by
Shelton et al[30],has predicted 2s state cross section
values which are in good agreement with the measured
n = 2 cross section.  Cross section values for 2p
state excitation are not available from their method.


2.                    SCATTERING PHENOMENA


2.1                        2p State

     The finite value of the 2p excitation cross section at
threshold was predicted by Damburg and Gailitis (1963) and
implied by the results of Chamberlain et al (1969) but shown
clearly by the results of Williams and McGowan (1968) and
McGowan et al (1969).  The latter group also confirmed the
presence of a large [1]P resonance, of width about 0.015 eV,
just above threshold at 10.2 eV as predicted in a close
coupling calculation by Taylor and Burke (1967).  The existence
of at least one, possibly two, further oscillations at about
10.4 and 10.6 eV was suggested by the measurements of McGowan
et al (1969).  A calculation by Marriot and Rotenburg (1968)
indicated that interference between background potential
scattering and resonance scattering could produce structure
in the [1]P and [1]D partial cross sections.  However recent data
by Williams and Willis[17] do not show any structure within the
improved error of ±1% of the data points, which is about a
factor of five better than that obtained by McGowan et al[16].
See figure 1.

     Further resonant states lying below the n = 3  and n = 4
levels have been observed in the 2p photon decay channel by
McGowan and Williams[16] as shown in figure 1.  These resonances
are discussed in the next section.


2.2                        2s State

     Only two studies have been made of the 2s state excitation
function near threshold.  Electron energy resolutions of
0.110 eV and 0.15 eV have been used by Koschmeider, Raible and

Kleinpoppen[22] and by Oed[23] respectively to study the threshold
and resonance behaviour of the 2s excitation function.  These
studies have not been inconsistent with, but also have not
shown clearly, the existence of the predicted[21] finite cross
section value, a resonant structure just above threshold and
further resonance structure below the n = 3 level.  These
phenomena are similar to those of the 2p excitation process.
However these features are shown clearly by recent measurements
by the author[24] as shown in figure 2 in which the data  of
Koschmeider  et al and Oed have been normalized at 10.8 eV to
the pseudo state close coupling values of Geltman and Burke[19].
The thick curve is the composite theoretical curve used by
Koschmeider et al.  From 10.2 to 10.7 eV the three-state-plus-
20- correlation terms values of Taylor and Burke[20] are used to
describe the threshold behaviour.  From 10.7 eV to 12.1 eV
the six state close coupling values of Burke et al[21] are used
to describe the resonance structure.  Within the limits of
experimental error the present measurements confirm the
predictions of theory in describing the threshold behaviour of
the 2s excitation process.


2.3                      Resonances below n = 3

     Figures 1 and 2 for the 2p and 2s photon decay channels,
respectively, both show that the experimental data contain
resonance structure at about those energies, 11.77, 11.81 and
11.91 eV which are the six-state close coupling predicted
locations of the $^1$S, $^1$D and $^1$P resonant states.  Further data
on these resonances have been provided by (a) Risley et al[26]
in a fast H$^-$ion beam collision experiment in the detached
electron spectra showing structure at 11.86 $\pm$ 0.04 eV and by
(b) Spence[27] who observed only one apparent resonance feature
at 11.860 $\pm$ 0.030 eV in a total cross section transmission
type experiment.  The preceding data do not indicate either
the orbital angular momentum, L, or the spin, S, of the
resonant states.

     Recently the author has studied[24] the inelastic electron
decay channel spectra, as shown in figure 4, by observing the
10.2 eV energy loss electrons.  Because of the low energies
(less than 2 eV) of the scattered electrons there is considerable uncertainty (up to 0.080 eV) in the energy scale.  The
overlapping nature of the resonances required that as good an
energy resolution as practical (0.016 eV) be used in the
incident and scattered analyzers which in turn produced a
small beam current and poor counting statistics.  However the
resulting angular differential energy loss spectra indicate
the presence of four resonant states at energies in the
vicinity of the predicted resonances.  The identification of

the states is not clear.  In order to improve the quality of
such data, the author's apparatus would need to be improved to
obtain better performance than atom beam number densities of
$10^{10}$ atoms $cm^{-3}$ and an electron energy resolution of 0.012 eV
at a current of $10^{-8}$ amps.


2.4            Electron-Photon Coincidence Studies

     Coincident detection of the prompt decay (2p) 1216 Å
photons with the 10.20 eV energy loss electrons offers a means
of separating the 2p state excitation process from the 2s state
excitation.  Efforts at Queen's University to study this process
have been hindered by a low atom beam density of $10^{10}$ atoms
$cm^{-3}$, a low signal-to-noise ratio of $10^{-2}$ and a high spurious
coincidence contribution from those electrons which have
excited the 2s state.  Recent preliminary data are presented
in figure 5 which shows the relative coincidence count rate as
a function of photon scattering angle for an incident electron
energy of 100 eV and electron scattering angles of $10^{0}$ and $15^{0}$.
The error bars represent 70% confidence limits.  The data have
been shown to be free of error from the usual sources of atom
and electron beam density variations, detector efficiency
changes and spurious electrons and electronic noise.  A
comparison is not made with any theoretical predictions of the
coincidence count rate because of the preliminary nature of
the data.  It is noted that the above coincidence type of
experiment does not permit[31] a measurement of the relative
phase of the $m_L$ components of the nP state as has been made
for helium by Kleinpoppen[32].  A further measurement is
required to separate the singlet and triplet scattering, for
example, by using spin polarized electrons and atoms.


2.5 Coherent Electron Impact Excitation of Atomic Hydrogen

     Mahan, Krotkov, Gallagher and Smith[5] have observed an
asymmetry of Balmer-alpha $6562^{0}$A radiation with respect to the
direction of a static electric field applied across the
interaction region along the axis of the primary electron
beam.  The asymmetry is interpreted as the result of the
coherent excitation of states of opposite parity and the same
orbital angular momentum component $m_L$ i.e. predominantly the
$P_{3/2} - D_{3/2}$, $S_{\frac{1}{2}} - P_{\frac{1}{2}}$ and $P_{3/2,\frac{1}{2}} - D_{3/2,\frac{1}{2}}$ pairs within the $n = 3$
shell.  Such observations are the first of their type for free
electron-free atom single collision processes and complement
the observations of similar collective processes in the beam-
foil experiments[5].  Smith's work was a preliminary study to
investigate, using time resolution, the 3S, 3P and 3D
excitation cross sections.

2.6                    Photon-photon coincidence studies.

The photon-photon coincidence detection of sequential cascading photons has been used, for example, in the measurement of atomic state lifetimes[33] and in the measurement of polarization correlations in the test of hidden variable theories of quantum mechanics[34] but for the hydrogen atom the method has been used only for the study of the quantum-efficiency of a 1216 Å photon detector[15]. The most recent experiment at Queen's University, Belfast, extends that method to determine the quantum efficiency of a 1216 Å detector composed of a Mullard type 318 BL channel electron multiplier preceded by a flowing oxygen filter enclosed with two magnesium fluoride windows. The method requires the simultaneous measurement of (a) the total $n = 3 \to 2$ photons decay rate and (b) the coincidence rate of the $n = 3 \to 2$ photons with the total $n = 2 \to 1$ photons. Then the product of detector efficiency, $\xi_\alpha$ and angular acceptance $\Omega_\alpha$ of the Lyman alpha detector is given by the ratio of the 6562 Å singles rate to the coincidence rate i.e. $\xi_\alpha \, \Omega_\alpha = N_{32} \, \xi_2 \, \Omega_2 \, \xi_\alpha \, \Omega_\alpha / N_{32} \, \xi_2 \, \Omega_2$. For the special case of atomic hydrogen with degenerate levels and a metastable 2S state, the application of an electric field in the direction of the incident electron beam will cause mixing of degenerate levels with an effective decrease in state lifetime such that all the $n = 3$ and $n = 2$ states will decay in the collision volume. Photon detectors placed at $\pm 54.7^\circ$ avoids problems with anisotropic photon distributions.

The main experimental problem arose from (a) those 1216 Å photons which were produced by direct excitation of the $n = 2$ levels rather than by cascade and (b) from the 1216 Å and 6562 Å photons produced from the hydrogen furnace. Such photons were the largest contributor to the spurious coincidence count rate. A value of $\xi_\alpha$ of $0.09 \pm 0.04$ (with 70% confidence limit) was obtained which was in agreement, within the experimental errors, with the value of $0.06 \pm 0.02$ obtained previously by Williams and Willis[17] using the method of Samson and Cairns[14]. Further applications of the photon-photon coincidence method to lifetime determinations and cascade effects are in progress.

## REFERENCES

1.   J.F. Williams, J. Phys. B8,1683(1975).
2.   R.L. Long, D.M. Cox and S.J. Smith, J. Res. Natl. Bur. Std. (U.S.) 72A, 521 (1968).
3.   W.E. Kauppila, W.R. Ott and W.L. Fite, Phys. Rev. A1, 1099 (1970).
4.   J.F. Williams, to be published in "Interactions of Electrons and Photons with Atoms" (H. Kleinpoppen and M.R.C. McDowell, Eds., Plenum Press, N.Y.) 1975.

5.    S.J. Smith, to be published in "Interactions of Electrons and Photons with Atoms" (H. Kleinpoppen and M.R.C. McDowell, Eds., Plenum Press, N.Y.) 1975.

6.    J.F. Williams, to be published.

7.    R.F. Stebbings, W.L. Fite, D.G. Hummer and R.T. Brackmann, Phys. Rev. 119, 1939 (1960).

8.    S.J. Smith, "Physics of the one- and two-electron atoms" (Ed. H. Blum and H. Kleinpoppen, North Holland) 574 (1969).

9.    J.R. Gibson and K.T. Dolder, J. Phys. 32, 741 (1969).

10.   D. Andrick, Advances in At. Molec. Phys., 9, 207 (1974).

11.   J.F. Williams and B.A. Willis, J. Phys. B8, 1641 (1975).

12.   K.G. Williams, Proc. 6th ICPEAC (ed. I. Amdur, M.I.T. Press, Cambridge, Mass), 731 (1969).

13.   W.L. Fite, R.F. Stebbings and R.T. Brackmann, Phys. Rev. 116, 356 (1959).

14.   J.R. Samson and R.B. Cairns, Rev. Sci. Inst., 36, 19 (1965).

15.   F. Christofori, P. Fenici, G.E. Frigerio, N. Molho and P.G. Sona, Phys. Letters 6, 171 (1963).

16.   J.W. McGowan, J.F. Williams and E.K. Curley, Phys. Rev. 180, 132 (1969).

17.   J.F. Williams and B.A. Willis, J. Phys. B7, L61 (1974).

18.   B. Moiseiwitsch and S.J. Smith, Rev. Mod. Phys. 38, 1 (1968).

19.   S. Geltman and P.G. Burke, J. Phys. B3, 1062 (1970).

20.   A.J. Taylor and P.G. Burke, Proc. Phys. Soc. 92, 336 (1967).

21.   P.G. Burke, S. Ormonde and W. Whitaker, Proc. Phys. Soc. 92, 319 (1967).

22.   H. Koschmeider, V. Raible and H. Kleinpoppen, Phys. Rev. A8, 1365 (1973).

23.   A. Oed, Phys. Letters, 34A, 435 (1971).

24.   J.F. Williams, J. Phys. B, to be published.

25.   S. Ormonde, J. McEwen and J.W. McGowan, Phys. Rev. Letters 22, 1165 (1969).

26.   J.S. Risley, A.K. Edwards and R. Geballe, Phys. Rev. A9, 1115 (1974).

27.   D. Spence, J. Phys. B8, L42 (1975).

28.   S. Geltman and M.B. Hidalgo, J. Phys. B4, 1299 (1971).

29.   B.N. Bransden, J.P. Coleman and J. Sullivan, J. Phys. B5, 546 (1972).

30.   W.N. Shelton, E.S. Leherissey and D.H. Madison, Phys. Rev. A3, 242 (1971).

31.   M.R.C. McDowell, L. Morgan and V. Nyerscoff, J. Phys. B (1975).

32.   H. Kleinpoppen, to be published in "Interactions of Electrons and Photons with Atoms" (H. Kleinpoppen and M.R.C. McDowell, Eds., Plenum Press, N.Y.) 1975.

33.   C. Camhy-Val and A.M. Dumond, Astron. Astrophys. 6, 27 (1970).

34.   C.A. Kocher and E.D. Cummins, Phys. Rev. Letters 18, 575 (1967).

THEORY OF LOW-ENERGY ELECTRON-ATOM COLLISIONS

AND RELATED PROCESSES

J. N. Bardsley*

Physics Department, University of Pittsburgh,

Pittsburgh, PA   15260   U.S.A.

Through the work of atomic physicists and quantum chem-
ists over the past fifty years there has developed an extensive technology by which one can construct wave functions for
bound states of many atoms and molecules, and thus study the
properties of these systems.  The major purpose of this paper
is to examine some of the ways that this wealth of technology
can be adapted or extended to the study of scattering phenomena.

One theory of this type is the algebraic expansion method
which was used by Schwartz[1] in his definitive work on low-
energy e-H scattering and which has been developed in recent
years by Nesbet and Callaway and their colleagues.  In this
approach one introduces a basis set of square-integrable ($L^2$)
functions which are chosen to permit an accurate representation of the short-range correlation between the electrons in
the system of projectile and target.  This set is supplemented
by a small number of continuum functions with the appropriate
asymptotic form to describe the initial and final states of
the collision process.

The R-matrix method also belongs to this class of theories.  Here one separates configuration space into an inner
and an outer region.  In the inner region the wave function
is expanded in terms of a basis set of $L^2$ functions.  In the

---

*Visiting Fellow, 1974-75, Joint Institute for Laboratory
Astrophysics, University of Colorado and National Bureau of
Standards, Boulder, Colorado.

outer region the effects of exchange are neglected and the
solution of the appropriate set of coupled differential equations presents no serious computational problems, provided
that the effect of long-range forces is treated with care at
energies close to thresholds.

Some of the recent calculations using these methods have
been mentioned already by McDowell, and I will not go into
more detail.   Instead I will discuss some of the techniques
by which one can avoid completely the explicit representation
of the asymptotic form and thus carry through a complete calculation using only $L^2$ functions.

First let us examine the method of coordinate rotation,
or dilatation, in which one introduces complex values of the
coordinates.   Each inter-particle distance, r, is replaced by
$re^{i\alpha}$, where   α   is a positive number.   Consider the effect of
such a transformation on the asymptotic forms of scattering
wave functions.   In outgoing waves   exp(ikr)   becomes
exp(-kr sinα + ikr cosα) which tends to zero as   r → ∞.
However, in incoming waves   exp(-ikr)   becomes
exp(+kr sinα - ikr cosα) which diverges as   r → ∞.   Hence we
obtain wave functions that vanish asymptotically, just like
bound state wave functions in real coordinate space, provided
that we tackle problems in which there are no incoming waves.
Two suitable problems are the determination of resonance energies and the calculation of photoionization cross sections.

Resonant states can be defined as solutions of the
Schrödinger equation that satisfy outgoing boundary conditions.   Thus if the coordinates are rotated through a positive
angle the wave functions vanish as   $|r|$ → ∞   and can be ex-
panded in a complete set of real square-integrable basis
functions, $\phi_n(r)$.   One can then obtain a variational estimate
of the energy and width of the resonant state from the secular
determinant

$$\left| <\phi_m| H(\alpha) - W|\phi_n> \right| = 0 \qquad (1)$$

where

$$H(\alpha) \equiv e^{-2i\alpha}T + e^{-i\alpha}V \qquad (2)$$

and     $$W \equiv E - \frac{1}{2} i\Gamma \quad . \qquad (3)$$

Here we have assumed that the potential energy   V   is composed
of a sum of Coulomb interactions, so that both the potential
energy and kinetic energy undergo simple transformations under
coordinate rotation.

The construction and solution of the secular determinant is straightforward, and results on the lowest $^1$S resonance in H$^-$ have been obtained by Bardsley and Junker,[2] by Doolen, Nuttall and Stagat,[3] and by Bain et al.[4] During the past year the discrepancies between the various theories concerning the position and width of this resonance appear to have been resolved, and there is now general agreement concerning their values (~9.557 and 0.048 eV, respectively).

Doolen[5] has presented a useful discussion of the dependence of the results upon the rotation angle. There is little variation for angles between 20° and 30° but significant change at larger and smaller angles. For values of $\alpha$ in the above range the results are also insensitive to small changes in the basis set.

Bain et al.[4] have shown how the dependence of the results upon $\alpha$ can be reduced by the introduction of complex basis functions and have applied their method to the dominant $^2$S resonance of He$^-$. However this extension leads to more difficult computation and raises some unresolved mathematical problems.

Extension of this approach to many-electron systems may be difficult unless pseudopotentials or model potentials are used to simulate the core electrons.

In addition to facilitating the accurate computation of resonance parameters for simple systems, the coordinate rotation method should help one to distinguish true resonances from computational artifacts. For example there has been some controversy concerning the existence of narrow resonances in e$^+$-H scattering above the threshold for positronium formation.[6] Papers on resonances in this system will be presented at this conference by Shimamura and by Nuttall and Yamani. The technique has also been used by Reinhardt and Bardsley in an analysis of some of the "bound states in the continuum" recently discussed by Stillinger and colleagues.[7,8]

The coordinate rotation method has also been applied by Rescigno, McCurdy and McKoy[9] in calculations of photoionization and photodetachment cross sections. The wave function describing the final state is the solution of the inhomogeneous differential equation

$$[H(r) - E_o - \omega]\psi^-(r) = \mu\psi_o(r) \tag{4}$$

in which $\mu$ is the dipole operator, $\omega$ is the photon frequency and $\psi_o(r)$ and $E_o$ represent the wave function and

energy of the initial state of the atom or molecule.  The
photoabsorption cross section  $\sigma(\omega)$  can then be expressed as

$$\sigma(\omega) = \frac{4\pi\omega}{c} \text{Im}[\alpha^-(\omega)] \tag{5}$$

in which the polarizability  $\alpha^-(\omega)$  is given by

$$\alpha^-(\omega) = \langle\psi_0|\mu|\psi^-\rangle \quad . \tag{6}$$

By rotating the coordinate frame we can express the
Green function for Eq. (9) in terms of a basis set of $L^2$
functions.  Let us assume that the Hamiltonian has been diagonalized within such a set leading to eigenfunctions  $\phi_n(r)$
with eigenvalues $E_n$.  For a system with N electrons the cross
section can then be computed from

$$\sigma(\omega) = \sum_n e^{3iN\alpha} \iint d^3r\, d^3r' \psi_0(re^{i\alpha}) \mu(re^{i\alpha}) \frac{\phi_n(r)\phi_n(r')}{E_n - E_0 - \omega}$$

$$\mu(re^{i\alpha})\, \psi_0(re^{i\alpha}) \quad . \tag{7}$$

This method has been applied to H$^-$ and He with very encouraging results.  However, the sensitivity of the results
to changes in  $\alpha$  has not been examined and it is unclear
whether the technique will prove feasible for many-electron
systems.

The second method for which considerably progress has
been made since the last conference is that of Stieltjes
Imaging.  This procedure was developed mainly by Langhoff and
colleagues[10,11] and has also been applied to photoionization
and photodetachment.

The spectrum of physical states that can be formed upon
photon absorption usually consists of both bound states and a
continuum.  However if one diagonalizes the Hamiltonian in a
set of square-integrable basis functions one obtains a finite
set of pseudostates with a discrete spectrum.  The lowest
pseudostates will often provide good approximations to bound
states of the target, and the oscillator strengths associated
with such states have obvious significance.

The basic idea in Langhoff's approach is that for those
pseudostates with energies that lie in the physical continuum
the oscillator strength gives a measure of the probability of
a transition into a band of states in the continuum.  If the
continuum oscillator strengths are considered to form an

integral then the energies and oscillator strengths associated
with the pseudostates are analogous to the quadrature points
and weights that permit the numerical evaluation of the integral.  The use of Stieltjes images enables one to implement
this physical idea in a very elegant manner.

A full description of this method has been given by
Langhoff and Corcoran,[10] and the theory has been applied to
atomic helium by Langhoff, Sims and Corcoran.[11]  Further applications of the method to H$^-$ and H$_2$ are reported at this
conference by Broad and Reinhardt, and by O'Neil and
Reinhardt, respectively.

The results of Broad and Reinhardt for photodetachment
of H$^-$ show signs of structure close to the threshold for the
production of H$^*$.  It is unclear whether this is a true
threshold effect or is due to the presence of a resonant
state.

It is not yet clear whether the occurrence of narrow
resonances in photoionization cross sections can be properly
described in terms of this method.  An alternative approach
which explicitly allows for such resonances has been proposed
by Doyle, Oppenheimer and Dalgarno.[12]  The coordinate rotation
method should also reveal the existence of resonances.

In regard to the calculation of cross sections for
electron-atom scattering little progress has been reported
in recent months using the Fredholm determinant method.  The
major advantage of the approach remains that one can treat
elastic scattering above the excitation threshold with allowance for the loss of flux into other channels.  However, application of the technique to inelastic collisions presents a
formidable challenge.  On the other hand, an interesting
application of the method to electron-molecule collisions has
been reported by Winter and Lane.[13]

Another approach by which one can compute cross sections
without explicit representation of the asymptotic forms is
through the projection of the interaction potential onto a
set of square integrable functions.  Rescigno, McCurdy and
McKoy[14] have published results on elastic e-H$_2$ scattering,
and at this conference will discuss calculations on electronic
excitation of H$_2$ by electron impact.  These latter results
look very promising, since there have been few reliable calculations of inelastic processes in low-energy electron-
molecule collisions.

I would like to conclude with a brief discussion of a
different topic.  Almost all calculations of scattering wave

functions are based on a variational principle that ensures that small changes in the wave function within the space spanned by a set of basis functions leads to second-order changes in the scattering cross sections. We do not know whether such changes would lead to a better or worse answer and we do not know how large would be the effect of expanding our set of basis functions.

It would seem to be very desirable to develop tests for the accuracy of trial wave functions used in variational calculations. One procedure is to compute other properties, such as the polarizability, using the same basis set. An alternative approach is through the study of the variance integral or variance sum

$$S \equiv \int \left| \omega(r)(H - E)\psi(r) \right|^2$$

in which $\omega(r)$ is a weighting function and $S$ is evaluated either as an integral over all configuration space or as a sum over a large number of points distributed throughout that space. It has been shown by Miller[15] and by Bardsley, Gerjuoy and Sukumar[16] that the minimization of $S$ leads to very ac-curate wave functions for potential scattering problems, and for certain potentials upper and lower bounds on the phase shift can be derived. Read and Soto-Montiel[17] have obtained very accurate results for e-H scattering by this technique. However, my principal interest in this method is that it might provide a way of choosing the best wave function from a set of simple trial functions, thus enabling one to get the best possible answer from a small computation. In this regard the technique has not yet been justified, and I have been disap-pointed by our own results.

In summary, various new and interesting techniques have been successfully applied to some simple problems. The scope of these methods for many-electron systems has not yet been established, but I hope that this situation will be clarified before the next conference.

## References

1. C. Schwartz, Phys. Rev. 124, 1468 (1961).

2. J. N. Bardsley and B. R. Junker, J. Phys. B 5, L178 (1972).

3. G. Doolen, J. Nuttall and R. W. Stagat, Phys. Rev. A 10, 1612 (1974).

4. R. A. Bain, J. N. Bardsley, B. R. Junker and C. V. Sukumar, J. Phys. B $\underline{7}$, 2189 (1974).

5. G. D. Doolen, J. Phys. B $\underline{8}$, 525 (1975).

6. Y. Hahn, Phys. Rev. A $\underline{10}$, 2512 (1974).

7. F. H. Stillinger and D. R. Herrick, Phys. Rev. A $\underline{11}$, 446 (1975).

8. F. H. Stillinger and D. K. Stillinger, Phys. Rev. A $\underline{10}$, 1109 (1974).

9. T. N. Rescigno, C. W. McCurdy Jr., and V. McKoy, to be published.

10. P. W. Langhoff and C. T. Corcoran, J. Chem. Phys. $\underline{61}$, 146 (1974).

11. P. W. Langhoff, J. Sims and C. T. Corcoran, Phys. Rev. A $\underline{10}$, 829 (1974).

12. H. Doyle, M. Oppenheimer and A. Dalgarno, to be published.

13. T. G. Winter and N. F. Lane, Chem. Phys. Lett. $\underline{30}$, 363 (1975); $\underline{32}$, 196 (1975).

14. T. N. Rescigno, C. W. McCurdy Jr., and V. McKoy, Phys. Rev. A $\underline{11}$, 825 (1975).

15. K. J. Miller, Phys. Rev. A $\underline{3}$, 607 (1971).

16. J. N. Bardsley, E. Gerjuoy and C. V. Sukumar, Phys. Rev. A $\underline{6}$, 1813 (1972).

17. F. H. Read and J. R. Soto-Montiel, J. Phys. B $\underline{6}$, L15 (1973); J. R. Soto-Montiel, Ph.D. Thesis, University of Manchester.

# ELECTRON SCATTERING BY LASER-EXCITED ATOMS

I.V. Hertel, H.W. Hermann, W. Reiland,

A. Stamatović and W. Stoll

Universität Kaiserslautern W.Germany

## INTRODUCTION

At the last ICPEAC in Belgrade, now only two years ago, we reported for the first time investigations of collision processes by laser excited sodium atoms in a crossed beam experiment[1]. The use of narrowband CW-dye lasers makes it possible to excite a substantial percentage of atoms to a short living state by tuning the laser frequency to the corresponding resonance transition and irradiating the atoms in the scattering region of an otherwise conventional crossed beam experiment. Up until now only a very limited number of researchers have made use of this possibility[2]. However, the growing interest in this field of research is quite apparent. For obvious reasons: A whole new variety of processes is now open for detailed investigation, even though the CW-excitation possibilities are at present still restricted to sodium: Thermal and suprathermal elastic differential cross sections of excited Na may be measured, inelastic processes induced by fast neutrals or ions may be studied as well as intramultiplett mixing, quenching of resonance radiation by diatomic molecules, ionisation and autoionisation phenomena. And of course the whole field of chemical reactions studies may profit from the new possibilities. It is, however, not only the new variety of questions which now may be asked but more important, the new quality of information which may be obtained in these studies.

The monochromacy and polarization of the exciting laser radiation allows to prepare the target atoms in well defined fine and hyperfine states with known orientation and alignment of the magnetic sublevels,which may be varied in a defined way with respect to the collisional system. Thus,information is revealed,hitherto hidden in the usual averaging procedures over unobserved state populations.

This latter quality makes even the comparetively simple e + Na* process with its known $1/r$ potentials a rather exhausting piece of work.

The present talk will strictly be limited to the electron collisions. However,the general considerations on the excitations procedure, the typical experimental difficulties and the principle of interpretation may - and I think should - be discussed in a completely analogue manner for any heavy particle collision involving laser excited atoms and atomic beams.

Thus, the strategy in our present electron collision experiment is twofold:

a.) it should provide detailed results on the electron collison dynamics and be compared with current e - Na calculations.

b.) the possibilities as well as the limitations and dangers of the method are studied in the electron scattering test case. Being the most simplest case the hope is justified that all phenomena occuring in this type of experiment may be understood and thus a basis will be provided for the interpretation of collision problems with a more complicated dynamics.

Considerable progress has been made during the last two years.  However,both goals are still far from being reached.

I will therefore try to review the basic experimental and theoretical methods encountered in the experiment for those who are not already familiar with the field. On the other hand I will try to illustrate the possibilities of the method at some experimental examples and I will hint at some of the typical difficulties encountered in the experiment.

## EXCITATION MECHANISM

The preparation of the excited atomic target
by the laser is certainly the crucial part of the
experiment.  Only its full understanding allows the
precise interpretation of the experimental data obtained.

The schematic of the experiment is given in
figure 1.

The sodium beam is intersected at right angles
by the exciting laser radiation in order to avoid
Doppler shift. The radiative life time of the atoms
is $\sim 10^{-8}$ sec, so the collision process has to occur
in the photon-atom-beam intersection region.

In order to obtain a high excited state density
the radiation density per unit frequency $\varrho_\nu$ has
to be high.  It can be made such that eq. 1 is near
to unity.  However, three aspects of the excitation
process have to be considered in more detail:

a.) Sodium atoms are not two level  systems and
    have a fine- and hyperfine splitting. The
    term scheme of the $3^2P_{3/2}$ excited and $3^2S_{1/2}$
    ground states is shown in figure 2.

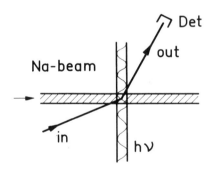

For two atomic
levels stationary:
$$\frac{n_{ex}}{n_{gr}} = \frac{1}{1 + \dfrac{8\pi h}{\lambda^3 \varrho_\nu}} \qquad (1)$$

Fig.1: Schematic of the experiment

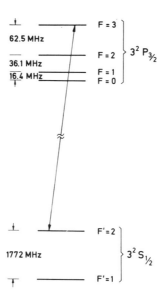

Fig.2: Sodium HFS term scheme.

Thus a defined hyperfine state may be excited
by the laser radiation.To avoid optical pumping to the other ground state only the $F'=2 \rightarrow$
$F= 3$     transition can be used, thus leading
to a limit of 31% of excited atoms with a
known F and J quantum number.

b.) The polarization of the laser light allows
only certain magnetic transitions $\Delta m = 0$ for
$\pi$ - and $\Delta m = \pm 1$ for $\sigma^+$ and $\sigma^-$ light, respectively. Thus an aligned or/and oriented
atom is produced.

c.) The atoms remain in the optical excitation
region for $\sim 10^{-6}$ sec. Thus each atom undergoes $\sim 100$ excitation processes and spontaneous
decays. The resulting optical pumping process
enhances the degree of non isotropic magnetic
sublevel population. The atoms are thus prepared with higher state multipole moments up
to multipole moments of the order $k_{max}=2 \cdot F =6$.
Their distribution may be calculated for
stationary conditions.However, more than
20 to 30 pumping cycles are needed before reaching this equilibrium.

Several other excitation schemes have been proposed[3] including two or more excitation frequencies

by making use of the Doppler shift in a multidirect-
ional excitation scheme or even incorporating a
magnetic field to remove unexcitable magnetic sublevels. The purpose is to excite other hyperfine
states and/or the $^2P_{1/2}$ level. We feel however, that
severe difficulties will be encountered in an attempt
to determine the magnetic sublevel population and
the degree of coherence among them.

Our excitation scheme has the great virtue of
producing a diagonal density matrix for the upper
magnetic  sublevel population which are only connected by spontaneous decay processes:

$$\rho(M_F, M_F') = \delta_{M_F, M_F'} \cdot w(M_F)$$

for one particular choice of the photon-frame co-
ordinate system. The latter is $\parallel$ to $\vec{E}$ for $\pi$-light
and $\parallel$ to the incident light direction for $\sigma$-light.
Besides, since the maximum multipole moment prepared
is $k_{max}$ = 2F the excitation of the F = 3 level will
reveal the most complete information on the scattering dynamics.

Let me, however, mention a few problems encountered in all excitation schemes:

a.) Since we are dealing with Gaussian light beams
the determination and maintenance of the excited state scattering volume and density is
non-trivial. Ideally a large expanded laser
beam should be used covering much more than
the scattering volume. This procedure is limited by the intensities available at the moment.

b.) High laser intensities are necessary for high
excited state population and fast pumping! Then
the spontaneous life time becomes large or at
least comparable to the induced: $\tau_{spont} \gg 1/B\rho_\nu$
This results in a saturation broadening

$$\Delta\nu_{sat} = \frac{1}{\hbar} \cdot e \cdot r_{12} \cdot |\vec{E}| \propto \sqrt{\text{laser intensity}}$$

which is typically of the order 100 MHZ at
1W/cm$^2$ laser intensity. As soon as $\Delta\nu_{sat} \approx$
$\Delta\nu_{laser}$ it is no longer $\rho_\nu$ which determines
the upper state density. This is favourable
for the excitation using poor laser band widths.
On the other hand, $\Delta\nu_{sat}$ is typically of the
order of the hyperfine splitting. Thus a destruction of the pumping process may occur,

resulting in a poorer degree of anisotropy
among the magnetic sublevels.
c.)A similar effect may be caused by radiation
trapping. The beam density therefore has to
be kept below $10^{10}$ atoms/$cm^3$.

Let us forget for the moment these difficulties,
which later on will show up as statistical uncertainties in our measurements, and turn to the experiment
itself.


## EXPERIMENT

The sodium beam is intersected in the scattering
region by the laser beam. The fluorescence is monitored and used to stabilise the laser frequency. If
an occasional mode hop occurs, the laser is scanned
automatically to search for maximum fluorescence
again. Meanwhile, the data accumulation is halted.

The electron scattering system is an otherwise
conventional system using hemispherical electrostatic
analysers. Electron gun and detector may be rotated
independently around the scattering region. So the
scattering angle $\vartheta_{col}$ may be varied, but also the
angle of incidence of the exciting photon beam. In
addition the polarization angle $\gamma$ can be rotated with
respect to the collision plane.

The experiment is controlled on line by a small
computer. Energy-loss spectra may be taken or the
dependence of the differential cross sections on the
polarization angle and direction of the incident
light are measured.

Figure 3 shows a typical energy-loss spectrum.

Without light the electron energy-loss spectrum
for scattering of the ground state is measured with
the most prominent 3 s ⟶ 3 p resonance excitation.
When a part of the atoms is ecited by the laser
light a variety of inelastic processes off the excited state are seen in addition. On an enlarged
scale the difference between light on and off is
shown (bottom right). Clearly the deexcitation process 3 p ⟶ 3 s is seen on the energy gain side.
We will later on mainly talk about this transition.

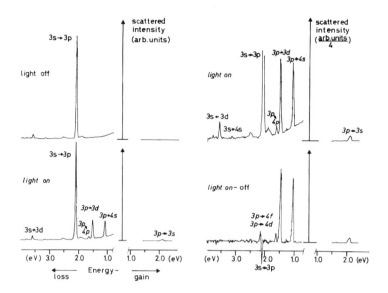

Fig.3:Energy-loss spectra for e+Na and e+Na,e+Na*.

The depletion of the ground state is demonstrated by
the "negative" peak at the position of the resonance
line 3s → 3p.   The most prominent feature of this
spectrum is however the large cross section for the
3p → 3d   and 3p → 4s excitation. One is tempted
to speculate from these observation and predict for
the ground state inelastic scattering 3s → 3p distinct resonances near the 3d   and 4s states.
Structures of this type have been observed by Eyb
and Hoffmann[4] in potassium where, however, unfortunately
the 4s and 3d   thresholds are close together.

    In the sodium case we are able to separate the
thresholds and preliminary measurements are shown
in figure 4.   Clearly a d-type structure is observed at or just below the 3d threshold, and an
additional feature at or below the 4s threshold.

    Let us however return to the excited state and
investigate the dependence of the 3p → 3s differenti-
al cross section $I(\vartheta_{col})$ on the polarization angle    ,
when using $\pi$ -light excitation. The light is incident in the scattering plane. Figure 5a.) shows a
typical experimental result demonstrating a clear
cos 2$\gamma$        dependence with a strong anisotropy as
we vary the polarization angle.

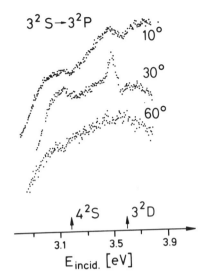

Fig.4: Excitation function for the 3s ⟶ 3p
transition in e+Na(ground state) near the
3d and 4s thresholds.

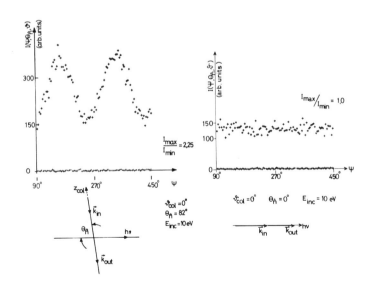

Fig.5 : Differential deexcitation cross section as
a function of the incident $\pi$ -light polarization angle $\psi$.
a.)for $\Theta_{\hat{n}}$ = 82° b.)for $\Theta_{\hat{n}}$ = 0°.

Now, if we change the incident light direction $\Theta_{\hat{n}}$ we observe a completely different picture (Fig. 5b): Only $\Theta_{\hat{n}}$ has been changed and no anisotropy is observed this time. This demonstrates clearly the importance of knowing the excitation conditions precisely.

The total information may be displayed in a more compact form by plotting maximum to minimum cross section $I(\Psi = 0°)/I(\Psi = 90°)$ as a function of the incident light direction $\Theta_{\hat{n}}$. Figure 6 gives such a plott for forward scattering together with a theoretical fit.  As the scattered electron is observed here in forward direction, this club has to be symmetric with respect to the electron beam direction $z_{col}$ as the light direction is changed.

Let us get a feeling what this shape here observed means: Figure 7a indicates the double pear like shape of the laser excited atom, which may be seen by the electron. As we turn the atom by rotating the polarization vector of the light, the electron experiences a completely different structure.

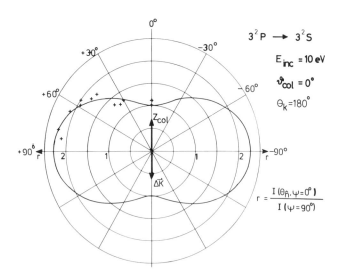

Fig.6:  Anisotropy of the collision cross section as a function of the incident light direction $\Theta_{\hat{n}}$

If we now change the incident light direction,
the electron may always  see the same projection of
the atom, independent of the polarization angle of
the light.(See figure 7b).

Thus the electron scattering current just reflects this shape and the club of figure 5 is really
an image of the interaction sphere as experienced
by the collision partners.

If the projectile is not scattered in the forward direction, it experiences a different interaction
sphere on its way and consequently the club should
be rotated, giving a different projection of the second  spherical harmonics. This is illustrated in
figure 8 for 3 eV $\vartheta_{col}$ = 20° together with a calculation using Moores and Norcross scattering amplitudes [5]. A good agreement between theory and experiment may be stated within the error limit of the
present measurements.

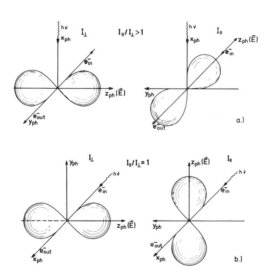

Fig.7: Symbolic view of the interaction sphere as
       experienced by the electron for different
       polarization angles of the light          .
       a.) for incident direction $\Theta_{\hat{n}}$ = 90°
       b.) for incident direction $\Theta_{\hat{n}}$ =  0° .

It should be noted, for later reference, that as seen here a larger angular momentum transfer rotates the observed club to smaller angles $\Theta_{\hat{n}}$ .

### THEORETICAL INTERPRETATION OF THE RESULT

Let me say a few words about the adequate theoretical description: In general one may describe any scattering process in terms of a set of scattering amplitudes $f_{ex\ M'}$ to excite an atomic state $\psi_{exM'}$ from an initial state $\psi$ o.

$$\psi_{col} = e^{i\vec{k}\cdot\vec{r}} \psi_c + \frac{1}{r} \cdot e^{i\beta_f r} \sum_{M'} f_{exM'} \psi_{exM'}$$

In a conventional scattering experiment the absolute squares of the scattering amplitudes summed over the various substates are measured. One way to derive more detailed information is to use an electron-photon coincidence technique. This method has been introduced by Kleinpoppen and Slevin et al. in Stirling and is now extensively used on Helium[6]. Ratios and phases for the scattering amplitudes in an s $\rightarrow$ p excitation can be measured in this manner, the now so called $\lambda$ and $\chi$ -parameters.

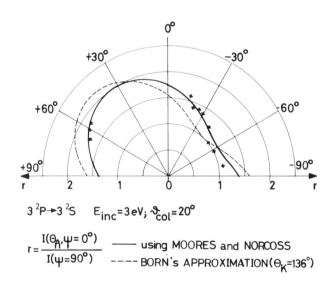

$3\,^2P \rightarrow 3\,^2S$   $E_{inc} = 3\,eV$; $\vartheta_{col} = 20°$

$r = \dfrac{I(\Theta_{\hat{n}}, \psi = 0°)}{I(\psi = 90°)}$   —— using MOORES and NORCOSS

---- BORN's APPROXIMATION ($\Theta_K = 136°$)

Figure 8 :  Anisotropy    of  the  collision  cross  section
as  a  function  of  the  incident  light  direction

The way in which we try to obtain similar information is to excite the atom in a known combination of M'states and then to investigate the
scattering process from this well prepared system.
Since only a special combination of the magnetic
sublevels is offered to the deexciting electrons, a
special combination of these amplitudes is observed.
It may be varied by using $\sigma$ or $\pi$ light excitation,
variing the angle of incidence and polarization.
Thus scattering amplitudes, phases and ratios
between them may be derived. You see that a great
variety of parameters may be used to obtain the most
complete picture. There are two ways to make life
for a collision physicist interesting: Either you
choose every day a different interaction potential
by changing target gas and projectiles, or you vary
the whole number of parameters available. It is the
latter possibility which keeps us busy. In figure 9
you see a general scheme of the geometry, displaying
the collision frame usually given by in-and out-going
electron in its relation to the photon frame here
given by linear polarized light.

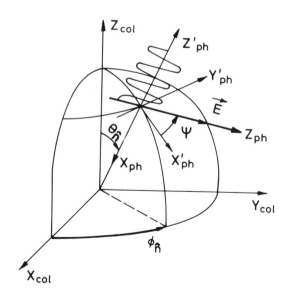

Fig. 9 :    Scattering geometry

Under such involved conditions the adequate language
for a description has to use irreducible tensors or
state multipole moments for
a.) the preparation of the atom by the light
b.) the interaction of the collision system.

This language has been introduced by Fano and
Macek[7] and has been applied to the scattering of
laser excited atoms by Macek and Hertel[8]. It dis-
entangles geometrical aspects i.e. the photoexcitation process, from the collision dynamics. The
differential cross section $I(\vartheta_{col})$, now also a
function of $\theta_{\hat{n}}$, $\phi_{\hat{n}}$ , $\gamma$     etc., can be written
simply

$$I(\vartheta_{col}) = C \cdot \sum_{k=0}^{k_{max}} \varpi(k) \cdot \langle \tau_o^{[k]}(Ph) \rangle$$

Where $\varpi(k)$ are the photon produced multipole
moments

$$\varpi(k) = \sum_M (-)^{k-F-M} W(M)(F-MFM|ko)$$

$\langle \tau_o^{[k]}(Ph) \rangle$ are the collisionally produced multipole moments of the inverse scattering process and
are related to the familiar scattering amplitudes
by:

$$\langle \tau_q^{[k]} \rangle \propto \sum_{MM'} (-)^M (J_{ex}-M J_{ex} M' | kq) f_{Mex} f_{M'ex}^*$$

Calculations are usually carried out in the
collision frame, so that a rotation has to be performed through the Euler   angles $\Theta$ and $\phi$ :

$$\langle \tau_o^{[k]}(Ph) \rangle \propto \sum_{q'} \langle \tau_{q'}^{[k]}(col) \rangle Y_{q'}^{[k]}(\Theta,\phi)$$

If you are an experimental physicist, you may
well think in terms of the double or multiple pear
shaped structure I showed earlier on, representing
these multipole moments (fig.7).

Of course, the method applies not only to the
electron deexcitation process but also to inelastic
and in principle to elastic scattering even in heavy
particle collisions.

One virtue of the technique is the possibility
to probe the nature of the interaction, without explicit calculations, just from the geometrical symmetries. If the colliding partners experience only
an interaction by the charge cloud, as usually assumed
in electron scattering, the angular dependence will
at most     show a second spherical harmonic shape,
as we see in figure 8 through 11.

In cases, however, where spin orbit interaction
enters explicitly into the collision dynamics,it is
no longer allowed to just project the LS coupling
scheme onto a j-j or j-F coupling and if $J_{atom}=3/2$
a third spherical harmonic dependence may be observed
when variing the incident direction of exciting
$\sigma$-light.

In heavy particle collisions this may well play
an important role. Even for the electron excitation
near threshold this may be the case as indicated by
photo-detachment measurements[9].

Let us discuss some less complicated investigation: In an intermediate energy range, we compare
Borns approximation with our experiment, lacking a
better theory.

Figure 10 shows the typical club for 6 eV and
$\vartheta_{col}$ = 10°. Born approximation, which is symmetric
around the momentum transfer vector, is seen to fail
substantially.

The experiment indicates a higher angular momentum transfer. For the excitation process 3p → 4s the
club is rotated to the other side, since here the
momentum transfer is reversed.

Going to 10eV incident energy Figure 11 a,b,c,d,
we observe a tendency toward a better agreement with
Born. For 5°,10°,15° and 20° measurements have been
performed,always indicating higher angular momentum
transfer than Born predicts.

This is similar to the observation by the Stirling group in He. It should be noted that only phases
andratios of amplitudes enter the present data and
to obtain a full set of information,the absolute
magnitudes have to be determined independently(e.g.[10])

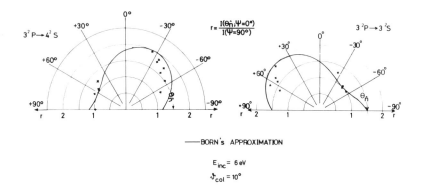

Fig. 10:   Comparison of Borns approximation and experiment for      a.) the 3p→3s and
b.) the 3p  ⟶  4s transition at 6eV,
$\vartheta_{col}$ = 10°.

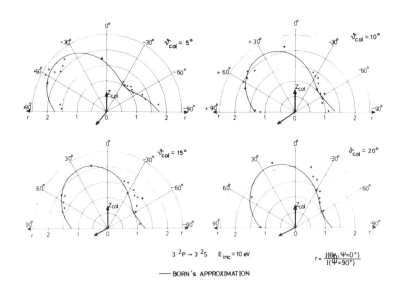

Fig. 11:   Anisotropy club similar to previous ones
for 10eV and a.) $\vartheta_{col}$ = 5°  b.) $\vartheta_{col}$ = 10°
c.) $\vartheta_{col}$ = 15° and d.) $\vartheta_{col}$ = 20°.

The statistical variations are at present still relatively large and may possibly be attributed to irregularities in the pumping process as discussed earlier.

A more sensitive test on Born approximation will be the measurement of the left right anisotropy for off plane excitation with $\sigma$ light, which is zero in Born or Glauber approximation. This encorporates major experimental changes, but results will be available within the next two months.

### STRONG FIELD LINEAR SUPERPOSITION OF STATES

Before ending this talk, let me mention one aspect of the experiment which still needs further clarification: In the experiments described here the spontaneous life time of the atoms is of the order of the induced life time or larger. Under these circumstances the atom can no longer be reguarded as an incoherent superposition of ground and excited state. Ground and excited states are in a partially coherent superposition, which is time dependent.

Thus a time dependent potential is encountered and the possibility of interference terms between ground and excited state scattering amplitudes arises in principle, especially for elastic scattering. In a forthcoming paper by Mittleman and Gertsen[11], a number of unexpected effects are predicted due to the possibility. Unfortunately only the e-$h\nu$-H problem is treated and nonumerical data are given. In a paper by Hahn and Hertel [12] we have tried to investigate this problem in a heuristic approach, too. Treating the photon-atom system in a rotating wave approximation and the collision problem in first Born we were able to show that such interference terms are of no importance as long as the experimental energy resolution is not extremely good. For higher order approximations e.g. including narrow resonances, this cannot be stated in general at present. So the need is felt for a more rigorous theoretical proof of the following theorem: An incoherent addition of the ground and excited atom cross-section is observed:

$$\sigma_{obs} = N_{gr}/N \cdot \sigma_{gr} + N_{ex}/N \cdot \sigma_{ex} \quad \text{as long as}$$

$$\Delta E_{exp} \gg h \cdot \Delta \nu_{sat} \qquad \text{ie. the experimental}$$

resolution in the scattering diagnostic is poor compared to the saturated line width $\Delta\nu_{sat}$ and if the collision time $t_{col} \ll 1/\Delta\nu_{sat}$.

## ACKNOWLEDGEMENT

I do not want to end this talk without gratefully acknowledging the many fruitefull discussion with  U. Fano and especially the invaluable cooperation with Joe Macek.

## REFERENCES

1.)   Hertel,I.V. and Stoll,W.VIIIth ICPEAC (Beograd 1973) Abstracts of papers pp 321, edited by Cobic,B.C. and Kurepa,M.V.,
Hertel,I.V. and Stoll,W. 1974 J.Phys.B:Atom molec.Phys.7,570-82 and 583-92
Hertel,I.V. 1974 in "Electron and Photon Interactions with Atoms", 1975  to be published by Plenum Comp. N.Y., edited by H.Kleinpopen and M.R.C. McDowell and in "Atomic Physics 4" 1975 Plenum Press NY and London edited by G.zu Putlitz, E.W. Weber and A. Winnacker  pp 381

2.)   Pritchard D.E. and Carter G.M., IXth ICPEAC (Seattle 1975) abstracts of paper,447, edited by J.S. Risley and R. Geballe
Carter,G.M., Pritchard,D.E., Kaplan,M. and Ducas,T.W. to be published
Düren,R., Hoppe,H.O. and Pauly, H., 50 Jahre MPI für Strömungsforschung (Göttingen 1975)and Düren,R., Hoppe,H.O. and Pauly,H. to be published

Further work is currently in progress in several groups e.g.: Bederson,B. et al.; van der Valk, Nienhuis et al.; Barat, Berlande,Pascale et al.

3.)   Gerritson,H.J. and Nienhuis,G., 1975,Appl.Phys. Lett. 26, 347-9
Carter,G.M., Pritchard, D.E. and Ducas,T.W.,to be published

4.)   Eyb,M., Dissertation 1974, Kaiserslautern

5.)   Moores,D. and Norcross,D., 1972   J.Phys.B.:
      Atom.molec.Phys.$\underline{5}$, 1482-505
      We are particular indebted to D.Norcross for
      communicating to us to the complex values of
      the inelastic scattering amplitudes

6.)   Kleinpoppen, H., Blum,K. and Standage,M.C.,
      Invited Progress Report IXth ICPEAC this volume
      see there for further references

7.)   Fano,**U**. and Macek J.H.,1973,Rev.Mod.Phys.45,553

8.)   Macek,J.H. and Hertel,I.V., J.Phys.B.$\underline{7}$,1974,2173

9.)   Lineberger,C., Invited Progress Report IX ICPEAC
      this volume

10.)  Shuttleworth,T., Newell,W.R. and Smith,A.C.H.,
      IXth ICPEAC (Seattle 1975) abstracts of papers,
      p 1109, edited by J.Risley and R. Geballe

11.)  Gertsen, J.I. and Mittleman,  to be published

12.)  Hahn,L. and Hertel,I.V. 1972, J.Phys. B $\underline{5}$,1975

ENERGY EXCHANGES BETWEEN TWO ESCAPING ELECTRONS

Frank H. Read[*]

Department of Physics, Schuster Laboratory

University of Manchester, U. K.

## Abstract

The final-state Coulomb interaction between three or more outgoing charged particles from a reaction can cause exchanges of energy and angular momentum between them.  Such exchanges may be present for example when the reaction products consist of two electrons receding from a positive ion.  The experimental evidence for the existence of such effects in threshold ionization and threshold autoionization are reviewed, and their possible theoretical descriptions are discussed.

[*]Visiting Fellow, 1974-75, Joint Institute for Laboratory Astrophysics, University of Colorado and National Bureau of Standards, Boulder, Colorado.

## 1.   INTRODUCTION

This talk will be concerned with the post-collisional exchanges of energy that may occur between the outgoing particles of a reaction, when these particles consist of two electrons receding from a positive atomic ion.  These outgoing particles may be produced by electron impact ionization of neutral or negatively ionized atoms, or by photo-double ionization or photo-double detachment of positive, neutral or negative atoms.

The monopole Coulomb interaction between the outgoing charged particles of a reaction has a special role to play in the time development of those final states which consist of three or more outgoing charged particles.  Even after these particles are outside the reaction zone in which their identities and internal energies are established the long-range Coulomb interaction continues to act in changing the partitioning of the available energy and angular momentum of the system, and the final kinetic energies and angular momenta of the individual particles are not fully determined until they are completely free from each other.  For example, in the case of two electrons receding from a heavy (and stationary) singly charged positive ion the Hamiltonian, once the electrons are further than a few atomic units from the ion, is

$$H = H^{(\text{internal})} - \frac{1}{2} \nabla_1^2 - \frac{1}{2} \nabla_2^2 - \frac{1}{r_1} - \frac{1}{r_2} + \frac{1}{|\vec{r}_1 - \vec{r}_2|} \tag{1}$$

where $\vec{r}_1$ and $\vec{r}_2$ are the position vectors of the two electrons.  Although strictly it is not possible to partition the initial potential energy terms between the two electrons, and therefore a definite total energy cannot be ascribed to each, there is nevertheless in some sense a continual exchange of energy between the two electrons through the last term in (1), and the final total energies of each are not known until this term becomes zero.  The last term also causes the forces acting on the electrons to be nonradial, leading to exchanges of angular momenta.  The final state Coulomb interaction therefore causes in effect a post-collisional exchange of energy and angular momentum in such reactions.

The situation is quite different when there are only two outgoing charged particles, since the total energy of each of them is fully determined as soon as they are outside the reaction zone.  Also in this case the monopole Coulomb interaction is unable to change the relative angular momentum of the two particles.  Even in the case of three or more outgoing charged particles the exchanges of energy would not usually be noticeable in most reactions, because the Coulomb energy between the particles, once they are otherwise free of each other, is usually so much smaller than their total energies that the effects of the continuing Coulomb interaction are not significant.

There are two cases, however, in which the final state Coulomb interaction may become more important. The first of these occurs near electron impact ionization thresholds or photon impact double-detachment thresholds, for which the total energy of the outgoing particles (apart from the internal energy of the residual ion) is approximately zero. In these reactions the slowly receding electrons may spend a considerable time interacting with each other and with the Coulomb field of the ion, and the resulting changes in the distribution of energy and angular momentum between them may become significant. The second example occurs in the near-threshold electron impact excitation of short-lived auto-ionizing states of atoms. In these reactions the inelastically scattered electron recedes slowly from the excited neutral atom, but at the time of autoionization of the atom into an electron and an ion, the Coulomb interaction energy between the scattered electron and these other two charged particles may still be comparable with its kinetic energy, again leading to observable changes in the final energies of both the scattered and ejected electrons. It is these two examples that will be discussed in this talk. Other examples, such as the near-threshold inner-shell ionization of atoms, followed by multiple Auger electron emission, may also exist.

Although for the present purposes we shall often be referring to the time evolution of the final state, and shall also sometimes be alluding to classical descriptions, it should be remembered that a more conventional approach, at least from the theoretical point of view, would be to consider the solutions of the time-independent Schrödinger equation which represents the reaction. It seems however that this latter approach, although undoubtedly the correct way to solve these scattering problems, is at the present time somewhat limited in its ability to take account of long-range final state interactions between more than two particles. It is this very point that provides the interest and motivation for the present studies.

## 2.   THRESHOLD ELECTRON-IMPACT IONIZATION

We shall primarily be considering the reaction

$$e + A^{m+} \rightarrow A^{(m+1)+} + 2e \tag{2}$$

near to its threshold, where  m ≥ 0.  The photo-double detachment reaction

$$hv + A^{(m-1)+} \rightarrow A^{(m+1)+} + 2e \qquad , \qquad (3)$$

where  m ≥ 0,  also provides the same products, but high resolution studies of such reactions have not yet been made.

These near-threshold ionization processes have been
treated theoretically by several authors, using a variety of
quantum mechanical,[1-9] semi-classical[10,11] and purely classical[12-17] methods, the main intent of most of this work being
to establish the exponent  n  of the power law dependence of
the total ionization cross section  σ  on the excess energy
E  above threshold,

$$\sigma \sim E^n \qquad . \qquad (4)$$

The value of  n  for single ionization of neutral atoms has
variously been found to range from 1.0 to 1.5, sometimes
being a function of  E  itself.

The earliest treatment, that of Wannier,[12] in which classical methods were used to obtain the value 1.127 for  n,  oc-
cupies a special place in this field.  The long-range final
state interactions are correctly included in this treatment,
as has been emphasized by Rau[6] and Fano and Lin,[18] but although it seems to be rigorous, apart from the assumption of
quasi-ergodicity in the initial conditions, it has not received general approval.  In the earliest quantum mechanical
treatment[1,2] the asymptotic wavefunction was taken to be the
product of two outgoing Coulomb waves, which leads to the
value 1.0 for the exponent n.  The effective ionic charges
$z_1$ and $z_2$ for the Coulomb waves are defined by

$$- \frac{1}{r_1} - \frac{1}{r_2} + \frac{1}{|\vec{r}_1 - \vec{r}_2|} = - \frac{z_1}{r_1} - \frac{z_2}{r_2} \qquad (5)$$

and although  $z_1$ and $z_2$  are dependent on the initial directions and momenta of the electrons they do not vary with time,
which implies that final state interactions of arbitrarily
long range are not included.  This therefore constitutes an
important difference between the two approaches, and we will
follow Rau[6] and Fano[19] in assuming that it is this difference
that is essentially responsible for the different values of n.

The Wannier exponent 1.127 has also been obtained more
recently by semi-classical (Peterkop[11]) and quantum mechanical
(Rau[6]) methods.  The quantum mechanical treatments of Temkin
and co-workers[3,4,8,9] and Kang and Kerch[5] give values of  n

somewhat greater than 1.0, while the classical trajectory integrations[14-17] give approximately 1.127, as indeed they must. Experimentally, it is difficult to differentiate between the exponents 1.0 and 1.127, but recent studies[20-24] favor the latter value. In the experiments of Marchand et al.[24] for example, n has been found to be 1.16 ± 0.03 for the near-threshold electron impact ionization of helium.

To understand why the final state Coulomb interactions increase the value of n it is helpful to consider what Rau[6] has described as "dynamic screening," or what in other words amounts to the changes with time of the effective charges $z_1$ and $z_2$ of equation (5). If two electrons start in "symmetric" orbits (by which is meant $r_1 = r_2$, $\dot{r}_1 = \dot{r}_2$ and $\theta_1 = -\theta_2$) they will retain this symmetry for all future times and will both eventually escape with the same kinetic energy $\frac{1}{2}$ E. If on the other hand the starting condition is nearly symmetric, but with electron 2 being slightly faster and further from the ion than electron 1, then it will be better screened from the ion than is electron 1 (i.e. $z_2$ will be smaller than $z_1$) and will therefore experience a lesser attraction, and its initial advantage in speed and distance will increase with time. Put in other words, the potential energy part of the Hamiltonian [Eq. (1)] is always a maximum at $r_1 = r_2$ (for all values of $\theta_{12}$), and therefore deviations from the symmetric orbit will increase with time.

This radial correlation instability is most effective when the two electrons have similar initial kinetic energies, and it tends to reduce the probability of the two electrons having the same or similar final kinetic energies. The instability leads in effect to an exchange of energy between the electrons, and in extreme cases electron 1 may finish with a total energy which is negative, remaining bound to the positive ion and thus removing flux from the ionization channel. The instability becomes more effective as the available excess energy E becomes smaller, since although the electrons always start with high kinetic energies when near the ion they must spend longer times and travel further distances from the ion before the Coulomb energy becomes sufficiently smaller than E for them to be free. When E = 0.1 eV for example, this stage is not reached until the electrons are further than about 200 Å from the residual ion, which takes a time greater than about $6 \times 10^{-13}$ s. The net effect of the dynamic screening is therefore to reduce the probability of ionization at small values of E, whilst having a smaller effect at higher energies. In other words the exponent n is made slightly greater than unity, actually giving a zero slope to the ionization cross section at the threshold itself. More complete and quantitative accounts of these effects can be found in the works of Wannier,[12] Rau,[6] Fano and Lin[18] and Fano.[19]

As well as the instability in the radial correlation there is an accompanying stability in the angle $\theta_{12}$ between the final directions of the two escaping electrons, since the electrostatic forces always tend to push this angle towards $\pi$, at the same time tending to impart equal and opposite angular momenta to the outgoing electrons. In terms of the excess energy $E$ the spread $\Delta\theta_{12}$ of angles about $\pi$ is of the order of $E^{1/4}$ in the Wannier model, which implies a maximum angular momentum of each electron of the order of $E^{-1/4}$ (where $E$ is measured in atomic units).

Although these various ideas and interpretations are not yet proven there is some support for them in the experiments of Cvejanović and Read,[25] who attempted to measure the correlations in energy and angle of the two escaping electrons. This was achieved by a time-of-flight coincidence technique, thereby exploiting the slow velocities of the electrons and avoiding the usual difficulties of the more conventional techniques. For small excess energies $E(\lesssim 1$ eV) in electron-helium ionization they were able to measure the probability distribution $P(E_1,E_2)$ for the partitioning of the energy $E$ between the two electrons, which is predicted to be uniform (i.e. independent of $E_1$ or $E_2$) by most theories of threshold ionization. In fact the experimental results are consistent with its being uniform over the whole range of values of $E_1$ and $E_2$, although for technical reasons this could not be definitely established for values of $E_1$ or $E_2$ less than about 0.05 eV. Also the angular correlation function $P(\theta_{12})$ was found to have a width $\Delta\theta_{12}$ which decreases as $E$ decreases, as predicted by the Wannier-Peterkop-Rau[12,11,6] theory.

These authors also obtained information of a quite different type through the development of a new threshold detection technique[26] which enabled electrons of nearly zero energy to be detected with high energy resolution $\Delta E(\lesssim 50$ meV) and high efficiency ($\approx 50\%$). If the only electrons to be detected are those having energies $E_{1,2} \leq \Delta E$, and if the probability distribution $P(E_1,E_2)$ is independent of $E_1$ and $E_2$, then the yield of detected electrons is proportional to $\sigma \times (\Delta E/E)$, which has the power law dependence $E^{n-1}$. Figure 1 shows a spectrum obtained by this technique. It can be seen that the value of the exponent $(n-1)$ above the ionization threshold is indeed consistent with that expected from the Wannier theory. An analysis of this and other spectra has shown[25] that the exponent for the yield above the ionization energy is $0.131 \pm 0.019$. This result does not however constitute a definite proof of the Wannier law since the uniformity of $P(E_1,E_2)$ for values of $E_1$ and $E_2$ less than $\Delta E$ has not been definitely established, as mentioned above.

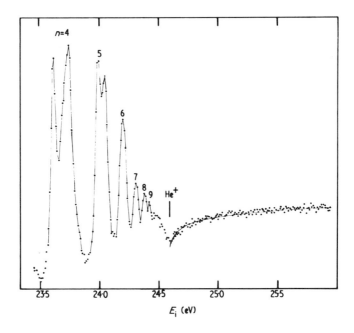

Figure 1.   A spectrum showing the measured yield of very low
            energy ($\lesssim$50 meV) electrons resulting from electron
            impact of helium.  The curve drawn through the
            points above the ionization threshold is propor-
            tional to $E^{0.127}$, where E is the energy excess
            above the ionization energy.  Reproduced from
            Cvejanović and Read, J. Phys. B $\underline{7}$, 1841 (1974),
            with permission.

     An equally important part of this spectrum, as pointed
out by Fano,[19] is the region below the ionization energy,
where peaks corresponding to the threshold excitation of
Rydberg states of helium can be seen.  These states have large
radii [$\sim n^2$ or $1/(-E)$] and can only be produced when it is possible for one electron to reach these large distances while
the other recedes to infinity, and therefore there is time for
the final state interaction to cause appreciable exchanges of
energy and angular momentum.  The net effect is that the probability of exciting these states near their thresholds is
reduced, and that those which are excited tend to have a large
range of orbital angular momenta.  In fact it can be seen from
the spectrum that as the ionization energy is approached these
peaks merge together and that a mean yield drawn through the
separate and merged peaks has a shape which approximately
mirrors that of the yield above the ionization energy.

In summary one may say that the decrease, or cusp, in the yield of very low energy electrons immediately above and below the ionization energy exists because the post-collisional exchange of energy between the outgoing electrons decreases the probability of their both having a small final energy. There should also be an associated increase in the range of angular momenta of the two electrons in this region,[19] as discussed above, but there is no experimental evidence for this yet.

More recently, the cusp at the ionization threshold has also been seen by Spence,[27] who used a modified version of the trapped well technique to detect inelastic and ionization electrons over a range of final energies. These experiments, as well as those of Cvejanović and Read,[25] are consistent with the Wannier law holding up to about 2 or 3 eV above the ionization energy, in the case of electron-helium scattering.

### 3. THRESHOLD EXCITATION OF AUTOIONIZING STATES

In the previous example we were concerned with the final state interaction between two electrons that leave the reaction zone (that is, the region of short-range interactions in which the identities and internal states of the outgoing particles are determined) at essentially the same time. In this second example we shall be concerned with two electrons of greatly different energies that leave the reaction zone at substantially different times, the slower electron leaving first and the faster one being delayed by its temporary capture as part of an autoionizing state of the target atom. As before, we shall adopt a simple description by emphasizing the time evolution of the final state and attributing the observed effects to final state interactions, since although this model may have weaknesses a more exact description does not yet exist (see Taylor and Yaris[28] and the discussion in section 4 below).

Consider the electron impact excitation of a double excited autoionizing (or pre-ionizing) state of helium

$$e + He \rightarrow He^{**} + e_1$$
$$\phantom{e + He \rightarrow} \lfloor \rightarrow He^+ + e_2 \qquad (6)$$

and suppose that the energy $E_1$ of electron 1 is less than a few eV. The lower lying states $He^{**}$ have lifetimes of $5 \times 10^{-15}$ s or longer and the ejected electrons $e_2$ have energies $E_2$ of 33.2 eV or more. In the case of the shorter lived states it is possible for $e_1$ to have traveled less than a few tens of Angstroms before the time of break-up of the autoionizing atom, and then the Coulomb energy between $e_1$ and $e_2$ at this instant may be significant compared with $E_1$. This may then lead to observable energy exchanges between the electrons, causing their final energies to be different from the nominal values $E_1$ and $E_2$.

There are two simple but equivalent ways of describing the origin of the energy exchanges, if $e_2$ can be regarded as being very fast compared with $e_1$ (that is, if the "sudden approximation" is assumed). Before the autoionization event $e_1$ is receding from the neutral atom as an expanding charge cloud. The ejected electron $e_2$ has an additional initial potential energy $\Delta E$ caused by its being immersed in the Coulomb field of this cloud and it retains this part of its total energy in traveling through the cloud, therefore having a final kinetic energy equal to $E_2 + \Delta E$. The detailed probability distribution of the values of $\Delta E$ depends on the lifetime $\tau$ of the autoionizing state, and on the nominal energy $E_1$ of the scattered electron, but clearly the magnitude of $\Delta E$ increases as $\tau$ and $E_1$ are decreased. Alternatively, from the point of view of the scattered electron, the neutral atom from which it is receding suddenly changes, at the moment of autoionization, into a positive ion, causing it to be retarded and to have the final energy $E_1 - \Delta E$. The total energy of the system must of course remain unchanged.

The distribution of energy exchanges $\Delta E$ can be found experimentally either by measuring the loss of energy of $e_1$ or the gain of energy of $e_2$. The latter method was in fact the means by which the effect was first noticed (Hicks et al.[29]) in electron-atom scattering. Some of their results are shown in Fig. 2. It can be seen that the energies of three of the four ejected electron peaks increase as the energy $E_1$ is decreased, the lifetime of the fourth state ($2s2p\,^3P$) being too long to show the effect. More recently (Spence[30]) the energy exchange $\Delta E$ has also been observed as a loss of energy of electron $e_1$.

An analogous effect in the excitation of autoionizing states of He by impact with $He^+$ ions of energy 1 to 4 keV had been noticed much earlier by Barker and Berry[31]; in these experiments the scattered ion, which is the analogue of the scattered electron $e_1$ and has a velocity of the same order, interacts with the ejected electron $e_2$ and causes it to have a decreased energy (the difference in the sign of $\Delta E$ being simply caused by the difference in sign of the charge of the scattered particle). Using a simple classical interpretation of a post-collision Coulomb interaction between the scattered particle and ejected electron they found that the probability distribution for the energy loss $\Delta E$ is given by

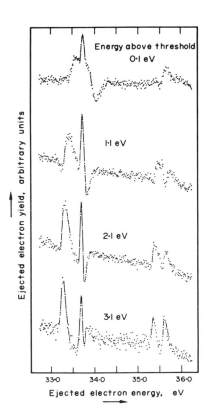

Figure 2.   Spectra of ejected electrons from the four lowest
            autoionizing levels of helium ($2s^2$ $^1S$, $2s2p$ $^3P$,
            $2p^2$ $^1D$ and $2s2p$ $^1P$), at an ejection angle of $70°$.
            The figures above the spectra show by how much the
            incident energy exceeds the energy of the auto-
            ionizing states.   The displacement of the three
            broader peaks to higher ejected electron energies
            can be clearly seen as the incident energy is de-
            creased.   Reproduced from Hicks, Cvejanović, Comer,
            Read and Sharp, Vacuum 24, 573 (1974), with per-
            mission.

$$P(\Delta E)d(\Delta E) = \frac{b}{(\Delta E)^2} \exp\left(-\frac{b}{\Delta E}\right) d(\Delta E) \qquad (7)$$

where

$$b = \frac{1}{v\tau} \qquad (8)$$

and where   v   is the velocity of the scattered particle.   The
resulting line shape of the ejected electron energy peaks is

asymmetric, with a long tail towards large values of $\Delta E$. The
most probable value of $\Delta E$ is

$$\Delta E_{peak} = \frac{1}{2} b \tag{9}$$

and the full width at half maximum of the distribution is
1.07b. Barker and Berry studied the $(2s2p)^1P$ state of He,
and found that the dependence of $\Delta E_{peak}$ on the ion velocity
v is as given by Eqs. (8) and (9), but with a value of $\tau$
somewhat smaller than the known lifetime of the state. These
classical expressions have also been used[29,32-34] to interpret
the energy shifts $\Delta E_{peak}$ seen in electron-helium scattering,[29] but again the fits to the experimental data have
required values of $\tau$ which are too small.

    In quantum mechanical terms the classical model of Barker
and Berry becomes the "shake-down" model.[35,36] This particular nomenclature has been chosen in analogy with the well-
known "shake-off" and "shake-up" events which occur in the
field of Auger electron spectroscopy when the outer electrons
of an atom have to adjust to the sudden removal of an inner
electron (see for example Carlson[37,38]). Before the autoionization event the scattered electron has the wave function
$\psi_{k\ell}$, where k is its momentum ($=\sqrt{2E_1}$, in atomic units) and
$\ell$ its angular momentum. This may be approximated by its
asymptotic form $Y_{\ell m}(\vec{r})\, j_\ell(kr)$ and it must be modified by the
appropriate factor[39] $[=\exp(-t/2\tau) = \exp(-r/2k\tau)]$ to allow for
the loss of amplitude as the autoionizing state decays. After
the autoionization event the final wavefunction for $e_1$ is the
analogous outgoing Coulomb wavefunction $\phi_{k'\ell}$, where k' is
the final momentum. The change in charge at the time of autoionization causes a "shake-down" of the free scattered electron from $\psi_{k\ell}$ to a continuum state $\phi_{k'\ell}$ of lower energy,
and in the sudden approximation (i.e. $E_1 \ll E_2$) the probability that it will find itself with the final momentum k' is
proportional to

$$P_{k'\ell}(k) = |q_{k'\ell}(k)|^2 \tag{10}$$

where the overlap integral is

$$q_{k'\ell}(k) = \int \phi_{k'\ell}^* \psi_{k\ell}\, d\vec{r} \quad . \tag{11}$$

There is no change in the angular momentum of the electron
since it experiences what is essentially a monopole perturbation, if the sudden approximation is valid. This shake-
down model is approximate in many respects,[35,36] but it is
nevertheless able to explain some features of the experimental

observations (see also below).  Recently for example the model
has been used[40] to obtain the probability distribution $P(\Delta E)$,
and it has been found that this is well approximated by the
classical expression (7), still being very asymmetric with a
long high energy tail, as can be noticed for example in some
of the peaks shown in Fig. 2.

A new development occurred in this field when it was
realized[32,33,41] that if the energy exchanges are large
enough, the scattered electron  $e_1$  may finish with a negative
total energy, becoming bound to the residual ion to form an
excited state of the neutral atom.  This process can be illustrated schematically as

$$e + A \rightarrow A^{**} + \boxed{e} \quad \begin{array}{l} \longrightarrow A^* \\ \longrightarrow \boxed{A^+} \;\; + e \end{array} \qquad (12)$$

and it constitutes, in effect, an extra exchange mechanism for
forming excited states of the atom  A  when the incident
energy is near the energies of autoionizing states of the
atom.  Smith et al.[33] tested this idea by studying the excitation cross sections of various Rydberg states of He at incident energies in the region of the first four autoionizing
states, and indeed found structures having a magnitude of a
few percent of the normal excitation cross section.  These
structures have positions that vary with the energy of the excited state being observed, and shapes that are unlike the
Fano-Beutler shapes of resonant structures.  Heideman
et al.[41,42] have also conjectured that this same process may
be responsible for certain structures seen in the optical excitation functions of helium at energies near 60 eV.  More
recently, a detailed study has been made[36] of the structures
in excitation cross sections of various states of He, Ne and
Ar:  Fig. 3 shows some of the results for He.

The shake-down model can readily be tested against these
new structures.  The scattered electron is shaken down from
its initial wavefunction  $\psi_{k\ell}$  to a bound state wave function
$\phi_{n\ell}$ of the neutral atom.  Since the excited states that are
populated in this way tend to be Rydberg states with high
values of n, King et al.[36] were able to approximate the  $\phi_{n\ell}$
by the appropriate hydrogenic wave functions, and they found
that the calculated values of the overlap integrals  $q_{n\ell}(k)$
have shapes that are similar to those of their experimentally
observed structures.  This is illustrated in Fig. 4, which
shows the functions  $q_{n0}(k)$, after convolution with a Gaussian
function of width (FWHM) 100 meV to simulate the finite energy
resolution of the experimental measurements.  It can be seen

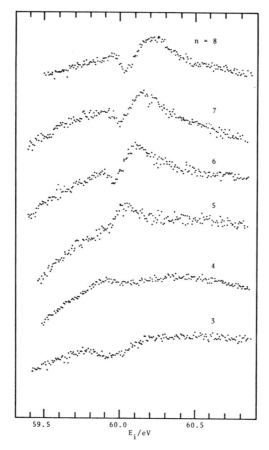

Figure 3.  Excitation functions at 0° for the states of helium
having the principal quantum number  n  from 3 to 8
(and including all possible values of the quantum
numbers  S, L and J  for each value of  n,  except
for the lowest spectrum, which is for $3^1P_1$ only).
The autoionizing state which is responsible for
these structures is thought to be the $2p^2$ $^1D$ state
at 59.90 eV.  Reproduced from King, Read and
Bradford, J. Phys. B, in press, with permission.

that the theoretical curves of Fig. 4 contain a sharp feature
(dip or peak) just above threshold, followed by a broader
feature of the opposite polarity (peak or dip), followed by
yet broader features at higher energies, and that the experimentally observed structures of Fig. 3 are similar, but with
the later and broader features being masked (if they exist)
by the energy dependent transmission function of the apparatus.  The fact that the observed structures are similar to
the overlap integral  $q_{n\ell}(k)$  rather than its square  $P_{n\ell}(k)$
is presumably caused by the new mechanism interfering

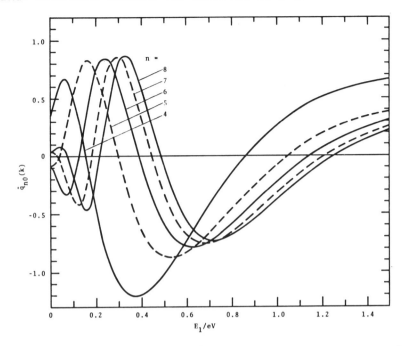

Figure 4.   Calculated values of the overlap integral $q_{n0}(k)$
as a function of $E_1 (=h^2k^2/2m)$, for n = 4 to 8, and
after convolution with an apparatus function having
a Gaussian shape and a width of 100 meV (FWHM).
The lifetime $\tau$ has been taken to be $9.1 \times 10^{-15}$ s,
(that of the $2p^2$ $^1D$ state of helium), and the
phase shift $\delta$ of the scattered electron has been
taken to be 0°.  The magnitude and sign of the normalization of these curves are arbitrary.  Reproduced from King, Read and Bradford, J. Phys. B, in
press, with permission.

coherently with a large direct contribution to the excitation
cross section.  A more detailed comparison[36] shows that the
shake-down model also agrees reasonably well with the observed
structures in other atoms, although many details (such as the
choice of phase shift of the scattered electron, and the
energy dependence of the excitation cross section of the
autoionizing state) have yet to be properly considered.

To summarize, the post-collisional interaction between
the scattered and ejected electrons causes an energy exchange
between them which has been observed in studying (i) the gain
in energy of the ejected electron $e_2$, (ii) the loss in
energy of the scattered electron $e_1$ when this electron has
a final total energy which is still positive, and (iii) the
formation of excited states when $e_1$ has a final energy which

is negative.  The shake-down model seems to give the magnitudes and distributions of these energy exchanges reasonably
well, but with some details remaining to be explained.  One
further type of investigation, namely that in which the electron  $e_1$  is detected when its final energy is nearly zero,
has yielded a threshold spectrum[29] that may contain additional
interference and resonance effects and therefore cannot yet be
used to obtain further information about the shake-down process.  More detailed information on these various effects and
their interpretation can be found in a recent review[35] of the
shake-down effect.

## 4.   SUMMARY

Two types of reaction have been discussed, and in both
there is some evidence for exchanges of energy between the
outgoing particles of the reaction after they have left the
inner zone in which their identities and internal states are
determined.  In the case of threshold ionization there are
also post-collisional exchanges of angular momentum.  Special
emphasis has been  placed on these exchanges, at the expense
of other aspects of the reactions, but before concluding it
is necessary to regain a more balanced view.

In a sense even the concept of a post-collision interaction does not exist, since a reaction or collision is not
complete until all the interactions have vanished.  A complete
theory would include all the final state interactions, of
whatever range, and there would then be no need to invoke any
further time-dependent processes.  Current theories indeed
attempt to do just this.  However, the phenomena that we have
been discussing involve such long ranges (hundreds of
Angstroms in the case of threshold ionization and tens of
Angstroms in the case of threshold autoionization) that it
must be difficult to do this in practice.

For example, Taylor and Yaris[28] have proposed that the
energy displacements seen in threshold autoionization should
be explicable in terms of resonances.  They suggest that the
effects seen near the energy of the $(2s)^2$ $^1S$ state of He could
be due to a series of resonances such as $(2s^2np)^2P$, the higher
members of which could have energies above that of the
$(2s)^2$ $^1S$ state.  These resonances may have an appreciable
overlap with the decay channel 1smpks  if  $n \approx m$,  and could
then lead to structures in the excitation functions of the
1smp states.  The overlapping of many such resonances could
give shapes that are different from the usual Fano-Beutler
shapes of isolated resonances.  In this model the autoionizing
state $(2s)^2$ $^1S$ is never formed as an intermediate step, although it is present as the core of the resonant state.  The

model might also explain the energy shifts and asymmetric profiles seen in ejected electron spectra. Although this model
may ultimately be the correct one to use, it seems probable
that it will give convergent results only if many levels of
several resonance series are included in the calculation. It
also seems possible that a superposition of the assumed resonant states, which could well have widths greater than their
spacings, might in effect look very similar to a single electron receding from a decaying $(2s)^2$ $^1S$ core, particularly
since this represents the dominant decay channel. The difference in the models would then become largely semantic.

In general it seems that long-range final state interactions between more than two particles cannot easily be included in current theoretical techniques, and that new methods, such as a flip-over to classical descriptions at large
distances, or a wave-packet approach with Fourier analysis to
regain specific energies, need to be developed to do this.

## Acknowledgments

I should like to acknowledge many useful conversations
with my colleagues in Manchester (in particular Drs. Comer,
Hicks and King) and with Profs. U. Fano and I. C. Percival.

## References

1.  S. Geltman, Phys. Rev. 102, 171-9 (1956).

2.  M. R. H. Rudge and M. J. Seaton, Proc. Roy. Soc. (London)
    83, 680-2 (1964).

3.  A. Temkin, Phys. Rev. Lett. 16, 835-9 (1966).

4.  A. Temkin, A. K. Bhatia and E. Sullivan, Phys. Rev. 176,
    80-9 (1968).

5.  I. J. Kang and R. L. Kerch, Phys. Lett. 31A, 172-3 (1970).

6.  A. R. P. Rau, Phys. Rev. A 4, 207-20 (1971).

7.  T. A. Roth, Phys. Rev. A 5, 476-8 (1972).

8.  A. Temkin and Y. Hahn, Phys. Rev. A 9, 708-24 (1974).

9.  A. Tamkin, J. Phys. B 7, L450-3 (1974).

10. R. Peterkop and A. Liepinsh, Abstracts of VIth ICPEAC
    (Cambridge USA: MIT Press), 212-4 (1969).

11. R. Peterkop, J. Phys. B 4, 513-21 (1971).

12. G. H. Wannier, Phys. Rev. 90, 817-25 (1953).

13. I. Vinkalns and M. Gailitis, Abstracts of Vth ICPEAC
    (Leningrad: Nauka), 648-50 (1967).

14.   D. Banks, I. C. Percival and N. A. Valentine, Abstracts of VIth ICPEAC (Cambridge USA: MIT Press), 215-6 (1969).

15.   R. Peterkop and P. Tsukerman, Abstracts of VIth ICPEAC (Cambridge USA: MIT Press), 209-11 (1969).

16.   R. Peterkop and P. Tsukerman, JETP $\underline{31}$, 374-7 (1970).

17.   P. Gruyić, J. Phys. B $\underline{5}$, L137-9 (1972).

18.   U. Fano and C. D. Lin, Atomic Physics IV (Plenum Press, New York), in press, (1975).

19.   U. Fano, J. Phys. B $\underline{7}$, L401-4 (1974).

20.   J. W. McGowan and E. Clarke, Phys. Rev. $\underline{167}$, 43-51 (1968).

21.   C. E. Brion and G. E. Thomas, Abstracts of Vth ICPEAC (Leningrad: Nauka), 53-5 (1967).

22.   C. E. Brion and G. E. Thomas, Phys. Rev. Lett. $\underline{20}$, 241-2 (1968).

23.   G. J. Krige, S. M. Gordon and P. C. Haarhoff, Z. Naturforschung $\underline{23a}$, 1383-5 (1968).

24.   P. Marchand, C. Paquet and P. Marmet, Phys. Rev. $\underline{180}$, 123-32 (1969).

25.   S. Cvejanović and F. H. Read, J. Phys. B $\underline{7}$, 1841-52 (1974).

26.   S. Cvejanović and F. H. Read, J. Phys. B $\underline{7}$, 1180-93 (1974).

27.   D. Spence, Phys. Rev. A $\underline{11}$, 1539-42 (1975).

28.   H. S. Taylor and R. Yaris, J. Phys. B $\underline{8}$, L109-12 (1975).

29.   R. J. Hicks, S. Cvejanović, J. Comer, F. H. Read and J. M. Sharp, Vacuum $\underline{24}$, 573-80 (1974).

30.   D. Spence, to be published (1975).

31.   R. B. Barker and H. W. Berry, Phys. Rev. $\underline{151}$, 14-9 (1966).

32.   F. H. Read, Atomic Physics IV (Plenum, New York), in press (1975).

33.   A. J. Smith, P. J. Hicks, F. H. Read, S. Cvejanović, G. C. M. King, J. Comer and J. M. Sharp, J. Phys. B $\underline{7}$, L496-502 (1974).

34.   G. Nienhuis and H. G. M. Heideman, J. Phys. B, in press (1975).

35.   F. H. Read, Radiation Research, in press (1975).

36.   G. C. King, F. H. Read and R. C. Bradford, J. Phys. B, in press (1975).

37. T. A. Carlson, The Physics of Electronic and Atomic Collisions (invited lectures and progress reports of VIII ICPEAC, Ed. B. C. Cobic and M. V. Kurepa, Institute of Physics, Belgrade, 1973).

38. T. A. Carlson, Radiation Research, in press (1975).

39. C. Bottcher, Electron and Photon Interaction with Atoms (Eds. H. Kleinpoppen and M. R. C. McDowell, Plenum Press, New York, 1975), in press.

40. P. J. Hicks, J. Comer and F. H. Read, to be published.

41. H. G. M. Heideman, G. Nienhaus and T. van Ittersum, J. Phys. B $7$, L493-5 (1974).

42. H. G. M. Heideman, T. van Ittersum, G. Nienhuis and V. M. Hol, J. Phys. B $8$, L26-8 (1975).

# ANGULAR CORRELATION OF ELECTRONS COMING FROM IONIZATION BY ELECTRON IMPACT

Derek Paul

Physics Department, University of Toronto, Toronto, Canada, M5S 1A7

K. Jung, E. Schubert and H. Ehrhardt

Fachbereich Physik, Universität Kaiserslautern, 675 Kaiserslautern, Postfach 3049, W. Germany

## ABSTRACT

The experimental work of Ehrhardt and coworkers on the ionization of inert gas atoms by electron impact, in which the full dynamics of the collisions are determined, is reviewed.

At the present time no theoretical calculations have been made which will explain all the experimentally observed features of these angular correlations, but the salient features of the 1s-ionization can be understood through a comparison with Born, especially Coulomb-projected Born approximations. One can gain insight into some differences between experiment and Coulomb-Born approximations for 2p-electron ionization by comparison with an impulse model, from which it is clear that electron-electron interaction at short range plays a major role in the emission of electrons in the direction of the momentum transfered by the fast electron.

Some measurements of the ionization of $H_2$ and $N_2$ are briefly discussed.

## INTRODUCTION

The experiments which we are about to describe were the first of the electron impact type in which the dynamics of all particles in an ionization event are fully determined. Fig. 1 shows the basic geometry of experiments of this sort in which

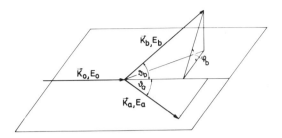

Figure 1

the incident electron is represented by a momentum vector $\vec{k}_o$ and kinetic energy $E_o$, and the outgoing particles by vectors $\vec{k}_a$ and $\vec{k}_b$ and corresponding energies $E_a$ and $E_b$. We say "outgoing particles" here rather than "outgoing electrons" because one of them could in principle be the recoiling ion. Whether or not it is the ion or the second outgoing electron which is detected, the dynamics are fully determined, including the state of excitation of the residual ion, provided $E_a$, $E_b$ and $E_o$ are all measured as well as the angles $\theta_a$, $\theta_b$ and $\phi_b$ for each event.

It is necessary to detect the outgoing particles in coincidence and it is therefore much more convenient to detect two outgoing electrons than to attempt to detect one electron in delayed coincidence with the recoil ion which is generally a much slower particle. Overall the counting rates in such experiments are such that one needs to exploit every technical advantage, such as good time resolution in the coincidence circuits, or one must sacrifice energy or angular resolution.

Fig. 2 shows schematically the experimental arrangement developed by the research group which has been located in Kaiserslautern since 1971. The arrangement is coplanar, and allows $\theta_a$ and $\theta_b$ to be independantly varied, but restricts $\theta_b$ to 0 or 180°. Sufficient details of the equipment and earlier mode of operation have been given by Ehrhardt et al. (1972c). The typical coincidence time resolution of our system was 10 ns full width at half maximum.

## Published data for ionization of inert gas atoms

We would like in this section to give you a rather rapid survey of the experimental data published in the period 1972-4 inclusive. The earliest paper (Ehrhardt et al. 1969) contained only two angular correlation curves, having somewhat poorer statistical accuracy than the later measurements.

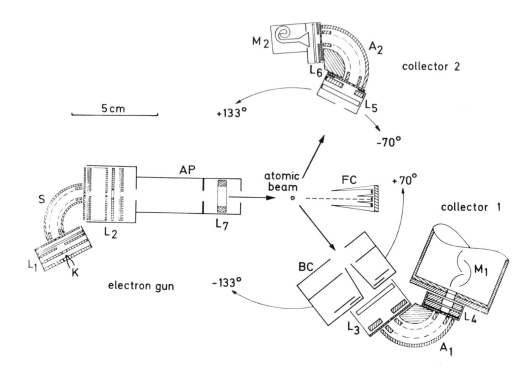

Figure 2.   Arrangement of electron gun and
analysers (A1, A2) within vacuum system

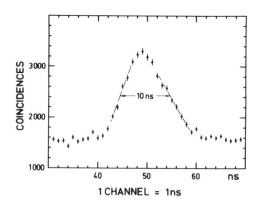

Figure 3.   Time resolution curve

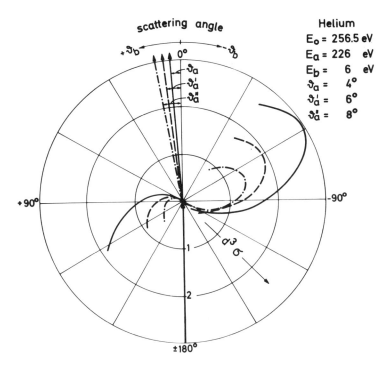

Figure 4.   Angular distributions of the slow electrons for
ionization in which the faster electron is ejected at 4°, 6°, 8°


        Figs. 4, 5, 6, 7 and 8 show respectively some results
for the ionization of helium with incident electrons of energies
256.5, 80.5, 50, 41 and 30.5 eV.  The data have been published
by Ehrhardt et al. (1972a, b) and are fully discussed in these
papers.  The solid curves in these diagrams are the experimentalists' smoothed versions of the experimental data and are
not derived from any theory.  We wished however, to present
the initial data without the bias that may be introduced by
theoretical knowledge.  One must bear in mind that the experiments were only able to operate in the angular ranges

$$-\quad 125° < \theta_b < \theta_a - 42°$$

and        $\theta_a + 42° < \theta_b < 125°$

where $\theta_a$ is taken as positive, and negative values of $\theta_b$ are
taken to indicate $\phi_b = 180°$ (see fig. 1).  Sometimes, therefore, a very interesting region of the angular correlation
could not be reached experimentally.  Theories, then, bring
in valuable clues as to the possible shapes of the missing
parts of the angular distributions, but can also be misleading

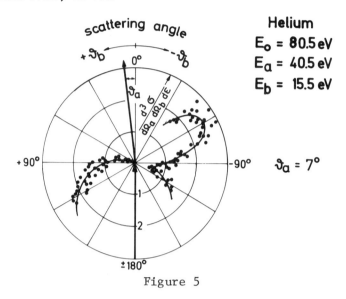

Figure 5

if they are incorrect.

In fig. 4 one sees at once that the distributions all
have two lobes, one of which is roughly in the direction $\vec{K}_{oa}$
where

$$\vec{K}_{oa} = \vec{k}_o - \vec{k}_a$$

and is the momentum transferred by incoming electron if one
considers $\vec{k}_o$ to have been changed into $\vec{k}_a$ by the impact. It
has been customary to call the lobe such that $\vec{K}_{oa} \cdot \vec{k}_b > o$ the
binary peak, and the opposite lobe the recoil peak.

In fig. 5 the distribution is similar, but whereas one
can imagine an axis of symmetry for the 256.5 eV data, such
an axis is already starting to look less plausible at 80.5 eV,
unless there is an entire missing lobe at an angle around
$\theta_b$ = o which could not be measured. The same can be said of
the data in figs. 6, 7 and 8 at lower primary energies, the
apparent asymmetry becoming larger as $E_o$ is decreased. One
can use arguments that the character of the distributions will
vary continuously with primary energy in these cases and, after
studying the experimental data in detail, would conclude that
the curves in figs. 5 - 8 are indeed asymmetric, and do not
have missing lobes in the forward direction. It would nevertheless be nice to be able to investigate the forward angles
experimentally.

For argon, 27 curves were measured over a prolonged
period and were published by Ehrhardt et al. (1974). These

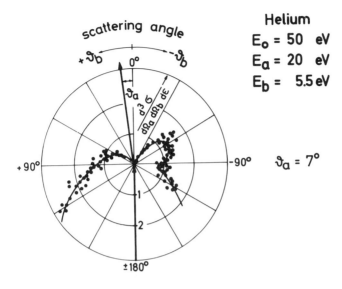

Figure 6

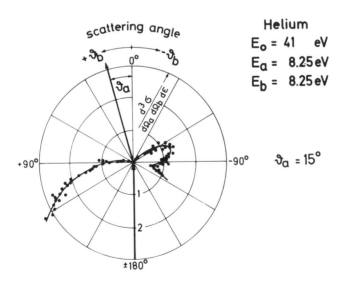

Figure 7

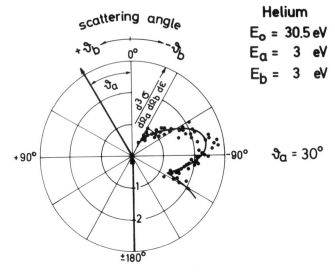

Figure 8

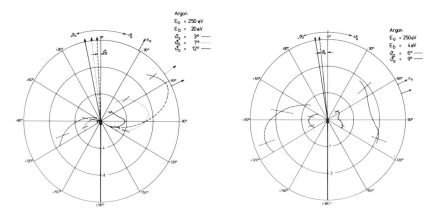

Figure 9

data are shown in figs. 9, 10 and correspond to primary electron energies of 250 and 100 eV respectively. In all cases
the outer 3p-electrons were ionized. The curves illustrated
are again all smoothed versions of the experimental data. The
statistical errors in typical individual points are indicated
by error bars, but the data points are so closely spaced and
so numerous that the statistical uncertainties in the smoothed
curves are much smaller except of course at the extreme ends
of each curve.

The argon data show in many cases the same characteristics

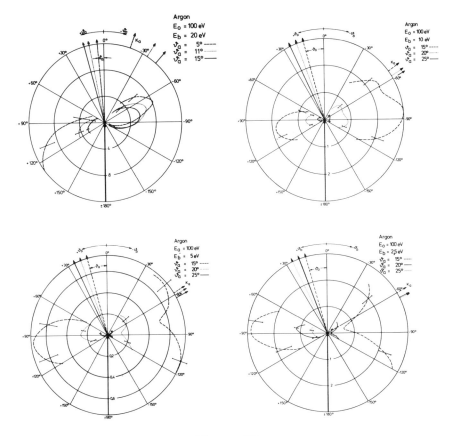

Figure 10

and tendencies as the helium data, in particular the binary
and recoil peaks, and the implied waist or minima between them.
However, some of the data reveal a conspicuous dip in the
direction normally associated, approximately at least, with
the binary maximum, namely $\vec{K}_{oa}$.  If one now attempts to
associate a symmetry axis with the direction of the binary
minima (where they occur) one finds  in all cases that the
distribution is not symmetric with respect to this axis except
where the statistics do not warrant any clear pronouncement on
the subject.

J.F. Williams (private communication) has recently made
a number of coincidence measurements on the ionization of
3p-electrons from argon for a primary electron energy of 250 eV.
His angular distributions are similar to ours where the other
parameters are also comparable.

THEORY

Born approximations

The earlier theories of the fully differential cross sections for the ionization of helium by electron impact have either been Born approximation, simplifications to such approximations, or, binary encounter theories (Vriens 1969, 1969a and 1970; Glassgold and Ialongo 1968). In fig. 11 the top left curve is the result of the Glassgold and Ialongo theory together with the experimental points for the corresponding parameters. This curve fails

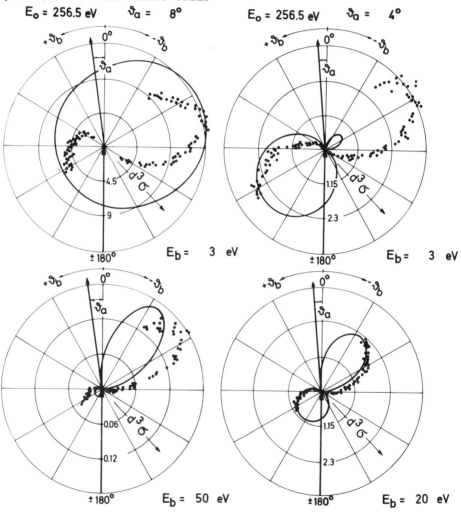

Figure 11.  Comparison of experimental data with Glassgold and Ialongo theory (upper left) and plane-plane Born approximation (right hand curves), and plane-Coulomb Born approximation (lower left).

entirely to produce the conspicuous minima in the experimental
data at angles of about $30°$ and $-150°$, because interaction of
the incident and emergent electrons with the helium ion is
entirely omitted.  The other three curves in fig. 11 are Born
approximation results in the sense that all the Coulomb interactions between electron and electron or electron and ion are
properly included.  These diagrams reveal fairly appropriate
shapes and, after normalising to the experimental points (say,
in the $\vec{K}_{oa}$ direction), quite good agreement with experiment in
a few cases.  Always, however, the binary maximum in the experiments is shifted to a larger angle than $\theta_{oa}$, between $\vec{K}_{oa}$ and
$\vec{k}_o$, than this theory predicts.  So far as one can see the
breadth of the binary maxima are about correctly predicted by
the theory, whereas the ratio of counts in the binary and recoil peaks is almost always wrongly predicted by the theory.

   Nevertheless the minima are present at, typically, $45°$
and $-135°$ in these theories, in reasonable agreement with
experiment.  The minima can be understood as follows.  In the
Born approximation one can set up the problem either in the
post or prior form, which means that the approximation is equivalent to one in which either the incident electron-ion interaction term (before the collision) drops out or the faster
emergent electron's interaction with the ion (after the collision) drops out.  This leaves in the matrix element only one
set of terms representing electron-ion interaction and one set
representing electron-electron interaction.  If we write the
appropriate matrix elements in the prior form

$$B \equiv \langle \psi^{(-)} | V_{12} | \phi_i \rangle$$

and $$R \equiv \langle \psi^{(-)} | W_{1,ion} | \phi_i \rangle$$

then the amplitude for the transition is  $f = B - R$.
The minima in the angular distribution of emitted electrons as
a function of $\theta_b$ is a direct result of the interference of the
B and R terms in f.  In the plane-plane[1] Born approximation
B and R are both real and the interference is often complete
at two angles symmetrically placed relative to $\vec{K}_{oa}$.  In most
of the diagrams in fig. 11 this interference is very marked
in the experimental data and comes near to the plane-plane
Born predictions.

   In order to explain the experimental data more fully Schulz
(1973) and Geltman (1974) have carried out complicated Born
computations in which one or both of the outgoing waves are
represented by Coulomb wave functions.  We usually refer to
these as plane-Coulomb and Coulomb-Coulomb Born approximations.

---

[1] i.e., in which ingoing and outgoing particles are all represented by plane waves.

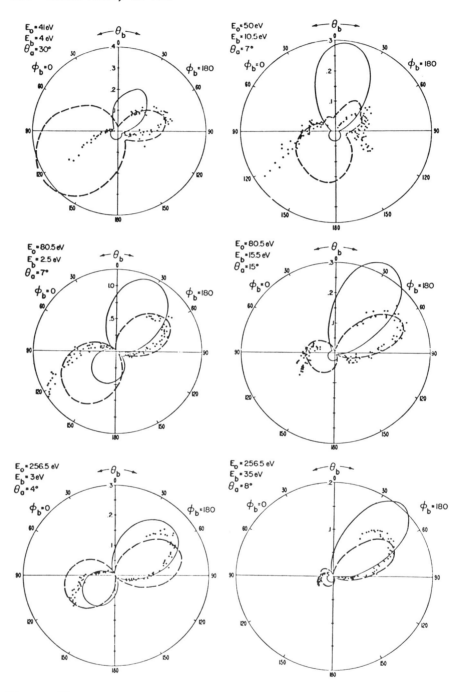

Figure 12.   Comparison of experimental data with Geltman's
plane-projected Born approximation (solid curves) and
Coulomb-projected Born approximations (dashed curves)

The plane-Coulomb approximations ("PP" or plane-projected in Geltman's paper) are given by the solid curves in fig. 12. For low incident energies these calculations fail very badly, but at 256.5 eV incident electron energy they often have the advantage of predicting a much better binary-to-recoil peak ratio than plane-plane Born computations. In the plane-Coulomb approximation the angular distribution is unfortunately still symmetric about $\theta_{oa}$.

In the Coulomb-Coulomb calculations ("CP" or Coulomb-projected approximation) Geltman used a Coulomb wave function also for the faster outgoing electron and included exchange with the incoming electron, which was omitted from the plane-Coulomb case. In the Coulomb-Coulomb calculation the target wave functions was taken as

$$\phi_o(\vec{r}_2, \vec{r}_3) = \frac{Z^3}{\pi} e^{-Z(r_2+r_3)}$$

with $Z = \frac{27}{16}$, the Hylleraas value, while the outgoing wavefunction for ion and slower electron was taken as

$$\phi_{kS}^-(\vec{r}_2, \vec{r}_3) = \frac{1}{\sqrt{2}} [u_k^-(\vec{r}_2)v_o(\vec{r}_3) + (-1)^S u_k^-(\vec{r}_3)v_o(\vec{r}_2)]$$

in which $v_o$ is the hydrogen-like ion ground state wave function and $u_k^-$ an outgoing Coulomb wave of effective charge Z. The outgoing faster electron wave function was also taken to have the same effective charge $Z = \frac{27}{16}$.

The results are shown by the dashed curves in fig. 12 revealing greatly improved agreement over the whole range of primary electron energies. Nevertheless the binary-to-recoil ratios are often seriously in error and the angular shifts of the binary peaks are only sometimes in agreement with experiment. The results shown are normalized to experiment in the binary direction.

Undoubtedly a difficult question in the Coulomb-Coulomb approximation is the choice of effective charge for the outgoing waves. It is very tempting to consider that the slower particle will see a larger effective charge than the faster one which is partly screened by the slower, and/or that the effective charges will depend on the directions and magnitudes of the vectors $\vec{k}_a$, $\vec{k}_b$. Schulz (1973) investigated this very thoroughly in a number of approximations, three of which were Coulomb-Coulomb and in one of these he set $Z = -1$ for both outgoing waves. In a second Coulomb-Coulomb approximation he allowed the effective charges to vary as a function of $\vec{k}_a$, $\vec{k}_b$; and in a third approximation an effective charge $\lambda_{b,m}$ was defined as the negative of the minimum value of

$$\left| \frac{1}{\lambda_b} \frac{B(\lambda_a = o, \lambda_b)}{R(\lambda_a = o, \lambda_b)} + 1 \right|$$

so that it aquires the value o in the recoil direction and -1 in the binary direction. An additional condition $\lambda_a = -1-\lambda_{b,min}$ is an attempt to simulate the mutual screening of the electrons. This approach led to the best agreement with experiment for the ratios of cross sections in the binary and recoil direction.

Fig. 13 illustrates some of the best fits obtained by Schulz, using the last-mentioned approximation. Although some of the fits are really good, there is still the problem that the angle shift and the binary-to-recoil ratio are poorly predicted overall when one allows a wide range of experimental parameters. Schulz concluded (private communication) that the Coulomb interactions at short range, namely the three body electron-electron-ion interaction, must be important in the parameter region which has been investigated experimentally, and that one might very well not trouble to take Born approximations much further.

Nevertheless the split binary lobes observed with argon gas deserve some theoretical attention, and since the problem of ionization of 3p electrons from argon is by no means the simplest to begin with theoretically, Knapp and Schulz (1974) investigated the ionization of 2p electron from the hydrogen atom and also the 2p electrons from neon. The ionization of neon has now also been investigated by Jung et al. (1975), so we shall restrict our discussion of the Knapp and Schulz paper to the theoretical results pertaining to neon. Because of the greater complication of calculating the ionization from the p-shell, the plane-Coulomb Born approximation was the most sophisticated which was attempted. It is necessary in this case, in order to obtain even fair agreement with experiment,

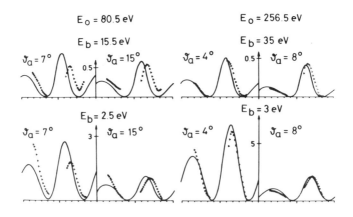

Figure 13.   Comparison of Schulz's approximation A8 (solid curves) with smoothed experimental data (triangles)

to orthogonalize the outgoing wave of the slow electron with
the wave function for the five remaining p-electrons. Also
exchange between the outgoing wave and all the p-electrons of
the ion is important. With these features included in the
calculation and exchange with the faster electron being neglected, the usual broad features of the experimental results
were predicted. We shall discuss the experiments further on.

Finally, regarding plane-Coulomb approximations, in addition to the predicted symmetry, etc., one expects the angular
correlations not to change their patterns with changing primary energy, provided $E_b$, the slower outgoing electron's energy,
and $K_{oa}$, the magnitude of the momentum transfer, are kept
constant.

The Impulse Model

To complete the discussion of p-electron ionization theory
we would like to mention a very simple impulse model which is
exactly the same in principle as a binary encounter model
(Vriens 1969, 1969a, 1970). Let us recall that the Born approximations fail to allow for electron-electron correlation in the
wave function, though the electron-ion interaction is of course
included. In fig. 14 one sees the opposite situation in which
the ionizing collision is treated as a purely electron-electron
collision. The ion is assumed to have a momentum $-\vec{k}_e$, but to
play no part in the collision, i.e., its instantaneous momentum
before the collision becomes its recoil momentum after the atom
has been ionized.

Vriens (1970) pointed out a basic difficulty with binary
encounter models, that they do not conserve energy unless one
adds suitable quantities $\Delta_j$ to the squares of the momentum
vectors. These $\Delta_j$ represent the additional free particle energies due to the presence of the atomic field. Initially it is
a matter of pure speculation whether one will obtain better
agreement with experiment by putting the $\Delta_j = o$ or choosing
values which ensure energy conservation in the classical sense.

The value of such a model is therefore very much in doubt,
but nevertheless one has the possibility of treating just those
effects which are neglected in the Born models of Knapp and
Schulz for neon, while one must neglect everything which is
included in the Born models except of course the electron-
electron potential.

In our model we selected experimental sets of values of
$\vec{k}_o$, $\vec{k}_a$ and $\vec{k}_b$ and transformed these vectors into the frame of
reference of the target electron. We then used the Mott cross
section appropriate to a scattering event in which the trans-
formed $\vec{k}_o'$ is the incident vector and $\vec{k}_a'$ is the scattered electron
direction. The solid angle elements and $dE_b'$ are then transformed into the laboratory coordinates.

à) Neon                    RECENT EXPERIMENTS

A number of improvements to the equipment have been
brought about in the last twelve months which are described in
full elsewhere (Jung et al. 1975b).  Here we mention only the
improved gas jet with inner and outter pumping regions (fig. 15),
and that a new electron gun was constructed which permits a
slightly wider range of angles for the analysers.  Most important perhaps was the attention paid to systematic errors.
Essentially all of the major sources of systematic error have
now been investigated by Jung et al. and we believe that all
such errors have been reduced to values which are certainly
below the statistics of recent experimental measurements.  That
is not to say that former measurements have been found to be
in error systematically – on the contrary our recent findings
if anything support the truth of former results – but the upper
limits which we had been able to set on certain systematic
uncertainties were formerly much greater.

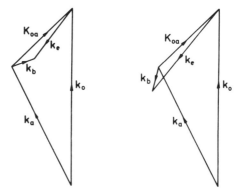

Figure 14.  Vector diagrams corresponding to no interaction with
the ion (binary encounter models).  Left: $k_e < K_{oa}$ and therefore
$k_b$ must lie in binary hemisphere.  Right:  $k_e > K_{oa}$ and $\vec{k}_b$ can lie
in any direction.

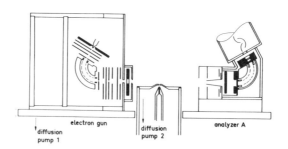

Figure 15.  Sectional view of new apparatus showing interior
arrangement of electron gun, and the gas jet forming nozzle and
cover.  Analyser B not shown.

With the rebuilt apparatus the curves of the figs. 16 - 18 were obtained for the 2p-ionization of neon and the 2s-electron ionization. The diagrams have been labelled as follows to facilitate simultaneous identification and grasp of the experimental parameters. We use the label $E_o/E_b/K_{oa}$ for any given run in which the 2p-electrons' ionization was studies, with $E_o$, $E_b$ in electron volts and $K_{oa}$ in atomic units, i.e., $a_o^{-1}$.

It is perhaps most fruitful to discuss these results first in comparison with the Born theory of Knapp and Schulz (1974). The obvious first question is how the experimental data and theory were normalized to each other. In figs. 16, 17 it was possible to normalize in the recoil direction for the lowest $K_{oa}$ value for each pair of values $E_o$, $E_b$. To change any of those experiments from $E_o/E_b/0.5$ to $E_o/E_b/0.8$ required only a change in the angle $\theta_a$ and no adjustments to either detector nor to the electron gun. Therefore no further arbitrary normalization was required when only changes of $K_{oa}$ were made in successive runs. Thus one can see at once from studying the vertical pairs of runs in fig. 16, or from the three p-electron ionization diagrams in fig. 17, how relatively good or bad the Knapp and Schulz recoil peak predictions are. We note that the recoil peak intensities seem to fit about correctly for the $250/20/K_{oa}$ series (fig. 17) and less well at lower $E_b$ values, but that the shape of the recoil peak at low $E_b$ can be in quite good agreement with experiment over a wide angular range (e.g., fig. 16, 250/5/0.5).

The general prediction of the simpler Born theories, that the form of the angular correlation is a function only of $E_b$ and $K_{oa}$, $f(E_b, K_{oa})$, is rather well borne out by the horizontal pairs of curves in figs. 16, 17, for which only the primary energy was changed. The agreement, while not very exact, is perhaps surprising having regard to the low incident energy in the right-hand diagrams. It could be that this is another of those examples of physical systems which accidentally display one property in more accurate agreement with theory than one has any strong reason to hope.

General features of the plane-Coulomb predictions of Knapp and Schulz are only fairly well borne out. The waist-lines in the directions almost normal to $\vec{K}_{oa}$ occur very much as predicted, though, as before with helium, one wishes one had more of the angular correlation curves available from the experiments. Where Knapp and Schulz predict a minimum in the binary peak such minima are invariably found in the experiments, but are generally sharper than the theory suggests. In those cases studied where no minimum was predicted minima were nevertheless found, a possible exception to this rule being run 250/20/0.5. In 250/20/0.8 or 250/10/0.5 the minimum is unquestionably present in the experiments, while in 100/10/0.5 it is clearly implied.

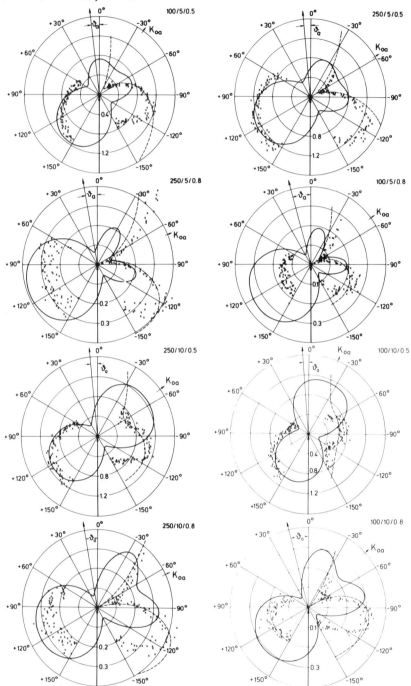

Figure 16.  Ionization of 2p-electrons in neon.  For notation
see text.

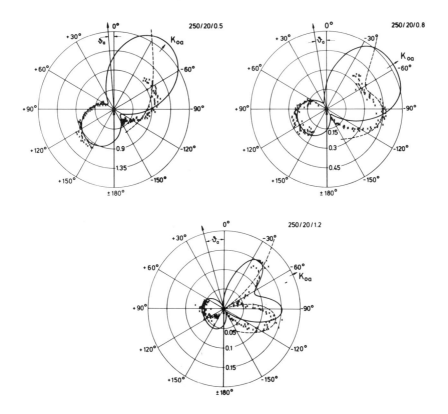

Figure 17.   Ionization of 2p-electrons in neon.   For notation
see text.

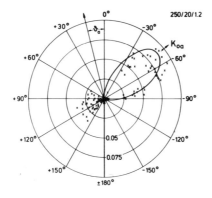

Figure 18.   Ionization of 2S-electrons in neon.

The consistency with which the minima are sharper and deeper in experiment than theory and the consistent tendency of the binary lobes to appear more spread-out in the experiments, and also larger than in the theory, led us to consider those electron interactions which are neglected in the Born theory as being to a large extent responsible for the differences.   The figs. 16, 17 therefore also have the impulse model results included, but these are only plotted in a narrow range of angles about $\vec{K}_{oa}$, since this approximation neglects too much to be of broad use.   Also each of the results is separately normalized to experiment at an angle somewhat greater than $\theta_{oa}$. Even so there is much that can be said from inspection of the diagrams.   Most obvious is the sharper minimum and more widespread lobes in the cases where the experimental results are also sharp, namely in fig. 16, and the 250/10/0.8 and 100/10/0.8 and 250/20/1.2 curves.   The impulse model shifts the angle of the minimum towards greater absolute values of $\theta_b$ than $\theta_{oa}$ by only a small amount in these cases, but then the experimental minima are not much shifted either.   For the rather flatter data such as 250/10/0.5, 100/10/0.5 and 250/20/0.8 our impulse model tends to shift the binary maximum to larger angles and may make it disappear, giving way to a forward scattering lobe. This feature is a property of the way in which the binary encounter theory is set up and does not occur similarly in the Vriens theory.   We have also applied Vriens' equations for the $\Delta_i$'s all equal to 0 and find analogous sharp minima in the $\hat{K}_{oa}$ direction when such minima occur in the experiments.   The agreement with experiment in the $\hat{K}_{oa}$ direction is a phenomenon arising from the distribution of electron momenta $P(\vec{k}_e)$ in the target atom.   Asymmetry about $\hat{K}_{oa}$ could be due to the forward peaking of the Mott scattering and the coordinate transformation, but the model is not capable of predicting this effect in a uniquely correct way.

The impulse model results normalize poorly to the experimental data in the sense that if one goes from $K_{oa}$ = 0.5 to 0.8 without changing $E_o$ or $E_b$ the ratio of predicted differential cross sections is badly in error.   Also the overall form of these cross sections is quite wrong, and it seems unlikely that the strong forward lobes predicted at small $\theta_b$ values will turn out to be so large in experiment.   However, we have up till now not been able to investigate small $\theta_b$ values experimentally.

With the parameters we have used the Mott and Rutherford scattering cross sections are almost equal, indicating that exchange is unimportant as far as the fast electron is concerned. However this need not be so and one needs only to go to larger momentum transfers $K_{oa}$ to bring about the situation in which, on the impulse model this exchange is at a maximum.   In our experiments we thus expect electron-electron correlation in the wavefunction to be by far the most important factor omitted from

the Knapp and Schulz (1974) models.  At the present we are not
aware of a simple, correct method of bringing the Born and
impulse theories together into one.

b.  Molecular hydrogen and nitrogen

     Immediately prior to the improvements in the apparatus
which were briefly described in the foregoing subsection a series
of measurements were made on $H_2$ and $N_2$.  These were the first
molecules to be investigated in this way, and the results show
many of the features of the earlier results on helium and argon.
This work has been described in a recent paper by Jung et al.
(1975a) and will not be dwelt on here.  The special features
were the binary minimum in the $\pi$-electron ionization of nitrogen[1] and the rather small recoil peaks in some instances.
The authors offered a possible explanation of the absence of
recoil peak (when they found none) by suggesting that the recoil ion had absorbed more vibrational energy in these cases
and therefore the slow electron had an energy $E_b$ which fell outside the analyzer's energy window.  For $H_2$ an attempt was made
to relate the energy of the outgoing electron to Frank-Condon
factors as determined by photo-ionization.  One of us
(E. Schubert) has recently measured the spectrum of outgoing
energies $E_b$ in coincidence with $E_a$ for a recoil direction from
$H_2$ and finds the spectrum to be similar to that in the binary
direction i.e., the vibrational energy loss distribution is the
same on the binary and recoil sides.  For nitrogen these measurements have not been made.

Conclusions

     This paper has merely given a survey of the work done to
date by one experimental group.  Much work carried out at
Flinders University, Adelaide, Australia by E. Weigold and
co-workers also involves coincidence experiments on ionization[2].
However, until very recently these authors were much more
interested in the cross sections at higher primary energies and
generally at larger momentum transfers.  The situation is probably just now being reached at which data, from new experiments
by the two groups, might be overlaid on the same plots for comparison.  In so saying we are quite mindful of the limitations
of all the work so far carried out in this field.  For example,
interesting findings are almost sure to result from the extension of the work described here for neon to higher momentum

---

[1] The minimum near $\theta_{oa}$ on the binary side was observed with
$E_o = 100$ eV, $E_b = 4$ eV and $\theta_a = 25^o$.

[2] See for example Weigold et al. (1975), Ugbabe et al. (1975).

transfers.  The split binary lobes become reunited into a single forward lobe at large momentum transfers, and true binary encounters (in the sense of the impulse model) can only result in events on the binary side.  Recoil scattering at such momentum transfers therefore provides a clean measurement of electron-ion interaction.  We note that the work of Weigold and coworkers though it has concentrated on the higher momentum transfers, has not probed this feature of ionization.  Another area of interest is ionization in the neighbourhood of threshold, when both outgoing electrons are slow.  The binary-recoil feature, of the angular correlation in such cases is expected to give way to a distribution in which the two electrons emerge predominantly in opposite directions.

## References

1.  Clementi E., 1965. in "Tables of Atomic Functions", a supplement to IBM J. Res. Develop. 9, 2.

2.  Ehrhardt H., Schulz M., Tekaat T., and Willmann K. 1969. Phys. Rev. Lett. 22 89-92.

3.  Ehrhardt H., Hesselbacher K.H., Jung K., Schulz M., Tekaat T., and Willmann K. 1971. Z. Phys. 244 254-67.

4.  Ehrhardt H., Hesselbacher K.H., Jung K., and Willmann K., 1972a. J. Phys. B: Atom. molec. Phys. 5 1559-71.

5.  Ehrhardt H., Hesselbacher K.H., Jung K., Schulz M., and Willmann K., 1972b. J. Phys. B: Atom. molec. Phys. 5 2107-16.

6.  Ehrhardt H., Hesselbacher K.H., Jung K., Schubert E., and Willmann K. 1974. J. Phys. B: Atom. molec. Phys. 7 69-78.

7.  Fano U. and Lin C.D., 1973. in "The Physics of Electronic and Atomic collisions" invited lectures and progress reports VIII ICPEAC, ed. B.C. Cobic and M.V. Kurepa (Inst. Phys. Beograd) pp. 230-7.

8.  Geltman S., 1974. J. Phys. B: Atom. molec. Phys. 7 1994-2002.

9.  Glassgold A.E., and Ialongo G., 1968. Phys. Rev. 175 151-9.

10.  Jung K., Schubert E., Paul D.A.L., and Ehrhardt H., 1975. J. Phys. B: Atom. molec. Phys. 8 1330-7.

11.  Knapp E.W. and Schulz M., 1974. J. Phys. B: Atom. molec. Phys. 7 1875-90.

12.  Schulz M., 1973. J. Phys. B: Atom. molec. Phys. 6 2580-99.

13.  Ugbabe A., Weigold E., and McCarthy I.E., 1975. Phys. Rev. A 11 576-85.

14.  Vriens L., 1969.  Physica <u>45</u> 400-6.

      "   "   1969a in "Case Studies in Atomic Collision Physics I" ed. E.W. McDaniel and M.R.C. McDowell 335-98.

15.  Vriens L., 1970. Physical <u>47</u> 267-76.

16.  Weigold E., Hood S.T., and McCarthy I.E., 1975. Phys. Rev. A <u>11</u> 566-75.

# ELECTRON-MOLECULE  SCATTERING

# THE THEORY OF LOW ENERGY ELECTRON-MOLECULE SCATTERING

KAZUO TAKAYANAGI

Institute of Space and Aeronautical Science

University of Tokyo

Komaba, Meguro-ku, Tokyo, Japan

Early theoretical works have been reviewed in many articles.[1,2,3]   Here, I discuss mainly the low energy electron scattering from hydrogen molecules and from polar molecules.

After some distorted-wave calculations have been reported, the first close-coupling (CC) calculations on electron-$H_2$ scattering with rotational transitions have been done by Lane and Geltman[4] with a semiempirical potential, and then by Lane and Henry[5,6] with more accurate interaction.   Another detailed calculation on the same system was done by Hara,[7] using the adiabatic approximation.   These calculations have shown the importance of the electron exchange and the target polarization effect in the low energy region studied.   When these effects, together with a reasonably accurate electrostatic interaction, are taken account, the resulting cross sections are in fairly good agreement with the electron beam experiment[8] and also with the electron swarm experiment.[9]

There exists, however, some difference between the rotational cross sections obtained by Henry and Lane and those of Hara.   This disagreement arises from the following three reasons :
1) Henry and Lane describe the collision problem in the one-center polar coordinates, while Hara's formulation is based on the two-center spheroidal coordinates.
2) Henry and Lane applied the CC method, while Hara's calculation is based on the adiabatic approximation.
3) The electron-$H_2$ interaction adopted in these two calculations are approximate and they are different from each other. (To some extent this is related to 1.)

Some comments on each of these will be instructive.

In the one-center treatment, the electron-molecule inter-
action is expanded in terms of the Legendre polynomial $P_\lambda(\hat{r}\cdot\hat{s})$,
while in the two-center approach, the same interaction is ex-
panded in $P_\nu(\eta)$, where $\eta= (r_a - r_b)/s$, $\vec{r}_a$ and $\vec{r}_b$ are the position vectors of the scattered electron relative to the nuclei
A and B of the molecule, $\vec{r} = (\vec{r}_a + \vec{r}_b)/2$, $\vec{s} = \vec{r}_a - \vec{r}_b$ is the
internuclear separation vector and finally $\hat{r} = \vec{r}/r$, $\hat{s} = \vec{s}/s$
are unit vectors.   In both approaches, the expansions  are
truncated and only the first two or three non-vanishing terms
are retained.   Different approximate potentials are thus ob -
tained.   The one-center expansion cannot represent the Coulomb
singularities at the position of the nuclei if only a few terms
are adopted.[1]  Recently, Darewych, Baille and Hara[10] compared
the one-center and two-center calculations for the positron
scattering from $H_2$ in the adiabatic approximation, and concluded
that the two calculations give  almost the same result in spite
of the truncations.   This is probably because the $H_2$ molecule
has a very small internuclear separation and the molecular electron cloud is nearly spherical and also because the low energy
positron cannot approach nuclei closely because of the Coulomb
repulsion.   Similar comparison for electron scattering from
other molecules would be interesting.

In the adiabatic approximation, the electron scattering
amplitude for the rotational transition of a rigid rotor from
$j_0 m_0$ state to j m is given by[1]

$$F(j_0 m_0, k_0 \to j\, m,\, k)$$

$$= \int Y_{jm}^*(\hat{s})\, f(k_0 \to k;\, \hat{s})\, Y_{j_0 m_0}(\hat{s})\, d\hat{s}, \qquad (1)$$

where $f(k_0 \to k; \hat{s})$ is the elastic scattering amplitude for
fixed molecular orientation $\hat{s}$ and $\vec{k}_0$, $\vec{k}$ are the initial  and
final wave vectors of the scattered electron.   Validity of
the approximation may be discussed in terms of the frame-transformation theory of Chang and Fano.[11]  The body frame (fixed to
the target molecule) is adopted at shorter distances where the
rotational Hamiltonian of the molecule can be disregarded  as
compared with the electron-molecule interaction, while the laboratory frame is used at larger distances.   If the electron
energy is much larger than the rotational level spacings, we
can neglect the rotational Hamiltonian even in the outer region, and then the adiabatic approximation is derived.   Chang
and Temkin[12] discussed the validity of the adiabatic approximation in a different way and suggested that for electron-
homonuclear diatomic molecule scattering the theory is applicable when the incident energy is above 1.65 times the thresh-

old energy.    It will be safer, however, to say that the approximation is valid when the effective collision duration is much
shorter than the period of molecular rotation and the electron
energy is much larger than the level spacings.    Although the
limitation of its applicability is not yet clearly shown, the
adiabatic approximation has many merits as I have discussed at
Atomic Physics Conference in Heidelberg last year.[13]

The close-coupling method, on the other hand, has no such
limitation, but the error introduced by truncation of the channel number is hard to estimate.

We now come to discuss the interaction potential.    To
take account of the electron exchange effect, both Henry-Lane
and Hara derived the integro-differential equations.    In deri-
ving these equations, the target molecule is assumed to be in
its ground state.    Henry and Lane used the simple Wang's function, while Hara used the five term SCF molecular orbitals   of
Kołos and Roothaan.    Henry-Lane's static potential decreases
exponentially with distance.    Thus they added the quadrupole
interaction which has the correct asymptotic form at large distances but modified rather arbitrarily at short distances   to
avoid singularity at r = 0.    Furthermore, both Henry-Lane and
Hara added the polarization potential to the static interaction
term.    The polarization potentials used in these two calcula-
ations are considerably different except in the asymptotic region.

It will not be an authentic way to include the polarization effect only in the static interaction term and leave the
exchange term unchanged.    The same problem arises in the theory of electron-atom scattering.    To improve the situation, one
can apply the variational method where the trial function contains the incident electron - atomic electron correlation terms.
Another method to include polarization interaction as well as
electron exchange effect is to solve coupled integr-differen-
tial equations taking account of a sufficient number of electronically excited states.    The third method is the use of
the polarized orbitals,[14] although the nature of this approach
is not so clear as the other two methods.    The polarized orbital method has been applied to the electron scattering from
a space-fixed $H_2^+$ ion by Temkin and his collaborators.[15]   However, in all the other papers on electron-molecule scattering,
the electron exchange and polarization effects are introduced
independently.

In spite of these differences and unsatisfactory or ambiguous features, the calculated cross sections for electron-$H_2$
system in both of these approaches agree fairly well with the
experimental data as mentioned already.    Similaly, the rota-

tional excitation cross sections of $D_2$ calculated by Henry and Lane[16] in the energy region below 1 eV are in accord with the electron-swarm experiment.[17] It is interesting to see that the calculated rotational cross sections for $D_2$ are almost identical with those of electron-$H_2$ scattering, except in the vicinity of the threshold. This result is what we expect in the adiabatic theory. Since Henry and Lane used the same interaction potential for e-$H_2$ and e-$D_2$ systems, the adiabatic theory gives exactly the same scattering amplitude for these two cases. Only difference in the cross sections thus comes from the ratio of final and initial wave numbers which should be multiplied to the absolute square of the amplitude.

The rotational excitation of $H_2$ by positrons has been studied in the Born,[18] the distorted-wave,[19] the adiabatic(one-center),[20] and the adiabatic (two-center)[21] approximations ( see also ref. 10). Unfortunately, the positron scattering experiments comparable with these calculations have just started recently[22] and it is too early to conclude anything definitely about the agreement between theory and experiment.

Burke and Chandra[23,24,25] have studied the electron scattering from $N_2$ in the adiabatic approximation. The interaction potential used was of the form of one-center expansion. A pseudo-potential method was introduced to avoid the complexity of the usual electron exchange term. Their approach presents a tractable way to study electron scattering from other molecules, including polyatomic ones. If the target molecule has some symmetry properties, these should be fully taken account. By decomposing the total wave function into partial waves corresponding to different irreducible representations of the symmetry group, we can reduce the whole set of coupled equations into sets of smaller dimensions.[26] However, when the target molecule is a large molecule it is no more appropriate to expand the interaction potential around single center. One of the possible approach in such cases is the variational method, where one can choose a multi-center trial function. Onda[27] presented a formulation along this line.

Vibrational excitations of $H_2$ will be discussed next. In early studies, the vibrational excitation of homonuclear diatomic molecules by electron impact was assumed to be due to the short-range interactions, and the calculated cross sections were too small to explain experimental data. In 1965, I pointed out that the polarization force could be important.[28] On the other hand, Bardsley, Herzenberg and Mandl[29] emphasized the resonance nature of the vibrational excitation of $H_2$, as in the case of $N_2$. The close-coupling method was applied to the vibrational excitation of $H_2$ by Henry.[30] Agreement of his calculations with experiment,[31,32] however, is semiquantitative.

This is probably because the interaction potential, especially its dependence on the internuclear distance, has not been chosen with sufficient accuracy.

A few years later, Henry and Chang[33] studied again the same problem. Their calculations, based on the frame-transformation theory,[11] are practically equivalent to the adiabatic approximation, which is obtained by the straightforward extension of (1). The interaction used is the same as that used in the close-coupling calculations except that a cut-off radius which depends on the internuclear distance is introduced for the long-range interaction. It is a surprise that with this small modification of the potential the calculated cross sections are now in much better agreement with experiment. Faisal and Temkin[34] independently studied the same problem in the adiabatic approximation and obtained a comparable result. In the low energy regions, the main contribution to the cross section comes from three partial waves: $s\sigma$, $p\sigma$, and $p\pi$. Of these, only the $p\sigma$ wave has the phase shift which is very sensitive to the internuclear separation. This may be regarded as a $^2\Sigma_u^+$ resonance.

It is an important task for theoreticians not only to calculate the absolute magnitude of a particular cross section, but also to explain the ratio (or, more generally, interrelation) of two related cross sections. Abram and Herzenberg[35] calculated as a function of the electron scattering angle the ratio of cross sections $R(1)$, where

$$R(n) = d\sigma/d\omega(v=0 \to n; \Delta j=0) \; / \; d\sigma/d\omega(v=0 \to n; j=1 \to 3). \quad (2)$$

They used the impulse approximation, where the scattering amplitude for vibrational excitation is first calculated for fixed molecular orientation and then the rotational transitions are taken account just as in the adiabatic approximation (1). Assuming that the main contribution to the vibrational excitation comes from

incident p wave $\to$ compound $^2\Sigma_u^+$ state $\to$ outgoing p wave,

They obtained the ratio $R(1)$, which was in fair agreement with experiment.[32] Their scattering amplitude consists of a product of the vibrational and rotational factors. Thus, we can expect $R(n) = R(1)$ for any possible n. According to the recent experiment by Wong and Schulz,[36] however, the ratios $R(1)$, $R(2)$ and $R(3)$ are all different. It is noted that Abram and Herzenberg's theory assumes the resonance process and ignores the direct process. Since the excitation of v=2 and 3 states is almost entirely due to the resonance process, it is reasonable that the theoretical ratio agrees better with $R(2)$ rather

than R(1).    However, the theory cannot explain that the ratio
R(3) is appreciably smaller than R(2).    The observed ratio
R(1), on the other hand, is fairly close to the theoretical
value of Henry and Chang[33] and that of Faisal and Temkin.[34]
In order to explain the observed behavior of the cross section
ratios, Temkin and Sullivan[37] took account of the dependence
of the vibrational wave function on the rotational quantum number j.    Applying the adiabatic approximation, they could suc-
ceed in explaining the experimental findings.    Chang[38] also
investigated the same problem.    He took account of the interference among the s$\sigma$, p$\sigma$, and p$\pi$ waves.    This interference
effect is large for v=0→1 but small for v=0→2 and 3.    Therefore, his calculation distinguishes R(1) from the others, but
does not explain the observed difference between R(2) and R(3).

    Let us now proceed to discuss the electron scattering
from polar molecules.    First, calculations for HD will be
briefly summarized.    This molecule has a dipole moment, but
its magnitude is so small that the transition j=0→1 due to the
dipole moment is expected to be very small.    The $\Delta j$=2 transitions are more important.[39]    If the fixed-nuclei scattering amplitude for the electron-$H_2$ system is known, the corresponding
amplitude for HD target is derived simply by multiplying a fac-
tor $\exp(-i\vec{K}\cdot\Delta\vec{R})$, where $\vec{K} = \vec{K}_0 - \vec{k}$ is the momentum change in
unit of $\hbar$ and $\Delta\vec{R}$ is the vector representing the shift of the
center-of-mass of HD from the midpoint of the nuclei.[39]    Using
this relation and applying the adiabatic approximation, Hara[40]
calculated some rotational cross sections in the 1 - 10 eV region.    It is interesting to see that the calculated cross
section $\sigma$(0→2) for HD is nearly equal to the corresponding
cross section for $H_2$, while $\sigma$(0→1) for HD is an order of magnitude smaller in this energy range.

    Another example of weakly polar molecule is CO.    Close-
coupling calculations[41] for this target show that, although
the dipole moment D of CO is only 0.11 debye, the dipole transition (j=0→1) is the largest among the partial cross sections
up to 0.1 eV.    This part of cross section is close to that
given by the Born approximation, which is of the form

$$\sigma(j \to j \pm 1) = \frac{8\pi}{k_0^2} \left(\frac{D}{ea_0}\right)^2 \frac{j_<}{2j+1} \log \frac{k_0+k}{|k_0-k|} . \quad (3)$$

Above 0.1 eV, the elastic cross section $\sigma$(0→0) becomes larger
than $\sigma$(0→1).    In the case of HCl, the dipole moment is much
larger (1.07 debye), so that even at 0.5 eV the cross section
$\sigma$(0→1) is still an order of magnitude larger than the elastic
one.[42]    The differential cross sections (DCS) for dipole
transitions $\Delta j$= ±1, generally, are strongly peaked in the forward direction which is due to the long-range nature of the

dipole interaction.     Because of this long-range nature, the
Born approximation is applicable to a good approximation to
this particular process.

We have seen in the electron scattering from non-polar
molecules that the adiabatic theory is applied quite satisfactorily.    In the case of polar molecules, however, care must
be taken of the long-range nature of the interaction.     In
particular, it has been shown by Garrett[43] that the total
scattering cross section diverges for the fixed polar molecule
whatever the short-range part of interaction may be.     Thus,
the frame-transformation at a certain distance is necessary
for obtaining a finite total cross section.     Chandra and
Gianturco[44] have proposed such a formulation where the transformation from the body frame to the laboratory frame is made
by means of a modified R-matrix method.

In relation to the electron scattering from more strongly
polar molecules, a large number of papers have been published
on the critical dipole moment and related topics.     Stimulated
by electron-swarm experiments, many people calculated the magnitude of the lowest dipole moment $D_c$ with which at least one
bound state of an electron exists in that field.     Experiments
have shown a systematic deviation from the Born preidtion for
the relation between the momentum-transfer cross section $\sigma_m$
and the dipole moment D of the target molecule.     Many people
thought of a resonance as an explanation of the experimental
findings.    We thought that a large cross section would be obtained at low collision energy when a new bound state is produced at zero energy as D is increased.     Because of the limited pages available for this article, it is impossible to tell
the whole story in details, which one can see in the review by
Garrett[45] and the references therein.     Garrett himself has
made many important contributions to clarify the problem.

Here, I shall summarize some interesting points.     First
of all, for a dipole field fixed to space, both for a point
dipole and for a finite dipole (charges ±q separated by a
distance R=D/q), no bound states exist for the dipole moment D
less than the critical value $D_c$ = 0.639 $ea_0$ (= 1.625 debyes).
For D > $D_c$, there are an infinite number of bound states.     In
the case of point dipole, especially, any negative energy E
can be an eigenvalue so that the energy spectrum is continuous, while no definite solution can be found in the positive
energy region for D > $D_c$ since there is no means to determine
from physical considerations the coefficients of linear combination of two independent radial functions.[46]     The ground
state, if exists, has the energy -∞.     For the fixed finite
dipole, we have no such difficulty for the positive energy
problem.     An infinite number of bound states are concentrated

in the infinitesmmal energy region just below E = 0.    The
ground state has a finite binding energy.

For a free (rotatable) finite dipole with a finite moment
of inertia $D_c$ is larger than for a fixed dipole and it depends
on the total angular momentum quantum number J of the whole
system (electron plus dipole).    The larger the quantum number is, the larger the critical dipole moment.    The number of
bound states is now finite.    The magnitude of $D_c$ depends also
on the charge separation R and on the moment of inertia I.    As
I is increased to infinity, the critical moment $D_c$ approaches
very slowly to the value for a fixed dipole.    For real elect-
ron-polar molecule scattering, the interaction is such that
the pure dipole field is observed asymptotically at large distances, but other multipole fields and polarization potential
become important as r decreases, and finally strong short-range
interactions appear at small distances.    The critical value $D_c$
depends also on these interactions.    In this case, additional
bound states may exist due to the attractive fields of shorter
ranges other than the dipole interaction.    In any case, the
number of bound states is finite.    When a new bound state is
produced at E=0 as we increase D, a peak appears in the low-
energy scattering cross section as a function of D.    The peak
or peaks which appear in the low-value region of D are due to
the elastic part of the scattering.    The position of the peak
depends on the incident electron energy.    However, roughly
speaking $\sigma_m$ is inversely proportional to the electron energy,
so that the contribution of the low energy electrons will be
emphasized in the electron-swarm experiments unless the mean
electron energy is high.    Since $D_c$ depends considerably on
I, R, and especially on the interactions other than the dipole
potential, Garrett[45] has concluded that for real polar molecules any correlation between the dipole moment and the cross
section is completely destroyed.    Nevertheless, various calculations tend to indicate that for real molecules $\sigma_m$ at low D
values is larger than the Born prediction for the point-dipole
interaction, while at high D values the true $\sigma_m$ is smaller
than the Born value.    The explanation is probably as follows.
The $\sigma_m$ in the Born approximation, originally derived by Alt-
shuler[47] on the basis of the adiabatic approximation, takes
account of the rotational transitions $j \rightarrow j \pm 1$ only.    For
smaller D, the elastic scattering $(j \rightarrow j)$, especially due to the
short-range interactions, is relatively more important and $\sigma_m$
becomes larger than the Born value.    For larger D, the Born
approximation, which is nothing but the first-order perturbation theory, is no more applicable even for    comparatively
distant collisions.    Under such condition, the perturbation
method often overestimates the transition probabilities.    This
explains why the Born approximation is much above the observed
cross section.    The electron-swarm experimental data are

consistent with this explanation of the deviation from the
Born prediction.

Recent beam experiments on electron scattering from strongly polar molecules (CsF, CsCl, KI) also indicate that $\sigma_m$ is considerably smaller than the Born value.   However, the energy
region studied (0.5 - 16 eV) is different from the energy range
relevant to the electron swarm experiments so that we cannot
compare them directly.   Stern and his collaborators[48,49,50]
applied the molecular-beam recoil technique and derived the DCS
from the observed data.   One of the most remarkable findings
is that the DCS is considerably smaller than the Born value
even in the relatively small scattering angles $(5^0-30^0)$ where
the long-range dipole interaction is expected to be of primary
importance.

Rudge[51] assumes that the cross sections are not strongly
dependent upon the details of the electron wave function in the
inner region of the target and simply replaces in the Born
cross section the incident and outgoing plane waves by zero inside a certain cut-off radius $r_c$.   By taking $r_c$ to be about
twice the internuclear distance of the molecule, he has obtained the DCS comparable with the experimental curves.   However,
his model is far from being realistic, so that it is hard to
say something definite out of his calculations.

Allison[52] has applied the CC aproach for the electron
scattering from CsF $(D = 3.1\ ea_0)$ which is initially in the
j=41 state.   At 1 and 2 eV, he has obtained the total cross
section slightly less than the Born value and $\sigma_m$ considerably
smaller than the Born prediction.   However, the calculated
cross sections are still larger than the experimenal values.
The accuracy of the experimental total cross section is somewhat uncertain since the DCS is less accurate at small angles
$(<5^0)$ where a considerable contribution to the integrated cross
section comes from.   The reason why the calculated $\sigma_m$ is
larger, by a factor of 2 or 3, is not certain.   In fact, this
result is against the simple expectation.   The interaction
adopted is the pure dipole interaction at larger distances, but
cut-off to zero in the inner region, so that the large-angle
scattering is expected to be suppressed.   Therefore, the theoretical cross section is expected to be less than the true
cross section, in contradiction to the reported results.

We have investigated the same problem in three different
methods: the Glauber approximation,[53,54] the adiabatic approximation (by K.Onda, unpublished) and the CC method (by Y.
Itikawa, unpublished).   In the first approach the simple point
dipole interaction has been used since this method has no
difficulty with the singularity at the origin.   In the second

approach, however, the singularity must be avoided to determine the wave function, so that the small hard core was introduced.   In the third method, the potential used was the dipole interaction with a cut-off factor with which the potential tends to zero at r = 0.   For 1 eV electron scattering from CsF, for instance, the Glauber results are in fairly good agreement with the CC results at small angles (< 50°), but the adiabatic approximation gives slightly larger values.   At larger scattering angles, the Glauber calculation is unreliable partly because this approximation is originally intended for high energy (relative to the interaction), small angle scattering, and partly because we have used the point-dipole interaction which is too strong at very small distances.   For small angles (<50°), however, this approximation gives the DCS  for e + CsCl at 4.77 eV which is in good agreement with the experimental values.   The DCS from the adiabatic approximation has been found to depend on the hard-core radius especially at large angles.   Evidently, a more realistic interaction should be used in future to make definite comparison between theory and experiment.

REFERENCES

1. K.Takayanagi, Prog. Theor. Phys. suppl. No.41, 216 (1967).
2. K.Takayanagi and Y.Itikawa, Adv. Atom. Mol. Phys. $\underline{6}$, 105. (1970).
3. D.E.Golden, N.F.Lane, A.Temkin and E.Gerjuoy, Rev. Mod. Phys. $\underline{43}$, 642 (1971).
4. N.F.Lane and S.Geltman, Phys. Rev. $\underline{160}$, 53 (1967); $\underline{184}$, 46 (1969).
5. N.F.Lane and R.J.W.Henry, Phys. Rev. $\underline{173}$, 183 (1968).
6. R.J.W.Henry and N.F.Lane, Phys. Rev. $\underline{183}$, 22 (1969).
7. S.Hara, J. Phys. Soc. Japan $\underline{27}$, 1592 (1969).
8. F.Linder and H.Schmidt, Z. f. Naturforschg.26a, 1603(1971).
9. R.W.Crompton, D.K.Gibson and A.I.McIntosh, Austral. J. Phys. $\underline{22}$, 715 (1969).
10. J.W.Darewych, P.Baille and S.Hara, J. Phys. B (Atom. Mol. Phys.) $\underline{7}$, 2047 (1974).
11. E.S.Chang and U.Fano, Phys. Rev. $\underline{A6}$, 173 (1972).
12. E.S.Chang and A.Temkin, J. Phys. Soc. Japan $\underline{29}$, 172(1970).
13. K.Takayanagi, Atomic Physics, vol.4 (Proceedings of the 4th International Conference on Atomic Physics, Plenum Press) in press.
14. e.g., R.J.Drachman and A.Temkin, Case Studies in Atomic Collision Physics, ed. by E.W.McDaniel and M.R.C.McDowell (North-Holland, Amsterdam, 1972), p.399.
15. A.Temkin and K.V.Vasavada, Phys. Rev. $\underline{160}$, 109 (1967); A.Temkin, K.V.Vasavada, E.S.Chang and A.Silver, Phys. Rev. $\underline{186}$, 57 (1969).

16. R.J.W.Henry and N.F.Lane, Phys. Rev. A4, 410 (1971).
17. D.K.Gibson, Austral. J. Phys. 23, 683 (1970).
18. K.Takayanagi and M.Inokuti, J. Phys. Soc. Japan 23, 1412 (1967).
19. S.Hara, J. Phys. B(Atom. Mol. Phys.) 5, 589 (1972).
20. P.Baille, J.W.Darewych and J.G.Lodge, Can. J. Phys. 52, 667 (1974).
21. S.Hara, J. Phys. B(Atom. Mol. Phys.) 7, 1748 (1974).
22. P.G.Coleman, T.C.Griffith and G.R.Heyland, Appl. Phys. 4, 89 (1974).
23. P.G.Burke and N.Chandra, J. Phys. B(Atom. Mol. Phys.) 5, 1696 (1972).
24. N.Chandra and P.G.Burke, J. Phys. B(Atom. Mol. Phys.) 6, 2355 (1973).
25. N.Chandra, J. Phys. B(Atom. Mol. Phys.) 8, 1338 (1975).
26. P.G.Burke, N.Chandra and F.A.Gianturco, J. Phys. B(Atom. Mol. Phys.) 5, 2212 (1972).
27. K. Onda, J. Phys. Soc. Japan 36, 826 (1974).
28. K.Takayanagi, J. Phys. Soc. Japan 20, 562, 2297 (1965).
29. J.N.Bardsley, A.Herzenberg and F.Mandl, Proc. Phys. Soc. 89, 305, 321 (1966).
30. R.J.W.Henry, Phys. Rev. A2, 1349 (1970).
31. G.J.Schulz, Phys. Rev. 135, A988 (1964).
32. H.Ehrhardt, L.Langhaus, F.Linder and H.S.Taylor, Phys. Rev. 173, 222 (1968). H.Ehrhardt and F.Linder, Phys. Rev. Letters 21, 419 (1968). see also ref. 8.
33. R.J.W.Henry and E.S.Chang, Phys. Rev. A5, 276 (1972).
34. F.H.M.Faisal and A.Temkin, Phys. Rev. Letters 28, 203 (1972).
35. R.A.Abram and A.Herzenberg, Chem. Phys. Letters 3, 187 (1969).
36. S.F.Wong and G.J.Schulz, Phys. Rev. Letters 32, 1089 (1974).
37. A.Temkin and E.C.Sullivan, Phys. Rev. Letters 33, 1057 (1974).
38. E.S.Chang, Phys. Rev. Letters 33, 1644 (1974).
39. K.Takayanagi, J. Phys. Soc. Japan 28, 1527 (1970).
40. S.Hara, J. Phys. Soc. Japan 30, 819 (1971).
41. O.H.Crawford and A.Dalgarno, J. Phys. B(Atom. Mol. Phys.) 4, 494 (1971).
42. Y.Itikawa and K.Takayanagi, VI ICPEAC, Abstracts of Papers (The MIT Press, Camb ridge, 1969) p.144; J. Phys. Soc. Japan 26, 1254 (1969).
43. W.R.Garrett, Phys. Rev. A4, 2229 (1971).
44. N.Chandra and F.A.Gianturco, Chem. Phys. Letters 24, 326 (1974).
45. W.R.Garrett, Molecular Phys. 24, 465 (1972).
46. For the solutions in $1/r^2$ potentials, see, e.g., L.D. Landau and E.M.Lifshitz, Quantum Mechanics (transl. from Russian, Pergamon Press, 1965); Wm.M.Frank, D.J.Land and R.M.Spector, Rev. Mod. Phys. 43, 36 (1971).

47. S.Altshuler, Phys. Rev. 107, 114 (1957).
48. R.C.Slater, M.G.Fickes, W.G.Becker and R.C.Stern, J. Chem. Phys. 60, 4697 (1974).
49. W.G.Becker, M.G.Fickes, R.C.Slater and R.C.Stern, J. Chem. Phys. 61, 2283 (1974).
50. R.C.Slater, M.G.Fickes, W.G.Becker and R.C.Stern, J. Chem. Phys. 61, 2290 (1974).
51. M.R.H.Rudge, J. Phys. B(Atom. Mol. Phys.) 7, 1323 (1974).
52. A.C.Allison, J. Phys. B(Atom. Mol. Phys.) 8, 325 (1975).
53. K.Takayanagi, Prog. The r. Phys. 52, 337 (1974).
54. O.Ashihara, I.Shimamura and K.Takayanagi, J. Phys. Soc. Japan 38, 1732 (1975).

ACKNOWLEDGEMENTS

The author is indebted to Drs. Y.Itikawa, S.Hara, I. Shimamura and K.Onda for valuable discussions.

# MULTIPLE SCATTERING APPROACH TO THE VIBRATIONAL EXCITATION OF MOLECULES BY SLOW ELECTRONS

G. Drukarev

Leningrad State University

Leningrad, USSR

## 1. INTRODUCTION

The subject of this report is the vibrational excitation of homonuclear two-atomic molecules by slow electrons (accompanied possibly by a rotational transitions). The most important feature of those processes are the broad resonances in cross sections. They are attributed customary to the temporarily capture of an electron and formation of a negative molecular ion in quasi-stationary state. After the decay of this state and release of the captured electron the molecule can go over to any vibrational and rotational state. The quasi-stationary states are described by the analytical continuation of the potential energy curves in the complex energy region /1/.

We would like to present here another approach based on the picture of multiple scattering of an electron inside the molecule.

In order to make the problem tractable we introduce some simplifications.

1. We will treat the interaction of an electron with the atoms forming the molecule as if they are simple potential field centers. Then we reduce our problem to the 3-body one.

This approximation is valid more or less if the energy of an electron is below the electronic excitation threshold of molecule.

2. We will use the adiabatic approximation for the transition amplitude:

$$F_{fi} = \int \varphi_f^*(\bar{R}) \, A(\bar{R}) \, \varphi_i(\bar{R}) \, d\bar{R} \tag{1}$$

Here $\varphi_i$, $\varphi_f$ — initial and final wave functions of the nuclear motion (in the ground electronic state of molecule). $A(\bar{R})$ — the scattering amplitude of an electron on two atoms in fixed positions at the distance $\bar{R}$.

The expression (1) can be deduced from Faddeev equations /2/ under the condition that the scattered electrons energy exceeds the binding energy of the molecule (in its ground electronic state). We will assume that it is the case.

3. Finally, to calculate the amplitude A we will use the zero range potential model. It is indeed a drastic approximation but it has an important advantage: it takes into account the multiple scattering effect explicitly.

Within the approximations outlined above we neglect completely all long range interactions including the polarization potential due to electric field of the scattered electron.

This will introduce no significant error as far as the vibrational excitation is concerned. The electronic cloud of the molecule is pushed by the field of scattered electron and returns back during the time interval much shorter than the period of vibration and one cannot expect any energy transfer to the vibrational degrees of freedom. However the polarization of molecule can affect strongly the elastic scattering.

In the case of pure rotational excitation it is well known since the work of E. Gerjuoy and S. Stein /3/ that the electric quadrupole moment of molecule plays a major role.

In 2. the expression for $A$ according to the zero range potential model is introduced and discussed.

In 3. we present the results of cross sections calculations for some transitions in $H_2, Li_2, Na_2, K_2$ made together with Dr. I. Yurova, and compare them with other calculations and experiment.

In 4. the relation between the multiple scattering approach and quasistationary negative ion picture is established.

## 2.  SCATTERING ON TWO FIXED CENTERS IN ZERO RANGE POTENTIAL MODEL

In this model the wave function of the scattered particle is subject to the boundary conditions at the centers

$$\frac{d}{d\rho_i}\left(\rho_i\,\psi\right)_{\rho_i=0} = -\frac{1}{a_i}\left(\rho_i\,\psi\right)_{\rho_i=0} \tag{2}$$

$$\rho_i = |\vec{r}-\vec{r_i}|$$

Here $a_i$ is the scattering length for the center $i$

Outside the centers the wave function is a proper solution of free motion Schredinger equation (without a potential). The conditions (2) together with the standart asymptotic conditions determine the wave function completely.

In the case of two centers located at the point $-R/2;\ R/2$ we put

$$\psi = e^{i\vec{\kappa}\vec{r}} + A_1\,\frac{e^{i\kappa\rho_1}}{\rho_1} + A_2\,\frac{e^{i\kappa\rho_2}}{\rho_2} \qquad \rho_{1,2}=|\vec{r}\mp\tfrac{1}{2}\vec{R}| \tag{3}$$

According to (2) we require that

$$\frac{d}{d\rho_1}\left(\rho_1\,\psi\right)_{\rho_1=0} = -\frac{1}{a_1}\left(\rho_1\,\psi\right)_{\rho_1=0} \quad \frac{d}{d\rho_2}\left(\rho_2\,\psi\right)_{\rho_2=0} = -\frac{1}{a_2}\left(\rho_2\,\psi\right)_{\rho_2=0} \tag{4}$$

Inserting (3) into (4) we obtain the equations for the coefficients
Let us consider now the asymptotic of (3) at $r\to\infty$
It has the standard form

$$\psi_{r\to\infty} \sim e^{i\vec{\kappa}\vec{r}} + A\,\frac{e^{i\kappa r}}{r} \tag{5}$$

where

$$A = A_1\,e^{i\frac{\kappa}{2}\vec{R}\vec{n}} + A_2\,e^{-i\frac{\kappa}{2}\vec{R}\vec{n}} \qquad \vec{n} = \frac{\vec{r}}{r} \tag{6}$$

Using the expressions for $A_1,\ A_2$ which follows from (4), one get for the amplitude

$$A = \frac{2\cos(\frac{1}{2}\bar{R}\bar{n})\cos(\frac{1}{2}\bar{R}\bar{n}_o)}{\frac{1}{a}+i\kappa+\frac{e^{i\kappa R}}{R}} + \frac{2\sin(\frac{1}{2}\bar{R}\bar{n})\sin(\frac{1}{2}\bar{R}\bar{n}_o)}{\frac{1}{a}+i\kappa-\frac{e^{i\kappa R}}{R}} \qquad (7)$$

$$\bar{n}_o = \frac{\bar{\kappa}}{\kappa}; \qquad \bar{n} = \frac{\bar{\tau}}{\tau}; \qquad a_1 = a_2 = a$$

(This expression was used by several authors in various scattering problems). The multiple scattering is represented in (7) by the term $\frac{1}{R}\exp(i\kappa R)$

Now we must take into account that the interaction of an electron with the atom depends on the total spin of the system "electron + atom". This can be done by including spin variables in the wave function of the scattered electron, and replacing the numbers $a_l$ by operators acting on the spin variables of the electron and of the target.

In the case when both atoms in the molecule have spin 1/2, instead of (4) we will write at the first center

$$\frac{1}{a_1} = \frac{1}{4a_t}(3+\bar{\sigma}\times\bar{\sigma}_1) + \frac{1}{4a_s}(1-\bar{\sigma}\times\bar{\sigma}_1) \qquad (8)$$

Here $\sigma$ and $\sigma_1$ are acting on the target and electronic spin variables, $a_t, a_s$ — the scattering length in the triplet and singlet states. Similar expression we will write at the second center. But in the molecule with the total spin 0 the projections of the spin of each individual atom are not fixed, and therefore we must average over all possible projections. The result is

$$\left\langle \frac{1}{a_1} \right\rangle = \frac{3}{4a_t} + \frac{1}{4a_s} \qquad (9)$$

It follows that in effect for the molecule with zero total spin one can still use the expression (7) with the average inverse scattering length given by (9).

Finally we should notice that neglecting the polarization of molecule we must use the numerical values of $a_t, a_s$ calculated in static approximation (without allowance for polarization).

### 3. EXCITATION AND TOTAL CROSS SECTIONS FOR $H_2$, $Li_2$, $Na_2$, $K_2$

To calculate the cross sections we insert (7) in (1) and use for $\psi_i$, $\psi_f$ the Morse potential wave functions. We reproduce here the results for the following transitions:

(a) in $H_2$

Pure vibrational transition $V = 0 \to 1$, $j = 0 \to 0$ (Fig.1);

Vibrational transition $V = 0 \to 1$ without change of $j$ ($\Delta j = 0$). The cross section is averaged over the distribution of rotational states at room temperature (Fig.2).

Differential cross section for $V = 0 \to 1$, $\Delta j = 0$ transition (Fig.3).

Ratio of the differential cross sections for transitions $V = 0 \to 1$; $\Delta j = 0$ and $V = 0 \to 1$, $J = 1 \to 3$ (Fig.4).

The results of calculations are compared with the close-coupling calculations /4/ and experiment /5/, /6/.

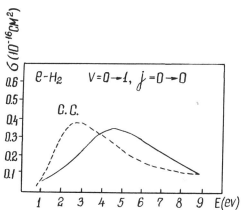

Fig.1   Calculated cross sections for $V = 0 \to 1$, $j = 0$ in $H_2$
      ---Close coupling /4/
      ——Present calculations

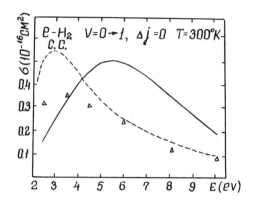

Fig.2   Cross sections for V=0→1, Δj=0 in H₂
---Close coupling /4/
— Present calculations
▲ ▲ Experiment /5/

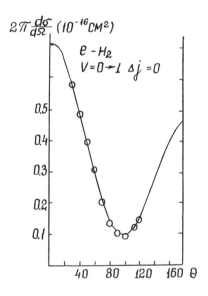

Fig.3   Differential cross section for V=0→1,
Δj = 0
—Present calculations
ₒₒₒ Experiment (arbitrary units) /6/

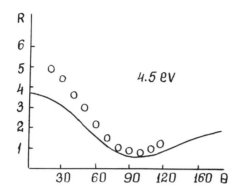

Fig.4    Ratio of differential. Cross sections
         for    Δ j=0 and j=1→3 ,    $V = 0 → 1$
         —Theory
         ∘∘∘Experiment /6/

(b) In    $K_2$  $Na_2$  $Li_2$    pure vibrational transitions
    $V = 0 → 1$,    $j = 0 → 0$.    (Fig.5).

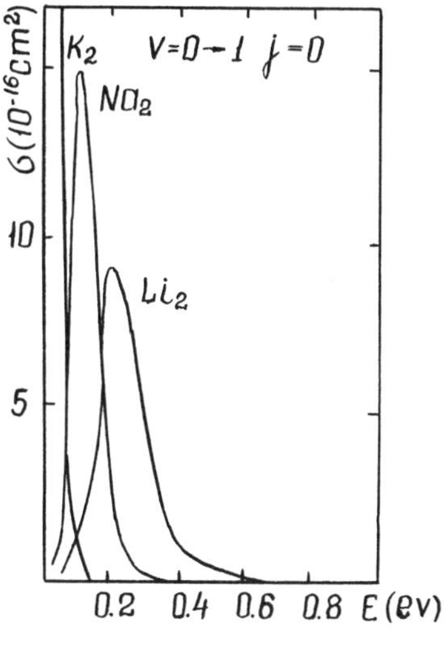

Fig.5    Calculated cross sections for $V=0→1$,
         $j=0$  in $K_2$, $Na_2$, $Li_2$

Using the optical theorem

$$\mathcal{I}_m \, F_{oo} \, (\bar{n}_o, \bar{n}_o) = \frac{K}{4\pi} \, \sigma_{tot} \tag{10}$$

we can calculate the total cross sections for all transitions including elastic scattering.

The result is

$$\sigma_{tot} = \int |\psi_o(\bar{R})|^2 Q(\bar{R}) \, d\bar{R} \tag{11}$$

where

$$Q = \frac{4\pi}{K^2} \left\{ \left[ 1 + \frac{(\frac{R}{a} + \cos KR)^2}{(KR + \sin KR)^2} \right]^{-1} + \left[ 1 + \frac{(\frac{R}{a} - \cos KR)^2}{(KR - \sin KR)^2} \right]^{-1} \right\} \tag{12}$$

The expression (12) is the cross section of the scattering on two fixed centers averaged over all directions of the initial momentum of the scattered particle.

Shown on Figs.6,7 are the calculated total cross sections in $H_2$, $Li_2$, $Na_2$, $K_2$ together with the experimental data.

We certainly underestimate the total cross section because of negligence of longrange interaction which affects the elastic part.

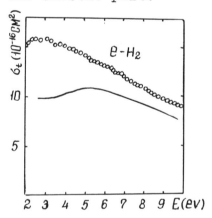

Fig.6   Total cross section for $H_2$.
          ooo Experiment /7/
          —— Present calculations

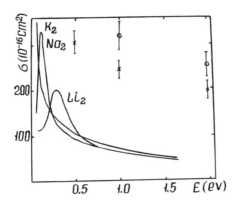

Fig.7   Total cross section for $K_2$, $Na_2$, $Li_2$

o $K_2$ /8/    x $Na_2$ /8/

—Present calculations

## 4.   CONCLUSION

The results of the calculations indicates that the multiple scattering determines the order of magnitude of the vibrational excitation cross sections in the energy region under consideration even if the zero range potential model is used.

The theory can be improved by using more realistic potentials which allows at least approximately to take into account multiple scattering effect.

Finally, we will establish the connection between the multiple scattering approach and quasistationary molecular ion picture.

Consider the denominator of amplitude in (7)

$$\mathcal{D} = \frac{1}{a} + iK - \frac{e^{iKR}}{R} \qquad (13)$$

It can become zero at certain imaginary or complex value of electronic momentum.

If

$$\mathcal{D}(i\gamma) = 0 \qquad (14)$$

then

$$\frac{1}{a} - \gamma - \frac{e^{-\gamma R}}{R} = 0 \qquad (15)$$

This relation is exactly the same which determines the electronic term in field of two centers. The complex zero of (13) corresponds to the analytical continuation of the electronic term in the complex energy region (quasistationary state).

REFERENCES

1.  J.Bardsley, F.Mandl. Reports on Progress in Physics, 31, part II, 472 (1968). G.Schulz. Rev. Mod. Phys. 45, 423 (1973).

2.  G.Drukarev. JETP, 40   (1974)

3.  E.Gerjuoy, S.Stein. Phys.Rev., 97, 1671; 98, 1848 (1955).

4.  R.Henry. Phys.Rev., A2, 1349 (1970).

5.  F.Linder, Abstracts VI ICPEAC, 141, (1969).

6.  F.Linder, H.Schmidt. Z.Naturfor., 26a, 1603 (1971).

7.  D.Golden, H.Bandel, J.Salerno. Phys.Rev., 146, 40 (1966).

8.  T.Miller, A.Kasdan. J.Chem.Phys., 59, 391/3 (1973).

# CLASSIFICATION OF FESHBACH RESONANCES IN ELECTRON-MOLECULE SCATTERING[*]

David Spence

Argonne National Laboratory

Argonne, Illinois 60439

The genealogy of Feshbach resonances, first introduced by Sanche and Schulz,[1] describes a positive ion core (which may be in either its ground or an excited state) as a grandparent state. Addition of an electron to the grandparent state may give rise to many Rydberg states of the neutral molecule which are referred to as parent states. Addition of another electron to the parent states gives rise to negative ion states, i.e., resonances. Such negative ion states are called Feshbach resonances if they lie energetically below the parent excited states.

Feshbach resonances are a general phenomenon which nevertheless form a subclass of the broader problem of understanding the dynamics of two excited electrons moving in the field of a positive ion core, i.e., electron correlation effects in Coulomb-like fields. Related phenomena in this broader category in which electron correlation effects play a dominant role (and are thus difficult to handle theoretically) include

[*] Work performed under the auspices of U.S. ERDA.

threshold electron excitation and ionization,[2,3] and doubly-excited states.[4,5]

For Feshbach resonances in molecules, the problem is further complicated since the ion core is nonspherical in shape and has internal structure (rotation, vibration, and perhaps electronic excitation).

Despite all these complications, however, spectra of Feshbach resonances are often strikingly simple, permitting straightforward analysis and configuration assignment. Further, despite the large number of all conceivable resonance configurations, the number of strong resonance features observed in electron-molecule scattering is fairly small, which must imply stability rules (some of which are known) for pairs of electrons in excited orbitals in Coulomb-like fields.     For instance, to get a quasi-bound state, one must put two electrons in orbitals of similar size ("dynamic screening" in Rau's words).[3]

Though many bands of resonances and isolated resonances have been located by electron scattering from atoms and molecules,[6] the assignment of electron configurations has, in general, required detailed angular distribution measurements of resonant electron scattering.[6] Though such measurements will produce great details of the resonant scattering processes, they are time-consuming and difficult. There have also been limited attempts to determine resonance electron configurations by comparison between resonance spectra and the spectra of neutral excited states in iso-electronic atoms and molecules.[7]

Studies of Feshbach resonances in electronically similar molecules, and comparisons between recent experimental and theoretical results for resonant electron-atom scattering, make it possible to derive certain systematics which often enable a simple determination of resonance configurations by a comparison between resonance spectra and known Rydberg-state spectra. These systematics will be discussed in three sections:

(i)   Determination of resonance grandparent states and the systematics of the binding energies of _pairs_ of Rydberg electrons in diatomic molecules;

(ii)  Systematics of the binding energies of electrons with various values of n and $\ell$ to parent Rydberg states;

(iii) Application of guidelines provided by the above systematics to new systems.

Binding of Pairs of Rydberg Electrons to Grandparent Position-Ion States

Weiss and Krauss[8] predicted that Rydberg states of diatomic molecules, unlike valence states, would support core-excited Feshbach resonances, the temporary negative ion thus formed consisting of two Rydberg electrons moving in the field of a positive ion core. Rydberg electrons are essentially non-bonding, and thus the nuclear motion of the positive ion core remains virtually unperturbed by the addition of two extra electrons. Indeed, the vibrational spacings and Franck-Condon factors (and hence the observed vibrational intensities) are sim-ilar for vibrational progressions associated with formation of both the _negative_ ion and the _positive_ ion grandparent. These properties enabled Sanche and Schulz[9] to verify experimentally the prediction of Weiss and Krauss. The spectrum of Feshbach resonances observed in NO by Sanche and Schulz[9] is shown in Fig. 1, and the comparison between the vibrational intensities and spacings of the $NO^-$ system, and the corresponding vibra-tional intensities and spacings in the $NO^+(X'\Sigma^+)$ system, is shown in Table 1. This table demonstrates convincingly that

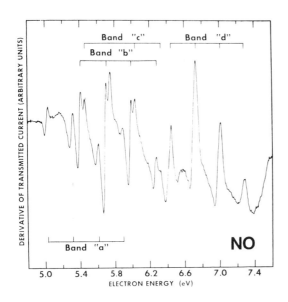

FIG. 1.--Derivative of transmitted current vs. electron energy in NO. The four vibrational bands of resonances consist of pairs of Rydberg electrons bound to the $X'\Sigma^+$ ground state of $NO^+$. [From Sanche and Schulz, Ref. 1]

the resonances of Fig. 1 are formed by addition of two non-bonding (Rydberg) electrons to the ground state positive ion core. Similar negative ion progressions associated with excited states of positive ion cores have been found by Sanche and Schulz,[1] an example of which is shown in the energy-level diagram of Fig. 2. Offsetting the positive ion spectra by 4.1 eV with respect to the negative ion spectra allows easy identification of each band or isolated resonance with a particular state of the positive ion core.

In such negative ion systems, the lowest-lying resonance associated with a particular ion core state consists of a pair of $ns\sigma$ electrons (where $n-1$ is the principal quantum number of the valence shell) bound to the ground state positive

Table 1.   Comparison of vibrational spacings and vibrational intensities for four bands of NO⁻ with appropriate values for NO⁺(X'Σ⁺) [Sanche and Schulz (1972)].

Vibrational spacings (meV)

| $\Delta v$ | NO⁻ band[a] | | | | NO⁺(X'Σ⁺) | |
|---|---|---|---|---|---|---|
| | "a" | "b" | "c" | "d" | Experimental | Theoretical |
| 0–1 | 286 | 290 | 292 | 283 | 290 | |
| 1–2 | 286 | 290 | 288 | 284 | 287 | |
| 2–3 | 282 | 284 | 286 | 275 | 283 | |
| 3–4 | | | | 275 | 278 | |
| 4–5 | | | | 275 | 273 | |

Vibrational intensities

| $v$ | NO⁻ band[a] | | | | NO⁺(X'Σ⁺) | |
|---|---|---|---|---|---|---|
| | "a" | "b" | "c" | "d" | Experimental | Theoretical F.C. Factors |
| 0 | 0.84 | 0.57 | --- | 0.66 | 0.7 | 0.48 |
| 1 | 1 | 1 | --- | 1 | 1 | 1 |
| 2 | 0.62 | 0.64 | --- | 0.7 | 0.7 | 0.92 |
| 3 | 0.16 | 0.24 | --- | 0.19 | 0.5 | 0.49 |

[a] "a", "b", "c", "d" represent designations of bands which start at 5.04, 5.41 5.46, and 6.44±0.05 eV, respectively.

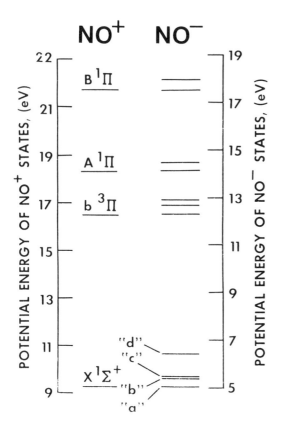

FIG. 2.--Energy level diagram of NO⁺ grandparent states and NO⁻ resonant states. The energy scales have been shifted by the binding energy of the lowest member of band "a" with respect to the X'Σ⁺ NO⁺ state. Each state of NO⁺ on the left gives rise to the NO⁻ states on the right (From Sanche and Schulz, Ref. 1)

ion core. In all the diatomic molecules studied by Sanche and Schulz,[1] the binding energy of a pair of $ns\sigma$ electrons to the positive ion grandparent state was always 4.1 ± 0.1 eV, irrespective of whether the ion core was in its ground state or an excited state. At first sight it might appear that the application of this value, 4.1 eV, would enable one to determine straightforwardly which resonances were associated with a particular ion core, even when comparison of vibrational intensity and spacings is impossible. However, a recent study[10] of the systematics of the binding energies of the lowest pair of $ns\sigma$ electrons to the positive ion core of molecular halogens (i.e., systematics along a column of the Periodic Table) has shown that the binding energy of a pair of $ns\sigma$ electrons to the ground

state positive ion core varies from 4.5 eV in $F_2$ to 3.5 in $I_2$.

The constancy of the binding energy observed by Sanche and

Schulz[6] results from the fact that all the diatomic molecules

they studied were composed of atoms from the first row of the

Periodic Table.

    The binding energy    (BE) of each Rydberg electron mov-

ing in the field of a positive ion core, can be schematically ap-

proximated by

$$BE = \frac{R}{(n - \mu)^2} + EA , \qquad (1)$$

where n is the principal quantum number of the Rydberg electron,

$\mu$ is its quantum defect, EA is the electron affinity of the neutral

Rydberg state, and R is the Rydberg energy.   In the molecular

halogens, the lowest excited electrons are in $ns\sigma$ orbitals,

where n takes the value 3, 4, 5, and 6 for $F_2$, $Cl_2$, $Br_2$, and

$I_2$, respectively.   The quantum defect ($\mu$) of a Rydberg state

is equal to $\pi^{-1}\delta$, where $\delta$ is the asymptotic phase shift of the

excited electron relative to an equivalent electron moving in a

pure Coulomb field.

    With calculated quantum defects based on an approxi-

mate atomic potential,[11] together with the reasonable value of

0.5 eV for the electron affinity of an excited state for an s

electron, Eq. (1) yields the binding energies of a pair of $ns\sigma$

Rydberg electrons in the molecular halogens shown in Fig. 3,

where the comparison with experimental binding energies is

made.

    The physical basis for the variation shown in Fig. 2 is

the increasing distance of the outermost loop of the ns radial

wavefunctions, as the shells progressively fill from $F_2$ to $I_2$.

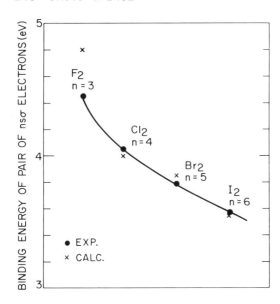

FIG. 3.--Calculated and experimental binding energies of nsσ electron pairs to ground state positive ion cores in the molecular halogens. (From Spence, Ref. 10)

The increasing distance is due to the orthogonality with the inner s electron wavefunctions.

## Systematics of Binding of Rydberg Electrons to Parent Rydberg States

Though this report is concerned primarily with electron-molecule scattering, it is instructive to examine certain systematic guidelines that can be obtained from simpler systems of electron-atom scattering. The most detailed comparison between experimental[12] and theoretical resonance[13] energies can be made for the e + O system, where observed resonances and their respective parent excited states and grandparent positive ion states are depicted in Fig. 4.

From this figure one can draw certain conclusions regarding the binding energy of an electron of given n and ℓ to a parent Rydberg state, namely, (i) pairs of Rydberg electrons are only likely to form bound resonant states when the principal quantum number n is the same for both excited electrons; (ii) the

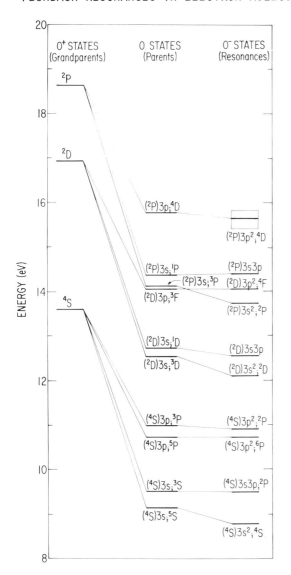

FIG. 4.--Energy level diagram of $O^+$ grandparent states, O parent states, and $O^-$ resonance states. (From Spence, Ref. 12)

electron affinity of a Rydberg state is larger for an additional electron whose angular momentum $\ell$ is the _same_ as that of the initial Rydberg electron; (iii) the electron affinities of Rydberg states, for resonance configurations $s^2$, $p^2$, and sp, is approximately of the order 0.5 eV, 0.01—0.10 eV, and < 0.01 eV, respectively.

These observations are in accordance with the requirements for minimized mutual screening of the positive ion core[3]
by the pair of Rydberg electrons which will occur when both
electrons are equidistant from the ion core. For illustration,
we shall now discuss resonance spectra recently obtained in
the hydrogen halides,[14] which are isoelectronic with rare gases.

## Application of Systematics to New Systems.    HBr

The apectrum of HBr is shown in Fig. 5. Most features
occur as doublets because of spin-orbit splitting of the positive ion core. Features associated with the $^2\Pi_{\frac{3}{2}}$ ion core are
marked 1, 2, 3, and 4, and those associated with the $^2\Pi_{\frac{1}{2}}$ ion
core are labeled 1A, 2A, 3A, and 4A. Bands of resonances "a"
and "b" are associated with excited states of the positive ion
core and will not be discussed here.

Interpretation of Fig. 5 is facilitated by comparison with
the Kr spectrum of Sanche and Schulz,[15] which is shown in Fig.
6. Noting that many features are separated by the $^2P_{\frac{3}{2},\frac{1}{2}}$ ion
core splitting allows the separation of resonances associated

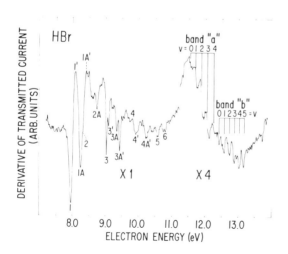

FIG. 5.--Derivative of
transmitted current vs.
electron energy in HBr.
(From Spence and
Noguchi, Ref. 14)

with each particular $^2P_{\frac{3}{2},\frac{1}{2}}$ ion core, as indicated by the vertical lines of Fig. 6. By <u>unambiguously</u> associating particular resonances with the appropriate positive ion core in this way, we can construct ladders of resonant states and neutral excited parent states of the type shown in Fig. 7.

For each $^2P_{\frac{3}{2}}$, $^2P_{\frac{1}{2}}$ ion core, there are four main sets of resonances, exactly the situation we observe in the hydrogen halides. The lowest resonance in each series in Fig. 7 occurs by addition of a pair of 5s electrons to the $^2P_{\frac{3}{2},\frac{1}{2}}$ ion core. The second lowest resonances are almost coincident energetically with the $^3P_1$ and $^1P_1$ neutral excited states, and can only occur by the substitution of a 5p electron for one of the 5s electrons, leading to resonance configurations $4p^5(^2P_{\frac{3}{2},\frac{1}{2}})5s5p$. This is the configuration tentatively assigned to these two resonances by Sanche and Schulz.[15] The next highest set of resonances

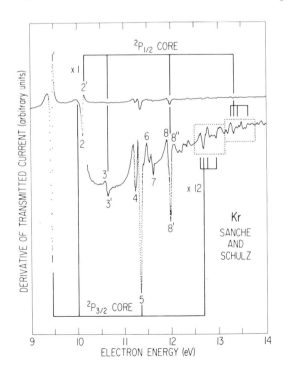

DERIVATIVE OF TRANSMITTED CURRENT (arbitrary units)

$^2P_{1/2}$ CORE

$^2P_{3/2}$ CORE

Kr
SANCHE
AND
SCHULZ

ELECTRON ENERGY (eV)

FIG. 6.--Krypton resonance spectrum of Sanche and Schulz (Ref. 15) showing the separation of resonances into two sets, each associated with the appropriate $X^2P_{\frac{3}{2},\frac{1}{2}}$ positive ion core. (From Spence and Noguchi, Ref. 14)

associated with the appropriate core configuration of Fig. 7
probably occurs by the addition of an extra 5p electron to the
Rydberg states of configuration $4p^5(^2P_{\frac{3}{2},\frac{1}{2}})5p$, leading to resonance configurations $4p^5(^2P_{\frac{3}{2},\frac{1}{2}})5p^2$.

No resonances associated with the 4d or 6s excited
states are observed. This is to be expected since the lowest
energy resonances associated with these excited states would
be $(^2P_{\frac{3}{2},\frac{1}{2}})5s4d$ and $(^2P_{\frac{3}{2},\frac{1}{2}})6s5s$ configurations which are unlikely to be bound because of the difference in principal quan-
tum numbers and angular momenta. We believe the highest set
of observed resonances in Kr to be associated with the 5d
excited states, the electron configurations being $(^2P_{\frac{3}{2},\frac{1}{2}})5d^2$,
though the lower resonances in this highest set may be as-
sociated with the 6p neutral excited states.

One should note that all the Kr resonances between
10.67 and 12.10 eV have been tentatively assigned the con-
figuration $4p^5(^2P_{\frac{3}{2},\frac{1}{2}})5s4d$ by Swanson et al.[7] These configurations are extremely unlikely for the following reasons.

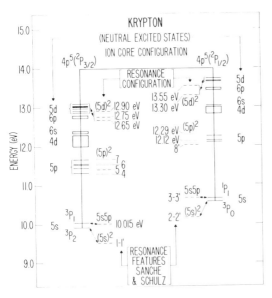

FIG. 7.--Energy level
diagram of neutral
excited states and
Feshbach resonances
in krypton (From Spence
and Noguchi, Ref. 14).

1) These assignments lead to values of the electron affinity of the Kr $4p^5(^2P_{\frac{3}{2},\frac{1}{2}})4d$ states for an s electron of between 0.6 and 1.3 eV. As the additional electron is bound mainly by Coulomb attraction, the electron affinity can hardly be greater than the prototype hydrogen 1s2s system where the electron affinity for an s electron is 0.64 eV.

2) The guidelines developed above show that the binding energy of an additional electron to a Rydberg state whose excited electron has different n and $\ell$ is likely to be very small, if, indeed, the state is bound at all. For Kr, the 4d orbital is 8.45 $a_0$ removed from the core, whereas the 5s is only 5.08 $a_0$,[16] thus providing screening of the core for the 4d electron, which is thus unlikely to be bound.

Having determined electron configurations for Kr, we may now transfer them to the corresponding resonances observed in the isoelectronic molecule HBr. Rydberg states observed in HBr by Ginter and Tilford,[17-19] by u.v. phosoabsorption spectroscopy, together with the resonance spectra observed[14] in HBr are shown in Fig. 8. The analogy of HBr to Kr makes identification of parent excited states a relatively straightforward matter, and is indicated in Fig. 8 by the dotted lines connecting resonance states with neutral states. The energetically highest observed resonances in HBr are clearly associated with neutral states of symmetry $[X^2\Pi_{\frac{3}{2},\frac{1}{2}}]\,5d\lambda$.

Similar assignments to those made in HBr have been made[14] to resonances observed in HCl and HI. The relative simplicity of the spectra, despite the very large number of possible electron resonance configurations and term values, implies strong underlying electron correlation stability rules.

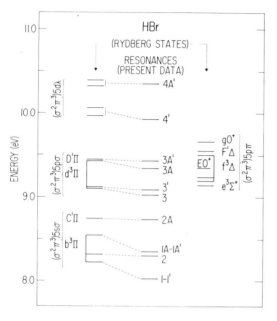

FIG. 8.--Energy level
diagram of Rydberg
states and Feshbach
resonances in HBr (From
Spence and Noguchi,
Ref. 14).

The precise nature of these correlation effects is not yet clear,
and must await theoretical elucidation.

## References

1.  L. Sanche and G. J. Schulz.  Phys. Rev. A 6, 69 (1972).
2.  G. H. Wannier.  Phys. Rev. 90, 817 (1953).
3.  A. R. P. Rau.  Phys. Rev. A 4, 207 (1971).
4.  R. P. Madden and K. Codling.  Astrophys. J. 141, 364
       (1965).
5.  U. Fano.  Atomic Physics (Plenum Press, N.Y., 1969),
       Eds. B. Bederson, V. W. Cohen, and F. M. J. Pichanick,
       p. 209.
6.  G. J. Schulz.  Rev. Mod. Phys. 45, 378 (1973).
7.  N. Swanson, J. W. Cooper, and C. E. Kuyatt.  Phys. Rev.
       A 8, 1825 (1973).
8.  A. W. Weiss and M. Krauss.  J. Chem. Phys. 52, 4363 (1970).
9.  L. Sanche and G. J. Schulz.  Phys. Rev. Lett. 27, 1333
       (1971).
10.  D. Spence.  Phys. Rev. A 10, 1045 (1974).
11.  J. L. Dehmer and R. P. Saxon.  Argonne National Laboratory
       Radiological and Environmental Research Division Annual
       Report, July 1972—June 1973. ANL-8060, Part I, p. 102.

12. D. Spence. To appear in Phys. Rev. A, August 1975.
13. J. J. Matese. Phys. Rev. A 10, 454 (1974).
14. D. Spence and T. Noguchi. J. Chem. Phys. 63, 505 (1975).
15. L. Sanche and G. J. Schulz. Phys. Rev. A 5, 1672 (1972).
16. J. L. Dehmer. Unpublished work.
17. M. L. Ginter and S. G. Tilford. J. Mol. Spectry. 34, 206 (1970).
18. M. L. Ginter and S. G. Tilford. J. Mol. Spectry. 37, 159 (1971).
19. M. L. Ginter and S. G. Tilford. Unpublished work.

MOLECULAR COLLISIONS
AND REACTIVE SCATTERING

# PROGRESS IN THE QUANTUM DYNAMICS OF REACTIVE MOLECULAR COLLISIONS[†]

Aron Kuppermann

Arthur Amos Noyes Laboratory of Chemical Physics,[‡] California Institute of Technology, Pasadena, California 91125

## ABSTRACT

Progress in accurate quantum mechanical calculations of reactive collisions is reviewed. The results of three-dimensional calculations are described and compared with those of approximate methods. Resonances in reactive scattering are discussed as well as electronically nonadiabatic chemical reactions.

## I. HISTORICAL INTRODUCTION

Given an electronically adiabatic potential energy surface for a triatomic system, it is possible in principle to solve the Schrödinger equation describing the motion of the nuclei on that surface and to obtain, from such solutions, state-to-state differential and integral reaction cross sections. These cross sections furnish very detailed information about the dynamics of the reaction and can also be used to calculate rate constants for bulk reactions. This problem is, however, computationally formidable, and as a result the first attempts to solve it were limited to collinear reactions in which the three

[†] Supported in part by the U. S. Air Force Office of Scientific Research, Grant No. AFOSR-73-2539.

[‡] Contribution No. 5208

atoms were constrained to lie on a straight line. This reduces the number of independent variables on which the wavefunction depends from six to two, greatly simplifying the problem. The first solution of this collinear problem for a realistic potential energy surface was obtained by Mortensen and Pitzer (1) for the H + H$_2$ exchange reaction and extended later (2) to some of its isotopic counterparts. The method used was a finite difference numerical solution of the two-variable time-independent Schrödinger partial differential equation coupled with an iterative procedure for imposing the appropriate reactive-scattering boundary conditions. A variation of the finite difference method in which the boundary conditions were imposed by a noniterative approach involving a sufficiently large number of linearly independent solutions of the Schrödinger equation was developed by Diestler and McKoy (3) and applied by Truhlar and Kuppermann (4) to H + H$_2$ and by Truhlar, Kuppermann, and Adams (5) to some of its isotopic counterparts. An interesting variation of the finite difference method was introduced by McCollough and Wyatt (6) who used it to solve the time-dependent Schrödinger equation for the H + H$_2$ exchange reaction, replacing thereby a boundary value problem by an initial value one.

The finite-difference approach in any of the variations mentioned above is computationally very inefficient and inappropriate for extension to a wide energy range or to problems of higher dimensionality. As a result, several other methods have been developed and used recently. These include the variational approach used by Mortensen and Gucwa (7) for collinear H + H$_2$, and the integral equation method developed by Sams and Kouri (8) and applied to several collinear systems by Adams, Smith, and Hayes (9). However, the most widely used approach for collinear collisions has been the coupled-equation (i.e., close-coupling) method, in one of its several forms. The basic method consists in choosing a set of two convenient variables, x and y, to describe the internal configuration of the system. These variables may be different in different regions of configuration space but satisfy the central property that for x equal to a constant $\bar{x}$ the potential energy function $V(\bar{x}, y)$ is bound. The wavefunction $\psi(x, y)$ is expanded in eigenfunctions of $V(\bar{x}, y)$, (which are called the vibrational basis set) and the resulting coupled ordinary differential equations in the x-dependent coefficients of this expansion are solved. Variations of this approach have been developed by Light (10), Kuppermann (11), Diestler (12), and Johnson (13) and applied to a variety of collinear systems (10-18). As a result, a significant amount of knowledge has accumulated about the reactive scattering properties of collinear reactive systems.

When the constraint of collinearity is
relaxed, the problem becomes significantly more difficult.
For coplanar triatomic reactions, the wavefunction depends
now on four variables (after the motion of the center of mass
is removed) and the potential energy function V on three. A
partial wave expansion reduces the problem to a set of
uncoupled partial wave Schrödinger equations depending on the
same three variables as V (19, 20). One of these is usually an
angle, and expansion of the wavefunction in terms of a complete
set of basis functions of that angle (the rotational basis set)
yields a set of coupled two-variable partial differential equations. For triatomic reactions in three-dimensional space,
the wavefunction depends on six variables, three of which can
be chosen to be the ones on which V depends (21, 22). It is
possible to eliminate the other three variables by a partial
wave expansion in terms of Wigner rotation functions, leading
to a set of coupled partial wave Schrödinger equations depending on the same three variables as V. A further expansion in
terms of the angular variable usually appearing in V leads
again to a set of coupled two-variable partial differential
equations, the number of which is usually appreciably larger
than for the coplanar case. There have been relatively few
studies of such noncollinear reactions. The coplanar $H + H_2$
system was investigated by coupled-equation techniques by
Light and co-workers (19), using a single vibrational basis
function, and by Kuppermann and co-workers (20), in a converged vibrational-rotational expansion approach. Baer and
Kouri (23) developed a coupled T-operator integral equation
technique and applied it to a simple three-dimensional model
atom plus diatom system in which reaction with only one end
is permitted. Wolken and Karplus (24) applied an integro-
differential equation method proposed by Miller (25) to the
three-dimensional $H + H_2$ reactive system using a one-
vibrational basis function approximation. Very recently,
Kuppermann and Schatz (21) have achieved accurate vibrationally and rotationally converged results by an extension of the
coupled-equation techniques previously used for collinear and
coplanar systems (11, 20), and Elkowitz and Wyatt (22) have
applied a different version of the coupled-equation method
using hindred rotor basis functions in the expansion (22, 26).
In a somewhat different vein, Micha (27) has developed a
Fadeev-equation approach to the problem of reactive scattering.

In this paper we focus attention on the accurate 3-D
results (21, 22) all of which have appeared since the previous
ICPEAC meeting was held in Belgrade in July of 1973. Coplanar and collinear results as well as other approximate
methods will be invoked mainly for comparison purposes.
However, some recent accurate collinear studies of reactive
scattering resonances and of electronically-nonadiabatic

chemical reactions will also be discussed.

## II.  POTENTIAL ENERGY SURFACES IN SYMMETRIZED HYPERSHPERICAL COORDINATES

In order to summarize the characteristics of the reactive scattering problem and the approaches used to solve it, it is useful to describe the nature of the interaction potentials under consideration.  For electronically adiabatic chemical reactions the Born-Oppenheimer separation approximation permits us to describe the motion of the nuclei during a chemical reaction as that due to a potential energy function V which depends on the relative positions of the nuclei but not on the electronic coordinates.  V is obtainable by assuming these nuclei to be fixed and solving the electronic motion problem.  The resulting electronic wavefunction and energy depend parametrically on the nuclear geometry.  Let us consider a system of three atoms $A_\alpha (\equiv A)$, $A_\beta (\equiv B)$, and $A_\gamma (\equiv C)$.  For notational purposes, let $\lambda \nu \kappa$ be any cyclic permutation of $\alpha \beta \gamma$, meaning that $A_\lambda + A_\nu A_\kappa$ represents any of the channels A + BC, B + CA, or C + AB.  Let $\underset{\wedge \lambda}{r'}$ and $\underset{\wedge \lambda}{R'}$ be respectively the vector from $A_\nu$ to $A_\kappa$ and from the center of mass of $A_\nu A_\kappa$ to $A_\lambda$.  For example, for $\lambda \nu \kappa \equiv \alpha \beta \gamma$ $\underset{\wedge \alpha}{r'}$ is the vector from B to C and $\underset{\wedge \alpha}{R'}$ the vector from the center of mass of BC to A.  The potential energy function can be considered to depend on any of the sets of variables $(r'_\alpha, R'_\alpha, \gamma_\alpha)$, $(r'_\beta, R'_\beta, \gamma_\beta)$, or $(r'_\gamma, R'_\gamma, \gamma_\gamma)$ where $\gamma_\lambda$ $(\lambda = \alpha, \beta, \gamma)$ is the angle in the $(0, \pi)$ range between $\underset{\wedge \lambda}{r'}$ and $\underset{\wedge \lambda}{R'}$.

It has recently been shown that a very convenient set of coordinates exists for mapping V (28).  These are the symmetrized hyperspherical coordinates $r = (r_\lambda^2 + R_\lambda^2)^{1/2}$, $\theta_\lambda = 2 \cos^{-1}(R_\lambda /r)$ and $\gamma_\lambda$, where $\underset{\wedge \lambda}{r} = (\mu_{\nu \kappa}/\mu_{\lambda, \nu \kappa})^{1/4} \underset{\wedge \lambda}{r'}$, and $\underset{\wedge \lambda}{R} = (\mu_{\lambda, \nu \kappa}/\mu_{\nu \kappa})^{1/4} \underset{\wedge \lambda}{R'}$ are scaled (29-31) distances.  The masses $\mu_{\nu \kappa}$ and $\mu_{\lambda, \nu \kappa}$ are respectively the reduced mass of $A_{\nu \kappa}$ and of the $A_\lambda + A_\nu A_\kappa$ pair, the angle $\theta_\lambda$ is in the $(0, \pi)$ range and r is independent of $\lambda$ (30, 31).  The factor 2 in the expression for $\theta_\lambda$ is crucial and makes the present hyperspherical coordinates differ from those suggested previously (30-32).  We now consider r, $\theta_\lambda$, $\gamma_\lambda$ to be the spherical polar coordinates of a point $P_\lambda$ in a three-dimensional internal arrangement

configuration space $OX_\lambda Y_\lambda Z_\lambda$.  In order for this space to be
completely scanned by those variables, we extend the range
of $\gamma_\lambda$ from $(0,\pi)$ to $(-\pi,\pi)$ by setting, for $\gamma_\lambda < 0$, $V(r,\theta_\lambda,-\gamma_\lambda)$
$= V(r,\theta_\lambda,\gamma_\lambda)$.  The mapping of V in this space (but not in the
space in which the factor of 2 just mentioned is omitted) has
two very important properties:  (a) an $r,\theta_\lambda,\gamma_\lambda \to r,\theta_\nu,\gamma_\nu$
transformation rotates equipotential surfaces around $OY_\lambda$
without distorting them and (b) the symmetry properties of
the map are the same as those of the reaction.  For example,
the mapping for the $H_3$ system has the symmetry properties
of an equilateral triangle and that for the $FH_2$ those of an
isosceles triangle.  These symmetry properties do not depend
on the choice of the arrangement channel $\lambda$ and permit us to
visualize the characteristics of V for all configurations in one
single representation, the $\alpha$ one for example.

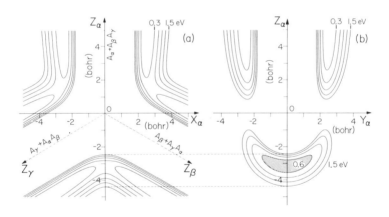

Figure 1.  Equipotential surfaces for $H_3$.  The cartesian coor-
dinates $X_\alpha, Y_\alpha, Z_\alpha$ are $r \sin \theta_\alpha \cos \gamma_\alpha$, $r \sin \theta_\alpha \sin \gamma_\alpha$, and
$r \cos \theta_\alpha$, respectively, with r, $\theta_\alpha$ defined in the text.  The
curves are intersections of $V(r,\theta_\alpha,\gamma_\alpha) = E$ surfaces with the
planes $OX_\alpha Z_\alpha$ (Fig. 1a) and $OY_\alpha Z_\alpha$ (Fig. 1b).  The origin of
measurement of E is the minimum of the $H_2$ diatomic potential
energy curve with the third atom removed to infinity.  The
values of E range from 0.3 eV to 1.5 eV in steps of 0.3 eV,
as indicated on top of figure.  All points on Fig. 1a and those
on the $OZ_\alpha$ axis of Fig. 1b correspond to linear configurations.
Those off the $OZ_\alpha$ axis on Fig. 1b correspond to perpendicular
(i.e., isosceles triangle) configurations in which $A_\alpha$ is the
odd atom.

In Fig. 1 we depict intersections of the equipotential surfaces with the $0X_\alpha Z_\alpha$ and $0Y_\alpha Z_\alpha$ planes for the Porter and Karplus (33) potential function describing the $H_3$ system. The $0Y_\alpha$ axis, perpendicular to the plane of Fig. 1a, is a three-fold axis of symmetry of V, due to the equivalence of the three H atoms and of the corresponding arrangement channels. The angle between any of the three axes $0Z_\alpha$, $0Z_\beta$, and $0Z_\gamma$ is 120°, rather than the usual 60° (34), due to the factor 2 in the definition of $\theta_\lambda$. It is this factor which permits the three arrangement channels $A_\lambda + A_{\nu\kappa}$ $(\lambda = \alpha, \beta, \gamma)$ to be represented equivalently. The lower part of Fig. 1b, in the negative $Z_\alpha$ half-plane, depicts in detail the "transition state" region of configuration space halfway between the $A_\gamma + A_\alpha A_\beta$ reactant and $A_\gamma A_\alpha + A_\beta$ product configurations. At any energy E, all classically allowed pathways leading from such reagents to such products must pass through the region enclosed by the corresponding equipotential. The hatched area on the bottom of Fig. 1b is enclosed by the E = 0.6 eV equipotential. The smaller E, the more confined is this region and the less can the intermediate reactive configurations deviate from col-linearity. The characteristics of these "passages" between reagents and products influence significantly the dynamical properties of V. For example, if they are narrow, the reac-tion is collinearly dominated.

With the help of these symmetrized coordinates we can describe graphically the nature of the atom-diatom reactive scattering problem and of the methods used for solving it. Collisions of $A_\alpha$ with $A_\beta A_\gamma$ correspond to configurations in Fig. 1 initially with $X_\alpha$ and $Y_\alpha$ of the order of the $A_\beta A_\gamma$ equi-librium internuclear distance and $Z_\alpha$ large with respect to that distance. After the collision has occurred, the system rebounds into that region for nonreactive collisions, or moves to regions in the vicinity of the $Z_\beta$ axis with large $Z_\beta$ for reactive collisions resulting in $A_\beta + A_\gamma A_\alpha$ products or the $Z_\gamma$ axis for $A_\gamma + A_\alpha A_\beta$ products. One must obtain scattering wavefunctions which behave accordingly in these different regions of configuration space. Using a time-dependent language, a wave packet approaching the origin O from the large $Z_\alpha$ direction is partially reflected and partially bifurcates into the $0Z_\beta$ and $0Z_\gamma$ directions due to reaction of $A_\alpha$ with either $A_\beta$ or $A_\gamma$ respectively. This bifurcation problem, which encompasses the competition between these two reactions, is conceptually a rather difficult one and has been at least partially responsible for the slowness of the progress in the reaction dynamics of noncollinear systems.

## III. COUPLED-EQUATION METHOD FOR 3-D REACTIVE SCATTERING

An approach we recently used in obtaining accurate solutions for the 3-D H + H$_2$ system (21) is summarized below. The Schrödinger equation for the triatomic system can be considered to be a function of the six variables $R_\lambda, r_\lambda, \gamma_\lambda, \theta_\lambda, \phi_\lambda, \psi_\lambda$. The first three have been defined in the previous section, $\theta_\lambda$ and $\phi_\lambda$ are the polar angles of $\underset{\sim}{R}_\lambda$ in a laboratory-fixed system of reference, and $\psi_\lambda$ is the "tumbling" angle between the instantaneous triatom plane and a fixed reference plane containing $\underset{\sim}{R}_\lambda$. We consider wavefunctions $\Psi^\lambda_{JM}$ of these six variables which are simultaneously solutions of the Schrödinger equation and eigenfunctions of the square of the total angular momentum and its component along a laboratory-fixed 0z axis. We expand $\Psi^\lambda_{JM}$ in terms of the Wigner rotation functions (35) $D^J_{M\Omega_\lambda} (\phi_\lambda, \theta_\lambda, 0)$ and the spherical harmonics $Y_{j_\lambda\Omega_\lambda} (\gamma_\lambda, \psi_\lambda)$. The resulting coefficients are functions of the two distances $R_\lambda, r_\lambda$ and satisfy a set of coupled differential equations. A final expansion is made in terms of local vibrational wavefunctions, which are cuts of the rotationally averaged potential along directions transverse to an appropriately chosen propagation coordinate, which varies from region to region of the $R_\lambda, r_\lambda$ configuration space. The resulting coupled ordinary differential equations are integrated into the interaction region from each of the three $\lambda$ arrangement channel regions, using the Gordon method (36). These solutions are then smoothly matched along three half-planes of the configuration space of Fig. 1, all limited by the 0Y$_\alpha$ axis of that figure and containing the negative half of the 0Z$_\alpha$, 0Z$_\beta$, and 0Z$_\gamma$ axes, respectively. This matching procedure contains built into it the solution to the bifurcation problem mentioned in the previous section. It involves using basis functions for the matching which are localized on those half-planes but when taken together are complete over the entire $\gamma_\alpha, \gamma_\beta$, or $\gamma_\gamma$ angular range. Finally, linear combinations of the smoothly matched solutions are used to obtain the reactance and scattering matrices for each value of J, from which the differential and integral cross sections for reactive and nonreactive processes are calculated.

## IV.   RESULTS AND DISCUSSION OF 3-D CALCULATIONS

The coupled-equation method described above was applied to the H + $H_2$ system (21) using the Porter and Karplus potential energy surface (33) depicted in Fig. 1. In addition to distinguishable atom reactive and nonreactive cross sections, calculations were made of antisymmetrized cross sections corresponding to scattering amplitudes which are the appropriate linear combinations of the direct and exchange contributions necessary to make the scattering wavefunction antisymmetric with respect to the exchange of any two hydrogen atoms. In the 0.40 to 0.70 eV range of total energy E, up to 30 rotational, 4 vibrational, and 100 total basis functions were found necessary for convergence of the results for each J to within 5%. Convergence of the reactive differential cross sections required all values of J from 0 to about 12.

The J = 0 reactive probabilities have a dependence on energy very similar to that of the corresponding coplanar and collinear ones, over several orders of magnitude of the probabilities. These curves are shifted towards higher energies by about 0.05 eV in going from 1-D to 2-D and from 2-D to 3-D probabilities, probably due to the zero point energy stored in the bending mode of the transition state. This behavior suggests how 3-D probabilities can be obtained approximately from 1-D ones for this collinearly dominated reaction (21).

In Fig. 2 we present the distinguishable atom integral reaction cross sections $Q_{0j}^R$ from the ground vibrational state and initial rotational state j (j = 0, 1, 2) of the reagent, as a function of E, for several calculations: the converged (21) (SK), the hindred rotor (22) (EW) and the quasi-classical (37) (KPS) ones. The latter agree quite well with the SK cross sections for j = 0 and 1, at energies above the effective quasi-classical threshold. This agreement was to be expected, since it had already been observed for the collinear reaction (38). As a consequence one obtains near equality in the SK quantum and KPS quasi-classical thermal rate constants at 600°K. At lower temperatures the effects of tunneling become very important, and the SK rate constants are larger than the KPS

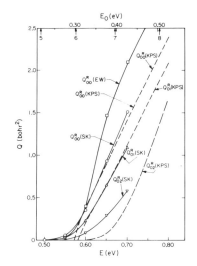

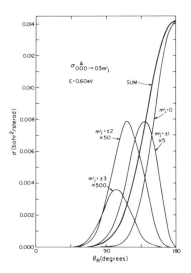

Figure 2. Comparison of integral reactive cross sections from several H + H$_2$ calculations.

Figure 3. Antisymmetrized differential cross sections and their sum in the helicity representation.

ones by factors of 3.2 and 18 at 300°K and 200°K respectively (21). The EW $Q_{00}^R$ results have been recently decreased, bringing them to within approximately 15% of the SK ones (39). The one vibrational basis function integro-differential equation results of Wolken and Karplus for the antisymmetrized $Q_{00 \to 01}^A$ cross section have an effective threshold about 0.05 eV lower than the SK ones and rise much more rapidly with energy. This is probably due to the severely truncated nature of their vibrational-rotational basis set. The Choi and Tang distorted wave calculation (40), done at a higher energy, seems to fall on an extrapolation of the SK $Q_{00 \to 01}^A$ curve.

In Fig. 3 we plot the antisymmetrized $\sigma_{000 \to 03m_j'}^A$ para → ortho differential cross sections as well as their sum, as a function of the scattering angle $\theta_R$, for $|m_j'| = 0$ through 3. We see that the $m_j' = 0$ cross section is backward peaked and that it dominates over the others. This highly nonstatistical behavior of the polarization of the products is the manifestation of an $m_j = 0$ to $m'_j = 0$ quasi-selection rule for low j which has been proposed (21) for collinearly dominated reactions, due to maximum atom-diatom collinear overlap for those values of $m_j$ and $m'_j$. In contrast, the degeneracy-averaged (i.e., summed over $m'_j$ and averaged over $m_j$) product rotational distributions can be fitted to temperature-like distributions to a high degree of accuracy, indicating a statistical-type behavior. The corresponding rotational temperature parameters increase nearly linearly with energy from 228°K at E = 0.45 eV to 446°K at 0.70 eV.

The degeneracy-averaged antisymmetrized para → para differential cross sections $\sigma_{00 \to 01}^A$, to which only exchange collisions contribute, are smooth backward peaked functions of the scattering angle. In contrast, the $\sigma_{00 \to 02}^A$ para → ortho cross sections are peaked sideways, due to the large contribution from direct processes, and display an oscillatory behavior as a function of the scattering angle, due to the interference between the direct and exchange scattering amplitudes, both of which contribute to this cross section. The backward-peaked shape of the $\sigma_{00 \to 01}^A$ differential cross section for the coplanar reaction at an energy E is essentially identical to the one for the 3-D reaction at an energy E + 0.05 eV over the entire energy range considered. The energy shift again suggests an effect of the bending energy of the transition state. The agreement in these angular distributions is not unexpected, since the same potential is sampled in both cases and the primary difference between the two calculations is the additional centrifugal coupling resulting from the tumbling of the three-atom plane. The existence of the strong product polarization effect mentioned above indicates that such coupling is weak compared with the potential coupling responsible for the linear geometry requirement. This suggests a close similarity between the 2-D and 3-D dynamics and conversion of 2-D results to 3-D ones promises to be an accurate approximate technique. Another promising approximation is the neglect of the weak coupling between different tumbling angular momenta, which reduces the numerical effort in a 3-D reactive scattering calculation to that of a 2-D one.


V.   RESONANCES IN REACTIVE SCATTERING

For the $H + H_2$ system, reactive scattering resonances have been observed in collinear (16), coplanar, and 3-D collisions (41). In Fig. 4 we depict the reaction probabilities for these three kinds of collisions.  As can be seen, a dip appears

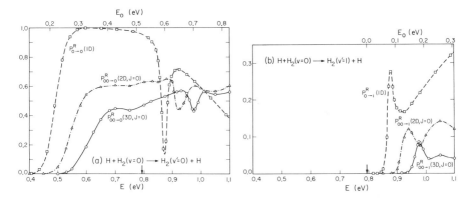

Figure 4. Collinear (1-D), coplanar (2-D), and three-dimensional reaction (3-D) reaction probabilities as a function of total energy E and translational energy $E_0$.  The noncollinear results have been summed over all product rotational states within the product vibrational manifold indicated.

in the **vi**brationally elastic reaction probability curves (left side panel) and a peak in the vibrationally inelastic ones (right side panel).  For the 1-D case, a time-delay analysis indicates (16) that this structure is due to interference between a direct mechanism and a compound state one associated with an internal excitation dynamical (Feshbach) resonance.  A vibrationally adiabatic analysis (42) of the collinear resonance indicates that in the region of the potential energy function saddle point and at the resonance energy the system has a 90% probability of being found in the first vibrationally excited state of the transverse (symmetric stretch) internally excited mode of motion.  This indicates that trapping of the energy in this mode is responsible for the resonance.  Estimates of the energy dependence of the integral cross section for the reaction from the ground vibrational-rotational state of the reagent to the first vibrationally excited state of the product, for 3-D collisions, is similar to that of the corresponding J = 0 reaction probability, depicted in the right panel of Fig. 4, and has a peak value of 0.05 bohr$^2$.    Therefore the partial wave and product rotational state sums involved in the calculation of this cross section do not wash out the resonances observed in collinear collisions, at least for the present collinearly dominated reaction.  This is the first time that a resonance has been predicted for a 3-D reaction whose potential energy function does not have an attractive well.  Since Feshbach resonances have also been detected in collinear calculations for the collinearly dominated $F + H_2$ ($D_2$, HD) (17) and $Cl + H_2$

(last of Ref. 18 papers) reactions (using potential energy
surfaces devoid of wells), it is quite possible that such resonances may exist for these reactions in the real world as well.
The shift in the position of the 1-D, 2-D, and 3-D resonances
of Fig. 4 is approximately equal, once more, to the zero-
point energy of the bending mode of the transition state. This
suggests how the position of 1-D resonances   could be used
to predict where the 3-D ones should lie for collinearly dominated reactions.  In going from the SSMK potential energy
surface used in the study of Ref. 16 to the Porter and Karplus
one used in Ref. 41, the width of the 1-D H + $H_2$ resonance
being considered decreases from about 0.05 eV to 0.022 eV,
indicating a great sensitivity of the properties of the resonance to the shape of the potential energy surface.  As a
result, resonances may prove a sensitive probe for the experimental characterization of potential energy surfaces and for
the development and testing of approximate reaction-dynamic
theories.

## VI.  ELECTRONICALLY NONADIABATIC
## CHEMICAL REACTIONS

Partly as a result of the possibility of developing
chemical lasers involving electronic state population inver-
sions, interest in the theoretical understanding of electronically chemiluminescent reactions has recently increased.  A
particular example is the Ba + $N_2O$ $\rightarrow$ BaO + $N_2$ reaction, in
which the BaO may be formed in an electronically excited
state (43-45).  No accurate theoretical calculations for such
electronically nonadiabatic reactive processes have yet been
attempted.  A classical trajectory-hopping scheme has been
developed and applied to the $H^+$ + $D_2$ $\rightarrow$ $HD^+$ + D reaction (46).
In addition, a semi-classical description of such reactions
based on the Feynman propagator has been formulated and
applied to this same reaction (47).  As a means of assessing
the validity of these approximate schemes, an accurate two-
potential energy surface collinear reactive scattering calculation method was developed and applied to a simplified model
of the Ba + $N_2O$ $\rightarrow$ BaO + $N_2$ reaction (48).  These calculations
indicate that a substantial fraction of the BaO product can
indeed, for this model, be produced in an electronically
excited state.  These accurate collinear calculations can now
be used to test the validity of approximate ones for the same
potential energy surfaces.

## VII. SUMMARY

Substantial progress has occurred over the last few years in the development of accurate 3-D quantum mechanical methods for calculating cross sections of electronically adiabatic reactions. Out of such calculations have emerged predictions of a product polarization quasi-selection rule and of the existence of Feshbach resonances for collinearly dominated reactions. Such predictive ability is indicative of the vigor of the field. Progress has also been made in the calculation of electronically nonadiabatic reaction cross sections.

## ACKNOWLEDGMENTS

I wish to thank my many collaborators who over the years have participated in much of the work described in this paper: D. G. Truhlar, J. T. Adams, J. M. Bowman, G. C. Schatz, M. Baer, R. Ling, and J. Dwyer. In particular, all of the 3-D work was done jointly with Schatz whose superb scientific ability and dedication to his work are greatly responsible for its successful completion. Thanks are also due to Ambassador College for generous use of their computational facilities.

## REFERENCES

1.    E. M. Mortensen and K. S. Pitzer, J. Chem. Soc. (London) Spec. Publ. 16, 57 (1962).

2.    E. M. Mortensen, J. Chem. Phys. 48, 4029 (1968).

3.    D. J. Diestler and V. McKoy, J. Chem. Phys. 48, 2951 (1968).

4.    D. G. Truhlar and A. Kuppermann, J. Chem. Phys. 52, 3841 (1970); ibid. 56, 2232 (1972).

5.    D. G. Truhlar, A. Kuppermann, and J. T. Adams, J. Chem. Phys. 59, 395 (1973).

6.    E. A. McCullough, Jr., and R. E. Wyatt, J. Chem. Phys. 51, 1253 (1969); ibid. 54, 3578, 3592 (1971).

7.    E. M. Mortensen and L. D. Gucwa, J. Chem. Phys. 51, 5695 (1969).

8.    W. M. Sams and D. J. Kouri, J. Chem. Phys. 51, 4809 (1969).

9.    J. T. Adams, R. L. Smith, and E. F. Hayes, J. Chem. Phys. 61, 2193 (1974).

10.   C. C. Rankin and J. C. Light, J. Chem. Phys. 51, 1701 (1969); G. Miller and J. C. Light, ibid. 54, 1635, 1643 (1971).

11.   A. Kuppermann, Proceedings of the Conference on Potential Energy Surfaces in Chemistry, University of California, Santa Cruz, W. A. Lester, Ed. (August 1970), p. 121; Electronic and Atomic Collisions, Proceedings of VII ICPEAC (North-Holland, Amsterdam, 1971), p. 3.

12.   D. J. Diestler, J. Chem. Phys. 54, 4547 (1971).

13.   B. R. Johnson, Chem. Phys. Letts. 13, 172 (1972).

14.   S. F. Wu and R. D. Levine, Mol. Phys. 22, 881 (1971).

15.   E. J. Shipsey, J. Chem. Phys. 58, 232 (1973).

16.   G. C. Schatz and A. Kuppermann, J. Chem. Phys. 59, 964 (1973).

17.   G. C. Schatz, J. M. Bowman, and A. Kuppermann, J. Chem. Phys. 58, 4023 (1973); 63, 674, 685 (1975).

18.   M. Baer and D. J. Kouri, Chem. Phys. Letts. 24, 37 (1974); M. Baer, J. Chem. Phys. 60, 1057 (1974); A. Persky and M. Baer, ibid. 60, 133 (1974); M. Baer, V. Halavee and A. Persky, ibid. 61, 5122 (1974); M. Baer, Mol. Phys. 27, 1429 (1974).

19.   R. P. Saxon and J. C. Light, J. Chem. Phys. 56, 3874, 3885 (1972); A. Altenberger-Siczek and J. C. Light, J. Chem. Phys. 61, 4373 (1974).

20.   A. Kuppermann, G. C. Schatz, and M. Baer, J. Chem. Phys. 61, 4362 (1974); J. Chem. Phys., forthcoming; G. C. Schatz and A. Kuppermann, J. Chem. Phys., forthcoming

21.   A. Kuppermann and G. C. Schatz, J. Chem. Phys. 62, 2502 (1975); G. C. Schatz and A. Kuppermann, J. Chem. Phys., forthcoming.

22.   A. B. Elkowitz and R. E. Wyatt, J. Chem. Phys. 62, 2504, 3683 (1975).

23.    M. Baer and D. J. Kouri, Chem. Phys. Letts. 11,
       238 (1971); J. Chem. Phys. 56, 1758 (1972); 57, 3991
       (1972).

24.    G. Wolken and M. Karplus, J. Chem. Phys. 60, 351
       (1974).

25.    W. H. Miller, J. Chem. Phys. 50, 407 (1969).

26.    S. A. Harms and R. E. Wyatt, J. Chem. Phys. 62,
       3162, 3173 (1975).

27.    D. A. Micha, J. Chem. Phys. 57, 2184 (1972); D. A.
       Micha and J.-M. Yuan, forthcoming.

28.    A. Kuppermann, Chem. Phys. Letts. 32, 374 (1975).

29.    D. Jepsen and J.O. Hirschfelder, Proc. Natl. Acad.
       Sci. U.S. 45, 249 (1959).

30.    L. M. Delves, Nucl. Phys. 9, 391 (1959); 20, 275
       (1960).

31.    F. T. Smith, J. Math. Phys. 3, 735 (1962).

32.    P. M. Morse and H. Feshbach, Methods of Theoretical
       Physics (McGraw-Hill, New York, 1953), p. 1729.

33.    R. N. Porter and M. Karplus, J. Chem. Phys. 40,
       1105 (1964).

34.    S. Glasstone, K. J. Laidler, and H. Eyring, The
       Theory of Rate Processes (McGraw-Hill, New York,
       1941), Chap. 3.

35.    A. S. Davydov, Quantum Mechanics (NEO Press, Ann
       Arbor, 1966), Chap. VI.

36.    R. G. Gordon, J. Chem. Phys. 51, 14 (1969).

37.    M. Karplus, R. N. Porter, and R. D. Sharma, J.
       Chem. Phys. 43, 3259 (1965).

38.    J. M. Bowman and A. Kuppermann, Chem. Phys.
       Letts. 12, 1 (1971).

39.    R. E. Wyatt, personal communication.

40.    B. H. Choi and K. T. Tang, J. Chem. Phys. 61,
       2462, 5147 (1974).

41.  G. C. Schatz and A. Kuppermann, Phys. Rev. Letts. in press.

42.  J. M. Bowman, A. Kuppermann, J. T. Adams, and D. G. Truhlar, Chem. Phys. Letts. 20, 229 (1973).

43.  Ch. Ottinger and R. N. Zare, Chem. Phys. Letts. 5, 243 (1970); C. D. Jonah, R. N. Zare, and Ch. Ottinger, J. Chem. Phys. 60, 4369 (1974); A. Schultz and R. N. Zare, ibid. 60, 5120 (1974).

44.  R. W. Field, C. R. Jones, and H. P. Broida, J. Chem. Phys. 60, 4377 (1974).

45.  C. J. Hsu, W. D. Krugh, and H. B. Palmer, J. Chem. Phys. 60, 5118 (1974).

46.  R. K. Preston and J. C. Tully, J. Chem. Phys. 54, 4297 (1971); J. C. Tully and R. K. Preston, ibid. 55, 562 (1971); J. R. Krenos, R. K. Preston, R. Wolfgang, and J. C. Tully, ibid. 60, 1634 (1974).

47.  W. H. Miller and T. F. George, J. Chem. Phys. 56, 5637 (1972); T. F. George and T.-W. Lin, J. Chem. Phys. 60, 2340 (1974); T.-W. Lin, T. F. George, and K. Morokuma, Chem. Phys. Letts. 22, 547 (1973); J. Chem. Phys. 60, 4311 (1974).

48.  J. W. Bowman, Theoretical Studies of Electronically Adiabatic and Nonadiabatic Chemical Reaction Dynamics, Ph.D. Thesis, California Institute of Technology, Pasadena, CA, 1975; J. W. Bowman and A. Kuppermann, forthcoming.

# NON-ADIABATIC EFFECTS IN COLLISIONAL VIBRATIONAL RELAXATION
# OF DIATOMIC MOLECULES

E. E. Nikitin

Institute of Chemical Physics, Academy of Sciences

Moscow, V-334, U.S.S.R.

## INTRODUCTION

There has been a renewed interest recently in processes of vibrational relaxation of diatomic molecules in collisions with atoms stimulated mainly by study of chemical reaction of excited species and of kinetic processes in gas lasers.[1,2] It has been found that very often            ational treatment[3-9] of the energy exchange by V-T (vibrational to translational) mechanism completely fails, and other mechanisms have been invoked. Among these are V-R or VRT (vibrational to rotational or vibrational to rotational and translational) energy exchange, scrambling ("chemical relaxation", i.e., deactivation or activation of molecules in atomic exchange reaction) and electronically non-adiabatic vibrational relaxation. The unique feature of the last process is that it cannot be interpreted in terms of motion of a three-atomic system — an impinging atom A and a diatomic molecule BC — on one potential surface corresponding to a fixed adiabatic electronic state. It has been pointed out[7,8] that this nature of a collision is responsible for very fast relaxation in some cases where other possible paths (VR, VRT or scrambling) could be definetely ruled out. It is the aim of this report to review the present status of the problem.

To simplify the discussion we limit ourselves to $0 \rightarrow 1$ or $1 \rightarrow 0$ vibrational transition in BC thus disregarding all effects connected with multiquantum jumps. Also, we stick, when possible, to semiclassical interpretation of the phenomena, leaving aside some subtle details which are, hopefully, not very important for the physical picture.

ELECTRONICALLY ADIABATIC VRT ENERGY EXCHANGE

Existing theory of vibrational activation (or deactivation) of a diatomic molecule with an atom in a collision process

$$A + BC(v = 0) \rightarrow A + BC(v = 1)$$

is essentially based on the concept of the potential energy surface which itself is valid if the adiabatic treatment of motion of electrons is justified.[7] If A is chemically inert then energy exchange proceeds <u>via</u> VRT process, two extreme limits of which are known as $\overline{VT}$ and VR processes.

The first mechanism is believed to be applicable for homonuclear diatomic molecules ($H_2$, $N_2$, $O_2$) or "almost homonuclear" molecules such as CO. The theory is essentially based on the celebrated Landau-Teller[10] approach with many corrections added during last year's development (see, e.g.[4,6]).

The second mechanism is valid when the intermolecular interaction is highly anisotropic, and dynamical pecularities of the system A-BC favor almost complete transfer of vibration to rotation or vice versa (the most salient examples refer to deactivation of hydrogen halides). The simple theory[3,11] again retains all main features of the Landau-Teller mechanism, the major difference being the interpretation of an effective mass which corresponds to a particular mode (called later effective) of a relative motion responsible for inducing the vibrational transition. In this case of the VR process the rotation of BC is effective, while for the VT process the radial relative motion of A and BC is effective).

To bridge the gap between these two limiting cases a simple VRT model has been suggested[7,12] which explicitly defines the active mode in terms of an intermolecular potential and atomic masses $m_A$, $m_B$, and $m_C$. At this point it should be stressed that the approach under discussion can be justified only for thermal systems in which translational and rotational motion are characterized by the Boltzman distribution with a common temperature T. In a sense the approach is a crude version of the combined phase-space/trajectory method discussed recently in connection with reaction dynamics.[13,14] This is the reason why we shall consider mainly the average transition probability $\langle P(v=1 \rightarrow v=0)\rangle = \langle P_{10}\rangle$ or $\langle P(v=0 \rightarrow v=1)\rangle = \langle P_{01}\rangle$ per one gas kinetic collision, but not corresponding cross sections.

According to this simplified model of the VRT process, we assume that the classical motion along some active coordinate $q = q(t)$ which partly defines the relative position of A and BC induces vibrational transitions among states of the oscillator BC caused by perturbation $V(q)$. The simple model which can be accepted would be a classical harmonic oscillator excited (or de-excited) by an external time-dependent force $F = F(q)$ acting during the collision time $\tau_c$. Though transition probabilities $P_{10}$ and $P_{01}$ are not readily defined in this model, there are semiclassical arguments relating $P_{01}$ and $P_{10}$ to certain classical quantities, such as the energy transferred to the oscillator either initially at rest $\Delta E_0$, or vibrating with the energy $E_V = \hbar\omega/2$.[15] The transition probability $P_{01}$ being roughly proportional to $\Delta E_0$, is given by the square of the Fourier-component of $F(t)$ corresponding to the vibrational frequency $\omega$:

$$P_1 = C \left| \int_{-\infty}^{\infty} F(t)\exp(i\omega t)dt \right|^2 \qquad (1)$$

where C is a coefficient dependent on masses and potential range. Now, on a quite general argument put forward by Landau and Teller,[10] we expect that at high frequencies, namely at $\omega\tau_c \gg 1$, the Fourier integral would be proportional to $\exp(-\omega\tau_c)$ with $\tau_c$ defined by the characteristic length $\ell$ of the interaction potential and the velocity $v_q$ before collision

$$P_{01} = A(\omega,\ \ell,\ v_q)\exp(-2\omega\ell/v_q) \qquad (2)$$

Here the pre-exponential factor A depends on $\omega$, $\ell$, $v_q$ but weakly compared to the exponent. When averaged over thermal velocity distribution, Eq. (2) yields the Landau-Teller formula which is often associated with the simple intermolecular interaction of the form $U \sim \exp(-\alpha q)$. For this interaction $\ell = \pi/2\alpha$, and $\langle P_{01} \rangle$ is

$$\langle P_{01} \rangle = B(\omega, T, \mu_q^*)\ \exp(-3[\pi^2\omega^2\mu^*/2\alpha^2 kT]^{1/3}) \qquad (3)$$

Here $\mu^*$ is defined via the reduced mass $\mu$ of a colliding pair, $\mu = m_A(m_B + m_C)/(m_A + m_B + m_C)$, the reduced mass M of the molecule, $M = m_B m_C/(m_B + m_C)$, and the properties of the contour line $R = R(\gamma)$ of the intermolecular potential $U = U(R, \gamma)$ for non-vibrating BC and the total energy $E = kT$. Explicitly,

$$\mu^* = \frac{\mu \cdot M r_e^2 (R^{*2} + R_\gamma^{*2})}{M r_e^2 (R^{*2} + R_\gamma^{*2}) + \mu R^{*2} R_\gamma^{*2}} \quad , \quad R_\gamma = \frac{dR}{d\gamma} \qquad (4)$$

where the star (*) denotes particular values of $R$ and $R_\gamma$ which make the rhs of Eq. (4) minimal, and $r_e$ is the equilibrium internuclear distance in BC.

Two limiting cases mentioned correspond to conditions $\mu \ll M r_e^2 / R_\gamma^{*2}$ or $\mu \gg M r_e^2 / R_\gamma^{*2}$ thus giving

$$\mu^* = \begin{cases} \mu & \text{for VT process} \\ M r_e^2 / R_\gamma^{*2} & \text{for VR process} \end{cases} \qquad (5)$$

with $M r_e^2 / R_\gamma^{*2} < \mu^* < \mu$ for the general VRT case.

Of course, this model is extremely crude, but it seems to explain the temperature dependence of $<P_{01}>$ and $<P_{10}>$ for collision of chemically inert species, and the dependence of probabilities on masses.[3,9] It is difficult to assess the accuracy of this simple model of VRT processes. A comparison between theory and experimental data cannot provide conclusive evidence because the interaction is often known insufficiently. The only way to this end seems to be a direct comparison between the model approach giving directly the probabilities and results of exact dynamical calculations giving cross-sections $\sigma_{v,j;v'j'}(u)$. These can be used to calculate the rate coefficient $k_{v,v'}(T)$ which in turn defines the average probability $<P_{v,v'}>$ via gas-kinetic cross-section $\sigma_{gas}$ and the mean velocity $\bar{u}$:

$$<P_{v,v'}(T)> = k_{v,v'}(T)/\sigma_{gas} \bar{u} \qquad (6)$$

This goal is being pursued now by many authors. We mention here only the most detailed study of $H_2$-He collisions. It has been found that the "breathing sphere" model corresponding to the pure VT mechanism, gives an order of magnitude lower relaxation rate than exactly combined dynamical[16] and relaxational[17] calculations though with similar temperature dependence of relaxation rate. It can be expected that this difference will be made essentially smaller if $\mu^*$ in Eq. (3) is identified with $\mu^*$ in Eq. (4) rather than with $\mu$ (which corresponds to the "breathing sphere" model). On this and other available grounds, we thus assume that the VRT theory, as represented by Eq. (3) and (4), provides an order of magnitude estimation of $\langle P_{10} \rangle$ and $\langle P_{01}. \rangle$  Equation (4) shows also what kind of information is needed about potential energy surface $U(R, \gamma, r)$ to carry out this rough estimation of $P_{vv'}$.

Figure 1 illustrates to what extent the simple correlation

$$\log \langle P_{10} \rangle = \text{const} - 3[\pi^2 \omega^2 \mu^*/2kT\alpha^2]^{1/3} \tag{7}$$

which follows from Eq. (3) is fulfilled.  The values of $\langle P_{10} \rangle$

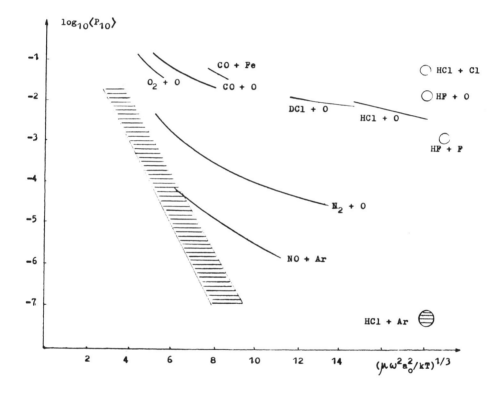

Fig. 1.  Mean transition probabilities for some electronically adiabatic and electronically non-adiabatic deactivation processes of diatomic molecules by atoms.

for the deactivation of $N_2$, $O_2$ and CO by inert gases believed to occur via the VT mechanism[18,19] fall in the shaded zone, the zone width being taken as a measure of the accuracy of Eq. (7).  Here $\log_{10}\langle P_{10} \rangle$ is plotted against a dimensionless parameter $(\omega^2 \mu a_o^1/kT)^{1/3}$, $a_o$ being the Bohr radius.  According

to Eq. (7), the slope of the shaded zone must have been
$2.2(\alpha a_0)^{-2/3}$ and the comparison with actual slope then gives
$\alpha = 4-6$ Å$^{-1}$ which is the usual value of the parameter $\alpha$ for
the VT processes.[3,9,18,19]  The experimental data[20] on deactivation of HCl(v=1) on Ar, also shown in this figure,
illustrate the large effect of rotation in inducing vibrational transition.  The value of $\mu^*$, which shifts the point
into the shaded zone is about 3-4.  This is considerably
lower than $\mu = 19$ and reasonably agrees with theoretical estimate.

## ELECTRONICALLY NON-ADIABATIC VRT ENERGY EXCHANGE

The approach discussed in the previous section fails if
the initial electronic state of a colliding pair is either
degenerate or  close to another state.  The degeneracy is
lifted by the intermolecular interaction, and this implies
that non-stationary electronic wave function $\Psi(\rho,t)$ ($\rho$ being
a set of coordinates of electrons) is described not just by
one function, but by a linear combination of functions $\varphi_k$
with different time-dependent exponents:

$$\Psi(\rho,t) = \sum_k c_k \varphi_k(\rho,q) \exp\left[-\frac{1}{\hbar} \int_{t_o}^{t} U_k(q) dt\right] \tag{8}$$

At the limits of free partners,  $q \to \infty$, $\varphi_k$ becomes just a
component of a degenerate electronic state of A + BC, and
$U_k(q)$ converges to the common limit taken here to be zero.

Consider again a perturbation $V(q)$ responsible for coupling of vibration of BC with a relative motion along the
active mode $q(t)$.  To calculate the force $F(t)$ acting on BC,
we form

$$\partial V/\partial r \big|_{r=r_e} \, ,$$

the first derivative of V with respect to the vibrational coordinate r of BC at $r = r_e$, and then average it using $\Psi$ from
Eq. (8).

$$F(t) = \int \Psi^*(\rho,t)\left(\frac{\partial V}{\partial r}\right) \Psi(\rho,t) d\rho \tag{9}$$

It follows from Eq. (8) that $F(t)$ has the form

$$F(t) = \sum_k |c_k|^2 f_{kk} + \sum_{k \neq m} |c_k c_m^*| \cos\left[\int_{t_o}^{t} \Omega_{km}(t) dt + \lambda_{km}\right]$$

$$= \sum_k F_{kk} + \sum_{k \neq m} F_{km} \tag{10}$$

with

$$f_{km} = \int \varphi_k{}^* \frac{\partial V}{\partial r} \varphi_m d\rho \ , \qquad \Omega_{km} = \frac{1}{\hbar} [U_k - U_m] \qquad (11)$$

The next important step is to realize that the time dependence of all $f_{km}(t)$ is qualitatively the same, and similar to $F(t)$ for electronically adiabatic VRT processes (EA-VRT). We can expect here a simple bell-shape form of $f_{km}$ with width $\tau_c$ and amplitude $f^0_{km}$ (see Fig. 2). If the coupling between $\varphi_k$ is

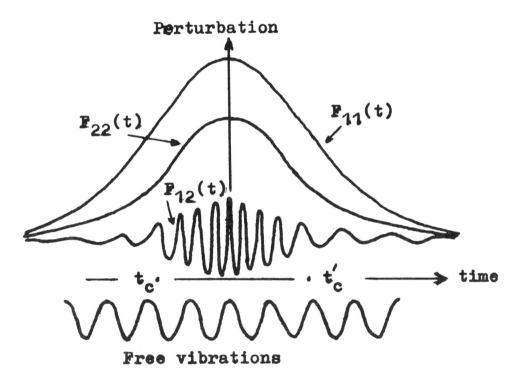

Fig. 2.  Adiabatic ($F_{11}$ and $F_{22}$) and non-adiabatic ($F_{12}$) contributions to the perturbation acting on the molecular oscillator.

not too strong, we do not expect an essential time-dependence of $c_k$. Thus the force $F(t)$ is a contribution of non-oscillating components $|C_k|^2 f_{kk}(t)$, and components

$$F_{km}(t) = |C_k C_m{}^*| f_{km}(t) \cos [\int^t \Omega_{km}(t) dt + \lambda_{km}]$$

oscillating with the time-dependent frequency ranging from zero (free partners) up to some maximum value $\Omega_{km}^o$ which corresponds to the maximum energy difference of adiabatic electronic terms $U_k$ and $U_m$. It is the latter feature of a perturbing force which provides the necessary conditions for a new type of VRT energy transfer via electronically non-adiabatic transitions (ENA-VRT).

However, for the ENA-VRT process to be effective sufficient conditions should be met to provide:

1.   not too small values of $f_{km}^o$ ,

2.   large enough values of $\Omega_{km}^o$ .

If $\Omega_{km}^o > \omega$ then at some time $t = t_c$ there will be a resonance condition $\omega = \Omega_{km}(t_c)$ which makes possible effective excitation or de-excitation of an oscillator via perturbation V.   Thus V must couple states $\varphi_k$ and $\varphi_m$ but still be sufficiently weak not to disturb the pattern of electronic terms at configurations of A and BC complex, at which $\Omega_{km} \approx \omega$. Three types of the interaction V of this kind have been previously considered:[21]

1.   Non-adiabatic coupling due to rotation of the instantaneous plane defined by three atoms A-B-C.   This interaction couples states characterized by different symmetry with respect to reflection of electron coordinates in this plane.

2.   Spin-orbital interaction which couples states with different or the same multiplicity.

3.   A small part of the Coulomb interaction which has been neglected for good reasons in constructing zero-order adiabatic functions $\varphi_k$.   This interaction couples states with the same quantum numbers of the point group $C_s$.

In all cases it is expected, of course, that $V_{km} \ll \hbar\omega$; otherwise a different formulation of the problem is needed.

Figure 2 gives a qualitative picture of three contributions $F_{11}$, $F_{22}$, and $F_{12}$ to $F(t)$ for a system A + BC with two-fold electronic degeneracy at $t \to \pm \infty$ (this model refers to process (17), vide infra).   Also shown are free vibrations of BC, and two resonance points $t_c$ and $t_c$.

Now, close to resonance we make a linear approximation to $\Omega_{12} = \omega + \Omega_{12}'(t-t_c)$ and arrive to the following contribution to the Fourier-component from the oscillating part of the interaction

$$\int f_{12}\cos\left(\int^{t}\Omega_{12}dt + \lambda_{12}\right)\exp(i\omega t) \approx f_{12}(t_c)[\Omega'(t_c)]^{1/2},$$

(12)

where $\Omega'_{12}$ is $\dfrac{d\Omega_{12}}{dt}$ with an estimate $\Omega'_{12} \approx \omega v_q / \Delta\ell$, $\Delta\ell$ being the range of the resonance region. This rough estimate where interference effects from two (or more) stationary phase points have been neglected leads to the expression for the transition probability $P_{10}$ of the ENA-VRT process, part of which is the probability for the EA-VRT process as given by Eq. (2):

$$P_{01} = A \exp\left(-\frac{2\ell\omega}{v_q}\right) + a \; \frac{\Delta\ell\omega}{v_q},$$

(13)

where all parameters have already been defined except a that stands for the strength of electronically non-adiabatic inter- action. Performing thermal averaging over $v_q$, we get finally

$$\langle P_{01}\rangle = B \exp\left[-3\left(\pi^2\omega^2\mu^*/2\alpha^2 kT\right)^{1/3}\right] + b \; \exp\left[-E_o/kT\right],$$

(14)

where B and b depend but weakly on all parameters, and $E_a$ is the activation energy needed for a system to reach a point where resonance conditions are achieved. Generally, $B \gg b$ and $E_a$ is expected to be rather low, in the range of $\hbar\omega$.[7,14] Physically, two terms in Eq. (14) represent near adiabatic ($\omega\tau_c \gg 1$), non-resonant transfer of vibrational energy into rota- tion and translation and resonant ($\omega \approx \Omega$), non-adiabatic transfer of vibrational energy into electronic energy followed by adiabatic transformation of electronic energy (i.e., the potential energy for atoms) into rotation and translation.

The crucial point now is that though $b \ll B$, the second term in Eq. (14) can successfully compete and often completely overweigh the first at low temperatures. Conversely, at higher temperatures, when the Landau-Teller exponent is not too low, the first term will give the main contribution. These general features of ENA-VRT processes are illustrated in Fig. 1 where full lines and points correspond to collision processes to be discussed now. To make reliable estimations of parameters B and b in Eq. (14), a quantal version of the approach is needed. The starting point for that is a "vibro- nic" adiabatic approximation which treats, on equal footing, the electronic and vibrational motion in the system.

Considering first a collision between atom A and molecule BC, both in non-degenerate electronic states, we assume that their relative motion is adiabatic with respect to zero-order vibrations of the molecule.  An intermolecular potential $U_{VV}(R,\gamma)$ can then be assigned to each vibrational level.  In the simple case of a harmonic oscillator these adiabatic potentials are obtained from the interaction potential between atom A and molecule BC in the absence of vibrations by shifting them by appropriate quantities $(n+1/2)h\omega$. Collision-induced changes in the vibrational state can be viewed as non-adiabatic transitions between such levels.  Such an interpretation of the single-quantum deactivation process is illustrated by Fig. 3.  One sees that the negligible transition probability expected for $\omega\tau_c \gg 1$ can be thought of as resulting from a tunnelling transition between non-intersecting adiabatic levels.

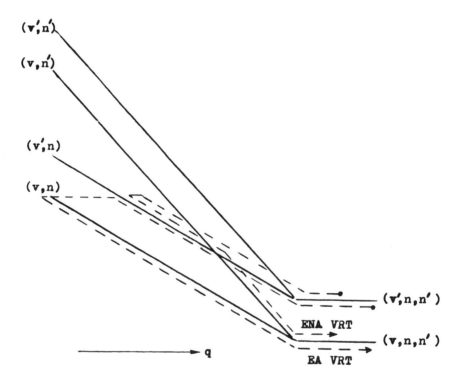

Fig. 3.   Quantum-mechanical interpretation of electronically adiabatic and electronically non-adiabatic VRT mechanisms.

Consider now the same collision, assuming that either A or BC are in a degenerate electronic state. Appropriate adiabatic vibronic levels must already be constructed in zero order. For a given value of the vibrational quantum number, the intermolecular interaction leads to a set of potential surfaces $U_{vv}(h)(R,\gamma)$ which split apart as the intermolecular separation decreases. Shifting this set of levels by successive amounts $(n+1/2)\hbar\omega$, one obtains a full set of vibronic levels exhibiting many intersections. If two crossing levels are of the same symmetry one should treat them as nearly crossing. Non-adiabatic transitions occur close to the points where vibronic levels cross or pseudocross; they are accompanied by changes in the vibrational and electronic states. Since such transitions are not really tunnelling processes, the corresponding probability decays much slower with increasing value of the parameter $\omega\tau_c$ than that for transitions between adjacent vibrational levels derived from the same electronic state. In a simple case the transition probability between vibronic levels can be computed according to the Landau-Zener formula or to the more sophisticated counterpart of it. It follows from the quantal treatment that the ENA-VTR transition probability is proportional to

$$| V_{vv'}^{nn'} |^2$$

i.e., to the square of the matrix element of non-adiabatic interaction calculated with zero-order vibronic wave functions at an A-BC arrangement which correspond to the resonance conditions. The vibronic function is written as a product adiabatic electronic function $\varphi_n$ already defined, and vibrational wave function $\chi_v^n$, corresponding to oscillation of BC in the nth electronic state of the complex A-BC with fixed intermolecular coordinates R and $\gamma$. In the Condon approximation $V_{vv}^{nn'}$ is factored into the electronic matrix element $V^{n,n'}(r_e)$ and an overlap integral $S_{v,v'}^{n,n'}$

$$S_{v,v'}^{n,n'} = \int \chi_v^n(r) \, \chi_{v'}^{n'}(r)dr . \qquad (15)$$

This finally gives the equation for ENA contribution to $\langle P_{10}\rangle$:

$$\langle P_{10}^{ENA} \rangle = \left\langle \left| V(r_e)_{10}^{nn'} \right|^2 \frac{\Delta\ell^{nn'} \omega}{v_q} \right\rangle$$

where $\langle \ldots\ldots \rangle$ means now averaging over all resonant pairs, over all resonance positions, and over thermal distribution of velocities. It is clear that $S_{v,v'}^{n,n'}$ and thus $\langle P_{v,v}^{ENA} \rangle$ critically depend on relative position of close lying potential surfaces $U_n(R,\gamma,r)$, and this makes the application of

the ENA-VRT theory more difficult than the EA-VRT theory. As
we have seen, for the latter case rather limited information
on a potential surface is sufficient for the interpretation
of absolute values and temperature dependence of probability
of vibrational deactivation.

## SOME REPRESENTATIVE ENA-VRT PROCESSES

Not many processes showing anomalously fast deactivation
rates have been studied to the extent of knowing the temperature dependence of $<P_{10}>$. Most thoroughly investigated are
the processes cited below, with indication of degenerate
electronic state, temperature interval, and references to
corresponding papers.

$$NO(^2\Pi) + Ar \qquad (600\text{-}3000K) \qquad [22, 23] \qquad (17)$$

$$CO \quad + Fe(^5D) \qquad (1600\text{-}2350K) \qquad [24] \qquad (18)$$

$$N_2 \quad + O(^3P) \qquad (300\text{-}4500K) \qquad [25,26,27,28] \qquad (19)$$

$$O_2 \quad + O(^3P) \qquad (1600\text{-}3200K) \qquad [29, 30] \qquad (20)$$

$$CO \quad + O(^3P) \qquad (1200\text{-}4000K) \qquad [31] \qquad (21)$$

$$HCl \quad + O(^3P) \qquad (200\text{-}400K) \qquad [32, 33] \qquad (22)$$

$$DCl \quad + O(^3P) \qquad (200\text{-}400K) \qquad [32] \qquad (23)$$

The experimental data are shown in Fig. 1 by curves. We add
also three points corresponding to deactivation processes

$$HCl + Cl(^2P) \qquad (294K) \qquad [34] \qquad (24)$$

$$HF \quad + O(^3P) \qquad (300K) \qquad [35] \qquad (25)$$

$$HF \quad + F(^2P) \qquad (300K) \qquad [35] \qquad (26)$$

It is seen from Fig. 1 that for all these processes experimental values of $<P_{10}>$ are from two to six orders of magnitude larger than expected in terms of the EA-VRT theory. This
and also the weak temperature dependence of $<P_{10}>$ are strong
indications that the ENA-VRT mechanism is operative. This is
relatively less important at higher temperatures when all
curves tend to merge with the Landau-Teller strip. Of course,
one must consider the possibility that the removal of vibrational excitation proceeds _via_ a chemical reaction. Scrambling is certainly occurring in (20) and (21); and a recent
estimate shows that the rate of the process (20) can be
accounted for by formulation and decomposition of an intermediate complex $O_3^*$.[36] However, a previous estimate[21,37]
gives non-negligible contribution of the ENA-VRT transfer due
to the high statistical weight of terms correlating with the

excited states of ozone.  Deactivation of HCl and DCl may
occur in the course of reaction,[32,33] but there are strong
indications[38] that deactivation happens before the reaction.
The same arguments apply probably to deactivation of HF by O
atoms.  As for process (26) scrambling is rejected on the
grounds of the rather high activation energy.[39]

Processes (17), (18), and (19) certainly include no
"chemistry", and thus will be taken to illustrate the ENA-VRT
theory.

Though actual calculation require knowledge of potential
surfaces $U_n$, qualitative insight can be gained by constructing
correlation diagrams of electronic terms at fixed internuclear
distance in the BC molecule.  Consider the case NO + Ar.  At
$R \rightarrow \infty$ the $^2\Pi$ term of NO splits into two components $^2\Pi_{1/2}$ and
$^2\Pi_{3/2}$ with a fine-structure interval $\Delta\epsilon$ due to the spin-orbital coupling.  Now, if the interaction between Ar and NO is
taken into account it has no effect on splitting for a linear
arrangement of atoms (point group $C_{\infty v}$), but induces a change
for non-linear configurations.  When electrostatic interaction
contributing to additional splitting is stronger than spin-
orbital interaction two emerging states are classified as $^2A'$
and $^2A''$ (point group $C_s$), i.e., as states with free spins
(decoupled from electronic orbital motion).  The correlation
diagram for two states $^2A'$ and $^2A''$ is shown in Fig. 4 for
fixed interatomic distances $r_{NO}$.  The dashed curve corresponds
to the fulfilled condition of resonance $\omega = \Omega(R,\gamma)$.  It is
seen that at $\Delta\epsilon \ll \hbar\omega$ (which is the case for NO) the spin
orbital coupling can be neglected in correlation diagrams at
configurations corresponding to the ENA-VRT transitions.  It
is clear, also, that there is an energetically favorable configuration at which the resonance occurs with least interaction energy.  Detailed calculations can be found in Ref. 40.
We mention only, that the rather small value of $\langle P_{10} \rangle$, as
compared to that for process (19) is due to weak non-adiabatic
coupling (small value of $V^{nn'}(r_e) \approx \Delta\epsilon_{so}$), to poor overlap
($S_{10}^{nn'} \ll 1$) and appreciable activation energy.  As a result,
ENA contribution to the overall deactivation process becomes
negligible at high temperatures.

For $N_2$ + O collision the correlation diagram shows that
there will be $^3A''$, $^3A'$ and $^3A''$ terms emerging from $^1\Sigma_g^+(N_2)$
and $^3P(O)$ which correlate with $^3\Sigma^-$ and $^3\Pi$ states of the linear
system.  Confining, for simplicity, to this configuration of
atoms, we see that repulsive terms $^3\Sigma^-$ and $^3\Pi$ eventually cross
the attractive $^1\Sigma^+$ term dissociating into $N_2(^1\Sigma_g^+)$ and $O(^1D)$
(see Fig. 5).  The intersection of diabatic terms $^1\Sigma^+$ and $^3\Pi$,
which is known to be responsible for thermal decomposition of

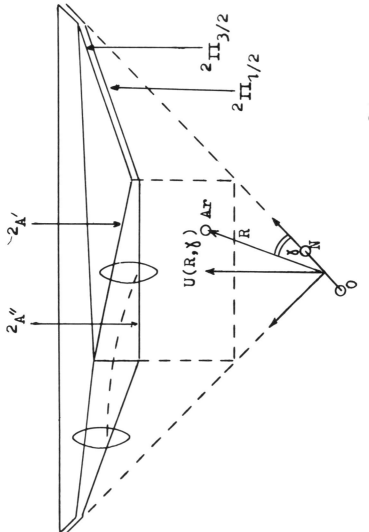

Fig. 4.   Correlation diagram of electronic terms for $NO(^2\Pi)$ + Ar system. Regions of ENA-VRT coupling are designated by ellipses.

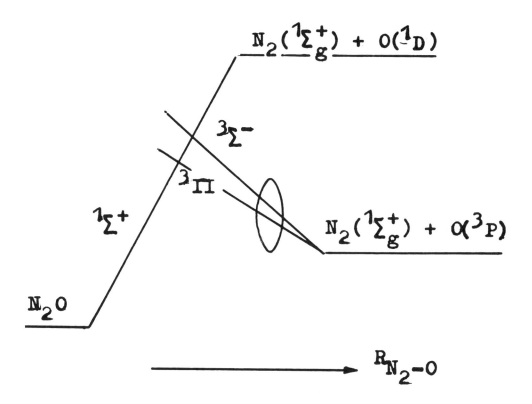

Fig. 5.  Correlation diagram of electronic terms for $N_2 + O(^3P)$ system for the linear arrangement of atoms.  Region of ENA-VRT coupling is designated by the ellipse.

$N_2O$[41] has been suggested[42] to provide an electronically non-adiabatic path or vibrational deactivation of $N_2$ by O.  We see, however, that there exists a much more energetically favorable path _via_ the ENA-VRT mechanism requiring less energy to reach the transition region.  The activation energy for the latter path is expected to be a fraction of $h\omega$ while for the former it would be about 1 e.v.  The Arrhenius plot of log $P_{10}$ versus 1/kT at low temperatures actually gives $E_a \approx 0.05$ e.v. which agrees with the ENA-VRT mechanism and rules out any appreciable contribution from the crossing of $^1\Sigma^+$ and $^3\Pi$ terms.  More details of calculation can be found in the original paper.[21,37]

REFERENCES

1.  C. B. Moore and P. F. Zittel, Science 182, 541 (1973).

2.  I.W.M. Smith, Accts. Chem. Res. (to be published).

3.  B. Stevens, Collisional Activation in Gases, Pergamon Press, 1967.

4.  D. Rapp and T. Kassal, Chem. Revs. 69, 61 (1969).

5.  Transfer and Storage of Energy by Molecules, Eds. G.M. Barnett and A. M. North, Vol 2; Vibrational Energy Transfer, Wiley-Interscience, London, 1969.

6.  D. Secrest, Ann. Rev., Phys. Chem. 24, 379 (1973).

7.  E. E. Nikitin, Theory of Elementary Atomic and Molecular Processes in Gases, Clarendon Press, Oxford, 1974.

8.  E. E. Nikitin, in Physical Chemistry: An Advanced Treatise, Vol. 6A, Ed. W. Jost, Academic Press, 1974, p. 187.

9.  V. N. Kondratiev and E. E. Nikitin, Kinetika i Mekhanizm Gazofaznykh Reaktzii, Moskva, Nauka, 1974.

10. L. Landau and E. Teller, Phys. Z. Sow. 10, 34 (1956).

11. C. B. Moore, J. Chem. Phys. 43, 2979 (1965).

12. E. E. Nikitin, Comm. At. and Mol. Phys. 2, 59 (1970).

13. R. N. Porter, D. L. Thompson, L. M. Raff and J. M. White, J. Chem. Phys. 62, 2429 (1975).

14. J. B. Anderson, J. Chem. Phys. 62, 2446 (1975).

15. W. H. Miller, J. Chem. Phys. 54, 5386 (1971).

16. H. Rabitz and G. Zarur, J. Chem. Phys. 61, 5076 (1974).

17. H. Rabitz and G. Zarur, J. Chem. Phys. 62, 1425 (1975).

18. R. C. Millikan and D. R. White, J. Chem. Phys. 39, 98 (1963).

19. R. C. Millikan and D. R. White, J. Chem. Phys. 39, 3209 (1963).

20. R. V. Steele and C. B. Moore, J. Chem. Phys. 60, 2794 (1974).

21. E. E. Nikitin and S. Ya. Umanski, Faraday Discuss. Chem. Soc. 53, 7 (1972).

22. G. Kamimoto and H. Matsui, J. Chem. Phys. 53, 3987 (1970).

23. J. C. Stephenson, J. Chem. Phys. 59, 1523 (1973).

24. C. von Rosenberg and K. L. Wray, J. Chem. Phys. 54, 1406 (1971).

25. W. D. Breshears and P. F. Bird, J. Chem. Phys. 48, 4768 (1968).

26. R. J. McNeal, M. E. Whitson and G. R. Cook, Chem. Phys. Lett. 16, 507 (1972).

27. D. J. Eckstrom, J. Chem. Phys. 59, 2787 (1973).

28. R. J. McNeal, M. E. Whitson and G. R. Cook, J. Geophys. Res. 79, 1527 (1974).

29. I. H. Keifer and R. W. Lutz, Symp. Combust. 11th, 67 (1967).

30. J. E. Breen, R. B. Quy and G. P. Glass, J. Chem. Phys. 59, 556 (1973).

31. R. E. Center, J. Chem. Phys. 58, 5230 (1973).

32. R.D.N. Brown, G. P. Glass and I.W.H. Smith, Chem. Phys. Lett. 32, 517 (1975).

33. D. Arnoldi and J. Wolfrum, Chem. Phys. Lett. 24, 234 (1974).

34. R. G. MacDonald, C. B. Moore, I.W.H. Smith and F. J. Wodarczyk, J. Chem. Phys. 62, 2934 (1975).

35. G. P. Quigley and G. J. Wolga, Chem. Phys. Lett. 27, 276 (1974).

36. M. Quack and J. Troe, Ber. Bunsenges. phys. Chem. 79, 170 (1975).

37. E. E. Nikitin and S. Ya. Umanski, Doklady Akademii Nauk SSSR 196, 145 (1971).

38.  D. Arnoldi, Dissertation, Georg-August Universität,
     Göttingen, 1975.

39.  S. V. O'Neil, H. F. Schaefer and C. F. Bender, Proc.
     Natl. Acad. Sci. USA $\underline{71}$, 104 (1974).

40.  A. A. Zembekov, S. Ya. Umanski, Khimiya Vysokikh Energii
     $\underline{7}$, 184 (1973).

41.  J. Troe and H. G. Wagner, Ann. Rev. Phys. Chem. $\underline{23}$, 311
     (1972).

42.  E. R. Fisher and E. Bauer, J. Chem. Phys. 1966 (1973).

# HEAVY PARTICLE COLLISIONS
# AT INTERMEDIATE ENERGIES

# SOME ASPECTS OF THE MOLECULAR APPROACH TO ATOMIC COLLISIONS

V. SIDIS[†]

Laboratoire des Collisions Atomiques

Université Paris-Sud, Bât. 220, 91405 Orsay, France

## 1. INTRODUCTION

### 1.1. Formal Analogy Between an Atom-Atom Collision System and a Diatomic Molecule

Let us consider a Quantum Chemist who has lately been involved in atomic collision problems. This man will soon find out that the study of an atom-atom collision problem can parallel that of a diatomic molecule in several respects since the two types of problems are *formally governed by the same hamiltonian.* Taking advantage of this analogy, he will first elect to treat the diatomic problem in a *molecular coordinate reference frame*[1] that rotates with the internuclear axis $R$ about the center of mass of the nuclei. Then, being familiar with the *Born Oppenheimer approximation*[2] in Molecular Physics, he will expand the total wave function of the system $\Psi(\{r\}, R)$ over a complete (orthonormal) basis set of functions $|\phi_n >$ that have the electronic coordinates $\{r\}$ and the internuclear separation R as variables :

$$\Psi(\{r\}, R) = \Sigma \int \chi_n(R) |\phi_n(\{r\}, R) > \qquad (1)$$

In analogy to the treatment of diatomic molecules he will further expand the functions $\chi_n$ over an orthonormal basis set of functions having as arguments the angular components of $R$ : $(\Theta, \Phi)$ ; the natural basis for him will be constructed from the *symmetric top wave functions.*[3] Once this is acheived, he will be able to bring the Schrödinger equation into the form of a set of *coupled equations* that determine the radial part of the $\chi_n$ functions. The resolution of these equations will enable

him to extract from the assymptotic behavior (R → ∞) of the
$\chi_n$'s the probability amplitudes for the states $|\phi_n(\{\underset{\sim}{r}\},R \to \infty)>$
when the nuclei are scattered in the direction $\Theta, \Phi$ : i.e. the
*scattering amplitudes*. For physical significance, and as well
in practice, to avoid convergence difficulties, he will require
the $|\phi_n (\{\underset{\sim}{r}\}, \infty) >$ to describe the non interacting eigenstates
of the electronic hamiltonian of the separated atoms. By
taking advantage of the symmetry properties of the various
operators appearing in the total hamiltonian he will generate
*symmetry adapted electronic functions* $|\phi_n >$. Our Quantum
Chemist already knows that by characterizing these functions
by the set of *molecular quantum numbers* $^{2S + 1}\Lambda^{\pm}_{g,u}(\Omega)$ [4] he will
be able to deduce a number of *selection rules*[5]. These will
helps him to determine which electronic transition is likely
to occur and will enable him to reduce somewhat the bulk of
his set of coupled equations. As to the explicit expression of
the many electron basis functions $|\phi_n >$ he is largely aware
of the important developments and progress that have been
acheived in Quantum Chemistry in the construction of various
types of orbitals involved in single configuration state
functions (CSF) and he knows much about configuration interaction
(CI) and multiconfiguration (MC) techniques...[6] Whatever the
computation facilities that can be made available for him, our
Quantum Chemist will be confronted by the problem of truncating
his basis set of $| \phi_n >$'s and he will look for some way of
minimizing the error produced by the neglect of a large number
of states. Turning to the literature he will find nowhere a
rigorous and general formulation of such a *minimum principle*.
At this stage he will rapidly discover that this situation
is due to the specific characteristics of atom-atom collision
problems. Indeed, contrary to the case of *bound states of
diatomic molecules* where the slow nuclear motion is confined
in a small range  of internuclear separations around the
equilibrium distance Re, in a collision problem the nuclei can
evolve in a *wide range of R*

$$0 \underset{\sim}{\leqslant} R_T \leqslant R \leqslant \infty \qquad (R_T \text{ being the classical turning point})$$

on a wide velocity scale and they can be found in a large
number of rotational states (a few tens to several thousands
of partial waves may be required to describe a collision
process). Let us then examine with our Quantum Chemist these
specific characteristics of atom-atom collision problems.


1.2. Nuclear Velocity and Spatial Extension of the Nuclear
Motion.


Qualitatively, at relative nuclear velocities much slower
than the electron velocities (say in a valence shell), a

molecular aspect appears and the atoms remain close to each other for sufficiently long to allow for the use of the Born Oppenheimer approximation and the related *variational procedures* on the eigensolutions of the electronic part of the hamiltonian : $H_{el}$. At these low nuclear velocities the system continuously adjusts (adiabatic behavior) to the R-varying interaction and the atoms remain on *one meaningful state* throughout the whole collision. This is just in the spirit of the original formulation of the Born–Oppenheimer approximation which thus only allows the treatment of purely *elastic scattering* at low energy[7]. This approximation successfully accounts for elastic rainbow scattering in e.g. the $H^+$-He, Ne, Ar[8] collisions as well as resonant charge transfer in symmetric systems at low energy.

On the other hand, at the largest nuclear velocities the molecular aspect is completely lost since the atoms may be considered as being submitted to a sudden perturbation that allows for a treatment based on *undistorted atomic states* (as e.g. the time dependent treatment of coulomb ionization[9]).

It is already seen that the study of the electronic part of an atomic collision problem, at moderate energies, is an awkward task  since the motion of the nuclei is sufficiently slow for a molecular aspect to manifest itself (adiabatic behavior), and it is sufficiently rapid to preserve some *atomic* characters of the system (non adiabatic behavior).

Moreover the probelm is complicated by the spatial extension of the nuclear motion. Indeed, in the course of an atom–atom encounter the system may traverse different regions characterized by the relative importance of the molecular axial field, the effects of the internuclear axis rotation and the spin orbit interactions. (fig. 1)

In each of these regions, appropriate states that minimize the static ($H_{el}$, $H_{so}$) or the rotational $v_0 b L_y / R^2$ coupling terms can be looked for. However, when going from one region to the other, the proper basis changes will cause mixings among the unperturbed states ; this may ultimately lead to electronic transitions. Of course, the accessible range of R can still be divided into a larger number of small domains and the system may be viewed as undergoing several step by step transitions[10]. Resonant charge exchange[11] and the Demkov[12] model illustrate this process when the system passes from the *atomic* region of interaction to the *molecular* region.

We will next examine some particular cases which demonstrate why the continuously adjusting adiabatic states do not always provide a convenient basis.

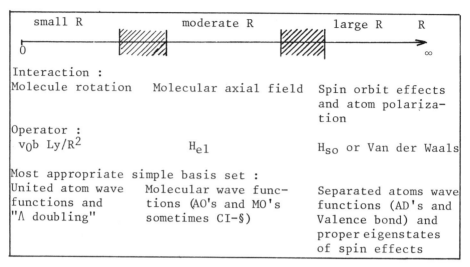

Fig. 1 : Schematic overall view of the different regions of interaction that can be traversed in the course of a collision. In the shaded regions the different interactions have the same importance and adiabatic states vary rapidly (effect of the $v_R$ d/dR operator).

## 2. SIMPLE EXAMPLES OF BASIS SET CHOICES. ADIABATIC STATES HAVING THE SAME SYMMETRY AND DIABATIC STATES HAVING UNDERLYING DIFFERENT SYMMETRIES.

### 2.1. Spin Orbit Coupling.

Let us take a system AB which at $R \to \infty$ is described by :

$$A(^2S_{1/2}) + B(^2P_{3/2} \text{ or } ^2P_{1/2})$$  (2)

(as $H + F$[13], $H + Kr^{+1}$[14], $I + Br$[15]) and consider only the two states (indexed 1,2) that come out from such a system and that have a zero projection of the total electronic angular momentum on the internuclear axis ($\Omega = 0^+$).[16] Three types of basis sets may be chosen (fig. 2).

(i) *At moderate R* the spin effects ($H_{so}$) are negligible in comparison to the axial field of the molecule ($H_{el}$) and one may then choose the two states :

$$|\phi_1 > = |^1\Sigma^+ >, \quad |\phi_2 > = |^3\Pi >$$  (3)

that diagonalize the $H_{el}$ operator. The coupling between these states, which is induced by the off diagonal matrix element of the $H_{so}$ operator, is such :

$$| <{}^1\Sigma^+ |H_{so}|{}^3\Pi > | << |E_{3\Pi} - E_{1\Sigma^+}| \qquad (4)$$

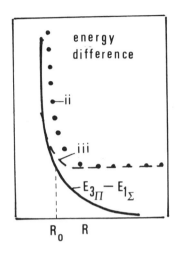

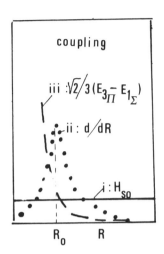

Potential energy difference                      couplings

Fig. 2 : Scheme of the potential energy differences and couplings corresponding to the different basis sets that can be chosen in the case of spin-orbit coupling.


(ii) Alternatively one may choose *for all R* the linear combinations :

$$|\phi_1'> = c(R)|{}^1\Sigma^+> + s(R)|{}^3\Pi>$$
$$|\phi_2'> = -s(R)|{}^1\Sigma^+> + c(R)|{}^3\Pi > \qquad (5)$$

that simultaneously diagonalize $H_{el} + H_{so}$. Nevertheless these states are coupled by the *radial matrix element* :

$$<\phi_1'|d/dR|\phi_2'> = -c^{-1}ds/dR \qquad (6)$$

This coupling *has a bell-shaped form* that usually maximizes around the distance $R_0$ where :

$$|E_{3\Pi} - E_{1\Sigma^+}| \approx 2|<{}^1\Sigma^+|H_{so}|{}^3\Pi >| \qquad (7)$$

This situation occurs at relatively large R (e.g. $R_0 \sim 8a_0$ in H Kr$^{+14}$). The radial matrix element is more or less peaked according to whether the energy difference $E_{3\Pi}-E_{1\Sigma^+}$ varies more or less rapidly. Since in a semi-classical treatment[17] the above d/dR matrix element is weighted by the radial nuclear velocity it may become important at moderate energies (far from the classical turning points) and will thus induce transitions between the two states considered. However this kind of coupling is somewhat cumbersome because :

- when a sharp peaking occurs, it requires accurate calculations of a large number of the c(R) or s(R) coefficients and it causes practical difficulties in solving the coupled equations for the nuclear motion.

- in a complete quantal treatment of the electronic transition induced by such a coupling one must also calculate the $<\phi_1'|d^2/dR^2|\phi_2'>$ matrix element since only the operator :

$$D = <\phi_1'|d^2/dR^2|\phi_2'> + 2<\phi_1'|d/dR|\phi_2'> d/dR \qquad (8)$$

is hermitian (i.e. $\int_0^\infty \chi_1 D\chi_2 \, dR = \int_0^\infty \chi_2 D\chi_1 \, dR$) (this difficulty does not appear in the semi-classical treatment)[17].

(iii) A third basis may be found *at large internuclear distance* when the total angular momentum J of the atoms is an almost good quantum number. This basis is constructed from the linear combinations of the $^1\Sigma^+$ and $^3\Pi$ states that diagonalize the $H_{so}$ operator at $R \to \infty$

$$|\tilde\phi_1, J_B = 3/2 > = 3^{-1/2}(\sqrt{2}|^1\Sigma^+ > + |^3\Pi >)$$
$$|\tilde\phi_2, J_B = 1/2 > = 3^{-1/2}(-|^1\Sigma^+ > + \sqrt{2}|^3\Pi>) \qquad (9)$$

At large R these states are weakly coupled by the samll energy difference between the $^1\Sigma^+$ and $^3\Pi$ states

$$|<\tilde\phi_1|H_{el}|\tilde\phi_2>| = \frac{\sqrt{2}}{3}|E_{3\Pi} - E_{1\Sigma^+}| \qquad (10)$$

The states of basis (i) or those of basis (iii) display *different symmetries*, related to the *almost good quantum numbers* $2S + 1\Lambda^\pm$ or J respectively. Such states are more convenient to use than the adiabatic states of basis (ii), except however, at the lowest nuclear velocities when the radial coupling induces no transitions at all. The 2 x 2 subspaces (i) and (iii) provide *diabatic states* in the sense defined by Smith[18]. These states vary with R and do not suffer from the criticisms[19] that have been raised against that definition.

Outside the critical region around $R_0$, of eq.(7), one may speak of the motion of the system on meaningful   potentials that correspond to weakly  interacting states. From these potentials it is possible to know "what are the actual forces acting on the nuclei"[7] and one is able to derive the semi classical phases of interest in studying the interference patterns[20] created at the transition region around $R_0$. To treat these transitions the use of basis (i) or (iii) is equally appropriate. However, if the set (i) is used, a basis change in the  asymptotic region (where $<{}^1\Sigma^+|H_{so}|{}^3\Pi>$ is constant) is required to recover the proper  asymptotic condition (in the "atomic" region).

## 2.2. Rotational Coupling

As a second example let us consider rotational coupling between two states, say $\Sigma^+$ and $\Pi$ that are widely separated at large R but approach each other closely with decreasing R (e.g. the $2p\sigma$-$2p\pi$ coupling in the $H^+$-H collision). Two sorts of basis sets may be chosen fig.3a

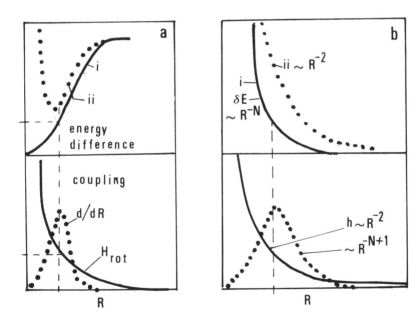

Fig.3 : Scheme of the potential energy differences and couplings in basis sets (i) and (ii) for rotational coupling at small R (a) and for rotational coupling at large R (b).

(i) the familiar states

$$|\phi_1> = |\Sigma^+> \quad \text{and} \quad |\phi_2> = |\Pi> \qquad (11)$$

that diagonalize the $H_{el}$ and $d/dR$ operators are coupled by the matrix element[17]

$$h = <\Sigma^+|H_{rot}|\Pi> = <\Sigma^+|L_y|\Pi> v_o b R^{-2} \qquad (12)$$

(ii) the adiabatic "$\Lambda$-doubling states" that diagonalize $H_{el}$ + $H_{rot}$ for every R

$$|\phi_1'> = c(R)|\Sigma^+> + s(R)|\Pi>, \quad |\phi_2'> = -s(R)|\Sigma^+> + c(R)|\Pi> \qquad (13)$$

In this case the states are coupled by the $d/dR$ operator[21]

$$<\phi_1'|d/dR|\phi_2'> = (hd\delta E/dR - \delta Edh/dR) / (\delta E^2 + 4h^2) \qquad (14)$$

where $\delta E = E_{\Sigma^+} - E_\Pi$. This basis is somewhat analogous to basis (ii) of §2.1. Again for the same reasons as those stated there, the $d/dR$ matrix element is difficult to handle and basis (i) is preferable.

Basis (i) has been widely used in most treatments of rotational coupling close to the united atom limit and to my knowledge nobody has ever claimed that the adiabatic states of basis (ii) should be used instead. Again, the pair of $\Sigma^+$ and $\Pi$ states obviously fulfil the diabaticity requirement $<\Sigma^+|d/dR|\Pi> \equiv 0$[18].

As a third example let us consider the opposite case where one considers a $\Sigma^+$ and a $\Pi$ state that are widely separated at small R but tend to the same level of the separated atoms as $R \to \infty$ (fig 3b). Here, in basis (i), it is assumed that *at small and moderate R* the rotational coupling h is negligible as compared to the energy separation $\delta E$. However, *at large R*, this coupling usually decreases slower than $\delta E$ (i.e. $h \sim R^{-2}$, $\delta E \sim R^{-N}$, $N \geqslant 3$)[22]. Thus it would seem more suitable to use basis (ii) at a sufficiently large R since with this basis

$$<\phi_1'|H_{el} + H_{rot}|\phi_1'> - <\phi_2'|H_{el} + H_{rot}|\phi_2'> \sim R^{-2} \qquad (15)$$

whereas the $d/dR$ coupling is of more shorter range

$$<\phi_1'|d/dR|\phi_2'> \sim R^{-(N-1)}, \quad N > 3 \qquad (16)$$

Actually in solving coupled equations, the two basis sets are equivalent since the integration must be carried out up to a sufficiently large R where the long range term $R^{-2}$, in either the potentials of basis (ii) or the coupling of basis (i) vanishes.

### 3. ON THE   TRACK OF DIABATIC STATES

In the above examples it has been shown to some approximation that two states ($^1\Sigma^+$, $^3\Pi$ or $^2\Sigma^+$, $^2\Pi$) may become degenerate (in the limits $R \to 0$, $R \to \infty$) or cross (finite R value). These degeneracies are allowed by the Wigner-Von Neumann (WVN) theorem[23] since the states display different symmetries that cancel out the d/dR matrix elements. When the remaining interaction between the states (e.g. $H_{so}$ or $H_{rot}$) is taken into account and is diagonalized, the degeneracies are lifted up and the resulting "adiabatic" states (of the same symmetry) may be strongly coupled by the d/dR operator. Analogous diabatic subspaces can be looked for when the states have the same spatial and spin symmetries.

### 3.1. Quasi-Diabatic States and Diabatic II Crossings

Consider the hamiltonian of a many electron system that has been cut into two terms :

$$H_{el} = H_0 + V \; ; \quad H_0 = \sum_i F_i = \sum_i (- \frac{1}{2} \Delta_i - \frac{Z_A}{r_{Ai}} - \frac{Z_B}{r_{Bi}} + w_i) \quad (17)$$

$$V = \sum_i - w_i + \sum_{j<i} 1/r_{ij}$$

Since $H_0$ appears as a sum over one-electron operators $F_i$, its eigenfunctions are antisymmetrized products (or Slater determinants $\phi$) of orbitals ($\gamma_i$). The eigenvalues ($\varepsilon_i$) of the $F_i$'s can *be called orbital energies.* The d/dR operator acting on such an antisymmetrized product affects one orbital at the time and can thus be considered as a sum over one electron operators $\sum (d/dR)_i$. Following Slater rules[24], if the orbitals are orthonormal, *the d/dR operator has no matrix element between two determinants $\phi$ and $\phi''$ that differ by at least two orbitals* $\gamma_1 \neq \gamma_1''$, $\gamma_2 \neq \gamma_2''$.

According to the WVN the two *configuration states,* $\phi$ and $\phi''$, are not prevented from crossing, even when they have the same symmetry. The two states indeed cancel the ($\partial H_0 / \partial R$) operator for all R as given by the Hellman-Feynman[25] theorem for the eigenstates of $H_0$ :

$$(\langle\phi|H_0|\phi\rangle - \langle\phi''|H_0|\phi''\rangle) \; \langle\phi''|\partial/\partial R|\phi\rangle = \langle\phi''|(\frac{\partial H_0}{\partial R})|\phi\rangle \quad (18)$$

The energy difference between the two states,

$$\langle\phi|H_0|\phi\rangle - \langle\phi''|H_0|\phi''\rangle = (\varepsilon_1 + \varepsilon_2) - (\varepsilon_1'' + \varepsilon_2'') \qquad (19)$$

can vanish when the orbital energies assume the behavior schematized in fig. 4.

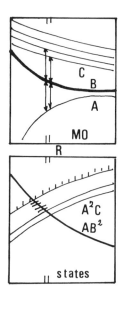

This situation always occurs in rare gas$^+$ + rare gas systems[26], [27,28,29,30] and gives rise to a set of crossings called DIABATIC II[31] between a *core excited state* and a *Rydberg series* e.g.

$1s\sigma_g 2p\sigma_u^2 - 1s\sigma_g^2 nl\sigma_g$ in $He_2^+$

$2p\sigma\,3d\sigma^2 - 2p\sigma^2\,nl\sigma$ in $HeNe^+$

$3p\sigma\,3d\sigma^2 - 3p\sigma^2\,nl\sigma$ in $HeAr^+$

$3s\sigma 3p\sigma^2\,3d\sigma^2 - 3s\sigma^2\,3p\sigma^2 nl\sigma$ in $HeAr^+$

Analogous examples can be found in Penning ionization problems (e.g.[32] $He^*(1s2s)$ + H → $He(1s^2)$ + $H^+$ : $\sigma\sigma'\sigma''-\sigma\,^2\Phi$), homogeneous perturbations in Molecular Spectroscopy[33] ($1\pi^3 2\pi^2 - 1\pi^4 3p\pi$ in NO) as well as in dissociative attachment problems[34].

Actually the potential curves are the mean values of the above $H_{el}$ operator. These potentials will still exhibit a curve crossing when the effect of the corrective terms $\langle\phi|V|\phi\rangle$ and $\langle\phi''|V|\phi''\rangle$ results only in slight shifts of the levels $\langle\phi|H_0|\phi\rangle$ and $\langle\phi''|H_0|\phi''\rangle$ respectively. In consequence the corrective term V should only play the role of a small perturbation. In practice, an adequate construction of $H_0$ and V may be realized by using the Hartree-Fock-Roothaan procedure [35] or any other judicious choice of one-electron pseudo potentials[36]. The only interaction between the states that differ by two orbitals is readily seen to come out from the *electrostatic interelectronic repulsions* $1/r_{ij}$ (electronic correlation).

Comparison between the couplings in the adiabatic and diabatic basis for $He_2^+$. In the $He_2^+$ system for the $1s\sigma_g 2p\sigma_u^2 - 1s\sigma_g^2 2s\sigma_g^2 2s\sigma_g$ diabatic crossing at $R_0 \simeq 1.5\ a_0$ one has : $\langle\phi|V|\phi''\rangle \simeq 2.75\,10^{-2}$ Hartree[37]. Using formula (14) with h = $\langle\phi|V|\phi''\rangle$ and $d\delta/dR|_{R_0} \simeq 1.65$ a.u[38] one obtains for the adiabatic states, that are not allowed to cross : $\langle\psi_1^a|d/dR|\psi_2^a\rangle \simeq 15.4\ a_0^{-1}$. This comparison shows that for all radial velocities $v_R$ such

$$v_R < \psi_1^a|d/dR|\psi_2^a > \gtrsim < \phi|V|\phi''> \qquad (20)$$

(i.e. for $v_R > 2.10^{-3}$ a.u. corresponding to a center of mass
energy of 0.2 eV) the diabatic basis has to be preferred
Indeed it allows to describe the system as undergoing elastic
scattering slightly perturbed by inelastic transitions[38]. That
justifies the use of two state approximations. On the contrast,
in the adiabatic basis, the important radial coupling has to
be considered at each pseudo crossing in the Rydberg series,
fig. 5b. One is then obliged to take the whole set of Rydberg
states as well as its associated continuum in order to simply
deal with elastic scattering and resonant charge exchange.

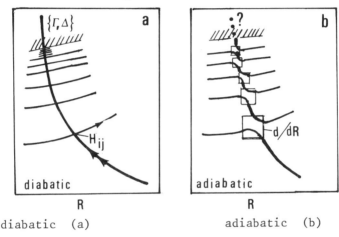

diabatic (a)                                          adiabatic (b)

Fig. 5 : At each crossing between diabatic curves schematized
in (a) the states are weakly coupled ; that justifies two
state approximations. At each pseudo crossing in the adiabatic
basis schematized in (b) the states are strongly coupled which
necessitates  the use of an infinite number of states.


## 3.2. Quasi-Diabatic States and Orbitals of Different Symmetries. Simplest Examples of Diabatic I Crossings

According to the WVN theorem when two orbitals *have
different symmetries* (e.g. $\sigma$-$\pi$  or  $\sigma_u$-$\sigma_g$...) their associated
energies $\varepsilon(R)$ are not prevented from crossing. If such an MO
crossing occurs, equation (19) shows that the energy curves of
the many-electron configuration states which differ by only
these orbitals also cross (for instance $(...\sigma_g^2)$ $^1\Sigma_g^+$ $-(...\sigma_u^2)$
$^1\Sigma_g^+$ or $(...\sigma^2)^1\Sigma^+-(...\pi^2)^1\Sigma^+)$. Examples of these crossings
called DIABATIC I[31] appear in the neutral symmetric rare-gas
combinations He$_2$[39], Ne$_2$,  Ar$_2$[31] as seen in fig. 6.

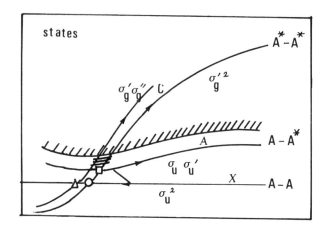

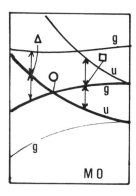

Fig. 6 :
Schematic picture of the behaviour
of some potential energy curves and MO
energies in the Ne$_2$ and Ar$_2$ systems :
○  Diabatic I crossing of potential
    energy curves reflecting the $\sigma_g$–$\sigma_u$
    MO crossing.

□  diabatic II crossings of section 3.1

△  diabatic II crossings generated by
   $\sigma_g$–$\sigma_u$ MO crossing.

Transition between such states are
again induced by the $1/r_{ij}$ interaction.

## 3.3. Quasi-Diabatic Subspaces and Incorrect Dissociation at Large R.

A drawback to the use of single configuration states built
from MO's is that in a number of cases they tend at large R
to interacting mixtures of separated atomic states[27,31,39,40].
This prevents from having a correct description of the proper
assymptotic conditions. It necessitates as well the consideration of a larger number of states than was necessary at
small and moderate R to describe a primary excitation mechanism
fig. 7.

As already mentioned in the case of diabatic states for
the spin orbit coupling, one may use the set of single configuration states up to a sufficiently large R, where the coupling
have stabilized to a constant value. A basis change to the
decoupled basis set of proper atomic states[40] has then to be

carried out.

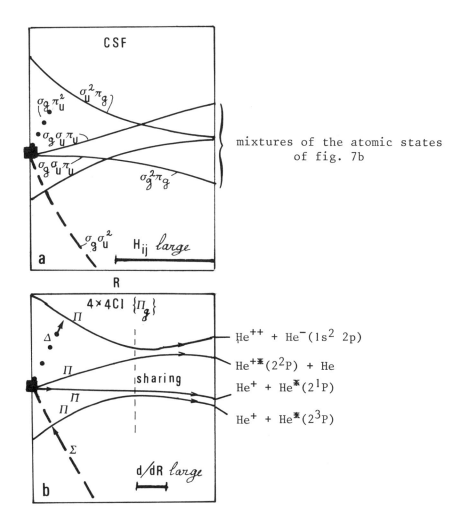

Fig. 7 : Scheme of the potential energy curves of the He$_2^+$ system in the single configuration description (a) (quasi-diabatic states) and after limited CI(b). In the diabatic basis the $<i|H_{el}|j>$ matrix elements are important at large R whereas in the adiabatic basis the $<i|d/dR|j>$ matrix elements peak between the molecular and atomic interaction regions.

This kind of basis change from a coupled diabatic molecular basis to an atomic uncoupled basis accounts for a population sharing between atomic states. This has been observed in the theoretical study of the process[40]

$$He^+ + He(1s^2) \rightarrow He^+ + He(1s\ 2p\ ^1P\ or\ ^3P)$$

This again shows that in addition to the molecular diabatic basis already discussed, there exists, at large R, another type of diabatic basis built up from simple valence bond type functions. This is just what has led O'Malley to suggest his "projected atomic orbitals" method[34]. This procedure, which consists in the use of both the LCAO-MO method and the valence bond approach, is presently used in the investigation of the near resonant charge transfer in $H^+ + Kr$ collisions[41]

## 3.4. An Extension of Quasi-Diabatic and an Insight to Diabatic III Crossings.

Coming back to single configuration states built from MO's, equation (19) shows that two states $|\phi>$ and $|\phi'>$ that differ by only one orbital $\gamma \neq \gamma'$ will cross when the associated orbital energies cross : $\varepsilon_\gamma = \varepsilon_\gamma'$. But according to the WVN theorem, orbitals having the same symmetry do not cross, in general. However, it has been found in the $(HeNe)^+$ system[29] that the states describing the configurations $(3\sigma 4\sigma n\sigma)^2\Sigma^+$ and $(3\sigma^2\ n\sigma)^2\Sigma^+$ do lead to a potential curve crossing, even though the $3\sigma$ and $4\sigma$ orbitals do not cross. In this case the crossing is produced by the diagonal matrix elements of the corrective operator V in equation (17) :

$$<\phi|H_{el}|\phi> - <\phi'|H_{el}|\phi'> = \varepsilon_{3\sigma} + <\phi|V|\phi> - (\varepsilon_{4\sigma} + <\phi'|V|\phi'>) \tag{21}$$

This kind of accidental crossing has been termed diabatic III[29]. It is also worth mentioning that of the two $^2\Sigma^+$ states that arise from the $(3\sigma 4\sigma n\sigma)$ configuration, the one which has a triplet *core* $(3\sigma 4\sigma)^3\Sigma^+\ n\sigma$ differs from the $(3\sigma^2\ n\sigma)^2\Sigma^+$ state by *two spin orbitals* and this is again an extension of quasi diabatic states.

## 3.5. Diabatic Orbitals : an Open Problem.

Apart from the previous case, the results of Hartree-Fock calculations on diatomic systems often show the existence of avoided crossings between orbitals that have the same symmetry.

Avoided crossings between adiabatic MO's are found when one studies quasi-symmetric systems. In these asymmetric species, the vanishing of the inversion symmetry lifts the u-g degeneracies[42]. Also, when one goes from a one electron system to a many electron system the particular symmetry of the pure two-center coulomb field[43] disappears and some crossings allowed in the one electron systems[44] are seen to be avoided in the many electron case. At these pseudo crossings the characters of the MO's may display a rapid variation with the internuclear distance. As a consequence the states built up from such MO's will display huge d/dR couplings. The situation is then analogous to those encountered in the above simple cases and there is again a need of states that conserve some characters as the internuclear distance varies. These states should thus involve slowly varying *diabatic orbitals* that run smoothly through the avoided crossings displayed by Hartree Fock MO's. The resulting crossings will be called diabatic $I^{31}$ as in § 3.2. However, the explicit determination of diabatic MO's remains as yet an open problem, although some attempts have been made to deal with it. To date, the various known methods to generate such orbitals can be classified as follows :

(i) Constant orbitals determined outside of the pseudo-crossing region ($R_c$), as is done in the valence bond approach ; but instead of using undistorted atomic orbitals, one uses the MO's obtained at a distance $R_c \pm \delta R$.

(ii) Orbitals obtained by removing from an LCAO (linear combination of atomic orbitals) expansion all basis orbitals that could cause a given MO to have a rapid change of character. An example can be found in the $Ar_2$ and $Ar_2^+$ systems[27] where, to describe the $\sigma_u$ 3p → $5f\sigma_u$ promotion, Rydberg orbitals have been deleted from the expansion basis. This is in the spirit of O'Malley's earlier suggestion of using Feschbach operators[45].

(iii) Orbitals obtained by a suitable R-varying linear combination of Hartree-Fock orbitals (adiabatic MO's) that minimizes the rapidly varying d/dR matrix elements as suggested by Smith[18] and realized by Taulbjerg and Briggs[42]. However this is an onerous procedure since it first consists in evaluating the d/dR matrix elements and then in looking for the proper basis change.

(iv) Node conserving orbitals obtained from a first order pseudo-potential separable in prolate spheroïdal coordinates as provided by the S.D.O. method proposed by Aubert[46] and used by Lesech and Aubert[47]. (Some nonorthogonality problems have yet to be solved before using this very promising method for systems that involve several electrons).

The latter method is the closest to the Barat-Lichten's[48] extension of the Fano-Lichten[50] promotion model. The success of this model has repeatidly been demonstrated in the qualitative interpretations of inner shell excitation phenomena where the interaction  is mainly governed by the two Center Coulomb field of the nuclei. It also works successfully for high lying Rydberg states which are submitted to an almost $H_2^+$-like field[49]. However, for valence shells, at large and moderate R, the validity of this approach is doubtful as it has recently appeared in the study of the He-Ar[51] and He-Ar$^+$[30] systems. Also some difficulties may have been encountered by the users of this model when     choosing the orbital energies of the separated atoms. Indeed, these orbital  energy levels may be very different in the initial and final states of the collision process and they can be very different in neutral and ionic systems[52].

Furthermore, even in the case of a pure two center coulomb field, avoided crossings between orbitals that have the same $(n_\xi)$ number of nodes are observed[44,53]. This situation, which is at the origin of the swapping of correlations in the Barat-Lichten[48] rules, can cause a rapid variation of the MO's with R[53]. *Thus even in the simplest systems the problem of defining diabatic MO's remains open.*

ACKNOWLEDGMENTS : The author would like to thank Dr. M. Barat and J.P. Gauyacq for their useful suggestions and their critical reading of the manuscript.

†Centre de Mécanique Ondulatoire Appliquée, 23 rue du Maroc 75019 Paris, France.

1. W.R. Thorson, J.Chem.Phys. 34, 1744 (1961)
2. M. Born and J.R. Oppenheimer, Ann. Phys. 84, 457 (1927)
3. M.E. Rose, Elementary Theory of Angular Momentum, Wiley (Interscience), New York, (1957)
4. G.H. Herzberg, Spectra of Diatomic Molecules, 2nd Ed., Van Nostrand, Princeton, New Jersey
5. H. Lefebvre-Brion, Winter College on Atoms, Molecules and Lasers  Trieste (Italy), (1973)
6. J.C. Browne, Advances in Atomic and Molecular Physics, Ed. D.R. Bates and I. Esterman, 7, 47 (1971)
   H.F. Schaefer, The Electronic Structure of Atoms and Molecules : A Survey of Rigourous Quantum Mechanical Results, Addison-Wesley, Reading, (1972)

7. T.F. O'Malley, Advances in Atomic and Molecular Physics, Ed. D.R. Bates and I. Esterman, 7, 223 (1971)
8. R.L. Champion, L.D. Doverspike, W.G. Rich and S.M. Bobbio, Phys. Rev. A, 2, 2327 (1970) and 4, 2253 (1971)
   H.U. Mittmann, H.P. Weise, A. Ding and A. Henglein, Z. Naturf., 26a, 1112 (1971)
   V. Sidis, J. Phys. B : Atom. Molec. Phys. 5, 1517 (1972)
9. N.F. Mott and H.S.W. Massey, The Theory of Atomic Collisions, 3rd Ed. Oxford Univ. Press, Great Britain, p. 794
10. R.G. Gordon, J. Chem. Phys. 51, 14 (1969) and in Meth. Comput. Phys. 10, 81 (1969)
11. W. Lichten, Phys. Rev. 131, 229 (1963)
12. Yu. N. Demkov, Sov. Phys. JETP 18, 138 (1964)
13. F.H. Mies, Phys. Rev. A 7, 942 (1973)
14. C. Benoit, C. Kubach, V. Sidis, J. Pommier and M. Barat, in this Conference.
15. G.A.L. Delvigne and J. Los., Physica 67, 166 (1973)
16. L.Landau and E. Lifschitz, Mecanique Quantique, Ed. Mir (Moscou), §85, (1967)
17. R. McCarroll, VIII ICPEAC, Invited Lectures and Progress Reports, Ed. B.C. Cobic and M.V. Kurepa, (Beograd - Yugoslavia), 71 (1973)
18. F.T. Smith, Phys. Rev. 179, 111 (1969)
19. B. Andresen and S.E. Nielsen, Molec. Phys. 21, 523 (1971)
   H. Gabriel and K. Taulbjerg, Phys. Rev. A 10, 741 (1974)
20. M. Barat, VIII ICPEAC, Invited Lectures and Progress Reports, Ed. B.C. Cobic and M.V. Kurepa, (Beograd-Yugoslavia), 44 (1973)
21. R.D. Levine, B.R. Johnson and R.B. Bernstein, J. Chem. Phys. 50, 1694 (1969)
   M. Oppenheimer, J. Chem. Phys. 57, 3899 (1972)
22. L. Landau and E. Lifschitz, Mecanique Quantique, Ed. Mir (Moscou), §89, (1967)
23. J. Von Neumann and E.P. Wigner, Physik Z., 30, 467 (1929)
   E. Teller, J. Chem. Phys. 41, 109 (1936)
   K.R. Naqvi and W.B. Brown, Int. J. Quant. Chem. 6, 271 (1972)
24. J.C. Slater, Quantum Theory of Molecules and Solids, 1, Mc Graw-Hill, New York
25. R.P. Feynman, Phys. Rev. 56, 340 (1030)
   W. Hobey and A.D. Mc Lachlau, J. Chem. Phys. 33, 1577 (1960)
26. W. Lichten, Phys. Rev. 131, 229 (1963)
   V. Sidis, J. Phys. B : Atom. Molec. Phys. 6, 1188 (1973)
27. V. Sidis, M. Barat and D. Dhuicq, J. Phys. B : Atom. Molec. Phys. 8, 474 (1975)
28. J.P. Gauyacq, work in progress
29. V. Sidis and H. Lefebvre-Brion, J. Phys. B : Atom. Molec. Phys. 4, 1040 (1971)
   M. Barat, D. Dhuicq, J.C. Brenot, J. Pommier, V. Sidis,

R.E. Olson, E.J. Shipsey and J.C. Browne, to be published

30. V. Sidis, M. Barat, J.C. Brenot, J. Pommier, O. Bernardini, D.C. Lorents and F.T. Smith, to be published (see also in this conference)

31. J.C. Brenot, D. Dhuicq, J.P. Gauyacq, J. Pommier, V. Sidis, M. Barat and E. Pollack, Phys. Rev. A $\underline{11}$, 1245 (1975)

32. W.H. Miller, C.A. Slocomb and H.F. Schaefer III, J. Chem. Phys. $\underline{56}$, 1347 (1972)

33. P. Felenbok and H. Lefebvre-Brion, Can. J. Phys. $\underline{44}$, 1677 (1966)

34. T.F. O'Malley, J. Chem. Phys. $\underline{51}$, 322 (1969)

35. C.C.J. Roothaan, Revs. Mod. Phys. $\underline{23}$, 69 (1951) and $\underline{32}$, 179 (1960)

36. J. Eichler and U. Wille, Phys. Rev. A $\underline{11}$, 1973 (1975)

37. J.N. Bardsley, Phys. Rev. A $\underline{3}$, 1317 (1971)

38. R.E. Olson, Phys. Rev. A $\underline{5}$, 2094 (1972)

39. G.H. Matsumoto, C.F. Bender and E.R. Davidson, J. Chem. Phys. $\underline{46}$, 402 (1967)
    D.R. Yarkony and H.F. Schaefer III, J. Chem. Phys. $\underline{61}$, 4921 (1974)
    J.P. Gauyacq, to be published (see also in this conference)

40. B. Stern, J.P. Gauyacq and V. Sidis, to be published (see also in this conference)

41. C. Kubach and V. Sidis, work in progress

42. K. Taulbjerg and J.S. Briggs, to be published

43. C.A. Coulson and A. Joseph, Int. J. Quant. Chem., $\underline{1}$, 337 (1967)
    S.P. Alliluev and A.V. Matveenko, Sov. Phys. JETP $\underline{24}$, 1260 (1967)

44. J.D. Power, Phil. Trans. Roy. Soc. of London $\underline{274}$, 663 (1972)
    M. Kotani, K. Ohno and K. Kayama, Handbook der Physik, Molecules II, Ed. S. Flugge, Springer Verlag (Berlin), XXXVII/2, 58 (1961)

45. T.F. O'Malley, Phys. Rev. $\underline{162}$, 98 (1967)

46. M. Aubert, N. Bessis and G. Bessis, Phys. Rev. A $\underline{10}$, 51 (1974) and A $\underline{10}$, 61 (1974)

47. M. Aubert and C. Lesech, to be published

48. M. Barat and W. Lichten, Phys. Rev. A $\underline{6}$, 211 (1972)

49. P. Lefebure, Thèse de Doctorat d'Etat, Université Paris VI (1975)
    J. Fayeton, M. Barat, J.C. Houver and F. Masnou-Seuws to be published (see also in this conference)

50. U. Fano and W. Lichten, Phys. Rev. Letters $\underline{14}$, 627 (1965)

51. J.C. Brenot, D. Dhuicq, J.P. Gauyacq, J. Pommier, V. Sidis, M. Barat and E. Pollack, Phys. Rev. A $\underline{11}$, 1933 (1975)

52. R. François, D. Dhuicq and M. Barat, J. Phys. B : Atom. Molec. Phys. $\underline{5}$, 963 (1972)

53. R.D. Piacentini, C. Harel and A. Salin, VIIe Colloque National sur la Physique des Collisions Atomiques et Electroniques, Université de Bordeaux I (France)

ASYMPTOTICALLY EXACT THEORY OF ELECTRON EXCHANGE IN DISTANT

COLLISIONS

Yu. N. Demkov[*]

Joint Institute for Laboratory Astrophysics,
University of Colorado
and Physics Department, Leningrad State University

In different atomic processes, such as resonant charge transfer ionization in a strong electric field, spin exchange, etc., the
electron penetrates a very broad potential barrier.

From the experimental point of view, such processes are interesting because they usually have very large cross sections and
therefore rather often they are of great importance -- for the
confinement of plasma in external fields, for the ion mobility in
gases, etc.

From the theoretical point of view these processes allow us to
use a rather simple theoretical treatment, because the only thing
one has to know in this case is the behavior of the wave function
of the electron -- far from the atom -- inside the potential
barrier -- in the region where this behavior is relatively simple.
The behavior of the wave function inside the atom usually is not
important and it can be replaced by the boundary condition at the
edge of the atom, i.e. by one or two numerical parameters which can
be treated as the adjustable parameters of the theory or can be extracted from the calculations of the wave function for an unperturbed atom. For this reason the theories of such deep underbarrier processes are relatively simple.

---

[*]Visiting Fellow, 1974-75 at the Joint Institute for Laboratory
Astrophysics of the National Bureau of Standards and University
of Colorado, Boulder, Colorado.

But there is another factor which complicates the same theories.   Indeed, when we are calculating such atomic parameters as
the energy of a state, correlation, fine structure, etc., we have
mainly to know the wave function inside the atom where this function is relatively large.   So in this case we do not need a very
accurate value of the wave function in the region where it is very
small, i.e., deep inside the potential barrier.   In the case of
tunneling processes the underbarrier part of the wave function is
of crucial importance for the theoretical description and therefore we must know this part to greater accuracy than in most of
the other cases.

So we need an asymptotically exact theory for the tunneling
processes which gives us an asymptotic expansion of the important
quantities approaching exact ones when the barrier becomes bigger
and bigger.   Such a theory now exists for a number of processes,
and is considered in the book of B. M. Smirnov [1], so we shall
emphasize here only the main points.

The first asymptotically exact calculation of $\Delta E$ -- the
splitting of $1s\sigma_g$, $2p\sigma_u$ states of the $H_2^+$ system for large internuclear distances R, which defines the $H^+ + H$ ground state resonance charge exchange, was performed by T. Holstein [2].   He found
the first two terms in the asymptotic expansion, which was later extended to four terms by Komarov and Slavyanov [3] and to nine terms
by Damburg and Propin [4]

$$\Delta E = 4Re^{-R-1}\left[1 + \frac{1}{2R} - \frac{25}{8R^2} - \frac{131}{48R^3} + 0\left(\frac{1}{R^4}\right)\right] , \qquad (1)$$

whereas the variational LCAO method gives

$$\Delta E = 4Re^{-R-1}\left[\frac{e}{3} - \frac{e}{2R^2} + \cdots\right] . \qquad (2)$$

Curiously enough the first terms of these expansions coincide within 9% accuracy although we do not know the theoretical reasons why
this difference is so small.

To understand better the nature of this discrepancy we can
consider the exchange integral

$$I = \int \psi_A U \psi_B (d\vec{r}) = \frac{1}{\pi} \int e^{-r_A - r_B} U(d\vec{r}) \qquad (3)$$

where   U   is a slowly varying Coulomb potential and   $\psi_A$, $\psi_B$   are
the hydrogen wave functions of atoms A, B.   The exponential factor
in the integrand remains constant on the surface of the ellipsoid

(see Fig. 1).  If  R  is large the main part of the volume of this
ellipsoid is outside the atomic region, and so the small changes of
$\psi_A\psi_B$ deep in the tunneling region can change the exchange integral
considerably.

Holstein [2], Firsov [5], Herring [6] and Landau & Lifshitz [7]
proposed a simple and effective formula expressing the splitting
$\Delta E$ through the integral over the median surface $\sigma$ (see Fig. 1)

$$\Delta E = \int_\sigma \chi_g \frac{d\chi_u}{dn} \, dS \quad , \tag{4}$$

where  $\chi_g$, $\chi_u$  are the exact  $1s\sigma_g$, $2p\sigma_u$ functions and  $d/dn$  is
the normal derivative to the  $\sigma$  surface.  Then the corrections to
$\chi_g$, $\chi_u$ can be found by the semiclassical method.  Such calculations
have been performed now for all the states of the $H_2^+$ system
[1,3,4].

Holstein [2] and Smirnov [1] pointed out that these calculations are valid for arbitrary $A^+ + A$ resonance charge exchange because outside the atom the wave function is hydrogen like.  The
most recent calculation (and the most exact one) of this sort has
been made by Bardsley [8], where two terms in the asymptotic expansion of  $\Delta E$  were calculated and good agreement between theory
and experiment was obtained.

The next problem proposed by Herring [6], then solved by
Gor'kov and Pitaevsky [3] and corrected by a numerical factor by

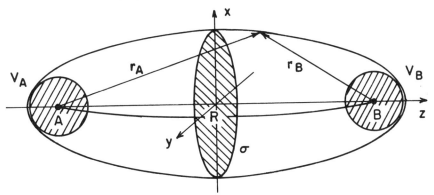

Fig. 1.   Illustration of the relative importance of the atomic
          vicinity regions $V_A$,$V_B$ and the interatomic one in the
          calculation of the exchange integral.  The volume of the
          ellipsoid essential for the integral is much larger than
          $V_A$,$V_B$ for large R.  The energy splitting can be expressed
          as a surface integral over the median plane $\sigma$.

Herring and Flicker [10] was the single-triplet splitting of the ground states of the $H_2$ molecule which defines the spin-exchange cross section in slow $H + H$ collisions.   The asymptotic formula for the splitting was found to be

$$\Delta E = 1.62 \ R^{5/2} e^{-2R} \ [1 + O(R^{-1/2})] \quad . \tag{5}$$

In this case the Heitler-London approximation gives a completely wrong pre-exponential factor

$$\Delta E = e^{-2R} \frac{4}{15} R^3 \ (4.09 - \ell nR + \dots) \tag{6}$$

which even interchanges the singlet and triplet states at very large distances.   The case of the singlet-triplet splitting for different atoms with spin $1/2$ was considered by Ovchinnikova [11] and Smirnov [1].

   Another two-electron problem is the splitting of the levels which defines the double resonance charge exchange $A^- + A^+$, $A^{++} + A$, etc.   At first sight the exchange integral which defines the g-u splitting of the levels must be proportional to the exponent

$$\exp\left(- \frac{1}{\hbar} \sqrt{2mI_1} \ R - \frac{1}{\hbar} \sqrt{2mI_2} \ R\right) = \exp(-\alpha_1 R - \alpha_2 R) \tag{7}$$

where   $I_1$, $I_2$   are the first and second ionization potentials of the atom (or ion) respectively.   But a more careful consideration shows that the real exponent should be

$$\exp\left(- \frac{2}{\hbar} \sqrt{2mI_1} \ R\right) = \exp(-2\alpha_1 R) \tag{8}$$

and is defined only by the first ionization potential.   The difference is especially large for $A^- + A^+$ exchange where   $\alpha_1$ and $\alpha_2$ can differ very considerably.   To understand better this difference we can consider Fig. 2 where the configurational space for both electrons is presented and only the motion of the electrons along the internuclear z axis is shown.   The double charge exchange process corresponds in this figure to the transition from point A to point C and the amplitude or the   $\Delta E$(g-u)   splitting also can be expressed as an integral over the median hyper-surface $\sigma_1$.   The overlapping of both localized wave functions at the origin is really defined by (7), but in the regions B and D corresponding to homopolar configuration of electrons this overlapping is much larger and is defined by (8).

   For the spin exchange $A + A$ process the corresponding transition on Fig. 2 is from region B to region D, and presenting the

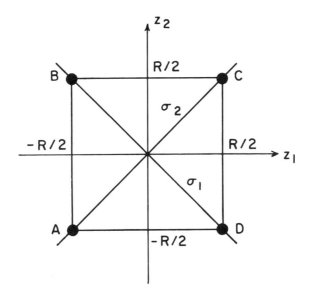

Fig. 2.    The simplified configurational space for a
double exchange process.  Transitions from
A to C and from B to D correspond to the
double charge exchange and to the spin exchange respectively.  The corresponding
energy splittings can be expressed as surface integrals over $\sigma_1$ or $\sigma_2$.

amplitude in the form of an integral over the surface  $\sigma_2$,  we can
easily see that on the whole surface between A and C the exponential order of magnitude for the overlapping is (8).   Thus we have
to calculate this integral over all the region between A and C.
Therefore the regions A and C become relatively less important and
we do not need to know the behavior of the wave function where both
electrons and the ion are all close together.   Just this point made
possible the above-mentioned $H_2$ calculation.

For a triple charge exchange, instead of a square and a two-
dimensional picture, we have to consider a cube and three $(z_1 z_2 z_3)$-
dimensional space (Fig. 3).   In the region of the thick line the
order of magnitude of the overlapping integral is

$$\exp[-(2\alpha_1 + \alpha_2)R] \qquad , \qquad (9)$$

and expressing the exchange probability through the surface integral we find that the maximum of the overlapping will be in the six
(a,b,c,d,e,f) regions of the surface.

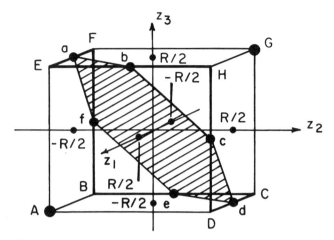

Fig. 3.  The configurational space for the triple charge
        exchange.  The initial state is A, the final
        one is G.  The line CDHEFBC defines the region
        where the overlapping has the order of
        $\exp[-(2\alpha_1+\alpha_2)R]$.  Expressing the energy split-
        ting as an integral over the median surface $\sigma$
        we have six regions a,b,c,d,e,f giving the
        main part of the splitting.

Correspondingly, the quadruple charge exchange will be defined by

$$\exp[-(2\alpha_1 + 2\alpha_2)R] \qquad (10)$$

and so on.  We can see that for a multiple resonance charge exchange we have even a different exponential behavior relative to
what one would expect from the simplest considerations.

The calculations for the double charge transfer including the
correct pre-exponential factor evaluation have been performed by
Komarov and Yanev [12].

All the problems considered above are connected with the
asymptotic behavior of the potential curves.  Therefore the further
application of these results to the collisional problems is possible only in the frame of adiabatic approximation, i.e., for slow
collisional velocities when the relative velocity of atoms is much
less than that of the atomic electrons involved in the transition.
Here we shall consider the simplest asymptotic calculation for a
dynamical problem of charge transfer where the result is applicable
for arbitrary velocities and the adiabatic approximation is not  .
used.

Let us consider two central potential wells $U_A$, $U_B$ moving in opposite directions with constant velocities $+\vec{v}/2$ $-$ $\vec{v}/2$ and with the impact parameter $\rho$. We introduce the coordinate system x,y,z according to Fig. 4. So the Schrödinger equation for the problem is

$$\left[ -\frac{1}{2} \nabla^2 + U_A\left(\left|\vec{r} - \frac{\vec{R}}{2}\right|\right) + U_B\left(\left|\vec{r} + \frac{\vec{R}}{2}\right|\right) \right] \psi = i \frac{\partial \psi}{\partial t} \qquad . \qquad (11)$$

We shall consider the transition of the electron between two states $\phi_A(\vec{r})$ and $\phi_B(\vec{r})$ which satisfy the equations

$$\left[ -\frac{1}{2} \nabla^2 + U_A(r) \right] \phi_A = E_A \phi_A \quad , \quad \left[ -\frac{1}{2} \nabla^2 + U_B(r) \right] \phi_B = E_B \phi_B \quad , \quad (12)$$

where

$$E_A = -\frac{\alpha^2}{2} \quad , \quad E_B = -\frac{\beta^2}{2} \qquad (13)$$

are the ionization potentials of atoms A and B. The asymptotic behavior of the functions $\phi_A$, $\phi_B$ at large r are

$$\phi_A \sim A \; r^{1/\alpha-1} \; e^{-\alpha r} \quad , \quad \phi_B \sim B \; r^{1/\beta-1} \; e^{-\beta r} \qquad . \qquad (14)$$

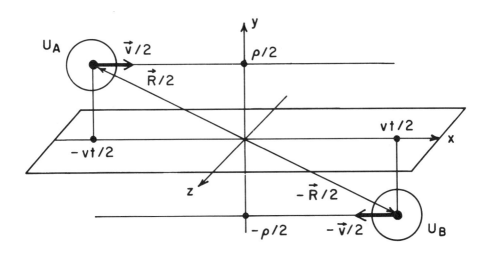

Fig. 4.   Impact parameter approximation to the charge exchange: the uniform motion along the straight line trajectory.

So we assume that the electron, at large r, is attracted to the $A^+$, $B^+$ positive ions by a Coulomb force, and keep only the leading term in the Coulomb wave function. The one-particle approximation considered here is not an essential one because for large r (the only region which is important for further considerations) the exact many-electron wave function of a neutral atom can always be presented as a product of the wave function of an ion and of an exponentially decreasing one-electron function (14). Thus the only parameters that characterize the initial and final states are the ionization potentials $-E_A$, $-E_B$ (or $\alpha$, $\beta$) and the normalization constants A, B, which can be obtained from the numerical calculations of the atomic wave functions, or can be considered as adjustable parameters. A simple expression for A, B can be found if we assume that $\phi_A$, $\phi_B$ are the exact Coulomb wave functions. For a detailed discussion about these constants see [1].

Considering the initial and the final states of the electron we have to take into account the uniform motion of the potential wells $U_A$ and $U_B$. The corresponding solutions of the equations

$$\left[ -\frac{1}{2} \nabla^2 + U_A\left( \left| \vec{r} - \frac{\overrightarrow{R(t)}}{2} \right| \right) \right] \psi_A = i \frac{\partial \psi_A}{\partial t}$$

$$\left[ -\frac{1}{2} \nabla^2 + U_B\left( \left| \vec{r} + \frac{\overrightarrow{R(t)}}{2} \right| \right) \right] \psi_B = i \frac{\partial \psi_B}{\partial t} \tag{15}$$

have the form

$$\psi_A = \phi_A\left( \left| \vec{r} - \frac{\overrightarrow{R(t)}}{2} \right| \right) \exp\left( i\, \frac{vx}{2} - i\, \frac{v^2}{8} + i\, \frac{\alpha^2}{2} t \right)$$

$$\psi_B = \phi_B\left( \left| \vec{r} + \frac{\overrightarrow{R(t)}}{2} \right| \right) \exp\left( -i\, \frac{vx}{2} - i\, \frac{v^2}{8} + i\, \frac{\beta^2}{2} t \right) . \tag{16}$$

(Here we used the Galilean transformation for the Schrödinger equation, see for instance [7, p.45].) So, to find the amplitude $f_{AB}$ of the electron transfer from the state $\psi_A$ to the state $\psi_B$ we have to solve equation (11) with the initial condition

$$\left. \Psi_A \right|_{t \to -\infty} = \psi_A \quad , \tag{17}$$

then consider $\Psi_A$ at $t \to +\infty$ and calculate the integral

$$f_{AB} = \lim_{t \to +\infty} \int \psi_B^* \Psi_A (d\vec{r}) \quad . \tag{18}$$

We can examine as well the inverse process and considering the solution of (1) $\Psi_B$ with the condition

$$\Psi_B \Big|_{t \to +\infty} = \psi_B \tag{19}$$

we have another expression for the same $f_{AB}$

$$f_{AB} = \lim_{t \to -\infty} \int \Psi_B^* \psi_A (d\vec{r}) \quad . \tag{20}$$

The expression

$$f_{AB} = \int \Psi_B^* \Psi_A (d\vec{r}) \tag{21}$$

does not depend on time and coincides with (18) and (20) at $t \to +\infty$ and $t \to -\infty$ respectively so it gives us another expression for $f_{AB}$. Due to the conditions (17), (19) for $t \to -\infty$ the main nonvanishing part of the integrand in (21) is in the upper semispace (y>0); for $t \to +\infty$ the main part is in the lower semispace (y<0).

Let us consider now the equations

$$\left( -\frac{1}{2} \nabla^2 + U_A + U_B \right) \Psi_A = i \frac{\partial \Psi_A}{\partial t} \quad ,$$

$$\left( -\frac{1}{2} \nabla^2 + U_A + U_B \right) \Psi_B^* = -i \frac{\partial \Psi_B^*}{\partial t} \quad , \tag{22}$$

multiply them by $\Psi_B^*$, $\Psi_A$ respectively, substract one from the other, and integrate over the lower semispace y<0. Then we have

$$-\frac{1}{2} \int_{y<0} \left( \Psi_B^* \nabla^2 \Psi_A - \Psi_A \nabla^2 \Psi_B^* \right) (d\vec{r}) = i \frac{d}{dt} \int_{y<0} \Psi_B^* \Psi_A (d\vec{r}) \quad . \tag{23}$$

Now we can transform the volume integral on the left-hand side into a surface one and integrate the whole expression over time from $-\infty$ to $+\infty$, taking into account that the integral on the right-hand side vanishes (up to the higher exponential terms) at $t \to -\infty$ and is equal to $f_{AB}$ at $t \to +\infty$. We obtain then the formula for $f_{AB}$

$$f_{AB} = \frac{i}{2} \int_{-\infty}^{+\infty} dt \int\int_{-\infty}^{+\infty} dx\,dz \left( \Psi_B^* \frac{\partial}{\partial y} \Psi_A - \Psi_A \frac{\partial}{\partial y} \Psi_B \right)_{y=0} \tag{24}$$

where $f_{AB}$ is expressed through $\Psi_A$, $\Psi_B$ and their derivatives on the x,z surface, i.e., in the deep underbarrier region (for large

$\rho$). Just like in the case of $H_2^+$ $1s\sigma_g$-$2p\sigma_u$ splitting we cannot replace $\Psi_A$, $\Psi_B$ in (24) with $\psi_A$, $\psi_B$ respectively, but we must calculate the distortion of $\Psi_A$, $\Psi_B$ due to the presence of the second potential well and find this distortion on the plane y=0. Consequently we present $\Psi_A$, $\Psi_B$ in the form

$$\Psi_A = S_A \psi_A \qquad , \qquad \Psi_B = S_B \psi_B \qquad (25)$$

where $S_A$, $S_B$ are the distortion factors which are equal to unity in the regions close to A and B respectively. $S_A$, $S_B$ are slowly varying functions and therefore, putting (25) into (11) we can neglect the second derivatives of $S_A$, $S_B$. Then we have the first order partial differential equations

$$-\frac{1}{2}\left(\frac{\nabla\psi_A}{\psi_A}\right) \cdot \nabla S_A + U_B S_A = i\,\frac{\partial S_A}{\partial t} \qquad ,$$

$$-\frac{1}{2}\left(\frac{\nabla\psi_B}{\psi_B}\right) \cdot \nabla S_B + U_A S_B = i\,\frac{\partial S_B}{\partial t} \qquad , \qquad (26)$$

with the boundary conditions

$$S_A\Big|_{\vec{r}-\vec{R}/2\to 0} = 1 \quad , \quad S_B\Big|_{\vec{r}+\vec{R}/2\to 0} = 1 \qquad . \qquad (27)$$

Replacing $U_A$, $U_B$ by a Coulomb potential, and $\psi_A$, $\psi_B$ by their asymptotic behavior, we can solve these equations. Then, using the corrected $\Psi_A$, $\Psi_B$, we can evaluate the integral (24) and find $f_{AB}$ in the form

$$f_{AB} \sim \frac{i}{v}\sqrt{\frac{\Pi}{32}}\,\frac{\alpha\beta}{\kappa^{5/2}}\left(\frac{\alpha\rho}{2\kappa}\right)^{1/\alpha-1}\left(\frac{\beta\rho}{2\kappa}\right)^{1/\beta-1}D\,e^{-\kappa\rho} \qquad (28)$$

where*

$$D = \left[\frac{\left(\alpha^2-\beta^2\right)^2 + v^2(\alpha+\beta)^2}{\left(\alpha^2-\beta^2\right)^2 + v^2\left(\alpha^2+\beta^2\right)}\right]^{1/\alpha+1/\beta}\exp\left[-\frac{2}{v}\arctan\frac{v}{\alpha+\beta}\right] \quad , \quad (29)$$

and

---

*Compared with the same formula in (13) this one is simplified and some obvious mistakes have been corrected.

$$\kappa = \frac{1}{2v} \sqrt{[(\alpha+\beta)^2 + v^2][(\alpha-\beta)^2 + v^2]} \qquad . \qquad (30)$$

The quantity $\kappa$ defines the exponential decrease of the exchange amplitude. In [14] the product $\kappa\rho$ was considered in detail and the exponential $e^{-\kappa\rho}$ factor was obtained from the steepest descent asymptotic calculation of the Born exchange amplitude. $\kappa\rho$ was called there a generalized Massey parameter and different limiting cases were considered. But the Born exchange amplitude giving a correct exponential factor cannot give us the correct pre-exponential factor due to the importance of the distortion of the wave functions, $\Psi_A\Psi_B$. In the course of the calculation it is important to include in (11) the time dependent term -- the interaction between the nuclei. Only then can we obtain agreement with the adiabatic limiting case when $v\to0$ ($v \ll |\alpha-\beta|$) considered by Smirnov and Bardsley [1,8].

Figure 5 gives the different theoretical calculations of f for the $H^+ + H$ case and for $v = 1$, i.e. for the intermediate velocity where both basic approximations, adiabatic and Born, are not applicable.

The asymptotic calculation (I) must give reasonable results for large $\rho$, and it is interesting that at $\rho=4$ it does not coincide with either the adiabatic approximation (II) or the Brinkman-Kramers one (IV). For the nonresonance case $Li+Na^+\to Li^++Na$

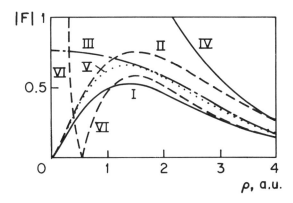

Fig. 5.    The amplitude of the resonant charge transfer
$p + H \to$ ($v = 1$ a.u.): I - asymptotically exact
amplitude of the present work; II - adiabtical
asymptotics; III - four-state Sturmian expansion
calculations of Gallacher and Wilets [15]; IV -
Brinkman-Kramers approximation; V - first asymptotical (large $\rho$) term of the Brinkman-Kramers
amplitude; VI - Born approximation.

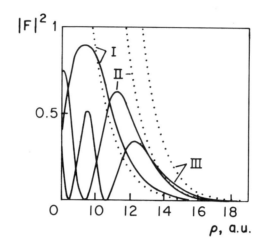

Fig. 6.   The non-resonant charge transfer probability
Li + Na$^+$ → Li$^+$ + Na, 0.25 eV for the inverse
velocities: I – v$^{-1}$ = 6 a.u.; II – v$^{-1}$ = 12
a.u.; III – v$^{-1}$ = 18 a.u.   Solid lines – two
state calculations of Melius and Goddard [16],
dots – the asymptotically exact probability
(this work).

(Fig. 6), the comparison of two theoretical calculations shows that
in the asymptotic region they are close to each other for different
velocities.

Using this calculation method we obtain the charge exchange
amplitude   only in the region where this amplitude is very small.
But this allows us to find the critical impact parameter $\rho_0$ which
divides the whole impact parameter region into two parts:   one with
a large probability of charge transfer and the other one with a
small probability.   Then the cross section for charge transfer can
be estimated by the formula

$$\sigma = \frac{\pi}{2}\, \rho_0^{\,2} \tag{31}$$

provided that there are no competing processes with comparable
cross section, such as detachment excitation, etc.   This approximation has been called in [14] a dense target approximation.   In
the H$^+$ + H case such a calculation is in good agreement with a much
more complicated calculation of Gallacher and Wilets [15].

It is also possible using these asymptotic formulae and the
eikonal method to find the angular dependence of the exchange amplitude at very small angles.

It is well known that the perturbation theory encounters some difficulties in the case of rearrangement collisions. In the Born approximation at high velocities the main part of the total cross section comes from small impact parameters, i.e., from the region where the asymptotic theory is inapplicable. Thus the total cross section calculated by the Born method and by the asymptotic one cannot be compared. However, considering the impact parameter method and comparing the asymptotic calculations with the Born expansion at large $\rho$, we probably shall be able to understand better the properties and the convergence of the Born expansion.

One of the important properties of this theory is that it provides the possibility to consider non-resonance processes, whereas all the previous calculations were confined to resonance processes, such as resonance charge exchange, spin exchange, and others.

The asymptotic theory considered here can be generalized in different directions. Particularly we can consider the asymptotic behavior of the exchange amplitudes in electron-atom scattering for large angular momenta, consider exchange processes for particles with comparable masses, consider the relativistic generalization, etc. A more detailed paper on this subject has been submitted for publication in JETP.

Part of this work was performed during my stay at the Joint Institute for Laboratory Astrophysics. I want to express my gratitude to the members of this Institute and especially to S. Geltman and J. N. Bardsley for stimulating discussions, and to L. Volsky for a careful check of the manuscript.

## References

1. B. M. Smirnov, Asimptoticheskie metody teorii atomnyh stolknovenii (Asymptotical methods in the theory of atomic collisions, in Russian), Atomizdat, 1972.

2. T. Holstein, Westinghaus Research Laboratories Reseach Report 60-94698-3-R9, 1955.

3. I. V. Komarov, S. Yu. Slavyanov, Sov. Phys. JETP $\underline{52}$, 1368, 1967; (Engl. transl. $\underline{25}$, 910, 1967).

4. R. J. Damburg, R. Kh. Propin, J. Phys. B $\underline{1}$, 681, 1968.

5. O. B. Firsov, Sov. Phys. JETP $\underline{21}$, 1001, 1951 (in Russian).

6. C. Herring, Rev. Mod. Phys. $\underline{34}$, 631, 1962.

7. L. Landau, E. Lifshitz, Quantum Mechanics, p. 291, Pergamon Press, 1965.

8.  J. N. Bardsley, T. Holstein, B. R. Junker, Swati Sinha, Phys. Rev. A 11, 1911, 1975.

9.  L. P. Gor'kov, L. P. Pitaevskii, Sov. Phys.-Doklady 151, 822 1963 (Engl. transl. 8, 788, 1964).

10. C. Herring, M. Flicker, Phys. Rev. 134, A362, 1964.

11. M. Ja. Ovchinnikova, Sov. Phys. JETP 49, 275, 1965　(Engl. transl. 22, 181, 1966).

12. I. V. Komarov, R. K. Yanev, Sov. Phys. JETP 51, 1712, 1966 (Engl. transl. 24, 1159, 1967).

13. Yu. N. Demkov, V. N. Ostrovsky, in Electronic and Atomic Collisions, Abstracts of Papers of the IXth ICPEAC, Univ. of Washington, Seattle, 1975, Vol. 1, p. 181.

14. Yu. N. Demkov, in Electronic and Atomic Collisions, Abstracts of Papers, VIIIth ICPEAC, Beograd, 1973, Vol. 2, p. 799; in The Physics of Electronic and Atomic Collisions, VIIIth ICPEAC Invited Papers and Progress Reports, Beograd, 1973, p. 26.

15. D. F. Gallacher, L. Wilets, Phys. Rev. 169, 139, 1968.

16. G. F. Melius, W. A. Goddard, Phys. Rev. A 10, 1541, 1974.

# OPTICAL EMISSION IN SLOW ATOMIC COLLISIONS

V. Kempter

Fakultät für Physik der Universität Freiburg

D 78 Freiburg i.Br., W. Germany

## I. INTRODUCTION

This report deals with experiments in which electronic excitation in ion-atom and atom-atom collisions is studied by means of optical methods. Both colliding particles are in their respective ground states before the collision. I would like to narrow the subject further to collisions in the energy range where the quasimolecular treatment of the collision can be adopted, e.g. to energies below approx. 1 keV.[1] Only outer-shell excitation will be considered. Collisions involving molecules will be excluded from the discussion which implies that chemiluminescence and dissociation with excitation will not be considered.[32]

The aim of this study is twofold:
(1) to present a collection of the new data in the field defined above;
(2) to discuss some new trends in the area by means of a few typical examples.

## II. EXPERIMENTAL METHODS AND RESULTS

Let us start by recalling the optical methods which yield information on excitation processes (see Table 1 (a) and (b)). The methods shown in Table 1(b) all involve a coincidence experiment between the scattered particle and the emitted photon; a separate review is dedicated to the determination of differential cross sections from

(a) Total cross sections

| Energy dependence of the total emission cross section $Q_{ij}$ | Properties of the potentials which are involved in the excitation process in the region of their interaction |
|---|---|
| (1) Absolute size | Crossing distance $R_c$ |
| (2) Shape | Coupling matrix element |
| (3) Threshold energy $E_{th}$ | Potential energy $V(R_c)$ at $R_c$ |
| (4) Structure | Long-range interaction between excited states |
| Spectrum S of the emitted radiation | Identity of the excited states, and relative cross sections for their population |
| Cross sections for population of different states arising from the same configuration | Relative population of the involved excited molecular states |
| Total polarisation P of the emitted radiation | Relative total cross sections for population of magnetic substates; Relative population of involved excited molecular states |

(b) Differential cross sections

| Differential polarisation $\Pi$ and Photon-particle angular correlation function | Differential cross sections for population of magnetic substates; Relative probabilities for population of the involved molecular states as function of impact parameter |
|---|---|

Table 1: Information on excitation processes from the study of the optical emission

| Projectile | Target | Studied process | Data | Reference |
|---|---|---|---|---|
| $Ne^+$, $C^+$, $Ar^+$ $N^+$, $O^+$ | He, Ne, Ar | a, b | S | 32 |
| $Mg^+$ | Cd | a, b | $Q_{ij}(rel)$; $E_{th}$ | 6 |
| $Cs^+$ | He, Ne | a, b | $Q_{ij}$; $E_{th}$ | 9 |
| $Cd^+$ | Cd | a, b | $Q_{ij}(rel)$; $E_{th}$ | 11 |
| $Rb^+$ | Ar | a | $Q_{ij}$ | 12 |
| $Li^+$, $Na^+$, $K^+$ $Rb^+$, $Cs^+$, $Mg^+$ $Ca^+$, $Sr^+$, $Ba^+$ | Cd | a, b | $Q_{ij}$; $E_{th}$ | 13 |
| $Mg^+$, $Sr^+$ | Na, K He, Kr | a, b | $Q_{ij}(rel)$ | 14 |
| $Cs^+$ | He, Ne, Ar | a | $Q_{ij}$; $E_{th}$ | 15 |
| $Mg^+$ | Rb, Cs | a, b | $Q_{ij}(rel)$; $E_{th}$ | 16 |
| $Li^+$, $Na^+$, $K^+$ | Na, K | a, b | $Q_{ij}$; $E_{th}$ | 17 |
| $Li^+$, $Na^+$, $K^+$ | K | a, b | $Q_{ij}(rel)$; $E_{th}$ | 18 |
| $Na^+$ | Ne | a, b | $Q_{ij}(rel)$; P | 19 |
| $N^+$, $O^+$, $Na^+$, $Mg^+$ | Ne | a, b | $Q_{ij}(rel)$; P | 20 |
| $He^+$ | He | b | $\Pi$; $\sigma(\vartheta)$ | 52 |

| Projectile | Target | Studied process | Data | Reference |
|---|---|---|---|---|
| $Li^+$, $Be^+$, $B^+$, $C^+$ $N^+$, $O^+$, $F^+$, $Ne^+$ $Na^+$, $Mg^+$, $Al^+$ | Ne | a, b | $Q_{ij}$(rel); P | 21 |
| $He^+$ | Ne | b | $E_{th}$; $Q_{ij}$(rel) | 22 |
| $He^+$ | Ne | a, b | $E_{th}$; $Q_{ij}$(rel) | 23 |
| $He^+$ | Ar | a, b | $E_{th}$; $Q_{ij}$(rel) | 24 |
| $He^+$ | He | a | $\sigma(\theta)$ II | 25 |
| He | He | a | $Q_{ij}$; $E_{th}$ | 7 |
| $H^+$ | Xe | b | $Q_{ij}$(rel) | 30 |
| Ne, Ar | Ne, Ar | a | $E_{th}$; $Q_{ij}$ | 5 |
| K | Ar, Kr, Xe | a | $Q_{ij}$(rel) | 33 |
| K | Ar, Kr, Xe | a | $Q_{ij}$; $E_{th}$ | 34 |
| K | He, Ne, Ar, Kr, Xe | a | P | 35 |
| K | He, Ar, Xe | a | $Q_{ij}$(rel); S | 36 |
| K | K, Na, Hg | a | $Q_{ij}$; $E_{th}$ | 37; 38; 39 |
| $H^+$ | Ne | b | $Q_{ij}$(rel); $\sigma(\theta)$ | 29 |

| Projectile | Target | Stu-<br>died<br>process | Data | Re-<br>ference |
|---|---|---|---|---|
| $H^-$ | He, Ar | b | $Q_{ij}(rel)$ | 31 |
| $Mg^+$ | He, Ne,<br>Ar, Kr, Xe | b | $Q_{ij}(rel)$ | 10 |

Process: (a) Direct Excitation (Projectile or target)
         (b) Charge transfer into excited states, or
             charge transfer with simultaneous exci-
             tation

Data: $Q_{ij}$ $(Q_{ij}(rel))$    Absolute (relative) line emission
                       cross section

      $E_{th}$         Threshold for level excitation

      P               Polarisation of line emission

      S               Spectrum of the emitted radiation

      $\sigma(\vartheta)$         Differential level excitation cross
                       section

      II              Differential polarisation

Table 2: Catalogue of available data


energy loss spectra.

Table 2 gives a catalogue of data obtained between
1971 and the end of 1974; data are only shown when they
do not already appear in Thomas' book[2] on electronic excitation in heavy particle collisions. Most of the data on
collisions between neutral atoms are already given in[43],
and are discussed there in some more detail.


III. NEW TRENDS

In this part of my report I'm going to discuss some
new trends in the area. A few experiments are presented
from which the author believes that they should be extended to other systems as well.

(i) Information on Excitation Mechanisms from
the Threshold Behavior of Excitation Cross
Sections

From the behavior of excitation cross sections in
the near-threshold region information on the location of
the region with nonadiabatic behavior may be obtained. In
general we have to distinguish three cases:

Case 1: A crossing or avoided crossing $R_c$ occurs at
$V(R_c)$, and $V(R_c)$ is smaller or equal to the endoergicity
$E_o$ of the excitation process. A sharp onset of the excitation cross section will be found at $E = E_o$. So far
such a behavior has only been found for collisions of alkali atoms with molecules.[43,3] In case 1 no information on
the location of the avoided crossing can be obtained from
the position of the threshold energy.

Case 2: If $V(R_c) > E_o$, the conservation of energy
classically gives a sharp onset of the cross section at
$E = V(R_c)$. In general a kink of the cross section is
found at this energy, but the cross section may actually
extend to much lower energies due to tunnelling between
the involved states at energies $E < V(R_c)$.[4] The tunnel
probability depends critically on the shape and the relative distance of the involved potential curves. The
strong rise of the cross section at $E = V(R_c)$, when the
transition becomes classically allowed, gives directly
the crossing energy $V(R_c)$. Examples for cross sections
where the threshold region seems to be influenced by
tunnelling may be found in[5,13].

Case 3: Rotational coupling without a crossing. One
expects a strong rise of the cross section when the collision energy is high enough to reach the region where
rotational coupling becomes efficient. No sharp onset of
the cross section will occur, but the cross section will
fall off almost exponentially towards lower energies.
Usually the apparent threshold is determined by the sensitivity of the detection system employed. The behavior
of the cross section at very low energies is again a very
sensitive test for the quality of the calculated potential curves in the region of strong nonadiabatic coupling.[6]

The excitation of $He(2^1P)$ in low energy collisions
between two groundstate He-atoms is entirely due to rotational coupling between the $X\ ^1\Sigma_g$-molecular ground
state and the lowest $^1\Pi_g$-state correlating with $He(2^1P)$
+ $He(1^1S)$.[51] Excitation of $He(2^1P)$ at low collision

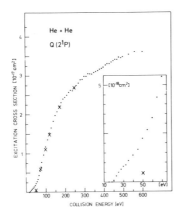

Fig. 1:  Total cross section for excitation of $He(2^1P)$ in
collisions between ground-state He-atoms: Solid
points, experimental results from[7]; crosses, theory from[6].

energies has been studied both theoretically[6] and experimentally.[7] Fig. 1 shows a comparison of the results;
since the absolute values of the measurements are only
accurate within a factor of three, both data sets have
been normalized arbitrarily at the highest energy which
is covered by both experiment and theory. As is pointed
out in[6], the disagreement at very low energies may partly be due to inaccuracy in the calculated $^1\Pi_g$-state
curve.

### (ii) Information on Excitation Mechanisms from Cross Sections for Population of Different States Arising from the Same Configuration

Let us turn to the population of close lying excited states which arise from the same configuration. In
order to explain the relative population of these states,
one has to ask how they correlate with the quasimolecular states which are populated by nonadiabatic transitions from the groundstate. Let us consider a simple example: the correlation diagram of the adiabatic states
arising from the two fine structure states of an alkali
atom when colliding with a ground state rare-gas atom, is
shown in Fig. 2. Transitions to $B^2\Sigma$ at $R_1$ would populate
$K(4^2P_{3/2})$ only, provided there are no non-adiabatic transitions between the excited molecular states. On the other

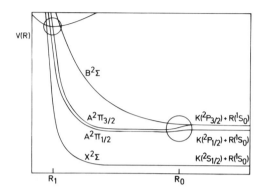

Fig. 2: Lowest adiabatic molecular states of the pot-
assium-rare gas pairs (schematically)

hand, exclusive and equal population of the two $^2\Pi$-states
would yield equal population of the two fine structure
states $^2P_{1/2}$ and $^2P_{3/2}$.

The interaction between the excited molecular states
is rather strong as turns out from fine structure mixing
experiments.[44] In the limit of high velocities one would
expect that the population of the fine structure states
could be obtained by using the sudden approximation in order to describe the transition from molecular to atomic
behavior at $R_o$.[46] The probability to populate $^2P_J$ is
given by

$$W_J = \Sigma'_{M_S, M_J} |\langle \Psi_{M_S} | JM_J \rangle|^2 \; ; \quad M_S = \pm 1/2$$

The state of the system between $R_1$ and $R_o$ is written as

$$\Psi_{M_S} = \{\Sigma_{M_L} c_{LM_L} |LM_L\rangle \exp(- i/\hbar \int_{t_o}^{t_1} V_{LM_L} dt)\} |SM_S\rangle \; ;$$

The excitation cross section $Q_J$ is obtained by integrating over all impact parameters which lead to excitation

$$Q_J = 2\pi \int db \cdot b \cdot W_J \quad ;$$

The relation between the states $|LM_L\rangle$ and $|JM_J\rangle$ is

$$|JM_J\rangle = \Sigma'_{M_L, M_S} \langle LSM_L M_S | JM_J \rangle |LSM_L M_S\rangle$$

and we obtain

$$W_J = \Sigma_{M_J, M_S} |\Sigma_{M_L} \ c_{LM_L} \langle LSM_L M_S | JM_J \rangle \cdot$$

$$\exp(-i/\hbar \int_{t_o}^{t_1} V_{LM_L} dt)|^2 \quad ;$$

One obtains interference terms of the form

$$Re\{c_{LM_L} \cdot c_{LM_L'}^* \} \cdot \cos \{\frac{1}{\hbar v} \int_{R_o}^{R_1} dR(V_{LM_L} - V_{LM_L'} )\} \quad ;$$

If these terms do not average out on integration over impact parameters, one obtains regular interference structure in the cross sections for population of the fine
structure states. This mechanism depending on long-range
interaction between excited states was used previously to
explain interference structure observed in total excitation cross sections for ion-atom collisions[45]: in all
these cases interference is thought to be due to interaction between the channels for direct excitation and
charge exchange.

Fig. 3 shows experimental results for K + Ar, Kr,
and Xe[33], and corresponding results for K + He, and Ne[47].
The gross structure in the energy dependence of the relative population for Kr and Xe was explained by the transition from adiabatic behavior at low energies to sudden
behavior at high energies. Regular structure is seen for
Ar, Kr, and Xe; indications are also visible for Ne. For
He the energy dependence could be interpreted as part of
regular structure with a rather low frequency.

Similar structure due to interaction between excited

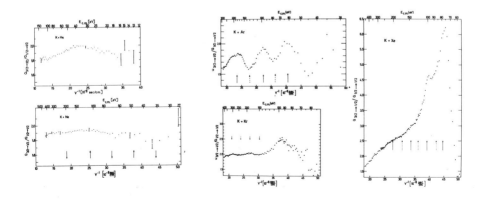

Fig. 3: Relative cross sections for population of the
potassium fine structure states in collisions
with rare-gas atoms (ref.[33] and [47])

molecular states leading to atomic states emerging from
the same multiplet was also observed for He + He', and
Ne + Ne, and Ar + Ar[5].

Cross sections for excitation of the neon resonance
states $Ne(2p^5[^2P_{1/2}]3s'$, J = 1) and $Ne(2p^5[^2P_{3/2}]3s$, J=1)
in collisions He[+] + Ne have been measured[23] between
the apparent thresholds and 10 keV[23]. The relative population of the two states is strongly energy dependent at
low energies. No explanation has been given yet in terms
of correlation diagrams between molecular and atomic
states. The regular structure is attributed to interaction between the channels for excitation with and without charge exchange.

I would like to mention shortly that a completely
new type of structure was recently discovered in the total cross section for excitation of H(2p) in collisions
of H[+] with Xe[30]. Until now no explanation at all was given for this effect.

(iii) Information on Excitation Mechanisms
from Cross Sections for Population of
Magnetic Substates

Let us now ask what can be learned from the popula-
tion of the magnetic substates $|Jm_J\rangle$ of **an individual**
state. Such states are degenerate without a magnetic
field. Relative cross sections for population of these
states can however, in simple cases, be obtained by measuring the polarisation of emission from the decay of
these states. Alternatively the angular distribution of
the emitted radiation can be measured in spectral regions
where no convenient polarisers are available.

The excited state can be represented by

$$\psi = \sum_{m_J} a_{m_J} |J m_J\rangle ;$$

small letters m refer to the space fixed primary beam
axis. The polarisation P of radiation from the decay of
this state can be expressed by the mean value of the
alignment component $A_o$ in the direction of the beam axis
as follows[40]

$$P = 3 h A_o/(4 + h A_o) ; \quad A_o = \overline{\langle \psi | 3J_z^2 - J^2 | \psi \rangle} ;$$

h depends on the angular momentum of the atomic state before and after decay, and on the dynamics of the radiation

process, and can be calculated from Eqs. (8) and (37) of[40].
$A_o$ can be expressed in terms of cross sections for population of the substates $|JM_J\rangle$

$$A_o = \Sigma_{m_J} Q_{m_J} [3m_J^2 - J(J+1)]/J(J+1)\Sigma_{m_J} Q_{m_J} \quad ;$$

In order to get information on the excitation mechanism from the $Q_{m_J}$, these cross sections must be related to the probability to populate the molecular states emerging from the combination of the excited atom with the other colliding particle. Between $R_1$ and $R_o$ the system is represented by the state vector $\psi = \Sigma_\Lambda c_\Lambda |\Lambda\rangle$ ; $\Lambda$ classifies the molecular states with respect to the internuclear axis. We introduce a density matrix $\rho_{\Lambda\Lambda'} = c_\Lambda c_{\Lambda'}^*$[41]; the probability to find the system at infinity in $|Jm_J\rangle$ is given by $\langle Jm_J |\rho|Jm_J\rangle$.

We first calculate the probability $\langle JM_J|\rho|JM_J\rangle$ to find the system in $|JM_J\rangle$, the internuclear axis still serving as quantization axis. The transition between $|\Lambda\rangle$ and $|JM_J\rangle$ is made by a transition matrix $P_{\Lambda,JM_J}$[40]

$$\langle JM_J|\rho|JM'_J\rangle = \Sigma_{\Lambda,\Lambda'} P_{\Lambda',JM_J} \cdot P_{\Lambda,JM_J} \cdot$$
$$\cdot \rho_{\Lambda,\Lambda'} \cdot \exp\{\frac{i}{\hbar}\int_{t_o}^{t_1}(V_\Lambda - V_{\Lambda'})dt\} \quad ;$$

where the exponential term accounts for the time evolution of the system from the time of excitation $t_1$ to $t_o$ where the transition to atomic behavior occurs.

The two basis sets $|JM_J\rangle$ and $|Jm_J\rangle$ are related to each other by a rotation of the molecule fixed frame into the space fixed frame

$$|Jm_J\rangle = \Sigma_{M_J} |JM_J\rangle \ D^{(J)}_{M_J,m_J} \quad ;$$

We finally obtain

$$\rho_{m_J,m_J} = \Sigma_{M'_J,M_J}\Sigma_{\Lambda',\Lambda} D^{(J)}_{M'_J,m_J} \cdot (D^{(J)}_{M_J,m_J})^* \cdot \rho_{\Lambda,\Lambda'} \cdot$$
$$\cdot P_{\Lambda',JM'_J} \cdot P^*_{\Lambda,JM_J} \cdot \exp\{\frac{i}{\hbar}\int_{t_o}^{t_1}(V_\Lambda - V_{\Lambda'})dt\} \quad ;$$

$\rho_{m_J,m_J}$ is still a function of impact parameter; $Q_{m_J}$ is obtained by integrating over all impact parameters $b$ leading to excitation: $Q_{m_J} = 2\pi \int db\ b \cdot \rho_{m_J,m_J}$ ;

For collisions between alkali and rare-gas atoms R $P_{\Lambda,JM_J}$ reduces to a Clebsch-Gordan coefficient, but more complicated transformations are needed f.i. for Ne + Ne[*]

to describe the transition from molecular to atomic behavior. For K + rare-gases the polarisation of the line
$K(4^2P_{3/2} \to 4^2S_{1/2})$ becomes[35] $P = 0.375\ D/(5\ B + 0.125 \cdot D)$

$$D = \int d\vartheta \sin\vartheta \cdot P_2(\cos\vartheta)\{|c_o(b(\vartheta))|^2 - |c_1(b(\vartheta))|^2\}\ ;$$

$$B = \int d\vartheta \sin\vartheta\{\tfrac{1}{3}|c_o|^2 + 2/3|c_1|^2\}\ ;$$

As was already shown in[41,42], P is non-zero only if the
angular distribution of the excited atoms is nonisotropic,
and if the molecular states correlating with $K(^2P_{3/2})$ +
+ $R(^1S)$ are populated nonstatistically. Complete information on the excitation mechanism is only obtained with
complete knowledge of the $c_\Lambda$ as function of the impact
parameter and the collision energy e.g. when the angular
distribution of the excited atoms is known. However when
the impact parameter dependence of $c_o$ and $c_1$ is not too
different, but only the energy dependence differs considerably, qualitative predictions on the excitation mechanism can still be made.[35]

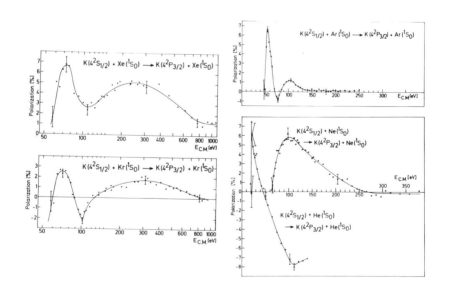

Fig. 4: Total Polarisation P of the line
$K(4^2P_{3/2} \to 4^2S_{1/2})$ for collisions of potassium
with rare-gas atoms (from[35])

35 <u>Fig. 4</u> shows the results for K + He, Ne, Ar, Kr, and
Xe.[35] The results require that
(i) for K + He, and Ne mainly $B^2\Sigma$ is populated at all stu-
    died energies
(ii) for K + Ar, Kr, and Xe $B^2\Sigma$ is populated preferentially
    at low energies, while at high energies the popula-
    tion of $A^2\Pi$ becomes comparable.

<u>Fig. 5</u> shows polarisation data for Na$^+$ + Ne[19,20,21];
measured was the radiation from the decay of states from
the Ne($2p^5 3p$) configuration. The strong oscillatory structure is attributed to long range interaction between direct excitation and charge exchange channels. The fact
that only the $m_J = \pm 1$ sublevels are populated was explained in the following way[19]: since the $m_J = \pm 1$ atomic
levels are populated, this implies that the molecular levels involved are the $\Omega = 1$ levels, where $\Omega$ is the projection of the total angular momentum on the internuclear
axis. Because $m_J$ refers to the space fixed beam axis, $\Omega$
to the rotating internuclear axis, this conclusion is only valid when the rotation of the internuclear axis during the collision can be neglected.

(iv) Information on Excitation Mechanisms from
        Photon-Particle Coincidence Measurements

Differential cross sections for excitation processes
are usually determined from measurements of energy loss
spectra of the scattered particles as a function of the

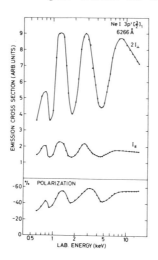

Fig. 5: Total polarisation of the line Ne($2p_5 \to 1s_3$) for
        collisions of Na$^+$ with Ne(ref.[21])

scattering angle.[51] This technique is limited in energy
resolution, and is usually not capable to separate energy
losses which are due to excitation of states with the
same main quantum number. A considerable improvement is
possible if the inelastically scattered particle is measured in coincidence with the emitted photon.[28]

When D. Jaecks gave his[28] report on this technique at
the last ICPEAC conference[28], all possibilities were not
fully utilized yet: coincidence studies were only performed in order to identify the process under study. However
this method also offers the possibility to measure differential cross sections for excitation of specific sub-
states $|Jm_J\rangle$: one either measures the differential pola-
risation $\Pi$, the polarisation of the emission which occurs
in delayed coincidence with the inelastically scattered
particle, or, alternatively, one determines the particle-
photon angular correlation function, the delayed coincidence rate for a fixed particle detector position as function of the angle between photon and particle detector.
The theory for both types of experiments can be taken
from[40,48,49,50] in which the coincidence rate is related
either to scattering amplitudes, or orientation and alignment parameters, or multipole states.

In order to gain information on excitation mechanisms
from differential cross sections for population of the
magnetic substates, the probability to populate these
states has to be correlated to the probability for population of the correlating molecular states. Let me explain this point in some more detail for measurements of
the differential polarisation $\Pi$. When the photon detector
is perpendicular to the scattering plane,

$$\Pi = 3\ h(A_0 - A_2)/\{4 + h(A_0 + 3A_2)\} \quad ;$$

The notation is the same as in (iii). $A_0$ and $A_2$ are alignment components in beam direction and in the plane which
is perpendicular to this axis. Since the scattering angle
is defined, no averaging over impact parameters as in
(iii) occurs, and

$$A_0 = \langle \psi | 3J_z^2 - J^2 | \psi \rangle =$$

$$= \Sigma\ \sigma_{m_J} [3m_J^2 - J(J + 1)]/J(J + 1)\ \Sigma\ \sigma_{m_J} \quad ;$$

$$A_2 = \langle \psi | J_x^2 - J_y^2 | \psi \rangle =$$

$$= \Sigma\ a_{m_J}\ a_{m'_J}^* \langle m_J | J_x^2 - J_y^2 | m'_J \rangle / J(J + 1)\ \Sigma\ \sigma_{m_J} \quad ;$$

As long the angular momenta are coupled to the internuclear axis, only alignment with respect to this axis
should occur; in a frame with the internuclear axis as
quantization axis $A_2$ must disappear. To be more specific,
let me discuss the results of[25]. The experiment is described in some more detail in[26,27]. $\Pi$ is measured for the
transition $He(3^3P \to 2^3S)$ in collisions $He^+ + He$. When the
internuclear axis is taken as quantization axis, $\Pi$ reduces to

$$\Pi = \frac{15(\sigma_o - \sigma_1)}{41\,\sigma_o + 67\,\sigma_1} \quad ;$$

If the photon intensity
is measured as function of $\alpha$, the angle between the po-
lariser axis and the internuclear axis, one obtains

$$I(\alpha) \sim 28\,\sigma_o + 26\,\sigma_1 + (15\,\sigma_1 - 15\,\sigma_o)\sin^2\alpha \quad ;$$

The measurements shown in Fig. 6 give a pronounced maximum in the direction of the internuclear axis indicating
that $\sigma_o \gg \sigma_1$. From this result it was concluded that the
primary excitation mechanism is a transition from the lowest $^2\Sigma_g$-state of $He_2^+$ to the next higher one of the same
symmetry which correlates asymptotically with $He^+$ +
+ $He(3^3P)$. For the same system $\Pi$ was also studied in[52];
while in[25] direct excitation was measured, excitation
with charge exchange was studied in[52].

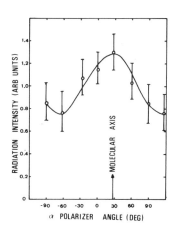

Fig. 6: Coincidence rate for the line $He(3^3P \to 2^3S)$ as
        function of the polariser orientation for
        $He^+ + He$ (ref.[25])

## IV. CONCLUDING REMARKS

In this report it was tried to point out the different possibilities to obtain information on excitation mechanisms for electronic excitation in slow atomic collisions by means of optical methods. Particular attention was paid to measurements of excitation cross sections in the threshold region, to high resolution studies allowing the separation of the multiplet components, and to measurements of the total and differential polarisation.

REFERENCES

1. R. McCarroll, VIII ICPEAC Beograd 1973, Invited Papers and Progress Reports, 71
2. E.W. Thomas, Excitation in Heavy Particle Collisions (N.Y., 1972, Wiley Interscience)
3. V. Kempter, W. Mecklenbrauck, M. Menzinger, G. Schuller, D. Herschbach, Ch. Schlier, Chem.Phys.Lett. 6 (1970) 97
4. M.Ya. Ovchinnikova, E.E. Nikitin, Sov.Phys. Uspekhi 14 (1972) 394
5. V. Kempter, F. Veith, L. Zehnle, IX ICPEAC Seattle 1975, Abstr. Papers
6. R.E. Olson, E.J. Shipsey, J.C. Browne, J.Phys. B, 8 (1975) 905
7. V. Kempter, F. Veith, L. Zehnle, J.Phys.B, 8 (1975) 1041
8. A.N. Zavilopulo, I.P. Zapesochnyi, O.B. Shpenik, Opt. Spectr. 32 (1972) 570
9. S. Bobashev, L. Shmaenok, VIII ICPEAC Beograd 1973, Abstr. Papers, 221
10. V.L. Ovchinnikov, O.B. Shpenik, I.P. Zapesochnyi, Sov.Phys.Tech.Phys. 19 (1974) 382
11. A.N. Zavilopulo, O.B. Shpenik, I.P. Zapesochnyi, Opt. Spectr. 32 (1972) 341
12. S.V. Bobashev, V.I. Ogurtsov, L.A. Razumovskii, Sov. Phys. Jetp 35 (1972) 472
13. O.B. Shpenik, A.N. Zavilopulo, I.P. Zapesochnyi, Sov. Phys. Jetp 35 (1972) 466
14. V.I. Ovchinnikov, G.S. Panev, A.N. Zavilopulo, I.P. Zapesochnyi, O.B. Shpenik, ICPEAC Beograd 1973, Abstr. Papers, 661
15. I.P. Zapesochnyi, S.S. Pop, Opt.Spectr. 33 (1973) 226
16. A.N. Zavilopulo, I.P. Zapesochnyi, G.S. Panev, O.A. Skalko, O.B. Shpenik, Sov.Phys.-Jetp Lett. 18 (1973) 245
17. V. Aquilanti, G. Liuti, F. Vecchio-Cattivi, G.G. Volpi, Entropie 42 (1971) 158
18. V. Aquilanti, G.P. Bellu, J.Chem.Phys. 61 (1974) 1618
19. N. Tolk, C.W. White, S.H. Neff, W. Lichten, Phys.Rev. Lett. 31 (1973) 671
20. T. Anderson, A. Kirkegaard Nielsen, K.J. Olsen, Phys. Rev.Lett. 31 (1973) 739
21. T. Anderson, A. Kirkegaard Nielsen, K.J. Olsen, Phys. Rev. A 10 (1974) 2174
22. B.M. Hughes E.G. Jones, T.O. Tiernan, VIII ICPEAC Beograd 1973, Abstr. Papers, 223
23. R.C. Isler, Phys.Rev. A 10 (1974) 2093
24. R.C. Isler, Phys.Rev. A 10 (1974) 117

25. G. Vassilev, G. Rahmat, J. Slevin, J. Baudon, Phys. Rev.Lett. 34 (1975) 444
26. G. Vassilev, J. Baudon, G. Rahmat, M. Barat, Rev. Scient.Instr. 42 (1971) 1222
27. G. Rahmat, G. Vassilev, J. Baudon, M. Barat, Phys. Rev.Lett. 26 (1971) 1411
28. D. Jaecks, VIII ICPEAC Beograd 1973, Inv. Papers and Progress Reports, 137
29. P.J. Martin, D.H. Jaecks, Phys.Rev. A 8 (1973) 2429
30. P.J. Martin, D.H. Jaecks, Bull.Am.Phys.Soc. 19 (1974) 1174
31. J.S. Risley, F.J. de Heer, C. Kerkdijk, J. Kistemaker, Bull.Am.Phys.Soc. 19 (1974) 1175
32. D. Brandt, Ch. Ottinger, J. Simonis, Ber.Bunsenges. 77 (1973) 648
33. V. Kempter, B. Kübler, W. Mecklenbrauck, J.Phys. B 7 (1974) 149
34. V. Kempter, B. Kübler, W. Mecklenbrauck, J.Phys. B 7 (1974) 2375
35. H. Alber, V. Kempter, W. Mecklenbrauck, J.Phys. B 8 (1975) 913
36. R. Düren, M. Kick, H. Pauly, Chem.Phys.Lett. 27 (1974) 118
37. V. Kempter, W. Koch, B. Kübler, W. Mecklenbrauck, C. Schmidt, Chem.Phys.Lett. 24 (1974) 117
38. V. Kempter, B. Kübler, W. Mecklenbrauck, W. Koch, C. Schmidt, Chem.Phys.Lett. 24 (1974) 597
39. V. Kempter, W. Koch, C. Schmidt, J.Phys. B 6 (1974) 1306
40. U. Fano, J. Macek, Rev.Mod.Phys. 45 (1973) 553
41. R.C. Isler, R.D. Nathan, Phys.Rev. A 6 (1972) 1036
42. R.J. Van Brunt, R.N. Zare, J.Chem.Phys. 48 (1968) 4304
43. V. Kempter, Molecular Beams (J. Wiley, 1975) 417
44. L. Krause, VII ICPEAC Amsterdam 1971, Inv. Papers and Progress Reports, 65
45. H. Rosenthal, Phys.Rev. A4 (1971) 1030
46. F. Masnou-Seeuws, R. McCarroll, J.Phys. B 7 (1974) 2230
47. W. Mecklenbrauck, Dissertation Freiburg (1975)
48. J.H. Macek, D.H. Jaecks, Phys.Rev. A4 (1971) 2288
49. J. Wykes, J.Phys. B 5 (1972) 1126
50. K. Blum, H. Kleinpoppen, J.Phys. B 8 (1975) 922
51. M. Barat, VIII ICPEAC Beograd 1973, Inv. Papers and Progress Reports, 43
52. W. de Rijk, F.J. Eriksen, D.H. Jaecks, Bull.Am.Phys. Soc. 19 (1974) 1230

# ELECTRON EJECTION IN SLOW HEAVY PARTICLE COLLISIONS

R. Morgenstern

Fakultät für Physik der Universität Freiburg/Br.

Hermann Herder-Str. 3     W.Germany

## 1 INTRODUCTION

In this report investigations of ionizing ion-atom
and atom-atom collisions by means of electron analysis
will be described. Only experiments at collision energies
up to several keV will be discussed. In this energy range
direct ionization is not an important process, and therefore electron ejection can always be regarded as due to
autoionization either from one of the collision partners
or from the quasimolecule formed during the collision. Although light emission from autoionizing states has been
observed in some cases[38], in general autoionization times
are very short ($\sim 10^{-14}$ sec) compared to optical decay
times. Therefore in most cases the fluorescence yield of
autoionizing states is negligible and light- and electron-
analysis are not competing methods to investigate a specific excitation but provide complementary information
about competing excitation processes occurring during the
collision.

As compared to the other methods of investigation
electron-spectroscopy has some specific advantages: (i) A
better energy resolution of the excited states can be obtained than in the analysis of the scattered heavy particles. (ii) Selection rules for autoionizing transitions
are not as restrictive as for optical transitions and therefore nearly all excited states lying in the ionization
continuum may autoionize. (iii) The collisionally induced population of autoionizing states is not altered by
subsequent cascading transitions, since autoionization

times are so short. (iV) Also due to the short transition times, autoionization may take place in the quasimolecule. In this case information about potential curves
and transition probabilities in the quasimolecule can be
extracted from the electron spectra.

Progress was made in several respects in the past
few years:

(i) Excitation of autoionizing states was investigated in a large number of collision systems. Agreement
of the experimental results with the predictions of the
Fano–Lichten-model was not always found to be satisfactory (chapter 2).

(ii) Electron emission from the quasimolecule was
found in several collision systems and was analyzed in
detail for some cases (chapter 3).

(iii) Angular distributions of ejected electrons
were measured – in some cases in coincidence with the
scattered heavy particles. An attempt to interpret this
in terms of collisionally induced alignment will be given
in chapter 4.

## 2 ELECTRONS DUE TO ATOMIC AUTOIONIZATION

Since the early investigations of ionizing alkali
ion/rare gas collisions by Beeck[1] and their interpretation
by Weizel and Beeck[2] it was suspected that autoionizing
states play an important role. The first discrete electron energy spectra – from $K^+$/Ne, Kr, Ar collisions –
were obtained by Moe and Petsch[3], but it was not yet realized that they were due to atomic autoionization. Berry[4] showed that atomic autoionization states were responsible for a large share of electrons in collisions of
rare gas ions and atoms. Since then a large number of
electron spectra from slow collisions of rare gas ions
and atoms[5-11] and of alkali ions and atoms[15-21] with rare
gas targets were obtained. Due to improved energy resolution it is often possible to give an identification of the
autoionization states the electrons are due to. At least
one can decide, which one of the collision partners is
ejecting the electrons in a certain peak by looking for
the Doppler-shift of the projectile-emitted electrons  at
different observation angles. These data can be compared
with the predictions of the promotion model, developed by
Fano and Lichten[22,23] and extended by Barat and Lichten[24].
The limitations of this model can be seen in several cases.

## 2.1 Collisionally Excited Autoionization States

The best agreement with the Fano Lichten model is found in collisions of small atoms such as He[+], He or Li[+] colliding on He[9,25,26]. In the single-electron-orbital correlation diagram promotion of two electrons — which is necessary for ionization — is always achieved by radial or rotational coupling of the $2p\sigma$ to the $2s\sigma$ or $2p\pi$ orbital. Since Li(1s) correlates to 1s of the united atom, always He electrons are promoted and therefore the same states should be excited in He[+]/He and Li[+]/He collisions. In a first approximation this is the case, as can be seen in Fig. 1. Differences are the broad structure at energies $> 36$ eV in the Li[+]/He spectrum, the origin of which is not yet clear, and the drastic kinematic broadening of the Li[+]/He-peaks which is due to the large recoil the He suffers at these low collision energies as a consequence of the unfavorable mass ratio. The most important contributions to both spectra are from $He(2p^2)^1D$ and $He(2s2p)^1P$. In the case of He[+] + He these states are excited via the molecular orbitals

$$He_2^+(1s\sigma_g\ 2p\sigma_u^2)^2\Sigma_g \rightarrow He_2^+(1s\sigma_g\ 2p\pi_u^2)^2\Delta_g,\ ^2\Sigma_g$$

which may either separate to $He^+(1s)$ and $He^{**}(2p^2)^1D$, or the system may proceed to

$$He_2^+(1s\sigma_g\ 3p\sigma_u\ 2p\pi_u)^2\Pi_g \rightarrow He^+(1s) + He^{**}(2s2p)^{1,3}P$$

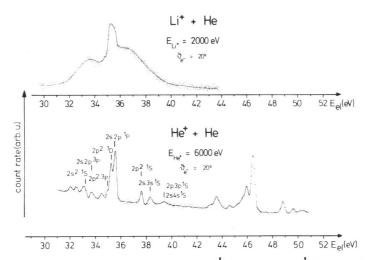

Fig. 1: Electron spectra from He[+]/He and Li[+]/He collisions. Mainly the same autoionization states are excited.[18,39]

In $Li^+$/He collisions the excitation may proceed similarly and in addition the $He^{**}(2s2p)^{1,3}P$ may be reached more directly due to the lack of u-g-symmetry.

In a recent $Li^+$/He collision experiment Stolterfoth et al.[27] observed electrons due to Li-autoionization states even at collision energies as low as 6 keV. This cannot be explained in the Fano Lichten model, but also the corresponding excitation probability was very low (< 0.05% of the He excitation).

Although the spectra resulting from He He collisions look very different since no He autoionization states are excited at low collision energies (Fig. 3), the excitation process is very similar in the first step. But instead of separating to $He + He^{**}$ - a state ∼ 60 eV above the incoming channel - the system rather separates to $He^- + He^+$, where target and projectile obtain one excited electron each, or to $He^+ + He^{-*}$.[9,28] These exit channels are strongly preferred since they lie only ∼ 40 - 45 eV above the incoming one. In a similar way double defect promotion in inner shells is observed to be symmetrical in symmetrical collision systems.[29]

Thus, keeping in mind that the correlation diagram for the molecular states has to be taken into account to see, how the promoted electrons are actually shared between the collision partners, the Fano Lichten model is in agreement with these experiments to a large extent. Similarly the symmetrical rare gas collisions can be described correctly within the Fano Lichten model, as was shown in great detail by Brenot et al.[30]

In Ne Ne collisions again as in He He the two-electron excitation is shared between the collision partners (or $Ne^+ + Ne^{-**}$ is produced), in all other cases rare gas autoionization states are excited.[9] Also it is in accordance with the model that in nearly symmetric collisions of alkali and rare gas atoms preferably the rare gas atom is excited, although the alkali partner contributes autoionization electrons to a perceptible extend.[26]

The situation is drastically different, when asymmetric collisions of rare gas atoms are investigated. In He on Ar, Kr and Xe collisions no He-autoionization was observed,[8] and Brenot et al.[31] in their energy loss measurements could not even observe single excitation of He, although in the diabatic MO diagram the He(1s) is promoted via the 3dσ, 4fσ and 4fσ in the respective correlations with Ar, Kr and Xe. Brenot et al.[31] "repaired" the promotion model by realizing that in the case, where two filled orbitals come close to each other, the system behaves adiabatically. The crossing is avoided and the outermost (looser bound) orbital is promoted. Since Ar3p, Kr4p and Xe5p in the collisions with He are the outermost

ones, their corresponding σ-orbitals are promoted instead of the He(1s).

Nearly the same occurs in other asymmetric rare gas
collisions. Čermák et al.[10] investigated Ar/Kr collisions
and observed electrons predominantly ejected from Kr.
Fetz and Scheidt[11] collided all kinds of rare gas ions on
all other rare gas targets and found out that always the
heavier collision partner is contributing most of the autoionization electrons. This reflects the fact that the
outer p-orbitals of the heavier rare gas partner are looser bound and therefore promoted according to the rule of
Brenot et al.[31]

In just the same way the spectra from collisions of
He with Cs, Rb and K may be explained. They are shown in
Fig. 2. All the electrons are due to alkali autoionization. Since the outer p-orbitals of Cs and Rb are looser bound than He(1s), and that of K is nearly degenerate
with it, one can again understand that they are promoted
instead of He(1s) – in disagreement with the original diabatic one-electron MO-model. However, Na(2p) is much
stronger bound than He(1s). In spite of that Na(2p) apparently is promoted, leading to Na autoionization states.

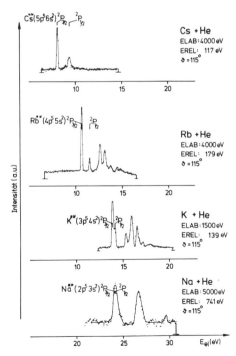

Fig. 2: Electron spectra from alkali/He collisions. Only
alkali autoionization is observed.[19]

This cannot even be understood in a "repaired" one-electron correlation diagram, but only in terms of total molecular states. Obviously interactions between these
states are strong enough to induce transitions between
them regardless if the corresponding one-electron MO's
are promoted or not.

Other systems were investigated at low collision
energies in the last years: Čermák et al.[32,33] collided
He on $O_2$, $N_2$, $N_2O$ and NO and observed autoionization
electrons from N and O. Gerber[39] collided $H^+$ and $H_2^+$ on
rare gases. Much progress was made in the analysis of
electrons from negative-ion atom collisions. Since there
is another report on this subject[34] it is not further discussed here.


## 2.2 Cross Sections and Excitation Functions

So far, accurate absolute cross sections for electron production in low energy neutral-neutral or positive
ion-neutral collisions have not been determined. Cross
sections for electron detachment in collisions of negative
alkali ions with He[35] and of $H^-$ on various gases[36] have
been measured but of course they are much larger (several
$Å^2$) than the former ones.

Some information about absolute cross sections may
be deduced from excitation functions, as was done in[9] for
Ar/Ar collisions. The cross section for excitation of
$Ar^{**}(3s3p^6 4s)$ could be fit to the Landau Zener theory and
from the fit-parameters a cross section of 1.14 $Å^2$ could
be deduced. This is believed to be in the right order of
magnitude.

Excitation functions in other cases were measured
(e.g. in[17]) and mostly did not have a Landau-Zener type
shape. This is not surprising because mostly the various
exit channels are not well separated but strongly interact with each other. A good example for this has been observed in collisions of $K^+$ on Ar, where a regular oszilla-
tion in the excitation cross section has been measured.[37]


## 3 AUTOIONIZATION OF THE QUASIMOLECULES

One influence of quasimolecular effects on ejected
electron spectra is the distortion of atomic autoionization.
E.g. an influence of the He collision partner on the angular distribution of electrons, resulting from $Ar^{**}$ autoionization[8] and from $H^-$ autodetachment[42] was observed.
An influence on position and width of autoionization peaks
due to the long range of the Coulomb potential is expect-

ed, if the other collision partner is an ion. This was discussed for $He^+$/He collisions by Barker and Berry.[5] Although Gordeev and Ogurtsov[48] showed that the peakwidths in the spectra, discussed in[5] were due to Doppler-broadening, the influence of the $He^+$ on the $He^{**}$ autoionization does exist, as was observed by Gerber et al.[7] Since valuable information on autoionization lifetimes can be deduced from this "Stark"-shift and -broadening, it is desirable to get rid of the kinematical peak-broadening. For this reason electron spectra were recently observed at $180^\circ$ with respect to the beam,[41] since here kinematical broadenings are strongly reduced.

Another influence of quasimolecular effects on the spectra is a type of autoionization, which can only occur at finite distances of the collision partners. This Penning type ionization was observed in Rb/Ar[43] and He He[9,39] collisions.

As also discussed in chapter 2, He/He collisions result in an excitation of both collision partners to various excitation states. The corresponding molecular states lie in the ionization continuum and may decay to $He + He^+ + e^-$. An electron spectrum is shown in Fig. 3. Indicated in the figure are electron energies which would be expected for a mutual ionization of the collision partners at infinite internuclear distances. For transitions at a range of somewhat smaller separations one expects peaks which - according to the variation of the involved potential curves - are shifted and broadened, as is actu-

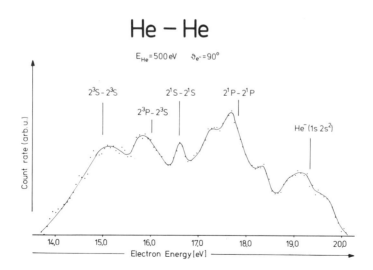

Fig. 3: Electron spectrum from He/He collisions due to Penning type ionization of $He^* + He^*$.[39]

ally observed. Since the peaks are mostly shifted to lower energies, the transitions appear to occur predominantly at distances, where the potential curve corresponding
to $He^* + He$ has a well; the curve for $He^+ + He$ is nearly
flat here. Classically one expects a singularity in the
electron spectrum at the minimum-energy of the difference potential. In a semiclassical theory[44] this singularity is replaced by the first maximum of a squared Airy-
function $Ai^2(-x)$ with

$$x = [ \hbar \, v(R^*) ]^{-2/3} \left( \frac{E_{el}^*{}''}{2} \right)^{-1/3} (E_{el} - E_{el}^*)$$

where $v$ is the radial relative velocity of the collision
partners, $E_{el} = V^*(R) - V^+(R)$ is the difference potential of excited and ionized state and $R^*$ and $E_{el}^*$ are internuclear
distance and electron energy at the minimum of the difference potential. At the classically expected singularity
this distribution reaches 44% of its maximum value. The
"width" of the distribution — defined here as separation
in energy between the maximum and the 44%-value at the
low-energy-slope of the peak — is increasing with the relative velocity of the collision partners as $v^{2/3}$. In
Fig. 4 this "width", as determined from the low energy
peak in the measured spectra, is plotted versus $v^{2/3}$. The
linear dependence on $v^{2/3}$ indicates that the ionization
process is reasonably described by the Penning theory. If
the difference potential near its minimum is approximated
by a parabola one can deduce its curvature at the minimum
from the slope of this linear $v^{2/3}$-dependence, and the
welldepth can be deduced from the energetical separation
of the 44%-value of the distribution from the difference
potential at infinity. The corresponding data obtained
in[39] for the $He(2^3S) + H(2^3S)$ potential curve are

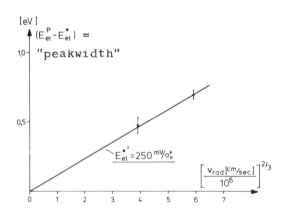

Fig. 4: Energy dependence of the "width" of the electron
peak due to ionization of $He(2^3S) + He(2^3S)$.[39]

$E_{el}^{*}{}'' = 250$ meV/$a_o^2$ for the curvature and 720 meV for the welldepth.

   A more advanced analysis of these measurements is given by Gerber and Niehaus.[45] They show that the low energy edge of an electron distribution is well described by an improved formula up to collision velocities corresponding to $\sim$ 500 eV for the He/He system. This formula has the $Ai^2(-x)$ function as its low velocity limit. In case of the $^1\Sigma_g^+$ potential curve dissociating to two He($2^3S$) atoms the analysis using the improved formula is consistent with the theoretical welldepth of 520 mV, and the theoretical curvature of $\sim$ 250 mV/$a_o^2$.[46] Alkali rare gas collisions should be predestinated for a Penning type ionization process, since a single excitation of the rare gas partner even in its lowest state is sufficient to ionize the alkali partner. For Rb/Ar collisions this has been observed by Bydin et al.[43] Spectra from more recent measurements[19] are shown in Fig. 5. An electron energy of 7.37 eV is expected for a mutual ionization of Rb + Ar at infinite internuclear distance. Again the observed peak is shifted to slightly lower energies and is broader than those from Rb autoionization states. From an analysis of

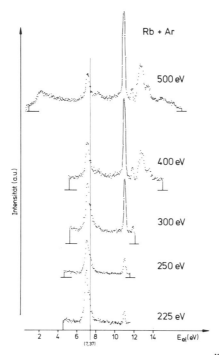

Fig. 5: Penning type ionization of Rb + Ar*, as observed in Rb/Ar collisions at various energies.[19]

this, one can deduce a welldepth of 0.45 eV for the $Rb/Ar^*$ potential curve with a curvature at its minimum of $5.5$ eV/$Å^2$. This seems to be reasonable when one compares with the interaction potentials of $K_2$, or $Rb_2$, which can be expected to have a similar shape as $Rb/Ar^*$.

Brenot et al.[31] suggested that also the structure at 16.1 eV, observed in electron spectra from He/Ne collisions, might possibly be due to this Penning type ionization. We believe that this structure is due to the formation of $He^+ + Ne^{-*}$ with subsequent decay of $Ne^{-*}$, especially since the peakwidth decreases with increasing collision energy. However in He/Ne collisions an intense continuous electron spectrum is underlying the discrete structure and the Penning type ionization probably plays an important role in He/Ne as well as in Ne/Ne collisions. But it seems likely that the transition probability is significantly higher than in $He^* + He^*$ and most of the transitions occur at smaller internuclear distances. This compares well with the observation of Brenot et al.[30] that the ratio of the cross section for ionization and double excitation is significantly higher in Ne/Ne collisions than in He/He collision.

Ionizing transitions at small internuclear distances which result in a continuous electron distribution at low energies are described theoretically by Demkov and Komarov[47], by Smirnov[48] and by Dalidchick.[49] A conclusive comparison of these theories with measured electron spectra has not yet been made due to the difficulties to measure continuous electron distributions at low energies without a disturbance from background electrons. Since this problem is not met in energy loss measurements of the scattered heavy particles it appears that they are better suited for a comparison with the theories.

## 4. COINCIDENCE MEASUREMENTS AND COLLISIONALLY INDUCED ALIGNMENT

Measurements of coincidences between electrons and scattered heavy particles provide additional information about the excitation process: (i) a correlation of energy loss and electron energy often clarifies if the system is electronically excited after the autoionization, (ii) background electrons can be discerned from a real continuous electron distribution at low energies, (iii) excitation functions for single states can be determined differential with respect to the scattering angle, and (iv) collisionally induced alignment can be observed without averaging about the scattering angle.

First measurements of coincidences were reported by

Gerber and Niehaus.[50,51,52]. In their experiment they used an ion spectrometer with rotational symmetry around the beam axis; an annular slit admitted ions, which were scattered at a certain angle $\Theta$ with respect to the beam, but at any azimuth angle $\Phi$. However, as was discussed by Gordeev and Ogurtsov[40], the electron energy varies dependent on $\Phi$ and thus by an electron-energy analysis, actually also the azimuthal scattering angle $\Phi$ of the autoionizing atoms is determined. If the electron detector is at an angle $\vartheta$ with respect to the beam and at an azimuth angle $\varphi$, the electron energy $E_e$ is given by

$$E_e = E_{eo} + A + B \cdot \cos(\varphi - \Phi) \tag{1}$$

where $E_{eo}$ is the electron energy in a coordinate system attached to the autoionizing atom, and A and B are functions of $\vartheta$, $\Theta$ and the velocities of electron and atom. The energy distribution of electrons emitted from atoms with a fixed scattering angle $\Theta$ and equally distributed $(\varphi - \Phi)$ consists of a symmetrical, doubly peaked structure with a peak separation 2B.

A measured spectrum of electrons from $He^+/He$ collisions, measured in coincidence with ions at $\Theta = 8°$ is shown in Fig. 6. Acutally the spectrum consists of two doubly peaked structures with an energy separation A due to electron emission from target and projectile. An important feature is the asymmetry of both of these structures, which may indicate a correlation between $\Phi$ and $\varphi$. Such a corre-

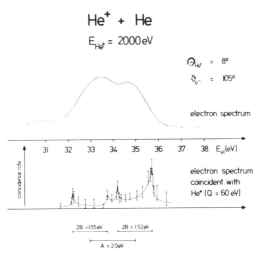

Fig. 6: Electron-ion-coincidences. Different electron energies correspond to different azimuth angles of ion-scattering. [39]

lation – and thus the measured asymmetry – may be due to
a collisionally induced alignment of the electron-emitting
atom.

Electron emission from aligned states has been observed and theoretically described by Cleff and Mehlhorn[53]
and by Risley and Geballe.[42] But there rotational symmetry
around the beam was assumed, which cannot be done in our
case, since a distribution dependent on $(\varphi - \phi)$ is to be
described. Therefore in our group some formulas were deduced[56] to describe measured distributions: The beam axis
is chosen as z-coordinate. Neglecting spin variables, the
initially excited state is described by

$$\psi_a = \sum_M a_M \mid \alpha\, L\, M \,\rangle \tag{2}$$

with complex scattering amplitudes $a_M$, which depend on
the scattering angles $\Theta$ and $\phi$ of the heavy particle as
$a_M(\Theta) \exp(-iM\phi)$. The final state of the system is written
as a product $\psi_f = \psi_I \cdot \varphi_k$ of an ion wave function

$$\psi_I = \sum_{M'} b_{M'} \mid \beta\, L'\, M' \,\rangle \tag{3}$$

and a plane wave for the ejected electron

$$\varphi_k = e^{i\vec{k}\vec{r}} = \sum_l \sum_m c_{lm}^{\vec{k}} \mid \gamma\, l\, m \,\rangle \tag{4}$$

Here $\alpha$, $\beta$ and $\gamma$ are quantum numbers describing variables
other than angular momenta and which are not further regarded. The number of electrons measured in coincidence
with scattered heavy particles is proportional to the
square of the matrix element $M_{af} = \langle \psi_f/A/\psi_a \rangle$ where A is
the autoionization operator. If A is approximated by a
tensor operator of zero[th] order, proper account is taken
of addition of angular momenta, and the Wigner-Eckart-
theorem is used, one obtains

$$M_{af}^o = \sum_{M=-L}^{L} \sum_{l=|L-L'|}^{L+L'} \sum_{m=-l}^{l} b_{M-m}^* c_{lm}^{\vec{k}*} a_M$$

$$\cdot R_{LL}^o(\delta,\alpha|L'1)\langle L'1(M-m)m|LM\rangle \tag{5}$$

Here $R_{LL}^o$ is a reduced matrixelement and $\delta$ combines properly the quantum numbers $\beta$ and $\gamma$. For an autoionizing
transition $^1P \rightarrow {}^1S$ the neglection of spin may be justified. For this case a simple formula for the angular dis-
tribution of coincident electrons $I \propto |M_{af}^o|^2$ is obtained
from (5). Using $a_1 = -a_{-1}$ (which follows from mirror symmetry in the scattering plane) and writing

$$|a_o|^2 = \sigma_o \qquad |a_1|^2 = \sigma_1 \qquad \mathrm{Re}\langle a_o a_1 \rangle = \sqrt{\sigma_o \sigma_1}\cos\chi \tag{6}$$

where $X$ is the relative phase between $a_o$ and $a_1$ one obtains

$$I \propto \sigma_o \cos^2\vartheta + 2\sigma_1 \sin^2\vartheta \cos^2(\varphi - \Phi)$$

$$- 2\sqrt{2\sigma_o\sigma_1} \sin\vartheta \cos\vartheta \cos(\varphi - \Phi) \cos X \qquad (7)$$

This formula has the same structure as that obtained by Macek and Jaecks[54] for the angular distribution of photons from a $^1P \rightarrow {}^1S$ transition, which are measured in coincidence with scattered particles, and which was used by Eminyan et al.[55] to interprete their corresponding measurements. A transformation to an energy scale using (1) gives

$$W(\epsilon) \propto 1/(1 - \epsilon^2)^{1/2} \cdot \left\{ \cos^2\vartheta + 2\sigma_1/\sigma_o \cdot \epsilon^2 \sin^2\vartheta \right.$$

$$\left. - 2^{3/2}\epsilon(\sigma_1/\sigma_o)^{1/2} \sin\vartheta \cos\vartheta \cos x \right\} \qquad (8)$$

with $\epsilon = (E_e - E_{eo} - A)/B$. With (8) one can describe asymmetric energy distributions of coincident electrons like the one shown in Fig. 6. Since at least an important contribution of He autoionization electrons is due to $He^{**}(2s2p)^1P$, autoionizing to $He^+(1s)^2S$, as can be seen from Fig. 1, the formulas deduced here may be applicable. Provided that additional checks are made to exclude experimental artefacts and that more measurements are made, coincidence measurements like these will enable us to obtain cross section ratios and relative phases for the population of magnetic sublevels and thus provide a full information on the excitation process.

I would like to acknowldege a large number of fruitful and elucidating discussions with Dr. Arend Niehaus.

1   O  Beeck, Ann.d.Phys. 6 1001 (1930)
2   W  Weizel, O  Beeck Z.f.Phys. 76 250 (1932)
3.  D E  Moe  O H  Petsch Phys.Rev. 110 1358 (1958)
4   H W  Berry  Phys.Rev. 127 1634 (1962)
5   B  Barker  H W  Berry  Phys.Rev. 151 14 (1966)
6   H W  Berry  Phys.Rev. A 6 1805 (1972)
7   G  Gerber  R  Morgenstern  A  Niehaus  Phys.Rev.Lett. 23 511 (1969)
8   As 7   but J.Phys. B 5 1396 (1972)
9   As 7   but J.Phys. B 6 493 (1973)
10  V Čermák M Smutek J Šrámek J.Electr.Spectr.2 1 (1973)
11  H Fetz H Scheidt pg.876 in$^{12}$ and priv.comm.
12  IX ICPEAC Seattle, Book of Abstracts (1975)
13  VIII ICPEAC Beograd, Book of Abstracts (1973)
14  VII ICPEAC Amsterdam, 1971 Book of Abstracts (1971)

15 V I Ogurtsov Yu F Bydin Sov.Phys.JETP Lett.$\underline{10}$ 85 (1969)
16 V I Ogurtsov Yu F Bydin Sov.Phys.JETP Lett.$\underline{30}$ 1032 (1970)
17 Yu F Bydin V A Vol'pyas V I Ogurtsov Sov.Phys.JETP Lett. $\underline{18}$ 322 (1973)
18 U Thielmann Dipl.-Thesis Freiburg 1974 (unpubl.)
19 R Morgenstern M Trainer pg. 872 in[12]
20 R Morgenstern M Trainer to be publ. in J.Phys. B
21 Yu F Bydin V A Vol'pyas S S Godakov pg. 867[12]
22 U Fano W Lichten Phys.Rev.Lett.$\underline{14}$ 627 (1965)
23 W Lichten Phys.Rev.$\underline{164}$ 131 (1967)
24 M Barat W Lichten Phys.Rev.$\underline{A6}$ 211 (1972)
25 D C Lorents G M Conklin J.Phys.$\underline{B5}$ 950 (1972)
26 R François D Dhuicq M Barat J.Phys.$\underline{B5}$ 963 (1972)
27 N Stolterfoth priv.comm.
28 R Morgenstern M Barat D C Lorents J.Phys.$\underline{B6}$ L330 (1973)
29 M E Rudd B Fastrup P Dahl F D Showengerdt Phys.Rev.$\underline{A8}$ 220 (1973)
30 J C Brenot D Dhuicq J P Gauyacq J Pommier V Sidis M Barat E Pollack Phys.Rev. $\underline{A11}$ 1245 (1975)
31 As 30 but Phys.Rev.$\underline{A11}$ 1933 (1975)
32 V Čermák J Šramek J.Electr.Spectr.$\underline{2}$ 97 (1973)
33 V Čermák J.Electr.Spectr.$\underline{3}$ 329 (1974)
34 A K Edwards Progress Report at this Conference
35 Yu F Bydin Sov.Phys.JETP $\underline{23}$ 23 (1966)
36 J S Risley R Geballe Phys.Rev.$\underline{A9}$ 2485 (1974)
37 as 21 but Phys.Lett.$\underline{50A}$ 239 (1974)
38 I S Aleksakhin G G Bogachev V S Bukstich I P Zapesochnyi Sov.Phys.JETP Lett.$\underline{18}$ 360 (1973)
39 G Gerber Ph.D.thesis Freiburg 1974 (unpubl.)
40 Yu S Gordeev G N Ogurtsov Sov.Phys.JETP $\underline{33}$ 1105 (1971)[12]
41 R Morgenstern A Niehaus U Thielmann pg. 870 in[12]
42 J S Risley R Geballe Phys.Rev.$\underline{A10}$ 2206 (1974)
43 Yu F Bydin V I Ogurtsov V S Sirazetdinov Sov.Phys.JETP Lett.$\underline{13}$ 240 (1971)
44 W H Miller J.Chem.Phys.$\underline{52}$ 3563 (1970)
45 G Gerber A Niehaus submitted to J.Phys.B 1975
46 B J Garrison W H Miller H F Schäfer J.Chem.Phys.$\underline{59}$ 3193 (1973)
47 Yu N Demkov L V Komarov Sov.Phys.JETP $\underline{23}$ 189 (1966)
48 B M Smirnov Sov.Phys.JETP $\underline{26}$ 812 (1968)
49 F I Dalidchik Sov.Phys.JETP $\underline{39}$ 410 (1974)
50 G Gerber A Niehaus Phys.Rev.Lett.$\underline{31}$ 1231 (1973)
51 G Gerber A Niehaus pg. 179 in[13]
52 G Gerber A Niehaus B Steffan J.Phys.$\underline{B6}$ 1836 (1973)
53 B Cleff W Mehlhorn J.Phys.$\underline{B7}$ 593 and 605 (1974)
54 J Macek D H Jaecks Phys.Rev.$\underline{A4}$ 2288 (1971)
55 M Eminyan K B MacAdam G Slevin H Kleinpoppen J.Phys. $\underline{B7}$ 1519 (1974)
56 A Niehaus R Morgenstern U Thielmann to be publ.

# INNER SHELL IONIZATION PHENOMENA

# EXPERIMENTAL STUDIES OF INNER-SHELL EXCITATION IN SLOW ION-ATOM COLLISIONS

Bent Fastrup

Institute of Physics, University of Aarhus

DK-8000 Aarhus C, Denmark

The main object of this invited talk is a discussion of inner-shell excitations in slow ion-atom collisions. The period up to 1972/73 has been covered by extensive review articles [1,2,3]. A contribution by Briggs [4] at this conference reviews the theoretical aspects of inner-shell excitation in slow ion-atom collisions. I shall therefore mainly discuss experimental data published since 1972/73. The limited time, however, will not permit a thorough discussion of all interesting aspects, and hence only such data which may illustrate various inner-shell excitation mechanisms will be considered.

Being close to the nuclei inner-shell electrons act as independent particles. The interaction with the other electrons is well accounted for through their effective screening of the nuclear fields. This means that single electron wavefunctions are adequate in the description, and that correlation effects ordinarily play an insignificant role. For the case of K electrons which are only slightly screened by the other electrons simple scaling properties exist which allow establishment of reduced, universal excitation cross sections [5,6].

In the following we shall deal with collisions for which $Z_1 \simeq Z_2$ and $v < v_{el}$, where $v$ is incident particle velocity and $v_{el}$ is orbital velocity of electron to be removed. This prescription ensures that quasi-adiabatic conditions prevail throughout the collision time. Hence, the molecular-orbital-promotion (MO) model [7,8] is expected to provide a good framework for our discussion of inner-shell processes. For the sake of comparison, some reference will also be made to the direct- or Coulomb excitation mechanism [5] which essentially applies when $Z_1 \ll Z_2$, i.e. for light incident particles.

Although the MO model and the Coulomb excitation model have their individuel and non-overlapping domains of applicability, there have recently been several attempts to modify both. By including typical low-velocity effects such as increased binding of target electrons due to the passing charged particle and certain kinematical effects, it has been possible to extend the applicability of the Coulomb excitation method considerably [9]. Such corrections tend to reduce the K excitation cross section. In the MO picture this situation corresponds to direct ionization from the demoted $1s\sigma$ which correlates with the high-Z collision partner.

In the next section I shall present and discuss some very recent attempts to deal with K excitation in medium and heavy collision systems with $Z_1 \sim Z_2$.

## INNER-SHELL EXCITATION MECHANISM

The Coulomb-excitation model yields in its BEA formulation, see, e.g., McGuire and Richard [10], a cross section $\sigma(V)$ for an inner-shell ionization,

$$\sigma(V) = NZ_1^2 \sigma_o G(V)/U^2 ,\qquad (1)$$

where N is the number of electrons in the shell from which the electron is removed, $Z_1$ is the projectile charge, $\sigma_o$ is a constant, U is the binding energy of the electron to be removed, and G(V) is a reduced cross section depending only on a reduced velocity, $V = v/v_{el}$ . The functional form of G(V) depends on the shell to be excited.

If we make the approximation $U = \frac{1}{2} Z_{2,\text{eff}}^2$ for K-binding energy, Eq.(1) displays simple scaling properties for K excitation. In the derivation of the formula, straight-line trajectories have been assumed which means that it does not account for kinematical effects, i.e., threshold behaviour of cross sections near onset.

In the MO model, inner-shell excitations are caused by strong interactions at crossings or between nearly degenerate MO's. This means that the complete collision process may be treated as a series of separable two-state interactions, an example of which is the production of a K vacancy in the high-Z partner where a $2p\sigma$-$2p\pi$ rotational coupling at small internuclear separation distances is followed by a $2p\sigma$-$1s\sigma$ radial coupling as the collision partners separate. On top of that, the Pauli principle brings in socalled exit-channel effects [6,7] to be discussed later, which are governed by the precollision-

al state of the outer shells. It is fairly evident that the resulting cross section for such a chain of processes does not possess simple scaling properties. If, however, we restrict our
attention to one specific interaction, such as the $2p\sigma$-$2p\pi$
rotational coupling leading to production of a K vacancy in
either partner, simple scaling properties may be reestablish-
ed. This was done originally by Briggs and Macek [6], who
gave the following scaling law for symmetric collision systems,

$$\sigma_K(Z,v) = N_o Z_s^{-2} \sigma_{rot}(v/Z_s) ,\qquad (2)$$

where $N_o$ is the $2p\pi$-vacancy number ($= 1/6$ of number of initial 2p vacancies), $Z_s$ lies between $(Z-1)$ and $(Z-1/2)$, and
$\sigma_{rot}$ is the $D^+$- D cross section for a K excitation assuming
a single $2p\pi_x$ vacancy. Very recently, Taulbjerg et al.[11]
have shown that a scaling law may also be established for

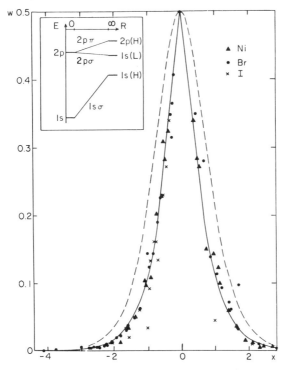

Fig.1 K-vacancy transfer probability W for Ni, Br, and I
beams for extended target and energy ranges versus x [12].

asymmetric collision systems. These scaling laws, which are
based on a detailed study of the $2p\sigma$-$2p\pi$ rotational-coupling

mechanism, are applicable only when this approximation is valid, i.e., at low velocities.

In his treatment of the K-vacancy-sharing process, which is related to a $2p\sigma$-$1s\sigma$ radial coupling, Meyerhof [12] very convincingly showed that a simple formula accounts for a large variety of experimental data on K-vacancy sharing, see Fig. 1. This formula, which is essentially based on Demkov's previous treatment [13] of the interaction between two parallel and nearly degenerate states, yields

$$R = e^{-2x} : 2x = \sqrt{2} \pi(I_H^{1/2} - I_L^{1/2})/v , \qquad (3)$$

where R is the ratio between the numbers of K vacancies produced in the heavy and light collision partners, respectively, and $I_H$ and $I_L$ are the corresponding atomic binding energies in a.u. The formula possesses simple scaling properties because $(I_H^{1/2} - I_L^{1/2}) = \frac{\gamma}{\sqrt{2}}(Z_H - Z_L)$, where $\gamma < 1$ is a screening constant.

The probability W of transferring a vacancy from $2p\sigma$ to $1s\sigma$ is given by $W = R/(1+R)$.

If, in Eq.(2), the vacancy number $N_o$ is zero, the cross section for a K excitation vanishes. For symmetric collision system, this corresponds to a situation where initially both collision partners have filled 2p shells like in Ne-Ne. For asymmetric collision systems where the number of $2p\pi_x$ vacancies is related to the number of 2p vacancies in the high-Z collision partner [6], a similar situation occurs when the 2p shell of this partner is initially filled as in N-Ne. This effect, which was referred to as the exit-channel effect, has been verified experimentally [1,14]. As has also been shown experimentally, the exit-channel effect is a typical low-velocity effect [14]. When the projectile velocity becomes comparable with a characteristic orbital velocity of outer-shell electrons, the exit-channel effect loses its significance. This has led to a modification [14] of the simple two-state rotational-coupling approximation,

$$\sigma_K = [N_o + N(v)]\sigma_{rot}(v) , \qquad (4)$$

where $N_o$ being a static $2p\pi_x$ vacancy number is 1/6 of initial number of 2p vacancies for symmetric systems, i.e., $N_o = (1/6)N_{2p}$, and 1/3 of initial number of 2p vacancies in high-Z collision partner for asymmetric systems, i.e., $N_o = (1/3)N_{2p}$. N(v) is a velocity-dependent component of the vacancy number and $\sigma_{rot}$ is the cross section of the transfer of a single $2p\pi_x$ vacancy into the $2p\sigma$ MO. It is obvious that $N(o) = 0$ in Eq. (4).

Although the modification seems purely phenomenological, some justification of it may be given. As $v$ is increased and becomes comparable with outer-shell orbital velocities, couplings, presumably of radial type, between $2p\pi$ MO and unoccupied higher MO's may produce additional $2p\pi_x$ vacancies (at most two). Such processes, which probably occur at fairly large internuclear-separation distances on the inward half of the collision, are well separated in time from the rotational coupling which takes place near the united limit, hence the product form of Eq.(4).

Unfortunately, no 'ab initio' theoretical studies have been carried out so far on these outer-shell processes determining $N(v)$.

Meyerhof [15] has proposed that since these processes occur at fairly large separation distances and presumably between close-lying MO's, the formula from the K-vacancy-sharing process, Eq.(3), may be used directly to estimate their strength and velocity dependence when appropriate atomic binding energies for the interacting MO's are applied. Although data on $N(v)$, which have been derived from Eq.(4) using experimental $\sigma_{KLL} \simeq \sigma_K$ and calculated $\sigma_{rot}$ [16] display a velocity dependence roughly in accordance with Eq. (3), there seems to be some ambiguity in selecting the proper atomic binding energies in Eq.(3).

Even though Eq.(4) represents a reasonable extension of the simple two-state rotational approximation, we shall see that other competitive processes may also be brought into the discussion of K excitations.

Two such approaches are due to Meyerhof [17] and Saris [18]. They both assume that direct ionization to the continuum is a plausible mechanism for K excitation when the exit channel is closed.

In contrast to the original and well-documented modification [9] of the BEA or PWBA approximation, where the passing point charge causes an increase in the K-binding energy of the target atom and thereby a decrease in the K-ionization cross section at low velocities, Meyerhof and Saris have noted that when $Z_1 \simeq Z_2$, molecular promotion of the $2p\sigma$ MO may effectively reduce the binding energy of the two K electrons correlating with the K shell of the low-Z collision partner. Hence, in the united limit, a direct ionization from the $2p\sigma$ MO is a more likely process than it would be if atomic K-binding energies were used. This approach does not contradict the binding-energy correction approach to the BEA model since this latter modification refers to the K ionization of the target atom or the high-Z partner. If, moreover, the $2p\sigma$ ionization probability is multiplied by W, the transfer probability of a vacancy from $2p\sigma$ to $1s\sigma$, the probability of producing a K vacancy in the high-Z collision partner results.

In his approach, Meyerhof [17] suggests that direct ionization from $2p\sigma$ is caused by rotational coupling to continuum $\mu\epsilon\pi$ states. Such a mechanism has been advanced before [3] to explain production of K vacancies in energetic Ar-Ar collisions. In the absence of theoretical calculations, Meyerhof has estimated a rotational-coupling cross section from appropriate results for $H_2^+$ system [19] by a scaling procedure similar to Eq.(2). The justification of this procedure is dubious.

In his treatment of K excitation when the exit channel is closed, Saris [18] proposes a mechanism which basically is a simple modification of the BEA method. Instead of using unperturbed atomic-binding energies, BEA(1s), he uses their united atom velues, i.e., $2p\sigma$ energy at united atom limit (low-Z partner) and $1s\sigma$ energy at the adiabatic limit in SCA [20] (high-Z partner). It is apparent that a smooth transition from the modified BEA to the BEA must take place as the collision velocity is increased. Hence, appropriate $2p\sigma$ and $1s\sigma$ binding energies must depend on velocity and tend towards their separated atom's values as velocity is raised. Introducing an effective atomic number, $Z_{eff}^2 = \frac{1}{2}(Z_1^2 + Z_1^2)$, the modified BEA yields

$$\sigma_K = Z_{eff}^2 \, G(E/\lambda \, U_{ua})/ U_{ua}^2 \,, \qquad (5)$$

where $U_{ua}$ is the united atom-binding energy and G is defined in Eq.(1).

When applying this formula to a series of symmetric and asymmetric systems for which the exit channel is closed, Saris claims to get an overall good agreement with experimental data. This agreement has even been extended to systems like N-Ne, for which the two-step process, Eq.(4), is more likely.

In summary, Fig.2 shows the lower MO's which may play a role in the K-excitation process. The various excitation mechanisms are depicted: (a), two-state rotational

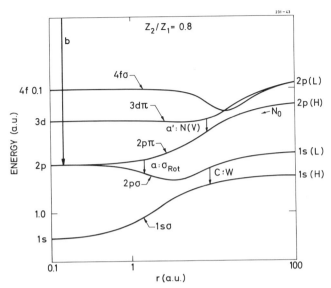

Fig. 2 One-electron calculated MO curves, $Z_1 = 1$ and $Z_2 = 0.8$ [16].

coupling, Eq.(2); (a'), two-step process, Eq.(4); (b), direct ionization from $2p\sigma$, Eq.(5).

So far, we have only discussed K-excitation processes. Excitations of other inner shells, L , M , etc., differ distinctively from the excitation of the K shell. While the main feature in the K-excitation mechanism is the $2p\sigma$-$2p\pi$ rotational coupling, highly promoted MO's are responsible for the excitation of the other inner shells [1]. $4f\sigma$ for the L shell, $6h\sigma$ for the M shell, etc. By crossing a multiplicity of excited MO's, these highly promoted MO's carry one or two electrons into an excited state, leaving one or two inner-shell vacancies, respectively. A large volume of experimental data exists on particular L-shell excitations. Since these data have been discussed in various review papers [1,2], we shall refrain from discussing them here. It might, however, be appropriate to mention that recent, very detailed studies [21] of rare-gas-atom collisions at low impact energies have revealed similar features for excitation of outer shells. For example, one observes L-shell excitation via $4f\sigma$ MO for Ne-Ne, M-shell excitation via $5f\sigma$ MO for Ar-Ar, see Fig.3, and N-shell excitation via $6f\sigma$ MO for Kr-Kr. All these excitations are characterized by a highly promoted MO, which carries one or two electrons up to a multiplicity of highly excited MO's.

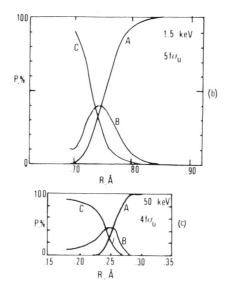

Fig. 3  Upper curve: outer M-shell excitation [21]. Lower curve: inner L-shell excitation [7]. No excitation (A), one electron excited (B), and two electrons excited (C).

Although in recent years, intense experimental and theoretical studies of inner-shell excitations have resulted in a deeper insight in inner-shell excitations in heavy ion-atom collisions, there still remain many unsolved problems. Let me mention a few here: Inner-shell excitation of L, M etc. has barely been touched theoretically. For the K-shell excitation where good agreement between experimental data and theory is obtained when the exit channel is open, i.e. when $(Z_1, Z_2) < (10, 10)$, there exist no detailed calculations when the exit channel is closed, i.e. when $(Z_1, Z_2) \geq (10, 10)$. For example, it is an open question when a two-step process is the leading process and when a direct ionization process is the leading process. It might also be mentioned that although we consider correlation effects or two electron transitions to be of negligible importance in inner-shell excitation, there seem to be some indication [22] that in rare cases such effects may become important.

## EXPERIMENTAL TECHNIQUES

In the investigation of inner-shell excitations in ion-atom collisions, several different methods have been employed: (a) Inelastic-energy loss measured in differential-scattering experiments; (b) Auger-electron emission yield, and (c) x-ray emission yield. Most data under (b) and (c) are total cross sections, i.e., inherently integrated over all impact parameters, although in a few cases also differential data have been obtained by measuring the emitted radiation in coincidence with the scattered projectiles, see, e.g., Fig.4.

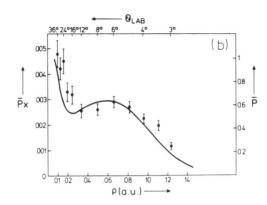

Fig.4 Experimental x-ray yields $\bar{P}_x$ versus impact parameter in 363-keV Ne$^+$- Ne collisions. Also shown is the theoretical vacancy-production yield $P$ [25]

Several review papers have described the various methods in detail [1,2,23,24]. Hence only a brief survey of interest for the forthcoming discussion of the data will be given here.

(a) <u>Inelastic energy-loss</u> data have provided detailed information on the excitation function [1,2]. This method, which has provided a convincing experimental verification of the MO model, is, however, restricted to collision systems for which excitation energy as well as excitation probability of a specific inner shell are appreciable so that isolated peaks occur in the energy-loss spectrum. This limitation implies that essentially only cases, for which a simple MO excitation mechanism is fully operative, can be studied in detail, i.e., K excitations when $(Z_1, Z_2) \leq (10, 10)$ , L excitation when $(12, 12) < (Z_1, Z_2) < (25\text{-}30, 25\text{-}30)$, etc.

Methods (b) and (c) are based on a measurement of the emitted radiation following the relaxation of an inner-shell vacancy. For light atoms, the predominant decay mode is an Auger process. Accordingly, the fluorescence yield, which is the probability that an inner-shell vacancy relaxes by x-ray emission, is low. For example, $\omega_K$ for Ne and Ar is 1.8% and 12%, respectively [26] For ionized species, the fluorescence yield is higher [27] and often not well known. This, combined with the lack of knowledge of the degrees of ionization of the collision products, makes x-ray emission-yield data less valuable for the determination of the primary excitation cross section. In such cases, Auger-emission yield data are preferable. For heavier collision systems where the fluorescence yield becomes appreciable, x-ray emission yields lead more directly to excitation cross sections.

(b) Auger-electron yield measurements are restricted to gaseous targets since the mean-free path of Auger electrons in solids ordinarily is too short to allow such targets to be used. This severely limits the applicability of method (b). Recently, however, the selection of appropriate targets has been extended by use of vapour targets [28]. Typical features of electron spectra are: (i) ~60-eV broad, unresolved peaked distribution superimposed on a continuum which decreases approximately exponentially with increasing electron energy [29]. The size of the continuum depends on both beam energy and atomic number of collision partners. This, in combination with a decrease in excitation probability when exit channels are closed, makes Auger-electron-yield determinations inadequate as a means of studying K excitations when $(Z_1, Z_2) > (11, 11)$. (ii) Kinematical or Doppler effects [23,29] which are: Doppler shift of projectile Auger electrons, Doppler broadening of projectile and target-electron spectra, and anisotropic emission in the laboratory frame. The Doppler broadening stems from the source particles being scattered into a cone around the beam direction, whereas the transformation of solid angles from source frame to laboratory frame invokes an anisotropy in the laboratory frame. At low and moderate energies, the Doppler broadening usually inhibits a resolving of discrete lines in the electron spectrum.

(c) X-ray emission yield. A major part of experimental data on inner-shell excitation stems from x-ray emission measurements, mainly because these measurements are less demanding in terms of equipment, but also because a larger variety of collision systems can be studied by this method. X-ray emission studies can be carried out with almost any projectile-target combination. This important feature allows systematic investigations to be carried out over extended ranges of Z values. An example of such investigations is the level-matching phenomenon [30].

While x-ray yields can be obtained with both solid and gaseous targets, the former constitute several tricky problems [31,32] when specific excitation cross sections have

to be evaluated from the measured yields. These problems,
known as 'solid' effects [31,32], stem essentially from two
features in connection with ion-solid interactions. One is related to the traversing projectile which, by having multiple
encounters with target atoms, achieves an unknown degree of
excitation/ionization. This means that we cannot specify the
initial state of a single event, and hence measured x-ray
yields represent some average over an unknown distribution
of such states. The effect is particularly significant if a previous collision event opens new exit channels for a subsequent
event. Such cases are known as 'double collisions' [33] and
are important in connection with production of MO x rays [34].
The other 'solid' effect comes into play when target-x rays
are measured. Energetic recoils, which are produced during
the slowing-down of the incident particles, create x rays in
collisions with target atoms. Based on a detailed theoretical
study, Taulbjerg et al.[31] concluded that recoils contribute
significantly to the generation of target-x rays when direct
collisions have a low probability of producing inner-shell
vacancies or when the incident ions are substantially heavier
than the target atoms, thereby producing energetic recoils.
Examples of such cases are N-Al and Ar-Al, respectively.

Projectile x-ray yields are less well-defined because of
multiple-collision effects which, by producing a multiplicity
of inner-shell vacancies, affect the fluorescence yield by an
unknown amount. In addition, the traversing projectile ion
achieves an excitation and charge state which may affect the
probability of producing an inner-shell vacancy strongly. Recent studies have indicated that emission from projectiles may
be significantly anisotropic [35], which means that angular
distribution must be known before total emission yields can be
derived.

## EXPERIMENTAL DATA

The impressive amount of experimental data on K excitation in light, medium, and heavy collision systems over a
broad range of energies, and the success with which quantitative theoretical studies on light collision systems have been
conducted [6,11,36] make these processes an extremely valuable tool to study collision dynamics in ion-atom collisions.
As we have already seen, the K excitation in either collision partner is determined by a few selective and localized
interactions of which the $2p\sigma$-$2p\pi$ rotational and $2p\sigma$-$1s\sigma$
radial couplings are the prominent. It is the aim of this experimental section to illustrate some characteristic features

of the K-excitation process by a few, typical examples taken from the literature. Since we are dealing with slow collisions, diabatic MO correlation diagrams will provide a sensible framework for our discussion. For the sake of simplicity, we shall assume that separated atomic orbitals have a 'regular' order, i.e., only collision systems which have no inner-shell swapping will be considered.

For the purpose of discussion, we shall divide collisions into three categories according to their atomic numbers:

I Light systems, $(Z_1, Z_2) < (10, 10)$
Open-exit channel. Process (a) at lower (v < 0.5 a.u.) and process (a') at higher velocities (v > 0.5 a.u.), see Fig.2. Mainly KLL Auger-emission data with gaseous targets, but also some K x-ray emission data with solid and gaseous tar. gets, do exist.

II Medium systems, $(10, 10) < (Z_1, Z_2) < (20-30, 20-30)$. Closed exit channel. Process (a') most likely at lower $(Z_1, Z_2)$ and low velocity. Process (b) may become important at higher $(Z_1, Z_2)$ and high velocity. Mainly K x-ray emission data but $KLL^2$ Auger-emission data with gaseous targets Ne [37] and Na [40] do also exist.

III Heavy systems, $(Z_1, Z_2) > (20-30, 20, 30)$
Closed-exit channel. Process (a') unlikely, especially at higher $(Z_1, Z_2)$. Process (b) more likely. Only K x-ray emission data on solid targets.

The K-vacancy-sharing process (c), see Fig.2, is expected to be the leading process in creating K vacancies in the high-Z collision partner. This applies to all three collision categories when $Z_1$ and $Z_2$ do not differ much. Although direct ionization of $1s\sigma$, Fig.2, leading to a vacancy in the high-Z partner is a possible process, we estimate this process to be of significance only when strongly asymmetric collision systems are considered. Such conditions are also close to the condition for which BEA is valid.

As most data listed under II and III have been obtained with solid targets and the systems have a closed-exit channel, double-collision processes [33] may play an important role in the production of a K vacancy. If, in the first collision event, a 2p vacancy is created, this vacancy may live long enough to be brought into the $2p\pi$ MO in a subsequent violent event. A $2p\sigma-2p\pi$ rotational coupling may then transfer the $2p\pi$ vacancy via the $2p\sigma$ MO to a K vacancy. The net effect of the double-collision mechanism is an effective

abolition of the closed-exit channel and hence a drastic increase in the observed cross section for producing a K vacancy.

By using Eq.(4) and calculated values of $\sigma_{rot}$ , we have
derived 2pπ vacancy numbers $N_0 + N(v) = \sigma_K/\sigma_{rot}$ and
$N(v)$. The $\sigma_{rot}$ values have been obtained by scaling appropriate one-electron results. Both symmetric and asymmetric systems have been considered. The scaling which has
been used is essentially analogous to Eq.(2) if $N_0 = 1$ and
$Z_s = Z_1 - 0.75$ (for the one-electron system, $Z_1 = 1$ and $Z_2$
varies between zero and unity). Exponentially screened Coulomb potentials have been used throughout. This has resulted in a very good reproduction of threshold behaviour of the
cross sections [37]. The scaled results agree very well
with 'ab initio' HF calculations on Ne + O by Taulbjerg and
Briggs [36], and also with low-velocity experimental data
when appropriate vacancy numbers $N_0$ are used [6], see,
e.g. Fig.5.

Values of $N_0 + N(v)$ and $N(v)$ have been derived for a
series of different light-collision systems (I). Following,
for example, the idea of Meyerhof [15], the data have been
plotted on a semilog plot versus $1/v$ . A straight-line dependence indicates a relationship $N(v) = A \exp(-c/v)$ , where
A and c are constants. In our last analysis of $N(v)$, we
have gone a step further. Assuming that a radial 3dπ-2pπ
coupling is the leading process responsible for bringing vacancies into 2pπ MO on the inward half of the collision, we
have derived a single-electron transition probability $t(v) =$
$= N(v)/N_{3d\pi_x}$, where $N_{3d\pi_x}$ is the number of $3d\pi_x$ vacancies which may couple to the 2pπ$_x$ MO. It is apparent that
since 3dπ correlates to 2p of low-Z collision partner,
$N_{3d\pi} = N_{2p}/3$, where $N_{2p}$ is the number of initial 2p vacancies in the low-Z partner.

I Light systems. Data are shown in Figs.5-8. Figure 5
displays two main features: The exit-channel effect of varying the charge state of the incident Ne particles from zero
to two [37] and the transition from MO interaction at low
velocity to Coulomb interaction at higher velocity. We also
notice that process (a) prevails at low velocity for the open-
channel cases (Ne$^+$, Ne$^{++}$-Ne), whereas process (a') presumably is the leading process for the closed-channel case
(Ne-Ne) and when the velocity exceeds 0.5 a.u. It should be
noted that none of the direct ionization processes from 2pσ ,
BEA(2pσ) and 2pσ-με$\pi$ , seems to reproduce the closed-
exit-channel data Ne-Ne very well.

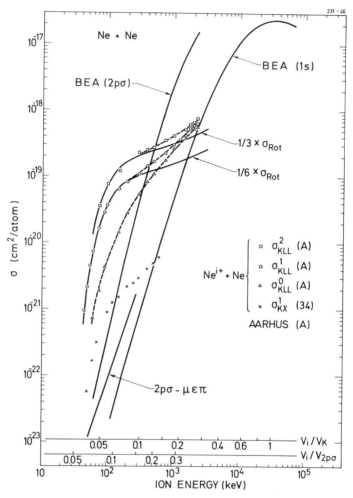

Fig.5   KLL  Auger  emission  cross  sections  $\sigma^i_{KLL}$  for  $Ne^{i+}$
+Ne [37] and  K  x-ray emission cross sections  $\sigma^1_{Kx}$  for
$Ne^+$ +Ne [29]. Theoretical curves, see text.

Figure 6 shows  $2p\pi_x$  vacancy numbers obtained for symmetric ($Ne^+$- Ne) and asymmetric ($Ne^+$- $N_2$) systems [37]. The
data at low velocity are in excellent agreement with static vacancy numbers  $N_o$ , as estimated by Briggs and Macek [6].

Figure 7 displays a single-vacancy transition probability
$t'(v)$ for a series of closed-exit-channel systems [37]. Within experimental uncertainty, the data agree with a simple formula, $t'(v) = \exp(-c/v)$, where  c  depends on  $Z_1$ ,  $Z_2$  being

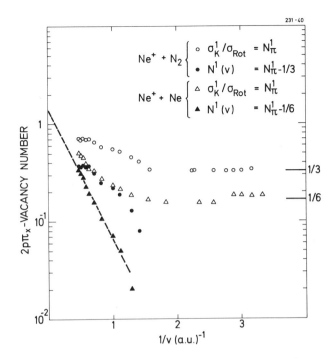

Fig.6 $2p\pi_x$ vacancy number, $N_o + N(v) = \sigma_K/\sigma_{rot}$, and its velocity-dependent component $N(v)$ for $Ne^+ + Ne$ and $Ne^+$ $Ne^+ + N_2$ [37].

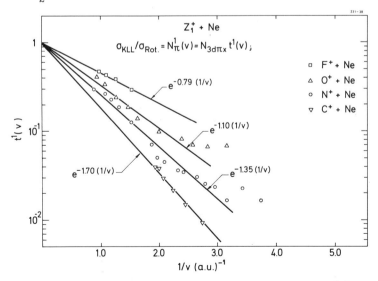

Fig.7 Single-vacancy transition probability $t'(v)$ into $2p\pi_x$.

kept fixed. We have been unsuccessful in relating $c/v$ to $2x$ in Meyerhof's formula, Eq. (3). It is, however, likely that $t(v)$ is determined by interactions between molecular states, see, e.g., Brenot et al.[21], where similar interactions have been discussed at length such that simple one-electron (MO) interactions are unrealistic. In this connection, it should be mentioned that the form $\exp(-c/v)$ for the transition probability may also result if a single passage through a pseudo-crossing take place.

In Fig.8 are shown K vacancy-sharing-ratio data for various systems over an extended velocity range. The data, which are based on measurements of KLL Auger emission yields in either collision partner [40,41] are in excellent agreement with Meyerhof's formula, Eq.(3), except for $B^+$-$CH_4$ [40], where presumably the single-electron picture is no longer appropriate.

II-III Medium and heavy systems. These systems, which all have closed-exit channels, have mainly been studied by x-ray emission with solid targets. Hence, data are hampered

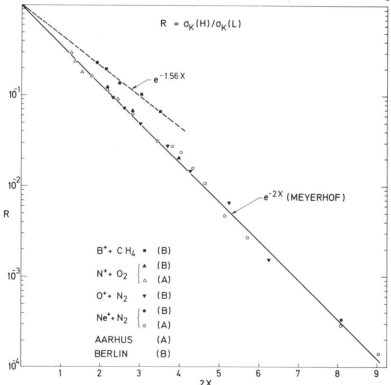

Fig.8   K-vacancy sharing ratio R .

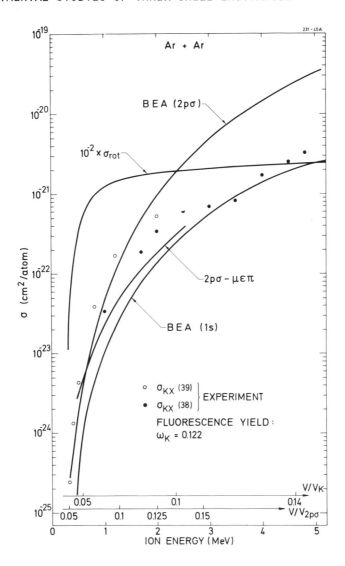

Fig.9 Ar-Ar K-vacancy-production cross section $\sigma_K =$
$= \sigma_{Kx}/\omega_K$ [38,39]. Theoretical curves, see text.

by double-collision processes [31,33] and, to some extent, also by the lack of knowledge of fluorescence yield [31]. There are only a few data where gaseous targets have been used [38,39,43,44]. By comparing solid- and gaseous-target data, one may obtain some information on the importance of the solid effects [43]. In Fig.9 are shown K x-ray emission

yields obtained in $Ar^+$- Ar [38,39] collisions. The data seem
to be in reasonable agreement with both BEA (1s) and
$2p\sigma$-$\mu\epsilon\pi$ results, whereas BEA ($2p\sigma$) is an order of magnitude higher at higher velocities. By comparing the data with
$\sigma_{rot}$, it is noted that the $2p\pi_x$ vacancy number is extremely
small ($\sim 10^{-2}$- $10^{-3}$) and presumably strongly velocity-dependent. Although the two-step process, Eq.(4), is a possibility,
the small value of N(v) may indicate that other, more direct,
mechanisms are responsible for a K excitation, e.g., rotational coupling to the continuum.

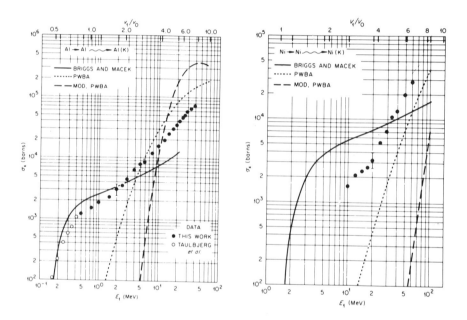

Fig.10  K x-ray-emission cross section $\sigma_x$(barns) [42].
The theoretical curves have been multiplied by neutral K-
fluorescence yields: 0.038(Al) and 0.41(Ni).

In Fig.10 are shown Al-Al and Ni-Ni x-ray data of Laubert et al.[42]. The solid curves represent $2p\sigma$-$2p\pi$ rotational-coupling results [6], assuming a single (on the average) $2p\pi_x$ vacancy created by ionization or by double-collision mechanism, i.e., $N_o$ = 1 in Eq.(2). The dotted curves
represent Coulomb-excitation results, whereas the dashed
curves include correction for increased binding, polarization,
and deflection in a Coulomb field [9]. The Al-Al data, which
have been extended up to 40 MeV or $v/v_K$ = 0.7, are in good
agreement with rotational-coupling results at lower energies
($\lesssim$5 MeV or $v/v_K \lesssim$ 0.25), whereas some deviation is noted

at higher velocities, presumably indicating that Coulomb excitation now becomes important in the K-excitation process.
Generally, the Al-Al data show very much the same feature
as Ne-Ne data, see Fig.5. The Ni-Ni data ($Z = 28$), which
have been extended up to 60 MeV or $v/v_K = 0.25$, are more
ambiguous. At low velocities, they are approximately three
times smaller than $2p\sigma$-$2p\pi$ rotational-coupling results, assuming $N_o = 1$ in Eq.(2). This indicates that in the case of
Ni-Ni, fewer $2p\pi_x$ vacancies are produced as a result of ion
-solid interactions. This is also expected since for Al-Al collisions a charge-state distribution with an appreciable amount
of charge states 4 and 5 may occur. Such cases have an open
-exit channel for a K excitation. In the Ni-Ni system, the
2p shell is an inner shell and only the less efficient double-
collision event may bring a vacancy via $2p\pi_x$ down into $2p\sigma$.

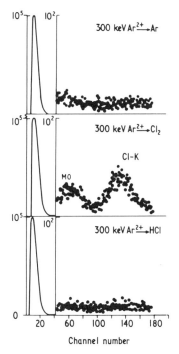

Fig.11   K x-ray spectra
for 300-keV $Ar^{2+}$ incident
on Ar, $Cl_2$, and HCl, normalized to same L x-ray
yield

    Although not identical systems, the Ni-Ni collisions in a
solid and the Ar-Ar collisions in a gas, see Fig.9, illustrate clearly the significance of solid effects. While the $\sigma_{rot}$
curve in the Ar-Ar case is 2-3 orders of magnitude higher
than experimental data, it is only 3 times higher in the Ni-Ni
case. At higher velocities, the Ni-Ni data are in reasonable
agreement with Coulomb-excitation results.

    The existence of the double-collision mechanism [33] has

been lent a very convincing experimental proof by Saris et al. [39], see Fig.11. The three spectra, which have been normalized to the same  L  x-ray yield, reveal only  K  x rays rays and MO x rays when a double-collision process is possible, i.e., when molecular $Cl_2$ target is used.


ACKNOWLEDGEMENTS

The author would like to thank J.Vaaben for providing him with unpublished one-electron calculations of $\sigma_{rot}$ and E.Bøving for his helpful assistance during the preparation of this manuscript.


References

1  Q.C.Kessel and B.Fastrup, Case Studies in Atomic Physics, 1973(North-Holland, Amsterdam) 3, 137

2  J.D.Garcia, R.J.Fortner, and T.M.Kavanagh, Rev.Mod. Phys. 45, 111 (1973)

3  B.Fastrup, Proc.Int.Conf.on Inner-Shell Ionization Phenomena and Future Applications, Atlanta, 1972 (USAEC Report No CONF-720404, Oak Ridge,Tennessee. 1973)

4  J.S.Briggs, 'Theory of Inner-Shell Excitation in Slow Ion-Atom Collisions', invited contribution, IX ICPEAC, Seattle, 1975

5  D.H.Madison and E.Merzbacher, 'Theory of Charged-Particle Excitation' in Atomic Inner-Shell Processes (B.Craseman, ed.) Academic Press, 1975

6  J.S.Briggs and J.H.Macek, J.Phys.B.: Atom-Molec. Phys. 6, 982 (1973); J.H.Macek and J.S.Briggs, ibid. 841 (1973); J.S.Briggs and J.H.Macek, J.Phys.B: Atom. Molec.Phys.5 , 579 (1972)

7  W.Lichten, Phys.Rev.164 , 131 (1967); M.Barat and W.Lichten, Phys.Rev.A 6 , 211 (1972)

8   W. Lichten, Proc. 4th Int. Conf. on Atomic Physics, Hei-
    delberg, 1974

9   G. Basbas, W. Brandt, and R. H. Ritchie, Phys. Rev. A 7
    1971 (1973); W. Brandt, Proc. Int. Conf. on Inner-Shell
    Ionization Phenomena and Future Applications, Atlanta,
    1972 (USAEC Report No CONF-720404, Oak Ridge, Ten-
    nessee, 1973)

10  J. H. McGuire and P. Richard, Phys. Rev. A 8, 1374 (1973)

11  K. Taulbjerg and J. Vaaben, Proceedings of Abstracts,
    IX ICPEAC, Seattle, 1975

12  W. E. Meyerhof, Phys. Rev. Letters 31, 1341 (1973)

13  Yu. N. Demkov, Soviet Phys. JETP 18, 138 (1964)

14  B. Fastrup, E. Bøving, G. A. Larsen, and P. Dahl, J. Phys.
    B: Atom. Molec. Phys. 7, L206 (1974)

15  W. E. Meyerhof, 3rd Int. Seminar on Ion-Atom Collisions,
    Gif-sur-Yvette (France), 1973

16  J. Vaaben, unpublished calculations of $\sigma_{rot}$.

17  W. E. Meyerhof, Phys. Rev. A 10, 1005 (1974)

18  C. Foster, T. P. Hoogkamer, P. Woerlee, and F. W. Saris,
    Proceedings of Abstracts, IX ICPEAC, Seattle (1975)

19  V. SethuRaman, W. R. Thorsen, and C. F. Lebeda, Phys.
    Rev. A 8, 1316 (1973)

20  J. Bang and J. M. Hansteen, Kgl. Danske Videnskab. Sel-
    skab, Mat.-Fys. Medd. 31 No 13 (1959

21  J. C. Brenot, D. Dhuicq, J. P. Gauyacq, J. Pommier, V.
    Sidis, M. Barat, and E. Pollack, Phys. Rev. A 11, 1245
    (1975)

22  V. V. Afrosimov, Yu. S. Gordeev, A. N. Zinoviev, D. H.
    Rasulov, A. P. Shergin, Abstracts of Papers, IX ICPEAC,
    Seattle, 1975

23  M. E. Rudd and J. H. Macek, Case Studies in Atomic Phys-
    ics, 1972 (North-Holland, Amsterdam) 3, 47

24 P. Richard, 'Ion-Atom Collisions' in Atomic Inner-Shell Processes (B. Crasemann, ed.), Academic Press, 1975

25 S. Sackmann, H. O. Lutz, and J. Briggs, Phys. Rev. Letters 32, 805 (1974)

26 W. Bambynek, B. Crasemann, R. W. Fink, H. U. Freund, H. Mark, C. D. Swift, R. E. Price, and P. V. Rao, Rev. Mod. Phys. 44, 716 (1972)

27 F. P. Larkins, J. Phys. B: Atom. Molec. Phys. 4, L29 (1971); C. P. Bhalla, Phys. Letters A 46, 185 (1973); M. H. Chen and B. Crasemann, Phys. Rev. A 10, 2232 (1974); E. J. McGuire, in Atomic Inner-Shell Processes (B. Crasemann, ed.) Academic Press, 1975

28 N. Stolterfoht, Z. Physik 248, 81 (1971)

29 N. Stolterfoht, D. Schneider, D. Burch, B. Aagaard, E. Bøving, and B. Fastrup, Phys. Rev. (in press)

30 F. W. Saris and D. Onderdelinden, Physics 49, 441 (1970); T. M. Kavanagh, M. E. Cunningham, R. C. Der, R. J. Fortner, J. M. Khan, E. J. Zaharis, and J. D. Garcia, Phys. Rev. Letters 25, 1473 (1970); H. Kubo, F. C. Jundt, and K. H. Purser, Phys. Rev. Letters 31, 674 (1973)

31 K. Taulbjerg and P. Sigmund, Phys. Rev. A 5, 1285 (1972); K. Taulbjerg, B. Fastrup, and E. Lægsgaard, Phys. Rev. A 8, 1814 (1973)

32 F. W. Saris, 'Solid Effects on Inner-Shell Ionization' in Atomic Collisions in Solids V (S. Datz et al., eds.) Plenum Press, 1975, pp. 343-63

33 J. Macek, J. A. Cairns, and J. S. Briggs, Phys. Rev. Letters 28, 1298 (1972)

34 F. W. Saris, W. F. van der Weg, H. Tawara, and R. Laubert, Phys. Rev. Letters 28, 717 (1972); F. W. Saris and F. J. de Heer, Proc. 4th Int. Conf. on Atomic Physics, Heidelberg, 1974; P. H. Mokler, S. Hagmann, P. Ambruster, G. Kraft, H. J. Stein, K. Rashid, and B. Fricke, Proc. 4th Int. Conf. on Atomic Physics, Heidelberg, 1974

35  E.H.Pedersen, S.J.Czuchlewski, M.D.Brown, L.D.
    Ellsworth, and J.R.Macdonald, Phys.Rev.A 11 , 1267
    (1975)

36  K.Taulbjerg and J.S.Briggs; J.S.Briggs and K.Taul-
    bjerg, J.Phys.B: Atom.Molec.Phys. (in press)

37  B.Fastrup, B.Aagaard, J.Vaaben Andersen, and E.
    Bøving, Proceedings of Abstracts, IX ICPEAC, Seattle
    1975, and unpublished data

38  H.O.Lutz, W.R.McMurray, R.Pretorius, I.J.van Heer-
    den, and R.van Reenen, unpublished results

39  F.W.Saris, C.Foster, A.Langenberg, and J.van Eck,
    J.Phys.B:Atom.Molec.Phys. 7 , 1494(1974)

40  N.Stolterfoht, P.Ziem, and D.Ridder, J.Phys.B:Atom.
    Molec.Phys. 7 , L409(1974); N.Stolterfoht, D.Schnei-
    der, and D.Ridder, Proceedings of Abstracts,
    IX ICPEAC, Seattle, 1975

41  B.Fastrup, B.Aagaard, E.Bøving, D.Schneider, P.
    Ziem, and N.Stolterfoht, Proceedings of Abstracts,
    IX ICPEAC, Seattle, 1975

42  R.Laubert, H.Haselton, J.R.Mowat, R.S.Peterson,
    and I.A.Sellin, Phys.Rev.A 11 , 135 (1975)

43  C.Foster, T.Hoogkamer, and F.W.Saris, J.Phys.B:
    Atom.Molec.Phys. 7 , 2563(1974)

44  L.Winters, M.D.Brown, L.D.Ellsworth, T.Chiao, E.
    W.Pettus, and J.R.Macdonald, Phys.Rev.A 11 , 174
    (1975)

THE THEORY OF INNER-SHELL EXCITATION IN SLOW ION-ATOM

COLLISIONS

J.S. BRIGGS

Theoretical Physics Division,

AERE Harwell, United Kingdom.

## 1.  INTRODUCTION

The results of observations of the production of
vacancies in inner-shells during slow ion-atom collisions has
been reviewed by Fastrup [1].  It has been shown that, largely
by use of the model due to Lichten and co-workers [2], the
seemingly complex interactions occurring when inner-shells
penetrate in slow collisions can be analysed in terms of the
simple properties of independent-electron (IE) molecular
orbitals (MO).  Furthermore, for inner-shell electrons the
behaviour of the relevant MO is dominated by the changing
nuclear field during the collision, the presence of other
electrons being taken into account only through their screening effect.  According to the Fano-Lichten model, excitation
of inner-shell electrons occurs with highest probability by
transition from an inner-shell MO into empty MO with which it
may become degenerate during the collision.  Hence the precise nature of the collision complex manifests itself through
the availability of such empty MO.  Most of the excitation
energy of the electrons is obtained from the nuclei by adiabatic motion and slight departures from adiabaticity at near
degeneracies are sufficient to excite electrons with high
probability at low impact velocity.

The task of this paper is to review the progress that has
been made in the calculation of cross-sections for excitation
of inner-shell electrons in slow collisions.  The cross-
sections have been calculated from an approximate solution of
the time-dependent Schrödinger equation by expansion in a
finite set of IE MO.  This expansion is not restricted to

those MO selected by the Fano-Lichten model as being strongly
interacting but, in principle, can be used to calculate the
magnitude of non-adiabatic transitions involving arbitrary
energy defects. Most of the results obtained so far [3-11]
have been restricted to consideration of those atoms whose
inner-shell electrons can be described adequately by non-
relativistic wavefunctions. This review also will be limited
to the non-relativistic regime.

## 2.   THE INDEPENDENT-ELECTRON MO MODEL

In most experiments involving ion-atom collisions which
result in inner-shell vacancies being created there is concomi-
tant excitation and ionisation of many outer-shell electrons.
The cross-section for inner-shell excitation that is measured
usually implies the total cross-section for preparation of all
final states having an inner-shell vacancy i.e. without speci-
fication of any particular final state of the many-electron
system. Auger electron and X-ray measurements under low resolution often do not differentiate between collisions producing
one or two inner-shell vacancies.

Since the energy required to excite an inner-shell electron is large compared to that needed to excite outer electrons,
the inner-shell excitation is characterised by transition to
a closely-spaced group of states all of which have an inner-
shell vacancy and are well-separated from other states.
Implicit in the independent-electron approximation is the
assumption that the aggregate probability for excitation of
all states of the group is just that for the transfer of a
single electron out of an inner-shell MO [12]. The only
effect of other electrons is to provide an effective field in
which an appropriate one-electron MO may be defined and to
limit the availability of final MO states via the Pauli
principle. It is a fundamental principle of the Fano-Lichten
ideas that the effective potential deciding the behaviour of
inner-shell electrons is dominated by the nuclear potential
and that the screening effect of other electrons does not affect
qualitatively the inner-shell MO. In addition the character of
inner-shell MO is assumed not to vary significantly with
changes in the excitation state of outer electrons. Under
these conditions it is appropriate to define a unique set of
inner-shell MO that describes the state of inner electrons
independently of the state of outer electrons. Because of
this, solutions of the time-dependent Schrödinger equation for
one electron in the field of two nuclei bear a close relationship to IE solutions of the scattering problems for inner-
shell electrons. With the classical treatment of nuclear
motion that is valid in all cases considered here, the fundamental quantity obtained from a scattering calculation is the

probability $P(b,v_I)$, as a function of impact parameter b and initial collision velocity $v_I$, that a transition is made between two spin-orbitals during a collision.

The Pauli principle can be accounted for simply by multiplying $P(b,v_I)$ by the factor $(q_i p_f)$ where $q_i, p_f$ are the probabilities that an <u>electron</u> is in the initial MO and a <u>vacancy</u> is in the final MO prior to the collision. Since most experiments measure excitation from initially filled subshells, $q_i$ is usually trivially unity. In this respect the Pauli principle acts to inhibit inner-shell excitation since any occupation of a particular group of degenerate final spin-orbitals reduces $p_f$ from its maximum value of unity.

Where both MO of opposite spin are emptied during a collision, the IE approximation assumes that the transitions are independent of each other. Since the one-electron transition operator conserves spin this requires that the probability of transferring both electrons from MO i to MO f is given by

$$P(i^2 \to f^2) \approx P(i^2 \to if) \; P(if \to f^2)$$
$$\approx P^2(i \to f). \tag{2.1}$$

The removal of one electron from an inner-shell MO causes a significant perturbation of other inner-shell MO, hence the approximation (2.1) may be suspect. However its validity has not been fully tested so far [13] because of lack of sufficient data on the separate cross-sections for excitation of one or two inner-shell vacancies.

Under the conditions that both MO i are occupied before the collision and $p_1$, $p_2$ are the probabilities that MO f contains 1 or 2 vacancies, the probability that only one of the spin-orbitals i becomes vacant after the collision is

$$P_1 = p_1 P + 2p_2 \, P(1-P). \tag{2.2}$$

The probability that both become vacant, from (2.1), is

$$P_2 = p_2 P^2. \tag{2.3}$$

The cross-section for formation of vacancies in the ith MO is then obtained by integrating the overall probability over impact parameter i.e.

$$\sigma_i(v) = 2\pi \int_0^\infty (P_1 + 2P_2) \, bdb$$
$$= 2\pi(p_1 + 2p_2) \int_0^\infty P(b,v_I) \, bdb. \tag{2.4}$$

The number $(p_1 + 2p_2)$ is the average number of vacancies in the MO f before the collision and will be called $N_f$.

## 3. INDEPENDENT-ELECTRON SCATTERING PROBLEM

The fundamental quantity to be calculated in the IE scattering problem is the probability $P(b,v_I)$ that a transition into a particular spin-orbital has been made. The derivation of appropriate statistical factors $N_f$ with which to construct probabilities relevant to the many-electron situation constitutes a separate problem.

The impact parameter version of the one-electron ion-atom scattering problem has been used widely to discuss charge transfer and excitation in real one-electron systems e.g. $H^+$-H collisions. The IE description of inner-shell electrons allows this formalism to be used in its entirety. An important difference is the strong influence of the shape of the nuclear trajectory on the very close collisions that are necessary for inner-shell excitation.

3.1 Coupled Equations. An approximate solution of the time-dependent Schrödinger equation

$$(h - i \frac{\partial}{\partial t}) \Psi(t) = 0 \tag{3.1}$$

in the form of a finite expansion in independent-electron wavefunctions $\chi_k$ i.e.

$$\Psi(t) = \sum_k a_k(t) \chi_k(t), \tag{3.2}$$

is obtained by solution of the coupled equations

$$i \underline{\underline{N}} \underline{\dot{A}} = \underline{\underline{M}} \underline{A}, \tag{3.3}$$

subject to appropriate boundary conditions. Here $\underline{A}$ is the column matrix of amplitudes $a_k$, $\underline{\underline{N}}$ is the overlap matrix with elements $N_{kj} = \langle \chi_k | \chi_j \rangle$ and $\underline{\underline{M}}$ is the coupling matrix with elements $M_{kj} = \langle \chi_k | (h - i \frac{\partial}{\partial t}) | \chi_j \rangle$. This last matrix is non-Hermitian when the basis set is non-orthogonal but satisfies the relation

$$\underline{\underline{M}} - \underline{\underline{M}}^\dagger = - i \underline{\dot{\underline{N}}} \tag{3.4}$$

which ensures unitarity of the solutions of (3.3). The probability that the k'th MO is occupied after a collision is obtained as

$$P_k(b,v_I) = \lim_{t \to + \infty} |a_K(t)|^2. \tag{3.5}$$

In the simplest case where the basis states $\chi_k$ are taken as eigenstates $\phi_k$ of the independent-electron operator h, the diagonal terms of $\underline{\underline{M}}$ are simply the MO energies $\varepsilon_k(t)$ and the off-diagonal elements involve the matrix elements

$$M_{kj} = - i \langle\phi_k|\frac{\partial}{\partial t}|\phi_j\rangle$$

$$= - i \ v_R \ \langle\phi_k|\frac{\partial}{\partial R}|\phi_j\rangle - i \ \dot\theta\langle\phi_k|\frac{\partial}{\partial\theta}|\phi_j\rangle$$

(3.6)

The radial and rotational coupling elements on the rhs of (3.6) connect different types of MO. A change in magnitude of the internuclear separation cannot alter the angular momentum of the electron and therefore the radial coupling connects states of the same angular symmetry i.e. $\sigma,\sigma$ or $\pi,\pi$ etc. The rotation of the internuclear axis does change the angular momentum of the electron and so the rotational elements connects states of differing angular symmetry e.g. $\sigma,\pi$.

As first pointed out by Bates and McCarroll [14], an expansion in an MO basis does not properly satisfy correct boundary conditions at infinite separation. This is because differential operations are performed with respect to an origin fixed at the centre-of-mass but the MO assume a constant asymptotic form only with respect to an origin fixed in the moving atom with which they are associated at infinite separation. A proper allowance for this motion of the electron at infinity is obtained by a Galilei transformation of the MO to a moving frame i.e. by using basis functions of the type

$$\chi_k = \phi_k \ e^{i p\underline{v}\cdot\underline{r} \ - \ \frac{1}{2}ip^2v^2t},$$

(3.7)

where $p\underline{R}$ is the distance from the centre-of-mass to the nucleus to which the MO $\phi_k$ correlates at infinite separation. Use of such a basis set has no effect on the diagonal matrix elements which still reduce to MO energies $\varepsilon_k$, but it modifies the coupling matrix elements such that the differential operator $\frac{\partial}{\partial t}$ is taken with respect to the correct origin. At this stage advantage has usually been taken of the low collision velocity at which the MO expansion is appropriate to expand the overlap off-diagonal and matrix elements to lowest order in $\underline{v}$.

$$\langle\phi_k \ e^{-i\underline{v}\cdot\underline{r}}\phi_j\rangle \approx \langle\phi_k|\phi_j\rangle$$

$$\langle\phi_k|e^{-i\underline{v}\cdot\underline{r}} \ \underline{v}\cdot\nabla_R|\phi_j\rangle \approx \langle\phi_k|\underline{v}\cdot\nabla_R|\phi_j\rangle.$$

The importance of retaining electron translation factors until differential equations have been performed in order to avoid the spurious asymptotic behaviour of coupling matrix elements is well-known and is illustrated in ref. [8].

The plane-wave translation factors are designed to give the correct asymptotic behaviour of the basis functions. However they are inconsistent with molecule formation at close distances and lead to a non-orthogonal basis. Alternative factors that are correct asymptotically but also lead to a better representation at short distances have been proposed by Schneidermann and Russek [15] and by Thorson [16] (see also ref. 10). The plane wave factors remove the awkward long-range couplings but they distinguish electrons for all R. It is evident that they may be inadequate for certain radial transitions occurring at very close distances.

## 4.   GENERATION OF MO FOR SMALL INTERNUCLEAR DISTANCES

In the generation of MO wavefunctions and energies for use in the IE model, three different types of approximation of increasing complexity have been employed. The first approximation is to ignore the effect of other electrons completely i.e. to solve the Schrödinger equation in the field of two bare nuclei. The next step consists of replacing the unscreened nuclear potential by a potential that allows for the varying screening as the internuclear separation is varied. The problem then consists of solving the one-electron Schrödinger equation in some effective one-electron potential. The third and most complicated method involves the solution of the many-electron Schrödinger equation in the Hartree-Fock approximation. This method generates a set of one-electron eigenvalue equations for each MO, with an effective potential that depends upon the interaction with electrons in all other occupied MO in a self-consistent way.

4.1  The one-electron molecular ions. The study of the states assumed by one electron in the field of two nuclei at varying separation forms the starting point of quantum chemistry and dates from the earliest days of quantum mechanics. However detailed calculations of MO energy levels and wavefunctions for the homonuclear $H_2^+$ problem did not appear until the work of Bates et al. [17] beginning in 1953. This was followed by calculations for various heteronuclear cases [18,19,20] and now MO correlation diagrams for any one-electron molecular ion are calculated routinely [6,10]. The one-electron Schrödinger equation in the field of two nuclei a distance R apart is

$$- \tfrac{1}{2}\nabla^2 \phi_k + (- Z_1/r_1 - Z_2/r_2)\phi_k = \varepsilon_k(R)\phi_k, \qquad (4.1)$$

where $r_1$, $r_2$ is the electron position with respect to nucleus 1 and 2 respectively and $\varepsilon_k$ is the kth eigenvalue. For the case of pure Coulomb potentials this equation can be separated in spheroidal coordinates and solved numerically to arbitrary accuracy. A particular example for $Z_1 = 1.0$, $Z_2 = 0.8$ is shown in fig. 1.

A contrast should be made in the use of one-electron correlation diagrams in the qualitative description of ion-atom collisions, as provided by the Fano-Lichten model, and their use in the performance of scattering calculations. The former use requires only the annunciation of rules to correlate separated atom (SA) and united atom (UA) levels, as done by Barat and Lichten [2], and the detailed behaviour of MO wave-functions and energies is not required.

Although outside the immediate scope of this article mention should be made of the advance of Müller et al. [21] by their numerical solution of the one-electron two-centre Dirac equation. This has opened the way for scattering calculations on slow ion-atom collisions where relativistic wavefunctions are required.

4.2  Effective one-electron potentials. In an effort to retain the simplicity of the one-electron Schrödinger equation (4.1) and yet represent more closely the binding energies of inner-shell electrons, Eichler and Wille have replaced the bare nucleus potentials by effective potentials to represent screening i.e. the Schrödinger equation is

$$(- \tfrac{1}{2}\nabla^2 + V_{eff}(r_1) + V_{eff}(r_2))\ \phi_k(R) = \varepsilon_k(R)\phi_k(R) \qquad (4.2)$$

Following a suggestion of Hund, Eichler and Wille [22] have proposed that a valid representation of the molecular potential may be obtained from a $V_{eff}$ of the atomic Thomas-Fermi form, but with parameters varied smoothly to interpolate between the SA and UA limits.

The deviation of the potential from Coulomb form intro-duces two main features that distinguish the effective-potential one-electron calculations. As in isolated atoms, the UA and SA subshells (of the same principal quantum number) are no longer degenerate. Furthermore, the extra constant of the motion that arises for a pure Coulomb potential is no longer present so that MO of the same symmetry are forbidden to become degenerate. This leads to the appearance of the avoided crossings, that are typical of adiabatic correlation diagrams (see fig. 2). It should be noted that the crossings appear in the bare nucleus correlation diagrams even though these are adiabatic, fixed nucleus calculations. These crossings are due to the separability, i.e. extra constant of

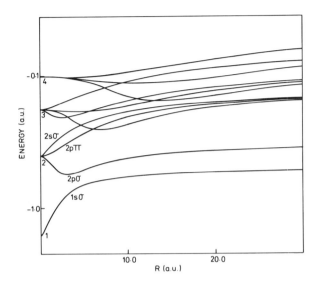

Fig. 1    One-electron MO correlation diagram for nuclear
          charge ratio $Z_1/Z_2=1.0/0.8$ (courtesy of J. Vaaben).

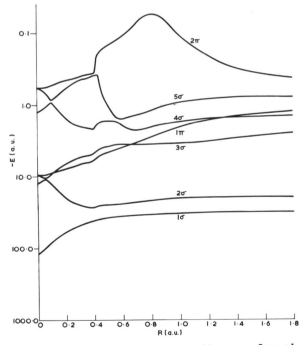

Fig. 2    Hartree-Fock MO correlation diagram for the NeO
          molecule (ref. 8).

the motion, that is present.

The effective Thomas-Fermi potential calculations appear
to be in good agreement with more sophisticated Hartree-Fock
calculations described below so far as the MO energies are
concerned.  However their ability to provide reliable wavefunctions to calculate coupling matrix elements between MO has
not been tested so far.  Their main attraction is that the
use of a statistical one-electron potential means that the
complexity of the calculation is independent of the number of
electrons, so that this method becomes advantageous for
heavier molecules where Hartree-Fock methods become prohibi-
tively expensive.

4.3  <u>The Hartree-Fock MO</u>.  The Hartree-Fock method constructs a many-electron wavefunction in determinantal form by
minimising the total energy of the molecule.  Such a variational principle leads to a set of coupled equations for the
MO wavefunctions $\phi_k$

$$[- \tfrac{1}{2}\nabla_i^2 - \frac{Z_1}{r_1(i)} - \frac{Z_2}{r_2(i)}] \phi_k(i)$$

$$+ 2 \sum_{\ell \neq k} <\phi_\ell(j)| \frac{1}{|\underline{r}(i)-\underline{r}(j)|} |\phi_\ell(j)> \phi_k(i) \qquad (4.3)$$

$$- \sum_{\ell \neq k} <\phi_\ell(j)| \frac{1}{|\underline{r}(i)-\underline{r}(j)|} |\phi_k(j)> \phi_\ell(i) = \varepsilon_k \phi_k(i).$$

The effective one-electron potential is then different for each
orbital and the set of equations (4.3) must be solved in a self-
consistent way.  However the eigenvalue $\varepsilon_k$ provides a good
approximation to the binding energy of an electron in the kth
MO.

In molecular calculations the Hartree-Fock equations are
usually solved by truncated expansion of the MO wavefunctions
in basis functions of Slater or Gaussian type.  A typical
result using the Hartree-Fock method is shown in fig. 2.  The
avoided crossings and irregular behaviour of outer orbitals
in fig. 2 should be contrasted with the smooth behaviour of MO
in fig. 1.  The first calculations of the Hartree-Fock type
were made by Larkins [23] and by Thulstrup and Johansen [24]
using only SA basis functions and therefore of limited accuracy.
More accurate calculations using both SA and UA basis sets have
been reported recently [25,26,27,8].

## 5.   K-SHELL EXCITATION

The only cases to be treated in detail so far by the coupled state IE method concern the excitation of K-shell electrons.   The Fano-Lichten model indicates that the dominant mechanism is the transfer of electrons from a promoted $2p\sigma$ MO into vacancies brought into the collision in a $2p\pi_x$ MO. The number $(p_1 + 2p_2)$ is the average number of vacancies in the $2p\pi_x$ MO and will be called $N_\pi$.   Since the $2p\sigma$ and $2p\pi$ MO are degenerate at $R = 0$ (fig. 1) it is the value of $N_\pi$ for small impact parameter collisions that is relevant.   It is assumed generally that $N_\pi$ is constant over the range of impact para- meters important for $2p\sigma$–$2p\pi$ coupling.   For the collisions of first-row atoms and ions Macek and Briggs [28] have derived the values of $N_\pi$ implied by the initial distribution of elec- trons in the separated atom and ion.   There is evidence that these values are only appropriate for impact velocities less that the outer-shell 2p electron orbital velocities.

5.1   Homonuclear Collisions.   The first case to be treated in detail by a numerical solution of the coupled equations (3.3) was the collision of $Ne^+$ ions with Ne atoms at energies below 500 keV.   In this case because of the additional symmetry of the homonuclear molecule only the $2p\sigma_u$ MO is coupled to the $2p\pi_u$ MO by the rotational operator $\dot{\theta}$ $<2p\sigma_u|\partial/\partial\theta|2p\pi_u>$, and (3.3) involve only two coupled equations.   The equations are identical to those used by Bates and Williams [29] and Knudson and Thorson [30] in the calculation of the probability of excitation of electrons to the 2p level in proton-hydrogen collisions.   In the $Ne^+$-Ne case the appropriate MO energies and rotational coupling matrix element were obtained partly from the calculations of Larkins [23] and partly by scaling from hydrogenic values.

It turns out that the shape of the nuclear trajectory has a decisive effect on this rotational transition.   In the $Ne^+$-Ne case an unscreened nuclear coulomb potential was used initially.   Then the total cross-section (fig. 3) exhibits a steep rise from an effective threshold in accord with the Auger electron data of Stolterfoht et al. [31].   The softer exponentially screened coulomb potential gives a closer agree- ment with the data in the threshold region.   This threshold effect is also seen in the probability $P(b,v)$.   This probabi- lity peaks for impact parameters less than the K-shell radius and the peak value rises rapidly in the threshold region (fig. 4).   Well above threshold a dramatic change takes place with the appearance of a narrow peak rising to unit transition probability at small impact parameter.   This peak was first explained by Bates and Sprevak [32] as corresponding to scatter- ing through 90° in the centre-of-mass frame.   The $2p\sigma$ and $2p\pi$

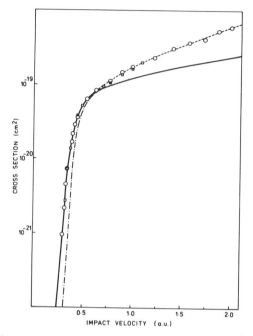

Fig. 3   K-shell ionisation cross-section in $Ne^+$-Ne collisions;
data points from ref. 31; ——— theory with screened
Coulomb internuclear potential ref. 13: —·—·— theory
with unscreened Coulomb potential ref. 3.

levels are degenerate at R = 0 and become the $2p_z$ and $2p_x$
atomic levels respectively.  Hence a simple rotation of the
internuclear axis, with no change in the electrons distribution, will effect the change from $2p_z$ to $2p_x$.  At the small
impact parameter appropriate to 90° scattering an electron in
the $2p\sigma$ MO first behaves adiabatically as the internuclear
distance shrinks, there then follows a sudden, completely non-
adiabatic rotation of the internuclear axis, after which the
electron adiabatically follows the changing nuclear potential,
now in a $2p\pi$ MO.  The rotation of the nuclear axis at zero
energy defect simply gives a probability $P(b) = \sin^2(\pi-\chi)$,
where $\chi$ is the final scattering angle.  For a Coulomb potential this probability has the form

$$P(b) = 4(b^2/K^2)/(1+b^2/K^2)^2 \qquad (5.1)$$

with $K = Z_1Z_2/Mv^2$ and M is the reduced mass.

This simple analytic form agrees well with the results of
coupled state calculations (fig. 5).  Lutz et al. have seen

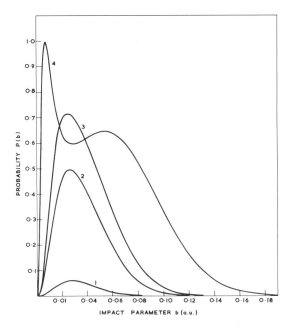

Fig. 4   2pσ-2pπ coupling probability in Ne$^+$-Ne collisions for
1, $v_I$ = 0.36 a.u.;   2, $v_I$ = 0.45 a.u.;   3, $v_I$ = 0.5
a.u.;   4, $v_I$ = 0.9 a.u. from ref. 13.

evidence of the onset of this peak in their differential X-ray-
ion coincidence experiments [33].

    For the Ne$^+$-Ne case, $p_1$ = 1/6, $p_2$ = 0 so that $N_\pi$ = 1/6.
This value gives a close agreement with experiment up to 200
keV but there is increasing discrepancy at higher energies
(fig. 3).   This is attributed to an effective increase in $N_\pi$
due to longer-range couplings amonst outer MO, as explained by
Fastrup [1].

    5.2  Hetero-nuclear collisions.   The essential simpli-
city of the K-shell excitation process in homonuclear collisions
is due to the strong coupling between just the 2pσ$_u$ and 2pπ$_u$
MO.   In heteronuclear collisions, the one-electron correlation
rules of Barat and Lichten suggest that the K-shell electrons
of the lighter partner are promoted along the 2pσ MO and couple
with vacancies in the 2pπ MO which emanates from the 2p shell
of the heavier partner.   Hartree-Fock calculations for a
typical example, the NeO molecule, (fig. 2) indicate the extra
complexity when the nuclear symmetry is relaxed.   There is a
clear avoided crossing between the 2σ (2pσ) and 3σ (2sσ) MO
that violates the one-electron correlation rules.   In addition

the $3\sigma$ and $1\pi$ levels may interact and the observation of
K-shell vacancies in the heavier partner indicates a finite
coupling of the $1\sigma$ MO. Hence the calculation must be
extended to include at least four MO.

One-electron calculations of hetero-nuclear correlation
diagrams do not show the avoided crossing between the $2s\sigma$ and
$2p\sigma$ MO (fig. 2). The first calculation, by Piacentini and
Salin [6] using only $2p\sigma$ and $2p\pi$ one-electron MO, with suitably
screened effective charges for the $Ne^+$-C molecule, obtained
reasonable agreement with experiment. Unfortunately an
unscreened coulomb internuclear potential was used which gives
inaccurate results in the threshold region. Taulbjerg et al.
[10], still ignoring the interaction with the $2s\sigma$ MO have shown
that one-electron calculations with a screened coulomb trajectory can provide reliable cross-sections for the $2p\sigma$-$2p\pi$
excitation.

The influence of the $2s\sigma$ MO has been investigated by
Briggs and Taulbjerg [9] using the Hartree-Fock MO for NeO.
The sudden change of character of the $2\sigma$ and $3\sigma$ MO at the
avoided crossing near $R = 0.0575$ a.u. gives rise to a large,
near singular radial coupling between them. This large coupling can be removed by a unitary transformation to a new basis
in which the MO retain their ellipsoidal nodal structure at
the crossing and are labelled according to their UA designa-
tion, $2p\sigma$ and $2s\sigma$, and so are 'diabatic' MO within the meaning
suggested by Lichten [2].

Within the diabatic Hartree-Fock basis the $2p\sigma$-$2p\pi$ coupling is again dominant and exhibits exactly the same characteristics as in the symmetric case (fig. 5). There is a
maximum of 5% reduction in P(b) for $2p\sigma$ occupation as a result
of coupling to the $2s\sigma$ MO, indicating that the neglect of this
coupling in the one-electron calculations is not serious.
For $Ne^+$-O collisions, $N_\pi = 1/3$ and the total cross-section for
oxygen-K vacancy formation using this value has been [9],
compared with the Auger data of Stolterfoht [34]. The overall agreement is good but again the value $N_\pi = 1/3$ appears to
be too low as the impact velocity is increased.

5.3   Excitation of the $1s\sigma$ MO. In hetero-nuclear collisions K-shell electrons of the heavier partner occupy the $1s\sigma$
MO and according to the Fano-Lichten model should not be
excited. However, experiment indicates a small but measurable
probability of excitation from the $1s\sigma$ MO. Calculations on
the $Ne^+$-O collision [9] indicate that this can be accounted
for solely by transfer of electrons from the $1s\sigma$ MO into
vacancies created in the $2p\sigma$ MO. The probability of this
coupling is decided entirely by $2p\sigma$-$1s\sigma$ radial coupling at

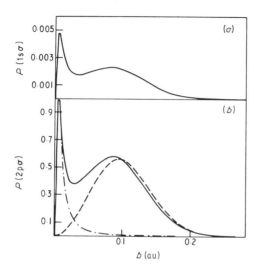

Fig. 5  a) The probability P(b) of a vacancy occupying the
1sσ MO after a Ne$^+$-O collision at $v_I$ = 1.0 a.u.
b) P(b) for the 2pσ MO in the same collision, ———
with screened Coulomb trajectory; --- with straight-
line trajectory; -·-·- eq. 5.1 of text.

separations where 2pπ-2pσ coupling, which must precede it,
is small. Accordingly the overall 1sσ probability may be
written approximately

$$P_{1s\sigma} = P(2p\pi \to 2p\sigma) \times P(2p\sigma \to 1s\sigma). \qquad (5.2)$$

Since P(2pσ-1sσ) varies slowly for impact parameters where
P(2pπ→2pσ) is important [35], the 1sσ probability 'borrows'
the shape of the 2pσ probability (fig. 5).

The calculated ratio [9] of the total cross-section for
neon-K vacancy formation in Ne$^+$-O collisions, to the oxygen-K
cross-section is independent of $N_\pi$ and very close to the experimentally measured values [34]. Under the conditions of the
approximation (5.2) the cross-sections for ionisation of heavy
(H) and light (L) collision partners are given by

$$\sigma_H = P(2p\sigma \to 1s\sigma) \, \sigma_L \qquad (5.3)$$

and where $P(2p\sigma{\to}1s\sigma)$ is calculated from 2-state equations
integrated along the <u>outgoing half of the</u> trajectory only, for
zero impact parameter. This probability will be called $S_o$.
On the assumption that the approximation (5.3) is valid,
Meyerhof [36] from a model solution of the two-state coupling
has derived a simple expression for the K-vacancy sharing
ratio

$$S_o = \exp(-K/v) \equiv \exp(-X),\qquad\qquad (5.4)$$

where K is a constant dependent upon the K-shell binding
energies. This formula gives good agreement with a great
variety of experimental data [36,37]. Using their one-electron
model Taulbjerg et al. [10] have investigated the K-vacancy
sharing in some detail. For a variety of collision partners
and impact velocities they have compared three ratios;

  i)   The cross-section ratio $R_K \equiv \sigma_H/\sigma_L$ from coupled
       $2p\pi$-$2p\sigma$-$1s\sigma$ MO calculations,

  ii)  The probability $S_o$ from $2p\sigma$-$1s\sigma$ MO calculations,

  iii) The Meyerhof expression (5.4).

     In terms of degree of approximation the results should
converge in the sense (iii) $\to$ (ii) $\to$ (i). However fig. 6
shows that this is not so and the Meyerhof expression appears
to be more accurate than its model assumptions should warrant.

     5.4 <u>Scaling of K-shell cross-sections.</u> It was recognised
from the outset [3] that K-shell excitation probabilities should
scale from one collision system to another. Since K-shell
radii scale inversely as the nuclear charge it is clear that
cross-sections decrease $\propto Z^{-2}$ as Z increases. Furthermore,
the adiabatic nature of the collision and hence the form of
the excitation probability is decided largely by the ratio $v/v_{e\ell}$
where $v_{e\ell}$ is the K-electron orbital velocity. Since $v_{e\ell}$ increases
$\propto Z$ then impact velocities must correspondly be scaled upwards
by a factor Z. Hence comparing excitation of K-shells of atoms
with nuclear charges 1 and Z one expects for given b and v,

$$P^Z(b,v) \approx P^1(Zb,v/Z)$$
$$\sigma^Z(v) \approx Z^{-2}\sigma^1(v/Z)\qquad\qquad (5.5)$$

     From a comparison of Hartree-Fock matrix elements and
energies for the $2p\sigma_u$-$2p\pi_u$ coupling with their counterparts in
the $H_2^+$ problem, Briggs and Macek [4] established the scaling
law for $2p\sigma_u$-$2p\pi_u$ coupling in symmetric collisions and obtained
reasonable agreement with a variety of experimental data.

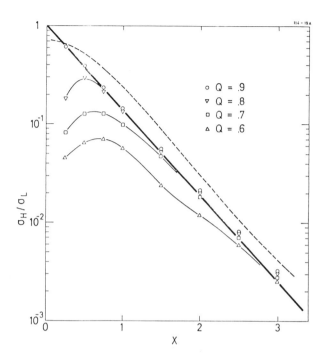

Fig. 6   K-excitation ratio versus the parameter X for
         selected values of $Q = Z_1/Z_2$.  Bold line, Meyerhof's
         formula ref. 36.  Marked points, the ratio $R_K$.
         Dashed curve, $S_0$ for $Q = 0.6$ (from ref. 10).

Koike et al. [7] have repeated the comparison recently.
Essentially the scaling law suggests that the cross-section for
$2p\sigma_u$–$2p\pi_u$ coupling in an ion-atom collision is the same as that
in the collision of a bare ion of charge $Z_S$ with a one-electron
atom with the same nuclear charge.  The effective charge $Z_S$ is
approximately equal to (Z-1), which allows for the screening in
real collision systems.

    By making explicit assumptions concerning the behaviour
of the $2p\sigma$–$2p\pi$ energy splitting, Taulbjerg et al. [11] have
promulgated a general scaling law for the $2p\sigma$–$2p\pi$ coupling
probability that incorporates the symmetric scaling as a special
case.  The aim of the scaling procedure is to establish a common
set of P(b,v) curves and a common total cross-section versus
impact velocity curve that obviates the need to perform expensive
Hartree-Fock calculations for each separate collision pair.

    Assuming a specific quadratic dependence for the $2p\sigma$–$2p\pi$
energy splitting in the region where this coupling is strong
and a Coulomb potential for the internuclear potential

Taulbjerg et al. suggest reduced units $(Z_\ell, Z_v^{-1})$ for length and velocity respectively in terms of the scaling parameters

$$Z_\ell = \left[ \frac{Z_1^2 Z_2^2 \, (Z_1 + Z_2/2)^4 M}{Z_A Z_B} \right]^{1/7}$$

$$Z_v = [Z_1 Z_2 (Z_1 + Z_2/2)^2 (Z_A Z_B/M)^3]^{1/7} \qquad (5.6)$$

Here $Z_A$, $Z_B$ are the full nuclear charges, $Z_1 = (Z_A - 1)$ and $Z_2 = (Z_B - 1)$, and M is the reduced mass of the collision system. The scaling law is strictly valid for a Coulomb potential trajectory but not for the screened Coulomb potential necessary to reproduce effective thresholds accurately. However it turns out that throughout the first row of the periodic table the screened potential does scale adequately enough except very close to the effective threshold. In addition, for collisions of first-row atoms and ions, it turns out that $Z_\ell$ and $Z_v$ are close to a mean value of the screened K-shell effective nuclear charges in accordance with the general considerations leading to (5.5).

Scaling of the results of a wide variety of Hartree-Fock and one-electron coupled-state calculations according to (5.6) does indeed reduce them to a set of common $P(b,v)$ and $\sigma(v)$ curves to within at least 10%. In particular the detailed shapes of the $P(b,v)$ curves are found to scale [11]. Figure 7 shows the universal $\sigma(v)$ curve compared with Hartree-Fock calculations. Well above threshold, where a straight-line trajectory is sufficient, the approximate analytic form $\sigma(v) = 10v^{2/3}$ in reduced units has been shown (fig. 7) to reproduce the universal cross-section [11].

## 6.   EXCITATION OF L-SHELLS

The $2p\sigma \to 2p\pi$ excitation represents the simplest example of the use of MO expansions in the IE approximation. The Fano-Lichten qualitative analysis has been applied to a large number of transitions involving electrons in L and higher subshells. Detailed calculations on such processes have so far not emerged.

The excitation of L shells by the $4f\sigma$ MO promotion is the classic example of Fano and Lichten, first discussed in connection with $Ar^+$-Ar collisions. However, this excitation mechanism is rather different from that operating for $2p\sigma$ excitation. In the latter case, there may be one avoided crossing to traverse followed by a single strong rotational coupling to the $2p\pi$ MO. In the case of $4f\sigma$ there are many

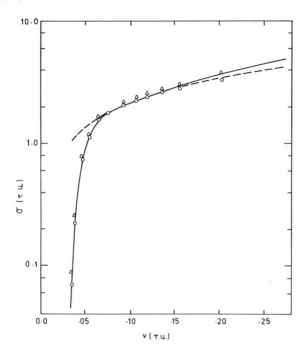

Fig. 7   Universal $\sigma(v)$ curve for $2p\sigma-2p\pi$ coupling in reduced
         units, ——;   Calculation with Hartree-Fock MO for
         $N^{+}$-N, $\Delta$;   for $Ne^{+}$-O, 0.   Dashed curve is the formula
         $\sigma(v) = 10.0\ v^{2/3}$ r.u. (from ref. 11).

avoided crossings and couplings with higher empty levels with
the result that the probability of occupation of the $4f\sigma$ MO
is dissipated amongst these levels during the collision.
Kessel [38] has explained the observed cross-section by a
steep rise from an effective threshold, corresponding to the
rapid promotion of the $4f\sigma$ MO as the 2p shells overlap, to a
geometric value.

A case of L-shell promotion analogous to the $2p\sigma-2p\pi$
K-shell promotion is the $3d\sigma$ promotion followed by $3d\sigma-3d\pi-3d\delta$
rotational coupling.   This situation often occurs for the
heavier L-shell in slightly asymmetric collisions and should
be amenable to a coupled-state calculation as for $2p\sigma-2p\pi$
coupling.

## 7.   UNSOLVED PROBLEMS

It appears that the IE MO expansion method can be applied
successfully when it is possible to isolate those MO which
couple strongly and whose initial occupation can be calculated
reliably.   Increasingly, experiments indicate the wide

occurrence of more complicated interactions contributing to inner-shell excitation. Fastrup [1] has discussed in detail the evidence for a velocity-dependent $N_\pi(v)$ in collisions of first-row atoms and ions. This number is an increasing function of velocity, leading to an increasing discrepancy between the theoretical $2p\sigma-2p\pi$ cross-section and experiment, as shown in fig. 3. This effect is particularly striking where the $2p\pi$ MO is initially full, i.e. $N_\pi = 0$, as in $Na^+$-Ne collisions. It appears that it can be explained in terms of emptying of the $2p\pi$ MO at large distances prior to strong $2p\pi-2p\sigma$ coupling. It is not clear at the moment to what extent the IEMO is applicable to this situation. At large distances the $2p\pi$ MO is an outer-shell MO for first-row atoms. Then the character of the MO and even the correlation diagram is not independent of the state of other electrons or of the initial and final ionisation states of the collision partners. The calculation of $N_\pi(v)$ is likely to receive increasing attention in the future.

For heavier atoms e.g. $Ar^+$-Ar, where the $2p\pi$ MO may properly be termed an inner-shell MO at all distances, two competing processes may lead to K-shell ionisation [1,39], a) Ejection of electrons from $2p\pi$ at large distances followed by strong $2p\pi-2p\sigma$ coupling as for first-row atoms, (b) direct coupling of the $2p\sigma$ MO to the many empty bound and continuum levels. At close distances, where the $2p\sigma$ MO has its minimum binding energy processes a) and b) cannot be distinguished in principle because of the strong coupling of the $2p\pi-2p\sigma$ MO.

In hetero-nuclear collisions two analogous processes exist for excitation from the $1s\sigma$ MO a) coupling to vacancies created in the $2p\sigma$ MO, b) direct coupling to the continuum. Process a) has been shown to be the dominant mechanism when $N_\pi$ is large, but for deep-lying $2p\pi$ MO this may not be the case.

No adequate, simple theory of direct coupling to the continuum in slow collisions of heavy ions under near-adiabatic conditions is available. Theories using atomic expansions valid for light projectiles, e.g. the Born approximation and its variants, are clearly inadequate. Direct excitation is characterised by coupling of a continuum of closely-spaced levels to a single inner-shell level well-separated from them, then, provided the transition probability to level n remains small it may be written

$$a_n(\infty) = \int_{-\infty}^{\infty} \frac{<\chi_n|\frac{\partial V}{\partial t}|\chi_o>}{(\varepsilon_n - \varepsilon_o)} \exp\{- i \int^t (\varepsilon_o-\varepsilon_n) \, dt'\}, \quad (7.1)$$

which is the appropriate first order solution of the equations
(3.3) in an orthogonal MO basis.  In (7.1) the convenient MO
identity [40,41] $\langle \chi_n | \frac{\partial}{\partial t} | \chi_o \rangle = \langle \chi_n | \partial V / \partial t | \chi_o \rangle / (\varepsilon_n - \varepsilon_o)$ has been
used.

When the energy defect $(\varepsilon_n - \varepsilon_o)$ is large the leading term
of (7.1) after integration by parts is

$$a_n(\infty) \approx \int_{-\infty}^{\infty} \langle \chi_n | V | \chi_o \rangle \exp\{- i \int^{t} (\varepsilon_o - \varepsilon_n) dt'\}$$

(7.2)

When the dominant contributions to the integral in 7.2
occur at large separations the distortion of the wavefunctions
may be neglected.  Then replacing $\chi_n, \chi_o$ by their SA functions
$\phi_n, \phi_o$ and assuming both are centred about target nucleus 2
say, (7.2) reduces to

$$a_n(\infty) \approx \int_{-\infty}^{\infty} \langle \phi_n | V_1 | \phi_o \rangle \exp\{-i \int^{t} E_o - E_n + \langle \phi_o | V_1 | \phi_o \rangle - \langle \phi_n | V_1 | \phi_n \rangle\}$$

(7.3)

where $E_o, E_n$ are atomic binding energies.  This is the amplitude
in the first-order distortion method [42].  Rudd and Macek [43]
have derived the same result from an LCAO expansion of MO and
have emphasised that it forms the link between an MO expansion
and the impact parameter Born approximation.  The latter is
obtained from (7.3) by neglect of the energy corrections in
the phase factor.  In an approximate way (7.3) has been used
for proton impact, where it has become known as the binding
correction [44].

Where a MO is promoted i.e. its binding energy is
decreased adiabatically as for the $2p\sigma$ MO it is likely that
the dominant contributions to (7.1) come from short distances.
Then (7.3) is inappropriate and the full distortion of MO as in
(7.1) or (7.2) should be retained.  Recently the apparent
dependence of direct excitation processes in heavy ion
collisions upon the UA binding energy has been demonstrated
[39,45].

In a series of important papers, Thorson and co-workers
[46] have discussed direct excitation of $1s\sigma_g$ and $2p\sigma_u$ MO in
$H^+$-H collisions.  This study may serve as a prototype for
heavier systems as was the case for $2p\sigma-2p\pi$ coupling.
Thorson et al. emphasized several difficulties connected with
direct excitation in a MO basis,

a)  Where strong rotational coupling occurs at short distances
it is not possible to make the first order expansion at all,
but all the initial strongly coupled manifold of states should

be included.    However since the coupling to continuum states
is weak, the strong-coupling problem may be solved separately.
For example in the case of $2p\sigma$-$2p\pi$ coupling $\chi_o\exp\{-i\int^t\epsilon_o\}$ in
(7.1) is replaced by

$$a_{2p\sigma}(t)\ \chi_{2p\sigma}\exp\{-i\int^t\epsilon_{2p\sigma}\} + a_{2p\pi}(t)\ \chi_{2p\pi}\ \exp\{-i\int^t\epsilon_{2p\pi}\}$$

$$(7.4)$$

where the a's are full solutions of the two-coupled state
problem.

b)    Final continuum states in which the electron moves slowly
with respect to one or other centre are difficult to define
in the two-centre basis.

c)    Matrix elements to continuum states are strongly dependent upon translational factors and the optimum choice of
such factors constitutes a major problem.

All of these difficulties remain for the collision of
slow heavier ions, the definition of final states being complicated by the variable screening of the nuclear charge.

## 8.    CONCLUSIONS

The cross-sections for excitation of the K-shells of both
collision partners in situations where a prior vacancy exists
in the $2p\pi$ MO can now be calculated reliably from independent-
electron coupled state MO calculations using curved trajectories.
The scaling properties of impact parameter dependent probabilities and total cross-sections has been established.    In con-
fronting theory with experiment the derivation of accurate
values of $N_\pi$, the average number of $2p\pi$ vacancies appearing at
close distances of approach remains a problem.    Ionisation
of the 2p shell at large distances may result in K-shell excitation when no 2p vacancies exist prior to the collision.    A
similar effect may arise from direct coupling to outer
unoccupied and continuum levels and the two processes are not
readily distinguished.    The calculation of direct excitation
processes in near-adiabatic collisions and the treatment of
the more complex excitation of L-shells, either by direct
excitation or MO promotion are the outstanding problems for
the future.

## REFERENCES

1.  B. Fastrup, IX ICPEAC Invited talks and Progress Reports (previous paper).

2.  U. Fano and W. Lichten, Phys.Rev.Letts. 14 627 (1965).
    W. Lichten, Phys.Rev. 164 131 (1967).
    M. Barat and W. Lichten, Phys.Rev. A6 211 (1972).

3.  J.S. Briggs and J. Macek, J.Phys., B5 579 (1972).

4.  J.S. Briggs and J. Macek, J.Phys. B6 982, 2484 (1973).

5.  T. Watanabe, M. Koike and F. Minami, J.Phys.Soc.Japan, 34 781 (1973).

6.  R.D. Piacentini and A. Salin, J.Phys. B7 L311 (1974).

7.  M. Koike, H. Sato and T. Watanabe, J.Phys.Soc.Japan, 38 216 (1975).

8.  K. Taulbjerg and J.S. Briggs, J.Phys. B8 1895 (1975).

9.  J.S. Briggs and K. Taulbjerg, J.Phys. B8 1909 (1975).

10. K. Taulbjerg, J. Vaaben and B. Fastrup (to appear in Phys.Rev.).

11. K. Taulbjerg, J.S. Briggs and J. Vaaben, (to appear in J.Phys.B).

12. J.H. McGuire and J.R. MacDonald, Phys. Rev., A11 146 (1975).

13. J.S. Briggs in Inner Shell Ionisation Phenomena ed. R.W. Fink et al. (USAEC:  Oak Ridge) p. 1209 (1972).

14. D.R. Bates and R. McCarroll, Proc.Roy.Soc. A245 175 (1958).

15. S.B. Schneiderman and A. Russek, Phys.Rev. 181 311 (1969).

16. W.R. Thorson, J.Chem.Phys. 42, 3878 (1965).

17. D.R. Bates, K. Ledsham and A.L. Stewart, Phil.Trans.Roy. Soc. A246, 215 (1953).

18. D.R. Bates and T.R. Carson, Proc.Roy.Soc. A234 207 (1956).

19. L.Y. Wilson and G.H. Gallup, J.Chem.Phys. 45 586 (1966).

20.   H. Hartmann and K. Helfrich, Theoret.Chim.Acta 10 406 (1968).

21.   J. Eichler and U. Wille, Phys.Rev.Letts. 33 56 (1974).

22.   B. Müller, J. Rafelski and W. Greiner, Phys.Letts. 47B 5 (1973).

23.   F.P. Larkins, J.Phys. B5 571 (1972).

24.   E.W. Thulstrup and H. Johansen, Phys.Rev. A6 206 (1972).

25.   R.S. Mulliken, Chem.Phys.Letts. 14 137 (1972).

26.   J.S. Briggs and M.R. Hayns, J.Phys. B6 514 (1973).

27.   V. Sidis, M. Barat and D. Dhuicq, J.Phys. B8 474 (1975).

28.   J. Macek and J.S. Briggs, J.Phys. B6 841 (1973).

29.   D.R. Bates and D.A. Williams, Proc.Phys.Soc. 83 425 (1964).

30.   S.K. Knudson and W.R. Thorson, Can.J.Phys. 48 313 (1970).

31.   N. Stolterfoht, D. Schneider, D. Burch, B. Aagaard, E. Boving and B. Fastrup, Phys.Rev. (to be published).

32.   D.R. Bates and D. Sprevak, J.Phys. B3 (1483) (1970).

33.   S. Sackmann, H.O. Lutz and J.S. Briggs, Phys.Rev.Letts. 32 805 (1974).

34.   N. Stolterfoht (private communication).

35.   J.S. Briggs, AERE Harwell report TP.594 1974 (unpublished).

36.   W. Meyerhof, Phys.Rev.Letts. 31 1341 (1973).

37.   J.D. Garcia, R.J. Fortner and T.M. Kavanagh, Rev.Mod. Phys. 45 111 (1973).

38.   Q.C. Kessel, Bull.Am.Phys.Soc. 14 946 (1969).

39.   C. Foster, T. Hoogkamer, P. Woerlee and F.W. Saris, Abstracts IX ICPEAC (1975).

40.   D.H. Madison and E. Merzbacher in Atomic Inner-Shell Processes ed. B. Crasemann (Academic Press:New York) 1974.

41.  W.R. Thorson and H. Levy, Phys.Rev. 181 230 (1969).
     H. Levy and W.R. Thorson, Phys.Rev. 181 244, 252 (1969).
     C.F. Lebeda, H. Levy and W.R. Thorson, Phys.Rev. A4
     900 (1971).
     V. SethuRaman, W.R. Thorson and C.F. Lebeda, Phys.Rev.
     A8 1316 (1973).

42.  M.R.C. McDowell and J.P. Coleman Introduction to the
     Theory of Ion-Atom Collisions (North-Holland:Amsterdam)
     p.123 (1970).

43.  M.E. Rudd and J. Macek, Case Studies in Atomic Physics
     3, 47 (1972).

44.  W. Brandt, R. Laubert and I. Sellin, Phys.Rev. 151 56
     (1966).

45.  J.U. Andersen, E. Laegsgaard, M. Lund and C.D. Moak,
     (to be published).

# EXPERIMENTAL STUDIES OF TARGET AND PROJECTILE X RADIATION IN HIGH VELOCITY ATOMIC COLLISIONS

James R. Macdonald[*]

Kansas State University

Manhattan, Kansas 66506 U.S.A.

The calculation in several approximations of inner-shell ionization of target atoms by the Coulomb field of swift projectiles has been reviewed by numerous authors.[1] In general, Coulomb ionization scales as the square of the ratio of projectile atomic number ($Z_1$) divided by the target inner-shell binding energy ($U_2$), and the cross sections increase monotonically with projectile velocity to a broad maximum at velocities comparable to the average orbital velocity of the inner-shell electron before decreasing to an $E^{-1}$ log E dependence at higher energy. Coulomb ionization theory is generally recognized as predicting results for ionization by light projectiles that is within a factor of two of experiment.

Experimental studies of the decay of states containing inner-shell vacancies formed by light ions at velocity below the peak of the ionization cross section have given results that fall below the predictions of first-order theory, as shown in Fig. 1(a). However, by considering the increased binding energy of inner-shell electrons in a perturbed-stationary-state treatment and the Coulomb trajectory of the projectile during the collision, Basbas et al.[2] have been able to obtain the agreement shown in Fig. 1(b) between theory and experiment. Additional systematic departure from the universal formulation for K-shell ionization of heavy atoms at low scaled velocity has been removed by Hardt and Watson[3] by considering relativistic effects on the scaled velocity in the parameterization. At velocities well above the peak of the

[*] Partially supported by the U. S. Energy Research and Development Administration.

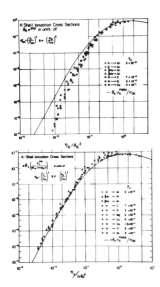

Fig. 1. K-shell ionization cross sections from hydrogen and helium impact. (a) scaled to include Coulomb deflection and increased binding as described in ref. 2. The quantities $\eta_K$, $\theta_K$, and $\varepsilon$ are parameters representing the beam energy, binding energy, and correction terms.

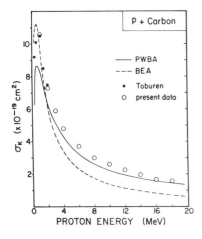

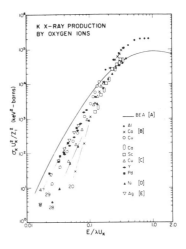

Fig. 2. Carbon K-shell ionization cross sections from proton impact. The solid points represent Auger measurements while the open circles are relative x-ray yields normalized to the Auger cross sections. (ref. 5)

Fig. 3. K-shell ionization cross sections for oxygen ion bombardment compared with the universal BEA prediction. (ref. 6)

ionization cross section, relatively little experimental work
has been done with K-vacancy production, although Jarvis et al.[4]
have studied the K-shell ionization of heavy targets by 160 MeV
protons and Burch[5] has studied the ionization of the carbon
K-shell with protons up to 18 MeV energy. The latter results
are shown in Fig. 2, and the conclusion has been made that the
experimental results are properly described at high velocity by
the asymptotic form of the plane wave Born approximation.

In recent years there has been an explosion in activity
in the study of x-ray production by high-energy heavy-ion impact.
There is a significant challenge in this field both to document
the many interesting phenomena and to establish a consistent
theory for these complicated atomic collision systems. Although the Coulomb ionization theories that have been so successful in describing inner-shell ionization by energetic protons have been formulated in a manner that contains a $Z_1^2$
scaling with the projectile atomic number, it is clear that the
conditions of many heavy-ion collision experiments violate
assumptions made in the theoretical derivation, in particular
the condition that $Z_1/Z_2 \ll 1$. In spite of this, the general
trend in the K-shell ionization cross sections for oxygen ion
bombardment[6] are seen in Fig. 3 to approach the scaled BEA
prediction for $Z_1/Z_2 \lesssim .2$.

In addition to total ionization cross section measurements
using light projectiles, experiments in which a coincidence between inner-shell vacancy decay and the scattering of the projectile at a particular angle have been performed by various
authors to measure the impact parameter dependence of inner-
shell ionization. For protons of energy from 1 to 2 MeV,
Laegsgaard et al.[7] have found that the impact parameter
dependence of the probability of K-shell ionization of Cu, Se,
and Ag is in good agreement with the semi-classical impact
parameter formulation of Coulomb ionization. Further, the
impact parameter dependence of the probability for ionization
of the Cu K-shell by energetic oxygen ions shown in Fig. 4 has
been found by Cocke and Randall[8] to agree with the calculated
dependence from the SCA model, although the results from the
BEA model underestimate the ionization probability at large
impact parameter. The theoretical models apparently describe
trends in the experimental heavy-ion data phenomenologically.
However, one must use caution in making detailed interpretation
of the agreement until the reasons for the extended limits of
validity of the theory are understood.

One of the complicating features of heavy-ion collisions
is the shift to higher energy of the characteristic x radiation
observed following inner-shell vacancy formation. High resolution spectra such as shown in Fig. 5 taken with a crystal

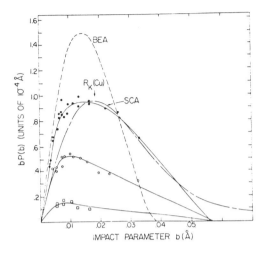

Fig. 4. Plot of bP(b) against b for K-shell ionization of Cu by oxygen ions. Solid points, 43 MeV; open circles, 35 MeV; open squares, 25 MeV. The curves through the data were used to calculate total cross sections for normalization to the theoretical curves. (ref. 8)

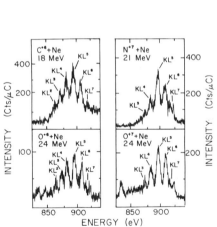

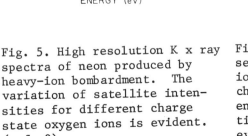

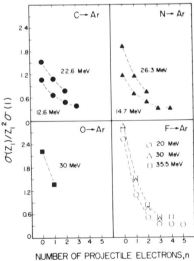

Fig. 5. High resolution K x ray spectra of neon produced by heavy-ion bombardment. The variation of satellite intensities for different charge state oxygen ions is evident. (ref. 9)

Fig. 6. K-ionization cross sections produced by heavy-ion bombardment in different charge states. The dependence on the number of projectile electrons, n=Z-q, is evident.

spectrometer[9] have shown that inner-shell vacancy production
by heavy ions is accompanied by multiple ionization of other
shells with a large probability. This leads to the formation
of a myriad of states that decay with dramatically different
energies, lifetimes, and fluorescence yields that in many cases
must be known in detail in order to deduce ionization cross
sections from x-ray yields. Although all the specific states
involved in the K x-ray satellite observations have not been
uniquely identified, successful interpretation of the binomial
distribution of line intensities has been made in terms of the
probability for independent L-shell ionization at small impact
parameter.[10]

Investigations of K x-ray production in heavy-ion collisions using gas targets have shown a strong dependence on the
initial charge state of the projectile both in target x-ray
spectra (Fig. 5) and in total cross section (Fig. 6). For the
ionization of neon, it is clear that a dramatic increase in
the mean fluorescence yield occurs for the highly ionized
states produced with increasing charge state.[11] However, for
argon K x rays the variation in fluorescence yield is not a
dominant factor and recent cross section measurements using
neon Auger electron yields[12] have confirmed that the vacancy
production depends significantly on the projectile charge
state. Explanations of this charge dependence of inner-shell
vacancy production have been proposed to include variations
in screening of the Coulomb field of the projectile, electron
capture by the projectile, as well as molecular processes
influencing these high velocity collisions. By measuring the
impact parameter dependence of the Ar K-shell ionization probability by fluorine ions of different charge states shown in
Fig. 7, Randall[13] has found an enhancement in vacancy production with the projectile charge state reaching ~ 25% for the
bare fluorine nucleus inside the argon K-shell radius. In
addition, the ionization probability falls off more slowly
with impact parameter than predicted by either the SCA Coulomb
ionization prediction or that of a Brinkman-Kramers calculation
of electron capture of argon K electrons. Attempts to explain
the observed impact parameter dependence using molecular
orbitals rotationally and radially coupled during the collision
have only been partially successful.

The importance of electron promotion processes during
high velocity collisions has been used for some time to explain
the oscillation of projectile inner-shell vacancy production
with target atomic number, $Z_2$, such as the work by Kubo et al.[14]
shown in Fig. 8. The rotational coupling of $\pi$ and $\sigma$ molecular
orbitals at small impact parameter is generally believed to
play an important role in producing the maxima in the oscillation as target and projectile energy levels match. In addition,

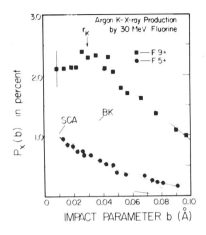

Fig. 7. Impact parameter dependence of Ar K x-ray production by fluorine nuclei and ions in charge state 5. The SCA Coulomb ionization values, and Brinkman-Kramers electron capture dependence are labeled. (ref. 13)

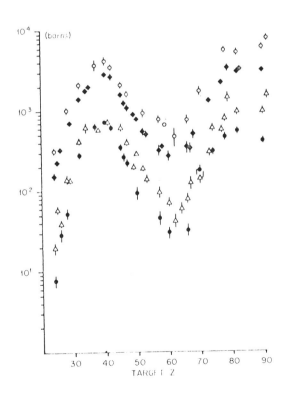

Fig. 8. Bromine Kα production cross section versus target Z for incident bromine energies of 110 MeV (open circles), 85 MeV (diamonds), 60 MeV (triangles), and 45 MeV (closed circles). (ref. 14)

the problem is complicated by the enhancement in the x-ray
cross section for the heavier collision partner in the near
symmetric collisions, and Meyerhof[15] has shown that vacancy
transfer between $1s\sigma$ and $2p\sigma$ by radial coupling, can account
for this enhancement. However, in applying the molecular model
to high velocity heavy-ion collisions one must be careful not
to ignore the importance of multistep excitation processes that
surely are changing the occupation numbers of less tightly
bound levels during the collision. For example, the charge
dependence of Cℓ K x-ray production in collisions of energetic
chlorine ions with hydrogen gas is the same as in heavier targets (Fig. 9). In particular, the increase in x-ray yield for
the four-electron incident ion (q=13) can only be explained if
2s excitation accompanies K-shell vacancy production in the
collision. Of course, the importance of excitation processes
accompanying inner-shell vacancy production changes the occupation numbers of relevant molecular orbitals during the collision. However, an examination of the total cross sections for
K x-ray production in symmetric collisions, such as the work
of Laubert et al.[16] with high energy aluminum ions shown in
Fig. 10, indicates that the rotational coupling model of Briggs
and Macek[17] describes the results to rather high projectile
energy. Further, the impact parameter results in Fig. 11 for
the K-shell ionization probability of energetic Cℓ ions in Cu,
Ti, and Al targets reported by Cocke et al.[18] are particularly
interesting. The almost symmetric Ti collisions are both more
efficient in producing K vacancies than the asymmetric collisions, and also hold the maximum ionization probability to
larger impact parameters for the symmetric collisions. This
plateau is not predicted by Coulomb ionization theory, and
suggests the importance of electron promotion mechanisms. No
detailed comparisons have yet been made with molecular models.

It is now generally recognized that the x-ray spectrum
from heavy-ion collisions consists of a high energy tail decreasing almost exponentially from the target or projectile
characteristic radiation. The continuous radiation has been
attributed to transitions between transient molecular orbitals
during the collision. A great deal of effort by numerous
groups has been invested in the experimental and theoretical
study of quasimolecular radiation and the present status of
this research will be reported by several other speakers in
this session. Another feature of the x-ray spectrum that has
been observed from a number of collision systems is a peak
resulting from radiative electron capture of target electrons
into inner-shell states of the fast projectile. The peak
energy of the photons is derived from the difference in binding energy of the initial and final bound states increased by
the translational kinetic energy of the target electron relative to the projectile, while the width of the peak is related

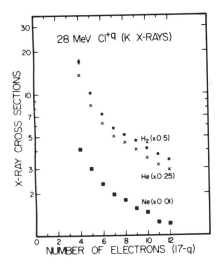

Fig. 9. Charge dependence of Cℓ K x-ray cross sections for 28 MeV Cℓ ions on $H_2$, He, and Ne.

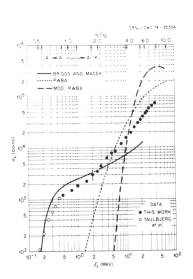

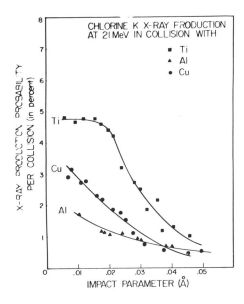

Fig. 10. Aluminum K x-ray cross sections for symmetric collisions of Al on Al. (ref. 16)

Fig. 11. Impact parameter dependence of Cℓ K x rays. (ref. 18)

to the momentum distribution of target electrons.  The present
status of this area of research will also be reported in this
session.

When fully stripped heavy ions participate in atomic
collisions there is a prolific yield of projectile characteristic x rays that clearly result from electron capture to excited
states.  The x-ray production cross section has been found by
Brown et al.[19] to fall with increasing projectile energy as
shown in Fig. 12 and the magnitude of the cross section is in
agreement with results of a Brinkman-Kramers calculation normalized to experimental total capture cross sections.  Indeed
for light targets ($H_2$ and He) essentially all of the electron
capture cross section by heavier nuclei is to excited states
and the x-ray cross section is within 20% of the total capture
cross section.  However, another feature of electron capture
theory is that cross sections are different for different substates and hence the excited one-electron ions formed in atomic
collisions are expected to be highly aligned if electron capture processes are important.  Pedersen et al.[20] have measured
the anisotropy of fluorine K x rays and obtained results shown
in Fig. 13 characterized by polarization fractions in excess
of 20%.  Czuchlewski et al.[21] have found that the polarization
fraction of the fluorine K x rays decrease to approximately
zero as electrons are added to the initial state of the
fluorine ion thus reducing the effect of electron capture.

Recently we have studied the alignment of projectile
states by directly measuring the polarization of oxygen Lyman
α radiation from foil-excited oxygen ions using a Bragg
crystal polarimeter.  The results are shown in Fig. 14, in
which the intensity of the Lyman α peak is shown for four
orientations of the polarimeter with respect to the beam axis.
The observed polarization of this K-shell radiation is about
16% and it confirms that there is significant alignment of
one-electron ions formed by complicated collision processes
encountered in foil excitation.

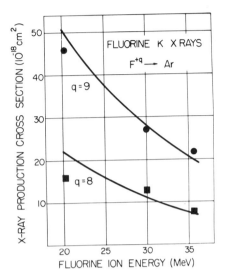

Fig. 12. Fluorine x-ray production cross sections F+q on
argon. The solid line through
the data with q=9 is proportional to the cross section
for electron capture to excited states calculated in a
Brinkman-Kramers approximation.
(ref. 19)

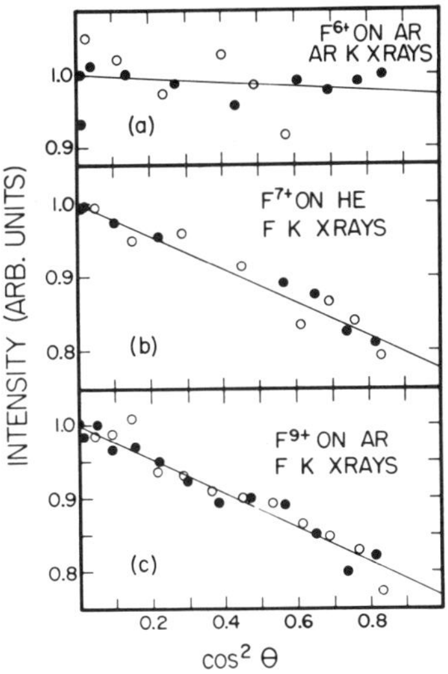

Fig. 13. Anisotropy of characteristic x radiation from projectiles. The solid and open
data points are taken at forward and backward angles respectively. The slope of the
plots against $\cos^2\theta$ gives
polarization fractions in excess of 20% for the projectile
radiation. K radiation from
the target is isotropic. (ref. 20)

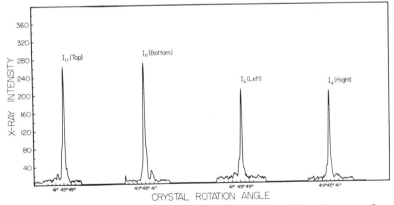

Fig. 14. The intensity of oxygen Lyman $\alpha$ radiation produced
in foil excited oxygen ions for four different orientations of
the Bragg crystal polarimeter. The intensities $I_{\parallel}$ and $I_{\perp}$ with
polarization parallel and perpendicular to the beam direction
are obtained independently in the different orientations.

## References

1. For a review and bibliography of theoretical approaches to inner-shell ionization see: D. H. Madison and E. Merzbacher, Atomic Inner-Shell Processes, Chap. 1, B. Crasemann, ed. (Academic Press, New York, 1975).

2. G. Basbas, W. Brandt, and R. Laubert, Phys. Rev. A $7$, 983 (1973).

3. T. L. Hardt and R. L. Watson, Phys. Rev. A $7$, 1917 (1973).

4. O. N. Jarvis, C. Whitehead, and M. Shah, Phys. Rev. A $8$, 2952 (1973).

5. D. Burch, submitted to Phys. Rev. A.

6. D. Burch, Proc. VII ICPEAC, Belgrade, B. C. Cobic and M. V. Kurepa, eds. (1974).

7. E. Laegsgaard, J. V. Andersen, and L. C. Feldman, Phys. Rev. Lett. $29$, 1206 (1972).

8. C. L. Cocke and R. Randall, Phys. Rev. Lett. $30$, 1016 (1973).

9. R. L. Kauffman, Ph.D Thesis, Kansas State University, 1975; R. L. Kauffman, C. W. Woods, K. A. Jamison, and P. Richard, Phys. Lett. $50A$, 75 (1974).

10. For a review and bibliography of multiple ionization phenomena in heavy-ion collisions see: P. Richard, Atomic Inner-Shell Processes, Chap. 2, B. Crasemann, ed. (Academic Press, New York, 1975).

11. D. Burch, N. Stolterfoht, D. Schneider, E. Wieman, and J. S. Risley, Phys. Rev. Lett. $32$, 1151 (1974).

12. C. W. Woods, Ph.D Thesis, Kansas State University, 1975; C. W. Woods, R. L. Kauffman, K. A. Jamison, N. Stolterfoht, and P. Richard, J. Phys. B $8$, L61 (1975).

13. R. Randall, Ph.D. Thesis, Kansas State University, 1975; R. Randall et al., submitted to Phys. Rev. A.

14. H. Kubo, F. C. Jundt, and K. H. Purser, Phys. Rev. Lett. $31$, 764 (1973).

15. W. E. Meyerhof, Phys. Rev. Lett. $31$, 1341 (1973).

16. R. Laubert, H. Haselton, J. R. Mowat, R. S. Peterson, and I. A. Sellin, Phys. Rev. A $11$, 135 (1975).

17. J. S. Briggs and J. Macek. J. Phys. B $5$, 579 (1972).

18. C. L. Cocke, R. Randall, and B. Curnutte, Proc. VIII, ICPEAC, Belgrade, B. C. Cobic and M. V. Kurepa, eds. (1974) p. 714.

19. M. D. Brown, L. D. Ellsworth, J. A. Guffey, T. Chiao, E. W. Pettus, L. M. Winters, and J. R. Macdonald, Phys. Rev. A $10$, 1255 (1974).

20. E. Horsdal Pedersen, S. J. Czuchlewski, M. D. Brown, L. D. Ellsworth, and J. R. Macdonald, Phys. Rev. A $11$, 1267 (1975).

21. S. J. Czuchlewski, L. D. Ellsworth, J. A. Guffey, E. Horsdal Pedersen, E. Salzborn, and J. R. Macdonald, Phys. Lett. $51A$, 309 (1975).

# STUDY OF K, L AND M INNER-SHELL IONIZATION

# BY PROTON IMPACT

V.S. Nikolaev, V.P. Petukhov, E.A. Romanovsky,

V.A. Sergeev, I.M. Kruglova and
V.V. Beloshitsky

Institute of Nuclear Physics, Moscow State
University; Moscow 117234, USSR.

The ionization of the inner shells by protons was studied at Moscow State University to make clear the difference between the ionization cross sections when electrons are ejected from different states. For this purpose we have measured the total cross sections for ionization of the K, L and M shells with electron binding energy $I = 2\text{-}10$ keV by protons with the energy $E = \cdot 150\text{-}500$ keV and have studied the ratios of the ionization cross sections for the $L_1$ and $L_2$ subshells of Pd, Sb and La atoms with the electron binding energy of about 4 keV. Theoretical calculations of the ionization cross sections for the ejection of electrons from the various states of the K, L and M shells have been made in the Born approximation and the connection of the specific features in the ratio of the ionization cross sections with the peculiarities of the electron initial states has been analyzed.

## 1. EXPERIMENTAL RESULTS

The cross section for the K, L and M shells ionization were determined by the detection of characteristic X-radiation / 1,2/. Foil targets 0.1 to 0.5 mm thick were used. X-ray quanta were detected by a proportional counter having a resolution of approximately 20 per cent. The mean error in determining the total ionization cross sections was 25 per cent.

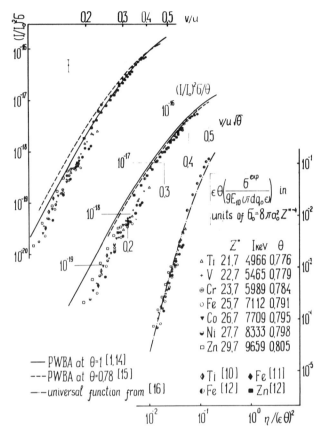

Fig. 1. The K shell ionization cross section.

When measuring the ratio of the ionization cross sections for the L subshells, X-rays of the lines $L_{\beta 1}$ and $L_{\beta 3}$ were analyzed by a diffraction crystal spectrometer with the Soller collimator and a crystal of lithium fluoride (2d = 4.2 Å) or quartz (2d = 6.66 Å). The spectrometer resolution was approximately 30 eV. The error in measuring the X-ray yields was not in excess of 5 per cent and the ratio of the ionization cross sections was determined to an accuracy of 20 per cent. The values of the fluorescence yields, the Coster–Kronig factors, radiation widths, stopping power and mass absorption coefficients are from refs. / 3–8/. The lacking values of the mass absorption coefficients were calculated using Jönsson's method /9/.

The results of measurement of the total K-shell
ionization cross sections are given in Fig. 1. The upper
part of the figure contains the ionization cross sections
$\sigma$ multiplied by the squared electron binding energy as
functions of the ratio of the proton velocity $v$ to the velocity $u$ determined from the electron binding energy I:
$u = v_o ( I / I_o )^{1/2}$ where $V_o = 2.19 \times 10^8$ cm/s, $I_o =$
13.6 eV. The binding energy values are taken from the
work by Bearden and Burr [13]. It should be pointed out
that for titanium, zinc and iron atoms our cross sections
agree with those obtained in the works [10-12]. Somewhat
below on this figure are presented the same data divided
by the outer screening parameter $\theta = I / I_o ( Z^*/n)^2$
against the reduced proton velocity $v/u \sqrt{\theta} \equiv v/(Z^*/n) v_o \theta$
( Here $Z^*$ is the effective nuclear charge and $n$ is the
principal quantum number). From the work by Merzbacher
and Lewis [1] it follows that the experimental data for various atoms should form a single curve in this case. It
is seen from the figure that the experimental-point scattering slightly exceeds the experimental errors, and in the
first variant which is not so well founded theoretically
this scattering is even somewhat smaller. In both variants
the experimental points in the low-velocity region lie below the curves calculated in the Born approximation. Calculations using the method of Basbas et al [16] which
accounts for the Coulomb deflection of the projectile in
the field of the target nucleus and for changes in the
binding energy of ejected electrons give a better agreement with the experiment.

The results of measurement of the ionization cross
sections for the L and M shells are presented in Figs. 2
and 3. Here $I = ( \sum I_i q_i )/ \sum q_i$ where $q_i$ is the number
in the $i$-th subshell of electrons and $I_i$ is the electron
binding energy. Our values of the reduced cross sections
$(I/I_o)^2 \sigma$ for the L shell of Pd atoms with Z = 46 agree
within the experimental error with the corresponding values for Ag atoms with Z = 47 from ref. [10] while the
ionization cross sections for the M shell of Bi atoms
( Z= 83) are consistent with the results of ref. [20]. The
conclusions drawn above concerning the K shell ionization of the hold for the L and M shells as well. However,
in contrast to the results for the K shell in the case of
the L – shell one observes an increase approximately by
a factor 2.5 in the reduced ionization cross sections
$(I/I_o)^2 \sigma$ and $(I/I_o)^2 \sigma/\theta$ at fixed values $v/u$ and $v/u\sqrt{\theta}$
with increasind Z from 45 to 74 and a small decrease in
the reduced cross sections in the case of the M shell
with increasing Z from 73 to 92. This may be associated
with errors in the theoretical values of the fluorescent

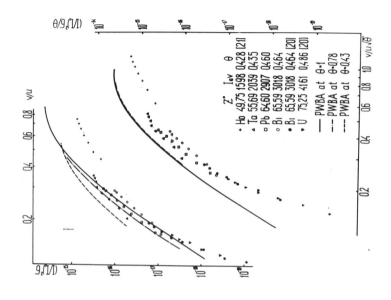

Fig.3. The M shell ionization cross section.

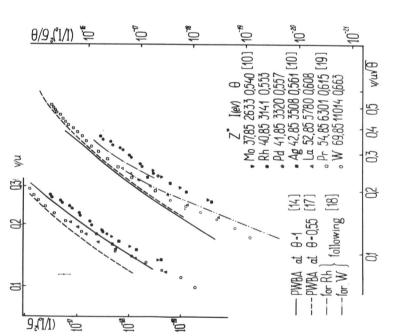

Fig.2. The L shell ionization cross section.

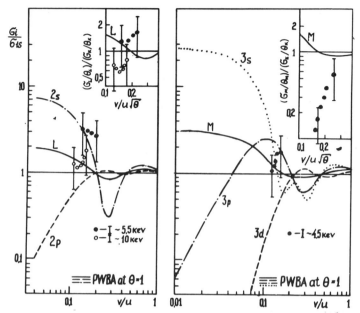

Fig. 4. Comparison of the K, L and M shell
ionization cross sections normalized to one electron at
equal electron binding energies.

yield /6/ we have used. Taking account of the deflection
of the proton trajectory and of changes in the binding energy of L  electrons due to their interaction with protons by
the method of Brandt and Lapicki /18/ leads to a better
agreement between theory and experiment.

     Noting that the reduced cross sections vary but
little with moderate variations in the electron binding energy I, we have compared the ionization cross sections for
the various shells. Fig. 4 gives the ratios of the total
ionization cross sections for the L and M shells to the
K-shell ionization cross section  $\sigma_i / \sigma_K$  ($i = L, M$) with
the same binding energy of electrons (when normalized to
one electron) for the electrons with binding energies of
about 5 and 10 keV as functions of $v/u$ , and the ratios
 $(\sigma_i/\theta_i)/(\sigma_K/\theta_K)$       as functions of $v/u\sqrt{\theta}$  are given
in the right upper parts of the figure. The lines represent
the results of the calculations in the Born approximation
for  $\theta = 1$. From the figure it is evident that when the
electron binding energy is about 5 keV the values $\sigma_i/\sigma_K$
for the L shell are much greater than those calculated in
the Born approximation and for the M shell they are equal
to the calculated ones. The values  $(\sigma_i/\theta_i)/(\sigma_K/\theta_K)$

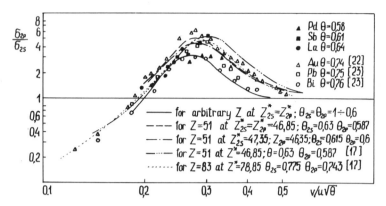

Fig.5. The ratios of the ionization cross sections for the $L_1$ and $L_2$ subshells.

are smaller than $\sigma_i/\sigma_K$ . For the L shell they are equal to the calculated values while for the M shell they are much lower than the calculated ones. When I is about 10 keV the experimental values $\sigma_L/\sigma_K$ agree within the experimental error with the calculated ones while the values $(\sigma_L/\theta_L)/(\sigma_K/\theta_K)$ lie somewhat lower.

The results of measurement of the ratio of the cross sections for the creation of vacancies in the 2s and 2p states for Pd, Sb and La atoms along with the results of similar measurements for Au, Pb and Bi /22,23/ are given in Fig. 5 as functions of the reduced proton velocity $v/u\sqrt{\theta}$ i.e., as functions of the ratio of the proton velocity $v$ to the mean orbital electron velocity $(Z^*/n)v_0$ times $\theta$ . The same figure gives the results of different variants of the Born calculations. From the figure it is seen that in all cases, in accordance with the results of the Born calculations / 14,17/ a pronounced maximum is observed in the values of the cross section ratio $\sigma_{2p}/\sigma_{2s}$ at the values $v/u\sqrt{\theta}\approx1/3$. This maximum, as was first shown in the works /14,24/, is determined by the node of the initial 2s electron wave function.

## 2. THEORETICAL INVESTIGATIONS

The results of the Born approximation calculations of the cross sections for ionization by protons of hydrogen-like ions in different states with an equal binding energy indicate that these cross sections are significantly different at proton velocities v smaller than the mean orbital velo-

city $(Z/n)v = u$ of the ejected electron (Fig.4). This comes from the fact that the ionization cross sections in the
region of small values $v/u$ are determined by value of
the generalized oscillator strength per unit energy transfer
$E_{if}$ in    rydbergs, $f_{if}(Q,K)$, for the ionization transitions
of an electron from the initial state i   to the final state f
in a relatively small region of the absolute values of the
momenta of the ejected electron $\hbar K$ and of the momentum
transfers $\hbar Q$ close to their minimum values $\hbar K_{min}=0$ and
$\hbar Q_{min}= I/v$ /25/. Therefore, with increasing proton velocity v, the ionization cross sections vary in accordance
with the change in the generalized oscillator strength
$f_{if}(Q,K)|_{K\to 0} \equiv f_{if}(Q)$ with decreasing momentum transfer
$\hbar Q \gtrsim I/v$. As regards the function $f_{if}(Q)$, it reflects, as
has already been noted /26,27/,the qualitative features of
the momentum distribution of electrons, in the initial state
$|\Phi_i(p)|^2$    (Fig.6), and most of all this takes place in
the case of the transitions from the s-states /26/. Therefore these qualitative features    exhibit themselves also

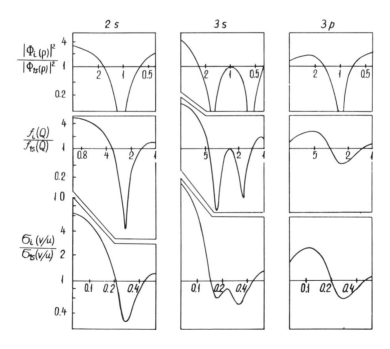

Fig.6. Comparison of the functions $|\Phi_i(p)|^2/|\Phi_{1s}(p)|^2$,
$f_{if}(Q)/f_{1s,f}(Q)$, $\sigma_i(v/u)/\sigma_{1s}(v/u)$for the hydrogen-like 1s, 2s, 3s
and 3p states of equal electron binding energies ( $\theta = 1$).
The momenta $\hbar p$ and $\hbar Q$ are presented in unities
$(Z^*/n)(\hbar/a_o)$

in the dependence $\delta_i/\delta_{1s}$ on $v/u\sqrt{\theta}$.

In order to understand the cause of such correspondence between the generalized oscillator strengths $f_{if}(Q)$ and the momentum distribution $|\Phi_i(p)|^2$ the generalized oscillator strengths were represented as the sum of the partial generalized oscillator strengths $f_{if}^\ell(Q)$ each of which corresponds to the transition of the electron to its final state with a definite orbital angular momentum $\ell$

$$f_{if}(Q) = \sum_{\ell=0}^{\infty} (2\ell+1)\, f_{if}^\ell(Q) \qquad (1)$$

when

$$f_{if}^\ell(Q) = \frac{E_{if}/I_o}{(Qa_o)^2} \sum_{\lambda=\ell-\ell'}^{\ell+\ell'} (2\ell+1) \left| \begin{pmatrix} \ell & \lambda & \ell' \\ 0 & 0 & 0 \end{pmatrix} \right|^2 |M_{if}^\lambda(Q)|^2 \qquad (2)$$

and   $M_{if}^\lambda(Q) = \int_0^{\infty} R_i(r)\, j_\lambda(Qr)\, R_f^{(-)}(r)\, r^2 dr$

where $R_i(r)$ and $R_f^{(-)}(r)$ are the radial parts of the corresponding wave functions and $\ell'$ is the orbital angular momentum of the electron in the initial state, $a_o = 5.29 \times 10^{-9}$ cm$^2$. The results of this expansion for the ionization transitions of the electron from the 2s and 3s states are shown in Fig. 7.

In the transitions from the s-states, the sum (2) is represented only by one term proportional to $|M_{if}^\ell(Q)|^2$. Therefore when the matrix element $M_{if}^\lambda(Q)$ changes

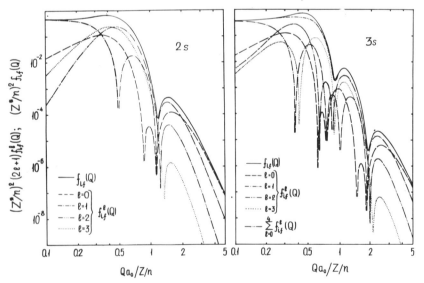

Fig. 7. Expansion of the generalized oscillator strenghts $f_{if}(Q)$ into the partial ones $f_{if}^\ell(Q)$ for the ionization transitions of the 2s and 3s electrons.

sign the values $f_{if}^{\ell}(Q)$ become zero, whereas in the total generalized strengths $f_{if}(Q)$, only minima occur.
An analysis of these matrix elements has shown that in
their value are reproduced the qualitative features of the
product of the radial parts of the wave functions for the
initial and final states $R_i(r)R_f(r)$ in the region of distances $r$ from the nucleus from zero to the values $\tilde{r}_i$
of the order of the most probable values $r_i$ for
the electron initial state. This correspondence in the case
of the electron transition from the 2s state to the p-state
of the continuum with zero energy is clearly seen in
Fig.8. This also leads to the correspondence between
the zeroes of the matrix elements $M_{if}^{\ell}(Q)$ and
the zeroes of the product of the radial parts of the wave
functions $R_i(r)R_f(r)$ in the region $r \leqslant \tilde{r}_i$ (Fig. 9).
Naturally, only those qualitative features of the partial generalized oscillator strengths which are only determined
by the electron initial state are presented in the total generalized oscillator strengths, $f_{if}(Q)$.

   In the case of the electron transitions from
the p-states the partial generalized oscillator strengths
are determined by two matrix elements of the sum (2)
   which change sign at different values $Q$ . Therefore
it is only minima that occur in the generalized oscillator
strengths $f_{if}(Q)$ and the qualitative features of
the initial state exhibit themselves in the total generalized
oscillator strength to a lesser extent / 27/.

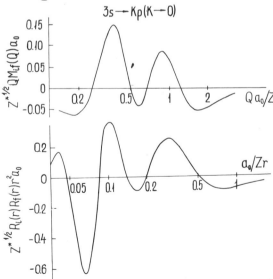

Fig.8. Comparison of the functions $M_{if}^{\ell}(Q)$ and
$R_i(r)R_f(r)r^2$ for the transition 3s $\rightarrow$ K$\ell$ when K=0, $\ell$ = 1.

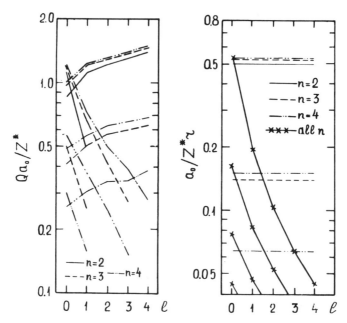

Fig. 9. The position of zeroes of the functions $M_{if}^{\ell}(Q)$ and $R_i(r)R_f(r)$ for the ionization transitions $ns \rightarrow K\ell$ when $K=0$ in the planes ( $\ell$ ; $Qa_0/Z^*$) and ($\ell$; $a_0/Z^*r$) respectively.

The structural features of the electron initial state are also reflected in the dependence of the ionization probability $W_i(\rho)$ on the impact parameter $\rho$ /28/. The values $W_i(\rho)$ were obtained by expansion of the proton inelastic scattering amplitude in the eigenfunctions of the operator of the proton orbital angular momentum $\ell_p$ and by the use of the quasiclassical relation: $\ell_p = Mv\rho$ where M is the reduced mass of colliding particles /28/. Earlier this method was used by Shift for calculating the probability of charge exchange of protons at various impact parameters /30/. The probabilities of ionization of the K shell and $L_1$ subshell with equal electron binding energies as functions of $\rho$ when normalized to one electron are given in Fig. 10.

At not too low velocities of protons the dependence of the ratio of these probabilities on $\rho$ qualitatively reproduces the dependence of the relations between electron densities in these states. The minimum of the electron density in the 2s state is especially pronounced at $v/u = 1/3$ when the minimum value of momentum transfer $Q_{min}$

$= 1.5(Z/n)(\hbar/a_0)$ is 1.5 times smaller than the values $Q = 2.35(Z/n)(\hbar/a_0)$ at which the function
$f_{2s}(Q)$ for the ionization of the 2s electron reaches its
minimum. At lower proton velocities, one observes inverse
relations between the probabilities of ionization for the 1s
and 2s electrons, which correspond to the relation between
the electron densities in the 1s and 2s states in the region
of distances from the nucleus smaller than the distance to
the node of the 2s electron wave function. This implies
that at low proton velocities electrons located at large
distances from the nucleus are not practically involved
in ionization processes, and ionization occurs due to the
interaction between a proton and electrons having great
momentum $p$ similarly to elastic reflection of a particle
from a heavy wall. It should be noted that these calculations of the probability of ionization at various impact parameters /31/ are reasonably consistent with the available
experimental data /33,34/ for K shell ionization of Cu, Se
and Ag atoms by 1 and 2 MeV protons, which corresponds
to the reduced proton velocity $v/u = 0.1\text{-}0.3$ (Fig. 11).

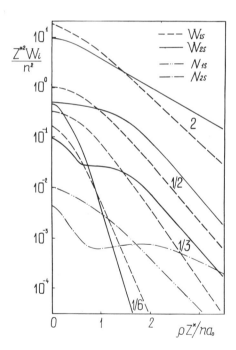

Fig. 10. The ionization probability $W_i$ and the two-dimensional electron density $N_i$ for the states i=1s and 2s
when $\theta = 1$ as a function of the impact parameter $\rho$.
The values $v/u$ are indicated by the curves ($N_i$ is given
in relative units).

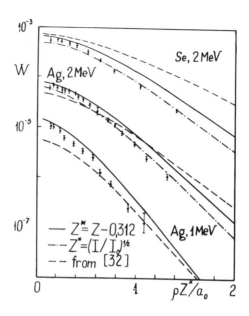

Fig. 11. The probability of ionization of the K shells of Se and Ag atoms by protons versus impact parameter.

REFERENCE S

1. E. Merzbacher and H.W. Lewis, Handbuch der Physik, v.34, p.166, Berlin, 1958
2. J.D. Garcia, R.I. Forther, T.M. Kavanagh, Rev.Mod. Phys. v.45, No.2, Part I, III (1973)
3. Int. Congress in X-ray Optics and X-ray Microanalysis, Paris, Herman, 1966
4. J.B. Woodhouse, ALA Fields and I.A. Bucklow, J.Phys. D: Appl.Phys. v.7, 483(1974)
5. A.H. Compton and S.K. Allison: X-rays in Theory and Practice (D. Van Nostrand Company Inc., Orincston, New Jersey,1960) p. 538
6. W. Bambynek, et al. Rev.Mod. Phys.44,1,716(1972)
7. L.C. Northcliffe, R.F. Schilling, Nucl. Data Tables, A7, 233 (1970)
8. J.H. Scofield, Phys.Rev. 179, 1 (1969)
9. E. Jönsson, Thesis, Uppsala (1929)
10. K. Shima, J. Makino, M. Sakisaka, J. Phys. Soc. Japan,30, 611(1971)
11. S. Messelt, Nucl.Phys. 5, 435 (1958)
12. R. Lear and T. Gray, Phys.Rev.A8,5,2469 ( 1973)
13. J.A. Bearden, A.F. Burr, Rev.Mod.Phys.39,1,125(1967)

14. V.S. Nikolaev, V.S. Senashenko, V. Yu.Shafer,
        J. Phys.B: Atom.Molec.Phys.$\underline{6}$,1779(1973)
15. G.S. Khandelwal, B.H. Choi, E. Merzbacher,
        Atom Data, $\underline{1}$,103 (1969)
16. G. Basbas, W. Brant, R. Laubert, Phys.Rev. A$\underline{7}$,3,
        983(1973)
17. B.H. Choi, E. Merzbacher, G.S. Khandelwal, Atom.
        Data, $\underline{5}$,291 (1973)

18. W. Brandt, G. Lapicki, Phys.Rev.A$\underline{10}$,2,474(1974)
19. F. Abrath and T. Gray, Phys.Rev.A$\underline{10}$,4,1157(1974)
20. P.B. Needham, Jr., and B.D. Sartwell, Phys.Rev.
        A$\underline{2}$,5,1686(1970)
21. J.M. Khan, D.L. Potter, R.D. Worley, Phys.Rev.$\underline{138}$,
        6A,1736 (1965)
22. S. Datz, J.L. Duggan, L.C. Feldman, E. Laesgaard,
        I.V. Andersen, Phys.Rev. A$\underline{9}$,1,192(1974)
23. D.H. Madison, A.B. Baskin, C.E. Bucz, S.M. Shafroth,
        Phys.Rev.A$\underline{9}$, 675(1974)
24. V.S. Senashenko, V.S. Nikolaev, V. Yu. Shafer,
        Abstracts of Papers VI ICPEAC, p. 740,
        Cambridge (1969); Phys.Lett.$\underline{31A}$,565 (1970)
25. I.S. Dmitriev, Ya.N. Zhilekin and V.S. Nikolaev,
        ZhETF,49, 500   (1965)
26. V.S. Nikolaev, anu I.M. Kruglova, Phys.Lett.$\underline{37A}$,
        315 (1971)
27. I.M. Kriglova, V.S. Nikolaev, V.A. Sergeev,
        Phys.Lett.43A, 375(1975)
28. V.V. Beloshitsky, V.S. Nikolaev, Phys.Lett.$\underline{51A}$,97(1975)
29. A.K. Kaminsky and V.S. Nikolaev, Abstracts of
        papers V All-Union CPEAC Uzhgorod, p.69 (1972)
30. H. Shiff, Can.J. Phys.$\underline{32}$,393(1954)
31. A.K. Kaminsky, V.S. Nikolaev and M.I. Popova,
        Abstracts of papers All-Union Seminar on theory
        of atoms and atomic spectra. Tashkent, p.45
        (1974)
32. J.M. Hansteen, O.P. Mosebakk, Nucl.Phys.A$\underline{201}$,
        541(1973)
33. E. Laegsgaard, I.V. Andersen, L. Feldman,
        Phys.Rev.Lett.$\underline{29}$,1206(1972)
34. D. Burch, VIII ICPEAC, Invited Lectures and
        Progress Reports, Belgrade, 1973.

THE IMPACT PARAMETER DEPENDENCE OF INNER SHELL EXCITATION

H.O. Lutz

University of Bielefeld

D-4800 Bielefeld, W-Germany

## I. INTRODUCTION

Off and on, I am asked (particularly from physicists not belonging to the inner shell community) "Why are you doing these cumbersome coincidence experiments?" Since these experiments are cumbersome indeed, what is their merit, aside from providing the fun of playing around with a complicated set-up? In this talk I shall try to give a brief account of recent developments in this area, and hopefully it will contain the answer to the above question. Since the topic of this talk is inner shell excitation, I shall restrict myself to target atoms heavier than hydrogen or helium.

If we consider inner shell excitation it appears that the relatively simplest system is that of a swift structureless point charge penetrating an atom which preferably consists of independent electrons in hydrogenic states. It is not surprising, therefore, that relatively early the theoreticians tackled just this problem, most closely represented by the ionisation of K-electrons by fast protons or $\alpha$-particles /1,2/. Thus, experimenters working in this field were in a fortunate position of having theories to compare to; at a relatively early stage they were already able to concentrate on correction to these theories, as binding energy and polarisation effects introduced by the projectile /3/.

In the early 1960s, the groups of Everhart and Afrosimov reported the first results on inner shell excitation at the other end of the spectrum /4/: the impact velocities were small and the projectiles certainly not structureless. These were

the famous Q-value experiments in slow heavy ion-atom collisions; the energy loss (Q-value) of Ar ions scattered from Ar
atoms through well defined angles showed discrete groups which
were attributed to the production of a distinct number of
vacancies in inner electron shells; these experiments were
thus the first impact parameter studies of inner shell excitation. They prompted the development of the molecular promotion model, and we all know the great impact this idea had on
inner shell excitation studies. The groups working in this
area were perhaps even more fortunate than the Coulomb-excitation community, since they did not have theories to compare
to; this left ample room for speculation and hand waving. In
their review paper, Kessel and Fastrup /4/ provided a very
detailed discussion of the differential energy loss experiments.
Since Bent Fastrup is also giving a review paper at this conference, I shall omit the Q-value measurements in my talk.
Furthermore, I shall also omit impact parameter studies of
molecular x-ray emission since there is an extra session at
this conference devoted to this phenomenon.

## II.   EXPERIMENTAL

The principle of the experimental technique is the same
in most of the work reported below (with one exception to be
discussed at the end of this section). Collimated ion beams
are injected into thin gaseous or solid targets. Ions scattered
through laboratory angles $\theta$ are detected in a more or less
annular detector to increase solid angle. This detector may
be moved along the incident beam axis to change $\theta$. The angle
$\theta$ can also be kept fixed, and the ion energy be varied. A
cylindrical electrostatic deflector may be used to analyse the
scattered ions according to their charge stated /5/. The impact
parameter $\varrho$ is related to $\theta$ through the interatomic potential.
The reaction channel is identified by the characteristic x-rays
(or Auger-electrons) emitted after the collision. In well
localized thin solid targets, the No. $N_c$ of x-rays coincident with
scattered ions, divided by the total number $N_s$ of ions scattered
through $\theta$, is directly proportional to the impact parameter
dependent inner shell excitation probability $P(\varrho)$ per collision (c.f., e.g., Ref. /7/). In gaseous targets, problems arise
from a not well defined scattering region /8/; in that case
the differential scattering cross section $\sigma$ must be known to
derive $N_s$, and the absolute value of $P(\varrho)$ is given by $P \cdot \omega =$
$c \cdot N_c / (\sigma \cdot N_x)$, where $N_x$ is the total number of the detected
x-rays, $\omega$ the fluorescence yield; the normalisation constant
$c$ can be derived by comparing $2\pi \int (P \cdot \omega) \cdot \varrho \, d\varrho$ to the
measured total x-ray cross section. In symmetric or nearly-
symmetric collision systems, scattered ions and knock-ons
cannot be distinguished; in that case, forward (f) and backward
angles (b) contribute to the x-ray yield, and one has

$(P_f \sigma_f + P_b \sigma_b) \cdot \omega = c \cdot N_c/N_x$, /8,9/. Since $\sigma$ falls off
rapidly with $\theta$, the contribution from backward angles is im-
portant only in a small region atround $\theta = 45^\circ$.

Even though the number of scattered ions is large, particularly at small $\theta$, coincidence events are a rather rare
occasion.  There is an upper limit to the number $N_{c,t}$ of true
coincidences as $N_s$ cannot be arbitrarily increased:  the
number of random coincidences $N_{c,r}$ is proportional to the
square of the incident beam intensity, $N_{c,r} \approx N_s \cdot N_x \cdot \tau$
($\tau$, coincidence time resolution).  Generally, it is only practical to increase $N_{c,t}$ up to $N_{c,r}$ . $N_x$ and $N_s$ are usually very
different in magnitude (typically, $N_s \approx$ 10-100 kHz, $N_x \approx$ 1-10 kHz,
and $N_{c,t}$ is quite low (typically only $\lesssim$ 1 Hz).  In light collision systems, Auger electrons may be used instead of x-rays to
identify the reaction channel.  Even though the detector solid
angle $\Omega$ will be smaller, the increase due to   will more than
compensate;  additionally, electrons can be detected by chan-
neltrons or open-ended multipliers which allow much faster
timing (reduction of $\tau$ by at least an order of magnitude) if
compared to proportional counters used for soft x-rays /10/.

In the technique described above, the scattering angle $\theta$
is used to identify the impact parameter.  Instead, the channeling effect provides a completely different approach /11/.
Energetic ions penetrating the open channels of a crystal undergo correlated small-angle collisions giving rise to well
defined oscillatory trajectories.  The amplitudes $y_m$ of these
oscillations are directly related to the energy loss in the
channel thus defining the impact parameters.  The entire energy
spectrum and thus the excitation probability over the whole
range of impact parameters is obtained in one run.  This technique has been demonstrated with the collision system I-Au
(Fig.1); it is particularly suited for measuring very small
values of the excitation probability (e.g., the experimental
data in Fig. 1 represent only a few hours beam time; furthermore, the I-Au system is a rather awkward one since the heavy
I-ions had to be measured in a time-of-flight set-up to obtain
the necessary energy resolution).

### III.   EXCITATION BY FAST STRUCTURELESS PROJECTILES

The terms "molecular excitation" and "Coulomb excitation"
serve a useful purpose as they separate in a qualitative way
regions of different theoretical methods, even though this
distinction often appears somewhat arbitrary.  An often helpful parameter is $\eta$ = v/u, where  v  is the ion velocity and
u  the orbital velocity of the excited electron.

At the "Coulomb-Excitation" end, mainly three types of

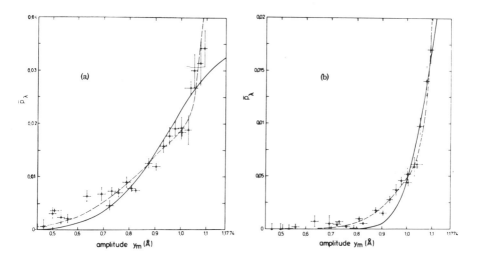

Fig. 1:   Characteristic x-ray production in 50 MeV I-Au
          collisions as a function of oscillation amplitude,
          $y_m$;   a) I-M, b) Au-M /11/.

theoretical methods are available:  the plane wave Born approximation (PWBA) /1/, the semi-classical approximation with or
without Coulomb deflection of the projectile (SCA) /2/, and the
binary encounter approximation (BEA) /3, 12, 13, 14/; SCA and
BEA additionally provide the impact parameter dependence $P(\rho)$
of the excitation process. Actually, the straight-line approximation seems to work well to quite low ion energies (e.g. in
SCA less than 10% difference compared to the Coulomb deflection
SCA) if the distance of closest approach is substituted for the
particular impact parameter (tangential approximation) /2/.
These theories could at first be tested experimentally only by
a comparison with total excitation cross sections.  These tests
were generally quite favorable for any of the above theories
depending on the ion energy range and the corrections introduced.
Much more stringent tests were possible when experimental data
on the impact parameter dependence of the Coulomb excitation
process became available, as those of Laesgaard and coworkers
on K excitation of Cu, Se and Ag by protons /6/.  The agreement
between the SCA /2/ and this experiment was considered to be
quite good (within about 30%), particularly since no fitting
parameters were employed.

    Recently, Amundsen and Kocbach /15,35/ introduced relativ-
istic wave functions into the SCA formulation, and simulated
the Coulomb deflection by a tangential approximation.  They

get excellent agreement with experimental data on K-excitation
by protons colliding with heavy target atoms, as obtained by
the Aarhus group (Fig. 2).

For the low velocity limit, Brandt and coworkers /16/
derived a simple universal expression, based on the non-relativistic SCA /2/, and proposed to plot a reduced excitation yield
Y vs. a reduced impact parameter y. The tendency of the experimental points is to fall on the universal curve if the binding
of the K-electron to the projectile is taken into account, and
the tangential approximation is used to account for the Coulomb
deflection (Fig. 3). However, if these data are plotted in a
more expanded form it becomes evident that the agreement between
this low-energy SCA formulation and the experiment is not quite
as good as suggested by Fig. 3. This is corroborated by other
experiments, most recently by Schmidt-Böcking and his group
/17/ in Frankfurt.

Since full SCA calculations /2/ are quite cumbersome,
McGuire /13/ extended the BEA to incorporate the impact parameter dependence of the excitation process. These BEA calculations are much simpler than the full SCA calculations, and
scaling laws allow application of tables for arbitrary projectiles and targets over a wide energy range. Generally, the

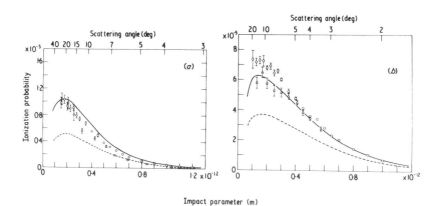

Fig. 2:   P($\varphi$) for K-excitation by protons in Ag; a) 1 MeV,
          b) 2 Mev; solid curve: relativistic SCA, broken
          curve: non-relativistic SCA /15/.

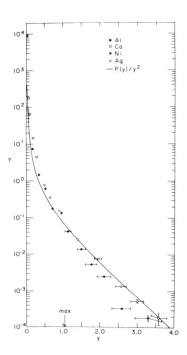

Fig. 3:   Reduced plot of the low energy SCA excitation yield
          Y vs. impact parameter y  for 2 MeV protons /16/.


shapes of the P($\varrho$) curves in the BEA and the SCA are fairly
similar in the entire energy range, and at high velocities the
BEA results coincide with the SCA results, as expected; there
are, however, no experimental P($\varrho$)-data to compare to in this
limit ( $\eta \gg 1$). At intermediate and low velocities ( $\eta \lesssim 1$),
particularly with very heavy target atoms, discrepancies arise.
It is at present difficult to assess the region of validity of
the different models; e.g., Hansen /14/ also found excellent
agreement between experimental P($\varrho$)-values by Laesgaard and
coworkers /6/ on K-excitation in $^{34}$Se by 2 MeV protons.  The
different scaling laws in the various BEA and SCA models become
apparent if very heavy collision systems are investigated.
Burch and coworkers /18/ found excellent agreement between the
simple BEA, PWBA and SCA theories, and experimental total K-
excitation cross sections in 50 to 100 MeV Cl → Pb collisions,
while, e.g., an improved version of the BEA by Hansen (including corrections for projectile deflection, retardation, and
relativistic effects) is much less sucessful.  When they investigated the corrections individually they found that the

good agreement with the simpler theories is largely due to a
fortuitous cancellation of these factors.  Again, much more
detailed information is provided by the $P(\rho)$-function, and a
comparison of SCA and BEA results clearly showed that for
50 MeV Cl $\rightarrow$ Pb the contributions in both models to the total
cross section stem from wholly different regions of $\rho$ : in
BEA, $P(\rho=0)$ is about 7 times as high as in SCA, but it falls
off about twice as quickly with $\rho$.  The experimental $P(\rho=0)$
values in the energy range investigated did not show good agreement with any of the above theories.  Cocke and coworkers /19/
obtained similar results using 0-, C-, and F-ions ($\eta \approx 0.5$).
The shape of the experimental $P(\rho)$ dependence is closely
represented by the SCA /16/, while the BEA gives larger contributions at smaller $\rho$.  For such cases the theoretical cross
sections may be nearly equal, while they are quite different
from the experimental values.

Recently, two groups independently studied the $P(\rho)$
dependence of K-excitation if the charge $Z_1$ of the projectile
is changed systematically.  Schmidt-Böcking and his group at
Frankfurt /17/ found that the deviation between the simple
SCA /2/ or BEA /13/ and their experimental data becomes
progressively larger for increasing $Z_1^2$ (Fig. 4).  In other
words, the simple $Z_1^2$ scaling is not fulfilled, as is well
known from total cross section data.  The low-energy SCA /16/
brought only slight improvement /17/, even though it takes
into account the increased binding of the K-electrons due to
the presence of the projectile.  Andersen and coworkers at
Aarhus /34/ obtained similar results, however, they found a
simple scaling procedure assuming a $\rho$-dependence of the increased binding: for a fixed value of $\eta$, $P(\rho E_B/\hbar v)$ has been
assumed to scale with the factor $v^6 Z_1^2/E_B^4$ ($E_B$ is the
$\rho$-dependent electron binding energy).  At small and intermediate
$\rho$, this procedure turns out to be quite effective.

An interesting feature is the strong dependence of the
excitation cross section on the projectile charge state particularly at large $\rho$ /19,34/.  This is especially noticeable
if the projectile K-shell is open.  No adequate theory exists
for describing such collision at intermediate velocities and
$Z_1/Z_2$-ratios; e.g., it was found that the Brinkman-Kramers
treatment of ls-ls charge exchange cannot quantitatively explain
the observed effect /19/.  This, of course, does not imply
that charge exchange is not operative; it just means that a
simple BA treatment is not sufficient.  Somewhat surprisingly,
there also appears to be an increasing disagreement between
theoretical (BEA) and experimental $P(\rho=0)$ values for targets
of decreasing $Z_2$ /20/.  For Al and Mg, the experimental data
consistently lie about a factor of 4 higher than the theory;
for Cu, Ti and Sc this factor does not exceed 1.5.  This

cannot be due to the less adiabatic nature of the collision in case of the lighter target atoms since the disagreement theory-experiment does not change substantially if the impact velocity is changed by about a factor of two for the same target. The authors suspect that the discrepancy arises from the difficulty of adequately describing the K-electrons in these atoms by a hydrogenic model.

Few experimental P($\wp$)-data is available on Coulomb L-excitation. Schmidt-Böcking and his group will report results on L-subshell excitation in Pb by 2 to 6 MeV protons ($\eta \approx 0.4$) at this conference. Neither SCA nor BEA seem to give consistent agreement in the entire energy range.

## IV.   EXCITATION AT LOW ION VELOCITIES

The work on the promotion model stimulated much interest in the electronic configuration of atoms which have undergone a collision. Thomson and coworkers /5/ detected in the first ion-electron coincidence experiment the energetic LMM-Auger electrons which were expected from the decay of L-vacancies in Ar-Ar encounters. Similar measurements with the asymmetric system P-Ar have been made by Dahl and Lorents /21/; they found pronounced structure in the coincident electron spectra and were able to associate it with electronic transitions involving L-vacancies. From non-coincident x-ray or electron spectra, it is a well known fact that these ion-atom collisions also result in outer shell excitation causing a strong   increase in the $L_{2,3}$ fluorescence yield /22/. Thomson and coworkers /23/ found evidence of outer shell excitation in the coincident electron spectra in L-vacancy producing Ar-Ar collisions; they showed that slow electrons from the excitation of the outer (M) shell may be ejected before as well as after the LMM-Auger event; nevertheless, no evidence has yet been found for molecular electron emission; in the systems investigated most inner shells are filled after the atoms have separated, in agreement with observations by Stolterfoht and coworkers for similar collision systems without the use of coincidence techniques /24/.

The first ion-photon coincidence experiments as performed in Jülich /7/ were prompted by the early investigations of Armbruster and coworkers /25/; they were designed to extend the work on the promotion model into the MeV range, and thereby to unambiguously identify the reaction channel. Results are shown in Fig. 5 for L x-ray production in the system I-Te. The systems were fairly adiabatic ($\eta \approx 0.2$).
As in Ar-Ar L-excitation, the main candidate for a promoted molecular level was $4f\sigma$. Nevertheless, the vacancy production showed a behavior different from that observed in Ar-Ar L-excitation: it varied only slowly with $\wp$ ; it was energy dependent;

and its maximum value was appreciably smaller than 2.  These
results could qualitatively be explained in the framework of
the promotion model, too.  Even though, one look at the correlation diagram makes it clear that the collision system
was far to complex to draw any quantitative conclusions as to

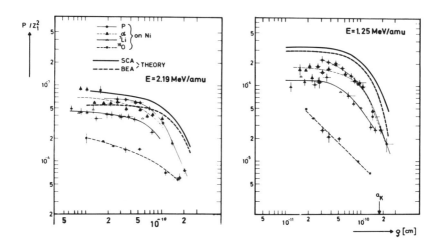

Fig. 4:  P($\varrho$) for Ni- K-excitation by various projectiles /17/.

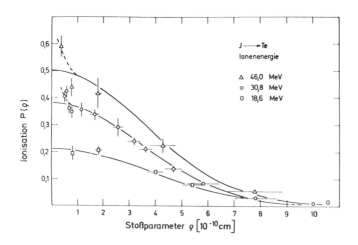

Fig. 5:  P($\varrho$) for L-excitation in I-Te collisions /7/.

the nature of the electronic states and transition involved in
the excitation process. One interesting feature is the additional steep rise of the $P(\rho)$-values at small $\rho$, indicated
by the dashed curves. This feature could not satisfactorily
be explained by any model of inner shell excitation, though it
was speculated that it may be due to the onset of direct Coulomb
excitation; however, a new analysis of recoil effects on the
$P(\rho)$-dependence in heavy ion collisions /26/ indicates that
x-ray excitation by target recoils may have been responsible
for this effect.

Up to this point, the electron promotion model was put on a
more qualitative basis. Much experimental information was
obtained on L-excitation in symmetric and asymmetric collisions;
however, because of the complexity of such collisions systems,
the electronic levels and transitions involved could not be
formulated quantitatively. Evidently, K-excitation is much
more suited for a quantitative analysis since (at least in
symmetric systems) the $2p\sigma - 2p\pi$ crossing is isolated, and
the states as nearly hydrogenic as one can get. This process
has been treated by Briggs and Macek /27/, and the experimental
data obtained by our group at Jülich and Bielefeld /8/ show an
almost embarrassingly good agreement with these calculations
(Fig. 6). In particular, the fairly flat adiabatic maximum at
large $\rho$ is definitely separated from the kinematic peak at small
$\rho$ allowing a clear dinstiction from the Coulomb excitation
mechanism. This is particularly gratifying since the total
cross sections would not provide that: plotting the Ne-Ne data
in reduced coordinates /3/, they are close to the Coulomb excitation curve regardless if the Ne-K or the united atom-L binding
energy is used (though the slopes of the curves are different).

A theoretical model can be tested very effectively by
investigating forbidden transitions. Fastrup and coworkers /28/
found a strong energy dependence of K-excitation in $Na^+$-Ne
collisions, even though this process has a closed exit channel
in the simple electron promotion model. They postulated a
velocity dependent vacancy occupation probability in the Na 2p
shell. Experimental $P(\rho)$-data /10,29/ for this system corrob-
orate this hypothesis (Fig. 7). The $Ne^+$-Ne and $Na^+$-Ne curves
look much alike, indicating that essentially the same mechanism
is operative. In a two-step process, a vacany is created in
the Na 2p-shell at large internuclear distances ($\hbar v/\Delta E \approx 1$ Å)
and then transferred to the Ne 1s-shell via the usual $2p\sigma - 2p\pi$
coupling. It should be noted that a straigth line in a "Demkov-
plot" ($\ln\sigma$ vs. $v^{-1}$) for a particular process does not prove
a Demkov coupling to be operative ! Plotting the total K-excitation cross section in $Ne^+$-Ne between 100 and 400 keV in this

manner also gives a straight line, as do, e.g., certain (and
rather long) sections of the total Coulomb excitation cross section.

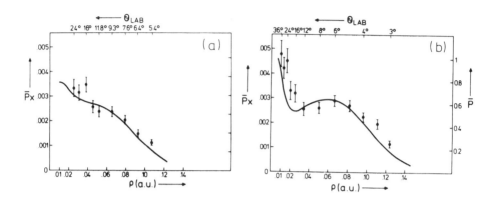

Fig. 6:   P($\varrho$) for K-excitation in slow Ne$^+$-Ne collisons /8/.

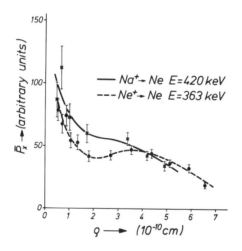

Fig. 7:   P($\rho$) for K-excitation in slow Na$^+$-Ne collisions /10,29/.
The Ne$^+$-Ne curve has been divided by 4.5 to give the
same $\int P \cdot \rho \, d\rho$ as for Na$^+$-Ne.

The experimental $Na^+$-Ne $P(\varrho)$-data show at small $\varrho$ in an even more pronounced way than $Ne^+$-Ne an enhancement over the theoretical curve. This is not likely to be due to the now effective $2s\sigma$ $-2p\sigma$ crossing (which is parity-forbidden in $Ne^+$-Ne). $P(\varrho)$-data for the system $Ne^+$-O /30/ show this deviation even more drastically (Fig. 8) even though the theory takes account of this crossing /31/. It may, therefore, be inferred that this deviation reflects the increasing importance of the coupling between the molecular 2p-states and higher unoccupied states. This is in agreement with total cross section experiments by Stolterfoht and coworkers /32/, it disagrees, however, with other total cross section experiments in similar systems (Al-Al, /33/) which suggest a much wider region of validity (up to 10 MeV) of the simple scaled D-D theory /27/.

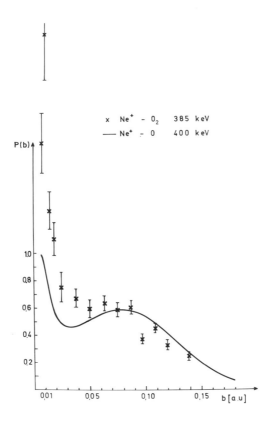

Fig. 8:  $P(\rho)$ for k-excitation in slow $Ne^+$-O collision /30/.
         The experimental data are normalized to the theory
         at $\rho$ = 0.073 a.u.

Unfortunately, these coincidence experiments are almost life-time experiments as they consume an awfully long time: each experimental point in Figs. 6-8 necessitated about 3 to 4 days beam time. Therefore, we started to use Auger electrons instead of x-rays to identify the reaction channel; even without improved coincidence timing, this increased the coincidence count rate by a factor of 4.

## V.   CONCLUSION

In most investigations, similar information can be obtained by either simple experiments followed by lengthy explanations, or by fairly complicated experiments giving short and clear-cut answers. The impact parameter dependence of inner shell excitation is no exception, even though the original stimulation for the development of the electron promotion model came from sophisticated differential (energy loss) measurements. A number of things have been pinpointed or at least indicated by studies of the impact parameter dependent inner shell excitation, such as the region of validity of various Coulomb excitation theories, the development of a quantitative theory of electron promotion, information on the decay mode of colisionally induced vacancies, etc. Such information could have probably been inferred from less sophisticated experiments, too, though most likely not as elegantly and unambiguously. Therefore, I believe (and this is corroborated by the increasing number of such papers at relevant conferences, e.g. ICPEAC) that such differential coincidence measurements will play an increasing role; this will hold particularly in investigations of processes in the no-man's land between Coulomb excitation and excitation by molecular promotion, and in the new and exciting field of molecular x-ray emission.

## REFERENCES

/1/   c. f. E. Merzbacher and H.W.Lewis, in: Handb.Phys. 34, 166 (1958), ed.: S. Flügge

/2/   J. Bang, and J.M.Hansteen, Mat.Fys.Medd.Dan.Vid.Selsk. 31, No.13 (1959); J.M.Hansteen, O.P.Mosebekk, Nucl.Phys. A201,541 (1973)

/3/   J.D.Garcia,R.J.Fortner, and T.M.Kavanagh, Rev. Mod. Phys. 45, 111 (1973)

/4/   c.f. Q.C.Kessel, and B.Fastrup, Case Studies in Atomic Physics 3, 137 (1973)

/5/   G.M.Thomson, P.C.Laudieri, and E.Everhart,
      Phys.Rev. $\underline{A1}$, 1439 (1970)

/6/   E.Laesgaard, J.U. Andersen, and L.C.Feldman, Phys.Rev.
      Lett. $\underline{29}$, 1206 (1972); Proc.Int.Conf.on Inner-Shell
      Ion.Phen., Atlanta, 1972

/7/   H.J.Stein, H. O. Lutz, P.H.Mokler, K.Sistemich, and P.
      Armbruster, Phys.Rev.Lett.$\underline{24}$,701 (1970); Phys. Rev.$\underline{A5}$,
      2126 (1972)

/8/   S.Sackmann,H.O.Lutz, and J.Briggs, Phys.Rev.Lett. $\underline{32}$,
      805 (1974);KFA-Rept.Jül-1154-KP, 1975

/9/   D.Burch, Phys. Lett. $\underline{47A}$, 437 (1974)

/10/  N.Luz,S.Sackmann, and H.O.Lutz, to be published

/11/  R.Ambros,H.O.Lutz, and K.Reichelt, Phys.Rev.Lett.$\underline{32}$,
      811 (1974); KFA-Rept. Jül-1047-KP

/12/  J.D.Garcia, Phys. Rev. $\underline{A1}$, 1402 (1970)

/13/  J.H.McGuire, Phys. Rev. $\underline{A9}$, 286 (1974)

/14/  J.S.Hansen, Phys. Rev. $\underline{A8}$, 822 (1973)

/15/  P.A.Amundsen, and L.Kocbach,J.Phys. $\underline{B8}$, L 122 (1975)

/16/  W.Brandt, K.W. Jones, and H.W.Kramer, Phys. Rev. Lett.
      $\underline{30}$, 351 (1973)

/17/  H.Schmidt-Böcking, R.Schulé, private communication,
      and paper contributed to this conference

/18/  D.Burch, W.B.Ingalls, H. Wieman and R.Vandenbosch,
      Phys. Rev. $\underline{A10}$, 1245 (1974)

/19/  C.L.Cocke, and R.Randall, Phys. Rev.Lett. $\underline{30}$, 1016 (1973)

/20/  J.F.Chemin, S.Andriamonje,B.Saboya,J.Roturier, and J.P.
      Thibaud, VI.Int.Conf.At.Coll.inSol.,Amsterdam, 1975.

/21/  P.Dahl, and D.Lorents, VII.Intern.Conf.Phys.El.Atom.
      Coll.,Amsterdam, 1971, pg. 395

/22/  R.S.Thoe, and W.W.Smith,Phys.Rev.Lett.$\underline{30}$, 525 (1973)

/23/  G.M.Thomson,W.W.Smith, and A.Russek, Phys.Rev. $\underline{A7}$, 168
      (1973); and paper contributed to this conference

/24/ N.Stolterfoth,F.D.Burch,and D.Schneider,VIII.Int.Conf. Phys.El.Atom.Coll.,Belgrade, 1973, pg. 731

/25/ c.f. H.J.Specht, Z. Physik 185, 301 (1965)

/26/ D.Burch, and K.Taulbjerg, to be publ.inPhys.Rev.A

/27/ J.S.Briggs, and J.Macek,J.Phys. 35, 579 (1972)

/28/ B.Fastrup,E.Boving,G.A.Larsen, and P.Dahl,J.Phys. B7, L 207 (1974)

/29/ H.O.Lutz, N.Luz, and S.Sackmann, IV.Int.Conf.Atomic Phys.,Heidelberg, 1974, pg.651; Verh.DPG (VI) 10, 31 (1975)

/30/ N.Luz, thesis, and to be publ.

/31/ J.S.Briggs, and K.Taulbjerg, AERE-Repts.T.P. 597 and 600, 1974

/32/ N.Stolterfoth,D.Schneider, D.Burch, B.Aagaard, E. Boving, and B.Fastrup, to be publ. in Phys. Rev.

/33/ R.Laubert, H.Haselton, J.R. Mowat, R.S.Petersen, and I.A. Sellin, Phys. Rev. A11, 135 (1975)

/34/ J.U.Andersen, E.Laesgaard, M.Lund, and C.D.Moak, Paper contributed to the Int.Conf.on At.Coll. Sol., Sept. 1975, Amsterdam.

/35/ Greenlees et al. in a paper contributed to this confer- ence found excellent agreement between the Amundsen and Kocbach relativistic SCA treatment and K-excitation by protons in heavy targets.  The final data showed by this group at this conference showed only a slight (and sys- tematic) deviation from this theory which might be attri- buted to an uncertainty in the total cross section.

# HIGH-RESOLUTION X-RAY AND AUGER ELECTRON MEASUREMENTS IN ION-ATOM COLLISIONS *

C. Fred Moore

Department of Physics, University of Texas,

Austin, Texas   78712

## ABSTRACT

Material presented in the last few years on Ion-Atom collision phenomena is summarized.  Many different aspects of the Ion-Atom interaction are discussed.  Work on collisions where both projectile and target have Z greater than 1 is stressed.

I.   Introduction

Although investigation of ion-atom collision phenomena has a long history, ninety-nine and forty-four one hundreths percent of all the data has been taken in only the last four years. Many different interesting phenomena have been observed.  A large number of these have been observed in experiments per-formed with low  resolution detectors; however, the explanation of these measurements has been delayed until good high-resolu-tion work offered detailed results to be compared with theory. The need for data with better resolution than is now possible will become apparent during the course of this talk.

Modern high resolution Auger spectroscopy has a serious experimental impediment:  that is, the ion-atom collision usu-ally creates such angular divergence that the doppler broaden-ing of the Auger electron emitted from the excited residual ion or atom is greater than the experimental resolution of a good electron spectrometer.  One means to reduce this limitation is to go to higher heavy ion energies.  Since, for a given impact parameter, the transfered momentum is inversely proportional to the incoming projectile velocity. To eliminate the kinematic broadening for a given incident ion energy, one will have to make coincidence experiments, detecting the recoiling projectile to uniquely determine its direction. High resolution X-ray spectroscopy, on the other hand, is primarily limited by the resolution obtainable using high efficiency Bragg crystal spectrometers. Consequently, it is unlikely that there will be a large number of experiments in the near future with resolution substantially improved over that obtainable using present-day technology reported in this talk.

A working definition of high resolution is data with $\Delta E/E$ better than 3%. In the case of X-ray measurements, this requires a Bragg crystal spectrometer or the equivalent. For electron data, better resolution is obtainable, but the solid angle suffers appreciably when one strives for resolution better than one percent.

As a part of the introduction, I should like to mention some of the scientists which contributed significantly with their experimental programs: Nikolaus Stolterfoht, Dieter Schneider (Berlin), Werner Mehlhorn, Arend Niehaus (Freiburg), M. O. Krause, Thomas Carlson, Bob Thoe, Randy Peterson, Ivan Sellin, David Pegg (Oak Ridge), David Burch (Seattle), Eugene Rudd, Dave Golden (Nebraska), Dennis Matthews, Richard Fortner (Livermore), Karl Groeneveld (Frankfurt), Patrick Richard, Lew Cocke (Kansas State), Tom Gray (North Texas State), Steve Shafroth (North Carolina), Rand Watson (Texas A & M), Hank Oona (Arizona), Larry Toburen (Battell NW), David Nagel, P.G. Burkhalter, A. P. Knudson, (Naval Research Lab.), Bent Fastrup (Aarhus), Quentin Kessel, W. Smith (Connecticut), H. Tawara (Fukuoka), Frans Saris (Amsterdam). F. Hopkins (Stoney Brook), and L. Feldman (Bell Labs.).

Finally, I should like to thank the fine scientists with whom I have had the pleasure of collaborating with in this program at Texas and who are responsible for the contributions from Austin: Wilfred Braithwaite, Joe Bolger, David Burch, Bill Hodge, Jerry Hoffmann, Forrest Hopkins, Brant Johnson, Bob Kauffman, Dennis Matthews, Joe McWherter, David Olsen, Patrick Richard, Dieter Schneider, Gene Smith, and Hermann Wolter.


II. Projectile Excitations

Much work in this area is in progress. A large portion of the work is known as beam-foil studies. In September there will be a conference in Gattlingburg, Tennessee, on this subject. There are two standard-beam-foil measurements; the first concerns the spectroscopy of the transitions in the ion; the second is the measurement of the lifetime of a state using a 1/e decay

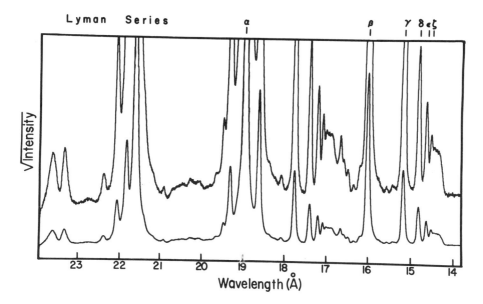

Figure 1:   An oxygen beam X-ray spectrum as viewed from the back of a carbon foil when bombarded by 20.53 MeV oxygen ions. A square root display is used to weight the data statistically for visual inspection and a scale change x8 is overlaid to show the finer details.


length method.  Figure 1 shows an example of an oxygen X-ray spectrum after a 20 MeV oxygen ion is transmitted through a thin carbon foil.  The hydrogenic Lyman series dominates the spectra. Also, two and three electron transitions are present.  Of particular spectroscopic   interest are the lines above the series limit for the He-like Lyman series at 16.84Å.  These are helium-like transitions whose initial states have neither electron in the K shell.  Recently, similar transitions have been found in a helium beam-foil experiment using an optical spectrometer, which is the standard beam-foil instrument.

Using the 1/e decay length procedure, we have measured in the two and three electron oxygen system several lifetimes of isomeric transitions and these agree well with theory.  An unusual feature was that the Lyman series was still observable down stream from the carbon foil.  The lifetime of the Lyman alpha line is four orders of magnitude shorter than the time-in-flight of the ion from the foil to the entrance of the spectrometer.  A careful examination of the series limit region shows that many states with high n are excited in the beam-foil collision.  Figure 2 shows this region of the spectrum

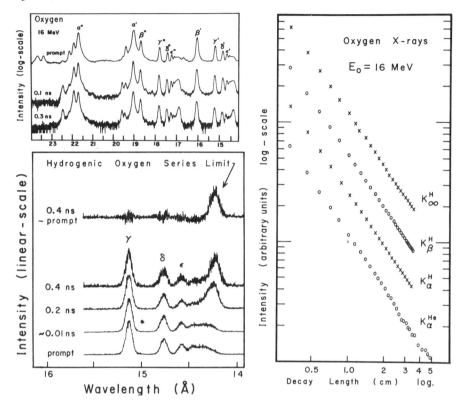

Figure 2:   An oxygen beam X-ray spectrum as viewed at the back
of a carbon foil and several down stream positions.   Isomeric
transitions are obvious in the time delayed spectra, for example
the $^3P$ - $^1S$ transition in orthonormal He-like oxygen at 21.88 Å
The lower figure is an excerpt from the region near the series
limit of Hydrogenic oxygen Lyman series.

Figure 3:   A log-log plot of the Intensity vs. position down
stream for $K_\infty^H$, a region of the spectra near the series limit in
hydrogenic oxygen (n≈9); $K_\alpha^H$ & $K_\beta^H$, the Lyman alpha and beta
transitions in hydrogenic oxygen; and $K_\alpha^{He}$, the Lyman-like alpha
transition in helium-like oxygen.


at 0.01, 0.2, and 0.4 nano-seconds after the ion traverses the
foil.   The maximum intensity is found at n = 13 at 0.4 nano-
seconds.   Since the lifetimes of the Lyman series go as $n^{-3}$,
it is reasonable to expect states populated with high n to
dominate the decay spectrum at long times, either by direct decay
to the ground state, or the more probable cascade through lower
n levels.   The decay curve for the Lyman α transition (as well
as for several other transitions) is plotted in Figure 3.   These

decay curves follow a power law rather than exponential function which is nearly:  Yield = $Y_0 t^{-1.5}$.  Several recent authors have calculations which reproduce these curves.  e.g.  F. Hopkins, P. von Brentano and R. M. Schectman.  The first of these authors assumed the initial population distribution decreased as $n^{-5}$ whereas Schectman used $n^{-m}$, with $2.2 < m < 4$.

Cascading from high n states should play an important role in beam foil 1/e lifetime measurements, and particularly in those of weak transitions.  Many authors have made measurements of lifetimes with low resolution detectors.  These measurements were made assuming that after a given length of time only one state persisted.  These results are surely in question.  One must have the resolution and also know the branches which feed the state to have a good measurement of the lifetime using the 1/e decay length method.

The use of a foil in measuring projectile excitations severely limits the scope of the physics which can be accomplished since reaction cross sections can not be measured.  The ideal situation is to measure beam-gas interactions under single collision conditions.  The information obtainable is not only the ionization cross section, or ionization cross section for a given charge state, but ultimately the cross section to a given final state of either (or even both) projectile and target. It has been reported (Johnson, 1975) that the prompt oxygen beam X-ray spectra differ very little between carbon foil excitation and nitrogen gas excitation.  More evidence is found in a similar measurement (Schneider, 1975) of the Auger electron decay of carbon, excited by a carbon foil and by neon gas. The carbon spectra are shown in Figure 4.  As can be seen, the foil-excited and gas-excited carbon spectra are qualitively the same.  The predominant electron transtions are from the three and four electron carbon ion system.  For comparison with previous work for heavier ions at Oak Ridge, a down-stream spectrum is shown.  The isomeric $^4P^o_{5/2}$ transition in the Li-like carbon ion dominates this spectrum.  The down-stream spectra are more easily obtainable since they are free of most of the electron background from the collision process.  Similar high resolution work is being carried out in Frankfurt and in Berlin. These measurements are needed for lifetime determination and line identification.

III.  Target Excitations

The study of various aspects of the excitation of atoms in gas phase or solid phase is one of the biggest interests in atomic physics.  High resolution work is essential to this endeavor.  The excitation of targets using ion beams has greatly increased our knowledge concerning ion-atom and ion-solid interactions.  A few of the phenomena of interest to us will be covered in the following sections.

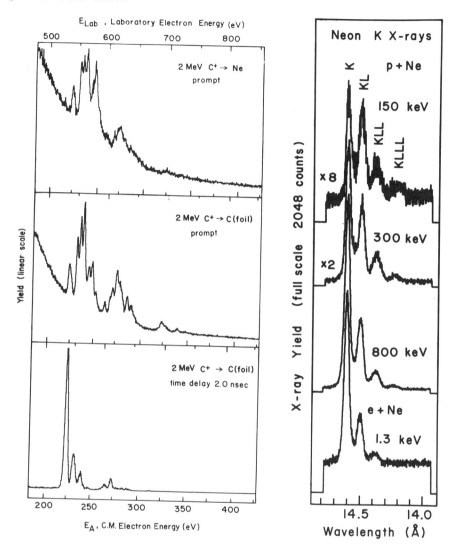

Figure 4:   Electron emission spectra from 2 MeV carbon ions as excited by a neon gas target under single collision conditions (upper), as excited by a carbon foil and viewed directly behind the back side of the foil (middle) and as viewed down stream from the back face of the carbon foil (lower).

Figure 6:   Neon X-ray spectra as excited by proton bombardment at 150, 300, and 800 keV and by electron bombardment at 1.3 keV under single collision conditions.   Notice the increase of the satellite intensity (KL, KLL, and KLLL) relative to the $K\alpha_{1,2}$ as the proton energy is decreased.

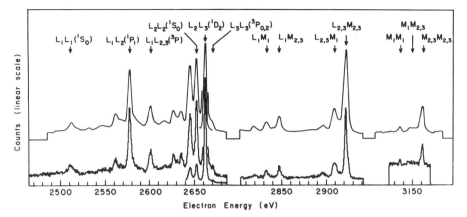

Figure 5:   Auger electrons from Argon KLL, KLM, and KMM transitions as excited by 6 keV electrons (Krause, 1975) and by 4 MeV protons. Note the similarity in the two spectra.   The vertical scales for the (electron vs proton) region are normalized (x1 vs x1) from 2430 to 2690 eV; (x2 vs x1) from 2800 to 2940 eV; and (x10 vs x3.7) from 3120 to 3170 eV.

1)   Electron and Proton Comparison

It has been established that for swift projectiles of electrons and protons, with the same velocities, the Auger electron spectra from Neon are nearly identical, (Stolterfoht, 1973).   Figure 5 shows a similar comparison between 4 MeV protons and 6 keV electron data taken be Krause (1975).   In this comparison there are some problems concerning the ratio KLL to KLM and to KMM which I believe must be an experimental error.   Otherwise, the similarity in the two spectra is remarkable.

2)   Excitation versus Projectile Energy

For projectile excitation, when a comparison is made between swift electrons and protons there is good agreement.   However, if the projectile velocity is decreased to a point near the threshold energy for the ionization of the particular electronic shell under investigation, a comparison is no longer meaningful.   The important parameter in the cross section calculation is the velocity. Low energy proton excitation functions are measured to study the behavior of atomic excitations where the velocities are small. Figure 6 displays several spectra which are part of such an excitation function.   An important feature to notice is that at low energies multiple ionization dominates.   That is, the ratio

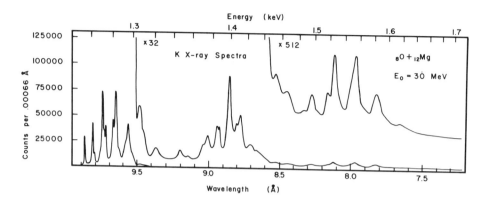

Figure 7:  Magnesium X-ray spectrum as excited by 30 MeV oxygen ions.  The target is a thick foil.  The intensity of the more energetic Mg X-rays is reduced above the K absorption edge for Mg (9.5122 Å).  Kα satellites are between 9.9 and 9.3 Å; both Kα satellites and Kβ hypersatellites are between 9.3 and 8.4 Å; and Kβ hypersatellites are between 8.4 and 7.5 A.

of satellite (multiple ionization) to diagram (single inner-shell ionization) lines is enhanced at lower collision velocities. Thus, low energy data is closely atuned to data obtained with heavier ions.  At higher energies multiple ionization is caused primarily by electron shake-off and the ratio of satellite to diagram lines is small.

3)   Heavy Ion Excitation

     The most dramatic feature of ion-atom scattering is the degree of multiple inner-shell ionization associated with heavy ion excitation.  Swift electron, proton and photon exci-tation predominately interacts with only a single electron in the atom.  The spectra associated with heavy ion excitation display an enormous number of transitions resulting from in-itial atomic states with a large degree of atomic electron re-arrangement and ionization.  A good example of an X-ray spec-trum from heavy ion bombardment is shown in Figure 7.  This spectrum displays satellite transitions and also hypersatellite lines from double K-shell ionization.  Although a Bragg crystal spectrometer is used, the resolution is usually far from ade-quate to resolve the many members of each multiplet, since there are hundreds of unresolved lines in the spectra.  A similar ef-fect for Auger electron spectra of neon is shown in Figure 8. Usually the members of multiplets are resolved in the Auger

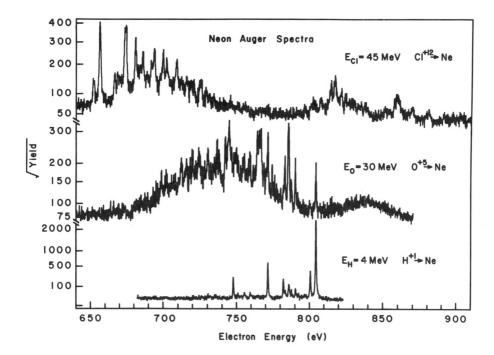

Figure 8:  Auger electron spectra resulting from K-shell
ionization of neon by protons, oxygen ions, and chlorine ions.

spectra, however, there are so many different lines that often
lines from one multiplet overlap with lines from other multi-
plets.  The normal neon Auger spectrum for proton excitation
is already complex; it is shown for comparison.  Again, the
resolution is inadequate; to completely resolve all lines,
however, the resolution for most of the lines approaches the
natural width of the levels themselves.  In heavy ion induced
spectra, one has to take the kinematical (Doppler) line broadening into account.  Thus, appreciable improvement in these
spectra will not be obtained in a singles experiment.  That is,
some coincidence technique is required which selectively gates
on transitions.

4)  Electronic Structure in Molecules.

Evidence for chemical bonding effects in the X-ray and
Auger electron decay of excited atoms can in part be predicted
due to the binding energy of the electrons in the valence shell.
Usually this is small in comparison with the energy required to

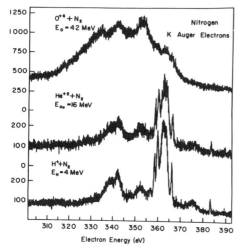

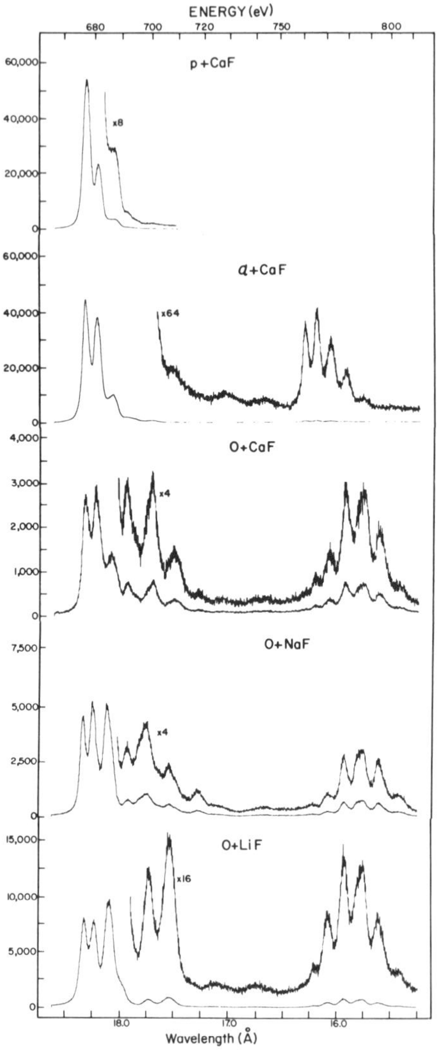

Figure 9:  Auger electron spectra of $N_2$ after proton, alpha particle, and oxygen ion bombardment.

Figure 10:  Florine X-ray spectra produced by proton, alpha particle, and oxygen ion bombardment.  Targets of LiF, NaF, and $CaF_2$ were used.  The satellite lines fall between 18.5 and 16.4 Å, while hypersatellite transitions fall in the region between 16.4 and 15.2 Å.  Chemical effects are observed in the satellite region but not in the hypersatellite region.

produce a K shell vacancy.  Auger electron spectra from electron excitation of oxygen differ greatly from that of a rare gas as reported by W. E. Moddeman (1971).  X-ray spectra from electron excitation do not yield such a dramatic result but have been intensely studied.

Heavy-ion excitation should enhance the chemical effects since multiple ionization will reduce the electron shielding of the nuclear charge and produce levels in non-neutral molecules.  These should be energetically different than those of a free ion.  Figure 9 displays the Auger electron spectra of Nitrogen as bombarded by protons, alpha particles and oxygen ions. The differences in the spectra are rather dramatic but unfortunately the resolution required must be improved by several orders of magnitude to resolve the many rotational, vibrational and disassociative states. Where individual lines are discernable in Figure 9, a factor of two in intensities is observed between proton and alpha particle excitation.

X-ray studies of heavy ion induced chemical effects have been reported by P. G. Burkhalter (1972) for Al and $Al_2O_3$ and by McWherter (1974) for Si and $SiO_2$. A rather dramatic presentation of the chemical effects of bonding fluorine in LiF, NaF and $CaF_2$ can be seen in Figure 10. Notice the complete dissimilarity in the $K\alpha$ satellite groups for the oxygen induced spectra for these three compounds of fluorine. Notice also, that the hyper-satellite lines of each are nearly identicle. This must derive from the disassociation of the fluorine in the collision process, although there is no hard experimental confirmation for this speculation. Also, these spectra are among the best examples of hypersatellite line excitations. In higher Z elements, the $K\beta$ transitions often mask the $K\alpha$ hyper-satellite lines since they have overlapping energies from sodium to vanadium.

As may be expected, theoretical interpretation of the data is tedious and a good description of these electronic structure effects from chemical bonding has not yet been achieved. These studies will provide a critical test for inter-atomic potentials and if experimental techniques are improved, open a new field of chemistry.

5)   Charge State Dependence

Early theoretical descriptions of ion-atom interactions had no particular expression relating.the charge state of the incident ion to the excitation of the target. Within the last couple of years, several different measurements have been made which show that the cross section changes by large factors depending on the charge state of the incident ion. The original X-ray experiments clearly showing these effects were done at Oak Ridge and at Kansas State. This effect was clarified by an experiment at Seattle showing the main contribution of the increase of the X-ray cross-section is the increase of the mean fluorescence yield caused by a higher degree of multiple inner shell ionization for more highly stripped ion induced reactions. The effects are largest when the ratio between the bare ion; one electron ion; and two electron ion excitation are compared. One of the major contributing mechanisms involved is electron capture by the incident ion when the ion is nearly fully stripped.

6)  Neon (X-ray: Auger Electron) comparison

The study of neon is particularly well suited to both experimental and theoretical investigations.  Measurements of Auger electron spectra can be analyzed to obtain relative intensities of the yield for each charge state of Neon.  The same can be done for X-ray measurements.  This information can be used to obtain a fluorescence yield for each individual charge state of neon.  Consequently, one can then use these branching ratios to determine cross sections from either X-ray or Auger data.  Of greater interest at the moment is an attempt to get experimental and theoretical measurements to agree, for these fluorescence yields.  The data is not easy to analyze because of overlapping lines, and caution must be taken not to misidentify transitions.  Consequently, many spectra are needed, taken under differing conditions so as to produce the  extremes  of the population of the various charge states of neon.  For this purpose projectiles of hydrogen(1+), helium(2+), carbon(4+, and 5+), oxygen(3+, 4+, 5+, 6+, 7+, and 8+), flourine(6+), chlorine(4+, 6+, 7+, and 12+) were measured, at various projectile energies, and the resulting X-ray and Auger electron decay of neon has been determined. Several of these spectra are shown in Figure 11 and 12.  One can see that the most probable electron population of neon after energetic chlorine bombardment and K-shell ionization is six vacancies in the L-shell, which is quite a contrast to no L-shell vacancies dominating after proton K-shell ionization.·

Several experimental and theoretical papers have been and are being written concerning the spectroscopic information in multiply ionized neon.  Also, the calculation of transition widths is a subject of current interest.  Recently, most of the descrepencies between measured results and theoretical calculations have been removed, since better calculations are now available by groups at Oregon (Craseman) and Kansas State (Bhalla).

7)  Statistical Population

In the absence of external electric and magnetic fields an atom in a specific state of total J, L, and S: $^{2S+1}(L)_J$ is degenerate with respect to the magnetic quantum number $M_J$. The number of different $M_J$ sublevels is 2J+1 and is called the statistical weight of the level.  The statistical weight of a term $^{2S+1}(L)$ (with all possible J) is (2S+1)(2L+1).  All theoretical calculations of cumulative transition probabilities for decay of atoms assume a statistical population for the determination of the fluorescence yields for specific charge states.  Experi-

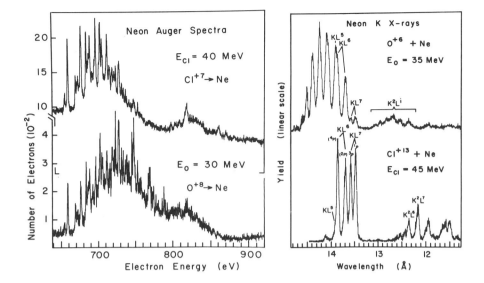

Figure 11:    Comparison of neon Auger electron spectra after chlorine and oxygen ion excitation.
Figure 12:    Comparison of neon X-ray spectra after chlorine and oxygen ion excitation.

ments with good resolution can in some cases resolve lines within a multiplet.  Recently, measurements of atoms excited by energetic particle bombardment have indicated this assumption is valid in some cases (Fortner, 1974) and false in others (Bhalla, Matthews, Moore, 1974).  Another investigation of the multiplet ($2p^5 - 2p^43s$) in chlorine does not support the previous finding that configurations $2p^43s^1$ ($^4P_{5/2,3/2,1/2}/^2D_{5/2, 3/2}/^2S_{1/2}$) have intensity ratios 18/10/2.  A spectrum of these data is shown in Figure 13.

Perhaps the nicest case which shows complete agreement with the usual assumption of statistical population is the intensity ratio of the $^3P_1$ and $^1P_1$ lines in the two electron neon X-ray spectra.  This ratio $^3P_1/^1P_1$ is 0.9:1.0, which is in complete agreement with the transition probabilites and branches as calculated.

The explanation of the mechanisms that give rise to these atomic excitations in the atom-ion collision is far from obvious. Various single channel models should predict results other than statistical population.  As well, two-channel approaches will give non-statistical populations.  One expects statistical population if the excitation process takes longer than the electron

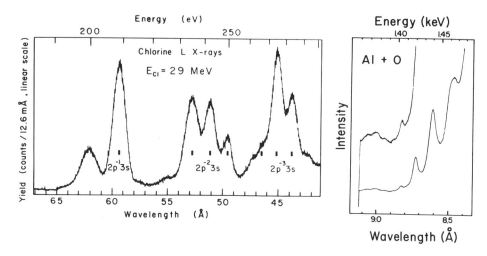

Figure 13:   Chlorine L X-ray spectra.   A chlorine beam was excited by a thin carbon foil.

Figure 14:   Aluminium X-ray spectrum in the energy region below the $K\alpha_2$ line.   The $K\alpha_2$ line falls at 8.34173 A and is off to the right of the figure.

rearrangement letting the system thermalize in a random distribution of states at the time of the collision.   This condition is obviously not satisfied, for the most part, in energetic heavy ion-atom collisions.

8)   Satellite Lines at Energies Below the $K\alpha_2$ Line.

Low Energy Satellite lines have been observed in heavy ion beam excitation of solid targets (Richard et al. 1973 and McWherter et al. 1973).   The portion of the spectrum below the $K\alpha_2$ of Aluminum is shown in Figure 14.   In this example a thick and then a thin aluminum foil were bombarded by 30 MeV oxygen ions.   The foil was positioned both perpendicular to the beam, and at a grazing angle to the beam.   The resulting spectra were in all respects the same.   These low energy satellite lines which appear in every heavy ion induced spectra below the $K\alpha_2$ diagram line have been observed in F(LiF), Na, Mg, Al (Al and $Al_2O_3$), Si (Si, SiO, and $SiO_2$), Ca, Sc, and Ti.   They generally have the features:   the line width is broader than that of the observed $K\alpha_{1,2}$ line, they have approximately even spacing, their intensities are about one percent of the $K\alpha_{1,2}$ line, and no more than four have been observed in any one spectra and their spacing increases with the Z of the target (in a manner similar

to the increase in spacing between the normal satellite lines).
In the case of magnesium and aluminum the energy spacing of
these satellites is about 1.3 times the energy of plasmon excitations as observed in electron scattering, and McWherter has
suggested that these excitations arise from a "local" plasmon
excitation (giant dipole states) occuring during the interaction
time when the heavy ion is still in the neighborhood of the
target atom.  The discrepancy with energy, and chemical effects
rule out the possibility of an identification of these with
either surface or volume plasmon excitation in the solid.

Due to the strong yield of excitation transitions, as
observed in the heavy ion excitation of argon and krypton, one
is tempted to identify these peaks as a radiative Auger transition.  This suspicion is further warrented in that the spectra
of Figure 14 contains structure between 1.36 and 1.39 keV, that
is, in the same region where T.Åberg observed radiative Auger
transitions (Utriainen, 1974).  As yet, however, there are no
published calculations which support this assignment.  However,
W. Hodge has preliminary results which in part substantiate
this claim; in his calculations the dominant transitions are
KLM Auger-radiative transitions rather than the KLL Auger-radiative transition dominant in the electron or photo excitation.

9)   Excitation Lines in Heavy Ion Induced Auger Spectra

It has been shown that bombardment of Argon by heavy ions
causes a tremendous increase in satellite lines in the Ar L
Auger spectra.  The structure between 205 and 210 eV in oxygen
and chlorine induced spectra, see Figure 15, is attributed to
Auger electrons whose initial states involve occupation of 4s
and 3d electrons.  The population probability of these electronic configurations in the ion-atom interaction is obviously
increased when high Z particles are used as projectiles.  The
lack of pupulation of normal L Auger lines in the chlorine
spectra is due in part to the larger fluorescence yield.

Figure 16 displays similar data for krypton.  However,
krypton gives the surprising result that there is very little
increase of satellite yield in oxygen ion bombardment.  Each
line in the proton spectrum can be related to a line in the
oxygen spectrum.  This is obviously due to a smaller ionization
probability for the N-shell than for the M-shell, since this
will reduce the $MN^m$ multiple ionization.  The broad structure
around 60 eV is attributed to Coster-Kronig transitions.  These
lines also coincide in position with satellite lines from M,
$M_{4,5}$, $M_{4,5}$ transitions.

462  C. FRED MOORE

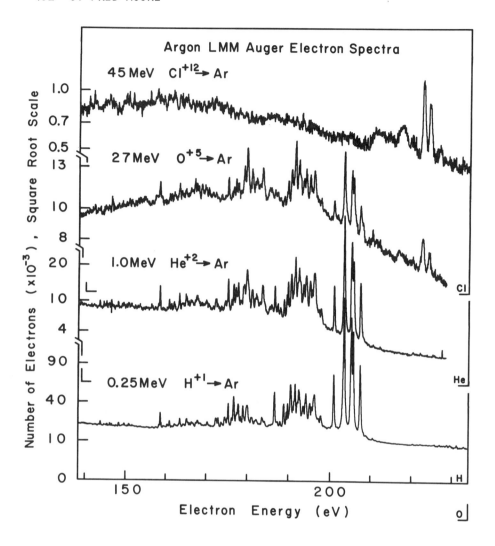

Figure 15: Argon L **Auger** electron spectra after bombardment
by proton, alpha particle, oxygen ion and chlorine ions.


IV.  Conclusion

I have only had time to discuss a few of the aspects of
high resolution data obtained from ion-atom collisions.  The
majority of present work is concerned with aspects of the
collision mechanism in the ion-atom collision.  Spectroscopy
using electron excitation is only necessary as a standard or
reference.  In almost    every case other than high energy

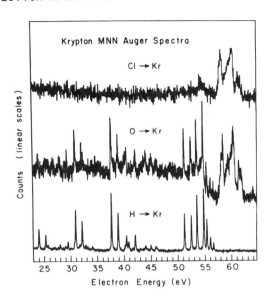

Figure 16: Krypton MNN Auger electron spectra after bombardment by proton, oxygen ion, and chlorine ions.

proton excitation, the spectra differ significantly from electron excitation due to the multiple inner-shell vacancies
created by the ion-atom interaction. The spectroscopy of
highly excited atoms is in itself interesting and must be
understood in order to get to the meat of the ion-atom interaction mechanism. At ion energies where the ion velocities
are comparable to or greater than the atomic electrons in the
shell under investigation in the measurement, the adiabatic
conditions normally used in a two center potential model do
not hold. This leaves us without a highly developed theory
(from basic principles) to predict energetic ion-atom collision
phenomena. By now, we have developed many sophisticated empirical guides for calculating ion-atom ionization cross sections,
but not to the extent where detailed information concerning
specific final states is concerned. The host of problems which
need to be solved is an enormous task. I expect that many new
and exciting discoveries lie ahead.

*) Supported in part by the Robert A. Welch Foundation, the
Air Force Office of Scientific Research and the Energy Research
and Development Administration.

V.   References and Bibliography of High Resolution Work since
ICPEAC VIII.

Bhalla,C. P., N. O. Folland and N. A. Hein, Phys. Rev. A8, 649,
(1973).

Bhalla,C. P. and P. Richard, Phys. Letters A45, 53 (1973).

Bhalla,C. P. and N. A. Hein, Phys. Rev. Letters 30, 39 (1973),
and C. P. Bhalla, N. O. Folland and N. A. Hein, Phys. Rev.
to be published.

Bhalla,C.P., D. L. Matthews, C. F. Moore, Phys. Lett. A46, 336,
(1973).

Biersack, J.P., U. Morfeld, D. Schneider, J. Phys. C9, Nr. 11-
12 34, (1973).

Bissinger, G. A., P. H. Nettles, S. M. Shafroth and A. W. Walt-
ner, Phys. Rev. A10 1932 (1974).

Bolger, J. E., D. K. Olsen, H. H. Wolter, and C. F. Moore, Z.
Physic 266, 173 (1973).

Bolger, J. E., W. J. Braithwaite, W. J. Courtney, and C. F.
Moore, Proceedings of the International Conference on Reactions Between Complex Nuclei, Volume I, (1974).

Braithwaite, W. J., D. L. Matthews, and C. F. Moore, Proceedings
of the International Conference on Reactions Between Complex
Nuclei, Volume I, (1974).

Breuckmann, B., V. Schmidt, Z Physik 268, 198 (1974).

Brown, M. D., J. R. Macdonald, P. Richard, J. R. Mowat, and I.
A. Sellin, Phys. Rev. A9, 1470 (1974).

Burch D. P. Richard, and R. L. Blake, Phys. Rev. Lett. 26, 1355
(1971).

Burch D., W. B. Ingalls, J. S. Risley, and R. Heffner, Phys. Rev.
Lett. 29, 1719 (1972).

Burch D., H. Wieman, and W. B. Ingalls, Phys. Rev. Lett. 30,
823 (1973).

Burch, D., N. Stolterfoht, D. Schneider, H. Wieman, J. S. Risley,
Phys. Rev. Lett. 32, 1151 (1974).

Burch, D., N. Stolterfoht, D. Schneider, H. Wieman, Phys. Rev.
Lett. (1974), to be published.

Burch, D., N. Stolterfoht, D. Schneider, H. Wieman, and J. S.
Risley, (Annual Report, University of Washington), (1974)
unpublished.

Burkhalter, P. G., A. R. Knudson, D. J. Nagal, and K. L. Dunning,
Phys. Rev. A6, 2093 (1972).

Cleff, B., W. Mehlhorn, J. Phys. B7, 593 (1974); ibid., 605
(1974).

Cocke, C. L., S. L. Varghese, J. A. Bednar, C. P. Bhalla, B.
Curnutte, R. Kauffmann, R. Randall, P. Richard, C. Woods,
and J. H. Scofield, Phys. Rev. A (to be published).

Crooks, J. B., and M. E. Rudd, Phys. Rev. A3, 1628 (1971).

Der, R. C., R. J. Fortner, and T. M. Kavanagh, Phys. Lett. A42,
337 (1973).

Crasemann, B., Atomic Inner Shell Processes, Acad. Press (1975).

Doyle, B. L., K. W. Hill, W. W. Jacobs, S. M. Shafroth, and J.
  Wu, Proceedings of the Fourth International Conference on
  Atomic Physics, 647 (1974).
Elliott, F.F. Hopkins, G. W. Phillips, and P. Richard, Z.
  Physik 269, 89-95 (1974).
Garcia, J. D., R. J. Fortner, and T. M. Kavanagh, Rev. Mod. Phys.
  45, 111 (1973), this is a general review of the ion atom col-
  lision phenomena.
Groeneveld, K. O., G. Nolte and S. Schumann, Phys. Lett., (to be
  published).
Groeneveld, K. O., R. Mann, G. Nolte, and S. Schmann, R. Spohr,
  Proceedings on the International Conference on Reactions Be-
  tween Complex Nuclei, Volume I, 197 (1974).
Hagmann, S., G. Herman, W. Mehlhorn, Z. Physik 266, 189, (1974).
Hansteen, J. M., and O. P. Mosebekk, Z Physik 234, (1970), and
  in Nucl. Phys., to be published.
Haselton, H. H., R. S. Thoe, J. R. Mowat, P. M. Griffin, J. P.
  Pegg, and I. A. Sellin, Phys. Rev. A11 468, (1975).
Hillig, H., B. Cleff, W. Mehlhorn, W. Schmitz, Z. Physik 268,
  225 (1975).
Hodge, B., R. Kauffmann, C. F. Moore, and P. Richard, J. Phys.
  B6, 2468 (1973).
Hodge, W., D. Schneider, B. M. Johnson, C. F. Moore, Proceedings
  of the Fourth International Conference on Beam Foil Spectroscopy
Hodge W., K. Roberts, D. Schneider, B. M. Johnson, W. Braith-
  waite, L. E. Smith, and C. F. Moore, submitted to J. Phys.
  B. Atom. Molec. Phys.
Hopkins, F., J. S. Eck, D. O. Elliott, and P. Richard, Phys.
  Rev. C8, 1721 (1973).
Hopkins, F., D. O. Elliot, C. P. Bhalla, and P. Richard, Phys.
  Rev. A8, 2952 (1973).
Hopkins, F. F., J. R. White, C. F. Moore, and P. Richard, Phys.
  Rev. C8, 380 (1973).
Hopkins, F., R. L. Kauffmann, C. W. Woods, and P. Richard, Phys.
  Rev. A9, 2413 (1974).
Hopkins, F., and P. von Brentano, Proceedings of the Fourth
  International Conference on Beam Foil Spectroscopy, (1975).
Jamison, K. A., C. W. Woods, R. L. Kauffmann, and P. Richard,
  Phys. Rev. A11, 505 (1975).
Johnson, B., M. Senglaub, P. Richard, and C. F. Moore, Z Physik,
  261, 413 (1973).
Johnson, B. M., D. L. Matthews, L. E. Smith, C. F. Moore, J.
  Phys. B: Atom. Molec. Phys. 6, L369 (1973).
Johnson, B. M., J. E. Bolger, James Whitention, K. Roberts, and
  C. F. Moore, Bull. Am. Phys. Soc. 19, 1174 (1974).
Johnson, B. M., D. Schneider, K. Roberts, J. E. Bolger, and C.
  F. Moore, accepted by Phys. Lett. (1975).
Johnson, B. M., D. Schneider, W. Hodge, C. F. Moore, Proceedings
  of the Fourth International Conference on Beam Foil Spectro-

scopy (1975).

Kauffman, R. L., C. W. Woods, F. F. Hopkins, D. O. Elliot, K. A. Jamison, and P. Richard, J. Phys. B6, 2197 (1973).

Kauffman, R. L., J. H. McGuire, P. Richard, and C. F. Moore, Phys. Rev. A8, 1233 (1973).

Kauffman, R. L., F. F. Hopkins, C. W. Woods, and P. Richard, Phys. Rev. Lett. 31, 621 (1973).

Kauffman, R. L., C. W. Woods, K. A. Jamison, and P. Richard, Phys. Lett. A50, 117 (1974).

Kauffman, R. L., C. W. Woods, K. A. Jamison, and P. Richard, J. Phys. B6, L335 (1974).

Kauffman, R. L., C. W. Woods, K. A. Jamison, and P. Richard, Phys. Rev. A11, 872 (1975).

Kessel, Q. C. and B. Fastrup, Case Studies in Atomic Physics 3, Eds, E. W. McDaniel and M. C. McDowell, p. 137 (1973).

Kessel, Q. C., Case Studies in Atomic Physics 1, 401 (1973).

Krause, M. O., Physica Fenninca 9, Si, 281, (1974).

Krause, M. O., Phys. Rev. Lett. 34, 633 (1975).

Krause, M. O., and J. G. Ferreira, J. Phys. B (to be published).

Langenberg, A. and J. van Eck, Phys. Rev. to be published.

Laubert, R., H. H. Haselton, J. R. Mowat, R. S. Peterson, J. S. Peterson, and I. A. Sellin, Phys. Rev. A11, 135 (1975).

Laubert, R., H. H. Haselton, J. R. Mowat, R. S. Peterson, and I. A. Sellin, Phys. Rev. A11, 1468 (1975).

Li, T. K., R. L. Watson, and J. S. Hansen, Phys. Rev. to be published.

Larsen, G. A., E. Boving, P. Dahl, and B. Fastrup, to be published.

Mackey, J. J., L. E. Smith, B. M. Johnson, C. F. Moore, and D. L. Matthews, J. Phys. B: Atom. Molec. Phys. 7, (1974).

Madison, D. H., K. W. Hill, B. L. Doyle, and S. M. Shafroth, Phys. Lett. A48, 249 (1974).

Matthews, D. L., W. J. Braithwaite, Claude Camp, and C. Fred Moore, Phys. Rev. A8, 1397 (1973).

Matthews, D. L., B. M. Johnson, J. J. Mackey and C. F. Moore, Phys. Rev. Lett. A45, 447 (1973).

Matthews, D. L., B. M. Johnson, J. J. Mackey and C. F. Moore, Phys. Rev. Lett. 31, 1331 (1973).

Matthews, D. L., B. M. Johnson, J. J. Mackey, L. E. Smith, W. Hodge, and C. F. Moore, Phys. Rev. A10, 1177, (1974).

Matthews, D. L., C. F. Moore, D. Schneider, Phys. Lett. A28, (1974).

Matthews, D. L., B. M. Johnson, G. W. Hoffmann, and C. F. Moore, Phys. Lett. A49, 195 (1974).

Matthews, D. L., B. M. Johnson, and C. F. Moore, Phys. Rev. A10, 451 (1974).

Matthews, D. L., B. M. Johnson, L. E. Smith, J. J. Mackey, and C. F. Moore, Atomic and Nuclear Data Tables 15, 41 (1975).

Matthews, D. L., W. J. Briathwiate, and C. F. Moore, Phys. Rev.

A, to be published.
Matthews, D. L., R. J. Fortner, D. Burch, B. M. Johnson and
C. F. Moore, Phys. Letters 50A, 441 (1975).
McWherter, J., J. Bolger, H. H. Wolter, D. K. Olsen, and C. F.
Moore, Phys. Lett. A45, 57 (1973).
McWherter, J., J. Bolger, C. F. Moore, and P. Richard, Z. Physik
263, 283 (1973).
McWherter, J., D. K. Olsen, H. H. Wolter, and D. F. Moore, Phys.
Rev. A10, 200 (1974).
Middleton, R., and C. T. Adams, Nucl. Istr. and Meth. 118, 329
(1974).
Moddeman, Q. E., T. A. Carlson, M. O. Krause, B. P. Pullen, W.
E. Bull, and G. K. Schweitzer, J. Chem. Phys. 55, 2317 (1971).
Moore, C. F., D. K. Olsen, J. McWherter, and P. Richard, J. Phys.
Soc. of Japan 34, 1020 (1973).
Moore, C. F., W. J. Braithwaite, and D. L. Matthews, Phys. Lett.
A44, 199 (1973).
Moore, C. F., W. J. Braithwaite, and D. L. Matthews, J. Phys. B
Atomic Molec. Phys. 6, 1592 (1973).
Moore, C. F., W. J. Braithwaite, and D. L. Matthews, Phys. Lett.
A47, 353 (1974).
Moore, C. F., J. Bolger, K. Roberts, D. K. Olsen, B. M. Johnson,
J. J. Mackey, L. E. Smith, and D. L. Matthews, J. Phys. B Atom.
Molec. Phys. 7, (1974).
Moore, C. F., J. J. Mackey, L. E. Smith, Joe Bolger, Brant M.
Johnson, D. L. Matthews, J. Phys. B:  Atom. and Molec. Phys.
7, L302, (1974).
Moore, C. F., W. Hodge, D. Schneider, and B. M. Johnson, Pro-
ceedings of the Fourth International Conference on Beam Foil
Spectroscopy, (1975).
Mowat, J. R., I. A. Sellin, D. J. Pegg, R. S. Peterson, M. D.
Brown, and J. R. MacDonald, Phys. Rev. Lett. 30, 1289(1973).
Mowat, J. R., R. Laubert, I. A. Sellin, R. L. Kauffman, M. D.
Brown, J. R. Macdonald, and P. Richard, Phys. Rev. A4, 1446,
(1974).
Mowat, J. R., P. M. Griffin, H. H. Haselton, R. Laubert, D. J.
Pegg, R. S. Peterson, I. A. Sellin, and R. S. Thow, Phys. Rev.
A10, 2198, (1975).
O. Keski-Rah Kouen, and M. O. Krause, AA. Date Tables 14, 137,
(1975).
O. Keski-Rah Kouen, and M. O. Krause, Phys. Fennica 9, S1, 261,
(1976).
Olsen, K., C. F. Moore, and P. Richard, Phys. Rev. A7, 1244,
(1973).
Olsen, D. K., C. F. Moore, Phys. Rev. Lett. 33, 194 (1974).
Pegg, D. J., H. H. Haselton, R. S. Thoe, P. M. Griffin, M. D.
Brown and I. A. Sellin, submitted to Phys. Rev.
Pegg, D. J., H. H. Haselton, P. M. Griffin, R. Laubert, J. R.
Mowat, R. Peterson, and I. A. Sellin, Phys. Rev. A9, 112 (1974).

Pegg, D. J., I. A. Sellin, R. Peterson, and J. R. Mowat, Phys. Rev. A8, 1350 (1973).

Richard, P., D. K. Olsen, R. L. Kauffman, and C. F. Moore, Phys. Rev. A7, 1437 (1973).

Richard P., R. L. Kauffman, J. H. McGuire, C. F. Moore, and D. K. Olsen, Phys. Rev. A7, 1437 (1973).

Richard, P., R. L. Kauffman, F. F. Hopkins, C. W. Woods, and K. A. Jamison, Phys. Rev. A8, 2187 (1973).

Richard, P., C. F. Moore, and D. K. Olsen, Phys. Lett. A43, 519 (1973).

Richard, P., Phys. Lett. A45, 13 (1974).

Richard, P., R. L. Kauffman, F. F. Hopkins, C. W. Woods, and K. A. Jamison, Phys. Rev. Lett. 30, 888 (1973).

Richard, P., C. L. Cocke, S. J. Czuchlewski, K. A. Jamison, R. L. Kauffman, and C. W. Woods, Phys. Lett. A47, 355 (1974).

Rutledge, C. H., and R. L. Watson, Atomic Data, to be published.

Rudd, M. E., B. Fastrup, P. Dahl, and F. D. Showengerdt, Phys. Rev., to be published.

Schectman, R. V., Phys. Rev. A, to be published.

Schneider, D., D. Burch, N. Stolterfoht, Proceedings of the Eighth International Conference on the Physics of Electronic and Atomic Collision, Beograd, 1973, pgs. 729.

Schneider, D., N. Stolterfoht, H. G. Gabler, Deutsche Physik-alische Gesellschaft (1973).

Schneider, D., D. Burch, N. Stolterfoht, to be published (1975).

Schneider, D., W. Hodge, B. M. Johnson, L. E. Smith, and C. F. Moore, submitted to Phys. Lett.

Schneider, D., B. M. Johnson, W. Hodge, L. E. Smith, C. F. Moore, Proceedings of the Fourth International Conference on Beam Foil Spectroscopy, (1975).

Schwietzer, J. Chem. Phys. 55, 2317 (1975).

Stolterfoht, N., D. Schneider, K. H. Harrison, Phys. Rev. A8, 2363 (1973).

Stolterfoht, N., D. Schneider, H. Gabler, Phys. Lett., A47, 271 (1974).

Stolterfoht, N., F. J. de Heer, and J. van Eck, Phys. Rev. Lett. 30, 1159 (1973).

Stolterfoht, N., D. Burch, D. Schneider, H. Wieman, J. S. Risley, Phys. Rev. Lett. 33, 59 (1974).

Stolterfoht, N., D. Schneider, P. Richard, R. L. Kauffman, Phys. Rev. Lett. 33, 1418 (1974).

Stolterfoht, N., D. Schneider, D. Burch, Proceedings of the Fourth International Conference on Atomic Physics, (1974).

Stolterfoht, N., D. Schneider, D. Burch, H. Wieman, J. R. Risley, Proceedings of the Fourth International Conference on Atomic Physics, (1974).

Stolterfoht, N., and D. Schneider, Phys. Rev. A11, 721 (1975).

Stolterfoht, N., D. Schneider, submitted to Phys. Lett. (1975).

Tawara, H., C. Foster, and F. K. de Heer, Phys. Lett. A43, 266-8 (1973).

Toburen, L. H., Phys. Rev. A9, 2505 (1975).
Toburen, L. H., and W. E. Wilson, Rev. Sci. Inst. to be published.
Utriainen J. and T. Aberg, Phys. Rev. A7, 1853 (1973).
Watson, R. L., and L. H. Toburen, Phys. Rev. A7, 1853 (1973).
Wilson, W. E. and L. H. Toburen, Phys, Rev. A11, 1303 (1975).
Winters, L. M., L. D. Ellsworth, T. Chiao, and J. R. McDonald,
    M. D. Brown, Phys. Rev. A7, 1276 (1973).
Woods, C. W., R. L. Kauffman, K. A. Jamison, C. L. Cocke, and
    P. Richard, J. Phys. B: Atom. Molec. Phys. 7, L474 (1974).
Woods, C. W., F. F. Hopkins, R. L. Kauffman, D. O. Elliott, K.
    A. Jamison, and P. Richard, Phys. Rev. Lett. 31, 1 (1973).
Woods, C. W., R. L. Kauffman, K. A. Jamison, N. Stolterfoht,
    and P. Richard, J. Phys. B. Atom, Molec. Phys. 8, L61 (1975).
Wuilleumier, F. and M. O. Krause, Phys. Rev. A10, 262 (1974).

QUASIMOLECULAR K X RAYS[*]

W. E. Meyerhof

Department of Physics, Stanford University

Stanford, California 94305

## INTRODUCTION

The discovery in 1972 by Saris et al.[1] of molecular-orbital (MO) L x rays in Ar+Ar collisions has stimulated much research on x-ray transitions between MO levels during a heavy-ion collision.[2] The motivation has been manifold: (1) attempts have been made to use the MO spectrum for tracing out the MO level spacing as a function of the internuclear separation,[3] (2) the generation of MO spectra is intimately related to the production mechanism of inner-shell vacancies and may aid to clarify the latter,[4,5] (3) detailed dynamical effects in the collision can be examined through the spectral shape[6] and through the anisotropy of the x-ray spectrum with respect to the projectile,[7] (4) in very heavy systems, such as U+U, important electrodynamic effects are predicted, including the possible emission of positrons,[8] (5) suggestions have been made that the properties of MO x rays can be used for the identification of superheavy elements.[9]

## MECHANISMS OF MO K X-RAY PRODUCTION

The present discussion is restricted to MO K x rays, i.e. to radiative transitions to the lowest MO level ($1s\sigma$). In contrast to outer-shell phenomena in which multielectron effects are very important, in inner-shell vacancy formation or decay the energy states can be treated a single-electron states. Multielectron effects can usually be taken into account by screening corrections.

Figure 1 shows the important transitions producing vacancies in the lowest states and possible radiative transitions filling the vacancies. Figure 1a sketches an atomic view and Fig. 1b a molecular view of the collision. In the atomic view one distinguishes projectile and target. In the molecular view one must think in terms of the higher-Z and the lower-Z collision partners.

For purpose of discussion, assume that the projectile is the higher-Z partner and that it carries a K vacancy produced in a prior collision. As Betz noted,[10] if the projectile velocity $v_1$ is larger than the intrinsic velocity $v_e$ of the jumping electron (transition from outer shells), radiative capture occurs. If $v_1$ is smaller than $v_e$ (transition from inner shells), MO radiation occurs and the molecular model is appropriate.[11] In either case the probability of a radiative transition occuring during the collision is of the order of the ratio of the collision time $t_{coll}$ to the projectile K-vacancy life time $\tau_x$, a ratio typically of the order of $10^{-2}$ to $10^{-3}$.

To produce MO K x rays a vacancy is needed in the $1s\sigma$ MO. Figure 1b indicates the three most important mechanisms to produce such a vacancy. (i) The vacancy can be brought into the collision through the 1s state of the higher-Z partner. This requires the projectile to be the higher-Z partner, and it must have undergone a prior, K-vacancy producing collision. (ii) If the projectile is the lower-Z partner and has undergone a prior, K-vacancy producing collision, its vacancy can be transferred to the $1s\sigma$ MO. The transfer probability has been computed and is found to decrease rapidly with increasing asymmetry of the collision.[12] (iii) The $1s\sigma$ vacancy can be produced by direct $1s\sigma$ electron excitation to vacant MO levels and the continuum. Although the excitation cross sections have not been computed for heavy-ion collisions, they can be extracted from experiment and systematized by means of scaling proposals.[13]

Since mechanisms (i) or (ii) require a prior projectile vacancy, they are called two-collision process: the vacancy formation and decay occur in separate collisions. Mechanism (iii) is called one-collisions process, since the $1s\sigma$-vacancy formation and decay occur in the same collision.

QUASISTATIC THEORY

Müller[6] will discuss the dynamic theory of MO x-ray production, which to date has been worked out mainly for the two-collision process. I will sketch an outline of the quasistatic

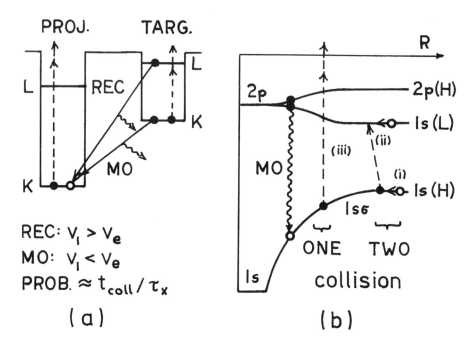

Fig. 1.   (a) Atomic and (b) molecular view of collision.

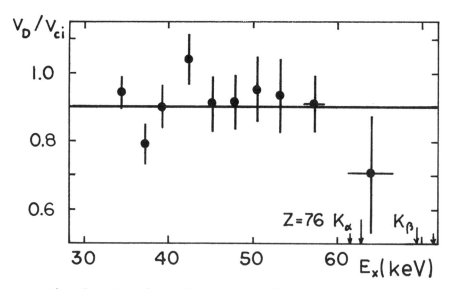

Fig. 2.   Experimental Doppler velocity for continuum spectrum from 200-MeV Kr+Zr versus x-ray energy.  Horizontal line is velocity expected for quasimolecular radiation. Arrows indicate UA transition energies.

theory, which is easily evaluated, but not correct in detail.[4]

The basic assumption of the quasistatic theory is that at every internuclear separation $R$ there is a probability $dt/\tau_x(R)$ of radiative decay, where $\tau_x(R)$ is the radiative lifetime of the $1s\sigma$ vacancy. Under certain approximations[4] the cross section for a one-collision process with a $1s\sigma$-vacancy formation probability $P(b)$ and subsequent decay is, per projectile

$$d\sigma^{(1)} = 2\pi b\, db\, P(b)\, dt/\tau_x(R) \quad . \tag{1}$$

Here $b$ is the impact parameter. For a two-collision process one has to consider that each projectile with a K vacancy has a probability $nv_1\tau_K 2\pi b\, db$ of making a second collision within the lifetime $\tau_K$ of the vacancy, where $n$ is the atomic target density. Calling $f$ the transfer probability of the projectile K vacancy to the $1s\sigma$ MO, the yield of MO radiation per projectile K vacancy is

$$dy^{(2)} = 2fnv_1\tau_K 2\pi b\, db\, dt/\tau_x(R) \quad . \tag{2}$$

The factor 2 has been added because there are two crossings of a given $R$ by a trajectory. Smith and Greiner have evaluated interference effects caused by the crossings, using the stationary phase approximation.[15] In most experiments the impact parameter is not determined, so that Eqs. (1) and (2) must be integrated over the range of impact parameters which can produce a given internuclear separation $R$.[4]

EXPERIMENTAL RESULTS

It is of interest to review the experimental evidence for MO K x rays as much as possible independently of any theory. Since MO K x rays form a continuum beyond the characteristic K lines, one must consider other continua with which MO x rays might be confused. Ordinary bremsstrahlung from electrons has been shown to lie several orders of magnitude below the experimental continua.[16] Radiative electron capture into projectile vacancies is an important continuum only close to the characteristic lines.[10] Nucleus-nucleus bremsstrahlung can form a significant background,[17] unless the projectile and target consist of identical isotopes. Fortunately, nucleus-nucleus bremsstrahlung can be computed absolutely.[18]

I shall now present the following pieces of experimental evidence. (1) The radiation typically assigned to MO K x rays is indeed radiated by the intermediate projectile-target quasimolecule and not by the target or projectile alone.

(2) The two-collision process does exist.  (3) The one-collision process does exist.  (4) Spectra have been measured as a function of impact parameter and I wish to discuss the expected shape.

   Important dynamic effects in the spectra are reviewed elsewhere in this conference, so I will only list them.  (1) As first shown by Mcdonald[19] and more recently by Greenberg,[20] the MO x-ray continuum is broadened beyond the united-atom (UA) limit, because of the change of level spacing with time. This has been  evaluated by Macek and Briggs.[21]  Müller will expand on new work.[6]  (2) As shown by Müller, Smith and Greiner,[22] there is an enhancement of the radiative decay probability due to the Coriolis effect.  Prediction of a characteristic anisotropy of the enhanced radiation has stimulated much experimental work on the anisotropy of the continuum x rays.  This is reviewed by Mokler.[7]  (3) The dynamic change from atomic to molecular orbitals and back again, which occurs during a collision, may give rise to interference effects in MO spectra.[23]  Experimental fluctuations have indeed been seen, but, according to Wölfli are due to discrete two-electron transitions,[24] rather than interference effects.

## The Radiating System

   The Doppler shift of the continuum radiation can be used to determine the nature of the radiating system.  If the Doppler velocity corresponds to that of the intermediate molecule, the radiation must originate there.  One can make use of the fore-aft symmetry of the continuum x rays to measure the Doppler velocity.  Since this is described elsewhere in the conference,[25]  I present only one experimental result (Fig. 2) for 200-MeV Kr-Zr.  The measured Doppler velocity $v_D$ is expressed in units of the initial velocity $v_{ci}$ of the center of mass of the intermediate molecule.  The solid horizontal line in Fig. 2 is the expected velocity if the radiation originates in the intermediate molecule, taking into account the slowing down of the projectile in the target material.  If the radiation originated in the projectile,  $v_D$ would be roughly twice as large, and if it originated in the target,  $v_D$  would be approximately zero.

## The Two-Collision Process

   The distinguishing feature of the two-collision process is that, _per projectile vacancy_, the MO cross section is proportional to the target density  n  [Eq. (2)], whereas the one-collision cross section per projectile is independent of

n [Eq. (1)]. Also, the two-collision process is most important for symmetric collisions [factor  f  in Eq. (2)].  To
check the  n  dependence we used a 30-MeV Br beam and bombarded  Br  containing targets in which the  Br  content was
varied.[4] The other constituents of the targets (K, Cl) were
chosen so as not to contribute appreciably to the MO yield.
In Fig. 3a the individual thick-target spectra are given.  The
relative  Br  density $n_{rel}$  is indicated, with 1.00 representing a pure, solidified  Br  target.  The x-ray energy
region of the characteristic  K  lines has been suppressed.
The predicted absolute quasistatic one-collision (0), two-
collision (T) and nucleus-nucleus bremsstrahlung (B) curves
are shown, as well as their sum (heavy lines).  For the most
diluted target, the spectrum is practically due to nucleus-
nucleus bremsstrahlung alone, but for a pure  Br  target it
is overwhelmingly due to the two-collision process.

In Fig. 3b the energy-integrated yields per projectile
K vacancy are given as a function of  $n_{rel}$.  After subtraction
of the nucleus-nucleus and small one-collision contributions,
the solid points are obtained.  The expected proportionality
to  $n_{rel}$  is found, showing that the two-collision process
exists and is dominant in 30-MeV Br+Br collisions.

### The One-Collision Process

The one-collision process is the only process expected to
be effective with a monoatomic gas target.[14,26] In solid
targets the process is generally dominant in sufficiently
asymmetric collisions, and also in symmetric collisions with
sufficiently high  Z  and high bombarding energy.[4] Figure 4
shows one example of each kind.  Betz kindly furnished me an
unpublished spectrum for 48-MeV S+Ne.[14] Here  $v_1/v_e$  is
nearly 0.6 for the  K  electrons of  S  and 1.0 for the  K
electrons of  Ne,  so that the molecular view of the collision
is barely applicable and the quasistatic theory not at all.
Nevertheless, the quasistatic one-collision theory gives the
right order of magnitude for the cross section far from the
static UA limit.  The 60-MeV Br+Ti and 82-MeV I+NaI spectra
were taken with solid targets.[4] The dominant role of the
one-collision process is apparent.  Work with high-Z systems
is very difficult because of very low yields.  This difficulty
can be overcome by increasing the bombarding energy, but then
Coulomb excited nuclear gamma rays may give an intense background in the MO x-ray region.

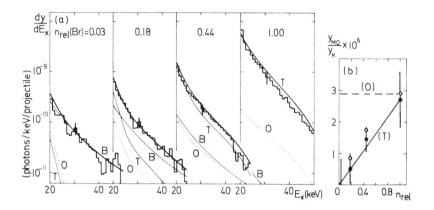

Fig. 3. (a) Continuum x-ray spectra from 30-MeV Br bombardment of various Br containing targets with relative atomic Br density $n_{rel}$. For pure Br, $n_{rel}$ = 1.00. The computed one-collision, two-collision and nucleus-nucleus bremsstrahlung spectra are denoted by O, T and B, respectively. (b) Integrated MO yield ($E_x$ = 27 to 50 keV) per projectile K vacancy versus relative Br density. Open symbols give the total MO yield, closed symbols are total minus one-collision yields. (O) and (T) are the relationships expected for one- and two-collision mechanisms, respectively.

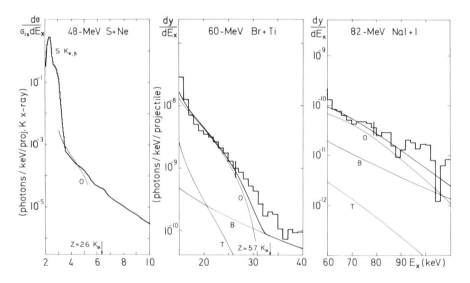

Fig. 4. Continuum x-ray spectra illustrating collisions in which the one-collision process is expected to be dominant. Gas target: 48-MeV S+Ne (Ref. 14); solid target asymmetric collision: 60-MeV Br+Ti, and high-Z, high-energy collision: 82-MeV NaI+I.

Impact Parameter Dependence

If a collision takes place with a given impact parameter, fine details of the interaction should be revealed.[6] At this conference the first measurements of continuum x-ray spectra in coincidence with the scattered particle are reported.[27,28] The experiments are very difficult and the experimenters are to be congratulated on their achievements. From theory[4,6] one expects an enhancement of the coincidence data toward the high x-ray energies, relative to the singles spectra. The quasistatic theory predicts a definite high energy cut-off for the spectrum (arrows on Fig. 5), which depends on the impact parameter. Actually, the dynamics of the collision smears out any cut-off, as first noted by Lichten.[29]

Figure 5 shows an x-ray spectrum from 60-MeV Cl+Pb, in coincidence with Cl ions backscattered at an average lab. angle of 162°.[26] The singles spectrum is also shown, as far as it could be determined above background. I am indebted to D. Burch for providing this data prior to publication. There is an enhancement of the coincidence spectrum with a shoulder near the expected cut-off. Only the one-collision process is expected to be effective in such an asymmetric collision. The predicted quasistatic coincidence and singles curves are shown. Both are low, but have the correct qualitative features.

The 32-MeV S+NaCl spectra in Fig. 5 are due to Schmidt-Böcking et al.[28] The enhancement of the coincidence spectra with respect to the singles spectra can be most clearly seen by forming the ratio of the two spectra. In this system one- and two-collision processes are effective. The solid lines are the predictions of the quasistatic theory, which can of course not reproduce the important dynamic effects. They are not in good agreement with experiment. The full dynamic theory is needed here.

CONCLUSIONS

The experimental evidence should leave no doubt that MO K x-ray spectra exist.[17] From an experimental point of view the challenge is to search for refined features in the MO spectra and to investigate other x-ray continua, particularly near the characteristic K lines where quantitative understanding is still lacking. [30] Betz has reviewed the situation at high projectile velocities ($v_1 > v_e$), where new radiative processes appear.[10] From a theoretical point of view the challenge is to explore the full dynamics of the collision process, and hopefully to end up with approximations easily applied to experiment. There is yet much to be done.

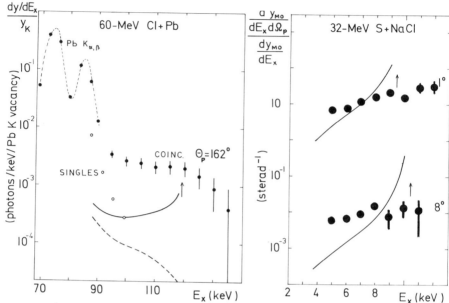

Fig. 5. Continuum x-ray spectra with defined impact parameter. 60-MeV Cl+Pb spectrum in coincidence with projectiles scattered through 162° (b = 1 to 4 fm). Singles spectrum given with open symbols. Coincidence and singles spectra are normalized to total Pb K-vacancy yields. Data from Ref. 27. Predictions of one-collision quasistatic theory for coincident and singles spectra are given by solid and dashed lines, respectively. 32-MeV S+NaCl spectra in coincidence with projectiles scattered through 1° (b = 630 to 790 fm) and 8° (b = 83 to 91 fm). Spectra are divided by singles spectrum at each x-ray energy. Data from Ref. 28. Solid lines are predictions of quasistatic theory. Arrows indicate static cut-off limits for $2p\pi$-$1s\sigma$ transition.

REFERENCES

* Supported in part by the National Science Foundation.

[1] F. W. Saris, W. F. van der Weg, H. Tawara, and R. Laubert, Phys. Rev. Lett. 28, 717 (1972).

[2] For references up to 1974 see P. H. Mokler, S. Hagmann, P. Armbruster, G. Kraft, H. J. Stein, K. Rashid, and B. Fricke, Proc. Fourth Int. Conf. on Atomic Physics, Heidelberg, 1974, to be published. Other references are given below.

[3] B. Knaf and G. Presser, Phys. Lett. 49A, 89 (1974).

[4] W. E. Meyerhof, T. K. Saylor, S. M. Lazarus, A. Little, B. B. Triplett, L. F. Chase, Jr., and R. Anholt, Phys. Rev. Lett. 32, 1279 (1974).

[5] K. Smith, B. Müller, and W. Greiner, J. Phys. B8, 75 (1975).

[6] B. Müller, this conference.

[7] P. H. Mokler, this conference.

[8] K. Smith, H. Peitz, B. Müller, and W. Greiner, Phys. Rev. Lett. 32, 554 (1974).

[9] P. H. Mokler, H. J. Stein, and P. Armbruster, Phys. Rev. Lett. 29, 827 (1972); B. Fricke and J. T. Waber, Phys. Rev. C8, 330 (1973); B. Müller, R. K. Smith, and W. Greiner, Phys. Lett. 53B, 401 (1975).

[10] H. D. Betz, this conference.

[11] J. S. Briggs and K. Dettmann, Proc. Fourth Int. Conf. on Atomic Physics, Heidelberg, 1974, to be published; M. Kleber and D. H. Jakubassa, to be published.

[12] W. E. Meyerhof, Phys. Rev. Lett. 31, 1341 (1973); J. S. Briggs, to be published; J. S. Briggs and K. Taulbjerg, to be published.

[13] W. E. Meyerhof, Phys. Rev. A10, 1005 (1974); W. E. Meyerhof, R. Anholt, T. K. Saylor, and P. D. Bond, Phys. Rev. A11, 1083 (1975); W. E. Meyerhof, T. K. Saylor, S. M. Lazarus, A. Little, R. Anholt, and L. F. Chase, Jr., to be published; C. Foster, T. Hoogkamer, P. H. Worlee, and F. W. Saris, this conference, paper 511.

[14] F. Bell, H. D. Betz, H. Panke, E. Spindler, W. Stehling, and M. Kleber, to be published.

[15] K. Smith and W. Greiner, Zeit. f. Physik, to be published.

[16] R. Anholt, unpublished.

[17] C. K. Davis and J. S. Greenberg, Phys. Rev. Lett. 32, 1215 (1974).

[18] K. Adler, A. Bohr, T. Huus, B. Mottelson, and A. Winther, Rev. Mod. Phys. 28, 432 (1956).

[19] J. R. Macdonald, M. D. Brown, and T. Chiao, Phys. Rev. Lett. 30, 471 (1973).

[20] J. S. Greenberg, C. K. Davis, and P. Vincent, Phys. Rev. Lett. 33, 473 (1974).

[21] J. H. Macek and J. S. Briggs, J. Phys. B7, 1312 (1974).

22 B. Müller, R. K. Smith and W. Greiner, Phys. Lett. 49B, 219 (1974).

23 R. K. Smith, B. Müller, W. Greiner, J. S. Greenberg, and C. K. Davis, Phys. Rev. Lett. 34, 117 (1975).

24 W. Wölfli, Ch. Stoller, G. Bonani, M. Suter, and M. Stöckli, to be published.

25 W. E. Meyerhof, T. K. Saylor, and R. Anholt, this conference, paper 308; P. H. Mokler, P. Armbruster, S. Hagmann, F. Folkmann, and H. J. Stein, this conference, paper 310.

26 Th. P. Hoogkamer, P. H. Worlee, C. Foster, and F. W. Saris, this conference, paper 300.

27 D. Burch, W. B. Ingalls, H. Wieman, and R. Vandenbosch, this conference, paper 306.

28 R. Schule, H. Schmidt-Böcking, I. Tserruya, G. Gaukler, and K. Bethge, this conference, paper 304.

29 W. Lichten, Phys. Rev. A9, 1458 (1974).

30 W. Frank, P. Gippner, K. H. Kaun, H. Sodan, W. Schulze, and Yu.P. Tretyakov, Joint Institute for Nuclear Research, Dubna, USSR, preprint E7-8616 (1975).

# RADIATIVE PROCESSES IN TRANSIENT QUASI-MOLECULES

B. Müller

Department of Physics, University of Washington

Seattle, Washington 98195 U.S.A.

The molecular model for the description of low energy ion-atom collisions was--after an early experimental result by Coates[1] in the 1930s--proposed by Fano and Lichten[2] in 1965. Since then it has become extremely useful in explaining a wide variety of experimental findings, and it is by no means clear to which limits the model can be ultimately pushed.

However, as for every model in physics, a crucial question is whether it just simply fits the available data  or whether it also has an independent physical meaning.  The decision between the two possible cases can only come from direct observations made of the ingredients of the model.  In the case of our interest this means that we should observe directly the conjectured (diabatic) quasi-molecular states formed during an ion-atom collision.[3]  The general technical tools to observe and identify states in atomic physics are radiative transitions. But it is not only a matter of philosophical interest to measure radiation from quasi-molecular states (MO-radiation):  it also provides us for the first time with a limit as to what the actual energies and wave functions are as a function of inter-nuclear separation.  These quantities have been calculated for a number of years, but experimental evidence so far was limited to the position of certain level-crossings.

And finally, and this is probably the most stimulating of reasons, the quasi-molecular states have as limits those of the so-called "united atom."  In the future, it may be hoped one will be able to extract the properties of superheavy atoms--atoms with nuclear charge beyond, say 110, which do not exist in either nature or laboratory--from data taken of superheavy

quasi-molecules.[4]  I should mention that the first steps into
this territory have been taken by Mokler et al.,[5] by the
Rochester group[6] and by Wölfli and collaborators.[7]

This last mentioned aim not only demands that we can successfully calculate a molecular radiative spectrum so that it
fits the experimental data, but also that we are able to extract information about the shape of the molecular levels from
a given spectrum.  An important step in this direction is the
confirmation of strongly frequency dependent anisotropy in the
MO radiation close to the united  atom limit of the spectrum.[8]
The detailed theory of this process has been recently reported
by K. Smith.[9]  To extract in this way information over the
united atom seems to be a promising perspective.

Today, however, I want to give a report of how well we
understand the shape and absolute cross-section of experimentally measured MO X ray spectra.  A number of results have
been published on radiation to the L resp. M shells of quasi-
molecules, but since the background problems are much more
severe for those, I shall concentrate  on non-characteristic
K radiation from heavy ion collisions.

In time dependent perturbation theory, the transition amplitude between two adiabatic molecular states--or more precisely between two electronic configurations in the quasi-
molecule--with emission of a photon $(\vec{k},\varepsilon)$ is:[10]

$$c_{fi,k\varepsilon} = -\frac{i}{\hbar} \int_{-\infty}^{\infty} dt\ a_{fi}(t) <f(R)|\vec{j}\cdot\vec{A}(k,\varepsilon)|i(R)>$$

$$\exp i[\int^{t} (\omega-\omega f_{i})dt] \quad .$$

Here $f(R)$ and $i(R)$ are the final and initial molecular states,
$\vec{j} = e\vec{\alpha}$ is the electromagnetic current operator and $\omega_{fi}(t) = -\frac{1}{\hbar}[E_{f}-E_{i}]$ is the instantaneous transition frequency between
the two states.  The evaluation of $\vec{j}$ depends somewhat on the
states that are chosen.  If they are solutions of the full
time dependent Hamiltonian,

$$H = H_{o}(R) - \vec{\omega}_{rot} \cdot (\vec{r} \times \vec{p} + \tfrac{1}{2}\hbar\vec{\sigma})$$

then one may write

$$\vec{j} = \frac{ie}{\hbar c} [H,r] + \frac{e}{c}(\vec{\omega}_{rot} \times \vec{r}) \quad ,$$

where the second part has been called induced current.[8a] Near
the united atom limit this contribution can be shown to be small
and it is more efficient to work with adiabatic two center
functions. Then

$$\vec{j} = \frac{ie}{\hbar c} [H_o, r] \sim - i\frac{e}{c} \omega_{fi} \vec{r}$$

in the long wave length approximation, where one simply sets

$A(k,\varepsilon) = (\frac{2\pi\hbar c}{kL^3})^{\frac{1}{2}} \vec{\varepsilon}.$ For a complete treatment of the radiation

process, at this point we would have to worry about how the
time-dependent amplitude behaves, especially in one-collision
processes.[11] Also care should be taken to include the rotation
between the intrinsic and the laboratory coordinate systems.[9]

Most calculations carried out so far have made use of the
static or stationary phase approximation[8a, 12] and a typical
spectrum is shown in Fig. 1. Even a superficial comparison
with the experimental spectrum (Fig. 2) taken by Greenberg et
al. shows that the shape comes out wrong. Even if an ad hoc
collision broadening is folded into the spectrum, no agreement
is reached.

Other calculations to mention are those by W. Lichten,[13]
utilizing the quasi-classical approximation, and by Briggs and
Macek[14] who calculated the numerical Fourier transform for the
system N+N, but no experimental results so far have been available for this system (see, however, Saris et al.[19]).

To be precise, let us make the approximation

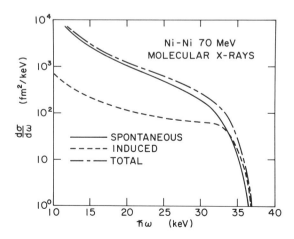

Fig. 1.  The static model

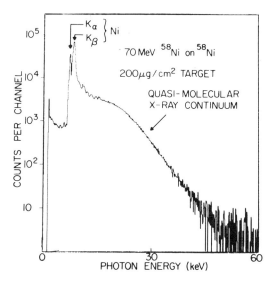

Fig. 2.   70 MeV Ni-Ni spectrum from Greenberg et al.

$$C_{fi} = -e \; a_{fi} \left(\frac{2\pi}{c\hbar k L^3}\right)^{\frac{1}{2}} <f|\vec{r}\cdot\vec{\varepsilon}|i>$$
$$\int_{-\infty}^{\infty} dt \; \omega_{fi}(t) \; e^{i\int^t dt(\omega-\omega_{fi})} \qquad .$$

Every time dependence--e.g. of $a(t)$ or the dipole matrix element--
can be folded in afterwards, if desired.  In order to have an
absolutely convergent integration, we separate between the
characteristic atomic transition frequency $\omega_S$ and the time dependent molecular contribution:

$$\omega_{fi}(t) = \omega_S + \omega'(t) \qquad .$$

Partial integration gives then for the integral

$$\tilde{C}(\omega) = 4\pi\delta(\omega-\omega_S)\omega_S \cos \int_0^\infty \omega'(t)dt \; +$$

$$+ \; \frac{\omega}{\omega-\omega_S} \int_{-\infty}^{\infty} \omega'(t) \; e^{i(\omega-\omega_S)t} \; e^{-i\int^t \omega'dt'} dt \qquad .$$

The first part constitutes the characteristic line with a factor which accounts for the phase shift during the collision and
the second part is what we call molecular radiation.  In order

to learn what its dependence on the photon frequency $\omega$ is, we have to investigate $\omega'(t)$. We will certainly have $\omega'(t) = \omega'(-t)$ and it will be a continuous, analytic function of time (Fig. 3). In order to analyze the Fourier transform further, we think of $\omega'(t)$ as a function of a complex variable. Mathematics tells us that it is either a constant or unbounded for $|t| \to \infty$ or it has singularities in the complex plane. The symmetry condition tells us that these singularities have to be symmetric with respect to the imaginary and real axes.

The important features to be represented are a flat slope at the time of closest approach, $t = 0$ and a point of inflection at the time $\tau$ corresponding to a characteristic internuclear distance. The simplest way to satisfy these requirements is to choose a single pole at $t = i\tau$ (and symmetrically at $t = -i\tau$), i.e. to write

$$\omega'(t) = \frac{-i\gamma(t)}{t-i\tau}$$

with a function $\gamma(t)$ that is regular in the upper half of the complex plane. Then also

$$\tilde{\gamma}(t) = \frac{\gamma(t) - \gamma(i\tau)}{t-i\tau}$$

is completely regular in this region, and we can evaluate

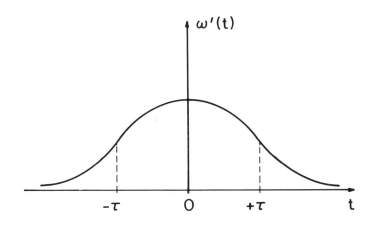

Fig. 3.  Typical shape of $\omega'(t)$

$$\int_0^t \omega'(t)dt \equiv -i\gamma_o \ln(\frac{t-i\tau}{i\tau}) -i \int_0^t \tilde{\gamma}(t)dt$$

with $\gamma_o = \gamma(i\tau) = i \underset{t=i\tau}{Res}\ \omega'(t)$.

The second part is a smoothly varying function everywhere.
With this we have

$$\tilde{C}_{mol}(\omega) = \frac{\omega}{\omega-\omega_s} \int_{-\infty}^{\infty} \frac{-i\gamma(t)}{t-i\tau} e^{i(\omega-\omega_s)t} (\frac{t-i\tau}{i\tau})^{-\gamma_o} e^{-\int_0^t\tilde{\gamma}(t)dt}\ dt.$$

Now perturbation theory tells us that $\omega'(t)$ vanishes at least like $R^{-2} \sim t^{-2}$ at $t \to \pm\infty$. Therefore an integration around an upper half-circle path of infinite radius will not contribute, and we may deform the integration path from the real time axis to the imaginary axis. The only restriction is that we may not cross any singularities. The remaining integral on the left side of the imaginary axis down to $t = i\tau$ turns around and runs

on the right side of the axis back to infinity (Fig. 4). With a simultaneous change of the integration variable we find:

$$\tilde{C}_{mol}(\omega) \sim + \sin(\gamma_o+1)\pi\ e^{-(\omega-\omega_s)\tau} \cdot \frac{\omega}{\omega-\omega_s} \tau^{\gamma_o}$$

$$\cdot \int_0^{\infty} dt\ e^{-(\omega-\omega_s)t} \frac{\gamma(it+i\tau)}{t^{\gamma_o+1}} e^{-\int^t\tilde{\gamma}(it+i\tau)dt}.$$

This integral, however, is not yet properly defined since it diverges at the lower limit. In order to make a meaningful expression out of the formal integral, we have to integrate n-times by parts, where n is the smallest integer $n > \gamma_o$. This yields:

$$\tilde{C}_{mol}(\omega) = \frac{\omega}{\omega-\omega_s} \frac{\Gamma(\gamma_o-n)}{\Gamma(\gamma_o)} \sin(\gamma_o+1-n)\pi\ e^{-(\omega-\omega_s)\tau} \tau^{\gamma_o}$$

$$\cdot \int_0^{\infty} dt\ \frac{e^{-(\omega-\omega_s)t}}{t^{\gamma_o+1-n}} [(\omega-\omega_s)-\frac{\partial}{\partial t}]^n (\gamma(it+i\tau)e^{-\int^t\tilde{\gamma}(it+i\tau)dt}).$$

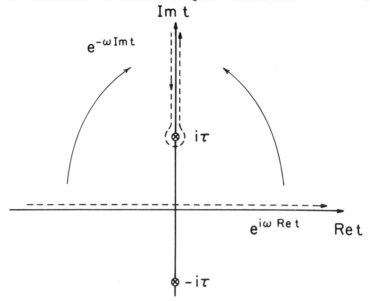

Fig. 4. Deformation of the Fourier integration path in the complex t-plane

The behavior of $|\tilde{C}_{mol}(\omega)|^2$ as a function of photon energy can
be readily determined from this expression (see Fig. 5):

(1)   Close to the characteristic atomic line it falls

off as $\dfrac{1}{(\omega-\omega_s)^2}$ , i.e. in the same way as the Lorentz

tail of the characteristic line does.

(2)   For high energies the exponential $e^{-2(\omega-\omega_s)\tau}$ will
prevail, thus giving the exponential fall off observed
in all experiments.

(3)   In the intermediate region, the integral is a polynomial of $(\omega-\omega_s)$ with degree n, exhibiting n zeros.
It also will contribute a factor $(\omega-\omega_s)^n$ to the high
energy tail of the spectrum.

The simplest possible choice of $\gamma(t)$ is[8a]

$$\gamma(t) = \frac{(\omega_c-\omega_s)R_o^2}{(\tau-it)v^2} ,$$

with which the integral for $\tilde{C}_{mol}$ can be evaluated explicity:[15]

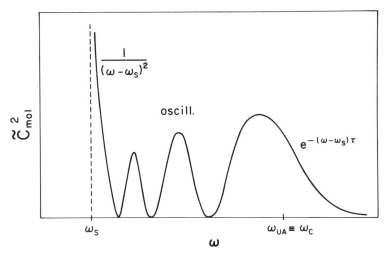

Fig. 5.   Schematical graph of a coincidence spectrum

$$\tilde{C}_{mol}(\omega) = 2\pi \; \gamma_o \; \frac{\omega}{\omega-\omega_s} \; k_{2\gamma_o}(\tau(\omega-\omega_s))$$   .

In this expression

$$k_\nu(x) = \frac{2}{\pi} \int_0^{\pi/2} \cos(x \tan\theta - \nu\theta)d\theta$$

is a Bateman function.[20]

In order to deduce a scaling law with respect to velocity v, we note that for all except the very smallest impact parameters b we may use the straight line approximation

$$R(t)^2 = b^2 + v^2t^2$$   .

Introducing $\rho = (R_o^2 + b^2)^{\frac{1}{2}}$ we may relate $\tau$ to the characteristic distance $R_o$ of the quasi-molecule:

$$\rho = v\tau$$   ,

and for the simple analytical model

$$\gamma_o = \frac{(\omega_c - \omega_s)R_o^2}{2\tau v^2} = \frac{(\omega_c - \omega_s)R_o^2}{2v\rho}$$   .

To obtain numerical values, $R_o$ was matched to a number of numerical calculations of molecular level diagrams to give

$$R_o \stackrel{\sim}{=} \frac{62100 \text{ fm}}{(\frac{z_1 + z_2}{2} - 1)}$$

which should not be used for highly asymmetric systems.  In Fig. 6, $\stackrel{\sim}{C}^2_{mol}(\omega)$ is shown for a Ni-Ni collision at two different impact parameters.  The oscillatory pattern in the molecular region is well represented and its variation with impact parameter is seen.

At this point our way of approach provides easy insight into how distant rearrangements in the quasi-molecule can affect the spectrum.  Such a process would correspond to the introduction of additional singularities at times $i\tau_R$ where $\tau_R \gg \tau$.  Straightforward application of our general results shows that this will introduce much more rapid oscillations into the spectrum which also fall off exponentially, but steeper, and thus should be detectable mostly in the vicinity of the characteristic line.

The fact that the spectrum extends beyond the classical "united atom" limit $\omega = \omega_c$ has been called collision broadening, and if we write the exponential fall off in this "tail" region as

$$\stackrel{\sim}{C}_{mol}(\omega) \propto e^{-\frac{\hbar\omega}{\Gamma}}$$

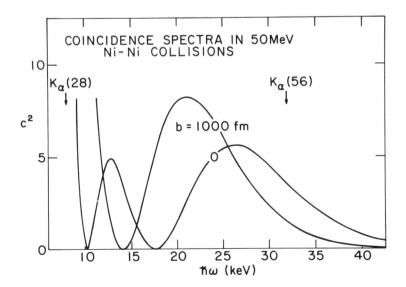

Fig. 6.  Coincidence apectrum at two impact parameters

we can define a typical width $\Gamma$.  In any analytical model that introduces a characteristic distance in the molecule, one has

$$\Gamma = \frac{hv}{\rho} = \frac{hv}{(R_o^2 + b^2)^{\frac{1}{2}}} \qquad ,$$

i.e. the broadening scales as v, for every impact parameter b. This result is in contrast to the scaling law $\Gamma \sim v^{\frac{1}{2}}$ given by Briggs and Macek,[14] which was obtained using an expansion $\omega'(t) = \omega_c - \dot{\omega}|t|$, which cannot be treated in an analytical way. However, it is obvious that at the distance of closest approach

$$\frac{d\omega'}{dt} = 0 \neq \dot{\omega} \qquad .$$

Therefore, for single impact parameter spectra their scaling law cannot apply.

The experimental results (except the one of D. Burch et al.[16] which he will report in this conference) have not discriminated between impact parameters.  For such a "singles" spectrum we have to calculate a cross section by folding $C_{mol}(\omega)$ with the Rutherford cross section:

$$\frac{d\sigma_{mo}}{d\omega} = 2\pi \int_0^\infty b \; db \; C_{mol}^2(\omega,b) \; \frac{k^2 L^3}{2\pi^2 c} =$$

$$= \frac{4}{3} \alpha \; a_{fi}^2 ||<f|r|i>||^2 \; \frac{\omega}{c^2} \int_{R_o}^\infty \rho d\rho \; \tilde{C}_{mol}^2(\omega,\rho) \qquad .$$

Even for the simplest model, the impact parameter integration cannot be performed analytically, but it may be done on a computer.  It is interesting to see which values of $\rho$ (or b) contribute for a given photon frequency $\omega$.  This is shown in Fig. 7 for a specific Ni-Ni collision, where one can see that from about $\hbar\omega = 26$ keV on the shape of the curves are almost unchanged with the main change being in absolute magnitude.  If there is any additional influence on the spectrum which depends on impact parameter (e.g. an ionization amplitude $a_{fi}(b)$) but does not discriminate between different b < 1500 fm (in this case) then the shape of the spectrum beyond 26 keV or so should remain practically unchanged.  Thus we can hope to represent-- at least in shape--many experimental results that originate in a much more complicated situation than our model was designed to deal with.

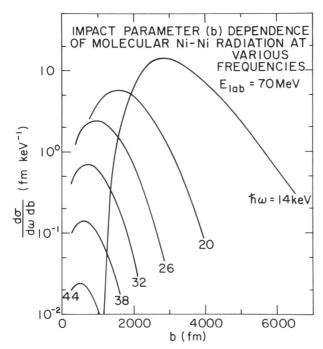

Fig. 7. Impact parameter dependence

The typical shape of an integrated spectrum is given in Fig. 8 for several Aℓ - Aℓ collisions. Characteristic is the long exponential fall off (tail) in the high frequency region, to which I shall come back in a moment. One also notices that roughly for photon energies below the united atom limit the cross section for radiation in one subsequent collision de-creases with increasing bombarding energy, whereas it grows beyond the united atom limit. The first part of this observa-tion is explained by the increasingly shorter collision time which causes the cross section to drop; the second part can be understood through the collision broadening which is bound to grow with bombarding energy. This effect reaches the full im-pact in the classically forbidden region.

This leads us quite naturally to investigate the dependence of the collision broadening $\Gamma$ on the projectile velocity v. If one fits the spectrum to a shape

$$\frac{d\sigma}{d\omega} = \text{const.} \times \omega\, e^{-\frac{\hbar\omega}{2\Gamma}}$$

(where a factor $\omega$ is added to describe effects from the radia-tion field alone), one can extract a spectral tail halfwidth

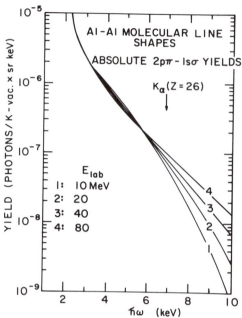

Fig. 8.   Singles spectra of Aℓ-Aℓ.

$$\Gamma_{\frac{1}{2}} = (2 \ln 2)\Gamma \quad .$$

The results are depicted in Fig. 9, together with the available experimental results.   Only the Ni-Ni, S-Aℓ, and the (preliminary) Aℓ-Aℓ data were given by the experimenters.   The others were read off the published spectra ₋nd thus have big error bars. With this in mind, the agreement is astonishingly good.   The most astonishing result--after the single impact parameter calculations--is that the broadening grows as $v^{\frac{1}{2}}$ and not like v. All numerical  results are very well fitted by

$$2\Gamma = \hbar \sqrt{\frac{(\omega_c - \omega_s)v}{3\pi R_o}} \sim 0.326\, \hbar \sqrt{\frac{(\omega_c - \omega_s)v}{R_o}}$$

which is in rather close agreement with the parametrized Briggs-Macek as used by several authors.   The good agreement is comfirmed by the (normalized) fit to spectral tails in Ni-Ni collisions of various energy obtained by Greenberg et al.[17] Fig. 10.

The second crucial test of the model is whether the experimentally obtained spectra are reproduced in absolute magnitude.   From the cross section in a single second collision

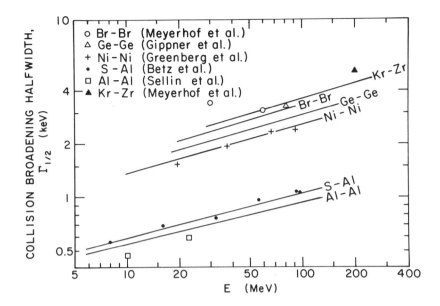

Fig. 9.   Dependence of tail broadening on bombarding energy

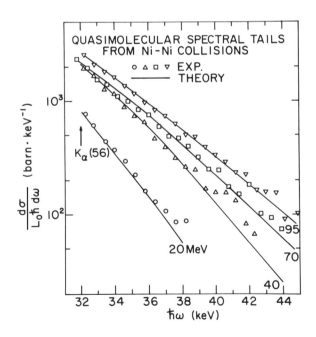

Fig. 10.   Tails beyond the UA limit in Ni-Ni (Greenberg et al.)

$\dfrac{d\sigma_{Mo}}{d\omega}$ the experimental yield per projectile K-vacancy is obtained by

$$Y_{Mo} = \frac{d\sigma_{Mo}}{h\,d\omega} \cdot \frac{v\tau_K}{d^3}\;f(Z_P,Z_T,v)\;.$$

Here $\tau_K$ is the lifetime of a projectile K-vacancy, d is the mean distance between two larger atoms, and $f(Z_P,Z_T,v)$ is the probability that the vacancy will end up in the 1s$\sigma$ molecular state. The good approximation f is given by Meyerhof's formula,[18] and certainly f = ½ for symmetric systems. If the yield is required per projectile K-X ray, $\tau_{KX}$ has to be substituted for $\tau_K$; and for a total experimental cross section $Y_{Mo}$ has to be multiplied by the projectile K-vacancy production cross section. For the actual computation of the yields, the dipole matrix element $||<f|r|i>||^2$ was chosen so that it reproduced the separated atom K-X-ray transition probability at $\omega = \omega_s$ and then scaled according to $\omega^{\frac{1}{2}}$. This should provide a reasonable approximation for the 2p$\pi$-1s$\sigma$ radiative transitions, which were only taken into account.

   Figures 11-13 show the calculations for various systems together with the experimental results. Greenberg's Ni-Ni data (Fig. 11) at four different bombarding energies are reproduced to within ±30% (the claimed experimental accuracy) except for the lowest photon energies. This may indicate either a breakdown of the severe approximations made in the model or it could be caused by some new process contributing to the experimental data. This can only be decided by exact molecular calculations for a few selected collision systems. Let me turn to the Br-Br and Br-Zr data obtained by Meyerhof et al.[12] which are shown in Fig. 12. The 30 MeV spectrum (12a) is very well represented. It should be mentioned that in this case almost the same agreement is obtained with Meyerhof's quasi-static calculation,[12] and that this is one of the few systems where the two-collision mechanism was established experimentally. For the 60 MeV Br-Br and Br-Zr spectra successively worse agreement is reached (12b,c). Whether this is due to the higher collision energy or competing (e.g. one collision mechanism) processes, remains to be seen. Anyway, the disagreement is never worse than a factor of 2, and the shape is always well reproduced. Finally, Fig. 13

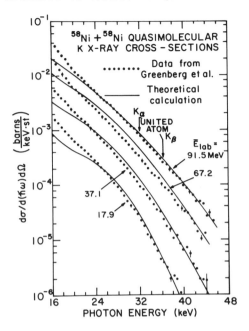

Fig. 11.   Absolute cross section for Ni-Ni spectra

gives a comparison with the Ge-Ge data from Gippner et al.[21]
No comparison of magnitude was possible in this case, because
only the number of events are given, but an estimate shows that
the theoretical calculation is not off by an order of magnitude.
In shape the spectrum beyond 20 keV is well described by theory.
This is the case also for the other systems investigated by
the Dubna group (Nb-Nb etc.).

Betz et al.[22] have given the molecular tail yield beyond
the united atom limit relative to the number of characteristic
K-X rays for various bombarding energies of s on Aℓ.  For energies up to 20 MeV there is fair agreement (within 40%) with
calculated fraction (Fig. 14).  But for higher bombarding energies the experimental yield increases almost exponentially,
whereas the calculated yield grows roughly like $E^{1/3}$.  (This
behavior is a result of various components, e.g. the decreasing molecular lifetime, the increasing broadening width, the
change in K-vacancy sharing, etc..)  Figure (15a) shows the
relative experimental tail abundance over theory.  Since the
theoretical Mo X-ray yield is proportional to the radiative
lifetime of a K-vacancy in the projectile, which in turn is a
function of the number of 2p electrons in the projectile, one

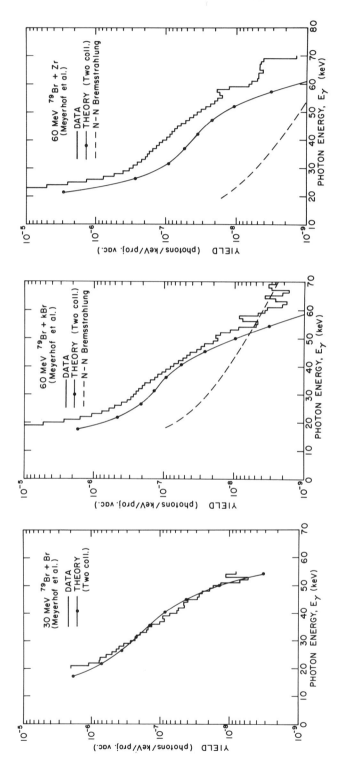

Fig. 12 Absolute cross sections for Br–Br (Zr) spectra

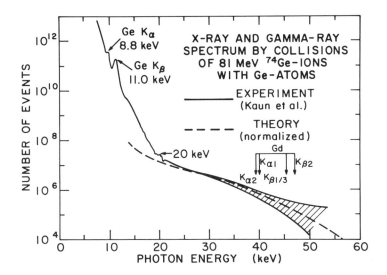

Fig. 13.   Comparison with the Ge-Ge data of Kaun et al.

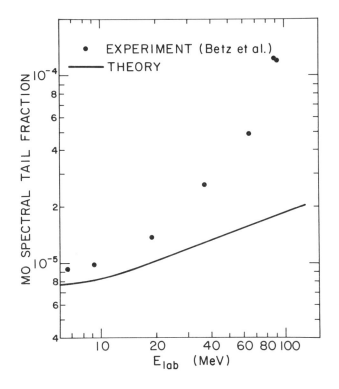

Fig. 14.   Mo tail fractions in S-Aℓ (Betz et al.)

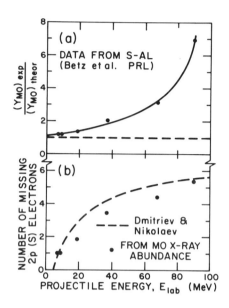

Fig. 15.   Zp-vacancies in S-projectiles Using data of Fig. 14

can deduce the latter from the comparison. Figure 15b shows the result of this, when the slightly scaled values of Bhalla et al.[23] for Ar are used. This can be compared to the statistical number of 2p electrons in a S ion travelling through a solid in equilibrium.[24] The two numbers agree roughly although the uncertainties are very great. Even if this rough estimate could be wrong by a factor of two, it certainly constitutes an effect which has to be considered in very energetic molecular collisions. It will definitely help to reduce the discrepancy between experiment and theory in this energy regime.[25]

The survey of presently available K-X-ray data of molecular origin has yielded a surprisingly good agreement between the experimental results and the two-collision theory. To appreciate this fully, one has to keep in mind that there is no adjustable parameter in the theoretical model. The approximation that was necessary to obtain a simple model certainly give error bars to the theoretical calculations, the size of which I estimate to be ±40%. This, however, only concerns the absolute magnitude and not the shape of the spectra. The slope of the Mo tail should be represented better than 15%. This has important consequences for a possible identification of collision systems. Figure 9, together with the relation $\Gamma \sim (Z_1 + Z_2)^{3/2}$,

indicates that from a precise measurement of Γ--at a single
convenient bombarding energy E--the combined charge of the two
colliding nuclei can be obtained within ±10% error.

To my mind this opens the promising perspective that,
after all, one can deduce some information about the quasiatomic system from molecular spectra. For a determination of
the united atom energies, one certainly will have to measure
the angular anisotropy of the radiation. If the present evidence is confirmed that this peaks at the united atom limit,
molecular K-X-ray spectroscopy will become the basic tool to
investigate intermediate superheavy systems.

## REFERENCES

1) W.M. Coates, Phys. Rev. 46, 542 (1934).
2) U. Fano and W. Lichten, Phys. Rev. Lett. 14, 627 (1965).
3) As with many ideas in physics, it is virtually impossible
   to trace back where it originated. The first experiments
   were reported at the Atlanta meeting (1972) by F. Saris
   and P.H. Mokler and their respective collaborators.
4) J. Rafelski, L. Fulcher, and W. Greiner, Phys. Rev. Lett.
   27, 958 (1971); P. Armbruster, P.H. Mohler, and H.J. Stein,
   Phys. Rev. Lett. 27, 1623 (1971).
5) P.H. Mohler, H.J. Stein, and P. Armbruster, Phys. Rev. Lett.
   29, 827 (1972).
6) F.C. Jundt, H. Kubo, and H.E. Gove, Phys. Rev. A10, 1053
   (1974).
7) W. Wölfli, Ch. Stoller, G. Bonani, M. Suter, and M. Stöckly,
   "Proceedings of the IX International Conference on the Physics of Electronic and Atomic Collisions," July, 1975,
   paper A1.
8) a) B. Müller and W. Greiner, Phys. Rev. Lett. 33, 469 (1974).
   b) C.K. Davis and J.S. Greenberg, ibid., p. 473.
   c) G. Kraft, P.H. Mokler, and H.J. Stein, ibid., p. 476.
9) K. Smith and W. Greiner, "Spontaneous and Induced Radiation
   from Intermediate Molecules," preprint, Frankfurt am Main,
   1975.
10) K. Smith, B. Müller, and W. Greiner, J. Phys. B8, 75 (1975).
11) W.R. Thorson,"Proceedings of the IX International Conference
    on the Physics of Electronic and Atomic Collisions," July,
    1975, p. 298.
12) J.H. Macek and J.S. Briggs, J. Phys. B7, 47 (1974); W.E.
    Meyerhof, T.K. Saylor, S.M. Lazarus, A. Little, B.B. Triplett,
    L.F. Chase, and R. Anholt, Phys. Rev. Lett. 32, 1279 (1974).
13) W. Lichten, Phys. Rev. A9, 1458 (1974).
14) J.S. Briggs and J. Macek, J. Phys. B7 , 1312 (1974)
15) This result has also been obtained by K. Smith.
16) D. Burch, W.B. Ingalls, H. Wieman, and R. Vandenbosch,

"Proceedings of the IX International Conference on the Physics of Electronic and Atomic Collisions," July, 1975, p. 306.

17) J.S. Greenberg, private communication.
18) W.E. Meyerhof, Phys. Rev. Lett. $\underline{31}$, 1341 (1973).
19) Th. P. Hoopkamer, P.H. Worlee, C. Foster, and F.W. Saris, "Proceedings of the IX International Conference on the Physics of Electronic and Atomic Collisions," July, 1975, p. 300.
20) Erdelyi et al., Higher Transcendental Functions, McGraw Hill, 1953.
21) P. Gippner, K.H. Kaun, F. Stary, W. Schulze, and Yu.P. Tretyakov, Nucl. Phys. $\underline{A230}$, 509 (1974).
22) H.D. Betz, F. Bell, H. Panke, W. Stehling, and E. Spindler, Phys. Rev. Lett. $\underline{34}$, 1256 (1975).
23) C.P. Bhalla, Phys. Rev. $\underline{A8}$, 2877 (1973).
24) I.S. Dmitriev and V.S. Nikolaev, Sov. Phys. JETP $\underline{20}$, 409 1965); H.D. Betz, Rev. Mod. Phys. $\underline{44}$, 465 (1972).
25) I acknowledge the helpful collaboration with D. Burch.

# EXPERIMENTAL EVIDENCE FOR ANISOTROPY OF

# NON-CHARACTERISTIC X RAYS

P.H. Mokler, P. Armbruster, F. Folkmann,

S. Hagmann, G. Kraft, and H.J. Stein

GSI Darmstadt, KFA Jülich, and Universität zu Köln

GSI, Postfach 541, 61 Darmstadt, Germany

During heavy ion-atom collisions both x-ray lines and x--ray continua may be observed.[1] The x-ray lines characterize the excited states of the atomic particles after a collision. Whereas, the x-ray continua may characterize the collision itself. As effects causing x-ray continua we have to consider: (i) x-ray transitions between orbitals of the transiently formed collision molecule, i.e., quasimolecular radiation (MO),[2,3] (ii) nucleus-nucleus bremsstrahlung,[4] (iii) bremsstrahlung of the ejected electrons ($\delta$-rays),[5] and (iv) radiative capture of target electrons into vacancies of the flying ions (REC);[6] Also (v) collisions of recoiling atoms may contribute to x-ray continua via the mentioned effects.[7] Which effect dominates depends on experimental conditions and on x-ray energy.

Especially effect (i) can give information on the transiently formed quasimolecule. Only this effect will be considered in this report giving main emphasis to the very heavy I-Au collision system. Generally, it can be stated, that effect (i) is the major process at quasiadiabatic collision velocities, i.e., $v/u \ll 1$ or at least $v/u < 1$, where $v$ is the collision velocity and $u$ the orbital velocity of the electrons of concern. All the measurements reported here will concern this velocity region. Nevertheless, the other effects mentioned can contribute to the x-ray continua and their influence must be considered on a case--by-case basis.

INFORMATION FROM SINGLE X-RAY SPECTRA

MO x-ray continua were first observed by Saris and coworkers for L-shell transitions in Ar-Ar collision molecules.[8]
Stimulated by this first experiment quasimolecular M radiation
was investigated for an extremely heavy collision system, the
I-Au system.[9] Measurements on quasimolecular K radiation
followed.[10] A brief review on the various collision systems
studied is given in Ref. 3. From a theoretical point of view
quasimolecular radiation from the heaviest possible collision
systems is of most interest;[11] this is especially true for K

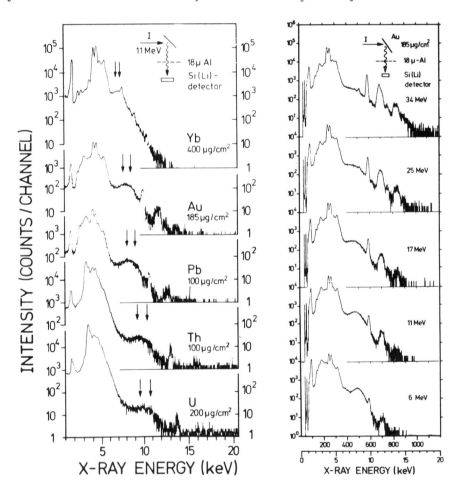

Fig. 1:   X-ray spectra for 11 MeV I bombardment on thin Yb,
Au, Pb, Th, and U-targets. The arrows give the Mα and Mβ
transition energies for the corresponding superheavy atoms.

Fig. 2:   X-ray spectra for 6 to 34 MeV I→Au collisions.

radiation. But, such experiments on quasimolecular K radiation from very heavy collision systems suffer appreciably
from extremely small cross sections, cf. Ref. 12. Moreover,
the corresponding x-ray spectra are steeply descending continua without endpoints. Hence, from an experimental point of
view quasimolecular M radiation from very heavy collision systems seemed to be more favourable.

In Fig. 1 x-ray spectra for 11 MeV I bombardment on thin
Yb, Au, Pb, Th, and U targets are shown. The broad peaks found
in the region below the arrows are of quasimolecular origin,
cf. Refs. 3 and 13. The arrows indicate the $4f(7/2,5/2) \rightarrow 3d5/2$
and $4f5/2 \rightarrow 3d3/2$ transition energies for a corresponding superheavy atom[14] with an atomic number $Z_1+Z_2$ = 123,132,135,143, and
145, respectively. This surprising good accordance of spectral
peak positions with the corresponding superheavy M-shell transitions led to speculations on even quasiatomic M radiation.[3,9]
It was extensively checked that the peak-like structure is real
and not caused by experimental effects as e.g. by absorber
effects.[3,15] The peak width increases about linearly with collision velocity; at higher impact energy the peak vanishes
within the underlying continuum, see Fig. 2. The x-ray production cross sections are extremely large for the quasimolecular
radiation: The x-ray production cross sections are given in
Table 1. Recently, the threshold behaviour was measured for
Xe-Au collisions.[16]

Within the electron promotion model[17] it seemed difficult
to understand the peak-like structure and the large cross sections for quasimolecular radiation. To overcome these difficulties it was primarily speculated on an induced emission mechanism.[9] Such a mechanism acting only at small internuclear distances, can qualitatively explain both points, cf. Ref. 18 and
3. To give a quantitative explanation both a complete theory
on an induced emission mechanism and realistic level calculations are necessary. A theory on rotationally induced transitions

| bombarding energy (MeV) | 6 | 11 | 17 | 25 | 57 |
|---|---|---|---|---|---|
| I L x rays | $1.2 \cdot 10^3$ | $3.5 \cdot 10^3$ | $1.0 \cdot 10^4$ | $2.5 \cdot 10^4$ | $9.5 \cdot 10^4$ |
| MO radiation | 9.7 | $2.7 \cdot 10$ | $7.1 \cdot 10$ | $1.2 \cdot 10^2$ | – |
| Au L x rays | 0.13 | 2.1 | $1.5 \cdot 10$ | $5.9 \cdot 10$ | $8.0 \cdot 10^2$ |

Table 1:   x-ray production cross section in barn for I-Au
collisions. The MO cross section concerns the x-ray energy
interval 6.5 to 9.2 keV.

was developed by B. Müller et al.[19] A relativistic many electron correlation diagram for the I→Au system was given by Fricke et al.[20] and is shown in Fig. 3. The quasimolecular radiation is assumed to be mainly caused by a vacancy carried in a second collision into the 3d 3/2 3/2 level. This is the highest dashed level in Fig. 3. This level shows for small internuclear distances a flat and an extremely wide minimum. For the proposed double collision mechanism this fact alone can possibly explain both points, cf. Refs. 3 and 20.

## EMISSION PATTERN AND ANISOTROPY

For I-Au collisions it is not quite clear what effect contributes mainly to the peak-like structure and large cross sections for the MO radiation. But, if an induced emission does exist there should be a non-isotropic emission because the emitting two center system may be allligned during the collision. For instance, for the rotationally induced emission mechanism an appreciable dipole contribution is calculated.[19] On the other hand, alone an unequal occupation of magnetic substates during collision may also give a non-isotropic spontaneous radiation, cf. Ref. 21. To study this effect is one reason to measure emission patterns and anisotropy distributions differential in x-ray energy. Also total cross sections will be influenced by a non-isotropic emission. Moreover, comparing spec-

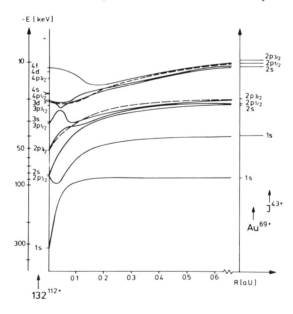

Fig. 3:   Relativistic multielectron level diagram from Ref. 20 for the I-Au system.

tra taken at different emission angles Doppler shift corrections
must be taken into account. From such shifts the velocity of the
emitting system, $v_E$, can be extracted. For the non-characteristic x rays the emitter velocity, $v_E$, will be a sensitive test
for the existence of a quasimolecule and, hence, a proof for
the existence of quasimolecular radiation. This is the other
reason for anisotropy measurements.

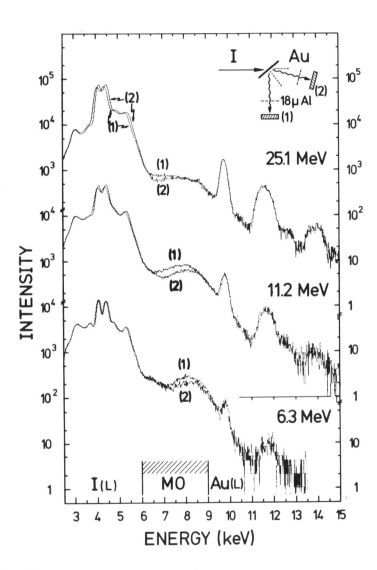

Fig. 4:    X-ray spectra for 6.3, 11.2, and 25.1 MeV I→Au
collisions. For each impact energy two spectra are compared
taken at 90° (1) and 15° (2).

We have done such investigations for the I-Au collision
system. The experiments were performed at the FN tandem van de
Graaff accelerator at the University of Köln. 6 to 45 MeV well-
-collimated I beams hit a 250-μg/cm² thin and selfsupporting
Au target. The x rays emitted were detected by a Si(Li) detector (80 mm², 220 eV FWHM at 5.9 keV, 1/2 mil Be window) mounted
on a turntable. X-ray detection angles θ between ± 15° and
± 146° were feasible. The band-like target chamber window was
covered by a vacuum tight 12 μm HOSTAPHAN foil. A simple x-ray
collimator on the detector nose defined a total angular resolution of Δ θ = 3.5° and prevented the detector from seeing non-
-target spot x rays. Also the target could be rotated around
the table axis in order to adjust it for two different angles θ
on equal target self absorption. To reduce electronic pile-up
effects an 18-μm thick Al-foil was inserted between target
chamber and detector.

For monitoring the primary beam intensity two methods were
used: (i) scattered I ions were registered by surface barrier
detectors and (ii) the x rays were counted by a fix-positioned
proportional counter. Both methods aggree within uncertainties
smaller than 3%. The whole experiment set-up was carefully adjusted to ensure that beam axis, and detection axis intersect
for all positions in one point of the target foil. This adjustment was experimentally tested by measuring right-left and forward-backward asymmetries using the characteristic lines emitted
during I-Au collisions. Taking all these points into account the
errors for anisotropies measured with this arrangement are certainly smaller than 4% including the normalization uncertainties.

In Fig. 4 three pairs of normalized spectra taken at 90°
and 15° are shown for 6.3, 11.2, and 25.1 MeV I impact, respectively. For corresponding spectra the target self absorption
is equal . The spectra show a pronounced anisotropy for the
quasimolecular radiation. The shape as well as the amplitude of
the anisotropy $I(90°)/I(15°)$ vary appreciably with impact energy. In contrast, the characteristic lines do not show a remarkable anisotropy. The small difference for the I L lines clearly
observable at the highest impact energy is due to the Doppler
shift caused by the flying I ions.

This Doppler shift can be determined more accurately by
comparing spectra taken at corresponding forward and backward
emission angles with the target surface perpendicular to the
beam axis. Assuming an x-ray emission pattern symmetric on 90°
we can extract the velocity of the x-ray emitter, $v_E$, from such
spectra. (This assumption may not be quite correct for the electron bremsstrahlung which is negligible in our case). The
emitter velocity found for various x-ray regions are given in

Table 2 for 25 MeV I impact. For various regions we find different $v_E$. Fitting the centroids of the characteristic x-ray
lines, the I L peaks show a shift corresponding to the full velocity of the bombarding I ions, $v_I$, while the Au-L peaks exhibit only a very small shift. Both facts are quite reasonable:
Considering the negligible energy loss in the thin Au target
the I ions must indeed show the full bombarding velocity throughout the target, whereas the emitting Au atoms should not move,
neglecting small recoil effects.

For steep continua, as e.g. the high energy part of the MO-
radiation the centroid method is not applicable. To determine
in such cases $v_E$ an exponential curve is fitted through the
data points. A Doppler velocity is defined by the shift of the
intersection point of the fitted curve with an appropriately
chosen intensity level. This shift depends on normalization,
energy interval, and detector solid angle correction which were
iteratively taken into account. The applicability of this method
is demonstrated for the characteristic Au-L radiation. The
emitter velocities extracted alone from the left or the right
wings of the lines aggree fairly well with $v_E$ determined by the
centroid method. Disregarding the small difference in right and
left values the results demonstrate that the applied method can
extract emitter velocities from not too flat continua. For the
MO-radiation we obtain from the high-energy wing an emitter velocity coinciding with the center of mass velocity, $v_E = v_{CM}$.

| x-rays | | region (keV) | $v_E/c$, exp. (%) | $v_E/c$, calc. (%) |
|---|---|---|---|---|
| I L α | , c. | ∿ 4.1 | 2.07 ± .09 | } 2.05 |
| I L β | , c. | ∿ 4.0 | 2.09 ± .09 | |
| I L + REC | r.w. | 5.6 – 6.1 | 1.70 ± .09 | —— |
| MO | , r.w. | 8.5 – 9.3 | 0.90 ± .15 | 0.81 |
| Au L α | , c. | 9.8 | 0.13 ± .02 | |
| | l.w. | 9.5 – 9.7 | 0.09 ± .02 | |
| | r.w. | 9.9 – 10.1 | 0.14 ± .02 | } ∿ 0 |
| Au L β | , c. | 11.6 | 0.13 ± .03 | |
| | l.w. | 11.1 – 11.3 | 0.08 ± .05 | |
| | r.w. | 11.9 – 12.2 | 0.14 ± .06 | |

Table 2:   Emitter velocities, $v_E$, for different x-ray regions
(c., centroid; r.w., right wing; l.w., left wing).

A double collision mechanism is assumed as the dominating mechanism for the production of MO-radiation:[9] The first collision gives a mean deflection of less than about 3° for the I ions, cf. Ref. 22, and an insignificantly changed I energy. Hence, in a second collision we will start with I ions having the full bombarding velocity directed within a narrow cone in beam direction. It is the same situation we have for a single collision mechanism. That means, independent on a single or double collision mechanism the result $v_E$ (MO) = $v_{CM}$ strongly supports the existence of both a quasimolecule and quasimole-

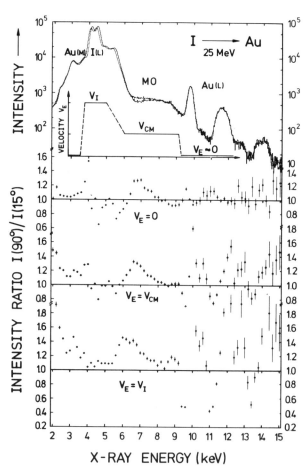

Fig. 5:   Anisotropy $I(90°)/I(15°)$ as a function of x-ray energy at 25 MeV impact energy. The anisotropies given are correlated on Doppler effects using emitter velocities, $v_E$, corresponding to 0, to the center of mass velocity, $v_{CM}$, and to the velocity of the I ions, $v_I$. Below the spectra at the top the different actual emitter velocities are indicated for the various x-ray regions.

cular radiation. Moreover, this result excludes several other
contributions to the 8 keV continuum, at least at the high-
-energy side. For instance, an appreciable admixture of characteristic line tails or REC to the MO-radiation should give
an emitter velocity between $v_I$ and $v_{CM}$. Such a $v_E$ inbetween is
indeed found for the steeply descending continuum around 6 keV,
where we expect such contributions.

Knowing now the emitter velocities from an experiment we
can evaluate the intrinsic anisotropies for the various parts
of the x-ray spectra. As example at 25 MeV I impact the anisotropy $I(90^\circ)/I(15^\circ)$ for the total spectrum is given in Fig. 5
using $v_E = 0$, $v_{CM}$, and $v_I$, respectively. Below the spectra in
top in Fig. 5 the anticipated $v_E$ is indicated for the various
x-ray regions. For the characteristic Au radiation we get
$v_E \sim 0$. Using $v_E = 0$ we find no significant anisotropy for the
Au L radiation. For the Au M radiation a small anisotropy (3%)
seems to be present, but the tremendously large absorber correction may influence this point. For the characteristic I L radiation we got $v_E = v_I$. Also in this case an isotropic emission
is found for the main lines of the characteristic radiation
within 3%. For one collision system a significant anisotropy
was reported for the characteristic radiation, but in this case
the adiabaticity condition is not fulfilled.[23] Hence, it may
be justified to assume isotropic emission for the characteristic
radiation in the quasiadiabatic collision region. Nevertheless,
this point should be investigated more thoroughly.

For the MO-radiation we got $v_E = v_{CM}$. Here, a pronounced
anisotropy is found varying significantly with x-ray energy.
Shape and amplitude of this differential anisotropy distribution
depends appreciably on $v_E$ used. For $v_E = v_{CM}$ we find a distribution with a maximum vanishing on both sides of the MO-band.

Using the correct emitter velocity $v_E = v_{CM}$ the differential
anisotropy distribution was determined for the quasimolecular radiation at impact energies between 6 and 45 MeV, see Fig. 6. Shape
and magnitude of this differential anisotropy vary considerably
with impact energy. For 11 MeV impact energy the anisotropy was
already measured for other detection angles $\theta$.[24,13,3] In Fig. 7
$I(\theta)/I(15^\circ)$ is given for $\theta = 30^\circ$, $45^\circ$, $75^\circ$, and $105^\circ$, respectively, using $v_E = v_{CM}$ for evaluation of the data. The curves
plotted should only guide the eye. The shape of the curve does
not change with $\theta$, only the amplitude varies roughly according
to $a+b\cdot\sin^2\theta$.[3,13] For the MO-radiation integrated over the range
between 7.1 and 9.5 keV this variation is demonstrated in Fig. 8.
A $\sin^2\theta$-function is fitted through the data points giving an integral dipole contribution around 20%.[24]

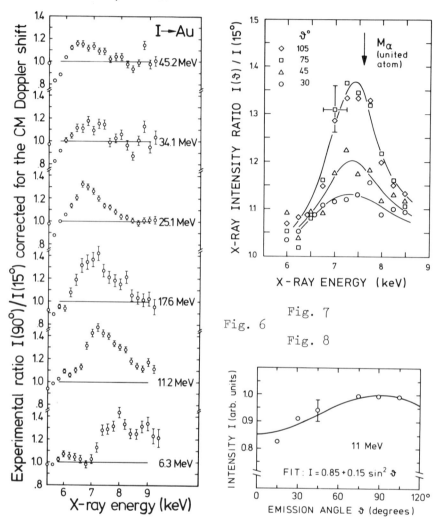

Fig. 6

Fig. 7

Fig. 8

Fig. 6:   Differential anisotropy distribution $I(90°)/I(15°)$ for the MO-radiation between 5.3 and 9.3 keV for impact energies between 6 and 45 MeV. The data are Doppler corrected for the center of mass velocity.

Fig. 7:   Differential anisotropy distribution $I(\theta)/I(15°)$ for the MO-radiation at 11 MeV impact energy. $I(\theta)/I(15°)$ is given for detection angles $\theta = 30°$, $45°$, $75°$, and $105°$. The data are Doppler corrected for the center of mass velocity.

Fig. 8:   Angular distribution for the MO x rays between 7.1 and 9.5 keV at 11 MeV I impact on Au. The data are corrected for the center of mass velocity.

Summarizing we may state for the I–Au collision molecule:

(i) The velocity for the emitter of the quasimolecular radiation coincides with the center of mass velocity.

(ii) The quasimolecular radiation is emitted non-isotropically with an appreciable dipole contribution prefering directions perpendicular to the beam.

(iii) The differential anisotropy distribution displays a maximum and vanishes approximately on both sides of the MO-radiation band.

(iv) The maximum reaches its highest value (∿45%) at 11 MeV impact energy. The maximum value of the differential anisotropy distribution decreases with increasing impact energy, see Fig. 9.

(v) The widths at half maximum of the differential anisotropy distributions do certainly not increase with increasing impact energy; the distributions are slightly asymmetric, cf. Fig. 6 and 10.

(vi) The position of the maximum in the anisotropy distribution shifts to lower x-ray energies with increasing impact energies reaching an asymptotic value at high impact energies, cf. Fig. 6.

Point (vi) is demonstrated once more in Fig. 10: The (negative) x-ray energy position of the maximum in the differential anisotropy distribution is plotted as a function of the reciprocal impact energy, i.e., as a function of the distance of closest approach in a head on collision. The bars in Fig. 10 are no error bars; they indicate the width at half maximum of the differential anisotropy distribution, cf. point (v). The representation chosen for Fig. 10 equals more or less the

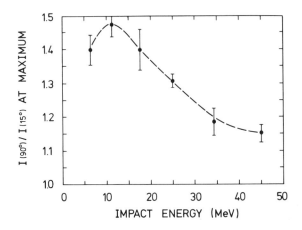

Fig. 9:    The maximum value of the differential anisotropy distribution as a function of impact energy for the quasi-molecular radiation form I→Au collisions.

usual representation of MO-levels: binding energy as a function
of internuclear distance. Hence, we may compare the resulting
curve with the 3d 3/2 3/2 level from the calculated level diagram[20] of Fig. 3, neglecting the initial binding energy of
the electron filling up the vacancy in the quasimolecule. This
comparison can only be of qualitative nature because only twenty
electrons were used for the level calculations, and because the
calculation procedure may be questionable for the few points at
small internuclear distance below 0.1 a.u. On the other side,
no weighing on internuclear distances is introduced for the experimental results. Nevertheless, the same trens of both curves
may indicate a close correspondence between peak positions in
the differential anisotropy distribution and transition energies at about the closest internuclear distances. For the rotationally induced emission mechanism, for instance, such a
behaviour is expected.[19] This and the other points will be discussed with the results on quasimolecular K and L radiation.


## RESULTS ON QUASIMOLECULAR K AND L RADIATION

Anisotropy measurements on quasimolecular L radiation are
done only recently. Wölfli et al. at Zürich investigated

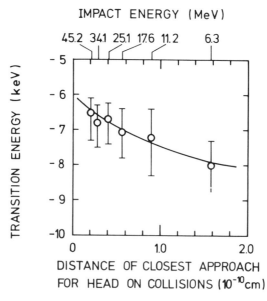

Fig. 10:  X-ray energy at the maximum of the differential
anisotropy distribution as a function of the distance of
closest approach in head on collisions. The bars indicate the
width of the distribution at half maximum for the quasimolecular (I+Au) radiation.

I→I, I→Ag, and Ag→Ag collision molecules for bombarding energies between impact energies of about 25 and 45 MeV.[25] Our
measurements on the I-Ag system aggree with the data of Wölfli
et al. at least at 17 MeV impact energy. The differential
anisotropy distributions of Wölfli et al. display up to three
maxima. The maxima positions seem to coincide with normal L
shell and L subshell transition energies each for the united
atom system. No shift of the maxima with impact energy is observed, because at the used impact energies, always extremely
small internuclear distances are feasible (asymptotic or run
way region). Also the widths of the distributions seem not to
depend significantly on impact energy.

For quasimolecular K radiation more collision systems are
investigated for their emission characteristic. The heaviest
system studied is the Nb-Nb system recently investigated at
Dubna.[26] The Kr-Zr system was studied at Stanford.[27,28]
The Ni-Ni collision molecule was investigated in Yale[29] and
more recently also in Dubna.[26] For all these heavy systems
differential anisotropy distributions are reported. All distributions display a maximum around the united atom $K_\alpha$ transition energy, neglecting the Nb-Nb case for background problems
and the low energy Ni-Ni result for its bad statistics. For the
Ni-Ni case the dependence on impact energy was measured. The
position of the maximum seems not to shift significantly with
impact energy (run way effect). Moreover, the width at half
maximum of this differential anisotropy distribution shows
only a very weak dependence on impact energy. For the Kr-Zr
system the emitter velocity was determined. It coincides with
the center of mass velocity indicating the very existence of a
quasimolecule and of quasimolecular K radiation.

For lighter systems quasimolecular K radiation was extensively studied by the Oak Ridge group.[21,30] They investigated
Si-Al, Al-Al, Al-C, O-C, and C-C collision systems at fairly
high impact energies ($v/u \lesssim 1$).  But only for Al-Al and partly
for Si-Al and Al-C the whole differential anisotropy distributions are communicated. For the Al-Al system also angular distributions are reported. For these light collision systems the
communicated differential anisotropy distributions do not show a
maximum at the $K_\alpha$ line of the united atom system. This fact may
possibly be caused by REC contributions dominating the spectra
at the high collision energies used.[31] An REC peak dominates
also the differential anisotropy distribution. Hence, the results from these light systems, especially from Si-Al and Al-Al
collisions, may not be representative for quasimolecular
radiation alone.

A brief summary of all the work done is given in Table 3.
Assuming only dipole contributions for the emission patterns of

Table 3:   Anisotropy studies done on quasimolecular radiation. (DAD, differential anisotropy distribution; $E_{imp}$, impact energy; $E_{x.max}$, x-ray energy at the postition of the maximum in DAD; x, yes)

| quasimolecular radiation | M | L | | | K | | | | | | | |
|---|---|---|---|---|---|---|---|---|---|---|---|---|
| collision molecule | I-Au | I-I | I-Ag | Ag-Ag | Nb-Nb | Kr-Zr | Ni-Ni | Si-Al | Al-Al | Al-C | O-C | C-C |
| (i) emitter velocity ($v_{CM}$, center of mass velocity) | $v_{CM}$ | — | — | — | — | $v_{CM}$ | — | — | — | — | — | — |
| (ii) emission pattern (d, dipole contribution) | d | — | — | — | — | — | — | d | — | — | — | — |
| (iii) DAD (m, maximum; a, asymptotic value) | m | m | — | m | ? | m | m | a | ? | a | ? | — |
| (iv) DAD maximum measured as function on $E_{imp}$ | x | x | x | — | — | x | x | x | x | — | x | x |
| (v) width of DAD (c, constant on $E_{imp}$) | ~c | ~c | ~c | — | — | — | ~c | REC | ? | — | — | — |
| (vi) $E_{x.max}$ correlated to close distances | x | x | x | x | ? | x | x | ? | — | — | x | x |

all investigated quasimolecular systems, we can plot all data
together in one diagram. In Fig. 11 $\beta$ coefficients are plotted
as a function of a generalized energy parameter $\eta$. $\eta$ is defined
by the ratio of the impact energy per nucleon ($E_{imp}/A_1$) over
the binding energy of concern in the united atom system ($E_{B u.a.}$)
multiplied by the electron-proton mass ration ($m_e/m_p$). Hence,
$\eta$ is the adiabaticity parameter squared $(v/u)^2$, u being the
orbital velocity of an electron in the corresponding united
atom shell. The coefficient $\beta$ describes the dipole contribution
for the emission assuming I $(\theta) = \alpha+\beta \sin^2\theta$, with $\alpha+\beta = 1$. I.e.,
the intensity distribution is normalized on $90^{\circ}$, $I(90^{\circ}) = 1$.
The $\beta$ values given in the graph correspond normally to the
maxima of the differential anisotropy distributions. Errors are
not given in this graph, but must often be taken into account.
Neglecting the REC effected data from the light systems, especially from Al-Al and Si-Al collisions, a general tendency may
possibly be extracted from this graph: The anisotropy characterized by $\beta$ tends to be maximal for low impact energies $\eta$. For
less adiabatic collisions the $\beta$ values seem to decrease.


## DISCUSSIONS

Because, in general, the differential anisotropy distribution seems to be maximal for x-ray transition energies at
small internuclear distances the theoretical model[19] on a rotationally induced emission was favoured.[21,24,25,27,29] In

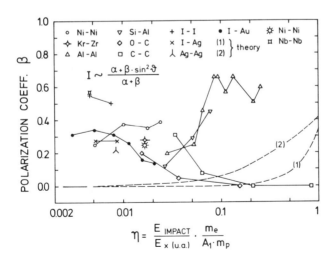

Fig. 11:    Anisotropy coefficient $\beta$ as a function on the gene-
ralized energy parameter $\eta$. ($\maltese$,$\divideontimes$ Ref. 26; $\Leftarrow$ Ref. 27; o Ref. 29; $\triangle$,
$\triangledown$,$\diamond$,$\square$,  Refs. 21 and 30; +, $\times$, $\lambda$ Ref. 25.)

this model the cross section ratio of induced over spontaneous emission was given by the squared ratio of transition frequency over rotation frequency of the internuclear axis, $(\omega_{rot}/\omega_{if})^2$. Applying a straight line path approximation and assuming all radiation as induced for $\eta = 1$ we get the dashed curve (1) in Fig. 10. This simple theory cannot at all explain the measured $\beta$ values. Later alignment was included in the theoretical calculations for the Ni-Ni case.[32,29] The dashed curve (2) in Fig. 10 corresponds to a more recent theory on rotationally induced emission.[33] In contrast to the first model[19] an explicit coupling between induced and spontaneous transitions is taken into account. Using similar simplifying approximations as above we get curve (2) for $2p\ 3/2\sigma \rightarrow 1s\ 1/2\sigma$ transitions. Alignment may alter the behaviour of curve (2) but the tendency may further hold true that the rotationally induced emission predicts only small $\beta$ values for small energy parameters $\eta$. Hence, rotationally induced emission may not explain the high anisotropy values at low impact energies. Also the more or less independence of the width of the differential anisotropy distribution on impact energy may be difficult to explain by models on induced emission demanding collision broadening.[19] Hence, other effects must contribute considerably to the observed anisotropies and their general features found at small $\eta$.

An unequal occupation of magnetic substates filling up the vacancy at close distance alone may often explain the observed anisotropies, cf. Ref. 21. For instance, applying an extreme population distribution $\beta$ values are possible twice as high as the ones found for the I-Au quasimolecule. With increasing $\eta$ the adiabaticity condition for the initial electron states get worser and worser and the MO model will not hold true further for these upper levels causing a more statistical occupation of magnetic substates. Such a mechanism can in principle explain the decrease in $\beta$ values observed with increasing $\eta$. Moreover, for symmetric collision systems also recoiling collisions may flatten out the intrinsic anisotropies at high collision velocities.[7] Concerning the light collision systems further arguments were given for the anisotropy decrease in C-C collisions.[21] On the other hand, the anisotropy increase for Al-Al collisions is consistently explained within a model proposed by Betz et al. including REC and induced emission.[31]

Considering all these facts, an induced emission mechanism should be studied at high $\eta$ whereas alignment effects seem to prevail at small $\eta$. Here, at small $\eta$, it should thoroughly be studied whether the general correspondence found between position of the maxima in the differential anisotropy distribution and x-ray transition energy at close distances is real, cf. Fig. 11 and point (vi) in Table 3. And questions arize:

Is the alignment alone responsible for this correspondence? What is the true mechanism causing this alignment at small internuclear distances and small η? What causes the rough independence of the width of the differential anisotropy distribution on impact energy, see point (v) Table 3? The results already available for the different quasimolecular shell transitions are very encouraging. Further work is necessary, especially at real low impact energies where no "run way effect" exists. But for each system to be investigated emitter velocity, angular distribution, differential anisotropy distribution, and its dependence on impact energy should be measured. At GSI in Darmstadt we are starting such investigations at the 1.4 MeV/amu point of the Unilac accelerator. A few runs have been made with heavy ion beams (Ar and Xe) of energies from 0.2 to 0.4 MeV/amu and in Fig. 12 is shown a preliminary spectrum with Xe on a Hg gas target testing the double collision model for producing the MO radiation, cf. also Ref. 16.

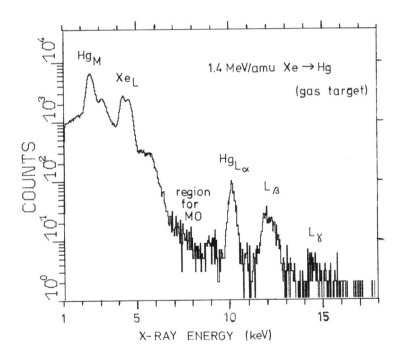

Fig. 12:   An x-ray spectrum from one of the recent first runs with the Unilac accelerator at GSI in Darmstadt, taken at the 1.4 MeV/amu point of the heavy-ion accelerator.

References

1) Q.C. Kessel, and B. Fastrup, Case Studies in Atomic Physics 3 (1973), 137; J.D. Garcia, R.J. Fortner, and T.M. Kavanagh, Rev.Mod.Phys. 45 (1973) 111.

2) F.W. Saris and F.J. de Heer, in Atomic Physics 4 (Plenum, New York, 1975).

3) P.H. Mokler, S. Hagmann, P. Armbruster, G. Kraft, H.J. Stein, K. Rashid, and B. Fricke, in Atomic Physics 4, p. 301 (Plenum, New York, 1975)

4) K. Alder, A. Bohr, T. Huus, B. Mottelson, and A. Winther, Rev.Mod.Phys. 28 (1956) 432.

5) E. Merzbacher and H.W. Lewis, in Handbuch der Physik 34 (1958) 166.

6) H.W. Schnopper, H.D. Betz, J.P. Delvaille, K. Kalata, A.R. Sohval, K.W. Jones, and H.E. Wegener, Phys.Rev.Lett. 29 (1972) 898.

7) K. Taulbjerg and P. Sigmund, Phys.Rev. A5 (1972) 1285; K. Taulbjerg, B. Fastrup, and E. Laegsgaard, Phys.Rev. A8 (1973) 1814.

8) F.W. Saris, W.F. van der Weg, H. Tawara, and R. Laubert, Phys.Rev.Lett. 28 (1972) 717.

9) P.H. Mokler, H.J. Stein, and P. Armbruster, Phys.Rev. Lett. 29 (1972) 827.

10) J.R. Macdonald, M.D. Brown, and T. Chiao, Phys.Rev. Lett. 30 (1973) 471.

11) J. Rafelski, L.P. Fulcher, and W. Greiner, Phys.Rev. Lett. 27 (1971) 958.

12) S.S. Hanna and W.E. Meyerhof, in Stanford University Progress Report 3/ 1/ 73 - 2/ 28/ 74 (1974).

13) P. Armbruster, G. Kraft, P. Mokler, B. Fricke, and H.J. Stein, Physica Scripta 10A (1974) 175.

14) B. Fricke and G. Soff, Gesellschaft für Schwerionen-forschung Report No. GSI - T1 - 74 (1974).

15) F.C. Jundt, H. Kubo, and H.E. Gove, Phys.Rev. A10 (1974) 1053.

16) H.O. Lutz, W.R. McMurray, R. Pretorius, I.J. van Heerden, and R.J. van Reenen, to be published.

17) U. Fano and W. Lichten, Phys.Rev.Lett. 14 (1965) 627; W. Lichten, Phys.Rev. 164 (1967) 131; M. Barat and W. Lichten, Phys.Rev. A 6 (1972) 211; W. Lichten, in Atomic Physics 4 (Plenum, New York, 1975).

18) H.J. Stein, in report of Gesellschaft für Schwerionen-forschung Darmstadt, GSI 73 - 11 (1973) 106.

19) B. Müller, R.K. Smith, and W. Greiner, Phys.Lett. 49B (1974) 219.

20) B. Fricke, K. Rashid, P. Bertoncini, and A.C. Wahl, Phys.Rev.Lett. 34 (1975) 243.

21) R.S. Thoe, I.A. Sellin, M.D. Brown, J.P. Forester, P.M. Griffin, D.J. Pegg, and R.S. Peterson, Phys.Rev.Lett. 34 (1975) 64.

22) H.J. Stein, H.O. Lutz, P.H. Mokler, and P. Armbruster, Phys.Rev. A5 (1972) 2126.

23) E.H. Pederson, S.J. Czuchlewski, M.D. Brown, L.D. Ells-worth, and J.R. Macdonald, Phys.Rev. A11 (1975) 1267; S.J. Czuchlewski, L.D. Ellsworth, J.A. Guffey, E.H. Pedersen, E. Salzborn, and J.R. Macdonald, Phys.Lett. 51A (1975) 309.

24) G. Kraft, P.H. Mokler, and H.J. Stein, Phys.Rev.Lett. 33 (1974) 476; G. Kraft, H.J. Stein, P.H. Mokler, and S. Hagmann, verhandl. DPG (VI) 9 (1974) 77; P. Armbruster, in XII Int. Winter Meeting on Nuclear Physics, Villars, Jan. 1974.

25) W. Wölfli, Ch. Stoller, G. Bonani, M. Suter, and M. Stöckli, to be published; The I→Ag system was also investigated by the authors at 6 and 17 MeV impact energy.

26) W. Frank, P. Gippner et al. to be published.

27) W.E. Meyerhof, R. Anholt, F.S. Stephens, and R. Diamond, in Abstracts of Contr.Papers to IV Int. Conf. on Atomic Physics (1974) 633.

28) W.E. Meyerhof, T.K. Saylor, and R. Anholt, to be published; W.E. Meyerhof, T.K. Saylor, and R. Anholt, Contr. to IX ICPEAC, Seattle (1975) 308.

29) J.S. Greenberg, C.K. Davis, and P. Vincent, Phys.Rev. Lett. 33 (1974) 473; J.S. Greenberg, C.K. Davis, and P. Vincent, Contr. to IX ICPEAC, Seattle (1975) 312.

30) R.S. Thoe, I.A. Sellin, J. Forester, P.M. Griffin, K.H. Liao, D.J. Pegg, and R.S. Peterson, to be published; R.S. Thoe, I.A. Sellin, R.S. Peterson, D.J. Pegg, P.M. Griffin, and J.P. Forester, Contr. to IX ICPEAC, Seattle (1975) 312.

31) H.-D. Beth, F. Bell, H. Panke, W. Stehling, and E. Spindler, Phys.Rev.Lett. 34 (1975) 1256.

32) B. Müller and W. Greiner, Phys.Rev.Lett. 33 (1974) 469.

33) R.K. Smith and W. Greiner, to be published; stimulating discussions with R.K. Smith are gratefully acknowledged.

# RADIATIVE AND NONRADIATIVE ELECTRON CAPTURE FROM AND INTO OUTER AND INNER SHELLS IN HEAVY ION-ATOM COLLISIONS

H.-D. Betz, M. Kleber, E. Spindler, F. Bell
H. Panke, and W. Stehling
Universität München, 8046 Garching, Germany

## SUMMARY

Nonradiative electron capture processes are discussed, based on the first Born approximation and adopted for use in heavy ion collisions. Competition of electron capture with direct target ionization and with other decays of projectile vacancies is indicated. Radiative electron capture (REC) is described theoretically; experimental REC spectra are shown along with calculated spectra. Discrepancies between experiment and theory are discussed and we comment on the possibility to use REC measurements to infer momentum distributions of target electrons.

## INTRODUCTION

Electron capture is an important process in heavy ion-atom collisions. Nonradiative electron capture occurs with large probability and cross sections may reach and exceed values of $\sim 10^{-15} cm^2$ and are, thus, often larger than the geometrical cross sections of projectile and target. Radiative electron capture (REC) is a relatively rare process with cross sections generally below $\sim 10^{-21} cm^2$. Only at very large collision velocities is it possible that REC exceeds nonradiative capture, but in this range both cross sections become very small. In the case of collisions between protons or hydrogen-like ions with atomic hydrogen the problem of calculating

cross sections is essentially under control for both
nonradiative and radiative capture processes.  For
heavy systems, however, considerable difficulties
arrise and there have been relatively few attempts
to cope with the problems.  In the past, the amount
of experimental data on electron capture in heavy
ion collisions remained limited; in recent years, an
increasing number of heavy ion accelerators became
available for atomic physics studies involving rearrangement processes in outer and inner shells,
and there is clearly urgent need to understand
quantitatively what is being observed in experiments.
In addition, progress in atomic structure calculations and availability of fast computers give a much
more favorable basis for extended theoretical work.
In this report, we discuss some basic features of
electron capture in fast heavy ion collisions and
we present experimental and theoretical results.
Nonradiative capture is treated briefly, and attention is given to REC processes.  Extensive and detailed descriptions on experiment and theory of REC will
be reviewed elsewhere[1] .

Significance of capture processes is manifold.
For example, when collision-induced inner-shell
ionization is studied, electron capture contributes
to electron rearrangement in projectile shells mostly before projectile x-rays are emitted, capture
into inner projectile shells destroys vacancies
produced in prior collisions and reduces the chance
for x-ray and Auger decays, capture from inner
target shells competes with other ionization mechanisms, and REC leads to contributions in x-ray spectra
which can be very intensive and may interfere with
other radiation features to be studied.

## NONRADIATIVE ELECTRON CAPTURE

As early as in 1930, Brinkman  and Kramers
calculated capture cross sections with an approximate
first Born approximation.  Capture proceeds due to
the attractive interaction between projectile nucleus
and target electron.  Brinkman  and Kramers' result
can be written in the following form:

$$\sigma_c = Aq^2 E_K^{4} E_i^{5/2} E_f^{3/2} \left[ E_K^{2} + 2E_K(E_i + E_f) + (E_i - E_f)^2 \right]^{-5},$$

$$(1)$$

where $A = \pi 2^{16} e^4/5 = 8.53 \times 10^{-16}$ keV$^2$cm$^2$, q is the ionic charge of the projectile, $E_K$ denotes the kinetic energy of an electron travelling with projectile velocity, and $E_i$ and $E_f$ stand for the binding energy of the electron before and after capture, respectively. In Eq.(1), all energies are to be inserted in units of keV. We note that in contrast to assertions in ref.2, Eq.(1) describes the capture cross section averaged over all initial hydrogen-like states within the principal target shell and summed over all hydrogen-like states in the principal projectile shell. In particular, contributions of capture into subshells with $\ell > 0$ are included.

When one takes into account the repulsion between projectile- and target nuclei capture cross sections will be reduced; in the range where electron velocities $v_e$ and collision velocity v are not too different, this reduction can be large and may reach a factor of $F \approx 10$. Jackson and Schiff[3], for example, treated this effect and derived an analytical expression for F. In addition, they calculated absolute contributions of capture into 2s and 2p states. Valuable calculations on those lines were also given by Mapleton[4] and summaries can be found in the work of McDowell and Coleman[5]. More recent references, and an evaluation of F for the limiting case of very high velocities are evident from ref.6.

For several reasons theoretical results such as Eq.(1) can not always be directly applied to heavy-ion encounters:

(1)   when the final principal shell is not completely empty it becomes necessary to calculate capture contributions into subshells.
(2)   A similar problem arrises for partially filled initial principal shells.
(3)   Inner target electrons to be captured do not always see the full projectile ionic charge because of screening effects of outer target electrons.
(4)   Inclusion of particle-nucleus interaction (factor F) becomes more involved.

In principle, these problems are well defined and can be solved, but considerable computational effort is required and it is probably not possible to derive simple analytical expressions. Moreover, points (3) and (4) require formulation of impact parameter

dependence of the capture process.  Let us briefly
discuss point (1).  Calculations for capture of 1s
electrons of atomic hydrogen into 3s, 3p, and 3d
proton subshells give values[4]  the ratio of which
varies from 41:52:7 at $v/v_e$ = 4.8 to 86:13:0.3 at
$v/v_e$ = 15 (for F=1).  Capture probabilities depend
on the overlap of electronic wavefunctions in
momentum space; since these functions drop more
rapidly with increasing $\ell$ it is evident that capture
into higher subshells becomes less likely especially for large $v/v_e$.  For $v/v_e \gg 1$ capture into ns
states dominates and follows the well-known $n^{-3}$ rule.
In heavy ion collisions, average ionization is
generally such that $v/v_e \simeq 1$ for the outermost projectile electrons, i.e. one is in a range where capture
contributions into s and p subshells of a principal
shell are roughly of comparable magnitude and where
capture into states with $\ell > 1$ seems to be relatively
small.

     In the following we illustrate quantitatively
some of the effects of capture processes mentioned
in the introduction.  Let us consider 92-MeV sulfur
ions which penetrate through various thin, solid
targets.  At this energy, the average ionic charge
is $\bar{q} \simeq 13$[7]  and K vacancies are abundant so that we
may select 1s2s2p as one of the typical projectile
configurations[8] .  In a carbon target, Eq.(1) shows
that only C-1s electrons have to be taken into
account.  When we take F=1, capture cross sections
into sulfur K,L,M, and N shell then amount to 0.42,
44,82, and $50 \times 10^{-19}$ $cm^2$, respectively, and decrease
approximately with $n^{-3}$ for still higher n.  When the
target is aluminum, the dominant contributions come
again from Al-1s and amount to approximately 36,177,
96, and $35 \times 10^{-19}$ $cm^2$ for capture into S-K,L,M, and
N shell, respectively.  In copper, 1s-1s transitions
occur with $\sigma_c \simeq 6 \times 10^{-19}$ $cm^2$, and Cu-L electrons fill
S-K and higher shells with cross sections of $45 \times 10^{-19}$
$cm^2$ and $\sim 10^{-16}$ $cm^2$, respectively.

     It is illustrative to use the values from above
for the following comparisons.  Radiative- and Auger
decay of the S-K vacancy proceeds with a "cross
section" of $\sim 2.7 \times 10^{-19}$ $cm^2$ [8] .  As a consequence,
the S-K holes decay mainly by Auger processes in C,
but mainly by electron capture in Al and heavier
targets.  As regards ionization of target atoms we
note that for Cu the 1s-1s capture is almost an order
of magnitude more likely than direct Coulomb

ionization[9] , whereas the two processes are about equally probable for Cu-L ionization. Finally, it is important to point out that highly excited projectile states are formed in abundance.

The accuracy of our numerical estimates is a factor of $\sim 2/F$ at best, but could be improved by more involved calculations. Trends are nevertheless apparent and similar estimates are possible for other collision systems.

## RADIATIVE ELECTRON CAPTURE (REC)

The REC process has been known for a long time, mainly from work in astrophysics (radiative recombination), but its first identification in heavy ion-atom encounters occurred very recently[10] . REC of a free electron by a nucleus has been treated, for example, by Bethe and Salpeter[11] . In heavy-ion collisions, REC processes are much more intricate and a satisfactory description is not worked out easily. We will outline some of the attempts which have been undertaken on those lines, new experimental results are presented, and problems presently unresolved will be discussed.

The mechanism of REC processes is easily understood. As a result of a collision, a target electron is captured into a projectile vacancy whereby a photon is emitted with an average energy $E_R \simeq E_f + E_K$. Thus, the energy of REC x-rays exceeds the one of characteristic projectile x-rays associated with the same vacancy. In case of capture of an electron with an initially sharp velocity a sharp REC x-ray would result. Target electrons, however, whether quasi-free conduction electrons or bound electrons, exhibit a well-defined velocity distribution which is reflected in the emitted REC x-ray spectra. In principle, it is thus possible to infer momentum distributions, or Compton profiles, of target electrons from REC measurements. Unfortunately, such a desirable procedure is not yet feasible for reasons discussed below.

Based on the results in ref.11 the differential cross section for REC into a 1s orbit is simply given by

$$\frac{d^2\sigma}{dE_x d\Omega} = \int d^3\vec{p} \frac{3}{8\pi} \sigma_R \sin^2\Theta |\phi_i(\vec{p}-\vec{p}_o)|^2 \delta(E'_f - E'_i), \quad (2)$$

with

$$\sigma_R = 9.1(\frac{\eta^3}{1+\eta^2})^2 \frac{\exp(-4\eta \ \text{arctg} \ 1/\eta)}{1-\exp(-2\pi\eta)} 10^{-21} \text{cm}^2. (3)$$

Here, $\phi_i$ signifies the initial momentum distribution of the electron to be captured, the delta function ensures energy conservation, $\Theta$ is the angle between $\vec{p}$ and the direction of the emitted photon, $\eta = v_e/v$ is known as Sommerfeld parameter, and $\vec{p}_o = m\vec{v}_o$ is the momentum of a target electron at rest, relative to the projectile.

Using the impact parameter method, Kienle et al.[12] calculated $\sigma_R$ in the Born approximation and formulated energy conservation $\delta(E'_f - E'_i)$. Initial momentum distributions were represented by hydrogen-like wavefunctions, averaged over a principal shell. The problems with this approach are essentially twofold: firstly, the Born cross section is less accurate than $\sigma_{BS}$ Eq.(3) especially when $\eta$ is large compared to unity; secondly, hydrogen-like 1s funct-ions are not satisfactory for most target electrons and involve screening estimates which can not be easily found without ambiguity. Some further com-ments on this method and Eq.(2) are given in ref.13 and ref.1.

For the present report and the extensive review[1] we have evaluated Eq.(2) with correct wavefunctions both in the Born approximation and, more exactly, with Eq.(3). The procedure was as follows: a relat-ivistic Hartree-Fock program[14] has been used to ob-tain wavefunctions $\Psi(x)$. Special care was taken to make $\Psi$ very accurate near $x=0$. A Fourier transform program was developed to obtain from $\Psi(x)$ the moment-um density $|\phi(p)|^2$. Then, Eq.(2) was evaluated for $\mathcal{F} = 90°$, where $\mathcal{F}$ is the angle between beam direction and observer. This procedure is free of any adjust-able parameter except that $E_f$ must be known. Since heavy projectiles are usually not fully ionized, the binding $E_f$ of projectile 1s holes is affected by presence of other electrons and is best determined by measuring energetic positions of characteristic projectile K$\alpha$ lines which are to be compared with Hartree-Fock transition energies for various approp-riate configurations. Some resulting REC profiles

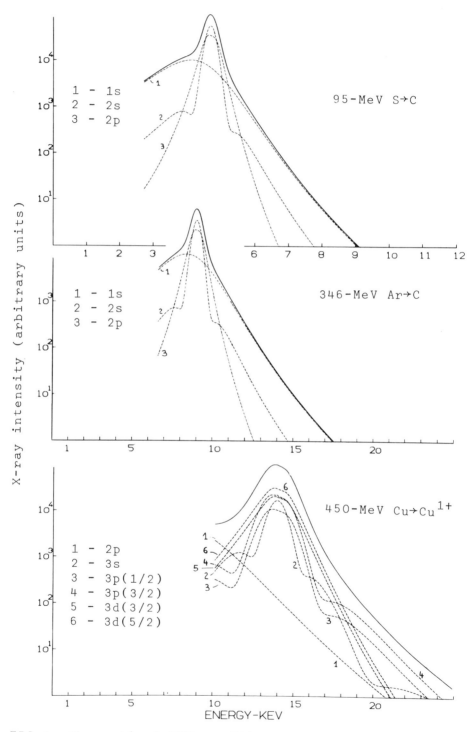

FIG.1. Theoretical REC profiles.

| $S$ MeV | in Carbon | | in Aluminum | | free |
|---|---|---|---|---|---|
|  | exact | Born | exact | Born |  |
| 8 | 735 | 92 | 982 | 189 | 368 |
| 16 | 408 | 111 | 678 | 212 | 191 |
| 32 | 226 | 122 | 489 | 235 | 93 |
| 55 | 140 | 117 | 275 | 226 | 52 |
| 95 | 81 | 100 | 157 | 194 | 28 |
| 110 | 66 | 90 | 130 | 179 | 23 |

Tab.1. Total REC cross sections $d\sigma/d\Omega$ at $90°$ for sulfur in C and Al, in barn. Results are given for the Born approximation, exact evaluation of Eq.(2), and in the free-electron approximation (see text).

are displayed in Fig.1, along with contributions of individual target electrons. Relative intensities are shown over 5 orders of magnitude, but the profile should not be taken too seriously for x-ray energies $E_x \gtrsim E_R + 2E_K$. In Tab.1 we list the integrated cross sections $d\sigma/d\Omega$ for $S = 90°$ obtained for collisions of 8- to 110- MeV sulfur on C and Al, using the procedures outlined above, along with the approximation $3\sigma_{BS}/8\pi$ for one free electron. It becomes obvious that the **free** approximation gives very reasonable total cross sections when scaled for the actual number of lightly bound target electrons.

Experimental REC spectra are shown in Fig.2 and correspond to the cases in Fig.1. Data with sulfur ions is from the tandem-accelerator laboratory in Munich[15], and the experiments with high-energy Ar and Cu ions were carried out at the ALICE accelerator in Orsay[16]. All spectra have been corrected for absorber effects and characteristic lines have been left out.

There is only partial agreement between experimental and theoretical REC profiles. The most obvious discrepancy concerns the peak widths which are experimentally wider than in theory. In view of very extensive further data and calculations not shown here, we indicate 3 possible reasons. Firstly, it appears that we have overestimated the contributions of the outermost target electrons; for example, if C-2p, Al-3s, and Cu-4s electrons are omitted from the REC summeration over target electrons, agreement

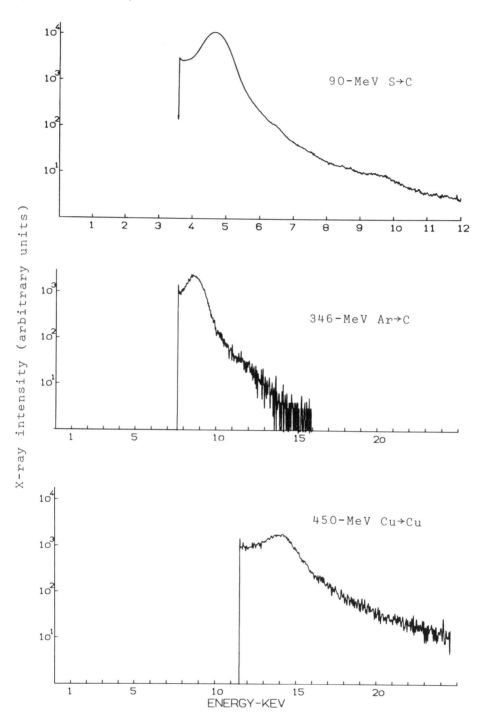

FIG.2. Experimental REC profiles.

becomes considerably better and is, in some cases, nearly perfect. Secondly, use of a sharp value of $E_f$ may not be justified since during capture the projectile vacancy may be screened by target electrons. This may cause a broadening of the REC peak which increases for heavier targets; the latter trend has indeed been observed[15]. Thirdly, a small broadening is caused by presence of different projectile states, i.e. in the experiment one averages over several discrete values of $E_f$. The tail on the high-energy side of the peak is reproduced quite well over at least one order of magnitude in intensity. Thus, there is definite hope that REC measurements can be used to infer momentum distributions of target electrons, but there are limitations and the questions outlined above must be clarified.

Finally, we note that a complete description of the REC process must include the impact parameter dependence; first attempts on those lines have been made[13]. Furthermore, angular distributions must be established. We point out that $\sin^2\theta \approx \sin^2\vartheta$ holds only in the vicinity of the REC peak and for angles which are not too far from $90°$; due to the vector addition $\vec{p}=\vec{p}_i+\vec{p}_o$, where $\vec{p}_i$ is the electron momentum, the angular distribution of REC may deviate significantly from $\sin^2\vartheta$ and does not decrease to zero at forward angles. Experimentally, such a trend has been well established[17]. It is also interesting to note that REC may also proceed to continuum states of the projectile and is then sort of Bremsstrahlung[18], and that capture of relatively fast electrons with $v_e \gg v$ leads to molecular-orbital x-rays. The latter effect, combined with dynamic collision broadening[19], becomes important for higher x-ray energies and dominates, for example, the spectra Fig.2 for the highest energies.

## REFERENCES

1. H.-D. Betz and M. Kleber, Radiation Effects (to be published in 1976).
2. H.C. Brinkman and H.A. Kramers, Proc. Acad. Sci. (Amsterdam) 33,973(1930).
3. J.D. Jackson and H. Schiff, Phys. Rev. 89,359(1953).
4. R.A. Mapleton, Phys. Rev. 126,1477(1962).
5. M.R. McDowell and J.P. Coleman, in "Introduction to the theory of ion-atom collisions", North-Holland, Amsterdam (1970), chapter 8.

6. M. Kleber and M.A. Nagarajan, J. Phys. B: Atom. Molec. Phys. $\underline{8}$,643(1975).

7. H.-D. Betz, Rev. Mod. Phys. $\underline{44}$,465(1972).

8. H.-D. Betz, F. Bell, H. Panke, G. Kalkoffen, M. Welz, and D. Evers, Phys. Rev. Lett. $\underline{33}$,807 (1974).

9. H.-D. Betz, F. Bell, and H. Panke, J. Phys. B$\underline{7}$, L418(1974).

10. H.W. Schnopper, H.-D. Betz, J.P. Delvaille, K. Kalata, A.R. Sohval, K.W. Jones, and H.E. Wegner, Phys. Rev. Lett. $\underline{29}$,898(1972); H.-D. Betz, in Proceedings of the Heavy-Ion Summer Study, Oak Ridge, Tennessee, 1972, ed. by S.T. Thornton, CONF-720669 (Oak Ridge National Laboratory, Oak Ridge, Tenn., 1972), p. 545 ff.

11. H.A. Bethe and E.E. Salpeter, Quantum Mechanics of One- and Two-Electron Atoms (Academic, New York, 1957), p. 320ff.

12. P. Kienle, M. Kleber, B. Povh, R.M. Diamond, F.S. Stephens, E. Grosse, M.R. Maier, and D. Proetel, Phys. Rev. Lett. $\underline{31}$,1099(1973).

13. M. Kleber and D. Jakubassa, Nucl. Phys. A (in print, July 1975).

14. J.P. Desclaux, private communication; for a more recent code see J.P. Desclaux, Comp. Phys. Comm. $\underline{9}$,31(1975).

15. H.-D. Betz et al. (to be published).

16. H.-D. Betz, C. Stéphan, B. Delaunay, E. Baron, E. Spindler, H. Panke, and F. Bell (to be published).

17. H.-D. Betz, F. Bell, H. Panke, and G. Kalkoffen, IV International Conference on Atomic Physics, Heidelberg 1974, ed. by J.Kowalski and H.G. Weber, p. 670ff.

18. H.W. Schnopper, J.P. Delvaille, K. Kalata, A.R. Sohval, M. Abdulwahab, K.W. Jones, and H.E. Wegner, Phys. Lett. $\underline{47A}$,61(1974).

19. H.-D. Betz, F. Bell, H. Panke, W. Stehling, E. Spindler, and M. Kleber, Phys. Rev. Lett. $\underline{34}$, 1256(1975).

# IMPORTANT PROBLEMS IN FUTURE HEAVY ION ATOMIC PHYSICS[*]

W. Betz, G. Heiligenthal, J. Reinhardt, R.K. Smith
and Walter Greiner[**]

Institut für Theoret. Physik der Johann Wolfgang
Goethe-Universität Frankfurt am Main, Germany.

## I.    INTRODUCTION

Fundamental problems are important problems, but not vice versa. The important problems can very often be found in the vicinity of fundamental questions; they are important to finally clarify the fundamental aspects.

In this sense we can see the following fundamental problems in heavy ion atomic physics:

1)  The verification of the predictions for quantum-electro-dynamics of strong fields; in particular the decay of the neutral electron-positron vacuum and the stability of the charged vacuum in overcritical electric fields.

2)  The possible spectroscopy of superheavy molecular (two-centre) orbitals.

3)  The atomic physics possible through naked (i.e. all electrons stripped off) nuclei.

4)  Atomic physics with heavy ions at ultra-high energies (1-2 GeV/Nucleon and more).

In our discussion of these problems we shall encounter a number of important effects which have to be studied. They are very often interesting in themselves and absolutely necessary to understand and appreciate the basic aspects.

*) Supported by BMFT and GSI
**) Invited speaker at the Int. Conf. on the Physics of Electronic and Atomic Collis., Seattle, July 1975.

II.   THE DECAY OF THE NEUTRAL VACUUM

The solution of the single-particle Dirac equation for electrons in Coulomb potentials of extended, non-compressed nuclei are meanwhile well known[1,2]. They are shown as a reminder in Fig. 1, where the "diving" of the 1s-level into the negative energy continuum appears at $Z_{critical} \approx 170$, that of the $2p_{1/2}$-level at $Z_{cr}^{(2)} \approx 184$ and that of the other bound levels correspondingly later. For point charges all the $j=1/2$ levels "dive" at $Z_{cr} = 1/\alpha = 137$, as shown in Fig. 2. The physical significance of this fundamental process becomes most transparent in Dirac's hole theory, where it is assumed that $E = -mc^2$ is the Fermi surface (all negative energy states filled up with electrons) and the corresponding state is called the vacuum $|0\rangle$. It is neutral and contains no currents, i.e. $\langle 0|j_\mu|0\rangle = 0$ for all $\mu = 1,2,3,4$, where $j_\mu$ denotes the current 4-vector.   Suppose now we are dealing with naked nuclei,

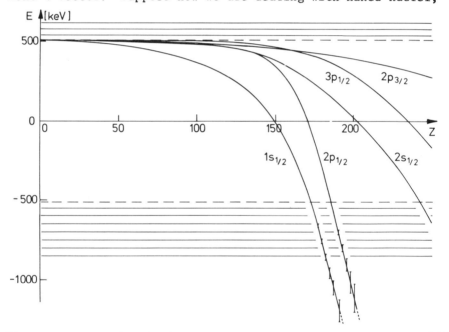

Fig. 1   Electronic binding energies for extended ($R=1.2$ fm·$A^{1/3}$) superheavy nuclei. Fermi charge distributions have been used. The diving points for the $1s_{1/2}$-, $2p_{1/2}$- and $2s_{1/2}$ levels are $Z_{cr} \approx 170$, 185 and 245 respectively. The influence of the other electrons are taken care of by a Thomas-Fermi distribution. The energy spreading of the states is shown by indicating the spreading width after the diving point (magnified by a factor 10). It is the same as the positron decay width of the bound state.

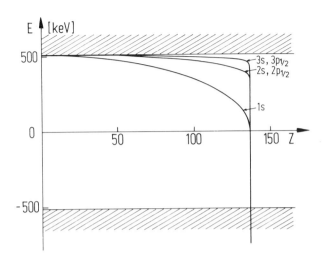

Fig. 2  Electronic binding energies for pointlike charges.
        All levels with $j = 1/2$ dive simultaneously at $Z = 137$.

i.e. all real electrons (that means those above the Fermi-level)
are stripped off, then at $Z = Z_{cr}$ an empty electron level dege-
nerates with the vacuum, energy-less electron-positron pair
creation becomes possible. We also say the vacuum autoionizes.
The created positrons, being in continuum states, move to infinity because they are repelled from the nuclei, while the
electrons are captured in bound states which now belong to the
vacuum: The new stable vacuum is charged by two electrons.
This process repeats itself at $Z_{cr}^{(2)} \approx 184$ where the $2p_{1/2}$-shell
dives, at $Z_{cr}^{(3)} \approx 240$ where the $2s_{1/2}$-shell dives, etc.
For pointlike (i.e. highly compressed) nuclei the diving of
the $n_{1/2}$ states appears perpendicular at $Z_{cr} = 137$. There the
vacuum becomes infinitely charged. This can – depending on
the selfshielding and on the compression energy of superheavy
nuclear system – yield isomers (stable charged vacua with highly
compressed kernels of nuclear matter) with effective charge
$Z - Z' = 137$. Here $Z$ is the proton number of the nucleus and
$Z'$ the charge of the vacuum.

    The fundamental significance of this effect may also be
revealed by pointing out, that the charge conjugation symmetry
for the vacuum state is simultaneously broken in overcritical
fields. This is illustrated in Fig. 3 by showing  a) the
undercritical and  b) the overcritical situation. The charge

534   W. BETZ, ET AL.

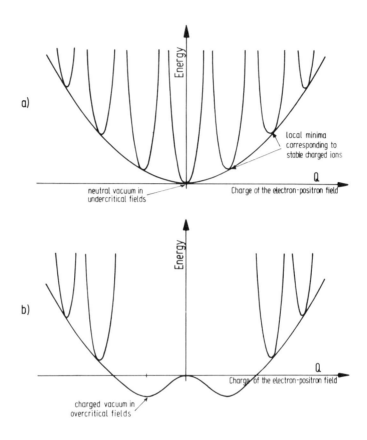

Fig. 3:   Schematic drawing of the energy of the system against
          charge, which is thought to be continous. The quan-
          tization of charge at integer values can be imaged
          to occur because of stabilization potentials.  In
          case  a) the vacuum (ground state) is symmetric under
          charge conjugation.  This symmetry is broken in over-
          critical fields (case b)) where the vacuum is charged.


is thought to be continous: The positive abzissa corresponds
to the charge of the electron-positron field of ordinary nuclei
built of protons and neutrons, while the negative abzissa
corresponds to that of antinuclei (built of antiprotons and
-neutrons). At the integer charge points we have stable atomic
states, which are schematically represented by potential minima as a function of the charge variable.  The vacuum, i.e.
the deepest of all these minima, is obviously charge-symmetric
in the undercritical case a) while it lost this symmetry in

the overcritical case b). This reflects the fact, that the
charge conjugation symmetry is spontaneously broken in overcritical fields. We mention briefly that this phenomenon of
changing vacuum state also occurs for other fields, which have
been discussed in literature after the overcritical electron-
positron vacuum: First, the pseudoscalar pion field may become overcritical; the changed vacuum is then called a pion-
condensate[3]. Second, the scalar σ-meson field may become overcritical; this has lead also to speculations on a type of
density isomers[4]. The basic physical processes in those cases
are the same as in overcritical quantum electrodynamics, but –
due to the strong interaction – less clean and different in
details.

### Experimental Possibilities

Overcritical fields can be generated in the collision of
very heavy nuclei, e.g. U-U. The basic processes in connection with the decay of the vacuum areillustrated in Fig. 4.
It shows schematically the superheavy molecular orbitals as
a function of the distance between the ions.

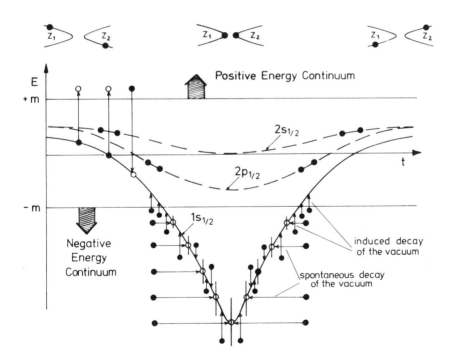

Fig. 4:   The dynamical process occurring in connection with
          the decay of the vacuum in overcritical fields.

At large distances of the approaching nuclei the Fourier-frequencies from the changing Coulomb field should give rise to ionization of the K-shells. Such K-vacancies can also be produced by a double collision process. At present it is unclear theoretically and experimentally, which of the two processes is contributing dominantly and how much. We assume here, because we do not know better at this time, that 1 K-hole will be created in hundred collisions. After the field has become overcritical at distances R < 36 fm in e.g. an U-U-collision, it will be populated by $e^+e^-$ -creations. The escaping positrons can then be observed experimentally as the indication for the decay of the vacuum. We shall study this process now more quantitatively.

In addition to the spontaneous positron creation (horizontal arrows in Fig. 4) there occur two effects due to the nonadiabaticity of the heavy ion collision: First, during the "diving"-process $e^+e^-$ -pairs are created in addition to the spontaneous (horizontal arrows) ones by induced autoionization (vertical small arrows). Second, before and after the diving a large number of positrons will be created by induced transitions (vertical small arrows) from the negative energy continuum to the $1s_{1/2}$-level. The latter transitions will occur even if there is no level-diving during the collision, but decreasing with smaller binding energy of the state.

The cross section for positron production is given by $d\sigma/dE_p d\Omega_{ion} = d\sigma_R/d\Omega_{ion} L_o W(E_p,\Theta)$ were $d\sigma_R$ is the differential Rutherford cross section, $W(E_p,\Theta)$ is the positron escape probability and $L_o$ the initial K-hole probability, which has been taken in all our calculations to be $L_o = 10^{-2}$ (see Refs. 5 and 6). This number has no justification. It is a simple guess. The present results of Meyerhof[7] and Saris[8] contradict this guess; they estimate from their data $L_o \approx 10^{-5}$. This in turn is in contradiction to results of Burch, Vandenbosch et al.[9], who estimate from their findings in Cl-U-U-collisions $L_o \approx 1/2$. Thus the situation is at present not quite clear, but exciting.

Fig. 5 shows the total ionization probability $W_T(E_p,\Theta)$. The full curves demonstrate that with decreasing ion energy $E_I$, the energies $\bar{E}_p$ for maximal positron cross sections are shifted to smaller values. This possibly allows to some extent a spectroscopy of the diving mechanism. For fixed ion energy and varying scattering angle $\Theta$, the energy maximum is only slightly shifted, as can be seen from the dashed curves. For different ion energies, $d\sigma/dE_p$ is shown for U-U in Fig. 6a). Compared with the purely spontaneous positron autoionization[10] the dynamical nonadiabatic effects lead to a "smearing out" of the sharply peaked excitation functions and

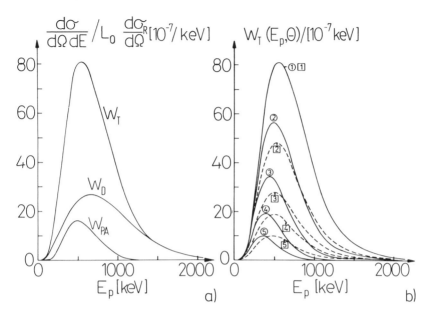

Fig. 5:  a) The probabilities $W_{PA}(E_p,\Theta)$, $W_D(E_p,\Theta)$ and $W_T(E_p,\Theta)$
for the system $U \to U$ in the case of central colli-
sion with $E_I$ = 812.5 MeV.

b) The full lines show $W_T(E_p,\Theta)$ at $\Theta = 180°$ for the
system $U \to U$ and its dependence on $R_{Min}$ (distance
of closest approach in fm), which corresponds to
different ion energies (denoted in the brackets,
units are MeV). (1) 15(815.5) (2) 20(609.4)
(3) 25(478.5) (4) 30(406.3) (5) 35(348.2).
In the last case $W_T(E_p,\Theta)\Theta = W_{PA}(E_p,\Theta)$ (no diving).
The dashed curves show for fixed ion energy
$E_I$ = 815.5 MeV the $\Theta$-dependence (1) $\Theta=180°$
(2) $\Theta=75°$ (3) $\Theta=50°$ (4) $\Theta=50°$ (5) $\Theta=30°$ .

to a considerable increase of the cross section by nearly two
orders of magnitude.

For the total positron cross section $\sigma$ one has to integrate
over the positron energy $E_p$. As a function of the ion energy
$E_I$, it is shown in Fig. 6b). It is considerably larger than
the corresponding one for purely spontaneous positron autoionization[10]. This is due to the induced transitions (non-
adiabatic effects). The vacuum thus decays - due to the nonadiabaticity introduced by the dynamics - as well spontaneously as by induction. In systems with higher total charge

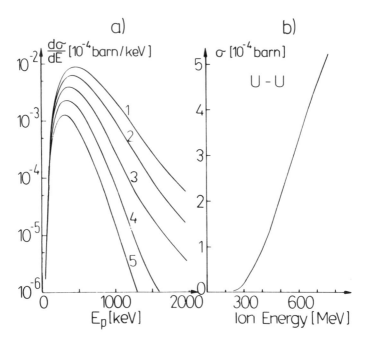

Fig. 6:   a) The positron production cross section $d\sigma/dE_p$ for the system U→U for various ion energies. The energies are the same as in Fig. 5 b).

b) The total positron cross section in dependence of the ion energy.

$Z_1 + Z_2$, the induced transitions to higher levels ($2p_{1/2}$, $2s_{1/2}$) which come close to diving or are just diving ($2p_{1/2}$) must be taken into account and will further increase the cross sections. A spectroscopy of the diving orbitals, i.e. of the structure of the charged vacuum, should be possible by either studying the positron spectra in their dependence on heavy ion energy and their change with the total charge $Z_1 + Z_2$ in the superheavy system. This, if done for lighter systems in a systematic fashion, should also yield information on the background effects. The vacuum structure can also be observed, perhaps even more accurately, by spectroscopy of the molecular X-rays of the superheavy, overcritical system.

## Background Effects

The background effects are for the vacuum-decay experiment most important. They have to be carefully identified and investigated, as far as possible. We mention here shortly in arbitrary order a few of them:

a) Coulomb pair creation: The time dependent Coulomb field may create directly ordinary electron-positron pairs (electron in either bound as in continuum state). The theory for this process is underway[11], but it seems that these cross sections are of the order 1/100 compared with the positron cross sections from vacuum decay.

b) Nuclear Bremsstrahlung can create an electron-positron background. Reinhardt and Soff[12] showed, that this process has a cross section of about $10^{-8}$ barns and can therefore be forgotten.

c) Internal conversion of γ-rays from Coulomb-excited nuclear levels: This turns out to be one of the most important background effects, which, in fact, deserves attention by itself. It has recently been investigated by Oberacker and Soff[13], who calculated the Coulomb excitation of the 35 lowest rotational and vibrational bands (which were described within the rotation-vibration model), folded this cross section with the various γ-decay-probabilities for those transitions giving γ-rays larger than 1 MeV and folded that in turn with the $e^+e^-$ -conversion probabilities. Their results are shown in Fig. 7 b) for unequal U-U collisions (e.g. $^{235}U$ – $^{238}U$ or - quite similar - $^{238}U$ – $^{232}Th$) and Fig.7 a) for equal, undistinguishable $^{238}U$ – $^{238}U$ collisions. In both cases the electron-positron cross section due to the induced and spontaneous decay of the vacuum is also shown. Obviously, at rather forward angles, the vacuum-decay-positrons outnumber those from the background by several orders of magnitude, so that even a decrease of the K-vacancy probability from $10^{-2}$ to $10^{-4}$ or $10^{-5}$ would make this observation possible. At backward ion angles, where the diving is stronger, however, the vacuum decay outnumbers the background only by a factor 5 – 7 ($L_0$ = 1% assumed). If, unexpectedly, there would be a number of 1⁻ states in Uranium around 2 MeV excitation energy, this would complicate the situation even more. Most important, however, is the fact, that a decrease of the K-vacancy probability to the estimated extrapolations of Meyerhof ($L_0 \approx 10^{-5}$) would make the observation of vacuum decay in U-U collisions impossible. The only possibility would then be to produce in high-energy heavy ion accelerators naked Uranium and bombard it onto Uranium-target (see next section). In that case $L_0$ = 1-2 and the positrons from vacuum decay would outnumber those of the discussed background by a factor

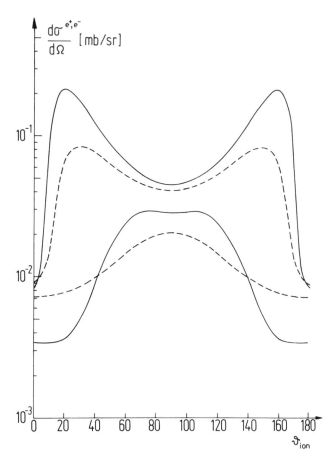

Fig. 7 a) Upper curve: Differential cross section for the
sum of spontaneous and induced positrons due to
vacuum decay in a U–U collision as function of
the heavy ion c.m. scattering angle $\vartheta_{ion}$. The
K-vacancy production probability is assumed to be
$L_0 = 0.01$. Lower curve: Corresponding differential
cross section for nuclear Coulomb excitation
positrons for the $^{238}_{92}U - ^{238}_{92}U$ system. The solid
lines show the results obtained for $E^{kin}_{CM} = 797$ MeV,
i.e. just at the Coulomb barrier, while the dashed
lines are for $E^{kin}_{CM} = 609$ MeV ($R_{min} = 20$ fm).

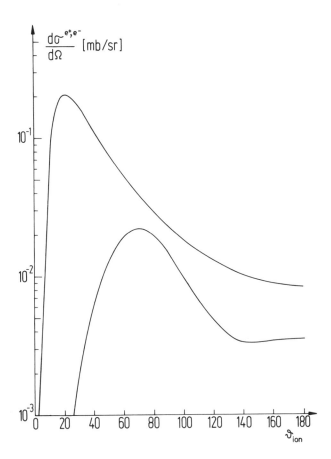

Fig. 7b): As in Fig.7a, but without symmetrization of the cross sections. $E^{kin}_{CM}$ = 797 MeV.

100 - 1000. An interesting side-result of the investigation of Oberacker and Soff is shown in Fig. 8. By multiplication of these Coulomb excitation cross sections with the pair conversion coefficient $\beta_{E2}$ ( $10^{-4}$) one obtains the corresponding pair production cross sections. This process occurs of course, only for such nuclei with first excited states separated from the ground state by more than 1 MeV. Obviously, in heavy ion bombardment, this amusing effect should be observable and would add to the understanding of positron-production in h.i. collisions, a field on the borderline between atomic and nuclear physics.

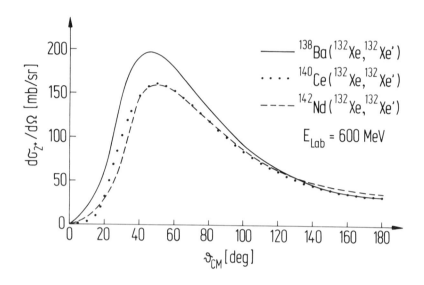

Fig. 8:   The angular distribution of the excitation cross
section for the first excited $2^+$ state of the nuclei
$^{138}$Ba, $^{140}$Ce and $^{142}$Nd is plotted.   In all cases
$^{132}$Xe projectiles with an energy $E_{lab}$ = 600 MeV have
been used.

## Naked Nuclei at High Energies

According to extrapolations of Blasche, Franzke and
Schmelzer[14], the U-ions can be stripped naked at energies
$\sim$ 300 MeV/N.   Such energies can be reached for Uranium at
the Bevalac within the next 3 – 5 years and plans at GSI in
this direction have also been worked out.   The atomic physics
with naked nuclei is indeed a very fascinating subject, since
it would enable us to study very heavy ions with only one or
two electrons around.   The Lambshift in very heavy atoms can
then directly be measured.   The decay of the vacuum can most
cleanly be studied. It is necessary, however, to deaccelerate
the completely stripped 300-400 MeV/N ions to energies below
the Coulomb barrier, i.e. down to about 7 MeV/N. Since these
ions are 90-times charged the deacceleration path can be relatively short. One might also think of other geometries like
splitting the fully stripped uranium beam and  arranging for
multiple collisions of the splitted beams. With the collision
angle one can arrange for the desired relative U-U energies.
A realistic geometry for such a set-up could consist in the
crossing of two colliding co-moving beams from storage rings

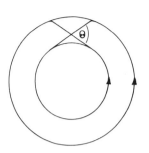

Fig. 9: Two heavy ion beams in storage rings are arranged to collide under the angle $\theta$. By changing $\theta$ their relative velocities can be adjusted.

(see figure 9)*.This seems to open a bright future for qualitatively new atomic physics.


III. POSSIBLE SPECTROSCOPY OF INTERMEDIATE SUPERHEAVY MOLECULES


The occurence of superheavy quasimolecules was predicted theoretically[15] and discovered experimentally by Mokler et al.[16] for I-Au collisions. The appearance of intermediate molecules, recognized by the molecular X-rays, was first found by Saris[17] for the L-shell, by Mokler,Armbruster et al.[16] and Gove et al.[18] for the M-shell, by McDonald et al.[19] for the K-shell; quasi-molecular physics was pioneered by Meyerhof et al.[20] for the Br-Br-system, by Greenberg et al.[21] for the Ni-Ni-system, by Kaun et al.[22] for Ge-Ge, Nb-Nb and other systems, by Groeneveld et al.[23] for Cu-Cu and Al-Al-systems and by Wölfli et al.[24] for Fe-Fe, Fe-Ni, Ni-Ni-systems. First ion-X-ray coincidence measurements were performed by Schmidt-Böcking et al.[25].The two-center Dirac orbitals are of fundamental importance for any of these investigations. It was first (and up to now only) solved for the single-particle two-center case by B. Müller[26]. The underlying Hamiltonian is

$$H_D(\vec{r},R) = c\vec{\alpha}\cdot\vec{p} + \beta mc^2 + V_1(r_1) + V_2(r_2)$$

where $r_1$ and $r_2$ are the distances of the electron from the corresponding centers. We show here systematically the Br-Br-levels (Fig. 10), the Br-Zr-diagram (Fig. 11), the Ni-Ni-levels (Fig.12), the I-Au-levels (Fig. 13), as renormalized for shielding effects by Fricke[27], the very asymmetric Cl-Pb-diagram (Fig. 14) needed by Burch et al.[28], and finally the U-U-diagram (Fig. 15). This gives us an impression of the manyfold to be investigated experimentally and of the new qualitative and quantitative features which appear (note e.g. the $2p_{3/2}-2p_{1/2}$ splitting in double uranium is about 900 keV!). The molecular

* This particular idea emerged in discussions with Ch. Schmelzer.

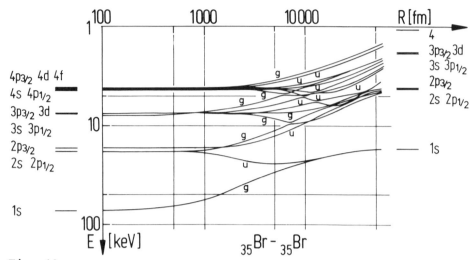

Fig. 10:
The 16 lowest $\sigma$-levels of the Br+Br ($Z_1 = Z_2 = 35$) molecular
system in double-logarithmic scale. States of opposite parity
are allowed to cross.

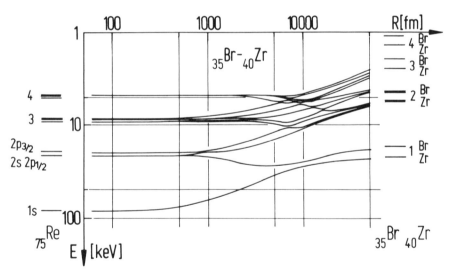

Fig. 11: The 13 lowest $\sigma$-states of the Br+Zr ($Z_1 = 35$, $Z_2 = 40$)
quasimolecule. All crossings are avoided in this
asymmetric system. The scale is double logarithmic.

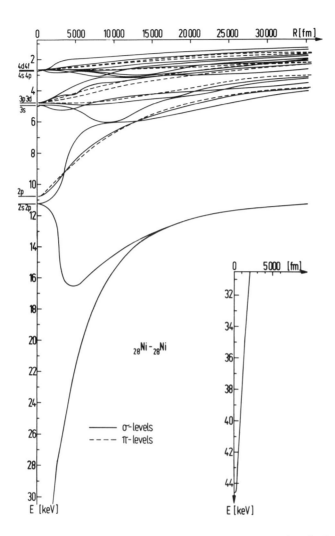

Fig. 12: The $\sigma$ - and $\pi$ - molecular levels of the Ni+Ni ($Z_1 = Z_2 = 28$) system in linear scale.

X-ray spectra are discussed by many colleagues at this conference. Let's therefore focus here on a possible spectroscopy
of the two-center levels. For that one needs some quantity
which peaks (or shows other observable features) at the distance of closest approach of the two nuclei. The most promising feature at present seems to be the peaking of the asymmetry of molecular X-rays as a function of the photon energy.
Such an asymmetry was predicted for induced radiation[29] which
should appear in addition to spontaneous radiation.

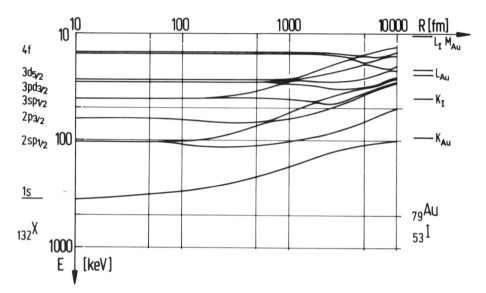

Fig. 13: The I+Au (Z = 53,Z = 79) quasimolecule in douple-
         logarithmic presentation. Eleven σ-states are shown.
         This is the first superheavy system investigated ex-
         perimentally.

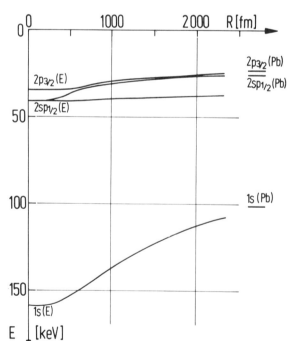

Fig. 14: The Cl+Pb
system $(Z_1=17, Z_2=82)$
in a linear scale. On-
ly the four lowest σ-
states are shown. The
system is extremely
asymmetric, so that up
to the M-shell the mo-
lecular states are for-
med without the parti-
cipation of the chlo-
rine atomic levels. This
puts the system at the
fringe of the molecu-
lar model.

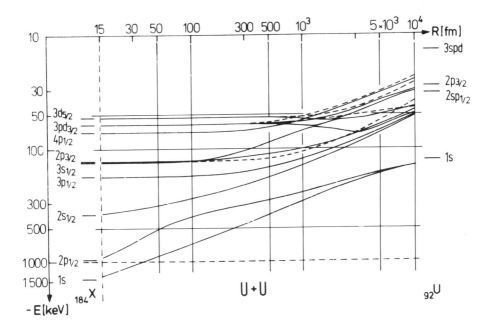

Fig. 15: The 10 lowest $\sigma$ -levels and four $\tau$ -levels of the su-
         percritical U+U ($Z_1=Z_2=92$) quasimolecule. The states
         are calculated with extended nuclear charge distri-
         butions. Note that the 1s$\sigma$ state reaches the lower
         Dirac continuum around R = 35 fm.

It was calculated in the quasistatic approach to give about 20%
asymmetry at the united atom limit for the Ni-Ni-system. The
possible alignment of electron levels was also considered in
this approximation, but seemed to give only constant (as a
function of photon-energy) contributions.

  R.K. Smith[30] recently carried out the full dynamical treatment of spontaneous and induced transitions by solving exactly the Fourier integral for the photon amplitudes

$$\left.\begin{array}{c} A(\omega) \\ B(\omega) \\ C(\omega) \end{array}\right\} = \int_{-\infty}^{\infty} dt\, d(t)\, a(t)\, e^{i\omega t}\, \frac{d}{dt}\, e^{-i\int_0^t \omega_{fi}dt'} \left\{\begin{array}{c} \cos\alpha_N(t) \\ \sin\alpha_N(t) \\ 1 \end{array}\right.$$

using a parametrization for the time dependence of the transition frequency $\omega_{fi}(t)$ and the dipole moment d(t). This lead
to the following surprising new results:

1) The induced radiation - contrary to our earlier belief -
does not increase substantially beyond the quasistatic estimate, but gives again a contribution of about 10-20% peaking

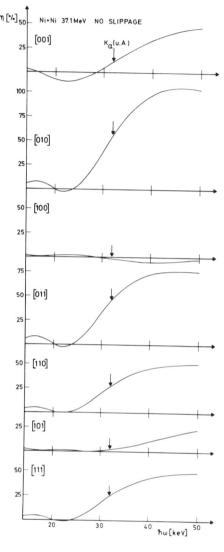

Fig. 16a: The X-ray-energy dependence of the angular anisotropy $\eta$ for the quasi-molecular K-shell radiation of the Ni-Ni-system at 37.1 MeV. The curves belong to different alignments of the 2p-levels. The numbers in square brackets denote the occupation probabilities of the states

$$\left[ {}^{2}p_{3/2}\pi \ , \ {}^{2}p_{3/2}\sigma \ , \ {}^{2}p_{1/2}\sigma \right] .$$

at the united atom limit. It alone ist not sufficient to explain the experiments.

2) The spontaneous radiation with alignment for the electrons (various unequal occupation probabilities for $\sigma$- and $\pi$-states respectively) leads also to a strong photon-energy dependent asymmetry, as shown in Fig. 16a. It shows characteristic features (rising, peaking) near the united atom limit. This comes from the fact, that the Fourier-integral takes the delay-times of the turning electric dipols during the collision into account.

This turning is fastest at the distance of closest approach (united atom limit), which transforms in Fourier space to those photon frequencies in the vicinity of the united atom limit.

3) Even closed electronic subshells lead to an asymmetry, because of the Fourier-integral (interference of photons from the way in and out). This is due to the break-down of the golden rule formula in this strongly time dependent process.

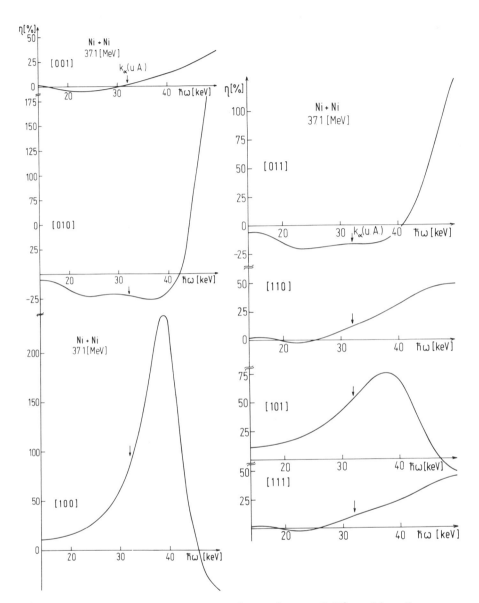

Fig. 16b: The curves correspond to those of Fig. 16a. Here, however, additionally the "slippage" (see below) is included to account for the nonadiabacity of the wavefunctions.

There is another effect of some importance, which we call slippage. It takes care of the fact, that the electronic mole-cular orbitals cannot follow the fast rotation of the system

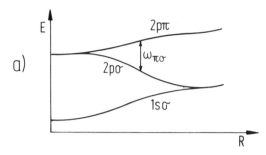

a)

Fig. 17a: A typical two center level-diagram showing the transitional frequency $\Omega_{\pi\sigma}$ between the 2p$\pi$ and 2p$\sigma$ states.

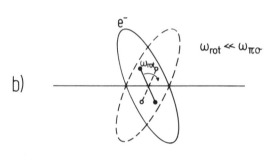

b)

Fig. 17b: For slow rotations ($\omega_{rot} \ll \Omega_{\pi\sigma}$) the electron wave function follows the changing molecular states.

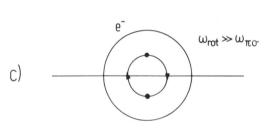

c)

Fig. 17c: For fast rotations ($\omega_{rot} \gg \omega_{\pi\sigma}$) the electron does not follow the molecular axis. This is taken into account by using the slippage probability p.

if $\omega_{rot} \gg \omega_{\pi\sigma}$, where $\omega_{rot}$ is the rotational frequency and $\Omega_{\pi\sigma}$ is the energy difference between the 2p$\pi$ and 2p$\sigma$ orbital (see Fig. 17). This is achieved by defining the slippage probability

$$p = \frac{\omega_{rot}}{\omega_{rot} + \omega_{\pi\sigma}}$$

and replacing the photon emission amplitudes A($\omega$), B($\omega$) and C($\omega$) by

$$\begin{bmatrix} \tilde{A}(\omega) \\ \tilde{B}(\omega) \\ \tilde{C}(\omega) \end{bmatrix} = \int_{-\infty}^{\infty} dt\, d(t)\, \tilde{a}(t)\, e^{i\omega t} [(1-p)\frac{d}{dt} e^{-i\int_0^t \omega_{fi} dt'} \begin{bmatrix} \cos\alpha_N(t) \\ \sin\alpha_N(t) \\ 1 \end{bmatrix} + \begin{bmatrix} p \\ ip\frac{\omega_{rot}}{\omega_{fi}} \\ p \end{bmatrix} \frac{d}{dt} e^{-i\int_0^t \omega_{fi} dt'}]$$

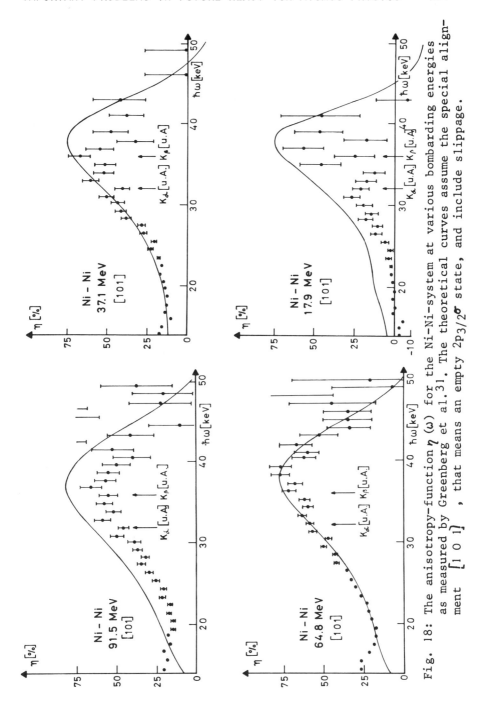

Fig. 18: The anisotropy-function $\eta$ ($\omega$) for the Ni-Ni-system at various bombarding energies as measured by Greenberg et al.31. The theoretical curves assume the special alignment [1 0 1], that means an empty $2p_{3/2}\sigma$ state, and include slippage.

This slippage effects also the asymmetry of the quasimolecular radiation in that the runway of the rotating dipole moments is prolonged. For various MO-transitions in the Ni-Ni-systems this is shown in Fig. 16a (without) and 16b (with slippage). The experiments of Greenberg et al.[31] on the asymmetry of the quasimolecular X-rays are now nearly quantitatively explained by theory (Fig. 18). Here the special alignment, that the $2p_{3/2}\sigma$-shell is empty, had to be assumed, to fit the experimental results. Obviously, there are some indications that this particular alignment changes somewhat with ion energy, which is plausible.

We would also like to mention here the very interesting recent experiments of Wölfli et al.[24]. He measured the MO-Xray asymmetry for Fe-Fe, Fe-Ni, Ni-Fe and Ni-Ni with high resolution (Figs. 19) and finds that the asymmetry peak scales indeed with $(Z_1+Z_2)^2$ where $Z_1$ and $Z_2$ are the proton numbers of the colliding ions, which eventually demonstrates the possibility to use the asymmetry for spectroscopy of quasimolecular orbitals. In the very accurately measured Ni-Ni-case, the asymmetry seems to oscillate with a frequency of about 2-3 keV and large amplitudes at the united atom limit. The envelope of these oscillations coincides with the Greenberg-measurement. It is not clear at this moment, whether these oscillations reflect a sudden rearrangement[32] or some other effect.

The next figure (20a) shows the prediction of the theory for the Mokler-Armbruster-experiment (I-Au) (Fig. 20b). Obviously as well the asymmetry as the quasimolecular peak itself can now be understood quantitatively and also in their ion energy dependence. Enhanced $\Delta m = 1$ alignment has to be assumed. This is, however, very reasonable. That the asymmetry peak moves with increasing energy theoretically towards higher photon energies, while the experiment shows the opposite trend, stems from the fact, that the calculations were carried out with an M-shell transition-frequency-dependence on the ion-ion distance R as shown in Fig. 21a while the exact two-center calculations yield a behaviour as in Fig. 21b, i.e. a decrease of the MO-transition-frequency for the M-shell at small distances. Again, this all demonstrates that the asymmetry peak can indeed be used for spectroscopy of the two center orbitals. This opens a very interesting future for spectroscopy of superheavy quasimolecular orbitals, which we consider as a rather fundamental contribution to physics.

However, here again a word of caution must be said, because of the disturbance of the quasimolecular spectrum in cases like U-U due to the nuclear X-rays (excited by Coulomb excitations). First calculations[34] of this effect indicate that the single MO-spectrum is severely disturbed (see Fig. 22).

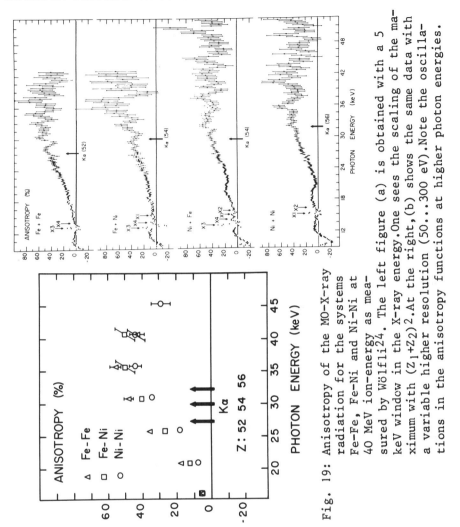

Fig. 19: Anisotropy of the MO-X-ray radiation for the systems Fe-Fe, Fe-Ni and Ni-Ni at 40 MeV ion-energy as measured by Wölfli[24]. The left figure (a) is obtained with a 5 keV window in the X-ray energy.One sees the scaling of the maximum with $(Z_1+Z_2)^2$.At the right,(b) shows the same data with a variable higher resolution (50...300 eV).Note the oscillations in the anisotropy functions at higher photon energies.

This can only be circumvented by investigating magic nucleus-nucleus collisions like Pb-Pb, because for them Coulomb excitation is expected to be very much decreased or by coincidence experiments.

We finally remark that with increasing accuracy of the measurements the nucleus-nucleus bremsstrahlung has to be identified and investigated. It should be recognizable through the interference effects between dipole and quadrupole bremsstrahlung, which we demonstrate in Fig. 23. These results were obtained by a semiclassical calculation[36].

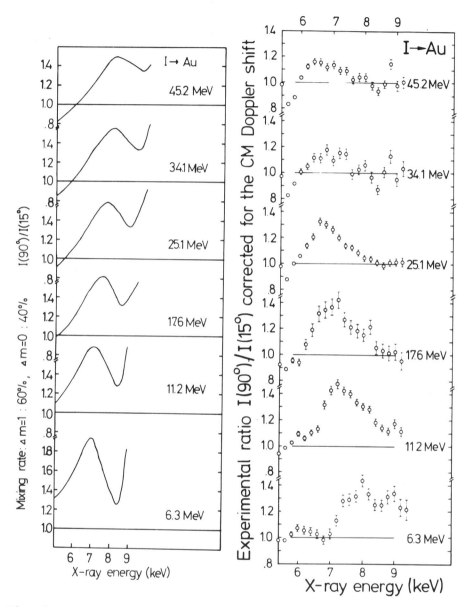

Fig. 20a: The theoretical anisotropy-function $\eta$ ($\omega$) for the
I-Au-system at various bombar-
ding energies. (Mixing rate
$\Delta$ m=1 : 60%, $\Delta$ m=0 : 40% )

Fig. 20b: The anisotropy $\eta$ ($\omega$)
for the I-Au-system as measured
by Folkmann et al[33].

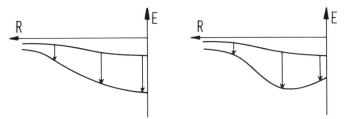

Fig. 21: Dependence of the transition frequency (arrows) on the
two-center distance. Whereas in the approximation,
which was used for the theoretical calculations (a),
the frequency increases monotonously with increasing
ion energy, the realistic two-center levels (b) show
the opposite trend.

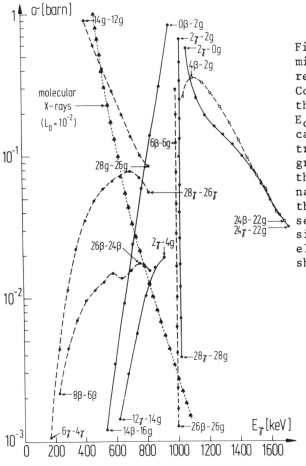

Fig. 22: The photon e-
mission cross-sections
resulting from nuclear
Coulomb excitation of
the $^{238}$U–$^{238}$U-system at
$E_{cm}$= 800 MeV are indicated for the various
transitions between the
ground state band and
the $\beta$- and $\gamma$-vibratio-
nal band. It is seen
that the nuclear X-rays
severely disturb the
single MO-spectrum for
electronic L- to K-
shell transitions[35].

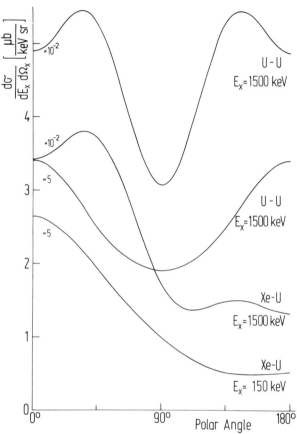

Fig. 23: The dipole + quadrupole radiation cross sections for the systems $^{132}Xe-^{238}U$ at 700 MeV and $^{238}U-^{238}U$ at 1400 MeV is shown for photon energies of 150 keV and 1500 keV. Note the effect of the dipole quadrupole interference for the asymmetric system.

## IV. ATOMIC PHYSICS AT RELATIVISTIC ION ENERGIES

Atomic physics with heavy ions can also be done at ultra-high ion energies like e.g. 2 GeV/N. To see the interesting and rather basic aspects of this field, let us shortly describe an experiment very recently performed by Raisbeck, Heckman and D. Greiner at the Berkeley Bevalac[37]. Its principle is shown in Fig. 24.

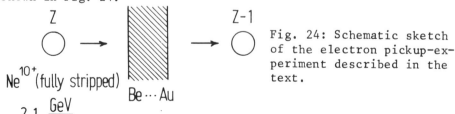

Fig. 24: Schematic sketch of the electron pickup-experiment described in the text.

They start off with completely stripped nuclei ( in the particular example naked Ne-nuclei ) with energies of about 2 Gev/N
impacting on various target layers (Be...Au) of varying
thickness. They observe the fraction of those ions which have
charge Z-1 i.e. which have picked up one electron while passing
the target. The target thickness has been varied. If it was
thicker than about 10 microns, the pickup-rate became independent of the thickness, which indicates an equilibrium process:
As many electrons are stripped off as are picked up again.
The typical equilibrium pick-up-rate is of the order $10^{-8}$.
So much about the experiment.

Its principle value - as we see it - lies less in the understanding of the pick-up and stripping mechanisms at ultra-
high ion energies (which is of course a very interesting subject in itself, because it tells the story of the high Fourier-
components of the electron wave function), but mainly in its astrophysical applications[38]: There occur certain elements in
cosmic rays which can only be formed via $\beta$-decay, like e.g.

$$^{59}Vd \rightarrow {}^{59}Ta \; , \; {}^{51}Cr \rightarrow {}^{51}Va \; , \; {}^{55}Fe \rightarrow {}^{55}Mg \; .$$

This decay proceeds only via pure $e^-$-capture. If at cosmic ray
energies (typically 1 GeV/N) all electrons would be stripped
off, these isotopes can stay stable over a long period of time.
However, under certain conditions of density of interstellar
gas or more importantly of their environment during their creation process, there is electron pick up possible for the mother-
nuclei, which then may undergo $\beta$-decay and hence lead to the production of these special daughter-nuclei. Hence one can learn
from such experiments and from the abundance of these special
isotopes about the environment and its density during the creation process of cosmic rays.

Concluding Remarks: We have thus seen in this lecture, that
heavy ion atomic physics may give the answer to a number of very
fundamental and interesting problems. It brings together nuclear
and atomic physics and one should not hesitate to turn down the
dividing line between the two fields and simply do very interesting, fundamental physical research. This should be a
challenge.

REFERENCES

1) W.Pieper and W. Greiner, Z. Phys. 218 (1969) 327;
   B.Müller, J. Rafelski and W. Greiner, Z. Phys. 257 (1972)
   82 and 163;
   P.G. Reinhard, W. Greiner and H. Arenhövel, Nucl. Phys.
   A166 (1971) 173;
   J. Rafelski, B. Müller and W. Greiner, Nucl. Phys. B68 (1974)
   585;
   B. Müller, R.K. Smith and W. Greiner, Proc. of the Int.
   Sympos. on Physics with Relativ. Heavy Ions, Ed. L. Schroe-
   der, LBL Berkeley, July 1974.
2) V.S. Popov, Soviet Phys. JETP 32 (1971) 526.
3) A.B. Migdal, Soviet Phys. JETP 34 (1972) 1184;
   J. Rafelski and A. Klein, Proceedings, Center for Theoretical Studies, Miami, Florida, Jan. 1974.
4) T.D. Lee and G. C. Wick, Phys. Rev. D9 (1974) 2291.
5) K. Smith, B. Müller and W. Greiner, Journal of Physics B8
   (1975) 75.
6) K. Smith, H. Peitz, B. Müller and W. Greiner, Phys. Rev.
   Lett. 32 (1974) 554.
7) W.E. Meyerhof et al., Contribution to the 4$^{th}$ Int. Atomic
   Physics Conf., Heidelberg, July 1974.
8) F. Saris et al. preprint from the Institute for Atomic and
   Molecular Physics, Amsterdam.
9) D. Burch, V. Vandenbosch et al., preprint from the University of Washington, Seattle (1974).
10) H. Peitz, B. Müller, J. Rafelski and W. Greiner, Lett.
    Nuovo Cim. 8 (1973) 37.
11) G. Soff and H. Peitz, Inst. f. Theoret. Physik der Univer-
    sität Frankfurt am Main.
12) J. Reinhardt, G. Soff and W. Greiner, preprint, June 1975
    Inst. F. Theor. Physik der Univ. Frankfurt am Main.
13) V. Oberacker, G. Soff and W. Greiner, preprint, June 1975
    Inst. f.Theor. Physik der Univ. Frankfurt am Main.
14) Blasche, Franzke and Ch. Schmelzer, priv. communication.
15) The idea of superheavy quasimolecules was discussed during
    various GSI-seminars in 1969..72 and first published in
    J. Rafelski, L. Fulcher and W. Greiner, Phys. Rev. Lett.
    27 (1971) 958.
16) P. Mokler,H.J. Stein, P. Armbruster, Phys. Rev. Lett.
    29 (1972) 827;
    P. Mokler, H.J. Stein, P Armbruster, Phys. Rev. Lett.
    33 (1974) 475.
17) F.W. Saris, W.F. van der Weg, H. Tawara, R. Laubert, Phys.
    Rev. Lett. 28 (1972) 717.
18) H.E. Gove et al., University of Rochester, to be published.

19) J. R. McDonald, M.D. Brown and T. Chiao, Phys. Rev. Lett. 30 (1973) 471.
20) W.E.Meyerhof, T.K. Saylor, S.M. Lazarus, W.A. Little, B.B. Triplett and L.F. Chasw, Phys. Rev. Lett. 30(1973)1279.
21) C.K. Davis and J.S. Greenberg, Phys. Rev. Lett! 32(1974)1215; J.S. Greenberg, C.K. Davis and P. Vincent, Phys. Rev. Lett. 33 (1974) 473.
22) P. Gippner, K.-H. Kaun, F. Stary, W. Schulz and Yu. P. Tretyakov, Nucl. Phys. A230 (1974) 509.
23) K.O. Groeneveld, B. Knaf, G. Presser, Proc. Int. Conf. on React. between complex Nuclei, Nashville (1974).
24) W. Wölfli, Ch. Stoller, G. Bonani, H. Suter and M. Stöckli, to be published; W. Wölfli, priv. comm. .
25) H. Schmidt-Böcking, I. Tserruya, R. Schulé, G. Gaukler, K. Bethge, IKF Jahresbericht, 1974, Univ. Frankfurt am Main.
26) B. Müller, J. Rafelski, W. Greiner, Phys. Lett. 47B(1973) 5; B. Müller and W. Greiner, The Two Center Dirac Equation, to be published in Phys. Rev. A.
27) B. Fricke, K. Rashid, P.Bertoncini and A.C. Wahl, Phys. Rev. Lett 34 (1974) 243.
28) D. Burch, W.B. Ingalls, H. Wieman and R. Vandenbosch, Phys. Rev. A10 (1974) 1245.
29) B. Müller, R.K. Smith and W. Greiner, Phys. Lett. 49B (1974) 219; B. Müller and W. Greiner, Phys. Rev. Lett. 33 (1974) 469.
30) R.K. Smith and W. Greiner, Spontaneous and Induced Radiation from Intermediate (Superheavy) Molecules,preprint, Univ. Frankfurt am Main; R.K. Smith and W. Greiner, to be published.
31) J-S. Greenberg, priv. comm. .
32) R.K. Smith, B. Müller, W. Greiner, J.S. Greenberg and C.K. Davis, Phys. Rev. Lett. 34 (1975) 134.
33) F.Folkmann, P. Armbruster, S. Hagmann, G.Kraft, P.H. Mokler and H.S. Stein, to be published in Phys. Rev. Lett. .
34) G. Soff, V. Oberacker and W. Greiner, to be published.
35) B. Müller, R.K. Smith and W. Greiner, Phys. Lett. 53B (1975) 401.
36) J. Reinhardt and W. Greiner, Proc, of the XIII Int. Winter Meeting, Bormio, Italy, 1975.
37) G.M. Raisbeck, H. Heckman and D. Greiner, priv. comm.
38) G.M. Raisbeck, C. Perron, J. Toussaint and F. Yiou. Proc. 13th Int. Cosmic Ray Conf., University of Denver, 1973; G.M. Raisbeck and F. Yiou, Phys. Rev. A4 (1971) 1858.

# PHOTON INTERACTIONS

# ATOMIC PHYSICS WITH SYNCHROTRON RADIATION

R. P. Madden

National Bureau of Standards

Washington, D.C.  20234

## INTRODUCTION

My intention in this talk is to review the contributions which have been made in the area of atomic physics
through the use of synchrotron radiation sources.  The
availability of these sources <u>and</u> their quality is increasing steadily.  The best possible use should be made of their
unique properties.

Synchrotron radiation was first observed and understood
in the late 1940's, but the first application of the soft x-
ray component was in 1956 when Tomboulian studied the transmission of metal foils using the then-existing 320 MeV
Cornell synchrotron.  Unfortunately, his use of this machine
only lasted two weeks.  The first synchrotron radiation
<u>facility</u> to be developed was SURF-I at NBS, beginning in
1961.  SURF-I utilized a 180 MeV synchrotron of modest
current.  The physics done with this machine throughout the
60's and into the 70's was largely atomic and molecular
photoabsorption.  Other synchrotron radiation sources were
developed in the 60's beginning with the Frascatti, Hamburg
and Tokyo synchrotrons and then the Stoughton Storage Ring.
In spite of the early atomic physics beginning, as these new
facilities were developed the emphasis was almost entirely
on solid state physics.  This momentum carried into the
70's; however by now a number of other atomic physics experiments have been accomplished utilizing synchrotron radiation.

Today I will not present many new results, nor will I

probe very deeply into the physics of the various experiments I will discuss.  Rather, I hope that today's overview
will leave you with an appreciation of the considerable
applicability of synchrotron radiation for experiments to
obtain data on atomic and molecular systems.

### ATOMIC PHOTOABSORPTION--

The first experiment attempted at the NBS synchrotron
radiation facility was the simplest--namely photoabsorption
by the noble gases.  This required a high resolution grazing
incidence spectrograph attached directly to the tangent arm
of the synchrotron as shown in Figure 1.  The spectrograph
was filled with the gas to be studied, the only leak being
the optically used portion of the entrance slit.  Spectra
were obtained for all the noble gases and a number of
molecular gases in this way.  The results for helium[1,2],
shown in Fig. 2, began a new era in the theoretical appreciation of the intricacies of configuration interactions.
In Fig. 2 we see only one strong Rydberg series of asymmetric resonances converging onto the n = 2 state of $He^+$
due to the simultaneous excitation of both electrons.  The
second series is extremely weak.  These are designated the
$sp,2n^+$ and $sp,2n^-$ series in an explanation by Cooper et
al.[3]  One sees also the higher lying $sp,3n^+$ and $sp,4n^+$
series converging onto the n = 3, and n = 4 states of $He^+$.

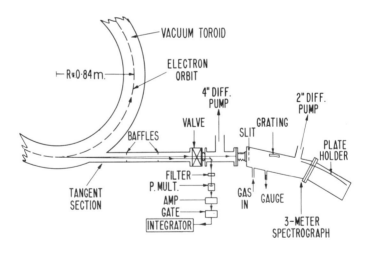

Fig. 1.  Early experimental arrangement for utilizing far
ultraviolet radiation from the NBS 180 MeV synchrotron (from
Madden and Codling, ref. 2).

A recent addition to this early work is the study of the profile of the sp,33$^+$ (3s3p) resonance by Ederer and Dhez[4] as shown in Fig. 3. This resonance has an asymmetry of opposite sign to the sp,22$^+$ resonance and interfers primarily with the sp,2ε$^+$ continuum. The parameters given for this resonance can be compared with others for helium given in reference 2.

Skipping over the work done on two-electron excitation and sub-shell excitation in Ne through Xe, let me dwell for a moment on the inner shell excitation of Xe first reported[5] by Codling and Madden in 1965. Here I refer to the excitation of the inner d electron requiring a photon energy greater than 65 eV. A more recent study of these resonance structures is that of Ederer and Manalis[6] where the resonance profiles were analyzed in detail as seen in Fig. 4. The important point here for further discussions is that only 4d → np type transitions are observed, the 4d → nf transitions being totally absent. Soon after the first report of this phenomena Ederer showed[7] that the d → f transitions held off until the ionized electron left with two Rydbergs of energy, causing a large maximum in the continuum cross section. This has been explained as a centrifugal barrier effect by Cooper[8]. More recently, Haensel et al.[9] using the DESY synchrotron as a background source studied Xe in the solid and gas phases in this region. Their result is shown in Fig. 5. This shows that the Ederer peak in the continuum for the gas exists with

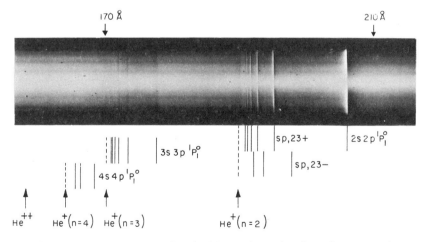

Fig. 2.  Resonances in the helium photoionization continuum due to two-electron excitation states (from Madden and Codling, ref. 2).

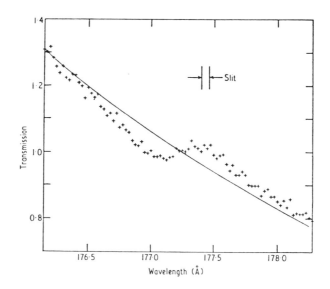

Fig. 3.   Profile of the 3s3p resonance in helium obtained with a high resolution scanning monochromator (from Dhez and Ederer, ref. 4).

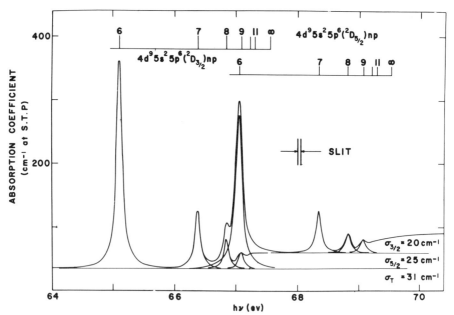

Fig. 4.   Study of the structure in the photoionization continuum of xenon due to transitions $4d^{10}5s^25p^6 \rightarrow 4d^95s^25p^6np$ (from Ederer and Manalis, ref. 6).

remarkably little change in the solid as well. This peak
has also been seen in $Sn^{10}$, $I^{11}$, and $Cs^{12}$. The case for
barium is interestingly different. The region of discrete
4d electron excitation in barium has recently been studied
by Connerade and Mansfield[13], Ederer et al.[14], and Rabe et
al.[15] Figure 6 shows the spectra reported in Ref. 14.

In ground state barium, as in Xe and Cs, the 4f orbitals
are unbound. However, if a 4d electron is excited, the
core becomes electrostatically stronger and the 4f orbital
pulls inside the potential barrier and becomes bound. This
is also true in La. The 4d $\rightarrow$ 4f transition has three J = 1
components $^3P_1$, $^3D_1$, $^1P_1$. The $^3P_1$ and $^3D_1$ are bound. Current
interpretations of Rabe et al.[15] and of Ederer et al.[14] are
in disagreement with recent calculations by Hanson et al.[16]
as to the position and strength of the $^1P_1$. The former
groups say that the large exchange interaction pushes the
$^1P_1$ component well above the d ionization limit--making a
major contribution to the broad maximum in the continuum
cross section. The latter group says that the $^1P_1$ is bound
and weak and that the broad maximum in the continuum above
the d ionization limit is due solely to the delayed onset
of the 4d $\rightarrow$ $\varepsilon$f transitions as in Xe and Cs. These differences
have yet to be resolved.

Before leaving the subject of atomic photoabsorption I
want to bring to your attention the strong effort in this
area which is currently underway at the synchrotron at U.
Bonn. This facility has an 0.5 GeV synchrotron and a 2.5
GeV synchrotron and atomic absorption spectra are being
taken on both machines. Many spectra of the alkalis and
alkaline earths have been taken and are under study. The
primary atomic workers at the U. of Bonn machines are J. C.
Connerade, M. Mansfield, K. Thimm, W. R. S. Garton and D.
H. Tracy.

## PHOTOELECTRON SPECTROSCOPY--

Photoelectron spectroscopy (PES) of gases was first
done with synchrotron radiation at the 300 MeV Glasgow
synchrotron by Mitchell and Codling from U. Reading. They
utilized the linear polarization of the radiation in the
orbital plane to obtain $\beta$ for the 3p excitation channel in
argon[17] in 1971. The $\beta$ I refer to is of course the param-
meter which describes the angular distribution in PES of
atomic systems, namely

$$N d\Omega = C[1 - \frac{1}{2}\beta + \frac{3}{2}\beta \cos^2\theta]d\Omega,$$

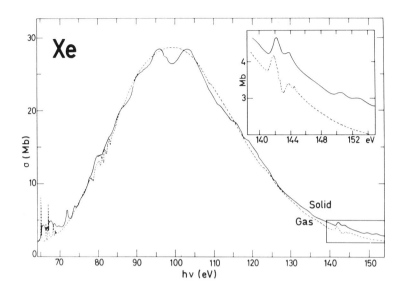

Fig. 5.   Study of the structure in the photoionization continuum of gaseous and solid xenon due to transitions
$4d^{10}5s^25p^6 \rightarrow 4d^95s^25p^6\varepsilon f$ (from Haensel et al., ref. 9).

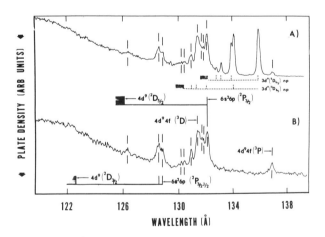

Fig. 6.   Discrete structure in the photoionization continuum
of barium vapor due to the excitation of one of the 4d
electrons (from Ederer et al., ref. 14).

where N is the number of photoelectrons emitted into solid
angle $d\Omega$, $\theta$ is the angle of emission relative to the electric vector of the plane polarized incident photon beam,
and C is a constant depending on the cross section.

Other gases were also studied at Glasgow, and the
Reading group then built an apparatus for continuing these
studies at the NINA synchrotron in Daresbury. This apparatus is shown in Fig. 7. Light is incident from the left
where some of it is intercepted by a sodium salicilate
screen which fluoresces into a monitor photomultiplier.
The remainder is incident on the interaction zone in the
center of the electron energy analyzer, labeled E in Fig.
7. This 127° analyzer can be rotated about an axis parallel
to the incident light direction, so that $\beta$ is uniquely
determined at each photon energy. It uses a channeltron
electron detector and a capillary array to focus the atoms.
The first results published in 1974 by Houlgate et al.[18]
are shown in Fig. 8. Here the experimental values for $\beta_{3p}$
in argon very nicely follow the theory of Amusia et
al.[19] (RPAE) but deviates from the Hartree-Fock (H-F)
calculations of Kennedy and Manson[20].

Recent new results by Houlgate and West[21] have extended
these experimental results up to a photon energy of 70 eV.
The extended results continue to agree very well with the
theory of Amusia et al.[19]

Houlgate et al.[18] also obtained and published the
partial cross section for excitation of the 3s electron in
argon, where the H-F and RPAE theories differ substantially
at threshold. Figure 9 shows the result. Again a quite
good comparison is achieved by the RPAE theory of Amusia et
al.,[22] although the predicted minimum appears too close to
the limit.

The recent work by Houlgate and West[21] also extended
these experimental results up to over 60 eV. In the extended
region the theoretical result of Amusia et al. remains
higher than the experimental result by a factor of two.

This is the region moreover where the RPAE and the H-F
theories give similar results.

Before leaving the field of photoelectron spectroscopy
I would like to mention an upcoming experiment. K. Thimm
of Univ. Bonn is cooperating with Siegbahn's Upsalla group
to put a new photoelectron experiment on the 2.5 GeV synchrotron at Bonn. The monochromator is completed and the
remaining apparatus is nearly ready. In this facility the

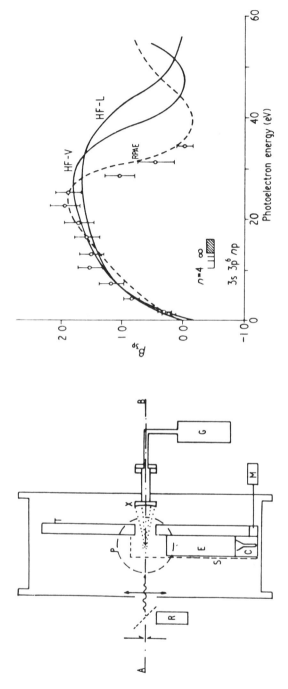

Fig. 7.  Schematic diagram of the 127°
analyzer.  C channeltron, E analyzer, G gas
supply, M motor drive and position monitor for
analyzer rotation, R reference photomultiplier
and sodium salicylate coated grid, s μ-metal
shield, T rotating table, X focusing capillary
array.  (from Houlgate et al., ref. 18)

Fig. 8.  β obtained for photoelectrons
from ionization of a 3p electron in argon.
H-F Hartree-Fock calculations of Kennedy
and Manson[20] RPAE random phase approximation with exchange by Amusia et al.[19]
(from Houlgate et al., ref. 18).

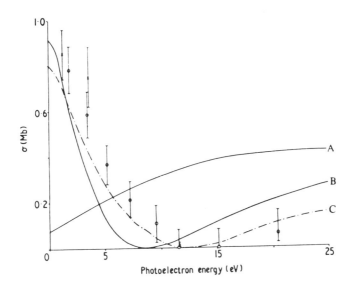

Fig. 9.   Cross section for photoionization of a 3s electron
in argon.   A Hartree-Fock calculations by Kennedy and
Manson[20], B RPAE theory by Amusia et al.[22] (from Houlgate
et al., ref. 18).

incident photon energy will be completely variable--allowing
a considerable extention of the Upsalla program in photoelectron spectroscopy of molecules.

<u>FLUORESCENCE</u>--

    The first attempt at atomic fluorescence using synchro-
tron radiation was a quite ambitious one.   Tracy and Roesler,[23]
using the Stoughton Storage Ring, succeeded in measuring
the photoionization excitation cross section to the $4\ell$
states of He$^+$ from observations of the He II 4S, P, D, F →
3S, P, D fluorescence complex at λ4686 Å.   The apparatus
they used is indicated in Fig. 10.   Synchrotron radiation
from the storage ring is transmitted from the left through
a polypropylene window into a cooled He cell.   The He$^+$
fluorescence is brought out and passes through a double
Fabry-Perot interferometer of high étendue and a filter to
a detector.   They determined the photoionization-excitation
cross section at threshold for the 4S state to be (3.2 $\pm$
1.3) x 10$^{-4}$Mb.   The 4P state excitation cross section was
estimated to be 1-3 times larger, and no direct excitation
of 4P and 4F was observed.

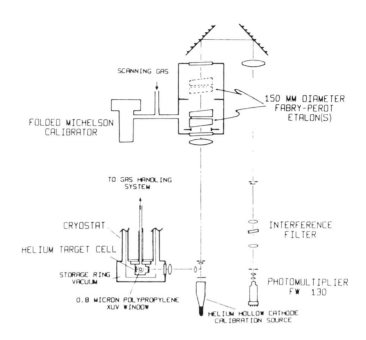

Fig. 10.   Apparatus for the high resolution observation of the
4S, P, D, F → 3S, P, D fluorescence complex in helium
excited by synchrotron radiation.   (from Tracy and Roesler,
ref.   23).

   Low counts were a serious problem at that time.   Since
then, the Stoughton ring has increased in current by an
order of magnitude, making experiments of this type much
more feasible.

   In the work just described, undispersed synchrotron
radiation was the primary light and the fluorescent radiation
was dispersed.   Alternatively Lee et al., used a grating
instrument to monochromatize the synchrotron radiation from
the Stoughton ring before allowing the primary light to
fall on the sample.   They collected the fluorescence over a
broad band.   The result for $O_2$ shown in Fig. 11 is an
example of their work with molecules[24].   Here the collected
wavelength band is 1050-1800 Å and the excitation cross
section of the fluorescence as a function of primary radiation wavelength is obtained.   Working with various filters
and available tabular data they were able to deduce that
most of the fluorescence came from an atomic O multiplet at
1300 Å.   The structure on the curve in Fig. 11 is due
largely to absorption of the primary radiation by $O_2$.

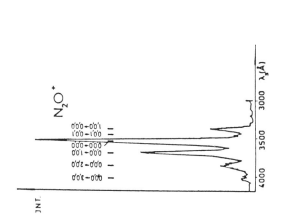

Fig. 12. Fluorescent yield as a function of the wavelength of the scattered radiation from $N_2O^+$ produced by irradiating $N_2O$ with 17.5 eV primary radiation. (from Sroka and Zietz, ref. 25)

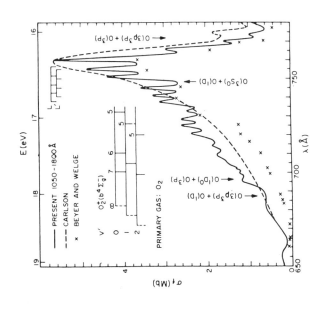

Fig. 11. Fluorescent excitation cross section for the production of fluorescence in $O_2$ in the wavelength band 1050-1800 Å as a function of primary radiation wavelength. (from Lee et al., ref. 24)

In another fluorescence experiment Sroka and Zietz[25] had sufficient intensity at DESY in Hamburg to use two monochromators. The primary one had a spectral band pass of 3.5 Å and the secondary one 24 Å. With this arrangement they could scan the fluorescence spectrum at any excitation energy or the excitation spectrum at any fluorescence wavelength. Figure 12 shows the emission of $N_2O^+$ obtained by irradiating $N_2O$ with 17,5 eV photons. This fluorescence is from the $A^2\Sigma^+$ state. Each peak consists of an unresolved doublet.

One can now ask what the excitation spectrum is for a particular component of the fluorescence spectrum, say for the 0 → 0 vibrational transition at 3550 Å. This is shown in Fig. 13 uncorrected for the primary light intensity distribution. They also observed fluorescence at higher photon energies (1300-1500 Å) which resulted from dissociative excitation of $N_2O$.

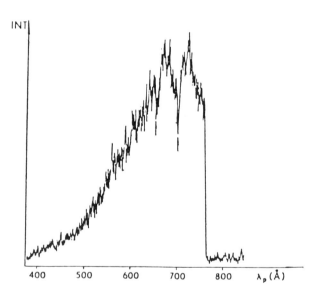

Fig. 13. Excitation spectrum for the 0 → 0 vibrational fluorescence transition of $N_2O^+$ at 3550 Å. The abscissa is the primary radiation wavelength. (from Sroka and Zietz, ref. 25).

Before leaving this subject I want to mention the recent study of the photodissociation of $H_2$ and $D_2$ using the ACO storage ring in Orsay, France which was reported at this meeting by Guyon and Mentall[26].

## MASS SPECTROMETRY--

The continuous wavelength coverage available with synchrotron radiation sources can be well utilized to obtain photoion yields in mass spectrometry. Such results are being obtained by Taylor's group at Stoughton who were the first to apply synchrotron radiation in mass spectrometry. The $CO_2$ results published by Parr and Taylor[27] will serve as an example. Figure 14 shows the ion yield of $CO_2^+$ as a function of wavelength of the incident radiation. Different thresholds are clearly discernable due to structure in the ground state for $CO_2^+$. The structure above the limits is due to the presence of autoionization states of $CO_2$ above the ionization threshold.

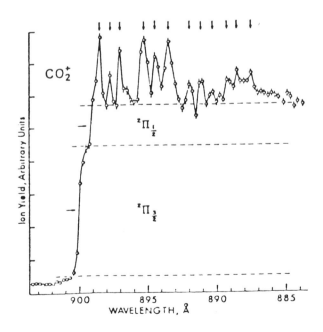

Fig. 14.  Ion yield of $CO_2^+$ as a function of primary radiation wavelength in the photoionization of $CO_2$. (from Parr and Taylor, ref. 27).

One can also examine dissociation by this method.    The production of O$^+$ from photodissociation of $CO_2$ is shown in Fig. 15 as a function of the wavelength of the primary radiation.

## TIME-RESOLUTION--

Storage rings have a little utilized property of considerable value for future applications.  That is the time structure of the radiation.  At Stanford's SPEAR, by placing only a single bunch of electrons in orbit, the pulse width can be reduced to 0.3 ns while the repeat period is 720 ns. At the ACO storage ring in Orsay, France the pulse width can be about 1 ns with a repeat period of 73 ns.   Such pulsed sources make the direct measurement of excited state lifetimes an attractive possibility.   Figure 16 shows an example of this application obtained by the LURE group at ACO[28].  Here we see the decay in emission from photo-excited fluorescene in aqueous solution.  The lifetime is easily determined from the slope of the log plot to be 4.1 ns.  The LURE group also was able to obtain the spectrum of the fluorescence at various times during the fluorescence.  This is shown for the same sample in Fig. 17.  Here we see a dramatic change in the spectral distribution in very short time intervals, since $t_3 - t_1 \approx 1$ ns.

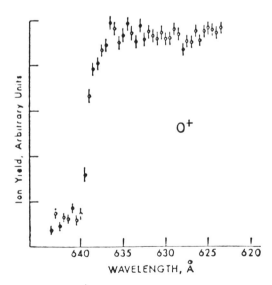

Fig. 15.   Ion yield of O$^+$ as a function of primary radiation wavelength in the photodissociation of $CO_2$ (from Parr and Raylor, ref. 27).

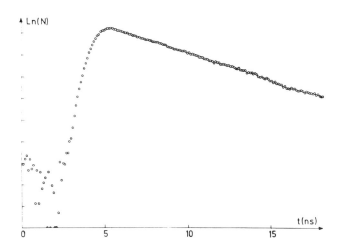

Fig. 16. Decay of fluorescence for photo-excited fluorescene in aqueous solution. The source of excitation is a
single pulse from the ACO storage ring at Orsay, France.
(from Dagneaux et al., ref. 28).

An example of lifetime determination when the lifetime
is short compared to the pulse width is shown in Fig. 18.
Here we see the fluorescence decay of pyrazine at 3240 Å
obtained[28] by the LURE group on the ACO storage ring. Also
shown is the ACO pulse profile used in this experiment,
indicating approximately a 3 ns width. When the ACO pulse
profile is convoluted with a decay profile having a 0.55 ns
time constant, the result agrees well with the experimental
observations. Using such techniques on Stanford's SPEAR
storage ring one might hope to measure lifetimes as short as
a few x $10^{-11}$ seconds.

## USE OF CIRCULAR POLARIZATION--

It is now well known to you all that polarized electrons
can be obtained by photoionizing alkali atoms with circularly polarized light. In fact a polarized electron injector
for SLAC is now operating successfully on this principle.
Heinzmann et al.[29] have demonstrated experimentally that
circularly polarized light can also produce polarized electrons (in thallium) when the photon energy is tuned to
autoionization resonances.

Near circularly polarized radiation is available from
synchrotron radiation sources. This occurs not in the

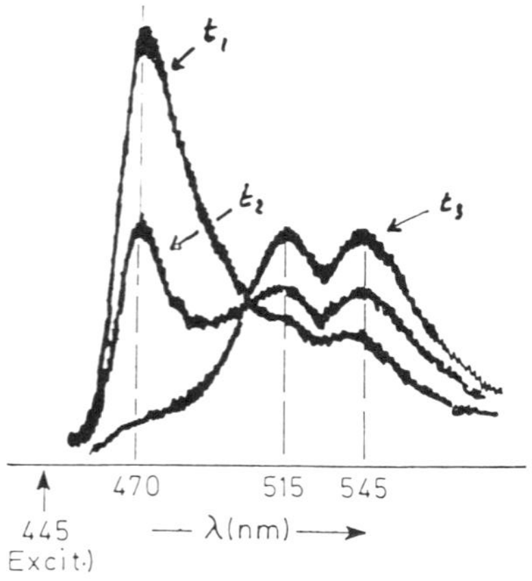

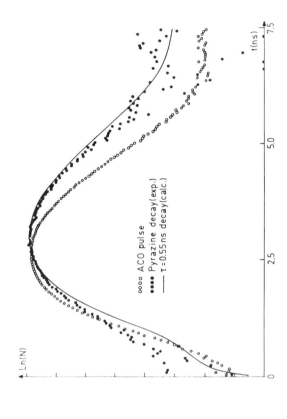

Fig. 18.  Decay of fluorescence of photo-excited
pyrazine compared with the time profile of the
exciting radiation.  The source of excitation is
a single pulse from the ACO storage ring at
Orsay, France.  (from Dagneaux et al., ref. 28)

Fig. 17.  Emission spectra at three
different times during fluorescence of
fluorescene in aqueous solution.  The
source of excitation was a single pulse
from the ACO storage ring at Orsay,
France.  $t_3 - t_1 \approx 1$ ns (from Dagneaux
et al., ref. 28).

orbital plane but above and below it where the perpendicular
and parallel (electric vector orientation to the orbital
plane) components approach the same magnitude, since these
two components are 90° out of phase.  Hence the near-circular
polarization available from synchrotron radiation sources
maybe advantageous for future experiments studying polarization effects.  Such experiments can extract great detail
about atomic wave functions.

<u>SUMMARY</u>---

As I have tried to demonstrate, many new experiments
and extentions of experiments in atomic physics are made
possible through the application of synchrotron radiation.
In many of the areas mentioned work is only just beginning
in the various synchrotron radiation laboratories around the
world.  At DESY in Hamburg, W. Germany, they have recently
obtained photoabsorption of the 2s and 2p electrons in
aluminum near 60-120 eV.  They will be concentrating on
photoabsorption of the transition metals in the near future.
The efforts at Univ. Bonn in photoabsorption and the upcoming
PES experiment were mentioned earlier in this talk.  At ACO,
they will soon have in operation an apparatus for analysis
of both electrons and ions in the photoionization of gases
and metallic vapors.

The news from the INS-SOR group in Tokyo is that they
have recently completed construction of a 300 MeV storage
ring.  While no atomic work was carried out at the old
synchrotron by this group, it is expected that the new
machine will be used for photoemission and Auger studies.
On the NINA synchrotron at Daresbury, England, they are
continuing to study angular distributions in PES and have
recently become interested in Auger electrons in the d
channels of Kr and Xe.  In the future they will be using
time-of-flight ion energy analyzers as well.  Perhaps not
all of you yet know that the English have committed themselves to building a new 2 GeV storage ring at Daresbury
which will be dedicated to synchrotron radiation experiments.

The use of synchrotron radiation for atomic physics is
also growing in the USSR.  Several synchrotrons (Lebedev -
650 MeV; Erevan - 6 GeV; Tomsk - 1.3 GeV) and the storage
rings at Novosibirsk (VEPP - 2 M; VEPP - 3) are now carrying
out or planning work in gas absorption spectroscopy, fluorescence, and ESCA.

Meanwhile in the US there are now three operating
storage rings.  SPEAR at Stanford is, as I have mentioned, a
very powerful tool for lifetime studies--however, I know of
no experiment planning to take advantage of this fact.
Great current attention at this facility is devoted to EXAFS
(extended x-ray absorption fine structure).  XPS at 8 keV
has also been achieved with a resolution of < 0.2 eV.  At
the University of Wisconsin storage ring at Stoughton,
Wisconsin the work on fluorescence yields of gases and the
photoion spectroscopy mentioned earlier is slated to continue.
This machine, a dedicated facility, continues to serve a
large number of outside users.

The newest storage ring in the US belongs to the oldest
synchrotron radiation facility, namely SURF-II at the
National Bureau of Standards in Washington, D.C.  This is a
250 MeV storage ring which first achieved beam last winter
and is just now approaching design performance.  At NBS, we
have already reactivated our transfer standard photodiode
calibration capability and have made available to outside
users a facility for the calibration of complete instru-
ments.  Next to be reactivated is an atomic metallic vapor
photoabsorption cross-section experiment designed to obtain
accurate cross sections and resonance profiles.

SURF-II has 11 possible beam lines each capable of
capturing the radiation from 50-60 milliradians of orbit.
As shown in Fig. 19 we currently have erected only three
beam lines having seven experimental stations capable of
simultaneous operation.  We have a number of potential
outside users who are discussing the use of this machine and
several of these experiments are atomic in nature.  I invite
you all to consider the use of this new facility yourselves
and would be glad to receive proposals for either independent
or cooperative experiments.  Thank you!

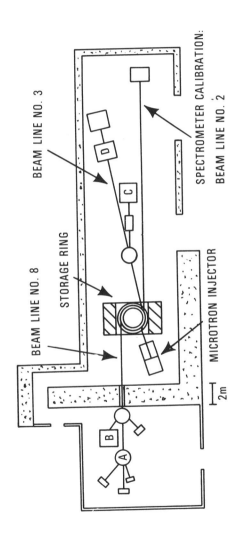

Fig. 19. The SURF-II facility at NBS in Washington, D.C. showing three of 11 available beam ports implimented. Stations A and D contain toroidal grating monochromators while B and C contain at 2.2 meter and 3 meter grazing incidence monochromator respectively.

REFERENCES

1.  R. P. Madden and K. Codling, Phys. Rev. Lett. 10, 516 (1963).

2.  R. P. Madden and K. Codling, Astrophys. J. 141, 364 (1965).

3.  J. Cooper, U. Fano, and F. Prats, Phys. Rev. Lett. 10, 518 (1963).

4.  P. Dhez and D. L. Ederer, J. Phys. B 6, L59 (1973).

5.  K. Codling and R. P. Madden, Phys. Rev. Lett. 12, 106 (1964).

6.  D. L. Ederer and M. Manalis, J. Opt. Soc. Am. 65, 634 (1975).

7.  D. L. Ederer, Phys. Rev. Lett. 13, 760 (1964).

8.  J. W. Cooper, Phys. Rev. Lett. 13, 762 (1964).

9.  R. Haensel, G. Keitel, P. Schreiber and C. Kunz, Phys. Rev. 188, 1375 (1969).

10. K. Codling, R. P. Madden, W. R. Hunter and D. W. Angel, J. Opt. Soc. Am. 56, 189 (1966).

11. F. J. Comes, U. Nielsen, and W. H. E. Schwarz, J. Chem. Phys. 58, 2230 (1973).

12. H. Peterson, K. Radler, B. Sonntag and R. Haensel, J. Phys. B 8, 31 (1975).

13. J. P. Connerade and M. W. D. Mansfield, Proc. R. Soc. A341, 267 (1974).

14. D. L. Ederer, T. B. Lucatorto, E. B. Saloman, R. P. Madden and J. Sugar, J. Phys. B. L3 (1975).

15. P. Rabe, K. Radler and H. W. Wolff, Proc. 4th Int.Conf. on VUV Radiation Physics, Hamburg, pub. Pergamon Veiweg, Braunschweig, W. Germany, pg. 247 (1974).

16. J. E. Hanson, A. W. Fliflet and H. P. Kelley, J. Phys. B 8, L127 (1975).

17. P. Mitchell and K. Codling, Phys. Lett. 38A, 31 (1972).

18. R. G. Houlgate, J. B. West, K. Codling and G. V. Marr, J. Phys. B 7, L470 (1974).

19. M. Ya Amusia, N. A. Cherepkov and L. V. Chernysheva, Phys. Lett. 40A, 15 (1972).

20. D. J. Kennedy and S. T. Manson, Phys. Rev. A 5, 227 (1972).

21. R. G. Houlgate and J. B. West, private communication.

22. M. Ya Amusia, Proc. 8th Int. Conf. on Physics of Electronic and Atomic Collisions, Belgrade--Invited Lectures and Progress Reports, p. 172 (1973).

23. D. H. Tracey and F. L. Roesler, Daresbury.

24. L. C. Lee, R. W. Carlson, D. L. Jusge, and M. Ogawa, J. Chem. Phys. 61, 3261 (1974).

25. W. Sroka and R. Zietz, Z. Naturforsch 28a, 794 (1973).

26. P. M. Guyon and J. E. Mentall, Proc. of the IX Int. Conf on the Physics of Electronic and Atomic Collisions (1975).

27. G. R. Parr and J. W. Taylor, J. Mass. Spec. and Ion

Phys. 14, 467 (1974)'.
28. P. Dagneaux, C. Depautex, P. Dhez, J. Durup, Y. Farge,
R. Fourme, P. M. Guyon, P. Jaegle, S. Leach, R. Lopez-
Delgado, G. Morel, R. Pinchaux,  C. Vermeil, and F.
Wuilleumier, Report on LURE activities (1974).
29. U. Heinzmann, H. Heuer and J. Kessler, Phys. Rev. Lett.
34, 441 (1975).

# PHOTODETACHMENT THRESHOLD PROCESSES

W. C. Lineberger*

Department of Chemistry, University of Colorado

and Joint Institute for Laboratory Astrophysics,

University of Colorado and National Bureau of

Standards, Boulder, Colorado  80302

## INTRODUCTION

The energy resolution attainable in photodetachment
studies has been substantially improved [1,2] to better than
1 meV by the use of continuously tunable laser light sources.
These techniques now permit determination of atomic electron
affinities with an accuracy of order  $10^{-4}$ eV, as well as a
thorough test of theoretical threshold laws.  In this paper
we give a brief summary of threshold laws for the case in
which a ground state neutral atom or molecule is produced in
the photodetachment process.  We also will look at cases in
which the resulting atom is left in an excited state, and the
two-electron ejection threshold resulting in a positive ion.

The threshold behavior of such processes is closely
related to electron-molecule scattering because of the fact
that photodetachment processes,

$$h\nu + A^- \leftrightarrows A + e^-(k,L) \tag{1}$$

where  k and L  are the electron linear and orbital angular
momentum with respect to the center of mass, can in some
sense be viewed as "half an electron-atom scattering process
having only a restricted number of partial waves accessible."
Within the constraints of this limitation one can use

---

*Alfred P. Sloan Foundation Fellow.

photodetachment to prepare electron-atom scattering systems
in total energy states with an energy resolution that is
essentially the optical resolution of the laser.  Thus, in
some cases we can investigate electron-atom scattering processes with an energy resolution that is several orders of
magnitude higher than currently obtainable in high resolution
electron beam scattering experiments.

The basic experimental technique employed [2] is to
intersect a  2 keV mass-analyzed negative ion beam with the
focused output of a pulsed, tunable dye laser.  The fast neutral atoms, produced in the photodetachment process, are then
detected by secondary electron emission on the first dynode
of a windowless electron multiplier.  The quantities measured
are the ion beam current, the photon flux, the neutral atom
signal, and the photon wavelength.  The result of this measurement is a relative cross section for the production of
neutrals as a function of photon wavelength.  Although to date
tunable laser photodetachment experiments have been performed
only in the photon wavelength region  4400-9000 Å, present
laser technology makes such experiments possible in the spectral region  2300Å-1 μm.

## GROUND STATE THRESHOLDS

The basic physics underlying photodetachment threshold
behavior has been known since  1948, when Wigner [3] showed
that the threshold energy dependence of a cross section for
processes involving two product particles depends only on the
long-range forces between the particles.  For the case of
atomic negative ion photodetachment, where the dominant long-
range final state interaction is the centrifugal potential,
the predicted behavior of the photodetachment cross section
near threshold, $h\nu_{thr}$, by photons of energy  $h\nu$  is given
by [3],

$$\sigma \propto h\nu (h\nu - h\nu_{thr})^{(L+1)/2} \propto h\nu \; k^{(2L+1)} \quad . \qquad (2)$$

If a  p-electron is detached, such as in  Se⁻, the outgoing
electron can be in an  s- or d-wave.  The theoretical threshold behavior is given by the  s-wave contribution  $\sigma \propto k$,
since the  d-wave cross section is suppressed relative to the
s-wave by the centrifugal barrier.  One of the interesting
questions that can now be addressed by a high resolution
photodetachment experiment is:  Over how large an energy
range does Eq. (2) give a satisfactory description of the
cross section?  The answer to this question can be complicated
by the fact that one actually observes a series of thresholds
corresponding to transitions from the various fine-structure
substates of the negative ion to fine-structure states of the
neutral atom.

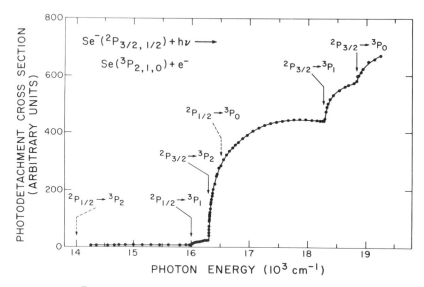

Fig. 1.   Se⁻ photodetachment cross section in the energy
          range 14000-19000 cm⁻¹. The individual fine struc-
          ture transition thresholds are labeled.

A study [2] of Se⁻ threshold photodetachment (Fig. 1)
shows six s-wave thresholds corresponding to transitions
from Se⁻ ($^{2}P_{1/2,3/2}$) to Se ($^{3}P_{2,1,0}$). The individual fine
structure thresholds labeled in the figure are obtained from
a knowledge of the neutral Se fine-structure energy separations. From a detailed investigation of a single fine-structure threshold, we find that the range of validity of the
Wigner threshold law is no more than 5 meV. One can qualitatively understand the departures from the threshold law in
terms of the low energy electron-atom elastic scattering
cross section and the electron atom polarization and quadrupole interactions, but only very limited success has been
achieved [2] in extracting such information quantitatively
from photodetachment data. In addition to these studies of
s-wave thresholds, a p-wave threshold has been studied in
some detail in the case of Au⁻ photodetachment [4]. As a
result of accidental cancellations in the higher order terms,
the Wigner threshold law was apparently valid over a much
larger energy region (approximately 50 meV) than in the Se⁻
s-wave case.

Molecular negative ion photodetachment threshold laws
were first derived by Geltman [5], who treated the diatomic
negative ion case by expanding the initial and final state
wavefunctions in a cylindrically symmetric basis and derived
threshold laws which depend upon the symmetry properties of
the electron being ejected. These predictions were subsequently verified by experiments. Although the Geltman

analysis did not explicitly depend upon the Wigner result,
Brauman and coworkers [6] have recently shown that the Geltman
result can be derived from the Wigner law, and have extended
the Wigner law to determine the form of the threshold energy
dependence of the photodetachment cross section for a molecular ion of arbitrary symmetry. Essentially, they note that
one can express the threshold energy dependence as a sum of
partial waves of the form of Eq. (2). Using the symmetry properties of the electron in the highest occupied molecular
orbital, together with electric dipole selection rules, they
present a straightforward group-symmetry argument to determine the L-value of the lowest nonzero term in the Wigner-
like expansion. They use this procedure to reproduce the
Geltman result, as well as to predict correctly the threshold
energy dependence in such polyatomic ions as $NH_2^-$ and $C_5H_5^-$.

The Wigner threshold laws assume that there are no $r^{-2}$
interactions other than the centrifugal potential, and that
the polarization potential ($r^{-4}$) is small in the relevant
portion of the asymptotic region. In the case of $OH^-$ photodetachment [7], the $OH$ permanent dipole moment gives rise to
an angle-dependent $r^{-2}$ interaction. One expects the threshold law to lie between $E^0$ (step function) and the Wigner
($E^{1/2}$) prediction, and to be $J$ dependent. Such behavior is
apparently observed in the $OH$ case [7], and, as seen in the
next section, these additional interactions may be very important for threshold processes leaving the neutral in an excited
state.

## EXCITED STATE THRESHOLDS

In the vicinity of the threshold for photodetachment to
an excited state of the neutral, the cross section may exhibit
features arising from interference between the ground state
channel and the excited state channel. In the case of an
excited state channel involving an s-wave free electron, this
interference manifests itself as a cusp in the open channel
cross section, having an energy dependence of the form
$A+B|E-E_n|^{1/2}$, where $E$ and $E_n$ are the system energy and the
threshold energy of the $n^{th}$ channel. The nature of these
so-called Wigner cusps has been pursued in substantial detail
in the study of scattering processes. Based on analogy with
electron-atom scattering, it is reasonable to expect similar
cusps in photodetachment cross sections near excited state
thresholds. Moores and Norcross [8] have examined theoretically the possibility of observing cusps in photodetachment,
and one in fact does experimentally see a cusp-like behavior
[9,10] in the total photodestruction cross section for the
light alkali negative ions at photon energies corresponding
to the opening of the lowest np state of the neutral.

In contrast to the behavior in the light alkali ions, photodetachment of the heavy alkalis (Rb⁻ and Cs⁻) at photon energies corresponding to the opening of the lower  np  final state is dominated by doubly excited states of the alkali negative ion, which lie very close to the  np threshold [10]. Figure 2 shows the  Cs⁻ photodetachment cross section in this energy region; the double-excited state of  Cs⁻ gives rise to a Fano profile in the photodetachment cross section, but the smooth development of phase across the resonance is interrupted by the opening of the inelastic channel.  This channel opening appears as a discontinuity in the resonance profile, as is seen in the inset in Fig. 2.   The  Rb⁻ cross section is qualitatively similar, but the resonances are significantly more narrow, the lowest energy resonance in  Rb⁻  being only 150 µeV wide, by far the most narrow feature observed to date in electron-neutral atom scattering processes.

Since, at each new photodetachment threshold, very low energy electrons associated with that threshold are produced, it is possible, if one can discriminate against the high energy electrons to look only at the low energy electrons and see the new threshold in detail.  We have recently completed modification of the photodetachment apparatus to permit observation of electrons with initial  KE less than some specified amount, in addition to the usual detection of neutral atoms. Following preliminary studies on  Te⁻ [12] to assure that

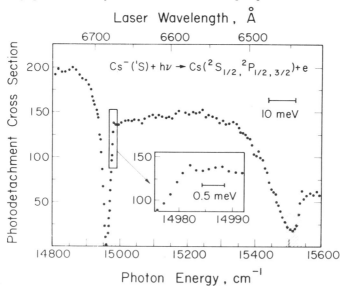

Fig. 2.   Cs⁻ photodetachment cross section  6800-6400 Å.  The shaded regions indicate the confidence limits on the opening of the  Cs $6^2P_{1/2}$ and $6^2P_{3/2}$ exit channels, as determined by photoelectron spectroscopy [11].

electrons with KE less than 50 meV were trapped with essentially 100% efficiency, a study was made of $K^-$ photodetachment near the K 4p ($^2P_{1/2}$, $^2P_{3/2}$) excited state exit channel thresholds. By proper selection of the trapping potential, the electron channel sees all electrons for which the final state is K $^2P$, and the neutral channel signal is proportional to the total ($^2S$ + $^2P$) photodestruction cross section. With appropriate normalization, the two cross sections can be subtracted to yield the elastic ($^2S$) and inelastic ($^2P$) contributions to the total cross section. The first preliminary data [13] are shown in Fig. 3. While at present the normalization between the total cross section and the $^2P$ partial cross section is only accurate to ±40%, a number of interesting features are still observable.

1) The $^2P_{1/2}$ cross section rises very sharply, from the threshold at ~17030 cm$^{-1}$, peaking less than 1 meV above threshold. The rate of rise is in fact instrumental, and it is possible that the rise is actually a step function.

2) The $^2P_{1/2}$ cross section decreases from the peak to the $^2P_{3/2}$ threshold, where it is quite small.

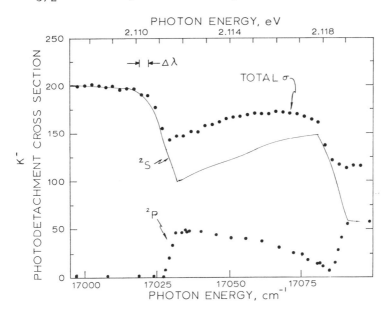

Fig. 3.  $K^-$ photodetachment cross sections near the $^2P_{1/2}$, $^2P_{3/2}$ channel openings. The total cross section is obtainable from neutral atom production, and the $^2P$ cross section is the cross section for producing electrons with KE ≤ 30 meV. The $^2S$ cross section is obtained from a subtraction of the two.

3) The $^2P_{3/2}$ cross section also rises sharply to a peak near threshold (also instrumentally broadened), but continues a slower rise above this peak.

4) The $^2S$ elastic channel is drastically affected by the $^2P$ channel openings. Independent of the normalization, the $^2S$ channel cross section increases between the $^2P_{1/2}$ and $^2P_{3/2}$ thresholds.

The $^2P$ channels should exhibit s-wave thresholds ($\sigma \propto \sqrt{E}$), and, based upon similar ground state thresholds, should peak several hundred meV above threshold. There should be Wigner cusps in the $^2S$ channel at the $^2P_{1/2}$ and $^2P_{3/2}$ thresholds, and in the $^2P$ channel at the $^2P_{3/2}$ threshold. Clearly, the observed behavior is much more complicated than such a simple picture.

The entire decrease in the $^2P_{1/2}$ cross section cannot be a Wigner cusp; the fact that the $^2P_{1/2}$ cross section must be positive above the $^2P_{3/2}$ threshold requires that the cusp-dominanted energy region be only ~100 μeV wide around the $^2P_{3/2}$ threshold. Another form of interference between the two channels must be invoked in order to explain the data. The effect on the $^2S$ cross section is equally profound, and implies that a very strong coupling exists between the three channels involved.

The most plausible reason for the extremely rapid rise in the $^2P$ channels is the presence of an interaction between the electron and atom strong enough to cause a step function threshold. Such steps can occur for the Coulomb interaction (e.g. photoionization of atoms), for "permanent" dipoles (e.g. $h\nu + H^- \rightarrow H(n=2) + e$), or possibly for very large polarizability. A hint to the interactions required is obtained by noting that the $^2P$ and $^2S$ channels are strongly coupled in the data, and are connected by optical transitions with large oscillator strength.

In a greatly oversimplified way, we note that a 1 meV electron can move 10 Å away from the atom in $50 \times 10^{-15}$ sec. If one views the resultant $^2P$ excited atom as in a coherent ($^2S + ^2P$) superposition state, the large oscillator strength permits large instantaneous dipole moments, and the period of the charge oscillation is $\sim 2 \times 10^{-15}$ sec. Even in such a crude picture, it is clear that the escaping electron can undergo substantial energy exchanges with the residual atom for considerable distances. Stated differently, the de Broglie wavelength of a 1 meV electron is ~400 Å, and the $^2S + ^2P + $ electron system must be regarded as closely coupled

over the time it takes a  1 meV electron to travel this distance,  ~2 × $10^{-12}$ sec.

Very preliminary measurements on  $Cs^-$ show [14] a behavior qualitatively similar to that of  $K^-$, and imply that the
$K^-$ behavior is not unique.  Further insight can be gained by
studying a system in which the excited and ground atomic states
are not optically connected; an experimentally accessible
example is the  $P^- \to P$ $^4S$, $^2D$ case.  It is undoubtedly true
that a detailed understanding of these phenomena awaits detailed calculations, including at least spin-orbit interaction
between the departing electron and the residual atom.

As our understanding of these processes increases, it
will be fun to continue to higher excited states, until one
reaches the two-electron ejection continuum, where the electron impact ionization threshold law can be tested.  If history is any guide, new and interesting surprises await there.

ACKNOWLEDGMENTS

All of the hard work and all of the new data reported
here are the fruits of the labors of J. M. Slater who will soon
complete his Ph.D., and F. H. Read, a JILA Visiting Fellow.
Throughout the course of this work, I have been fortunate to
have such outstanding coworkers as H. Hotop, A. Kasdan, T. A.
Patterson, and theoretical guidance from D. Norcross.  Support
from the National Science Foundation is gratefully acknowledged.

REFERENCES

[1]  W. C. Lineberger and B. W. Woodward, Phys. Rev. Lett.
     24, 424 (1970).

[2]  H. Hotop, T. A. Patterson, and W. C. Lineberger, Phys.
     Rev. A 8, 762 (1973).

[3]  E. P. Wigner, Phys. Rev. 73, 1002 (1948).

[4]  H. Hotop, and W. C. Lineberger, J. Chem. Phys. 58,
     2379 (1973).

[5]  S. Geltman, Phys. Rev. 112, 176 (1958).

[6]  K. J. Reed, A. H. Zimmerman, H. C. Andersen, and J. I.
     Brauman, J. Chem. Phys. (in press).

[7]  H. Hotop, T. A. Patterson, and W. C. Lineberger, J. Chem.
     Phys. 60, 1806 (1974).

[8]  D. L. Moores and D. W. Norcross, Phys. Rev. A 10, 1646
     (1974).

[9]   T. A. Patterson, Ph.D. Thesis, University of Colorado, Boulder (1974).

[10]  T. A. Patterson, H. Hotop, A. Kasdan, D. W. Norcross and W. C. Lineberger, Phys. Rev. Lett. 32, 189 (1974).

[11]  A. Kasdan and W. C. Lineberger, Phys. Rev. A 10, 1658 (1974).

[12]  J. M. Slater and W. C. Lineberger, to be published.

[13]  J. M. Slater, F. H. Read and W. C. Lineberger, to be published.

[14]  J. M. Slater and W. C. Lineberger, to be published.

# MULTIPHOTON PROCESSES WITH POLARIZED LIGHT[*]

P. Lambropoulos

Physics Department
Texas A&M University
College Station, Texas 77843, USA

## INTRODUCTION

It is not longer possible to review the subject of multiphoton processes in one lecture. Owing to the large amount of experimental as well as theoretical work that has been performed during the last three years or so, we are now faced with a mushrooming field which overlaps with a number of conventional areas of atomic, molecular and radiation physics. In this lecture, I shall discuss a selected number of topics with emphasis on problems that are being actively pursued at this time. In all cases, the discussion shall be limited to the essential aspects of the process under consideration without entering into theoretical or experimental details. I have chosen to concentrate on questions connected with the polarization of the light mainly because many of the relatively recent results have been stimulated by the recognition that the light polarization is responsible for a number of interesting effects. This choice does not limit the variety of the phenomena one can discuss since the more general aspects, such as resonance effects, photon correlation effects, etc. are present in any case.

The present discussion is limited to processes which can be understood within the framework of radiation perturbation theory including corrections for shifts and widths induced by the laser. The vast majority of the existing experimental results certainly fall in this category.

ANGULAR MOMENTUM IN PHOTON ABSORPTION AND EMISSION

The interaction of electromagnetic radiation with electrons involves not only the exchange of energy but also of
angular momentum.  In the dipole approximation, each photon
carries one unit of angular momentum (in units of $\hbar$).  This is
the intrinsic angular momentum of the photon.[1]  For higher
order multipoles, the photon carries additional angular
momentum which can be understood as orbital.  In an electronic
transition in which one photon is absorbed or emitted, in the
dipole approximation, one unit of angular momentum is transferred between the electromagnetic field and the electron.
This transfer occurs according to the rules of the vector
addition of angular momenta.

In a multiphoton transition, a final state is reached
from an initial state via the absorption or emission of more
than one photon.  Consequently, even in the dipole approximation, more than one unit of angular momentum is transferred.
This means that, in the case of absorption for example, the
second photon interacts with an electron whose state of angular
momentum has been altered by the absorption of the first photon;
and so on.  The angular momentum state of the photon is related
to its state of polarization.  A circularly polarized dipole
photon is a simultaneous eigenstate of the total intrinsic angular momentum S as well as of the projection $S_{\vec{k}}$ of S on the
direction $\vec{k}$ of propagation of the photon.  Such a photon is
said to have definite helicity[1] whose value is +1 or -1
depending on whether the photon is right or left circularly
polarized, respectively.  A linearly polarized dipole photon
on the other hand, although an eigenstate of S, is not an
eigen-state of $S_{\vec{k}}$.  As a result, the absorption of a circularly
polarized photon alters the state of the electron differently
from a linearly polarized photon.  This implies that a second
photon will see different states of the atom in the two cases.
So does a third photon, and so on.  It is then evident that
multiphoton absorption rates will depend on the state of polarization of the photons even if the initial electronic state is
totally unpolarized.  The same is true for stimulated multiphoton emission.  By way of contrast, note that the photon
polarization also affects single photon transitions.  But in
that case, it is only the differential cross section - for example, the angular distribution of photoelectrons - and not the
total transition rate that depends on the light polarization.

Qualitatively, one can understand this light polarization
effect by considering a succession of transitions and using
the dipole selection rules.  For light linearly polarized along
the z-axis the selection rules are $\Delta J = \pm 1,0$ (with 0 0 forbidden) and $\Delta M = 0$.  For light circularly polarized and propagating

along the z-axis the selection rule on M becomes $\Delta M = \pm 1$, the (+) corresponding to right and the (−) to left circularly polarized light.  In a simple hydrogenic model, the channels open to a multiphoton transition from an S initial state are shown in Fig. 1.  Clearly, light linearly polarized has more channels available to it, and this is the result of the selection rules on M.

## DEPENDENCE OF MULTIPHOTON ABSORPTION ON LIGHT POLARIZATION

The first obvious implication of the previous discussion is that the total transition rate of multiphoton processes depends on the light polarization.  Consider first a bound-bound transition; say 2-photon absorption from the ground state $|nS\rangle$ of an alkali atom. Owing to the selection rules, the absorption of two circularly polarized photons (see Fig. 1) must lead to a $|n'D\rangle$ state.  If, however, the frequency of the light is such that $E_{ns} + 2\hbar\omega = E_{n's}$ the process $|nS\rangle + 2\hbar\omega \rightarrow |n'S\rangle$ can only take place either with light linearly polarized or with two photons of opposite circular polarization.  This property has recently been used in a number of experiments to achieve Doppler-free excitation[2] of atomic states via the absorption of two photons propagating in opposite directions. This represents a significant advance in high resolution spectroscopy.

In multiphoton ionization, the final state is in the continuum and hence it can be written as a superposition of partial waves[3] i.e. states with well defined angular momentum. Unlike a bound-bound transition, for a given continuum final state energy, – which is determined by the number of photons needed to ionize – all angular momenta are available.  How much each of the partial waves contributes is determined by the light polarization, according to the rules illustrated in Fig. 1.  Thus, for example, 3-photon ionization with circularly polarized light leads to a photoelectron with orbital angular

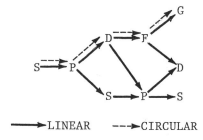

━━━▶LINEAR    ---▶CIRCULAR

Fig. 1:  Channels available for four-photon ionization with linearly or circularly polarized light.  Angular momentum channels are indicated in spectroscopic notation.

momentum $\ell=3$ (F channel); with linearly polarized light, it leads to a photoelectron whose state is a superposition of a P- and an F-wave. The light polarization will therefore influence not only the angular distribution, but also the total photoionization rates. This was first noticed experimentally by Fox et al.[4] and has subsequently been observed in a number of other experiments.[5-7] The theoretical interpretation of such experiments requires the detailed calculation of the relevant transition rates, which has been the subject of references 8-11.

To briefly review the basic aspects of the theory, recall that, for example, the rate of 3-photon ionization[3] is given by $W_3 = \hat{\sigma}_3 I^3$ where the generalized cross section $\hat{\sigma}_3$ can be written as

$$\hat{\sigma}_3 = 2\rho\alpha^3 \frac{m}{\hbar} \omega^3 K \left| \sum_m \sum_n \frac{\langle K|\varepsilon\cdot r|m\rangle\langle m|\varepsilon\cdot r|n\rangle\langle n|\varepsilon\cdot r|1\rangle}{(\omega_m-\omega_1-2\omega-i\Gamma_m)(\omega_n-\omega_1-\omega-i\Gamma_n)} \right|^2 \tag{1}$$

where $\alpha$ is the fine structure constant, m the electronic mass, $\omega$ the photon frequency, $\vec{\varepsilon}$ its polarization vector, and $\vec{r}$ the position operator of the electron. The states occuring in the matrix elements are atomic states with $|1\rangle$ being the initial and $|\vec{K}\rangle$ the final state of the outgoing electron with wave vector $\vec{K}$. Atomic energies are given by $\hbar\omega_1$, $\hbar\omega_n$ etc., and the double summation extends over all atomic states (including the continuum in principle). The light intensity I is measured in number of photons per $cm^2$ per sec; and $\hat{\sigma}_3$ is measured in $cm^6 sec^2$. The resonance denominators in Eq. (1) contain the widths of the intermediate states and such widths are in many cases[12] functions of the light intensity. Strictly speaking, there is also an intensity dependent shift of the intermediate level that should be included. If the photon frequency is such that one of the energy differences in the denominators of Eq. (1) vanishes, or is very small compared to $\Gamma$, the summation to which this term belongs reduces to essentially one term which gives the major contribution to the sum. Then the process is referred to as "resonance multiphoton ionization". Obviously, for 3-photon ionization, one can have two simultaneous resonances.

For N-photon ionization we have a transition rate given by $W_N = \hat{\sigma}_N I^N$ where the generalized cross section is

$$\hat{\sigma}_N = \frac{(2\rho\alpha)^N}{4\pi^2} \frac{m}{\hbar} K\omega^N \left| M_{K1}^{(N)} \right|^2 \tag{2}$$

and $M_{K1}^{(N)}$ is an (N-1) fold summation of the type of Eq. (1)
involving N atomic matrix elements and a product of (N-1)
resonance denominators of the form $(\omega_{n_{N-1}}-\omega_1-(N-1)\omega-i\Gamma_{N-1})\cdots$
$(\omega_{n_2}-\omega_1-2\omega-i\Gamma_2)$ $(\omega_{n_1}-\omega_1-\omega-i\Gamma_1)$ where $|n_1\rangle$, $|n_2\rangle$, ... $|n_{N-1}\rangle$
are the intermediate atomic states over which the summation is
performed. In N-photon processes, one can have (N-1) simultaneous resonances. But, unless there happens to be an unusual coincidence, in an experiment with one laser one expects
to have only one resonance for a given frequency. Three important exceptions must however be noted. First, a double
resonance can be readily achieved by using two lasers; and a
number of such experiments have been performed.[13,14] Second,
multiple resonances, or near-resonances, may occur in very
high order ionization (say N>20) with a single laser where the
last few photons induce transitions in the dense part of the
spectrum. Third, multiphoton processes in molecules will often
exhibit multiple resonances.

From the above basic equations it is evident that, as a
function of frequency, multiphoton ionization transition rates
will exhibit peaks and valleys depending on whether the photon
frequency in one of the denominators is near resonance or not.
Note that the absolute square in Eqs. (1) and (3) is taken
after the summation is performed and as a result there will
usually be considerable interference for photon frequencies
between resonances. The influence of the light polarization
comes from the matrix elements in the numerator. Thus, even
if the photon frequency is in near-resonance with an intermediate state, the transition probability will be small if the
light polarization is such that the transition to the resonant
state is forbidden. Consider, for example, 3-photon ionization from the ground state 3S of Na with a laser whose fre-
quency $\omega$ is such that $2\hbar\omega = E_{5S}-E_{3S}$ which means there is a
2-photon resonance in one of the denominators. If the light
is circularly polarized the resonance has no effect because
the transition $|3S\rangle+2\hbar\omega^+|5S\rangle$ is not allowed with circularly
polarized light of a single sense. The transition rate will
then be small since the only contribution will come from the
off resonant D-state. If on the other hand the light is
linearly polarized, the rate will be enhanced due to the resonance with the 5S state to which the 2-photon transition
is now allowed. The case of a 2-photon resonance with an
intermediate D-state (say $E_{4D}-E_{3S} = 2\hbar\omega$) is less obvious be-
cause then the 2-photon transition $|3S\rangle+2\hbar\omega^+|4D\rangle$ is allowed
for both circular and linear light polarization. Yet the
rates in the two cases will be different owing to angular
momentum considerations. In fact, the rate for circular
polarization will in this case be larger than for linear
polarization, although the latter has more channels available!

To discuss the general case, a more elaborate analysis
is necessary. The basic result is that the ratio of the rate
of multiphoton ionization for linearly polarized light to
that for circularly polarized is in general different from
unity, and is a function of the photon frequency. For low
order N-photon ionization (say, N=2,3,4), and for certain
ranges of photon frequencies, circularly polarized light will
give an ionization rate higher than linearly polarized. The
fact that linear polarization has more channels available is
in these cases counterbalanced by the strength of the channels
available to circular polarization. For large N however,
linearly polarized light will in most cases give higher
transition rates, because the number of channels open to
linear polarization simply outweighs the strength of circular
polarization.

Independently of the above general conclusions, it must
be always kept in mind that the ratio of the rates for circular to linear can be a very sensitive function of photon frequency and therefore each case must be examined in detail.
This ratio has been measured in several experiments[4-7] of
2- and 3-photon ionization of alkalis. The results as well as
calculated values of the ratio are given in Table I.

TABLE I

| Atom | Order of Process | Photon Wavelength | Ratio $\hat{\sigma}^{circ.}/\hat{\sigma}^{linear}$ | | Exper. Reference |
|------|------------------|-------------------|---------------|--------------|------------------|
| | | | Theory* | Experiment | |
| Cs | 2-photon | 3471 $\overset{o}{A}$ | 1.14 | 1.25±0.28 | 4 |
| Cs | 3-photon | 6942 $\overset{o}{A}$ | 2.23 | 2.15±0.4 | 4 |
| K | 3-photon | 6942 $\overset{o}{A}$ | 2.18 | 2.34±0.22 | 5 |

* Calculated values by P. Lambropoulos and M. R. Teague[11]

For the experiments of references 6 and 7 it is not
possible to calculate ratios which can be compared with the
experiment without knowing the detailed spectrum of the laser.
Calculations for a range of wavelengths averaged over typical
line-shapes yield reasonable agreement with experiment.

In general, the theoretical results are within the experimental errors.  Such measurements are interesting as they provide one test of theoretical calculations and there is a definite need for such tests.  At this point the available experimental data are neither extensive nor accurate enough to be used for distinquishing between various sets of calculations.

For near-resonance N-photon ionization in which the first N-1 photons bring the electron to near-resonance with an excited state, the ratio of the rate for circular to linear depends on the ratio of the two bound-free matrix elements that lead to the two allowed channels in the continuum.  For example, in 3-photon ionization from an S-state the two allowed channels are the F and P.  The ratio of the total rates can thus be used to obtain the absolute value of the ratio of the two bound-free matrix elements.[15]  Combined with absolute measurements of the generalized cross sections, it can also be used to obtain absolute values of the matrix elements themselves.  Moreover, the sign of the matrix elements can be obtained from measurements of angular distributions of photoelectrons for light of various polarizations.  Angular distributions[16] of photoelectrons in multiphoton ionization are a very interesting topic and offer attractive possibilities as a tool for the exploration of excited atomic and molecular states.  At this point, however, there are very few experimental data on this subject.[16]

## SPIN-ORBIT COUPLING EFFECTS

Since light circularly polarized has definite helicity, multiphoton ionization in the presence of spin-orbit coupling can lead to photoelectrons with definite spin-polarization along the direction of propagation of the photon.  This will usually occur when there is a near-resonance with an intermediate atomic state whose fine structure splitting is larger than the laser line-width.  How this comes about can be seen qualitatively by examining the channels available to the process.  Consider, for example, 3-photon ionization of an alkali atom.  For light right circularly polarized we have the following channels:

$$S_{1/2}(\tfrac{1}{2}) \longrightarrow P_{3/2}(\tfrac{3}{2}) \longrightarrow D_{5/2}(\tfrac{5}{2}) \longrightarrow F_{7/2}(\tfrac{7}{2}) \uparrow$$

$$S_{1/2}(-\tfrac{1}{2}) \begin{cases} \nearrow P_{3/2}(\tfrac{1}{2}) \longrightarrow D_{5/2}(\tfrac{3}{2}) \longrightarrow F_{7/2}(\tfrac{5}{2}) \uparrow\downarrow \\ \searrow P_{1/2}(\tfrac{1}{2}) \longrightarrow D_{3/2}(\tfrac{3}{2}) \longrightarrow F_{5/2}(\tfrac{5}{2}) \uparrow\downarrow \end{cases}$$

The numbers inside the parentheses are the $m_j$ values of the
states.  The arrows at the right hand side indicate the photoelectron spin polarization, with up and down indicating spins
in and opposite to the direction of propagation of the photon,
respectively.  Evidently, the first channel leads to photoelectrons 100% spin-polarized up.  However, even if the light
frequency is such as to be in resonance with the $P_{3/2}$ state,
there will also be an admixture of the second channel.  This
channel leads to a mixture of polarizations.  If all photoelectrons are collected, the net spin polarization is

$$P = \frac{N_\uparrow - N_\downarrow}{N_\uparrow + N_\downarrow}$$

where N indicates number of photoelectrons.  Thus the net
photoelectron polarization will depend on the extent to which
the two channels mix, which is determined by the Clebsch-Gordon
coefficients appearing in the equations.  Similarly, if the
photon frequency is in resonance with the $S_{1/2} \rightarrow P_{1/2}$ transition, only the last channel contributes; but it leads to a
mixture of polarizations.  In this case, the mixture is such
that the net polarization is about -70%, where +100% polarization corresponds to electrons totally polarized in the
direction of the photon propagation.  If the photon frequency
is between the $P_{3/2}$ and $P_{1/2}$ states, the admixture of the
channels is also influenced by the energy-difference denominators (see Eq. (1)).  For a frequency corresponding to about

$$E_{P_{1/2}} - E_{S_{1/2}} + \frac{2}{3}\left(E_{P_{3/2}} - E_{P_{1/2}}\right)$$ the spin-polarization is

+100%.  The calculation of the polarization as a function of
photon frequency requires some angular momentum algebra as
well as the calculation of a number of matrix elements.
One can then obtain a curve representing the polarization as
the laser is tuned around the $P_{1/2}$ and $P_{3/2}$ levels.  Such
calculations have been presented in references 11, 17 and 18.
More recent calculations[11] have shown that non-zero spin polarizations in 2- and 3-photon ionization of Cs (and to a lesser
extent in Rb) can also be obtained relatively far from resonance with a P-doublet; for example, between the 6P and 7P in
Cs.  This results from interference between contributions of
different P-doublets.  However it occurs near the minimum of
the generalized cross section which makes its observation
more difficult.  On the other hand, data in such cases provide
a rather severe and valuable test of theoretical calculations.
Although the above discussion has been concerned with the
alkalis, photoelectrons emitted in multiphoton ionization of
any atom will in most cases be spin-polarized to some degree
if the laser line-width is smaller than the fine structure
splitting of a level participating in the process.  Unlike

single-photon ionization, spin-polarized photoelectrons are the rule rather than the exception in multiphoton ionization.

At this point, there are only preliminary results on the observation of spin-polarized photoelectrons from multiphoton ionization. The experiments[14] have been performed in Na with two lasers, one tuned around the 3S → 3P transition and the other around the 3P → 4D. Polarization has been observed but it has been found to depend on laser power which can be explained on the basis of resonance effects (see the relevant section of this article and also ref. 14). Refinements of the experiments are presently underway.

## OPTICAL PUMPING EFFECTS

Multiphoton processes with circularly polarized light will in some cases be affected by optical pumping effects[14,19] between fine or hyperfine structure substates. Consider, for example, resonance 2-photon ionization via the transition $nS_{1/2} \rightarrow n'P_{1/2} \rightarrow \vec{K}$. For light perfectly (100%) right circularly polarized, only the induced transition $S_{1/2}(-1/2) \rightarrow P_{1/2}(1/2)$ can take place. But the $P_{1/2}(+1/2)$ can also decay spontaneously to the $S_{1/2}(+1/2)$ by emitting a linearly polarized photon. If this part of the spontaneous lifetime is shorter than the laser pulse duration, atoms are pumped into the $S_{1/2}(+1/2)$ state out of which they cannot be taken with right circularly polarized light. Consequently, the relative populations of the two substates $S_{1/2}(\pm1/2)$ change with time during the laser pulse. The end result, in the above case, is that the total photoelectron production is less than one would have predicted by ignoring this effect. In other words, the "effective" pulse duration of the laser is smaller than the actual duration.

If the light is not exactly 100% circularly polarized, but has a small admixture of the opposite circular polarization, optical pumping will be taking place in both directions; i.e. $S(+1/2)$ is pumped into $S(-1/2)$ and vice versa. The point is that for large laser intensities even a very small percentage of the opposite polarization can be sufficient to saturate the respective transition. In such cases, one must obtain expressions for the populations of the various substates as a function of time, and use these time dependent populations to find total photoionization rates, photoelectron polarizations, etc. This problem is of particular importance when two lasers of different frequencies merged into one beam are used.[14] For instrumental reasons, it is then rather difficult to have both 100% circularly polarized. It should also be noted that optical pumping effects depend on the ratio of the bound-free matrix elements which contribute to the process.

THE DEPENDENCE OF MULTIPHOTON IONIZATION ON LIGHT INTENSITY

## Photon Correlation Effects

The most simplistic approach to N-photon ionization leads to a transition rate of the form $W_N = \hat{\sigma}_N I^N$, i.e. proportional to the Nth power of the light intensity. This result rests upon two basic assumptions, not always stated as such in all papers. First, the electric field vector is assumed to be well represented by a perfectly amplitude-stabilized monochromatic wave of the form $\vec{\varepsilon}(t) = \vec{\varepsilon}_0 \cos\omega t$ where $\varepsilon_0$ is constant. The equivalent assumption when the field is quantized is that the initial state of the field is a pure coherent Glauber state.[20] Second, it is either assumed that the widths and shifts of the intermediate atomic states are negligible - as is the case in off-resonance processes - or, if not negligible, that they are independent of the laser intensity. Both of the above assumptions are not necessarily valid in many multiphoton ionization experiments.[14] It is therefore important to have a detailed theoretical understanding of this more general situation if one is to interpret experiments correctly.

The amplitude of a high power laser usually is not well stabilized. If nothing else, such lasers more often than not operate in more than one mode. And it is very important whether these modes are correlated, and if so, to what extent. From the point of view of quantized radiation theory, the physical picture is as follows: In multiphoton absorption, the atom absorbs more than one photon in going from the initial to the final state. Consequently, it is not only the average rate of arrival of photons that matters, but also whether they arrive in groups; and if so, the details of how they are grouped (or bunched, as it is usually called) can have a substantial effect on the rate. It is evident that the rate of absorption will be higher if the photons arrive bunched than if they arrive "single file", so to speak. This is referred to as the effect of photon statistics or correlations. A more careful analysis[21-23] shows that N-photon ionization is proportional to the Nth order correlation function of the radiation field, provided there are no resonances with intermediate atomic states. Denoting this correlation function by $G_N$, the transition rate is written as $W_N = \hat{\sigma}_N G_N$. The correlation function $G_N$ can be related to the intensity (which is the first order correlation function) by an equation of the form $G_N = \beta_N I^N$, where the coefficient $\beta_N$ depends on the state of the light source. If the light is in a pure coherent Glauber state, then $\beta_N = 1$. If it is in a chaotic state[20-23] (which is the state of light emitted by thermal sources), then $\beta_N = N!$ Clearly, a chaotic source, having the same intensity as a pure coherent source, will give a rate higher by a factor of $N!$, which can be a very large number. Note for example that $15! \cong 1.3 \times 10^{12}$ and that multiphoton ionization of order higher than 15 has been observed in noble gases.[24]

A typical high power pulsed laser usually is in neither of the above states and moreover the theoretical representation of its state is not a trivial matter, if feasible at all. Experimentally, one can at least conduct measurements under a variety of conditions and thus demonstrate the effect of photon correlations. In general, a single-mode laser (especially a cw laser) operating well above threshold can be assumed to have a well stabilized amplitude. Its photon statistics is then very close to that of a coherent state. The light of such a laser can then be converted to something resembling chaotic light, by introducing some randomization. Without going into details, we quote two experiments, one by Shiga and Imamura[25] and another by Kransinski et al.,[26] in which it was found that 2-photon absorption with a randomized laser gave a rate higher by about a factor of 1.8 than before randomization. A completely chaotic laser should have given a factor of 2.

A much more dramatic demonstration of this effect has recently been given by Lecompte et al.[27] Using a Nd-glass pulsed laser in single-mode and multimode operation, they observed 11-photon ionization rates in Xe differing by many orders of magnitude. In this experiment, it appears that the single-mode laser can be considered to have the statistics of a pure coherent state.

The enhancement factor N! given by a chaotic source is not the maximum photon statistics factor one can obtain. Chaotic sources have Gaussian fluctuations and that is what gives the N!. A laser, especially in multimode operation, can have fluctuations stronger than Gaussian which will result in a larger enhancement factor. Obviously, knowledge of the coefficient $\beta_N$ is necessary if one is to extract the magnitude of $\hat{\sigma}_N$ from measurements of $W_N$. Ideally, one would want to perform such experiments with well stabilized, single mode lasers for which $\beta_N = 1$. This has not always been possible thus far, and uncertainty in the value of $\beta_N$ is probably one of the most serious causes of discrepancy between theoretical and experimental[28] values of $\hat{\sigma}_N$.

## Resonance Effects

As already noted, the equations for $\hat{\sigma}_N$ involve in the denominators energy differences with imaginary parts (see Eq. (1)). When the photon frequency is in resonance with an intermediate atomic state, the real part vanishes and only the imaginary part remains in the denominator. For weak light intensities, such imaginary parts are determined by the inverse of the spontaneous lifetime of the resonant excited state.

For higher light intensities, the lifetime of the resonant state is no longer determined by spontaneous decay alone, but induced processes[12] contribute as well; and their contribution may be the dominant one.

Consider, for example, 2-photon ionization of an alkali atom via a resonance with a P-state. If the initial state is $|nS>$, and the photon frequency $\omega$ is such that $E_{nS}+\hbar\omega = E_{n'P}$ where $|n'P>$ (with $n'>n$) is an excited state, the second photon will ionize the $n'$P-state. Typically, when the photon flux reaches about $10^{19}$ photons/cm$^2$ sec (the exact value depending on the particular transition) the stimulated lifetime becomes comparable to or larger than the spontaneous lifetime of $|n'P>$. In that case the width $\Gamma_{n'}$ (as well as the associated shift $S_{n'}$) occuring in the denominator will depend on the light intensity and its spectral composition; and usually not in a simple way.

For a somewhat different situation, consider 3-photon ionization of an alkali atom with a 2-photon resonance. Let the photon frequency be such that $E_{nS}+2\hbar\omega = E_{n'D}$ where $|n'D>$ is an excited D-state. The spontaneous lifetime of the D-state is determined by its transitions to the lower P-states. For stimulated emission to become significant, the light intensity must be such that the 2-photon stimulated emission $|n'D> \rightarrow |nS> + 2\hbar\omega$ become comparable to or larger than the single-photon spontaneous decay of the D-state. For this to occur, a higher light intensity than in the previous case is required; because we now have a 2-photon process which is of higher order. Clearly, for N-photon resonances (with N>2) in higher order multiphoton ionization, even larger light intensities will be necessary for stimulated emission to become significant. Of course in the 3-photon ionization case mentioned above, the light intensity will probably enter when the ionization rate of $|n'D>$ becomes larger than its spontaneous decay. Thus each case must be examined in detail. The general consequence, however, is that the presence of intensity-dependent quantities (including the shifts) in the denominators will cause the transition rate to depart from the $I^N$ dependence on intensity. And this departure will change with changing intensity. Such effects have been seen in a number of experiments.[14,24]

One word of caution: These effects should be differentiated from "instrumental saturation" which occurs when the intensity is so large that all atoms in the interaction region are ionized. This type of saturation can be avoided by reducing the time of interaction. For some of the experiments in which the departure from the $I^N$-dependence has been observed, it is not clear at this time which of the two types of saturation is responsible for the observed behavior.

## ELECTRIC QUADRUPOLE CONTRIBUTIONS

It has been shown recently[29] that under certain conditions the dipole approximation is not sufficient for the interpretation of multiphoton ionization. For certain photon frequencies, there will be substantial contributions from quadrupole transitions which will be larger than the dipole by several orders of magnitude. As an example, consider 2-photon ionization of an $|nS>$ state with a photon frequency $\omega$ such that $E_{nS} + \hbar\omega = E_{n'D}$. The dipole nS $\rightarrow$ n'D transition is forbidden but the quadrupole is allowed. Detailed calculation shows that in this case the dominant 2-photon ionization will proceed via a quadrupole transition to the $|n'D>$ state and a dipole transition to the continuum.

For higher order multiphoton ionization, there can be a variety of combinations of quadrupole and dipole transitions. Thus for 3-photon ionization, and by choosing two appropriate laser frequencies $\omega_1$ and $\omega_2$, one can observe the process

$$nS \xrightarrow{\omega_1} n'P \xrightarrow[(Q)]{\omega_2} n''P \xrightarrow{\omega_2} \vec{K}$$

where the quadrupole transition is indicated by (Q). This process, as well as the process nS $\rightarrow$ n'P $\underset{(Q)}{\rightarrow}$ n''F $\rightarrow\vec{K}$ have recently been observed[13] in Na. Experiments of this type offer very interesting possibilities for the measurement of quadrupole transitions between excited states. Moreover, quadrupole contributions to multiphoton ionization will make the dominant contribution to the transition rate for photon frequencies corresponding to the deep valleys of the generalized cross section. Recall that such valleys are caused by interference and cancellation between the dipole contributions.

## CONCLUDING REMARKS

This has been a rather sketchy review of a few topics on multiphoton ionization. There has been considerable surge of experimental as well as theoretical activity in the last three years. Although there is general qualitative agreement between theory and experiment, we are only beginning to see quantitative comparisons. Presumably, in the next two years or so, we will be able to gain a much more detailed understanding of the observed facts and in the process improve our predictive capability. Recent experiments[30-34] − some of which are still underway − will provide very valuable data for comparison with perturbation theory as well as more recently proposed approaches.[35,36]

The list of references quoted in this review is only intended to be representative and not exhaustive. For a more complete list, the interested reader is referred to a recent review by Bakos[37] and to a forthcoming article by the author.[38] The article by Bakos contains a fairly complete list of Soviet references up to 1972.

## ACKNOWLEDGMENT

The author gratefully acknowledges the hospitality of the Joint Institute for Laboratory Astrophysics of the University of Colorado during the preparation of this article.

## REFERENCES

* Work supported by a grant from the National Science Foundation No. MPS74-17553.

1. See, for example, J.J. Sakurai, Advanced Quantum Mechanics, (Addison-Wesley Publishing Co., Reading, Mass. 1967) p. 31

2. N. Bloembergen, M.D. Levenson and M.M. Salour, Phys. Rev. Letters 32, 867 (1974); see also T.W. Hänsch, K.C. Harvey, G. Meisel and A.L. Schawlow, Optics Comm. 11, 50 (1974), and H.T. Duong, S. Liberman, J. Pinard and J.L. Vialle, Phys. Rev. Lett. 33, 339 (1974).

3. H.B. Bebb and A. Gold, Phys. Rev. 143, 1 (1966).

4. R.A. Fox, R.M. Kogan and E.J. Robinson, Phys. Rev. Lett. 25, 1416 (1971); also Bull. Am. Phys. Soc. 16, 1411 (1971).

5. M.R. Cervenan and N.R. Isenor, Optics Sommun. 10, 280 (1974).

6. P. Agostini and P. Bensoussan, Appl. Phys. Lett. 24, 216 (1974).

7. P. Agostini, P. Bensoussan and M. Movssessian, Phys. Lett. A, 53, 89 (1975).

8. P. Lambropoulos, Phys. Rev. Lett. 28, 585 (1972); and 29, 453 (1972).

9. S. Klarsfeld and A. Maquet, Phys. Rev. Lett. 29, 79 (1972).

10. Y. Gontier and M. Trahin, Phys. Rev. A 7, 2069 (1973).

11. For more recent and detailed calculations on the alkalies, see, P. Lambropoulos and M.R. Teague (submitted to J. Phys. B).

12. P. Lambropoulos, Phys. Rev. A 9, 1992 (1974); Y. Gontier and M. Trahin, Phys. Rev. A 7, 1889 (1973); C.S. Chang and P. Stehle, Phys. Rev. Lett. 30, 1283 (1973).

13. Melissa Lambropoulos, S.E. Moody, S.J. Smith and W.C. Lineberger, Phys. Rev. Lett.

14. For a review of other two-laser experiments, see, P. Lambropoulos and Melissa Lambropoulos, Atoms in Intense Electromagnetic Fields, Invited paper presented at the International Symposium on Electron and Photon Interactions with Atoms in Honor of Ugo Fano, University of Stirling, Stirling, Scotland, July 16-19, 1974 (to be published by Plenum Press).

15. P. Lambropoulos, J. Phys. B: Atom. and Molec. Phys. $\underline{6}$, L319 (1973).

16. Melissa Lambropoulos and R.S. Berry, Phys. Rev. A $\underline{8}$, 855 (1973); also S. Edelstein, M. Lambropoulos, J. Duncanson and R.S. Berry, Phys. Rev. A $\underline{9}$, 2459 (1974).

17. P. Lambropoulos, Phys. Rev. Lett. $\underline{30}$, 413 (1973); also, J. Phys. B: Atom. and Molec. Phys. $\underline{7}$, L33 (1974).

18. P.S. Farago and D.W. Walker, J. Phys. B: Atom. and Molec. Phys. $\underline{6}$, L280 (1973).

19. P. Lambropoulos, presented at the Fourth International Conference on Atomic Physics, Heidelberg, July 1974. Also P. Lambropoulos and K.M. Stuart (to be published).

20. R. Glauber, Phys. Rev. $\underline{131}$, 2766 (1963).

21. P. Lambropoulos, C. Kikuchi and R.K. Osborn, Phys. Rev. $\underline{144}$, 1081 (1966); also P. Lambropoulos, Phys. Rev. $\underline{168}$, 1418 (1968).

22. B.R. Mollow, Phys. Rev. $\underline{175}$, 1555 (1968).

23. G.S. Agrawal, Phys. Rev. A $\underline{1}$, 1445 (1970).

24. P. Agostini, C. Barjot, J.F. Bonnal, G. Mainfray, C. Manus and J. Morellec, IEEE J. Quantum Electronics $\underline{QE-6}$, 782 (1970).

25. F. Shiga and S. Imamura, Phys. Lett A $\underline{25}$, 706 (1967).

26. J. Krasinski, S. Chudzynski, W. Majewski and M. Glodz, Optics, Commun. $\underline{12}$, 304 (1974).

27. C. Lecompte, G. Mainfray, C. Manus and F. Sanchez, Phys. Rev. Lett. $\underline{32}$, 265 (1974).

28. B. Held, G. Mainfray and J. Morellec, Phys. Lett A $\underline{39}$, 57 (1972).

29. P. Lambropoulos, Gary Doolen and S.P. Rountree, Phys. Rev. Lett. $\underline{34}$, 636 (1975).

30. E.H.A. Granneman and M.J. van der Wiel, J. Phys. B: Atom. and Molec. Phys. (to be published); also contributed

paper No. 471 of this conference.

31. J.E. Bayfield and P.M. Koch, Phys. Rev. Lett 33, 258 (1974); also contributed paper No. 473 of this conference.

32. G. Mainfray et al. (private communication).

33. Melissa Lambropoulos and S.E. Moody (private communication).

34. N.B. Delone, P.N. Lebedev Physical Institute, Academy of Sciences of the USSR, Preprint No. 146 (1974).

35. S. Geltman and M.R. Teague, J. Phys. B:  Atom. and Molec. Phys. 7, L22 (1974).

36. M.H. Mittleman, Phys. Lett. 47A, 55 (1974); also J. Gersten and M.H. Mittleman, Phys. Rev. A 10, 74 (1974).

37. J.S. Bakos, in Advances in Electronics and Electron Physics, Vol. 36, pp. 58-152 (1974).

38. P. Lambropoulos, to be published in the Advances in Atomic and Molecular Physics.

# LASER PHOTODISSOCIATION OF HYDROGEN MOLECULE IONS

# WITH FRAGMENT KINETIC ENERGY ANALYSIS

J. DURUP

Laboratoire des Collisions Ioniques*,

Université de Paris-Sud, 91405-Orsay, France

## 1. INTRODUCTION

Laser photodissociation of molecular ions is a powerful tool for studying the dissociation processes of simple ions in specific vibronic levels. It has been developed up to nowadays in two different and complementary ways:
- (i) measurement of total cross sections at variable excitation wavelength,
- (ii) registration of kinetic energy distributions of fragment ions formed at fixed wavelength.

Experiment (ii) permits a straightforward identification of the vibrational levels populated in the parent ions, and furthermore may be applied e.g. to triatomic ions to yield the vibrational distribution of the diatomic dissociation fragment. But except for accidental resonances experiment (ii) is of no value in the case of photo-predissociation, which is likely to be the main channel in the photodissociation of most ions of interest.

Experiment (i) on the contrary is best suited for studying discrete-to-discrete transitions, i.e. photo-predissociation, but gives no detailed information on the photodissociation process.

*Associated to the C.N.R.S.

It appears now possible to combine both methods as will be shown in section 3.   Future increased
availability of powerful tunable lasers operating
in the uv and vuv may then bring this technique to
a complete spectroscopy of dissociating and predissociating states of molecular ions.

The present review will be concerned only with
type (ii) experiments, or 'ion photofragment spectroscopy' (IPFS), which are presently performed in
Amsterdam, the Netherlands, and in Orsay, France,
and have been essentially run on $H_2^+$, $HD^+$, and $D_2^+$
ions.   Photodissociation of these ions through the
$1s\sigma_g \rightarrow 2p\sigma_u$ transition can be completely worked out
theoretically for any vibrational-rotational initial
level, so that this process is well suited for a
testing of the possibilities of quantitative IPFS.

Some of the relevant transitions in $H_2^+$ are
shown in fig. 1 with the potential curves of the

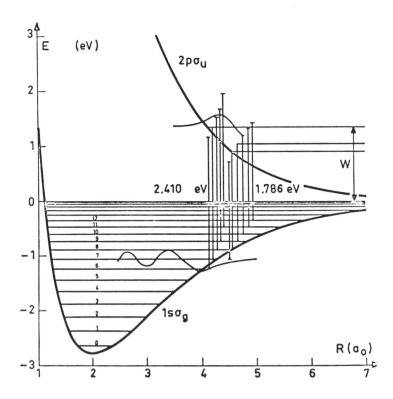

Fig. 1   Potential curves of $H_2^+$, $1s\sigma_g$ and $2p\sigma_u$,
with some of the observed transitions.

$1s\sigma_g$ and $2p\sigma_u$ states. The total relative kinetic
energy W of the fragments is shown for one particular transition. The wavy line sketches the relevant parts of the vibrational wavefunctions of the
corresponding initial and final states.

    The workers who have developed the laser photo-
dissociation IPFS experiment in Amsterdam[1-3] are
J. Los, J. G. Maas, and N. P. F. B. van Asselt.
The Orsay IPFS experiment[4-6] is performed by J.-
B. Ozenne, M. Tadjeddine, C. Pernot, A. Tabche-
Fouhaille and J. Durup, with earlier collaboration
of Pham D. and more recently of R. W. Odom.

    The present author is much indebted to Profes-
sor J. Los and his coworkers for sending him published and unpublished material for this review. Earlier reviews of ion photodissociation spectroscopy
were given by J. Durup[7] and by W. C. Lineberger[8].


## 2. <u>EXPERIMENTAL</u>

    Both IPFS experiments were done with fast ions,
typically 2 keV in Orsay and 10 keV in Amsterdam.
This is a general requirement for measuring accurately the relative kinetic energy of fragments
issuing from a parent species initially formed with
a non-negligible thermal energy spread. Thus for
$H_2^+$ an initial momentum spread of each proton equal
to 1.76 a.u. FWHM (corresponding to a maxwellian
distribution at 400 K) will be narrowed to 3.86 x
$10^{-3}$ a.u. after acceleration of the parent ion  to
an energy $T_0$ = 2 keV, and to 1.72 x $10^{-3}$ a.u. after
acceleration to 10 keV. For comparison, measured
W values correspond to moments of the order of 10
a.u. for each proton.

    However, from the formula connecting the lab-
oratory energy T of the ejected hydrom with the
total separation energy W in the c.o.m. frame,

(1)   $T = \frac{T_0}{2} + \cos\theta.\sqrt{T_0 W} + \frac{W}{2}$ (for symmetric molecules),

where $\theta$ is the ejection angle in the c.o.m. frame
with respect to parent ion flight direction, it
appears that since the relative momentum of the fragments is proportional to the second term in the r.
h.s. of (1) (which under usual conditions is much
larger than the third  term), the bandwidth in the

momentum measurement will be proportional to the uncertainty on T, which in any device is proportional to $T_0$.

Therefore, there will be for each apparatus some optimal value of $T_0$, which will depend on the temperature of the ion source, on its potential dispersion, on the resolving power of the momentum analyzer, and of course on the $T_0$-dependence of the signal intensity, which in turn is connected with the $T_0$-dependence of the ion beam current intensity in the acceptance angle.

In the IFPS experiments under consideration a stringent requirement is to achieve a high angular resolution along with the best possible alignment of secondary and primary beams, to ensure that $\theta$ in eqn (1) remains close to 0 or $\pi$. A misalignment of 3 mrad would shift the peaks by typically 0.1 momentum a.u. at 2 keV, or 0.3 a.u. at 10 keV, to be compared with peak separations e.g. between the v = 8 and v = 9 peaks of 0.6 or 0.8 a.u., respectively. Similarly, the width of the angular definition will broaden the peaks by corresponding amounts.

A detailed discussion of the factors controlling the resolving power in measurements of momentum distributions of dissociation fragments of fast molecular species was given by Fournier[9].

We shall now describe in more detail the experimental setups. The apparatus built in Amsterdam by Professor Los and his coworkers[1][2] is represented in fig. 2, the one we built in Orsay[4][5] in fig. 3. We shall in the following refer to these apparatuses and experiments by symbols (A) for Amsterdam and (O) for Orsay.

The hydrogen molecule ions are produced in a monoplasmatron in (A) and in a Nier-type 100-eV electron-impact ion source in (O). Acceleration electrodes, electrostatic lenses, deflecting plates and collimating holes are provided for preparation, adjustment and angular definition of the ion beam. In experiment (A) a primary mass analysis is performed before the interaction of the ions with the photons.

The laser beam crosses at right angle the ion beam. The lasers used were a continuous argon ion

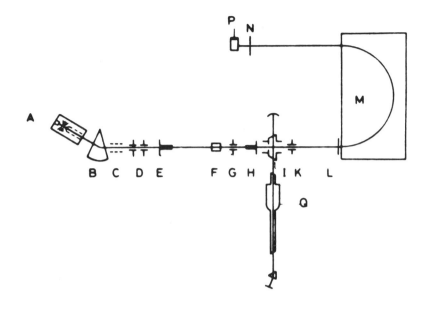

Fig. 2   The laser photodissociation apparatus of J.Los
and coworkers in Amsterdam (from refs (1)(2) ).

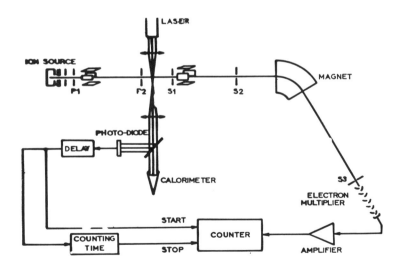

Fig. 3   The laser photodissociation apparatus of J.-
B. Ozenne and coworkers in Orsay (from ref.(5) ).

laser with selection of each of the seven main lines
in (A), and a relaxed ruby laser in (O); some experiments were performed in (O) with a Q-switched ruby
laser and with a flash-pumped tunable dye laser.
The ion beam in (A) is outside the laser cavity,
which increases the available photon intensity by a
factor of 15 to 20.

Most experiments were performed with light polarization parallel to the ion beam; thus, since the
$\sigma_g \to \sigma_u$ transition has its dipole moment along the
internuclear axis, the molecule ions which absorb
the light are mainly oriented parallel to their
flight direction, so that the $\theta$ distribution is
peaked at 0 and $\pi$. In (O) some additional experiments were performed at different polarization directions of the light with respect to the ion beam.

Momentum analysis of the fragment ions is performed by magnets. In (A) a pair of deflecting
plates located after the interaction region allow
for a scanning of the fragment ejection angle $\theta$. In
both experiments the ions are collected on electron
multipliers and counted.

The only noticeable background is due to collision-induced dissociation of the hydrogen molecule
ions interacting with the residual gas. To minimize
this effect ultra-high vacuum conditions (5 x 10$^{-8}$
to 1 x 10$^{-9}$ torr) are achieved in (A). The (O)
experiment is performed under usual high vacuum conditions, but the ions are counted only during the
time corresponding to the laser flashes.

In spite of large differences between the two
experiments, the signal-to-noise ratios and the
resolving powers obtained are comparable. However
the time necessary to scan a spectrum in (A) is ten
times shorter than in (O), which allows for a better
adjustment of the beam. Thus the theoretically
symmetrical forward-backward pattern which requires
perfect alignment was achieved in (A) but not in (O).

## 3. RESULTS AND DATA TREATMENT

Typical momentum spectra from both experiments
are shown in figs (4-8). Others may be found in the
original references[2][3][5][6][10]. A spectrum of
comparable intensity was obtained also with a tunable

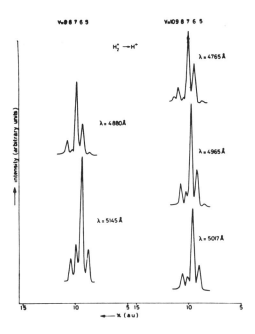

Fig.4 $H_2^+$ photofragment spectra from the (A) experiment.[2]

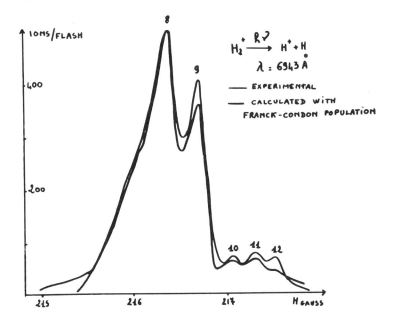

Fig.5 Experimental (upper curve) and computed (lower curve) $H_2^+$ photofragment spectra from the (0) experiment.

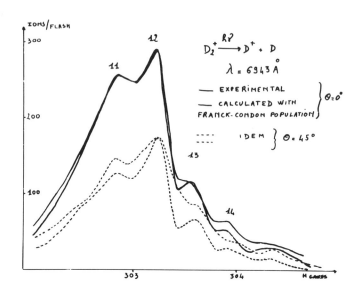

Fig.6 Experimental (upper curves) and computed (lower curves) $D_2^+$ photofragment spectra from the (0) experiment.

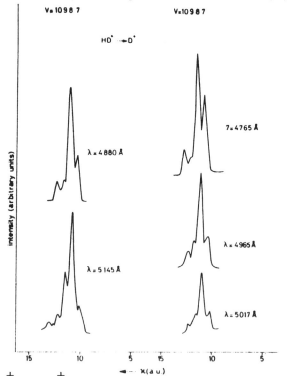

Fig. 7  $HD^+ \longrightarrow D^+ + H$ photofragment spectra from the (A) experiment (from ref. (3) ).

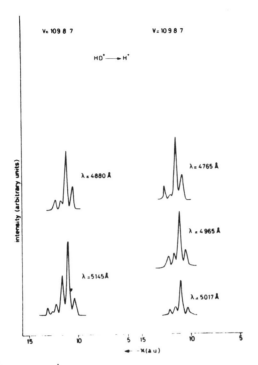

Fig. 8   HD$^+$ $\longrightarrow$ H$^+$ + D photofragment spectra from the
(A) experiment (from ref. (3) ).

dye laser at 6000 Å in (0).  The vibrational structure of the initial state is clearly resolved, and
the positions of the peaks agree very well with the
expected values (precise comparisons   are given in
ref.(2)(3)).

The differences in heights of the various peaks
arise from a combination of (i) the initial vibrational populations, (ii) the photodissociation
cross sections at a particular wavelength, and
(iii) the apparatus collection efficiency.  Factors
(i) and (iii) vary smoothly with the vibrational
quantum number.  Factor (ii) depends mainly on the
Franck-Condon factors (FCF) for photodissociation
as illustrated in fig. 1; for any given wavelength
the FCF's (which are known exactly from theory)
vary in a rather erratic way from one level to the
next one, and the general pattern of each spectrum
reflects well this variation(1-3)(5).

Momentum spectra of H$^+$ and D$^+$ from HD$^+$ are very
similar to each other as apparent on figs   7 and 8

618  J. DURUP

(3). This was expected since any level of the vibrational continuum of the $2p\sigma_u$ state will have
almost equal chances to give rise to $H^+ + D$ or
$D^+ + H$.

The main unknowns which may be extracted from
the experimental results are the vibrational populations in the incident hydrogen molecule ion
beams, which vary negligibly during the ion time
of flight.

Photodissociation cross sections for each vibrational-rotational level may be calculated theoretically with a high accuracy since Schrödinger equation
is exactly solvable for one-electron molecules in
elliptical coordinates. Precise potential curves
for the $1s\sigma_g$ and $2p\sigma_u$ states of $H_2^+$ were calculated
long ago by Bates et al[11]; from these curves vibrational wavefunctions in the Born-Oppenheimer
approximation are easily computed. The oscillator
strength for the $1s\sigma_g \to 2p\sigma_u$ transition was calculated at various internuclear distances by Bates[12].
From such data the photodissociation cross sections
of all vibrational levels of $H_2^+$ and $D_2^+$, with rotational quantum numbers K = 1 for $H_2^+$ and K = 2
for $D_2^+$, were calculated by Dunn at 25 Å intervals
of wavelengths[13]. Following Dunn's method Tadjeddine and Parlant[14] computed the photodissociation
cross sections of all (but the highest one) vibrational levels of $HD^+$, with rotational quantum number K = 2, at the wavelengths used in the IPFS
experiments of Orsay and Amsterdam.

Once the individual photodissociation cross
sections are known, derivation of the vibrational
populations from the experimental spectra requires
a full account of the apparatus effects and transformation Jacobians. In both the (A) and (O) experiments the successive peaks overlap each other, so
that any deconvolution procedure would be of doubtful significance. Therefore for both apparatuses
convolution procedures were used, where the full
spectra were reconstituted ab initio from arbitrary
but reasonable vibrational populations, using computer programs taking into account all experimental
features.

The arbitrary populations chosen were the Franck-
Condon populations, i.e. the populations which would
be generated if ionization took place according to

FCF's between the ground vibronic state of the parent molecule and the various levels of the $1s\sigma_g$ state of the ion.

Two further assumptions were made in the ab imitio computation of spectra: the orientation of the internuclear axis of the incident ions was taken as isotropic and the dissociation was assumed to take place in a very short time with respect to rotation. The latter approximation will be removed in further calculations but we do not expect the results to change significantly. The former assumption is more critical. A study of the angular dependence of the fragment ion ejection was performed in (A) for $H_2^+$ and $D_2^+$ and a significant anisotropy was reported[2]. Possible causes of anisotropy are (i) the angular dependence of the ionization cross section, and (ii) occurence of collisions in the ion source during the acceleration of the ions. The latter effect may be noticeable in the (A) experiment but not in the (O) one. The former effect may be known in principle from theory.

Finally, saturation effects have to be taken in consideration in the (O) experiment. If the orientations of the ions were fixed during the time they cross the laser beam ($2 \times 10^{-8}$ sec), those which have their axis close to the light polarization direction would be largely depleted if in the vibrational levels with high photodissociation cross sections, such as v = 9 in $H_2^+$ at 6943 Å, so that saturation effects would occur. Actually, spectra computed using this assumption are in sharp disagreement with experiment . In fact most of the ions do rotate during their flight  through the laser beam (only 10% of the $H_2^+$ ions under the relevant conditions have zero angular momentum), so that saturation effects were ignored in a first approximation. However, under stronger laser power an alignment of hydrogen molecule ions by selective photodissociation is possible as discussed earlier by Dehmelt et al[15].

Figs  5 and 6 show  comparisons between experimental and computed spectra from the (O) group. The general agreement appears very good. The effect of varying the direction of polarization of the  light is seen in fig. 6.

4.   VIBRATIONAL POPULATIONS OF $H_2^+$, $HD^+$, AND $D_2^+$

FORMED IN ELECTRON-IMPACT IONIZATION

Vibrational populations of $H_2^+$ and $D_2^+$ formed by impact of electrons of energy either between 106 and 141 eV or about 200 eV were deduced by von Busch and Dunn[16] from a photodissociation experiment with total cross section measurement as a function of wavelength (as designated by type (i) experiment in our Intro- duction).  At each wavelength the total cross sec- tion is a sum of factors including the population of each level and its individual cross section as known from theory[13].  Thus von Busch and Dunn could derive vibrational populations by a least- square fit of the ensemble of data.

Vibrational populations in photoionization ex- periments were obtained by various authors using photoelectron spectroscopy at 584 Å[17,18] or 304 Å [19], or threshold photoelectron spectroscopy[20]. We shall here use for comparison the values of Berko- witz and Spohr[18], which are among the most accurate, and which besides were used for comparison in an ab imitio computation of photoionization cross sections by Itikawa[21], who found good agreement between theory and experiment.

Table I compares relative vibrational populations of $H_2^+$, $HD^+$, and $D_2^+$ obtained from IPFS in Amsterdam (2)(3) with those derived by von Busch and Dunn[16], along with photoelectron spectroscopical results and with Franck-Condon factors.  All data were norm- alized to the peaks v = 7 in $H_2^+$, v = 8 in $HD^+$, and v = 10 in $D_2^+$.

No populations are given here from the  Orsay IPFS experiment.  Figs  5 and 6 and similar data for $HD^+$[10] show that for $H_2^+$, v = 8 to 12, $HD^+$, v = 9 to 12, and $D_2^+$, v =11 to 14 there is no signifi- cant differences between observed relative popula- tions and relative FCF's, within statistical uncer- tainty and possible corrections of a few percents of the main peaks to account for saturation effects. As may be seen from Table  I neither the Amsterdam IPFS results nor the data derived by von Busch and Dunn do show any systematic departure from Frank- Condon populations for these high levels.  However, the relative populations of the lower levels are significantly higher in the Amsterdam IPFS experi-

Table I

Relative vibrational populations of $H_2^+$, $HD^+$ and $D_2^+$ from various sources

| Vibrational level | FCF[a] | Photoelectron Spectroscopy Experimental (ref(18)) | Theory (ref(21)) | Photodissociation (ref(16)) | Electron Impact IPFS (ref(2))(3) |
|---|---|---|---|---|---|
| $H_2^+$ | | | | | |
| 5 | 2.03 | 1.85 | 1.84 | 2.03 | 2.39 ± 0.25 |
| 6 | 1.44 | 1.51 | 1.37 | 1.41 | 1.53 ± 0.01 |
| 7 | 1.00 | 1.00 | 1.00 | 1.00 | 1.00 |
| 8 | 0.69 | 0.77 | 0.72 | 0.65 | 0.67 ± 0.04 |
| 9 | 0.47 | 0.44 | 0.52 | 0.43 | 0.46 ± 0.01 |
| 10 | 0.32 | | 0.37 | 0.32 | 0.41 ± 0.03 |
| $HD^+$ | | | | | |
| 7 | 1.40 | 1.54 | | | 1.72 |
| 8 | 1.00 | 1.00 | | | 1.00 |
| 9 | 0.71 | 0.83 | | | 0.66 |
| 10 | 0.50 | 0.61 | | | 0.66 |
| $D_2^+$ | | | | | |
| 8 | 1.83 | 1.80 | 1.74 | 1.75 | 2.26 ± 0.05 |
| 9 | 1.36 | 1.27 | 1.33 | 1.42 | 1.50 ± 0.04 |
| 10 | 1.00 | 1.00 | 1.00 | 1.00 | 1.00 |
| 11 | 0.73 | | | 0.71 | 0.78 ± 0.03 |
| 12 | 0.53 | | | 0.51 | 0.55 ± 0.03 |
| 13 | 0.39 | | | 0.32 | 0.55 ± 0.08 |
| 14 | 0.28 | | | 0.27 | 0.30 ± 0.01 |

(a) Ref(22) for $H_2^+$ and $D_2^+$, (23) for $HD^+$. Accurate calculations of FCF's for vibrational-rotational levels of $H_2^+$ and $D_2^+$ by Villarejo(24) agree well with refs (22) and (23).

ment than the relative FCF's.

No detailed discussion of this tendency and of the comparison between electron and photon impact will be given here. Following factors are to be considered: (i) the electron ejection energy distribution; (ii) the magnetic quantum number distribution of the ejected electron; (iii) from (i) and (ii) the dependence upon internuclear distance of the transition moment[21]; (iv) possible orientation amistropies arising from (ii); (v) autoionization.

## 5.   CONCLUSIONS

The ion photofragment spectroscopy (IPFS) data here reviewed show the possibilites offered by this method. More thorough experiments with tuneable lasers may permit a complete description of the ionization and photodissociation of hydrogen molecules. Population analyses of the hydrogen molecule ions from any plasma source may be performed; anisotropies in ion beams may be detected.

Extension of such experiments to other simple ions by using tuneable lasers will give rise to an absorption                of discrete-to-continous and discrete-to-quasibound transitions, with a deter ination of the symmetry of such transitions connecting ground and metastable states with dissociating and predissociating ones.

## Acknowledgements

The present review was completed during a stay in Trent University, Peterborough, Ontario. The reviewer wishes to thank Professor R. E. March for his kind hospitality and Ms. J. Burrett for preparing the text for publication.

## REFERENCES

(1) N.P.F.B. van Asselt, J.G. Maas and J. Los, Chem. Phys. Lett. (1974) 24, 555.
(2) N.P.F.B. van Asselt, J.G. Maas and J. Los, Chem. Phys. (1974) 5, 429.
(3) N.P.F.B. van Asselt, J.G. Maas and J. Los, submitted to Chem. Phys.

(4)  J.-B. Ozenne, D. Pham and J. Durup, Chem. Phys.
     Lett. (1972) 17, 422.
(5)  J.-B. Ozenne, D. Pham, M. Tadjeddine and J.
     Durup, 4th Int. Symp. Molec. Beams, Cannes,
     France (1973), 591.
(6)  J.-B. Ozenne, J. Durup, R.W. Odom, C. Pernot,
     A. Tabche-Fouhaille and M. Tadjeddine, Colloque
     National des Collisions, Bordeaux, France (1975).
(7)  J. Durup, 21st Annu. Conf. Mass Spectrom., San
     Francisco, Calif. (1973), 109.
(8)  W.C. Lineberger, in Chemical and Biochemical
     Applications of Lasers, vol. I, Acad. Press (1974).
(9)  P. Fournier, Colloques Intern.C.N.R.S. (1973)
     217, 169.
(10) J.-B. Ozenne, J. Durup, R.W. Odom, C. Pernot, A.
     Tabche-Fouhaille and M. Tadjeddine, to be
     published.
(11) D.R. Bates, K. Ledsham and A.L. Stewart, Phil.
     Trans. Roy. Soc. (1953) A246, 215.
(12) D.R. Bates, J. Chem. Phys. (1951) 19, 1122.
(13) G.H. Dunn, Phys. Rev. (1968) 172, 1; JILA
     Report No. 92, Univ. of Colorado (1968).
(14) M. Tadjeddine and G. Parlant, to be published.
(15) H.G. Dehmelt and K.B. Jefferts, Phys. Rev. (1962)
     125, 1318; C.B. Richardson, K.B. Jefferts and
     H.G. Dehmelt, ibid. (1968) 165, 80 (and 170, 350).
(16) F. von Busch and G.H. Dunn, Phys. Rev. A (1972)
     5, 1726.
(17) D.W. Turner and P. May, J. Chem. Phys. (1966) 45,
     471; R. Spohr et al, Zts Naturf. (1967) 22a, 705;
     D.C. Frost et al., Proc. Roy. Soc. (1967) 296A,
     566; Chem. Phys. Lett. (1970) 5, 486; D.W.
     Turner, Proc. Roy. Soc. (1968) 307A, 15; L.
     Åsbrink, Chem. Phys. Lett. (1970) 7, 549; A.
     Niehaus et al., ibid. (1971) 11, 55.
(18) J. Berkowitz and R. Spohr, J. Electron Spectr.
     (1973) 2, 143.
(19) T.H. Lee and J.W. Rabalais, J. Chem. Phys. (1974)
     61, 2747.
(20) D. Villarejo, J. Chem. Phys. (1968) 48, 4014.
(21) Y. Itikawa, J. Electron Spectr. (1973) 2, 125.
(22) G.H. Dunn, J. Chem. Phys. (1966) 44, 2592.
(23) J.M. Peek, quoted in ref. (18).
(24) D. Villarejo, J. Chem. Phys. (1968) 49, 2523.

# TIME-DEPENDENCE AND ANISOTROPY
## OF COLLISION PRODUCTS

# ALIGNMENT AND ORIENTATION IN ATOMIC COLLISIONS

Joseph Macek

Behlen Laboratory of Physics
The University of Nebraska
Lincoln, NE  68508

## INTRODUCTION

Excitation of an atom by collisions leaves it generally in an anisotropic state.  Until recently, measurements of the anisotropy of collision excited states has concentrated on the alignment of atoms by a beam of electrons or heavy particles without regard to the direction of scatter of the incident ions.  Such sources of light have cylindrical symmetry so that the magnetic substates, with the beam axis as the axis of quantization, are incoherently populated.  The corresponding radiation can only exhibit linear polarization.  Polarization measurements have been made since the thirties,[1] although only in 1958 was the theory of polarization of impact radiation including fine and hyperfine structure correctly formulated by Percival and Seaton.[2]  It is not my intention to review the data on linear polarization measurements of the type considered by Percival and Seaton, although that is an area of continuing interest, rather I will discuss more recent techniques which examine impact radiation under conditions where the geometry has a lower degree of symmetry.

A collision geometry loses its axis of symmetry if another axis besides the incident beam axis is introduced.  Examples are, photon-scattered particle coincidence measurements where the momentum of the outgoing particle represents a second axis,[3,4] beam-foil collisions with tilted foils where the normal to the foil represents a second axis,[5,6] scattering from optically pumped atoms where the final electron momenta represents a second axis[7] and measurements incorporating spin analysis of either target or projectiles.[8]  A variety of formalisms have been developed to treat these measurements. I will first briefly review the relation between them in the context of atomic collision theory and their connection with "coherence". A few formulas pertinent to recent measurements will then be written down and finally I will indicate some preliminary conclusions which one can draw from the present fragmentary data. The conclusions will be rather general, but will serve to indicate future directions for experiment and theory.

## DESCRIPTION OF ATOMIC STATES

Atomic collision theory concentrates on calculating the scattering amplitudes $a_m(ab)$ for exciting a particular atomic magnetic substate m while the rest of the system undergoes a transition a → b. When we have complete knowledge of the entire collision system we can write a wave function for the excited atom

$$\psi = \sum_m a^{(ab)}_m \psi_m \tag{1}$$

where $\psi_m$ is an atomic eigenfunction. If all observed atoms are described by the same wave function $\psi$ the system is said to be in a pure state, or a state of maximal information. The state represented by Eq. (1) is also seen to be a coherent superposition of magnetic substates. Usually our knowledge of a and b is incomplete. This more general mixed state is described by a weighted average of the various pure state and is characterized by the density matrix $\rho_{m'm}$ defined by[9]

$$\rho_{m'm} = W_a^{-1} \sum_{ab} a^{(ab)*}_m a^{(ab)}_{m'} . \tag{2}$$

The density matrix describes both pure and mixed states so it represents an improvement over the language of amplitudes.

Fano[9] pointed out that certain linear combinations of density matrix elements offered considerable advantages for the description of atomic systems. These parameters, called state multipoles, are defined by

$$\rho^{[k]}_q = \sum_{m'm} \rho_{m'm}(-1)^{k-j-m'} (j'm'j - m|kq) , \tag{3}$$

where $(j'm'j-m|kq)$ are Clebsh-Gordan coefficients. The state multipoles are seen to be irreducible components of the density

matrix and therefore have more simple transformation properties
under coordinate rotations than do the non-reduced components
in equation (2). The theory of angular correlations is con-
ventionally expressed in terms of the state multipoles of
equation (3).

The language of state multipoles is entirely adequate for
describing atomic systems, but it is rather abstract and,
since it is framed in terms of matrix elements, is closely
married to the quantum theory. Accordingly, it does not make
direct contact with intuitive physical quantities and possible
alternative fundamental atomic theories such as the classical
theory. An alternative language in terms of mean values of
a set of operators offers some conceptual advantages.

Physical quantities, that is quantities measured in an
actual experiment relate to the mean value of some operator O
given by

$$<0> = \sum_{m',m=-j}^{j} \rho_{m'm} O_{mm'} = Tr\rho 0. \tag{4}$$

If the mean values of a set of $N = (2j+1)^2$ different operators
$O_1, O_2 \cdots O_N$ are known from measurements, then equation (4)
represents a set of N linear equations for the N density matrix
elements $\rho_{m'm}$ and can be inverted provided the set of N equations are linearly independent. The set of operators is said
to be complete[3] if the corresponding equations (4) are independent. Alternatively one could use the mean values of a
complete set of operators rather than the density matrix to
characterize atomic states.

For a multiplet of levels characterized by angular momentum quantum numbers a complete set of operators $T^{[k]}q$ are
constructed from the components of the angular momentum operator $\vec{J}$. The mean values of $T^{[k]}q$ equal, to within a normalization constant, the state multipoles $\rho^{[k]}q$. This is not to
say that the parameters are equivalent. By no means! The
mean values of angular momentum components represent a more
general language. For example, in those theories[4] which represent atomic states as ensembles of classical orbits, the
mean values of $T^{[k]}q$ can be calculated by averaging the tensors constructed from the electronic angular momentum components over an ensemble of classical orbits, whereas construction of the density matrix elements requires a representation
in terms of eigenstates, i.e., it is a language specific to
the quantum theory. Such a classical calculation could prove
useful in the limit of large principle quantum numbers near
the ionization threshold. In any event the mean values of

$T^{[k]}q$ tell us something about angular momentum transfer in atomic collisions.

## THEORY AND MEASUREMENTS

The main application of these concepts concerns collision induced fluorescence. The intensity of electric dipole radiation measured by a light detection system incorporating polarization analysis is

$$I = C \sum_{m_f} <(i'|\hat{\varepsilon}\cdot\vec{r}|f)(f|\hat{\varepsilon}^*\cdot\vec{r}|i)> \tag{5}$$

where $\hat{\varepsilon}$ is the polarization vector selected by the detector and C is a constant incorporating such factors as the total cross section, solid angle and detection efficiencies.

Eq. (5) has the form of a mean value of an operator $S = \vec{r}'|f)(f|\vec{r}$, but this operator is not in reduced form, nor is it constructed from angular momentum components. To express I in terms of the basic source parameters $<T^{[k]}q>$ we (a) write S in terms of its irreducible components $S^{[k]}q$ in a coordinate system defined by the detector, (b) use the Wigner-Eckart theorem to replace the mean values of $S^{[k]}q$ by the mean values of $T^{[k]}q$ and (c) transform to a preferred frame where the mean values take on their simplest form. The result is[10]

$$I = \frac{1}{3}CS\{1 - h^{(2)}(j_i,j_f)A_0^{det} + \frac{3}{2}h^{(2)}(j_i,j_f)A_{2+}^{det} \cos 2\beta$$

$$+ \frac{3}{2}h^{(1)}(j_i,j_f) 0_0^{det} \sin 2\beta\} \tag{6}$$

where $h^{(2)}(j_i,j_f)$ is a ratio of 6-j symbols, $\beta$ is a parameter specifying the degree of elliptical polarization selected by the detector ($\beta = 0$ represents linear polarization and $\beta = \pi/4$ circular polarization), and S is the line strength.

Upon transforming to a frame defined by the collision with z axis along the incident beam and x-axis in the scattering plane we have

$$0_0^{det} = 0_{1-}^{col} \sin\theta \sin\phi$$

$$A_0^{det} = A_0^{col} \frac{1}{2}(3 \cos^2\theta-1) + A_{1+}^{col} \frac{3}{2} \sin 2\theta \cos\phi \tag{7}$$

$$+ A_{2+}^{col} \frac{3}{2} \sin^2\theta \cos 2\phi$$

$$A_{2+}^{col} = A_0^{col} \frac{1}{2} \sin^2\theta \cos 2\psi + A_1^{col} \{\sin\theta \sin\phi \sin 2\psi$$

$$+ \sin\theta \cos\theta \cos\phi \cos 2\psi\}$$

$$+ A_{2+}^{col} \{\frac{1}{2} (1 + \cos^2 \theta) \cos 2\phi \cos 2\psi$$

$$- \cos\theta \sin 2\phi \sin 2\psi \}$$

where $\theta$, $\phi$ are the polar coordinates of the detector axis in the collision frame and $\psi$ is the angle between the xz plane and the axis of the linear polarization analyzer. The source parameters are defined in terms of mean values of angular momentum components as

$$O_1^{col} = <J_y>/j_i(j_i + 1)$$

$$A_0^{col} = <3J_z^2 - J^2>/j_i(j_i + 1)$$

$$A_{1+}^{col} = <J_x J_z + J_z J_x>/j_i(j_i + 1)$$

$$A_{2+}^{col} = <J_x^2 - J_y^2>/j_i(j_i + 1)$$

In general we see that four real parameters specifying the source anisotropy are relevant to angular correlation measurements involving electric dipole radiation when the source has a plane of symmetry. For sources with cylindrical symmetry $O_1^{col} = A_{1+}^{col} = A_{2+}^{col} = 0$ and we have to do with only one parameter $A_0^{col}$.

Eq. (6) pertains to observations with sharp spectral resolution. Collision experiments almost always employ low spectral resolution so that the light observed pertains to an entire fine structure multiplet in low Z atoms or a hyperfine multiplet in more complex atoms. Then the effect of the fine or hyperfine structure interaction is to perturb the initial alignment and orientation. Consider, for example, the fine structure interaction in light atoms where it is sufficiently weak that it plays no role in the collision excitation. The spin orbit interaction results in the reversible transfer of angular momentum from orbital motion to spin angular momentum. This exchange is readily accounted for in Eq. (6) by defining the alignment and orientation parameters as mean values of operators constructed from components of $\vec{L}$ rather than $\vec{J}$ and replacing $h^{(k)}(j_i,j_f)$ by $h^{(k)}(L_i, L_f)G^{(k)}(t)$

$$G^{(k)}(t) = \sum_{JJ'} \frac{(2J' + 1)(2J +1)}{2S + 1} \left\{ \begin{matrix} J'J\ k \\ L_iL_iI \end{matrix} \right\}^2 \cos W_{J'J}t \qquad (8)$$

The modulation of the alignment predicted by Eq. (8) has been observed in beam-foil collisions for many systems. A particularly nice illustration is provided by Wittmann et al.'s[11] measurement of the fine structure oscillation of 3 P He lines in Fig. (1). This data is a more refined version of Andrä's first observation which established that beam-foil excited atoms are sometimes aligned.

In most observations that I will discuss here the time resolution is sufficiently low that $G^{(k)}(t)$ in Eq. (8) is to be averaged over the exponential decay. Then the effect of the fine structure is to smear out the angular distribution.

The beam-foil source normally has cylindrical symmetry about the beam axis. It loses this symmetry when the foil is tilted with respect to the beam. A surprising result discovered by Berry[13] and co-workers was that atoms excited by tilted foils are both aligned and oriented implying that the foil surface plays a significant, but not clearly understood, role in excitation. The initial measurements of the alignment and orientation

**HELIUM-4  $3^3P_1 - ^3P_2$  BEATS  658 MHZ**

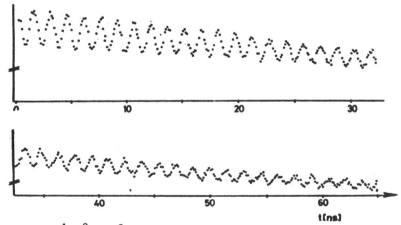

Fig. 1.   $He^4$ $3^3P_1$-$3^3P_2$ fine structure oscillations.  Data of Wittmann et al.[11]

TABLE I. Alignment and Orientation parameters for the He $3^1P$ level excited at 130 keV beam energy. Data of Berry et al.[5]

| Tilt Angle | $A_0^{col}$ | $A_{2+}^{col}$ | $A_{1+}^{col}$ | $O_{1-}^{col}$ |
|---|---|---|---|---|
| 0° | -0.090(36) | 0.016(9) | ... | ... |
| 20° | -0.081 | 0.012 | -0.024(7) | -.013(11) |
| 30° | -0.072 | 0.008 | -0.021 | -.038 |
| 45° | -0.054 | -0.002 | -0.040 | -.040 |

parameters vs. foil tilt angle are given in Table I. Substantial orientation is produced with increasing tilt angle. The values of the alignment and orientation parameters also indicate that the foil excited state is a mixed state. This might be expected since the atoms interact with a macroscopic system, the foil, whose state is not observed.

Beam-foil collisions represent one extreme in terms of the complexity of the excitation mechanism, a complexity mirrored to some extent in the mixed nature of the excited states. At the other extreme are the electron-photon coincidence measurements of Eminyan et al.[4] on the 2 $^1P$ and 3 $^1P$ levels in He. To the extent that spin orbit coupling plays no role in the collision, the excited state must be a pure state.[14] Intermediate in complexity are the recent coincidence measurements of Jaecks et al.[20] on the alignment of He($3\ ^1P$) states excited by charge transfer of $He^+$ with $H_2$. Here the helium atom interacts with a moderately complex system, namely the $H_2$ molecule. Because of the large number of alternative final states of $H_2^+$ (including vibrational and rotational states) one expects to find the helium atoms in a mixed state. Somewhat surprisingly this does not happen in all circumstances. At 1.5 keV and 1.5° scattering the data indicate that the helium atoms are in pure state with $M_L = 0$ along some axis in the scattering plane. This in turn implies that the vibrational and rotational degrees of freedom or alternatively the orientation of the molecular axis, play only a minor role in the collision.

The four source parameters in Eq. (7), measurable by coincidence techniques, serve to characterize j = 0,1 state completely, but are insufficient for j > 1. A technique introduced by Hertel and Stoll[7] provides sufficient information to determine all multipole moments and thereby completely characterize collision excited states. They prepare an atomic beam in a particular excited state by optical pumping. Inelastic and superelastic electron scattering cross sections are measured as a function of the incident light polarization and direction. In principle every multipole moment of the state excited in the time inverse collision can be measured.[14]

The time inverse scattering process populates a particular level $n_i$ according to the density matrix $\rho_{M_F,M_F'}$ while optical pumping prepares the same level according to the density matrix $W(M_F)$, diagonal in photon frame. The photon frame has its z axis along the light axis for circularly polarized light, and in the plane of polarization for linearly polarized light. The scattered electron flux is proportional to the overlap of the two alternatively prepared states

$$I = C \sum_{M_F} W(M_F)\rho_{M_F,M_F} \tag{9}$$

We can write this in the form of Eq. (4) by introducing the projection operator

$$\tau = \Sigma_{M_F} |FM_F)(FM_F|$$

so that Eq. (9) becomes[14]

$$I = c \, <(i|\tau|i')> \tag{10}$$

Again we write $\tau$ in terms of its irreducible components $\tau^{[\ell]}q$, use the Wigner-Eckart theorem to replace mean values of $\tau^{[\ell]}q$ by mean values of $T \quad q$ and transform to the collision frame. The result is

$$I = C \sum_{k=0}^{2F} \sum_{qp=0}^{k} W(k)v^{(k)}(F)(4\pi/2k+1)^{-\frac{1}{2}} <T^{[k]}qp(col)>$$

$$\times Y^{[k]}qp(\theta,\phi) \tag{11}$$

where

$$W(k) = \Sigma_{M_F} (-1)^{k-F-M_F} \, W(M_F)(F - M_F \, F \, M_F|kq)$$

$$\tag{12}$$

$$v^{(k)}(F) = 2^k(2k + 1)^{\frac{1}{2}}[(2F - k)!/(2F + k + 1)!]^{\frac{1}{2}} .$$

Here $\theta,\phi$ are the coordinates of the light frame's z-axis in a frame defined by the collision. The $p = \pm1$ index on the irreducible tensor components and spherical harmonies indicates that real quantities which transform into themselves under reflection in a plane are used in preference to the more common complex quantities. Owing to reflection symmetry we have

It follows from Eq. (11) that every multipole moment of the collision excited state i can be determined from measurements with suitable combinations of light polarizations and direction.

Let us now consider what has been learned from the few measurements of alignment and orientation parameters in collision experiments. These parameters are chosen to emphasize the central role of angular momentum transfer in the interpretation of angular correlation data.

In our first example, electrons scattered in the forward direction impart no angular momentum component parallel to the beam axis. If the target atom is initially in a S ground state the state that is excited must have $\langle L_Z \rangle = 0$ so that $A_0{}^{col} = -1$. A photon-scattered electron coincidence experiment was suggested by Imhoff and Read[15]      to test this prediction and was later successfully applied.[16] King et al.[16] did indeed find $A_0{}^{col} = -1$. Hertel and Stoll [7] obtained an equivalent result with their optical pumping technique.

Kleinpoppen and co-workers[4]      have carried out an extensive series of measurements of collision induced fluorescence in coincidence with scattered electrons exciting the 2 P state of helium at a variety of incident energies and scattering angles. They found non-zero orientation at impact energies up to 200 eV and scattering angles greater than 16° (Fig. 2). A significant feature of the curves in Fig. (2) is the close agreement with the distorted wave calculations of Madison and Shelton.[17] The Born approximation, which treats excitation

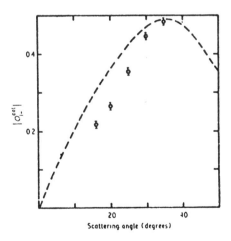

Fig. 2. $O_1{}^{col}$ for $3^1P$ He excited by 80 eV electron impact. Data of Eminyan et al.[4]

as an impulsive transfer of momentum $\vec{q}$ by the incident electron, predicts no orientation. The distorted wave approximation, which incorporates some distortion of the scattered wave, appears to account for the observed transfer of angular momentum neglected in the Born approximation.

Not all distorted wave approximations share this feature, however. The Glauber approximation, in particular, predicts zero orientation. This prediction is the consequence of an unphysical symmetry of the Glauber transition operator. A modified version,[18] which does not have the additional symmetry, does predict orientation.[19] Fig. (3) shows $O_{1-}^{col}$ calculated by Gau and Macek for $e^- + H$ collisions using a slightly modified version of the Glauber approximation. Note that $O_{1-}^{col}$ vanishes at 0 and 180° as required by symmetry. Most other approximations such as the close-coupling approximation and various types of distorted wave approximations also predict non-zero orientation. A future task will be to bring theory and experiment pertaining to orientation and alignment by electron excitation into detailed comparison.

Heavy ion collisions[3] represent a fruitful area for coincidence studies. In the low incident energy range ($\sim$ 1 keV) where the molecular eigenstate expansion applies alignment and orientation measurements indicate the type of coupling, rotational or translational, involved in excitation.[13]

Jaecks et al.'s[20] recent measurement of the excitation of He($3\ ^3P$) by $He^+ + He$ charge exchange collisions has identified a particularly simple mechanism for transfer of angular momentum. The detected photons from the $3\ ^3P \rightarrow 3\ ^3S$ decay in

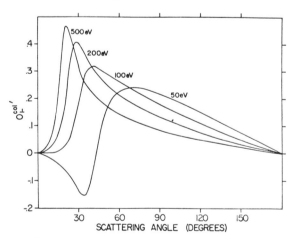

Fig. 3. $O_{1-}^{col}$ for the 2p level in hydrogen calculated in a modified Glauber approximation.

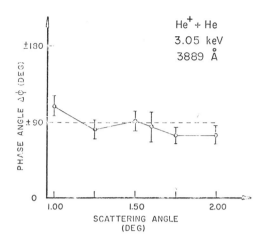

Fig. 4. Relative phase of the $M_L$ = 0 and 1 amplitudes for $3^3P$ He excited by charge transfer collisions of $He^+$ with He at 1.5 keV. Data of Jaecks et al.[20]

coincidence with the scattered neutral atoms with polarization analysis. They thereby obtain the ratio $|a_1|/|a_0|$ of scattering amplitudes and their relative phase. The relative phase in Fig. (4) was found to be nearly constant for the range of scattering angles involved, even though the ratio of amplitudes varied considerably.

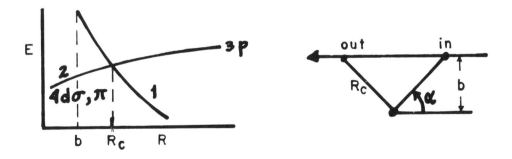

Fig. 5. Schematic molecular energy level diagram for the proposed mechanism for excitation of He p-states. Right hand picture defines the collision parameters $\alpha$ and b.

An explanation of this constancy follows from the Fano-Lichten   molecular eigenstate representation, schematically illustrated in Fig. (5).  The excitation is assumed to proceed via a Landau-Zener crossing between the 1s $\sigma_g(2p\sigma_a)$   and (1s $\sigma_g$)  $4d\sigma_g$ curves near $R_c \simeq 1$ a.u.  Such a transition takes place on both the inward outward passage of the crossing.  Because the $4d\sigma$, $4d\pi$ and $4d\Delta$ curves are nearly degenerate for small internuclear separations, the $4d\sigma$ wave function maintains the direction of the internuclear axis at the time of excitation as the axis of quantization throughout the collision. The two $4d\sigma_g$ wave functions, one at the in and one at the out position coherently superimpose to form the final wave function separating into the 3p atomic function.  To superimpose the two wave functions, we must take into account the change in orientation of the internuclear axis and the phase difference between the alternative paths given by

$$A = \int_0^{t_c} (E_2 - E_1)dt/h. \tag{13}$$

A transformation of the $4d\sigma$ functions to a frame with Z axis along the final beam direction, taking into account that the $4d\sigma g$ and $4d\pi g$ correlate with $3p(m_\ell = 0)$ and $3p(m_\ell = 1)$ atomic orbitals respectively gives the amplitudes $a_0$ and $a_1$ that the atom is in the $m_\ell = 0$ or $m_\ell = 1$ magnetic substate. We find

$$a_0 \propto 2^{-\frac{1}{2}} (3 \cos^2 \alpha - 1) \cos A$$

$$\tag{14}$$

$$a_1 \propto i \, 3^{\frac{1}{2}} \sin \alpha \, \cos \alpha \, \sin A.$$

The phase difference $\Delta\Phi$ equals 90° in good agreement with the data.  Since $0_{1-}$   is given by

$$0_{1-}^{col} = -2\sqrt{2}(|a_1|/|a_0|) \, \sin\Delta\Phi/(1 + 2|a_1|^2/|a_0|^2) \tag{15}$$

we see that $\Delta\Phi = 90°$ implies maximum orientation for a given ratio $|a_1|/|a_0|$.  This mechanism transfers  angular momentum from relative to internal motion very effectively.

A similar measurement by Vassilev et al.[21] found zero orientation at lower energy.  This would hold if A or $\alpha$ in Eq. (14) is small.

It is interesting to note that the phase distortion represented by A is a key quantity in this mechanism for angular

momentum transfer. The distorted wave approximation for electron collisions takes account of a similar phase distortion,
but in a less transparent way. This same mechanism also seems
to be involved in the formation of spin polarized hydrogen
atoms by charge transfer in Xe, another example of transfer of
angular momentum from relative to internal motion.[22]

    To briefly summarize the situation we now have available
various experimental techniques for measuring orientation and
alignment in atomic collisions. Furthermore collisions of
great complexity such as beam-foil collisions with tilted foils
produce aligned and oriented atoms. Interpreting these results
theoretically and identifying the underlying physical mechanisms
for angular momentum transfer represents a continuing task for
atomic theory. At present the results are fragmentary and no
comprehensive picture has emerged.

# REFERENCES

1.  H. W. B. Skinner and E. T. S. Appleyard, Proc. Roy. Soc.
    Lond. A117, 224 (1927).
2.  I. C. Percival and M. J. Seaton, Philos. Trans. R. Soc.
    London. A251, 113 (1958).
3.  D. H. Jaecks, D. H. Crandell and R. H. McKnight, Phys.
    Rev. Lett. 25, 491 (1970).
4.  M. Eminyan, K. MacAdam, J. Slevin and H. Kleinpoppen.
    Phys. Rev. Lett. 31, 576 (1972).
5.  H. G. Berry, L. J. Curtis, D. G. Ellis and R. M. Schectman,
    Phys. Rev. Letters 32, 751 (1974).
6.  D. A. Church, W. Kolbe, M. C. Michel and T. Hadeishi,
    Phys. Rev. Letters 33, 564 (1974).
7.  I. V. Hertel and W. Stoll, J. Phys. B:  Atom. Molec.
    Phys. 7, 570 (1974) and 583 (1974).
8.  J. Kessler, Rev. Mod. Phys. 41, 3 (1969).
9.  U. Fano, Rev. Mod. Phys. 29, 74 (1957).
10. U. Fano and J. Macek, Rev. Mod. Phys. 45, 553 (1973).
11. W. Wittmann, K. Tillman, H. J. Andrä and P. Dobberstein,
    Z. Physik 257, 299 (1972).
12. H. J. Andrä, Phys. Rev. Lett. 25, 325 (1970).
13. J. Macek and D. H. Jaecks, Phys. Rev. A4, 2288 (1971).
14. J. Macek and I. V. Hertel, J. Phys. B 7, 2173 (1974).
15. R. E. Imhof and F. H. Read, J. Phys. B 4, 450 (1971).
16. G. C. King, A. Adams, and F. H. Read, J. Phys. B 5, L254
    (1972).
17. D. A. Madison and W. N. Shelton, Phys. Rev. A7, 449 (1973).
18. F. W. Byron, Phys. Rev. A4, 1907 (1971).
19. J. N. Gau and J. Macek, Phys. Rev. A10, 522 (1974).
20. D. Jaecks, F. Eriksen, W. de Rijk and J. Macek, submitted
    to Phys. Rev. Letters.

21. G. Vassilev, G. Rahmat, J. Slevin and J. Baudon, Phys. Rev. Letters 34, 444 (1975).
22. R. Shakeshaft and J. Macek, Phys. Rev. A6, 1876 (1972).

# ANALYSIS OF INELASTIC ELECTRON-PHOTON COINCIDENCE EXPERIMENTS

H. Kleinpoppen, K. Blum[*] and M.C. Standage

Institute of Atomic Physics
University of Stirling
Stirling, Scotland

## I.  INTRODUCTION

At the VIIIth ICPEAC conference in Belgrade Jaecks[1] surveyed the applications of coincidence studies between photons and scattered particles in atomic collision processes.  Since then considerable progress has been made in such studies, particularly in connection with experimental investigations of inelastic electron-photon angular correlations[2-6].  Furthermore the theory of the measurement of electron-photon coincidences has been significantly extended.  Previously Macek and Jaecks[7], Rubin et al[8] and also Wykes[9] related electron-photon coincidence rates to "collision parameters" such as inelastic scattering amplitudes, partial sub-level cross sections and phase differences of scattering amplitudes.  Recently Fano and Macek[10], Macek and Hertel[11], and Blum and Kleinpoppen[12] have related the electron-photon angular correlation to "target parameters" such as orientation and alignment parameters and also to multipole moments of the atom excited during the collisional excitation process.  The progress both on the experimental and theoretical side, namely the extraction of collision amplitudes and the detailed analysis of target parameters, has brought us to a much deeper understanding of the electron-atom collision process.  This state of affairs is similar to the progress achieved in spin analysis scattering experiments[13-16] which have provided information on direct and exchange scattering amplitudes.

In this paper we review the recent progress in the research field under consideration.  This includes a report on the theoretical and experimental analysis of polarization and coherence of coincident photon radiation[28,30].

## II.   OBSERVATIONAL SCHEME FOR ELECTRON-
## PHOTON COINCIDENCE EXPERIMENTS

We assume that both electrons and atoms are unpolarized and we neglect spin orbit interactions between the electron and the atom.  Fig.1 shows the experimental geometry so far used in electron-photon coincidence experiments.  A well collimated monochromatic beam of electrons is cross-fired with an atomic beam target.  Inelastically scattered electrons, having excited atoms in the atomic beam, are detected in co-incidence with photons emitted in the de-excitation process. Experimental details of the coincidence experiments have been described extensively in the literature[2-6].  We, therefore, focus our discussion on some of the basic theoretical concepts involved.  We choose a fixed right-handed orthogonal xyz-coordinate system in such a way that the z-axis coincides with the direction of the incident electrons and the fixed momentum of the scattered electrons lies in the xz-plane (collision system).  The emitted photons may be observed in the direction of the unit vector $\underline{n}$ with polar angles $\theta,\phi$.  It is essential, that by fixing the directions of incident and scattered elec-trons a scattering plane is selected.  It is with respect to this plane that the polar angles $\theta,\phi$ can be defined in a physically reasonable way and the dependence of photon intensity and polarization on these angles can be observed. Without observing the scattered particles only a single axis (beam axis of the incident electrons) is defined.  The first consistent theory of this type of experiment with regard to anisotropic photon emission was given by Percival and Seaton[17].  Due to the rotational symmetry of the system about the electron beam axis the intensity of the line emission is still isotropic in a plane perpendicular to this axis.  Thus, one obtains only an average over the $\phi$- dependence and detailed information on the excitation process is lost in traditional excitation experiments with axial symmetry.

In the first electron-photon coincidence experiment by King et al[18] only electrons scattered in the forward direction were detected.  In this case, axial symmetry is restored which requires the selection rule $\Delta m_L = 0$.  In order to excite atomic sublevels with $m_L \neq 0$ it is necessary to detect the scattered electrons outside the forward direction.  Of course, the most general case for the electron-photon coincidence measurement is connected with any arbitrary direction of propagation of the photon and the scattered electron.  However, from an experimental point of view, the most obvious cases for electron-photon correlation measurements are the two cases

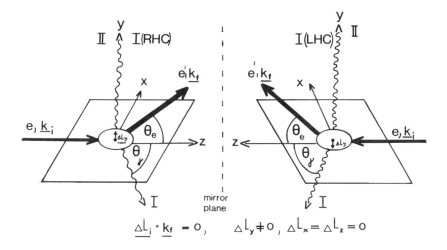

Fig.1 :  Scheme of the electron photon coincidence experiments including mirror symmetry operations.

where either the photon observation direction is parallel (case I in Fig.1) or perpendicular (case II in Fig.1) to the scattering plane.  Before we describe the results of these measurements we discuss some simple consequences derived from the geometry and invariance principles involved in the collision process.

In a classical picture we assume knowledge of the position of the atom as the centre of the scattering force in the scattering plane defined by the incoming and the scattered electron.  Orbital angular momentum transfer is then only possible perpendicular to the scattering plane.  However, this restriction on the orbital angular momentum transfer applies in a more general sense for any scattering processes which obey parity invariance in the following way.  A simple relation which follows from parity invariance for a given process can always be obtained by extracting those physical quantities from the process under consideration which allows one to form a scalar product of an axial and polar vector (pseudoscalar).  In our case, such a product can be formed by $\underline{\Delta L}_i \cdot \underline{k}_f$ where $\underline{k}_f$ is the wave vector of the outgoing electron and $\underline{\Delta L}_i$ is the net orbital angular momentum transfer with respect to the coordinate axis i = x,y or z.  For any arbitrary $\underline{k}_f$ parity invariance of the excitation process requires

$\Delta L_i \cdot k_f = 0$ and thus the only possible non zero component of the orbital angular momentum transfer is $\Delta L_y$.

Of course the argument based upon parity invariance can also be extended to show that any orientation of the atom (note that orientation is directly connected to the orbital angular momentum transfer to the atom, see chapter III) in the excited state is only possible perpendicular to the scattering plane. Similarly, it follows from symmetry arguments that any orientation of the atom or any circular polarization of a photon emitted normal to the scattering plane reverse their sign if the scattering of the electrons is switched from a given scattering angle to that with opposite sign (see Fig.1).

### III. RESULTS AND INTERPRETATION OF
### ELECTRON-PHOTON CORRELATIONS

#### 1.  Collision Parameters

The first electron-photon angular correlation from atomic excitation processes were measured in the scattering plane (case I in Fig.1). The theoretical description of this process has been extensively treated in literature[7-12]. The basic physics underlying electron-photon correlations is most easily demonstrated for singlet transitions of helium where spin and spin-orbit effects are negligible. We will discuss in detail the excitation/de-excitation process $^1S \rightarrow {^1P} \rightarrow {^1S}$ in helium which is based on the assumption of coherent excitation of the sublevels[7]. The state vector $\psi(^1P)$ of the excited $^1P$ state is given by:

$$\psi(^1P) = a_o|10> + a_1|11> + a_{-1}|1-1>  \qquad (1)$$

with $a_o$, $a_1$ and $a_{-1}$ as excitation amplitudes for the three magnetic sublevels $m_L = 0$ and $m_L = \pm 1$. Mirror symmetry of the scattering process with respect to the scattering plane requires that $a_1 = -a_{-1}$. The wave function of eq. (1) can be normalised so that the amplitudes $a_o$ and $a_1$ are related to partial cross sections for exciting the magnetic sub-levels: $|a_o|^2 = \sigma_o$, $|a_1|^2 = \sigma_1$ with $\sigma = \sigma_o + 2\sigma_1$ as the total differential excitation cross section. The amplitudes $a_o$ and $a_1$ are in general complex numbers defined only up to an overall phase factor. If $a_o$ is assumed to be real and positive, the relative phase between $a_1$ and $a_o$ is then defined by $a_1 = |a_1|e^{i\chi}$.

The state vector $\psi(^1P)$ is completely described by $\lambda = \sigma_o/\sigma$ and $\chi$ ($-\underline{\underline{\pi}}< \chi <\underline{\underline{\pi}}$) together with the total differential cross section $\sigma$ for given electron energy and scattering angle.

In the experimental study[2-6] of electron-photon angular correlations for the excitation process under consideration the electron scattering angle $\Theta_e$ is kept fixed whereas the photon observation angle $\Theta_\gamma$ (measured with respect to the direction of the incident electron beam) is varied in the scattering plane. For this case the angular correlation function N can be derived from theory[2,7] as follows:

$$N=\lambda\sin^2\Theta_\gamma+(1-\lambda)\cos^2\Theta_\gamma-2\{\lambda(1-\lambda)\}^{\frac{1}{2}}\cos\chi\sin\Theta_\gamma\cos\Theta_\gamma \qquad (2)$$

Based upon this theoretical expression the angular correlation function can be used to extract the parameters $\lambda$ and $|\chi|$. Fig.2 gives typical examples for such correlation data determined by experiment whereby predictions of the Born approximation and also also a least square fit of the theoretical correlation function (eq.2) to the experimental data are included. Note the excellent fit of the experimental data to the theoretically required form of the correlation function and also the discrepancy between the prediction of the Born approximation and the experimental data, particularly for larger electron scattering angles. Based upon the validity of the Born approximation a selection rule $\Delta m_L = 0$ should hold with respect to the momentum transfer direction. In other words, the line transition $^1P \rightarrow {}^1S$ should be completely polarized parallel to the momentum transfer which means that no photon emission would occur in that direction. However, it is

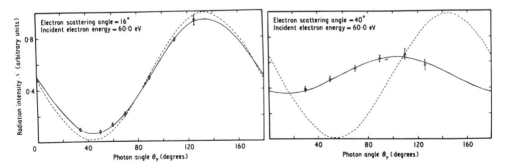

Fig. 2: Experimental data points[2] of electron-photon angular correlations from the helium excitation/de-excitation process $1^1S \rightarrow 2^1P \rightarrow 1^1S$ for two different electron scattering angles at 60eV. One rms error. Full curve, least squares fit of eq. 2 to the experimental data; dashed curve predictions of first Born approximation.

**Born approximation :**
$\Delta m_L = 0$ for $\Delta p$

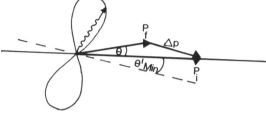

**as observed :**

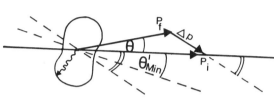

Fig. 3:  Momentum analysis ($P_i$ initial momentum of electron
travelling in z direction, $P_f$ final momentum of electron after
collisional excitation, $\Delta P$ momentum transfer to atom) and
photon angular distribution curve (arrow〜⇛) of 1S→1P→1S
excitation/de-excitation process of helium.

evident from Fig.2 that the angle at which minimal photon intensity is observed does not coincide with the Born prediction
and does not decrease to zero intensity.  Fig.3 demonstrates
this behaviour of the photon angular distribution.

Based upon the above model of coherent excitation of the
magnetic sub-levels the two parameters $\lambda = \sigma_0/\sigma$ and $|\chi|$ can be
extracted from the observed angular correlation by fitting the
data to eq.2.  Fig.4 and 5 present such data for the helium
$2^1P \to 1^1S$ and $3^1P \to 1^1S$ transitions together with the
$3^1P \to 2^1S$ transition at 80eV in comparison with theoretical
predictions.  The theoretical prediction of Madison and
Shelton[19] (distorted wave approximation) is in remarkably good
agreement in $\lambda = \sigma_0/\sigma$ with the experimental data whereas the
agreement with the phase parameter $|\chi|$ is only qualitatively
satisfactory.  Similarly, the eikonal distorted wave Born
approximation with Glauber distorting potentials qualitatively
approaches the angular dependence of $|\chi|$ at 80eV (Joachain and
Vanderpoorten[20]).  This approximation is also in good quan-
titative agreement with experimental data for $|\chi|$ at 200eV but
it shows some discrepancy with $\lambda$.

A typical variation of the angle $\theta_{min}$ for the minimum
photon intensity (see definition of $\theta_{min}$ in Fig.3) observed in
the scattering plane for fixed electron scattering angles is
demonstrated in Fig.6.  Note the considerable tilt between the

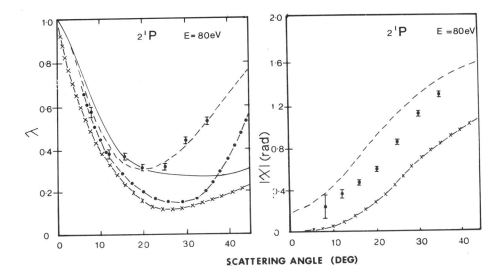

Fig. 4:  Variation of $\lambda$ and $|\chi|$ with electron energy and scattering angle for $2^1P$ excitations at 80eV; data points from the experiments of Eminyan et al[2], error bars represent one standard deviation.  ——first Born approximation, ---- distorted wave theory (Madison & Shelton[19] –·–·– Many-Body theory (Csanak et al[23]), x-x- eikonal distorted wave Born approximation (Joachain & Vanderpooten[20]).

momentum transfer direction (Born approximation) and the direction of minimum photon intensity at larger scattering angles.

Recently Arriola, Teubner, Ugbabe and Weigold[5] reported electron photon angular correlations from the excitation of the J = 1 levels in argon at 11.62 and 11.84eV and the 1066 and 1048Å photons which result from the decay of these levels to the J = 0, $^1S_0$ ground state.  The excited J = 1 levels are $^3P_1$ and $^1P_1$ states in the LS coupling scheme.  The experiment of the above authors was similar to the original one of Eminyan, MacAdam, Slevin and Kleinpoppen[2].  However, due to insufficient energy resolution of the scattered electrons the electron photon angular correlations from the excitation/de-excitation of the $^1P_1$ and $^3P_1$ could not be separated from each other.  Since the excitation to the $^{1,3}P_1$ states should be described by the same type of picture for the differential cross section with $\sigma = \sigma_0 + 2\sigma_1 = |a_0|^2 + 2|a_1|^2$ for both the singlet and triplet P excitation, the angular correlation function corresponding to eq. (2) can be derived as follows:

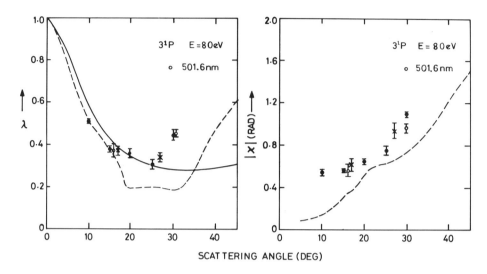

Fig. 5:  $\lambda$  and  $|\chi|$  data versus electron scattering angle for:
He $3^1P$ excitation at 80eV[3,28]. $\downarrow$ 53.7 nm and $\frac{\circ}{\downarrow}$ 501.6 nm
angular correlation data.  $\ddagger$ polarization data (Chapter III.3).
One rms error.  Solid curves represent first Born approximation.
Dashed curves are the multichannel eikonal predictions of
Flannery and McCann[24].

$$N = \sigma^s \left\lfloor \overline{\cos}^2\theta_\gamma - \lambda_1\cos2\theta_\gamma - \{\lambda_1(1-\lambda_1)\}^{\frac{1}{2}}\sin2\theta_\gamma\cos\chi_1\overline{\phantom{/}}\right\rfloor \qquad (3)$$

$$+ \sigma^t \left\lfloor \overline{\cos}^2\theta_\gamma - \lambda_3\cos2\theta_\gamma - \{\lambda_3(1-\lambda_3)\}^{\frac{1}{2}}\sin2\theta_\gamma\cos\chi_3\overline{\phantom{/}}\right\rfloor$$

where the subscripts 1 and 3 refer to the singlet and triplet
J = 1 excitations, respectively.  In order to extract  $\lambda$  and  $\chi$
parameters from the measured angular correlations it was
necessary to estimate the relative cross sections for the
excitation of the triplet state to the singlet state ( $\sigma^t/\sigma^s$ ).
From the theory and corresponding ratios in neon for  $\sigma^t/\sigma^s$
Arriola et al[6] assumed a value of approximately 0.2 for this
ratio at the energy and angles employed in their experiment.
Table 1 shows the data for  $\lambda$  and  $|\chi|$  obtained by these authors.
The results were, within the experimental error, consistent
with  $\lambda_1 = \lambda_3$  and  $\chi_1 = \chi_3$

Recently, several authors have investigated the theory of
electron-photon (Lyman- $\alpha$ ) coincidence measurements for the 2P
state of atomic hydrogen.  Theoretical predictions for this
process were given for  $\lambda = \sigma_o/\sigma$ , for the coincidence rates,

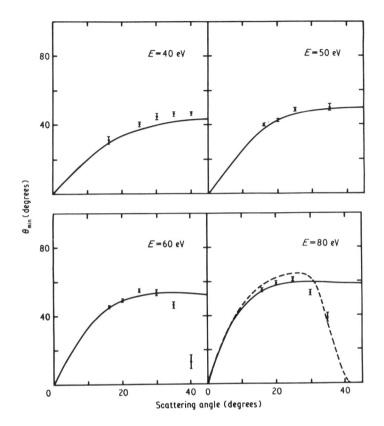

Fig. 6: Variation of $\theta_{min}$ with electron energy and scattering angle for $2^1P$ excitation of helium, Eminyan et al[2]. Legend as in Fig.4.

and for the target parameters (orientation and alignment para-meters, see sub-chapter III.2.) in various approximations. Morgan and McDowell[21] have used a distorted wave polarized orbital approximation. Morgan and Stauffer[22] applied Born, Born-Oppenheimer, Coulomb projected Born, Coulomb projected Born exchange (CPBE) and the generalised CPBE approximations. Both Morgan and McDowell[21], as well as Morgan and Stauffer[22] presented very interesting two-dimensional plots for the co-incidence rate as a combined function of electron scattering and photon emission angles. Morgan and McDowell[21] also showed that such electron-photon coincidences do not allow a measure-ment of the relative phase of the $m_L$ components of the excit-ation amplitudes for the hydrogen nP states unless the singlet

| $E_o$ | $\Theta_e$ | $\lambda$ | | $\chi$ (deg) | | $\Theta_{min}$ (deg) | |
|---|---|---|---|---|---|---|---|
| (eV) | (deg) | Expt | BA | Expt | BA | Expt | BA |
| 80.4 | 8.5 | 0.22±0.01 | 0.286 | 37±1 | 0 | 65 | 57.7 |
| 113.4 | 9.0 | 0.23±0.01 | 0.162 | 42±2 | 0 | 65 | 66.3 |

| $E_o$ | $\Theta_e$ | Best fit parameter | | | |
|---|---|---|---|---|---|
| (eV) | (deg) | $\lambda_1$ | $\lambda_3$ | $|\chi_1|$ (deg) | $|\chi_3|$ (deg) |
| 80.4 | 8.5 | 0.22±0.01 | 0.23±0.03 | 37±1 | 35±5 |
| 113.4 | 9.0 | 0.23±0.01 | 0.26±0.04 | 42±2 | 38±11 |

Table 1: Experimental results (Expt. [6]) and predictions by Born approximation (BA) for $\lambda$, $\chi$ and $\Theta_{min}$ from electron-photon angular correlations for the excitation of the first excited ($^1P_1$ and $^3P_1$) state of Ar. One rms error.

and triplet scattering can be separated from each other by using spin polarized electrons and atoms.

## 2. Target Parameters

The interpretation of the electron-photon correlations given in II/1 provides collision parameters like partial sub-level cross sections and phase differences between excitation amplitudes. Alternatively, one can express the measured quantities in terms of "target parameters". These parameters are expectation values of angular momentum quantities which characterize the angular momemtum state of the atom. It is, therefore appropriate to consider first the quantities which characterize the orbital angular momenta transfer to the atom for our case study $^1S \rightarrow {}^1P \rightarrow {}^1S$ of helium. As discussed in Chapter I, orbital angular momentum can only be transferred to the atom perpendicular to the scattering plane. The orbital angular momentum quantities $<L_i>$ and $<L_i^2>$ (orbital angular momentum quantities with components parallel to i = x, y and z axes) can be connected with the collision parameters $\lambda$ and $\chi$ as follows (in units of $\hbar$):

$$<L_x> = 0, \quad <L_y> = -2\{\lambda(1-\lambda)\}^{\frac{1}{2}}\sin\chi, \quad <L_z> = 0,$$

$$<L_x^2> = \lambda, \quad <L_y^2> = 1, \text{ and } <L_z^2> = 1-\lambda$$

Instead of using these quantities it is more convenient to characterize the atomic state in terms of irreducible tensors which have simpler transformation properties under rotation.

For the following we use the Fano-Macek notation. The Fano-Macek alignment tensor (tensor of rank two) for our case is connected with $\lambda$ and $\chi$ as follows:

$$A_O^{col} = \tfrac{1}{2}<3L_z^2 - L^2> \qquad = \qquad \tfrac{1}{2}(1-3\lambda)$$

$$A_{1+}^{col} = \tfrac{1}{2}<L_xL_z + L_zL_x> \qquad = \qquad \tfrac{1}{2}\{\lambda(1-\lambda)\}^{\frac{1}{2}}\cos\chi \qquad (4)$$

$$A_{2+}^{col} = \tfrac{1}{2}<L_x^2 - L_y^2> \qquad = \qquad \tfrac{1}{2}(\lambda-1)$$

Similarly the Fano-Macek orientation vector of the target atom can be calculated from $\lambda$ and $\chi$:

$$O_{1-}^{col} = -\{\lambda(1-\lambda)\}^{\frac{1}{2}}\sin\chi = \tfrac{1}{2}<L_y> \qquad (5)$$

The electron-photon angular correlation measurement in the scattering plane gives $\lambda$ and $|\chi|$. Accordingly, $|O_{1-}^{col}|$ can be calculated from the knowledge of $\lambda$ and $|\chi|$.

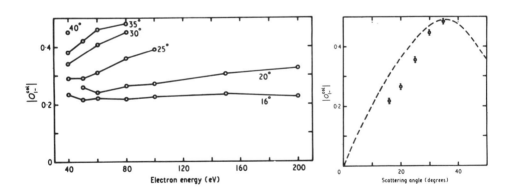

Fig. 7: Variation of the Fano-Macek orientation parameter with electron energy and different scattering angles (left) and with the scattering angle and fixed energy (80eV) for the helium $1^1S \to 2^2P$ excitation process. The left hand diagram only includes the experimental data [2] whereas the right hand diagram also displays theoretical predictions (dashed curve) for 78eV by Madison and Shelton [19].

Fig.7 shows data for $|O_{1-}^{col}|$ from the experiment and from the
predictions of the distorted wave theory.  Notice that the
maximum possible value of 0.5 (complete orientation of the atom
parallel or antiparallel to the normal of the scattering plane)
is almost reached for certain scattering angles and energies.
Morgan and McDowell[21] and also Gau and Macek[25] calculated $O_{1-}^{col}$
for the excitation of the 2P state of atomic hydrogen based
upon the distorted wave polarized orbital and the Glauber
approximation, respectively.  Table 2 presents some of their
results for 100eV:

| $\Theta$(deg) | $\|O_{1-}^{col}\|$ | $\|O_{1-}^{col}\|$ |
|---|---|---|
| | (a) | (b) |
| 80 | 0.243 | 0.208 |
| 100 | 0.211 | 0.158 |
| 130 | 0.120 | 0.0919 |
| 160 | 0.081 | 0.0353 |
| 170 | – | 0.0176 |

(a)   Morgan and McDowell[21]

(b)   Gau and Macek[25]

## 3.   Photon Polarization and Degree of Coherence:

## Photons Observed Perpendicular to the Scattering Plane.

The above analysis of electron-photon coincidences was
based on the assumption of coherent excitation of the magnetic
sub-levels.  Although coherent excitation is expected to be
correct for our case ($1S \rightarrow 1P \rightarrow 1S$ in helium) it is important to
test this assumption experimentally as far as possible.  In
order to do this we first investigate the coherence properties
of emitted photon radiation.  We will present some first results
which show that the light emitted perpendicular to the scattering plane is completely coherent.  These results indicate
the radiation is emitted by a coherently excited ensemble of
atoms.  The experimental data from the electron-photon coincidences in the scattering plane can satisfactorily be fitted to
the angular correlation function (eq. 2) derived on the
assumption of coherent excitation.  There is certainly a high
degree of probability that the interpretation based upon the
model of coherent excitation is correct.  It must, however, be
admitted that the observational data from the angular correlations in the scattering plane (see sub-chapter III) do not

provide an unambiguous proof for the model of coherent excitation. E.g. the angular correlation as observed in the scattering plane can also be fitted to an angular correlation which
results from the incoherent super-position of two harmonic
oscillators, oscillating incoherently parallel to the main axes
of the polarization ellipse. Thus, to obtain more conclusive
information, it is necessary to observe the photons perpendicular
to the scattering plane.

In order to obtain information about the coherence properties of the emitted light is is necessary to introduce
quantities which completely characterize the quantum mechanical
state of the emitted light. Such quantities are the elements
of the polarization density matrix of the photons or, equivalently, the Stokes parameters. These quantities have been
discussed extensively in the literature (see e.g. Born and
Wolf[26].

$$
\begin{aligned}
I &= I(\alpha = 0^{o}) &+& \quad I(\alpha = 90^{o}) \\
IP_1 &= I(\alpha = 0^{o}) &-& \quad I(\alpha = 90^{o}) \\
IP_2 &= I(\alpha = 45^{o}) &-& \quad I(\alpha = 135^{o}) \\
IP_3 &= I(RHC) &-& \quad I(LHC)
\end{aligned}
\qquad (6)
$$

With regard to photons observed perpendicular to the
scattering plane $I(\alpha)$ is the linearly polarized intensity component measured at an angle $\alpha$ to the incident electron beam.
I is the total intensity; $P_1$ and $P_2$ are the linear polarizations
with reference to the incident electron beam direction or $45^{o}$
against this direction whereas $P_3$ is the circular polarization
of the photons. From the Stokes parameters some quantities can
be derived which characterise the "degree of coherence" of the
emitted light. Following Born and Wolf[26], a "correlation
factor" (i.e. the ratio of the off-diagonal component to the
square root product of the diagonal components of the photon
density matrix) for partially coherent quasi-monochromatic light
can be determined from the measurement of $P_1$, $P_2$ and $P_3$:

$$
\mu_{xz} = |\mu_{xz}| e^{i\beta}{}_{xz} = \frac{P_2 + iP_3}{\sqrt{1-P_1^2}}
\qquad (7)
$$

$\mu_{xz}$ is effectively a measure of the correlation between the x
and z components of the electric vector of the radiation,
$|\mu_{xz}|$ is a measure of the "degree of coherence" of the light

and the phase $\beta_{xz}$ is the "effective phase difference" between the two components of the electric vector. It can be shown that $|\mu_{xz}| \leq 1$; the equality sign holds if and only if the photon beam is completely coherent.

An additional quantity which can be used to characterise the degree of coherence of a photon beam can be obtained as follows. The three Stokes parameters $P_1$, $P_2$ and $P_3$ transform under rotation as components of a three-dimensional vector, $\vec{P} = (P_1, P_2, P_3)$. This polarization vector characterises the polarization state of a photon beam in the same way as the spin-polarization vector characterises the polarization of a beam of spin-$\frac{1}{2}$ particles[27]. The magnitude of the vector $\vec{P}$ is defined as the "degree of polarization" of a photon beam:

$$|\vec{P}| = \sqrt{|P_1|^2 + |P_2|^2 + |P_3|^2}$$

Analogous to the case for spin-$\frac{1}{2}$ particles, a photon beam is completely coherent, if, and only if the beam is completely polarized; that is $|\vec{P}| = 1$.

Standage and Kleinpoppen[28] determined $\mu$ and $\vec{P}$ for the He, $3^1P \rightarrow 2^1S$ (501.6 nm) transitions with photons observed perpendicular to the scattering plane. The relevant measurements of the linear and the circular polarization are shown in Fig.8 for an incident electron energy of 80eV. Note the significant behaviour of $P_3$; unlike the linear polarizations $P_1$ and $P_2$ the circular polarization $P_3$ changes sign as the electron scattering angle is changed from positive to negative. This result is an outcome of the parity invariance argument discussed in Chapter II. From the polarization data $|\mu_{xz}|, \beta_{xz}$ and $|\vec{P}|$ can be determined (Fig.8 and 9). Note that the experimental values for the degree of polarization and the degree of coherence are unity. Thus, the radiation emitted perpendicular to the scattering plane is completely coherent within the experimental error. As mentioned above, it is this result which indicates that the radiation is emitted by coherently excited atoms (in order to prove this unambiguously, $\mu$ and P should be measured in two different directions[30]). The non-zero phase differences $\beta_{xz}$ also change sign as the electron scattering angle goes from positive to negative (Fig.9). We would emphasise that these results for the degree of coherence and the effective phase difference of the coincident photon radiation are obtained independent of any model for the excitation process of the atom.

Based upon the assumption of coherent excitation of the

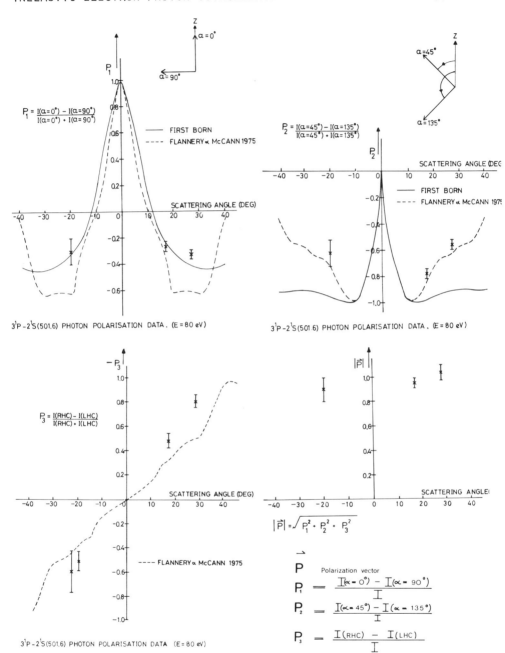

Fig. 8: Experimental data for $P_1$, $P_2$, $P_3$ and $|\vec{P}|$ (Standage and Kleinpoppen[28]) of the $3^1P \rightarrow 2^1S$ (501.6 nm) coincident radiation of helium at 80eV incident electron energy compared with first Born and multi-channel eikonal predictions of Flannery and McCann[24].

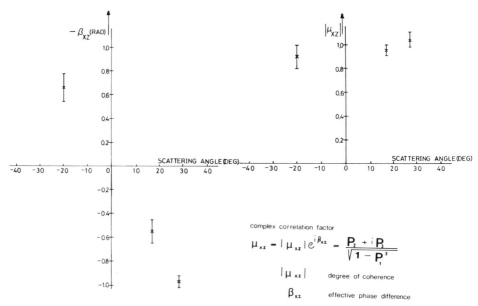

Fig. 9:   $|\mu_{xz}|$ and $\beta_{xz}$ for $3^1P \rightarrow 2^1S(501.6$ nm) coincident radiation of helium at 80eV incident electron energy.

degenerate magnetic sub-levels of the He, $n^1P$ states, the components of the polarization vector $\vec{P}$ can be expressed in terms of the parameters $\lambda$ and $\chi$ of the atomic radiation source (Blum and Kleinpoppen[12]) as follows:

$$P_1 = 2\lambda-1, \quad P_2 = -2\sqrt{\lambda(1-\lambda)}\cos\chi, \quad P_3 = 2\sqrt{\lambda(1-\lambda)}\sin\chi \quad (8)$$

Several conclusions can be drawn from eq.(8). In general, non-zero values for $P_2$ and $P_3$ depend upon non-random phase differences $\chi$ between the excitation amplitudes $a_0$ and $a_1$. (Note that in traditional electron atom excitation experiments with axial symmetry $P_2$ and $P_3$ are zero). The complex correlation factor becomes $\mu_{xz} = -e^{-i\chi}$ with $\beta_{xz} = -\chi$. We emphasise that this result directly relates the phase difference $\chi$ of the excitation amplitudes $a_0$ and $a_1$ to the phase difference $\beta_{xz}$ between the x and z components of the electric vector of the coherent radiation observed perpendicular to the scattering plane. Comparison of Figs.5 and 9 shows that values for $|\chi|$ extracted from the angular correlation data are in reasonable agreement with the measured phase differences $|\beta_{xz}|$. It also follows from eq. (8) that $|\vec{P}| = |\mu_{xz}| = 1$ is independent of the impact energy and the scattering angle. Comparison of eqs. (4) and (8) show that the Fano-Macek alignment and orientation parameters may also be expressed in terms of the components of the polarization vector as follows.

$$A_0^{col} = -(1+3P)/4, \quad A_{1+}^{col} = -P_2/2, \quad A_{2+}^{col} = (P_1-1)/4, \quad O_{1-}^{col} = -P_3/2.$$

Based upon the model of coherent excitation of the $^1P$ state of helium the linear and circular polarizations provide a direct measure of the Fano-Macek parameters.

It is interesting to compare our results for the measurements of $P_1$, $P_2$ and $P_3$ of the 501.6 nm helium line with those of a beam foil experiment by Berry et al[29]. These authors found partial elliptical polarization of the 501.6 nm line which depends upon the tilt angle of carbon foil relative to the helium ion beam axis. Applying the above polarization analysis to this beam foil experiment $|\vec{P}|$ and $|\mu|$ are substantially smaller than one and their values depend on the beam foil direction. This may indicate that the light of the 501.6 nm helium line was emitted by a partially coherent excited atomic source.

Finally, we would like to remark that the helium $3^1P$, $\lambda$ and $\chi$ data [3,5] obtained from the angular correlation measurements in the scattering plane are consistent with the polarization data measured perpendicular to the scattering plane.

## REFERENCES [+]

1. D.H. Jaecks, Invited Lectures and Progress Reports, VIIIth ICPEAC, Belgrade, 1973, p.137, edited by B.C. Copic and M.V. Kurepa.

2. M. Eminyan, K.B. MacAdam, J. Slevin and H. Kleinpoppen, Phys. Rev. Lett. 31, 576, 1973 and also J. Phys. B, Vol.7, No. 12, 1519, 1974.

3. M. Eminyan, K.B. MacAdam, J. Slevin, M.C. Standage and H. Kleinpoppen, J. Phys. B, in press.

4. M. Standage and H. Kleinpoppen, in Book of Abstract of the IXth ICPEAC Conference, 1975.

5. J. Slevin, M.C. Standage, H. Kleinpoppen, M. Eminyan and K.B. MacAdam, in Book of Abstracts, IXth ICPEAC Conference, 1975.

6. H. Arriola, P.J.O. Teubner, A. Ugabe and J. Weigold, J. Phys. B, 8, 1275, 1975.

7. J.H. Macek and D.H. Jaecks, Phys. Rev. A4, p.2288, 1971.

8. K. Rubin, B. Bederson, M. Goldstein and R.E. Collins, Phys. Rev. 182, 201, 1969.

9. J. Wykes, J. Phys. B, 5, 1126, 1972.

10. U. Fano and J.H. Macek, Revs. Mod. Physics, Vol.45 No. 4, p.553, 1973.

11. J. Macek and I.V. Hertel, J. Phys. B, 7, 2173, 1974.

12. K. Blum and H. Kleinpoppen, J. Phys. B, 8, 922, 1975, and in "Electron and Photon Interactions with Atoms" edited by H. Kleinpoppen and M.R.C. McDowell, to be published by Plenum Press 1975.

13. R.E. Collins, B. Bederson and M. Goldstein, Phys. Rev. 3, A 1976, 1971.

14. D. Hils, M.V. McCusker, H. Kleinpoppen and S.J. Smith, Phys. Rev. Lett. Vol.29, 398, 1972.

15. D.M. Campbell, H.M. Brash and P.S. Farago, Phys. Rev. 36A, 449, 1971.

16. G.F. Hanne and J. Kessler, Phys. Rev. Lett. 33, 341, 1974 and in "Electron and Photon Interactions with Atoms", 1975 to be published by Plenum Company, New York, edited by H. Kleinpoppen and M.R.C. McDowell.

17. I.C. Percival and M.J. Seaton, Phil. Trans. Roy. Soc. Lond. Ser. A, No. 990, Vol. 251, p.113, 1958.

18. G.C.M. King, A. Adams and F.H. Read, J. Phys. B, 5,1.245, 1972.

19.  D.H. Madison and W.N. Shelton, Phys. Rev. A7, 449, 1973.

20.  D.J. Joachain and R. Vanderpoorten, J. Phys. B, 7, L528, 1974.

21.  L.A. Morgan and M.R.C. McDowell, J. Phys. B, 1073, 1975.

22.  L.A. Morgan and A.D. Stauffer, J. Phys. B, 1975, in press.

23.  G. Csanak, H.S. Taylor and D.N. Tripathy, J. Phys. B, 6, 2040, 1973.

24.  M.R. Flannery and K.J. McCann, to be published 1975.

25.  J.N. Gau and J.H. Macek, Phys. Rev. A10, 522, 1974.

26.  M. Born and E. Wolf in "Principles of Optics", Perganon Press, 3rd edition, 1965.

27.  W.H. McCaster, Revs. Mod. Physics, 33, 8, 1961.

28.  M.C. Standage and H. Kleinpoppen, to be published.

29.  H.G. Berry, L.J. Curtis, D.G. Ellis and R.M. Schectman, Phys. Rev. Lett. 32, 751, 1974.

30.  K. Blum, H. Kleinpoppen and M.C. Standage, to be published.


+    Note the Book of Abstracts of the IXth ICPEAC which includes further applications of electron-photon correlations of atomic excitations:

   A)  P.J.O. Teubner, H. Arriola, A. Ugbabe and E. Weigold (on He and Ne excitations).

   B)  J.W. McConkey, K.-H. Tan, P.S. Farago and P.J.O. Teubner (on He and Ar excitations).

   C)  M.R. Flannery and K.J. McCann (on $\lambda$ and $\chi$ parameters of He).

*   K. Blum, present address, Department of Applied Mathematics and Theoretical Physics, Queen's University, Belfast.

ELECTRONIC STATE ALIGNMENT, ORIENTATION, AND COHERENCE

PRODUCED BY BEAM-FOIL COLLISIONS

D. A. Church

University of California
Lawrence Berkeley Laboratory
Berkeley, California 94720

## INTRODUCTION

The beam-foil collision is the basic excitation means for the light source used in Beam-Foil Spectroscopy (BFS). BFS is the study of electronic level parameters and ionic structure through observation of the spectra of fast beam ions charge-changed and excited by passage through thin, self-supporting foils. Micro-Ampere beams of ions are accelerated to a known fixed energy above 10 keV, charge-to momentum analyzed, and directed through the foil to a Faraday cup. The ion-foil interaction occurs in about $10^{-14}$ sec., and the uniform ion velocity distributes the subsequent radiative decays of the excited levels in space downstream from the foil relative to this well-defined origin. Some of this radiation is collected, spectrally analyzed, and detected using photon-counting techniques. The radiation is generally found to have a non-isotropic spatial distribution, or alternatively to be partially polarized: both are indicative of anisotropic excitation. The spectroscopy aspects of BFS have recently been critically reviewed (1-3), and the whole field has been the subject of a continuing series of international conferences (4-6). We are concerned here with the ion-foil collision process itself; particularly those aspects which result in preferential population of certain magnetic sublevels of particular electronic levels, and the coherence effects which depend for their observation on this anisotropic excitation. The excitation is coherent when non-equal amplitudes of excitation for each sublevel have well-defined phase relations. Alternatively, the cross-sections for population of specific magnetic substates may be favored, producing incoherent non-isotropic population.

In either event, a non-stationary coherence may be induced in
the wave-function subsequent to the collision by known internal or external interactions. This non-stationary coherence
then results in time-dependent intensity modulations of particular polarization components of the optical decay radiation
intensity. The measurement of the relative amplitudes of
these intensity modulations, called quantum beats, provides a
convenient method for the measurement of the excited level
anisotropy. Non-isotropic population distributions in electronic levels are conveniently described by tensor multipole
moment components (7). Because of the electric dipole transition selection rules $\Delta m=0,\pm 1$, only effects from tensor moments
of the first three orders can be directly observed in such
radiation. These moments are called respectively the line
strength S, the orientation O, and the alignment A. The
zeroth order moment S is a measure of the spherically symmetric excitation of the level, while the possible existence
of dipole or quadrupole components follows from the particular
symmetry properties of the collision geometry and the interaction (7) (See Fig. 1). The orientation and alignment components are independent, and their magnitudes are characteristic
of the type and strength of the interaction producing them.
The ion-foil interaction is currently not understood, so these
magnitudes are measured rather than predicted. Excitation
anisotropies of the outer electronic states have been studied
in visible and uv radiation only for ions with incident energy
less than a few MeV; the following discussion is limited to
this regime. Several distinct, but related anisotropic excitation techniques will not be discussed here. They include
laser excitation of fast ions (8), orientation of atoms by
capture of polarized electrons during channeled passage
through a magnetized crystal (9), electronic orientation of
ions by hyperfine coupling to oriented nuclei (10), and angular distributions of characteristic x-ray emissions (11).

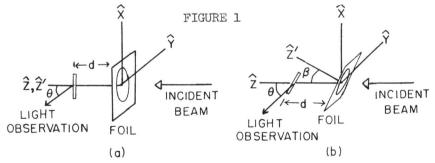

FIGURE 1

XBL 756-1611

Collision and detection geometries for (a) cylindrically symmetric, and (b) reflection symmetric beam-foil collisions. A
magnetic field may be applied along $\hat{x}$ or $\hat{y}$ in (a) or $\hat{x}$, $\hat{y}$, or
$\hat{z}$ in (b) to induce wavefunction coherence.

## THE FOIL

The self-supporting foils usually used in beam-foil measurements are made of carbon and are thought to be polycrys-
talline (3). The carbon is evaporated in vacuum onto deter-
gent coated microscope slides. The coating thickness is optically determined (12) and is customarily expressed as a sur-
face density, with $1 \ \mu g/cm^2 \approx 50 \overset{\circ}{A}$. Typical foil surface densities fall in the range 5-20 $\mu g/cm^2$. The non-carbon surface
density component is about 1.6 $\mu g/cm^2$ independent of thickness
(12). After mounting, the foil surface may have visible deviations from a plane, which change with bombardment time.

A typical beam-foil measurement is performed at pressures
in the low $10^{-6}$ Torr range, resulting in a low degree of surface cleanliness. The excitation characteristics of surfaces
freshly evaporated in vacuum initially differ from those of
the typical "dirty" carbon (13) but all materials investigated
relax to the "dirty" carbon values in times reciprocally related to pressure. Under ion bombardment, sufficient excited
foil atoms are sputtered forward to produce observable spectra
(14), up to 100 electrons per ion may be driven forward (15),
and continuum photons are emitted by the foil (16). The foils
exhibit certain aging characteristics after prolonged use (17)
and eventually break.

The ion-foil collision differs essentially from an ion-
atom collision in that the final ion state evolves from multiple interactions, with the final interaction possibly preceded
by several different degrees of ionization and excitation
while the beam particle is in the foil. The primary final observables (18) of the ion-foil collision are properties of the
ion: its energy, direction, charge, excitation, and excitation anisotropy. The characteristics of the first four of
these observables are described in the literature (19). It
appears probable that the observed excited states of transmitted ions are created either at the final surface of the
foil, or in the last few atomic layers of the bulk material.
This we will denote by "the final surface interaction". Fig.
2 compares the excitation of two levels of fast He atoms using
foil and gas collisions.

## EXCITATION ANISOTROPY

### Collisions with Cylindrical Symmetry

When an ion beam passes through a foil with surface normal along the beam direction, the mean collision is cylindrically symmetric, and possible alignment of the excited levels

is described by the tensor alignment component $A_O^{col} \propto <3J_z^2 - J^2>$
(7). The initial density matrix elements corresponding to
this alignment component are diagonal, characteristic of incoherent excitation. The linear polarization of radiation
emitted from a level so aligned is described in terms of the
radiation intensities emitted perpendicularly to the beam,
polarized respectively parallel $(I_{\parallel})$ and perpendicular $(I_{\perp})$ to
the beam direction. From these intensities, a polarization
fraction $P = (I_{\parallel} - I_{\perp})/(I_{\parallel} + I_{\perp})$ may be calculated for each
transition (20). If $\theta$ is the angle of observation relative
to the beam axis, the angular distribution of the radiation is
$I(\theta) \propto (1 - P\cos^2\theta)/(1 - P/3)$. In terms of the alignment parameter, the intensity of linearly polarized light is given by

(7) $I_{\ell p} = \frac{1}{3}CS\{1 - \frac{h^{(2)}}{4} A_O^{col}(3\cos^2\theta - 3\sin^2\theta\cos 2\psi - 1)\}$     [1]

where C and S are constants, $\psi$ is the angle of the linear
polarizer axis relative to the beam axis, and $h^{(2)}$ is a ratio
of 6j symbols determined by the angular momenta of the initial
and final levels, the order of the moment, and the photon
angular momentum. From this equation, one finds to first order that $P \simeq 3h^{(2)}A_O^{col}/4$.

A measurement technique which establishes coherence between the sub-levels is useful to minimize errors due to instrumental polarizations and cascade effects from higher populated levels. Such coherence is a consequence of any interaction that begins suddenly subsequent to the collision, has
a different axis of symmetry, and removes the sublevel degeneracies. Definite phase relations are then established between the sublevel wavefunctions, and interference terms with
difference frequencies characteristic of the unresolved sublevel splittings produce quantum beat modulations in the emitted light intensity. The interaction may be internal, such
as the fine-structure or hyperfine-structure interactions, or
it may be external, such as that produced by a uniform magnetic field directed perpendicularly to the beam direction.
The phase uncertainty of the ensemble is limited only by the
relative time interval uncertainties between creation of the
level at the foil and detection of the emitted light.

If the field strength H is fixed, the intensity of the
exponential decay of a foil excited level with time constant
T is then periodically modulated in time (space) $(t - t_0) =$
$(z - z_0)/v$ according to (21):

$I(t - t_0) = Ae^{-(z - z_0)/vT}(1 + P \cos(2\gamma_J H(z - z_0)/v))$     [2]

If the time relative to excitation $(t - t_0) = (z - z_0)/v \equiv d/v$
is fixed, by observing light omitted at a fixed distance d
from the foil while the magnetic field is swept, the modulation appears on a constant background (22). Equation [1]

exhibits this same form when $\theta=\pi/2+\omega t$ is substituted, to account for the precession of the moment. In either event, to first order, the polarization fraction P is obtained directly from the relative beat amplitude. It has been demonstrated that the period of these beats is not significantly affected by cascades (23,24), but the relative amplitude may vary with distance from the foil in the case of severe cascading.

Alignment by foil excitation is observed only when $L \geq 1$, characteristic of spin-independent excitation. Results from magnetic-field-generated beat measurements appear in Table 1. One sees that the polarization fraction is positive, and the alignment is negative, for almost all levels investigated; indeed, only certain p states have exhibited a positive alignment. The fractional alignment is generally $\lesssim 20\%$. There is a trend for the highest alignment to occur in the lower charge states, and for a given quantum number n in a specific charge state the highest L levels are generally more highly aligned at a given energy. Such tendencies are consistent with simple models for charge capture alignment (25). These trends are of course subject to the energy dependence of the alignment, which typically is not identical to that of the excitation (Fig. 2), and may be quite pronounced.

Similar coherence effects are observed in the absence of external fields when incoherently aligned L sublevels are subjected to internal interactions, such as fine-and hyperfine-structure couplings. These quantum beats appear superposed on the exponential decay of the radiated light intensity, as a function of the distance from the foil (time relative to excitation). A cosinusoidal modulation at frequencies corres-

FIGURE 2

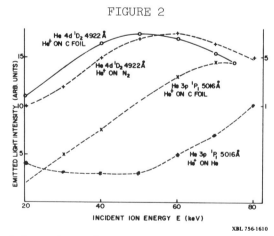

Comparison of ion-atom, ion-molecule, and ion-foil exitation for two levels of helium. The data are from Refs. 42 and 43.

TABLE I

Zeeman beat measurements of polarization fractions P of levels.

| Ion | Upper Level | J | Beam Energy (KeV) | P(%) | $A_o^{col}$ |
|---|---|---|---|---|---|
| $O^+$ | $3p\ ^2P^o$ | 3/2 | 500 | $-2.4\pm0.2$[a] | $+0.032$ |
| " | $3p'\ ^2F^o$ | 5/2 | " | $4.2\pm0.3$[a] | $-0.064$ |
| " | $3p'\ ^2F^o$ | 7/2 | " | $4.1\pm0.3$[a] | $-0.073$ |
| " | $3d'\ ^2G$ | 7/2 | " | $5.5\pm0.5$[a] | $-0.096$ |
| " | $3d'\ ^2G$ | 9/2 | " | $5.5\pm0.5$[a] | $-0.107$ |
| $O^{+2}$ | $3p\ ^3D$ | 2 | 800 | $2.0\pm0.2$[a] | $-0.027$ |
| " | $3p\ ^3D$ | 3 | " | $2.0\pm0.2$[a] | $-0.033$ |
| " | $3d\ ^3F^o$ | 3 | " | $3.5\pm0.2$[a] | $-0.058$ |
| " | $3d\ ^3F^o$ | 4 | " | $3.7\pm0.2$[a] | $-0.07$ |
| $O^{+3}$ | $3p\ ^4D$ | 7/2 | 1100 | $2.0\pm0.3$[a] | $-0.036$ |
| " | $3d\ ^4F$ | 7/2 | " | $1.9\pm0.3$[a] | $-0.034$ |
| Ne | $2p9\ ^2D$ | | 450 | $12.\ \pm2.0$[b] | $-0.20$ |
| $Ne^+$ | $3p\ ^2D$ | 5/2 | 1000 | $3.9\pm0.4$[c] | $-0.06$ |
| $Ne^{+2}$ | $3p'\ ^1F$ | 3 | " | $6.0\pm0.4$[c] | $-0.10$ |
| $Ar^+$ | $4p\ ^2P^o$ | 3/2 | 500 | $3.2\pm0.3$[d] | $-0.034$ |
| " | $4p\ ^2D^o$ | 5/2 | " | $3.4\pm0.1$[d] | $-0.052$ |
| " | $4p'\ ^2F^o$ | 5/2 | " | $4.8\pm0.8$[d] | $-0.073$ |
| " | $4p'\ ^2F^o$ | 7/2 | " | $6.5\pm0.2$[d] | $-0.116$ |
| $Ar^{+2}$ | $4p'\ ^3F$ | 4 | " | $2.2\pm0.2$[d] | $-0.041$ |

TABLE II

Beat measurements of P with relations to cross-sections.

| Ion | Upper Level | Beam Energy (KeV) | P(%) | Sub-level Cross-sections |
|---|---|---|---|---|
| H | $2p\ ^2P$ | 1000 | $-45.0$[e] | $\sigma_0\approx0.38\sigma_1$ |
| He | $3p\ ^1P$ | 70 | $10.0\pm1.5$[f] | $\sigma_0=(1.25\pm0.15)\sigma_1$ |
| " | $4d\ ^1D$ | 40 | $9.4\pm1.4$[g] | $\sigma_2/\sigma_0=0.385(1+0.85\sigma_1/\sigma_0)$ |
| " | $5d\ ^1D$ | " | $12.0\pm1.8$[g] | $\sigma_2/\sigma_0=0.357(1+0.80\sigma_1/\sigma_0)$ |
| " | $6d\ ^1D$ | " | $4.3\pm1.9$[g] | $\sigma_2/\sigma_0=0.445(1+0.94\sigma_1/\sigma_0)$ |
| " | $3d\ ^3D$ | " | $3.6\pm1.5$[g] | $\sigma_2/\sigma_0=0.406(1+0.89\sigma_1/\sigma_0)$ |
| " | $4d\ ^3D$ | " | $2.9\pm1.6$[g] | $\sigma_2/\sigma_0=0.423(1+0.91\sigma_1/\sigma_0)$ |
| " | $5d\ ^3D$ | " | $4.0\pm1.6$[g] | $\sigma_2/\sigma_0=0.398(1+0.87\sigma_1/\sigma_0)$ |
| " | $6d\ ^3D$ | " | $5.0\pm1.8$[g] | $\sigma_2/\sigma_0=0.375(1+0.83\sigma_1/\sigma_0)$ |
| $Be^+$ | $4d\ ^2D$ | 300 | $7.8\pm2.0$[h] | $\sigma_0+0.85\sigma_1\approx2.7\sigma_2$ |
| " | $5d\ ^2D$ | " | $3.4\pm1.0$[h] | $\sigma_0+0.90\sigma_1\approx2.3\sigma_2$ |
| " | $4f\ ^2F$ | " | $7.5\pm2.0$[h] | $\sigma_0+1.35\sigma_1-0.6\sigma_2\approx3.5\sigma_3$ |
| " | $4f\ ^2F$ | " | $4.6\pm2.0$[h] | $\sigma_0+1.45\sigma_1-0.4\sigma_2\approx3.2\sigma_3$ |
| $Be^{+2}$ | $2p\ ^3P$ | 600 | $1.0\pm0.5$[h] | $\sigma_0=(1.2\pm0.1)\sigma_1$ |

[a]Ref. 44; [b]Ref. 22; [c]Ref. 45; [d]Ref. 46; [e]Ref. 28; [f]Ref. 30; [g]Ref. 47; [h]Ref. 48.

FIGURE 3

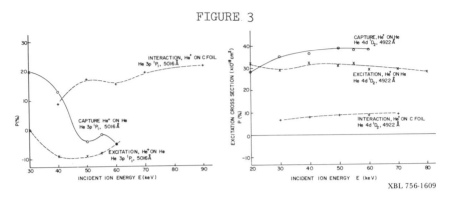

XBL 756-1609

Comparison of polarization fraction results from ion-foil and
ion-atom collisions.  Data from Refs. 30, 31, and 41.

ponding to the separation of different J or F levels is predicted and observed (19)  The initial phase of zero has been
verified (26).  Other formalisms (27) to analyze the data express the alignment in terms of the polarization fraction or
quantum beat amplitudes, and directly relate them to the
cross-sections for the population of particular $m_L$ sublevels.
Examples of data so obtained and analyzed are displayed in
Table II, where the cross-sections for population of sublevels
with magnetic quantum numbers $|m_L|$ are denoted $\sigma_{m_L}$.

The dependence of the alignment on the incident ion
energy has been measured for only a few levels:  the $2p$ (28) &
$3p$ levels (29) of H, and the $3p$ $^1P$ level (30) of He.  One notes
particularly the similarity of the H(2p) and H(3p) alignments,
which exhibit broad peaks with superposed structure.  In Fig.
3, some polarization fraction results from He$^+$-atom (31) collisions are compared with those of foil excitation.

### Tilted-Foil Collisions

When the plane of the foil surface is tilted at an angle
$\beta$ to the beam direction (see Fig. 1-b) the collision has at
most reflection symmetry in the x-z plane.  Excited level moments of order $k \leq 2j$ are then in principle possible but the
anisotropy directly detectable by light emission is completely
described by the orientation and alignment parameters (7):
Orientation $O_{1-}^{col} \propto <'|J_y|>$; Alignment $A_{0}^{col} \propto <|3J_z^2-J^2|>$;
$A_{1+}^{col} \propto <'|J_x J_z + J_z J_x|>$; $A_{2+}^{col} \propto <'|J_x^2-J_y^2|>$.  Orientation of
a level is defined for $J>\frac{1}{2}$; it is characterized by a non-zero
component of angular momentum, a net magnetic moment, a population distribution among the magnetic sublevels dependent on
$m_L$, and the emission of circularly polarized light.  Alignment

components are defined for levels with $J \geq 1$; they are described by quadratic forms of the angular momentum components, no net magnetic moment, a population distribution dependent on $|m_L|$, and the emission of linearly polarized light. The orientation and alignment parameters are related to the density matrix components of the level (32). With the exception of $A_o^{col}$, the parameters are described by off-diagonal elements, and consequently are coherent superposition states.

It is not obvious that on the microscopic level, the gross symmetry properties represented by the tilting of the foil should affect the collision. Nevertheless, orientation manifested by a net emission of elliptically polarized radiation was observed (33) for the 3p $^1$P level of He, and subsequently the effect was demonstrated to be quite general (34-36). Two observation techniques have been used to study these phenomena. The static measurement technique involves the measurement of polarized light intensity at angles of 90 degrees and 56 degrees relative to the beam direction as a function of the foil tilt angle and the beam energy (33, 37). The light was collected from a verticle beam segment, not parallel to the foil surface. The measurements are analyzed in terms of the Stokes parameters, which completely describe the polarization of the light. The dynamic measurement technique relies on coherence effects produced by an external uniform magnetic field (34, 35, 38). Quantum beats are observed as before, when the field is applied perpendicular to the beam, but the excitation coherence now permits the observation of beats when the field is applied parallel to the beam as well. The emitted light intensity is collected from a spatial region parallel to the foil surface to preserve this initial coherence.

General equations for the polarized light intensity emitted by any level (7) express static measurements in terms of the orientation and alignment parameters, and by substitution of the phase of the Larmor precession also describe the results of beat measurements (38). When a magnetic field is applied in the observation (y) direction (see Fig. 1), no orientation beats occur in circularly polarized light, demonstrating that the symmetry axis for the orientation coincides with the foil tilt axis (7, 34). All orientation and alignment parameters for the $4_d{}^1D_2$ level of He were separately determined for 40 KeV incident ion energy at the 30 deg. foil tilt angle (38). Similar measurements for the $3_p$ $^1P_1$ level of He were performed using the static measurement technique (33).

Table III shows the results of orientation measurements at particular foil tilt angles for several transitions of various ions. One sees that even with small foil tilt angles, the polarization fraction $P = (I_{\sigma+}-I_{\sigma-})/(I_{\sigma+}+I_{\sigma-})$, written

TABLE III

Orientation of atom and ion levels produced by tilted-foil
   collisions

| Ion | Upper Level | Beam Energy (KeV) | Foil Tilt Angle (degrees) | Orientation $O_{1-}^{col}$ |
|---|---|---|---|---|
| He | $3p^1P_1$ | 130 | 30 | $0.027^a$ |
| " | $3p^1P_1$ | 260 | 25 | $0.053^b$ |
| " | $4d^1D$ | 40 | 30 | $0.008^c$ |
| " | " | " | 45 | $0.016^c$ |
| " | " | " | 60 | $0.024^c$ |
| $O^+$ | $3p'\ ^2F_{7/2}$ | 540 | 25 | $0.014^b$ |
| $Ar^+$ | $4p\ ^2P_{3/2}$ | 675 | 25 | $0.02^b$ |
| " | $4p'\ ^2F_{7/2}$ | 675 | 25 | $0.022^b$ |
| $Ne^{+2}$ | $3p'\ ^1F_3$ | 1000 | 30 | $0.008^d$ |
| " | " | " | 45 | $0.0165^d$ |
| " | " | " | 60 | $0.024^d$ |

[a]Ref. 33; [b]Ref. 35; [c]Ref. 38; [d]Ref. 36. The orientation is
calculated from the data in these references.

in terms of right and left circularly polarized light intensities emitted perpendicular to the beam, are comparable to or
larger than the alignment of the same levels with an untilted
foil (Tables I, II).

    Measurements of the dependence of the magnitudes of the
anisotropy parameters on the beam energy, foil tilt angle, or
final surface material provide information about the dynamics
of the interaction. Several theories for possible interactions
have been suggested. A general torque interaction is expected
to produce a variation of the orientation proportional to sin
β, (37, 38). This is not generally observed. The torque
theory of Fano (32) applies to systems which are not distorted
by the interaction. In this theory, alignment is transformed
to orientation by the action of an external electric field
gradient. Static electric fields can also produce such a transformation (39, 40). Eck has proposed and worked out for a $^1P_1$
level a theory in which the Stark effect produced by electric
fields near the foil surface produce the orientation (39). He
predicts an orientation $O_{1-}^{col}$ varying as sin 2β. The total polarization fraction of the emitted light  is expected to remain constant as β is varied. None of these predictions are
bourne out by experiment (36, 38). Also, the orientation of
the He 3p $^1P_1$ level is found to reach a maximum when the alignment M/I drops to zero (37), but any relationships, or lack
thereof, between these parameters are otherwise undefined.

FIGURE 4

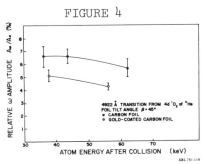

Dependence of the relative orientation signal on the material of the final surface of the foil. This dependence is found to depend on the amount of evaporated material. No reduction is found when the gold layer is on the <u>upstream</u> side of the foil.

The orientation interaction definitely depends on the final surface material of the foil, as Fig. 4 shows, where a thin gold layer evaporated onto the carbon foil causes a reduction of orientation (34, 41). A similar layer of aluminum produces a barely distinguishable effect, however (34, 37). Since the surfaces are presumed to be contaminated, any such observed differences may be associated with an interaction in the final foil layers, and indeed, the effect depends on the amount of evaporated material (41). The general surface field or field gradient interactions which follow are then associated with the surface contamination.

SUMMARY

The cylindrically symmetric beam-foil collision produces excitation and alignment of atom and ion levels similar, but not identical, to that resulting at comparable energies from ion-atom or ion-molecule collisions. When the foil is tilted, the macroscopic change acts on the microscopic scale to produce coherent alignment and orientation of the excited levels. The maximum beam energy range bounding this interaction has not yet been defined. The dynamic interaction which produces these effects is currently not predicted by any theory, although the dynamics of the ions subsequent to the collision are well understood. Refinement of current experimental technique can be expected to better define the final foil surface. The beam-tilted-foil collision promises to be useful in the study of ionic structure via quantum beat, radio-frequency and level-crossing spectroscopy techniques, and may provide a useful probe for certain surface interactions.

## REFERENCES

1.  S. Bashkin, "Beam-Foil Spectroscopy", Progress in Optics XII, E. Wolf, ed.: North-Holland, (1974), p. 288.

2.  H.J. Andrä, Physica Scripta 9, 257 (1974).

3.  I. Martinson and A. Gaupp, Physics Reports 15, 113 (1974).

4.  S. Bashkin, ed., "Beam-Foil Spectroscopy", Gordon and Breach, New York (1968).

5.  I. Martinson, J. Bromander, and H.G. Berry, ed., "Proc. Second Int. Conf. on Beam-Foil Spectroscopy", Nuc. Instrum. Meth. 90, (1970).

6.  S. Bashkin, ed., "Proc. Third Int. Cont. on Beam-Foil Spectrosopy", Nuc. Instrum. Meth. 110, (1973).

7.  U. Fano and J.H. Macek, Rev. Mod. Phys. 45, 553 (1973).

8.  H.J. Andrä, A. Gaupp, and W. Wittmann, Phys. Rev. Letters 31, 501 (1973); H.J. Andrä, A. Gaupp, K. Tillmann, and W. Wittmann, Nuc. Instrum. Meth. 110, 453 (1973).

9.  M. Kaminsky, Phys. Rev. Letters 23, 819 (1969).

10.  G.D. Sprouse, R. Brown, H.A. Calvin, and H.J. Metcalf, Phys. Rev. Letters 30, 419 (1973).

11.  E.H. Pedersen, S.J. Czuchlewski, M.D. Brown, L.D. Ellsworth, and J.R. Macdonald, Phys. Rev. A11, 1267 (1975).

12.  J.O. Stoner, Jr., J. Appl. Phys. 40, 707 (1969).

13.  K. Berkner, I. Bornstein, R.V. Pyle, and J.W. Stearns, Phys. Rev. A6, 278 (1973).

14.  See e.g., H.G. Berry, I. Martinson, and J. Bromander, Phys. Letters 31A, 521 (1970); S. Bashkin, Nuc. Instrum. Meth. 90, 3 (1970).

15.  C.F. Moore, W.J. Braithwaite, and D.L. Mathews, Phys. Letters 47A, 353 (1974).

16.  S. Bashkin, D. Fink, P.R. Malmberg, A.B. Meinel, and S. G. Tilford, J. Opt. Soc. Am. 56, 1064 (1966).

17.  See e.g., J.H. Brand, C.L. Cocke, B. Curnutte, and C. Swenson, Nuc. Instrum. Meth. 90, 63 (1970).

18.  H.G. Berry, J. Bromander, and R. Buchta, Nuc. Instrum. Meth. 90, 269 (1970).

19.  See reviews, Refs. 1-3, for general references.

20.  I.C. Percival and M.F. Seaton, Phil. Trans. Roy. Soc. (London) 251, 113 (1958).

21.  C.H. Liu and D.A. Church, Phys. Rev. Letters 29, 1208 (1972); D.A. Church and C.H. Liu, Nuc. Instrum. Meth. 110, 147 (1973).

22.  C.H. Liu, S. Bashkin, W.S. Bickel, and T. Hadeishi, Phys. Rev. Letters 26, 222 (1971).

23.  C.H. Liu, M. Druetta, and D.A. Church, Phys. Letters 39A, 49 (1972).

24.  M. Dufay, Nuc. Instrum. Meth 110, 79 (1973).

25.  R.H. Hughes, p. 103 of Ref. 4.

26.  D.J. Burns and W.H. Hancock, Phys. Rev. Letters 27, 370 (1971); J. Opt. Soc. Am. 63, 241 (1973).

27.  See the reviews, especially Ref. 2.

28.  H.J. Andrä, P. Dobberstein, A. Gaupp, and W. Wittman, Nuc. Instrum. Meth. 110, 301 (1973).

29.  D.J. Lynch, C.W. Drake, M.J. Alguard, and C.E. Fairchild, Phys. Rev. Letters 26, 1211 (1971).

30.  H.G. Berry and J.L. Subtil, Phys. Rev. Letters 27, 1103 (1971); Nuc. Instrum. Meth. 110, 321 (1973).

31.  F. J. DeHeer, L. Wolterbeek-Muller, and R. Geballe. Physica 31, 1745 (1965).

32.  U. Fano, Phys. Rev. 133, B828 (1964); Rev. Mod. Phys. 29, 74 (1957).

33.  H.G. Berry, L.J. Curtis, D.G. Ellis, and R.M. Schectman, Phys. Rev. Letters 32, 751 (1974).

34.  D.A. Church, W. Kolbe, M.C. Michel, and T. Hadeishi, Phys. Rev. Letters 33, 565 (1974).

35.  C.H. Liu, S. Bashkin, and D.A. Church, Phys. Rev. Letters 33, 993 (1974).

36.  H.G. Berry, L.J. Curtis, and R.M. Schectman, Phys. Rev. Letters 34, 509 (1975).

37.  H.G. Berry, S.N. Bhardwaj, L.J. Curtis, and R.M. Schectman, Phys. Letters 50A, 59 (1974).

38.  D.A. Church, M.C. Michel, and W. Kolbe, Phys. Rev. Letters 34, 1140 (1975).

39.  T.G. Eck, Phys. Rev. Letters 33, 1055 (1974).

40.  M. Lombardi and M. Giroud, Compt. Rend. B266, 60 (1968).

41.  D.A. Church and M.C. Michel, (to be published).

42.  W.S. Bickel, K. Jensen, C.S. Neuton, and E. Veje, Nuc. Instrum. Meth. 90, 309 (1970).

43.  C.E. Head and R.H. Hughes, Phys. Rev. 139, A1392 (1965).

44.  D.A. Church and C.H. Liu, Physica 67, 90 (1973); Nuc. Instrum. Meth. 110, 267 (1973).

45.  M. Druetta and A. Denis, Nuc. Instrum. Meth. 110, 291 (1973).

46.  D.A. Church and C.H. Liu, Phys. Rev A5, 1031 (1972).

47.  J. Yellin, T. Hadeishi, and M.C. Michel, Phys. Rev. Letters 30, 1286 (1973), Phys. Rev. Letters 30, 417 (1973).

48.  O. Poulsen and J.L. Subtil, Phys. Rev A8, 1181 (1973); J. Phys. B7, 31 (1974).

RECENT THEORETICAL ADVANCES

# ON GLAUBER AND GLAUBER-RELATED METHODS IN ATOMIC PHYSICS

F. W. Byron, Jr.

Department of Physics and Astronomy, University
of Massachusetts, Amherst, Mass. 01002

The purpose of this article is to present a brief summary of recent developments in the field of Glauber and
Glauber-related approximations.  Let me begin by reminding
the reader that since the original work by Franco [1] on the
elastic scattering of electrons by atomic hydrogen, the
Glauber approximation has been used extensively.  It gives a
simple form for the scattering amplitude as follows:

$$f(k,\vec{\Delta}) = \frac{k}{2\pi i} \int e^{i\vec{\Delta}\cdot\vec{b}} <n_f| e^{-\frac{i}{k}\int_{-\infty}^{\infty} V(\vec{b},z;\vec{r}_j)dz} - 1|n_i> d^2b \quad (1)$$

where the integral in the phase is along a z-direction taken
perpendicular to the momentum transfer, $\vec{\Delta}$.  With this choice,
the first Born approximation is obtained in the high energy
limit.  In Eq. (1), $\vec{r}_j$ stands for the internal coordinates
of the bound target system whose initial and final states
are labelled by $n_i$ and $n_f$ respectively, and k is the wave
number of the incident electron.  Equation (1) is very tractable for simple atoms and by now the basic elastic and excitation processes in $e^- + H$ and $e^- + He$ scattering have
been thoroughly studied by various authors.  A very complete
review of these studies has been given by Gerjuoy and Thomas
[2].

The Glauber approximation suffers from a number of difficulties:

   a. the cross-sections for $e^- + A$ scattering is the same
      as that for $e^+ + A$ scattering, where A is any atom;

   b. in each order of perturbation theory, the Glauber

amplitude is either purely real or purely imaginary, alternating from order to order (this actually implies a.);

c. for <u>inelastic</u> processes, the choice of trajectory at small angles is rather unphysical since the z-direction along which the phase is integrated is nearly perpendicular to the incident direction rather than nearly parallel, as one would expect.

A number of years ago, this author suggested [3] that the Glauber "wavefunction amplitude" might solve all these problems in one shot:

$$f_w(k,\vec{\Delta}) = -\frac{1}{2\pi} \int e^{i\vec{\Delta}\cdot\vec{r}} <n_f|V(\vec{r},\vec{r}_j)e^{i\chi(\vec{r},\vec{r}_j)}|n_i> d^3r \qquad (2)$$

where

$$\chi(\vec{r},\vec{r}_j) = -\frac{1}{k} \int_{-\infty}^{z} V(\vec{b},z',\vec{r}_j)dz' , \qquad (3)$$

the z-direction lying along the incident direction. One can easily see that this form deals in a non-trivial way with the three difficulties just mentioned. Whether or not it deals correctly with them remains to be seen.

Recently Gau and Macek [4] have directed their attention to this form in an attempt to remedy the large-angle behavior of the Glauber approximation for $s \rightarrow p$ transitions where it is found experimentally (by, for example, Suzuki and Takayanagi [5]) that $1^1S \rightarrow 2^1S$ and $1^1S \rightarrow 2^1P$ transitions in helium have very nearly equal differential cross-sections at large angles, whereas the Glauber approximation gives a $1^1S \rightarrow 2^1P$ differential cross-section at least an order of magnitude smaller than that for $1^1S \rightarrow 2^1S$ in the intermediate energy region.

Although this problem has not yet been completely analyzed, a few words of caution may be in order. If one applies the form of the Glauber amplitude given by Eqs. (2) and (3) to potential scattering by, say, a Yukawa potential, the results are not at all encouraging as is seen in Fig. 1. We see that the wide-angle behavior of the second Born term is indeed modified considerably but in a very undesirable way. For Im $f_{B2}$ the modified eikonal amplitude is, except at small angles, worse than the traditional eikonal amplitude which in fact agrees with the exact Im $f_{B2}$ at all angles. The approximation to Re $f_{B2}$ given by the modified eikonal amplitude is completely out of keeping with the exact Re $f_{B2}$. The point here is that it is very difficult to put back what is missing in an impact parameter formalism by kinematical

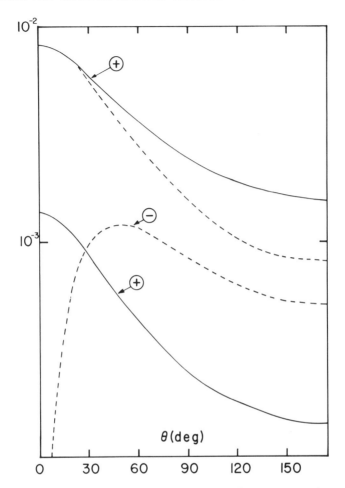

Fig. 1. The imaginary (upper curves) and real (lower curves) parts of the second Born amplitude for scattering by the Yukawa potential $V(r) = 0.5\exp(-r)/r$. The solid curves give the exact results, while the dashed curves give the results corresponding to Eqs. (2) and (3) applied to potential scattering. The incident wave number is $k=3$.

modifications. The elegant work of S. J. Wallace [6] and the ingenious perturbation summation method of A. R. Swift [7] seem rather conclusive on the need for additional phase functions to obtain the terms lost in the lowest order expansion of the Green's function.

However, at small angles in inelastic scattering, particularly for transitions involving non-spherically symmetric states, this modified eikonal amplitude may indeed offer an

improvement which will be well worth having. Figure 2 illustrates one of the difficulties encountered by the Glauber approximation for electrons exciting atomic hydrogen to the 2p0
state at 200 eV. For the 2p+ state the Glauber approximation
does very well at small angles, but for the 2p0 state, as
shown in Fig. 2, this is not the case. Figure 2 shows just
the real part of the second Born term, but presumably similar
effects are present in higher order terms, in particular the
third order term which is very important in the Eikonal-Born
series method. The ingenious work of Gau and Macek makes the
modified Glauber amplitude tractable for the first time, and

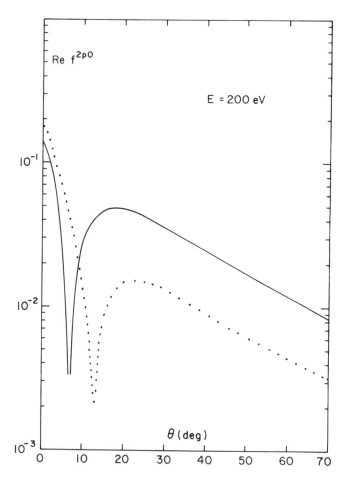

Fig. 2. The real part of the second Born amplitude for
the excitation of atomic hydrogen to the 2p0 state by electron bombardment at 200 eV. The solid curve is the result of
reference 8, using closure and making no kinematical approximations. The dotted curve is the Glauber result.

thus offers a possible remedy to the difficulty illustrated in Fig. 2, at least at small angles.

Turning now to the Eikonal-Born series (EBS) method of Byron and Joachain [9], let us begin by recalling the basic idea of this approximation. It consists in looking for the consistent leading corrections (in powers of $(1/k)$) to the first Born approximation. It turns out that this means writing, for example in the case in which both the initial and final states are spherically symmetric,

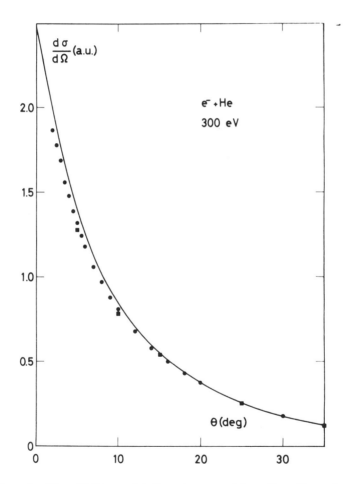

Fig. 3. The differential cross-section for the elastic scattering of electrons by helium at 300 eV. The solid curve gives the Eikonal-Born series results of reference 9. The circles represent the experimental values of Bromberg [11], while the squares show the experimental values of Jansen et al [12].

$$f^d = f_{B1} + \text{Re } f_{B2} + i \text{ Im } f_{B2} + \text{Re } f_{G3} \qquad (4)$$

$$f^{ex} = g_{Och}. \qquad (5)$$

Here $g_{Och}$ is the Ochkur approximation to the first order exchange amplitude, $f_{B1}$ and $f_{B2}$ are the first and second Born
terms, respectively, and $f_{G3}$ is the <u>Glauber</u> approximation to
the third Born term. The terms Re $f_{B2}$ (missing from the
Glauber approximation) and Im $f_{B2}$ (given well by the Glauber
approximation for transitions between spherically symmetric
states except at very small angles) are readily obtained by
using the closure approximation along with the exact inclusion
of those intermediate states which dominate at large angles
(the elastic intermediate states). In addition, because of
the importance of intermediate p-states in giving the small-
angle behavior of $f_{B2}$, these states are put in with special
care at small angles where their analytic behavior is very
simple.

By now the basic processes in atomic hydrogen have been
studied by Byron and Joachain [9] (elastic scattering) and
Byron and Latour [8] (1s → 2s and 1s → 2p excitation); in
helium, elastic scattering [9] and $1^1S → 2^1S$ excitation [10]
have been investigated by Byron and Joachain. Figures 3 and 4
illustrate the situation for elastic electron-helium scattering and for the sum of 1s-2s and 1s-2p excitation in electron-
hydrogen scattering, respectively. In the former case, the
most striking feature is the rapid increase of the differential cross-section at small angles. Here the effects of polarization contained in Re $f_{B2}$ are of paramount importance. At
large angles Re $f_{B2}$, which of course interferes with $f_{B1}$ in
computing the differential cross-section, is again very important. The agreement between the EBS results and experiment
[11,12] is excellent. For the case of excitation to the n = 2
state of atomic hydrogen, where recent experimental results
of Williams and Willis [13] are available, we see that outside
the small-angle region the difference between the Glauber and
the EBS results is striking. This is a result of the poor
behavior of the 1s-2p Glauber amplitude in the large-angle
region, which we have mentioned earlier. The agreement between the experimental results and the EBS theory is rather
good, although the experimental error bars are fairly large.

A few comments on the strengths and weaknesses of the
EBS method may be in order at this point. The weaknesses are
threefold:

   a. It is a perturbative method, and clearly only a few
      orders of perturbation theory are tractable.

b. For non-spherically symmetric systems (such as 1s-2p excitation in atomic hydrogen) the term $f_{G3}$ in Eq. (4) is suspect, as we noted in connection with the work of Gau and Macek [4].

c. At wide angles, the first Born term is neglible for inelastic scattering at intermediate and high energies so we get only the leading wide-angle term. The additional terms needed for a leading correction to $f_{B2}$ include the second order exchange amplitude at large angles.

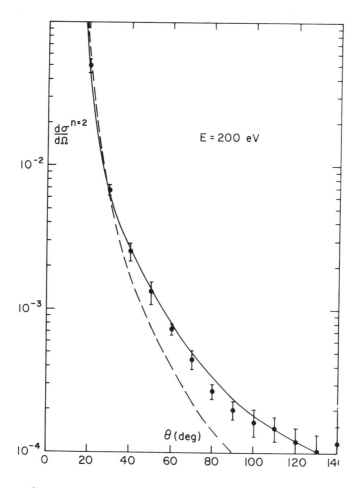

Fig. 4. The differential cross-section for the excitation of the n = 2 state of atomic hydrogen by electron impact at 200 eV. The solid curve is the EBS result of reference 8; the dashed curve is the Glauber result. The experimental points are those of Williams and Willis [13].

Among the advantages should be mentioned:

    a. It gives a consistent expansion in powers of $k^{-1}$, so one knows fairly well what one has got.

    b. It has coupling to all states.

    c. It has considerable pedagogic value. One can use this approach to get an idea of what other methods will do well or poorly.

To illustrate this last point we may give several examples. Firstly, in studying the perturbation series one finds no trace of contributions which would modify the first Born approximation differential cross-section for momentum transfers of the order of one or less in the high energy limit. Such modifications have been found by McDowell, Morgan and Myerscough [14] in their static coupling, distorted wave method; such a result should, we feel, be viewed with caution. Secondly, at intermediate energies one finds that in second order both initial and final state static distorting potentials make important contributions to the scattering amplitude. To include just initial or just final state distortion in a DWBA calculation at intermediate or high energies would be incorrect.

To conclude this report, let us briefly discuss another area in which the Glauber approximation has proven to be useful, namely, in the optical potential method for the elastic scattering of electrons and positrons from complex atoms. If one wants to look at more complicated systems than hydrogen and helium, it will be profitable to try to abstract the key features out of the ab initio method just discussed and put them into a more general context, even though this may mean giving up some fine details. Clearly, for heavier atoms, one expects the central, charge-cloud potential, $V_{static}$, which varies like $-Z/r$ at small distances, to dominate strongly the large momentum transfer behavior of the differential cross-section, while at small momentum transfers polarization and absorption effects should be very important.

The polarization potential can be obtained either by a study of intermediate p-states in the second order term of perturbation theory discussed above or by using the "double eikonalization" method of Joachain and Mittleman [15,16]. These give a dynamic polarization potential of the form

$$V_{pol}(r) = -\tfrac{1}{2}(\alpha/r^4)f(r), \tag{6}$$

where $f(r)$ is a function which tends to unity when r becomes greater than $a \simeq k/\Delta$ and falls rapidly to zero when r is less than a. Here k is the incident wave number, and $\Delta$ is an average target excitation energy similar to that used in the EBS method. The energy dependence of the cutoff parameter, a, means that as one goes to higher and higher incident energies the effect of polarization is forced to smaller and smaller scattering angles.

Similarly, absorption effects can be treated in a number of ways; the Glauber approximation, since it gives Im $f_{B2}$ well at all angles except the very smallest, is a natural candidate. Byron and Joachain [17] have studied electron-neon

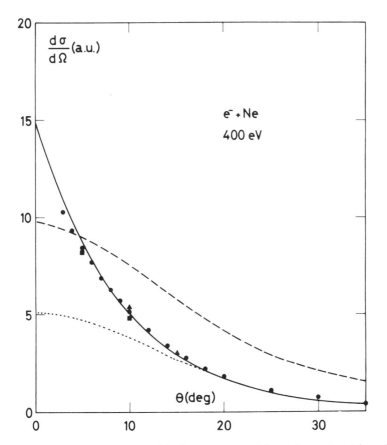

Fig. 5. The differential cross-section for elastic electron-neon scattering at 400 eV. The solid curve represents the optical model results of reference 17, the upper dashed curve is the first Born approximation and the lower dashed curve shows the result of an optical model calculation using only the static potential. The circles are the experimental results of Bromberg [11], the squares are those of Jansen et al [12] and the triangles are those of Gupta and Rees [19].

elastic scattering using a Glauber second-order absorption potential, modified at large distances to eliminate undesirable small-angle effects. Electron (or positron) scattering by the resulting optical potential

$$V_{Opt} = V_{static} + V_{pol} + iV_{abs} + V_{exch} \qquad (7)$$

is readily handled in the partial wave formalism. Note in Eq. (7) that we have also included an exhange pseudopotential [17,18] to take into account the lowest order effects of the Pauli Principle. $V_{exch}$ would be omitted for positron-atom scattering, and the sign of $V_{static}$ would be reversed. Figure 5 shows electron-neon scattering at 400 eV, compared with the experimental data [11,12,19], with the first Born approximation and with an approximation which takes simply $V_{Opt} = V_{static}$. Clearly at small angles the inclusion of polarization and absorption effects is extremely important but is much less important at larger angles. Even at an energy as high as 400 eV the first Born approximation does very poorly.

## REFERENCES

1. Victor Franco, Phys. Rev. Letters, 20, 709 (1968).
2. E. Gerjuoy and B.K. Thomas, Reports on Prog. in Phys. (1975).
3. F.W. Byron, Phys. Rev. A4, 1907 (1971).
4. J.N. Gau and J. Macek, Phys. Rev. A10, 522 (1974).
5. M. Suzuki and T. Takayanagi, Abstracts of Papers of VIII ICPEAC (Belgrade, 1973).
6. S. J. Wallace, Phys. Rev. Letters 27, 622 (1971); Ann. Phys. (N.Y.) 78, 190 (1973); Phys. Rev. D8, 1846 (1973).
7. A. R. Swift, Phys. Rev. D9, 1740 (1974).
8. F. W. Byron and L. J. Latour, Phys. Rev. (to be published).
9. F. W. Byron and C. J. Joachain, Phys. Rev. A8, 1267 (1973).
10. F. W. Byron and C. J. Joachain, J. Phys. B Lett. (to be published).
11. J. P. Bromberg, J. Chem. Phys. 61, 963 (1974).
12. R. H. J. Jansen, F. J. de Heer, H. J. Luyken, B. van Wingerdin and H. J. Blaauw, J. Phys. B (to be published); R. H. J. Jansen, thesis, University of Amsterdam (1975).
13. J. F. Williams and B. A. Willis, J. Phys. B8, 1641 (1975).
14. M. R. C. McDowell, L. Morgan and V. Myerscough, J. Phys. B8, 1053 (1975).
15. C. J. Joachain and M. H. Mittleman, Phys. Rev. A4, 1492 (1971).
16. F. W. Byron and C. J. Joachain, Phys. Rev. A9, 2559 (1974).
17. F. W. Byron and C. J. Joachain, Physics Letters, 49A, 306 (1974).
18. M. H. Mittleman and K. M. Watson, Ann. Phys. (N.Y.) 10, 268 (1960).
19. S. C. Gupta and J. A. Rees, J. Phys. B8, 417 (1975).

# VARIATIONAL PRINCIPLES, SUBSIDIARY EXTREMUM PRINCIPLES,

# AND VARIATIONAL BOUNDS

LARRY SPRUCH

Department of Physics, New York University

4 Washington Place, New York, N.Y.    10003

Ignoring many details, and studiously avoiding the temptation to state results in their most general context, we review some very recent developments in the construction and use of variational principles (V Ps), variational bounds (V Bds), and related techniques. These developments include: an approach that enables one to construct a V P for almost any well posed problem in mathematical physics by the use of a Lagrange undetermined multiplier (which can be a constant $\lambda$, a function L, an operator $\Lambda$, . . .) to account for each constraint; subsidiary extremum principles which provide a very useful means for the approximate determination of such multipliers; variational identities; upper and lower V Bds on relatively arbitrary bound state matrix elements; an upper V Bd on the scattering length A when the target ground state wave function $\phi$ and energy $\varepsilon_1$ are only imprecisely known; and a lower V Bd on A which uses the adiabatic approximation.

## 1.   INTRODUCTION

Though we will be concerned primarily with work of the past two years, it would be well to recall that what would now be appropriate would be not the bicentennial but roughly the bimillenial celebration of the work of Hero of Alexandria which, at least roughly, introduced the concepts of the variational principle ( V P) and the stationary or variational bound (V Bd) into mathematical physics. Hero – appropriately named – noted that a light ray reflected from a plane surface took the shortest possible path.

The advantages of V Ps and of V Bds are well known.  The error of a V P estimate of the quantity of interest is of the order of the square of the error made in the choice of various trial functions.  The V P approach thereby has the charming feature of largely masking the ignorance of the person making the estimate.  V Ps work beautifully in estimates of energies of bound systems and of scattering parameters.  In the estimation of arbitrary matrix elements, one must pay the price in performing V P estimates that one must introduce and develop approximations for the so-called Lagrange undetermined multipliers, in addition to the functions of direct physical interest. The V Bd has the further advantage that the error in the estimate is of definite sign.  However, V Bds are not always as simple to obtain, and can be more difficult to use, than V Ps.  V Ps are more commonly used than is generally appreciated.  Those approximation methods that are most heavily used - the Rayleigh-Ritz, the Born, and the close-coupling, to cite only three examples - give better answers than might have been expected (and have therefore won out against competing methods) precisely because they _are_ V Ps (and in some cases V Bds).

Super V Ps are estimates with an output error which is of third or higher order in the input error.  If they could be put into a useful form they would introduce a new era in the computational aspect of mathematical physics.

It will now be useful to catalog some of the concepts and techniques that have proved most fruitful in the development and use of V Ps and V Bds.  Some of the points made will be elaborated upon later.  i)  Essentially every properly posed problem of mathematical physics can be attacked variationally.  Some folklore to the effect that V Ps cannot be generated if the system is not time reversible, or invariant under rotation, simply isn't true.  ii)  The Rayleigh-Ritz (RR) principle for the ground state energy $\varepsilon_1$, that

$$\varepsilon_{1t} \equiv (\phi_t, h\phi_t)/(\phi_t, \phi_t), \tag{1.1}$$

where h is the hamiltonian and $\phi_t$ is a trial ground state wave function, is extremized by the choice of the exact ground state wave function $\phi$ for $\phi_t$, serves as an extraordinarily powerful tool for the determination of variational parameters contained in $\phi_t$, and should serve as a prototype: the approximate determination of any function required for the analysis should almost always proceed via the introduction of variational parameters which are to be evaluated with the aid of an extremum principle --- if no such extremum principle arises naturally, a "subsidiary extremum principle" designed for that purpose can normally be devised.  iii)  Operators of the form

$\mathcal{H}$ - E often appear, where $\mathcal{H}$ is a hamiltonian and E an energy. If $\mathcal{H}$ - E has a negative eigenvalue, it will often be useful to introduce a new non-negative operator, $\hat{H}$ - E, where $\hat{H}$ is a modified hamiltonian. This can serve different purposes. Firstly, assume that the real function X to be estimated is defined by

$$BX = f, \tag{1.2}$$

where B is a non-negative real symmetric operator and f is a known real function. Consider the functional

$$M(X_t) \equiv (X_t, BX_t) - 2(X_t, f). \tag{1.3}$$

Writing $X_t = X + \delta X$, one finds $M(X+\delta X) = M(X) + (\delta X, B\delta X) \geq M(X)$. $M(X_t)$ achieves its minimum value for $X_t = X$ and can therefore be used as an extremum principle for the evaluation of any variational constants contained in $X_t$. iv) A second advantage of the introduction of non-negative operators concerns the development of V Bds. To obtain a V Bd on a quantity Q of interest, one would normally not proceed directly. Rather, one constructs a variational identity (V Id) for Q, of the form $Q = Q_{var} + s$, where $Q_{var}$ is an explicit calculable V P for Q and where s is a term of second order, involving unknown functions of first order. To obtain a V Bd on Q one need not obtain a V Bd on s; a simple bound, that is, a bound on s which is of the same (second) order as s, will do. If, for example, s is of the form $s = (\delta X, B\delta X)$, where B is non-negative with respect to a class of functions which contains $\delta X$, we have $s \geq 0$ and $Q \geq Q_{var}$, that is, $Q_{var}$ is in this case not merely a V P for Q but a lower V Bd. v) Easily the most significant remark that can be made about the development of V Ps and V Bds for parameters characterizing scattering processes is that one can often reduce a scattering problem to a bound state problem and that there can be great advantage in doing so. A concrete example is the development of an upper V Bd on the scattering length A for scattering of a positron or electron by an atom whose ground state wave function and associated ground state energy $\varepsilon_1$ are only imprecisely known.[1,2] The standard[3,4] upper V Bd on A, which is explicit when $\phi$ and $\varepsilon_1$ are known, can be written in the form $A \leq F(\psi_t)$, where F is an explicit functional and $\psi_t$ is a trial scattering function. The asymptotic boundary condition that $\psi_t$ must satisfy requires a knowledge of $\phi$. If $\phi$ is only imprecisely known, the upper V Bd becomes a formal one. However, considering the positron case for simplicity, we can integrate - in actuality, not formally - over the positron coordinate, and arrive at $A \leq m(\phi)$, where m is a bound state matrix element. The determination of an upper V Bd on the scattering parameter A has thereby been reduced to the determination of an upper V Bd on a bound state matrix element $m(\phi)$ for $\phi$ imprecisely known.

Before closing this general discussion let me note that limitations of space have forced me to largely restrict myself to work of the N.Y.U. group and various collaborators. Let me further note that in variational studies it is important to extremize the high quality of the people one works with, and it has been my pleasure and good fortune these past two years to do just that. Formally, we have

$\delta$(high quality of I. Aronson, R. Blau, our chairperson E. Gerjuoy, Y. Hahn, K. Kalikstein, C. J. Kleinman, R. Mueller, A. R. P. Rau, my coworker for many many years L. Rosenberg, and R. Shakeshaft) = 0.

## 2. SYSTEMATIC CONSTRUCTION OF VARIATIONAL PRINCIPLES

The determination of the extremum values of a function whose arguments are constrained normally proceeds indirectly, with the constraint equations accounted for through the use of Lagrange undetermined multiplers. This suggests[5] that we can construct a V P for a functional $Q(\phi)$ by accounting for each of the equations that define[6] $\phi$ via a Lagrange multiplier, which can be a constant $\lambda$, a function L, an operator $\Lambda$, ... Assume, for example, that the quantity Q of interest is $\langle W \rangle \equiv (\phi, W\phi)$, where W is a real symmetric operator and where the real non-degenerate ground state wave function $\phi$ is defined by $(\phi, \phi) - 1 = 0$, which represents one constraint, and by $(h - \varepsilon_1)\phi = 0$, a constraint at each point in space. (The Schroedinger equation therefore requires a Lagrange undetermined multiplier at each point in space, that is, a Lagrange undetermined function.) The ground state energy $\varepsilon_1$ is not known exactly but can be estimated variationally using (1.1). We write[7,5]

$$(\phi, W\phi)_{var} = (\phi_t, W\phi_t) + \lambda_t[(\phi_t, \phi_t) - 1] + 2(L_t, [h - \varepsilon_{1t}]\phi_t). \quad (2.1)$$

We clearly have $(\phi, W\phi)_{var} = (\phi, W\phi)$ for $\phi_t = \phi$, and we choose $\lambda_t$ and $L_t$ to be approximations to $\lambda$ and L, respectively, with $\lambda$ and L chosen so that the difference $\delta(\phi, W\phi) \equiv (\phi, W\phi) - (\phi, W\phi)_{var}$ vanishes to first order in $\delta\phi \equiv \phi_t - \phi$. Neglecting the second order terms $(\delta\phi)^2$, $(\delta\phi)(\delta\lambda)$, $(\delta\phi)(\delta L)$, and $\varepsilon_{1t} - \varepsilon_1$, we have

$$\delta(\phi, W\phi) = 2(\phi, W\delta\phi) + 2\lambda(\phi, \delta\phi) + 2([h - \varepsilon_1]L, \delta\phi) = 0.$$

Since $\delta\phi$ is arbitrary, this demands $(h - \varepsilon_1)L + W\phi + \lambda\phi = 0$. Taking the inner product with $\phi$ gives $\lambda = -(\phi, W\phi)$, and L then satisfies

$$(h - \varepsilon_1)L = (\phi, W\phi)\phi - W\phi \equiv q(\phi). \tag{2.2}$$

L is made unique by (arbitrarily) specifying $(L, \phi)$.

With appropriate adaptations, the above procedure should normally be applicable to just about every well posed problem of mathematical physics[5]; ingenuity should not be necessary. We will return in Sec. 5 to the implementation of the above approach, that is, to methods of obtaining an adequate estimate $L_t$. Before doing so, it will be useful to exceed our two year historical limit in Sections 3 and 4 in order to examine some well known V Bds which can serve as models for the development of other V Bds.

### 3. TWO STANDARD VARIATIONAL BOUNDS

#### A. The Ground State Energy $\varepsilon_1$

The classic quantum mechanical example of a V Bd is the RR estimate of the ground state energy $\varepsilon_1$. With respect to quadratically integrable functions (this will often be understood in what follows) we have $h - \varepsilon_1 \geq 0$. It follows immediately that $\varepsilon_{1t}$ of (1.1) represents an upper V Bd on $\varepsilon_1$. Formal proofs which involve the expansion of $\phi_t$ in terms of the eigenfunctions $\phi_n$ of h --- we have been writing $\phi$ for $\phi_1$ --- are necessary to satisfy requirements of rigor but with regard to a comprehension of the origin of the V Bd they only confuse the issue.

#### B. The Scattering Length A - No Composite Bound State[3]

Consider a positron of zero initial kinetic energy incident on a target in its ground state. As always, the target ground state is characterized by $h(\vec{r})$, $\phi(\vec{r})$, and $\varepsilon_1$. $\vec{r}$ denotes the totality of electron coordinates. The hamiltonian for the full system is $H(\vec{r}, \vec{\rho}) = h(\vec{r}) + T(\vec{\rho}) + V(\vec{r}, \vec{\rho})$, where $\vec{\rho}$ is the positron coordinate, $T(\vec{\rho})$ the positron kinetic energy operator, and $V(\vec{r}, \vec{\rho})$ is the positron-atom interaction. The true scattering wave function satisfies $(H - \varepsilon_1)\Psi = 0$ and the boundary condition

$$\Psi(\vec{r}, \vec{\rho}) \sim (m/2\pi\hbar^2)^{\frac{1}{2}} \phi(\vec{r})(A - \rho)/\rho, \qquad \rho \sim \infty$$

or the appropriate modification for $e^+$-ion scattering. The only condition imposed upon the trial function $\Psi_t$ is that it satisfy a boundary condition of the same form as $\Psi$, but with A replaced by $A_t$. The V Id $A = A_{var} + s$ follows readily, where

$$A_{var} = A_t + (\Psi_t, [H - \varepsilon_1]\Psi_t), \quad s = -(\delta\Psi, [H - \varepsilon_1]\delta\Psi), \quad (3.1)$$

and where $\delta\Psi \equiv \Psi_t - \Psi$.   If the $e^+$ and the atom cannot form a composite bound state, $\varepsilon_1$ is at the bottom of the spectrum so that $H - \varepsilon_1 \geq 0$ with respect to normalizable functions <u>and</u> functions which approach constant multiples of $\phi/\rho$, as does $\delta\Psi$, and we are in a situation very similar to that which obtains in a RR analysis; we have $s \leq 0$ and therefore $A \leq A_{var}$, so that $A_{var}$ is not simply a VP estimate of A but an upper V Bd on A.

The above result is terribly restrictive in that $\phi$ is known only for H, $He^+$, $Li^{++}$, etc. We will show later how this crucial difficulty can be bypassed. On the other hand, the restriction to $e^+$ was entirely unnecessary; to consider $e^-$ scattering, we need merely antisymmetrize $\Psi$ and $\Psi_t$.

## 4.   THE CONSTRUCTION OF NON-NEGATIVE OPERATORS $\hat{H} - \varepsilon_1$

### A.   The Scattering Length - One Composite Bound State[4]

If the incident paritcle and the atom can form one and only one composite bound state, with energy $E_B < \varepsilon_1$ and with a normalized wave function $\Psi_B(\vec{r},\vec{\rho})$, we no longer have $H - \varepsilon_1 \geq 0$ and the approach which led in Sec. (3B) to an upper V Bd on $\bar{A}$ is inapplicable. We can however proceed as follows.

We assume once again that the target properties $\varepsilon_1$ and $\phi$ are known. (We will show later that this requirement can be bypassed.) $E_B$ and $\Psi_B$ are only imprecisely known, but we can formally introduce the projection operator $P_B \equiv |\Psi_B><\Psi_B|$. Since $1 - P_B$ annihilates $\Psi_B$, we have $H(1 - P_B) - \varepsilon_1 \geq 0$. Equivalently, since $P_B H = E_B P_B$, we have $H - (HP_B H/E_B) - \varepsilon_1 \geq 0$. This result is a purely formal one. However, choosing a normalized trial bound state wave function $\Psi_{Bt}$ good enough to give binding, that is, such that $E_{Bt} \equiv (\Psi_{Bt}, H\Psi_{Bt}) \leq \varepsilon_1$, and defining $P_{Bt} \equiv |\Psi_{Bt}><\Psi_{Bt}|$, it can be shown, surprisingly, that

$$\hat{H} - \varepsilon_1 \equiv H - (HP_{Bt}H/E_{Bt}) - \varepsilon_1 \geq 0. \quad (4.1)$$

The effect of the composite bound state has been extracted, in the sense that we have constructed a non-negative operator $\hat{H} - \varepsilon_1$ even though $\Psi_B$ is not known. (This "extraction" is crucial to the development of a number of V Bds.) The spectrum of $\hat{H}$ includes $E_B$ and a continuum starting at $\varepsilon_1$. $H - (HP_B H/E_B)$ and $H$ have identical spectra, continua starting at $\varepsilon_1$, the $E_B$ eigenvalue of $H$ having been raised to zero; thus, we have $[H - (HP_B H/E_B)]\Psi_B = 0$ and $\hat{H} \Psi_{Bt} = 0$.

It is simple to show that the V Id    $A = A_{var} + s$ can be recast in the form $A = \hat{A}_{var} + \hat{s}$, where the modified quantities, those with a caret, are obtained from (3.1) by replacing H by $\hat{H}$. $\hat{s} \leq 0$ follows from (4.1), and we have an upper V Bd for the one composite bound state case, $A \leq \hat{A}_{var}$. The extension to more than one composite bound state is trivial.

### B.    $\varepsilon_1$ and $\phi$ Imprecisely Known

In subsection A, we had an incident particle, and a target of known ground state energy $\varepsilon_1$ and known wave function $\phi$, with the energy of the entire system $\varepsilon_1$, and we constructed $\hat{H}$ such that $\hat{H} - \varepsilon_1 \geq 0$ even though there was a composite bound state with energy $E_B$ below $\varepsilon_1$, and $\Psi_B$ and $E_B$ were imprecisely known. The system was not in its lowest energy level. We now turn our attention to a system which is bound and which _is_ in its lowest state, its ground state, with energy $\varepsilon_1$, but we no longer take $\varepsilon_1$ or $\phi$ to be known. We clearly have, formally, $h - \varepsilon_1 \geq 0$, but in practice it will be necessary to work with $h - \varepsilon_{1t}$, which, unfortunately, has the negative eigenvalue $\varepsilon_1 - \varepsilon_{1t}$. An operator inequality that will prove very useful is[8]

$$\hat{h} - \varepsilon_{1t} \equiv h - (hp_t h/\varepsilon_{1t}) - \varepsilon_{1t} \geq 0, \qquad (4.2)$$

where $p_t \equiv |\phi_t><\phi_t|$ is an approximation to $p \equiv |\phi><\phi|$; we must satisfy the condition    $\varepsilon_{1t} \leq \bar{\varepsilon}_2 \leq (\varepsilon_1/\varepsilon_{1t})\varepsilon_2$. $\bar{\varepsilon}_2$ represents the energy of the first excited state if a second bound state exists and the threshold of the continuum otherwise. $\bar{\varepsilon}_2$ has been chosen using bounds on $\varepsilon_1$ and $\varepsilon_2$. We also have $\hat{h} - \bar{\varepsilon}_2 > 0$. The condition can be rewritten as $\varepsilon_{1t} \leq -(\varepsilon_1\varepsilon_2)^{\frac{1}{2}}$, and should be rather simple to satisfy. It will now be shown that the inequality (4.2) enables us to construct an extremum principle for the evaluation of $L_t$.

### 5.    EXTREMUM PRINCIPLE[9] FOR $L_t$

Variational parameters $c_i$ in $\phi_t$ can be readily determined, especially if they appear linearly, by minimizing $(\phi_t, h\phi_t)$. For $\phi_t$ sufficiently accurate, one can of course estimate any quantity which depends upon $\phi_t$, such as $(\phi_t, W\phi_t)$, but hopefully it will be more effective to proceed via the V P, Eq. (2.1). The roadblock in the past has been the lack of a broad-based method for the determination of a reliable estimate $L_t$ of L; one was restricted, for example, to the choice of a $\phi_t$ which was the eigenfunction of some trial hamiltonian.[10,11] An $L_t$ obtained from the "obvious" replacement of Eq. (2.2) by (2.2)', namely, $(h - \varepsilon_{1t})L_t = q(\phi_t)$, can be awful. To quote some economists in a

different context, authors who chose that route to determine
the $c_i$ were "caught with their parameters down". Indeed, it
was pointed out[10] that $(\phi_t, q(\phi_t)) = 0$ and therefore that
$(\phi_t, [h - \varepsilon_{1t}] L_t) = 0$; for $(\phi_t, \phi_t) = 1$, (2.1) gives the ridiculous
result $(\phi_t, W\phi_t)$! The vaunted V P isn't even variational, let
alone vaunted. What happened to the V P? The problem was recently reanalyzed, and recast in a trouble-free form. Thus, L
is defined by (2.2) only to within a multiple of $\phi$. The choice
of any value, c, for the multiple causes $(h - \varepsilon_1)^{-1}$ to be well
defined even though $\varepsilon_1$ is an eigenvalue of h. But (2.2)' uniquely
defines $L_t$; one finds $L_t = (h - \varepsilon_{1t})^{-1} q(\phi_t)$. As $\phi_t \to \phi$, we have
$\varepsilon_{1t} \to \varepsilon_1$ so that $(h - \varepsilon_{1t})^{-1}$ becomes nearly singular; the $\phi$ component of $L_t$ and therefore of $L_t - L$ becomes enormous, whereas
the derivation of the V P assumed that $L_t - L$ was of order $\delta\phi$.

One way[12] to avoid the near-singularity is to introduce $\hat{h}$.
Thus, make the particular choice

$$c \equiv (\phi, L) = (\phi, W\phi)/\varepsilon_1 \qquad\qquad (5.1)$$

and subtract $\varepsilon_1 pL = (hph/\varepsilon_1)L = (\phi, W\phi)\phi$ from both sides of (2.2);
one finds

$$(h - [hph/\varepsilon_1] - \varepsilon_1)L = -W\phi. \qquad\qquad (5.2)$$

There are no solutions of the homogeneous equation associated
with (5.2), which thereby uniquely deines L. Indeed, the inner
product of (5.2) with $\phi$ reproduces the boundary condition (5.1).
If we define $L_t$ to be the solution of the "obvious" replacement
of (5.2),

$$(\hat{h} - \varepsilon_{1t})L_t = -W\phi_t, \qquad\qquad (5.3)$$

we're in excellent shape, essentially because $\hat{h} - \varepsilon_{1t}$ is positive definite, an operator property with three very useful consequences:

i)   The homogeneous equation associated with (5.3) has no
solutions, so that (5.3) defines $L_t$ uniquely. A boundary condition on $L_t$ need not be specified; indeed, using $\hat{h}\phi_t = 0$, the
inner product of $\phi_t$ with (5.3) leads to $c_t \equiv (\phi_t, L_t) =$
$(\phi_t, W\phi_t)/\varepsilon_{1t}$, the natural approximation to (5.1).

ii)   Eq. (5.3) for $L_t$ goes smoothly to Eq. (5.2) for L as
$\phi_t \to \phi$, so that $L_t$ approaches L smoothly; there are no near singularities.

iii)   We have immediately at our disposal a simple but
powerful extremum principle for the estimation of $L_t$. We need

merely identify the entities in (5.3) with the entities in (1.2), $L_{tt}$ with $X_t$, and $\delta L_t \equiv L_{tt} - L_t$ with $\delta X$. Since $M(L_{tt})$ achieves its minimum for $L_{tt} = L_t$, one determines any variational parameter in $L_{tt}$ by minimizing $M(L_{tt})$, and thereby obtains a reasonable estimate of $L_t$ which, in turn, for a reasonable choice $\phi_t$, provides a reasonable estimate of $L$.

It will take a number of studies to properly test the power of the above approach, but studies[13] of $(\phi, W\phi)$ for $\phi$ the ground state wave function of helium and $W = r_1^2 + r_2^2$ (the diamagnetic susceptibility) or $\exp(i\vec{k}\cdot\vec{r}_1) + \exp(i\vec{k}\cdot\vec{r}_2)$ (the form factor) are encouraging.

The above discussion of an auxiliary extemum principle for the estimation of a Lagrange function concerned the bound state case; it should at least be mentioned that similar procedures are available for continuum problems.[14]

## 6.   VARIATIONAL BOUNDS ON MATRIX ELEMENTS

We now turn our attention to the determination of upper and lower V Bds on bounds state matrix elements[1,2,15], limiting ourselves to the diagonal case. It will first be necessary to develop a V Id for $\phi$ and to consider matrix elements which contain one rather than two unknown functions.

### A.   A Variational Identity for $\phi$

Proceeding along the same lines as in Sec. 2, one can construct a V P for $\phi$. The starting point would be[16]

$$\phi(\vec{r}) = \phi_t(\vec{r}) + L_t(\vec{r})[(\phi_t,\phi_t) - 1] + \int \Lambda_t(\vec{r},\vec{r}')[h(\vec{r}') - \varepsilon_{1t}]\phi_t(\vec{r}')d\vec{r}'.$$

We will however, proceed differently.[1] The essential point is to avoid arriving at an equation of the form $(h - \varepsilon_1)X = \ldots$, with the attendant difficulty associated with the inversion of $h - \varepsilon_1$, or rather of $h - \hat{\varepsilon}_{1t}$. We bypassed that difficulty in Sec. 4B by working with $\hat{h} - \varepsilon_1$, and we will do the same here. Thus, it is an immediate consequence of $(h - \varepsilon_1)\phi = 0$ and of the definition of $\hat{h}$ that

$$(\hat{h} - \varepsilon_1)\phi = -(hp_t h/\varepsilon_{1t})\phi = -Nh\phi_t, \tag{6.1}$$

where $N \equiv (\varepsilon_1/\varepsilon_{1t})S$ and $S \equiv (\phi_t,\phi)$. Since $\hat{h} > \bar{\varepsilon}_2 > \varepsilon_1$, it follows that $\hat{g}(\varepsilon_1) \equiv (\varepsilon_1 - \hat{h})^{-1}$ is non-singular and we have

$$\phi = N\hat{g}(\varepsilon_1)h\phi_t = N\hat{g}(\varepsilon_1)(p_t + q_t)h\phi_t = S\phi_t + N\hat{g}(\varepsilon_1)q_t h\phi_t,$$

where $q_t \equiv 1 - p_t$ and we have used $p_t h \phi_t = \varepsilon_{1t} \phi_t$ and $g(\varepsilon_1)\phi_t = (1/\varepsilon_1)\phi_t$, the latter a consequence of $\hat{h}\phi_t = 0$. Introducing a trial estimate $\hat{g}(\varepsilon_1)$, defining $\delta\hat{g}(\varepsilon_1) \equiv \hat{g}_t(\varepsilon_1) - \hat{g}(\varepsilon_1)$, and using $q_t h \phi_t = (h - \varepsilon_{1t})\phi_t$, we arrive at

$$\phi = \phi_{var} - N\delta\hat{g}(\varepsilon_1)(h - \varepsilon_{1t})\phi_t \equiv \phi_{var} + s, \qquad (6.2)$$

$$\phi_{var} \equiv S\phi_t + N\hat{g}_t(\varepsilon_1)q_t h\phi_t. \qquad (6.3)$$

(6.2) is a V Id for $\phi$, with $\phi$ written as the sum of a V P and a second order term s, which is explicit if only formal. It will be useful to record the identity, obtained by multiplying $\hat{g}_t$ by a unit operator,

$$\delta\hat{g}(\varepsilon_1) = -[\hat{g}_t(\varepsilon_1)(\hat{h} - \varepsilon_1) + 1]\hat{g}(\varepsilon_1). \qquad (6.4)$$

## B. Variational Bounds on $(F,\phi)$

We now obtain upper and lower V Bds on $(F,\phi)$, where F is a known function. Using (6.2) and (6.4), we have

$$(F,\phi) = (F,\phi_{var}) + N\Delta,$$

$$\Delta \equiv (\hat{g}(\varepsilon_1)J, [h - \varepsilon_{1t}]\phi_t), \quad J \equiv [(\hat{h} - \varepsilon_1)\hat{g}_t(\varepsilon_1) + 1]F.$$

With $||f|| \equiv (f,f)^{\frac{1}{2}}$ the norm of f, and with $\bar{\varepsilon}_2$ defined as above so that $|\hat{g}(\varepsilon_1)| \leq (\bar{\varepsilon}_2 - \varepsilon_1)^{-1}$, we find

$$|\Delta| \leq (\bar{\varepsilon}_2 - \varepsilon_1)^{-1} ||J|| \times ||(h - \varepsilon_{1t})\phi_t|| \equiv \Delta^{(+)}.$$

$\Delta^{(+)}$ is a simple upper bound on $|\Delta|$, that is, it is of the same order as $|\Delta|$, but since $|\Delta|$ is of second order, that will suffice. It follows that $|(F,\phi)| \lesseqgtr (F,\phi_{var}) \pm N\Delta^{(+)}$. In the above analysis, $\varepsilon_2$, S, and $\varepsilon_1$ --- other than when it appears in $h - \varepsilon_1$ in a situation in which the treatment of $h - \varepsilon_1$ is "delicate" --- and therefore also N, are carried along as if they are known; at the very last step, they must be replaced by the appropriate upper and lower bounds or V Bds.

## C. Upper and Lower V Bds on Bound State Matrix Elements

With $q \equiv 1 - p$, we start from $p\phi_t = \phi_t - q\phi_t$, and use $p\phi_t = S\phi$; we have $\phi = (\phi_t - q\phi_t)/S$, which leads to $(\phi, W\phi) = [2S(W\phi_t, \phi) - (\phi_t, W\phi_t) + (q\phi_t, Wq\phi_t)]/S^2$. We have just seen that we can obtain upper and lower V Bds on $(W\phi_t, \phi)$; we need merely identify $W\phi_t$ with F. The determination of upper and lower V Bds on $(\phi, W\phi)$ has therefore been reduced to the

determination of <u>simple</u> bounds on the second order quantity
$(q\phi_t, Wq\phi_t)$. Simple bounds can be obtained for a wide range of
operators W, by methods[17] which are straightforward but labori-
ous and dull, and you will be spared the details. W can be
any operator of the form of sums and products of the coordinates, relative coordinates, momenta and relative momenta of
the electrons, or any operator which can be bounded by such a
form.


D.   Generalizations

     We have throughout restricted ourselves to real functions,
to real symmetric operators, and to diagonal ground state
matrix elements. In fact, it is relatively straightforward to
generalize the results to complex functions, to hermitian
operators, and to off-diagonal bound state matrix elements.
Scattering problems at zero incident energy are in many ways
quite similar to bound state problems, and, as indicated above,
many results have been obtained for that case. Scattering at
non-zero incident energy is an entirely different matter.
Recently, upper and lower V Bds were obtained[18] on elements of
the K matrix for scattering of positrons by atoms in their
ground state, when $\phi$ and $\varepsilon_1$ are only imprecisely known. The
results are much more difficult to apply, however. Furthermore, the extension to electron scattering will by no means
be trivial. (The exchange integrals will be very troublesome,
much more so than at zero incident energy.)


## 7.   THE ADIABATIC APPROXIMATION

     The adiabatic approximation, in which the effective inter-
action $V_{ad}(\vec{R})$ between two systems at a separation $\vec{R}$ is taken
to be that in which the centers are frozen at that separation,
has two extremely interesting properties. Firstly, one can
often obtain quite accurate numerical solutions in that approximation. Secondly, speaking schematically, the approximation
tends to <u>overestimate</u> the attractiveness of the effective
interaction, $V_{eff}(\vec{R})$, that is, $V_{ad}(\vec{R}) \leq V_{eff}(\vec{R})$, because the
particles have an infinite time to adjust themselves to the
state of lowest energy, whereas in actuality they have only a
brief time.[19,20] This is as opposed to the static or close
coupling approximations, for example, in scattering, or to the
RR approach to the ground state, where one attempts to solve
the real problem with a restricted form of trial function and
<u>underestimates</u> the attractiveness of the effective interaction.
The adiabatic approximation therefore provides the <u>other</u> more
difficult bound on the ground state energy --- it provides a
necessary rather than sufficient condition for the existence

of a bound state, and can be used to prove the non-existence
of bound states[19] --- and on the scattering length A.  Thus
$V_{ad}$ is reasonably well known for a proton and a He atom[21], or,
equivalently, for a positron and a He atom.  This immediately
provides a rather accurate lower bound on A for e -He scattering.[20]  If one uses the adiabatic approximation indirectly,
to bound the second order error term in a V Id for A, one obtains a lower V Bd on A for positron-atom scattering.[22]

## 8.  VARIATIONAL PRINCIPLES AS AN ANALYTIC TOOL

V Ps are useful in at least three different connections.

i)  V Ps normally provide the most elegant formulation of
the laws of physics; $\delta \int L dt = 0$, with L the Lagrangian, provides
a classic example.

ii)  V Ps and V Bds are among the most powerful of all
numerical tools.  This use of V Ps and V Bds was, of course,
the main burden of the present paper.

iii)  It is perhaps not so well known that V Ps and V Bds
are also useful as analytic tools, as in the determination of
the form of an expansion.  Thus, for example, perhaps the
earliest derivation[23] of effective range theory (the expansion
of k cot $\eta$ in powers of $k^2$, where $\eta$ is the phase shift) used
a V P.  I strongly suspect that a very compact derivation of
Low's theorem on low energy bremsstrahlung[24], in which a knowledge of the associated radiationless cross section can be used
to determine the two leading terms in an expansion of the bremsstrahlung cross section in powers of the photon momentum,
could be given in the same spirit; a relatively poor trial
function inserted in a V P would give the desired result.  We
limit ourselves to one final example.  It had long been believed
that the high energy limit of Green's functions and cross sections is dominated by the Born approximation, but this is not
the case.  At very high energies the charge exchange cross
section for proton-hydrogen scattering behaves as $v^{-11}$ in the
second Born approximation[25] but only as $v^{-12}$ in the first
Born approximation, where v is the velocity.  It does not
follow that the second Born dominates.  The v dependence of
each higher term is known to be dominated by $v^{-11}$, but it is
not clear if the sum of higher terms converges; the total contribution from beyond the second Born term could conceivably
dominate the $v^{-11}$ contribution.  The situation is somewhat
similar for cross sections for exchange of light particles for
short range interactions; contributions in the first and second
Born approximations go as $v^{-20}$ and $v^{-19}$ , respectively.[26]
Again, the contribution of each of the higher terms is known[26]

to be dominated by the second Born term, but it is not known whether the sum converges. In the short range case, however, treating the problem in the impact parameter approximation so that the problem becomes time dependent, and using upper and lower V Bds on transition amplitudes[27], it has been shown[28] that the total contribution from beyond the second Born term is indeed dominated by the $v^{-19}$ contribution.

I take pleasure in acknowledging the various courtesies extended to me at the Aspen Center for Physics where this manuscript was completed.

†Work supported by the Office of Naval Research, under Contract No. N00014-67-A-0467-0007, and by the National Science Foundation under Contract No. MPS-7500131

[1]R. Blau, L. Rosenberg, and L. Spruch, Phys. Rev. A 10, 2246 (1974).
[2]R. Blau, L. Rosenberg, and L. Spruch, Phys. Rev. A 11, 200 (1975).
[3]L. Spruch and L. Rosenberg, Phys. Rev. 116, 1034 (1959).
[4]L. Rosenberg, L. Spruch, and T.F. O'Malley, Phys. Rev. 118, 184 (1960).
[5]E. Gerjuoy, A. R. P. Rau, and L. Spruch, Phys. Rev. A 8, 662 (1973).
[6]A point that has often been missed is that the functions that appear in Q($\phi$) must be sufficiently well defined to make Q uniquely defined. If, for example, Q = $(\phi_1, W\phi_2)$, where $\phi_1$ and $\phi_2$ can be complex, we must at least specify the relative phase of $\phi_1$ and $\phi_2$.
[7]C. Schwartz, Ann. Phys. (N.Y.) 2, 156 (1959); 2, 170 (1959). A. Dalgarno and A. L. Stewart, Proc. R. Soc. A 257, 534 (1960). J. C. Y. Chen and A. Dalgarno, Proc. R. Soc. 85, 399 (1965). L. M. Delves, Proc. Phys. Soc. Lond. 92, 55 (1967). J. O. Hirschfelder, W. B. Brown, and S. T. Epstein, Adv. Quant. Chem. 1, 255 (1964).
[8]W. Monaghan and L. Rosenberg, Phys. Rev. A 6, 1076 (1972).
[9]E. Gerjuoy, A. R. P. Rau, L. Rosenberg, and L. Spruch, Phys. Rev. A 9, 108 (1974).
[10]Useful results can still be obtained, not only for He, as in J. B. Krieger and V. Sahni, Phys. Rev. A 6, 919 (1972), A 6, 928 (1972), and references therein, but for atoms Li through A. (See ref. 11.) The point is, however, that once such a $\phi_t$ has been chosen, there is a limit to the accuracy that can be obtained.
[11]R. O. Mueller, A. R. P. Rau, and L. Spruch, Phys. Rev. A 10, 1511 (1974).
[12]One can make other choices of c. More significantly, one can construct positive-definite operators other than $\hat{h} - \varepsilon_1$. See E. Gerjuoy, L. Rosenberg, and L. Spruch, J. Math. Phys.

$\underline{16}$, 455 (1975); the possibility had been suggested by S. Aranoff and J. Percus, Phys. Rev. $\underline{166}$, 1255 (1968). See also S. Borowitz and E. Gerjuoy, U. of Colorado JILA Report No. 36, 1965 (unpublished).

[13]L. Rosenberg, R. Shakeshaft, and L. Spruch, submitted for publication. An attempt is also being made to go beyond some work in which, assuming the auxiliary function has been obtained sufficiently accurately, an expression was obtained for the second order error - see E. Gerjuoy, J. Math. Phys. $\underline{16}$, 761 (1975) - and to obtain a useful super V P. A brief comment on super V Ps is contained in ref. 16.

[14]L. Rosenberg, Phys. Rev. D $\underline{9}$, 1789 (1974); L. Rosenberg and L. Spruch, Phys. Rev. A $\underline{10}$, 2002 (1974).

[15]Earlier but limited results are given in R. Blau, A. R. P. Rau, and L. Spruch, Phys. Rev. A $\underline{8}$, 131 (1973).

[16]E. Gerjuoy, A. R. P. Rau, L. Rosenberg, and L. Spruch, J. Math. Phys. $\underline{16}$, 1104 (1975).

[17]The results obtained in ref. 1 were obtained by an adaptation of methods introduced in S. Aranoff and J. K. Percus, Phys. Rev. $\underline{162}$, 878 (1967).

[18]R. Blau, L. Rosenberg, and L. Spruch, Phys. Rev. A, to be published.

[19]I. Aronson, C. J. Kleinman, and L. Spruch, Phys. Rev. A $\underline{4}$, 841 (1971), and references therein.

[20]Y. Hahn and L. Spruch, Phys. Rev. A $\underline{9}$, 226 (1974).

[21]J. D. Stuart and F. A. Matsen, J. Chem. Phys. $\underline{41}$, 1646 (1964); L. Wolniewicz, J. Chem. Phys. $\underline{43}$, 1087 (1965).

[22]L. Rosenberg and L. Spruch, Phys. Rev. A (to be published), and Y. Hahn (submitted for publication).

[23]J. S. Schwinger, Phys. Rev. $\underline{72}$, 742 (1947) A.

[24]F. E. Low, Phys. Rev. $\underline{110}$, 974 (1958).

[25]R. M. Drisko, thesis (Carnegie Institute of Technology, 1955, unpublished).

[26]K. Dettmann and G. Liebfried, Z. Phys. $\underline{218}$, 1 (1969); K. Dettmann, Springer Tracts Mod. Phys. $\overline{58}$, 119 (1971).

[27]L. Spruch, in Lectures in Theoretical Physics - Atomic Collisions, edited by S. Geltman, K. T. Mahanthappa, and W. E. Brittin (Gordon and Breach, New York, 1969) Vol. XIC, p. 77. (See also p. 55)

[28]R. Shakeshaft and L. Spruch, Phys. Rev. A $\underline{8}$, 206 (1973) and Phys. Rev. A $\underline{10}$, 92 (1974). (The latter paper is concerned with the possibility of developing V Ps and V Bds when one uses a Sturmian basis in the impact-parameter method.) Shakeshaft has subsequently shown that it will be difficult to use the V Bd of ref. 27 to obtain accurate estimates of charge exchange cross sections, because of the complexity of Coulomb functions.

HIGHLY EXCITED RYDBERG STATES

# IONIZATION OF HIGHLY EXCITED ATOMS BY ATOMIC PARTICLE IMPACT

B. M. Smirnov

I.V. Kurchatov - Institute of Atomic Energy

Moscow, USSR

We consider the ionization of a highly excited atom by a collision with an atom or molecule. The aim of this paper is to present the theory of these processes[1] and to compare the theoretical results with experimental data.

In the cases considered, the ionization process involves a detachment of a weakly bounded electron and is realized through the scheme:

$$A^{**} + B \rightarrow A^{+} + B + e \tag{1}$$

where $A^{**}$ is an excited atom, and B is an atomic or molecular particle. The excited electron in the atom possesses an orbit with a large principal quantum number n. There are many transitions between excited states initiated by the collision between the highly excited atom and the other atom or molecule. Hence we may treat the problem using the semiclassical approximation. In consequence of the collision process, the weakly bound electron is scattered on another colliding particle and it leads to ionization. The classical method is valid, if the distance between the electron and the colliding particle during the impact is small compared with the dimensions of the highly excited atom.

Let us consider the ionization of a highly excited atom by a collision with an atom. Given the straight line trajectory of the incident atom, the ionization probability for the highly excited atom equals

$$W_i = \int dt |\psi(\vec{R})|^2 |\vec{v}-\vec{v}_a| \int_{\Delta E \geq J} d\sigma$$

$$= \int \frac{dz}{v_a} |\psi(\vec{R})|^2 |\vec{v}-\vec{v}_a| \int_{\Delta E \geq J} d\sigma \ . \tag{2}$$

$v_a$ is the relative velocity of the nuclei, $v$ is the electron velocity, $d\sigma$ is the differential cross section for elastic scattering of the valence electron on the incident atom and the inner integral must be given for scattering angles at which the energy exchange $\Delta E$ between the electron and the atom exceeds the ionization potential $J$ of the highly excited atom. $dz$ is the element of trajectory.  $dt$ is the element of time. $|\psi(\vec{R})|^2$ is the electron density at a given point of trajectory, so that $|\psi(\vec{R})|^2 |\vec{v}-\vec{v}_a| \int d\sigma$ is the ionization probability per unit time of the incident atom in a given point of space.

Making the integration in Eq. 2 over the impact parameter, we obtain the cross section for ionization by the highly excited atom in collision with an unexcited atom:

$$\sigma_{ion} = \int W_i d\vec{\rho} = \int d\vec{\rho} dz |\psi(\vec{R})|^2 \frac{|\vec{v}-\vec{v}_a|}{v_a} \int_{\Delta E \geq J} d\sigma = < \frac{|\vec{v}-\vec{v}_a|}{v_a} \int d\sigma > \tag{3}$$

where $d\vec{\rho}$ is the element of the impact parameter and the bracket denotes averaging over the electron distribution function in the atom.

Equation 3 can be derived in another way.  The ionization probability per unit time of a highly excited atom by collision with ground state atoms has the form

$$N_a < |\vec{v}-\vec{v}_a| \int_{\Delta E \geq J} d\sigma >$$

where $N_a$ is the incident atom density and the bracket refers to averaging over spatial distribution of weakly connected electrons.  Dividing this expression with the flux of the incident atoms $N_a v_a$, we receive Eq. 3 for the ionization cross section of an excited atom.

Let us calculate the ionization cross sections of Eq. 3. As a result of collision with an atom the change of electron energy can be written

$$\Delta E = \frac{P^2}{2\mu} - \frac{(\vec{P}-\Delta\vec{P})^2}{2\mu} = \vec{v}_a \Delta\vec{P} - \frac{\Delta P^2}{2\mu} \tag{4}$$

with the $\vec{P} = \mu \vec{v}_a$ being the nuclear momentum before the collision, $\mu$ is the reduced mass of the nuclei, $\Delta \vec{P}$ is the change of electron momentum after the collision. This quantity is equal to $\Delta P = 2m |\vec{v} - \vec{v}_a| \sin \frac{\Theta}{2}$, where m is the electron mass, and $\Theta$ is the electron scattering angle on the atom. We consider now the ionization of an atom in the range of the threshold $\mu v_a^2 \gg J$. Since $\Delta E \sim J$ and $\Delta P \sim J/v_a$, it follows $\Delta P^2/\mu \sim J^2/\mu v_a^2 \ll J$, that is the second term in Eq. 4 is small compared to the first one and we find

$$\Delta E = \vec{v}_a \Delta \vec{P}. \tag{5}$$

Provided the nuclei are moving on straight line trajectories we shall study two limiting cases for collision velocities of atoms. These cases are with respect to two limiting relations between nuclear collision velocity $v_a$ and characteristic electron velocity in the atom $e^2/\hbar n$ (n is the principal quantum number of electron in the excited atom). Ionization of the atom accompanies detachment of the electron with typical momentum $P \sim J/v_a \sim me^4/\hbar^2 n^2 v_a$. It follows in the case of small atomic velocities $v_a \ll e^2/\hbar n$, that the ionization process is determined by electrons with large momentum compared with the characteristic electron momentum in the atom, $P \gg me^2/\hbar n$. Under these circumstances the distribution function of weakly connected electrons in the field of a Coulomb centre, which is averaged over electron moments, has the form[2] ($\int f(v)dv = 1$)

$$f(v)dv = \frac{32dv}{\pi n^2 v^6} \frac{e^{10}}{\hbar^5}, \quad v \gg \frac{e^2}{\hbar n} \quad . \tag{6}$$

Including the angle $\Theta$ between vectors $\vec{v}_a$ and $\vec{P}$ and averaging the cross sections over the energy changes $\Delta E = v_a \Delta P \cos\Theta \geqslant J$,

$$\int_{\Delta E \geqslant J} d\sigma \frac{d\cos}{2} = \int_{v_a \Delta P \geqslant J} (1 - \frac{J}{v_a \Delta P}) d\sigma \quad .$$

We now use the assumption that the relative collision velocity between a weakly connected electron and an atomic particle is small compared with the characteristic atomic velocities. It fulfills well. Under this condition the scattering is isotropic, that is $d\sigma/d\cos\Theta = \sigma_o/2$, where $\Theta_o$ is the scattering angle and $\sigma_o$ is the total elastic cross section for the electron scattered by the atom. Using this relation we receive

$$\int_{\Delta E \geqslant J} d\sigma = \int_{v_a \Delta P \geqslant J} (1 - \frac{J}{2mvv_a \sin\theta/2}) d\sigma = \sigma_o (1 - \frac{J}{2mvv_a})^2, v \geqslant \frac{J}{2mv_a} >> v_a.$$

It follows for the ionization cross section

$$\sigma_{ion} = \int f(v)dv\frac{v}{v_a} \int_{\Delta E \geqslant J} d\sigma = \sigma_o \frac{2^{13}}{105\pi} (\frac{\hbar v_a n}{e^2})^3 =$$

$$= 2.5\sigma_o (\frac{\hbar v_a n}{e^2})^3, \quad v_a << \frac{e^2}{\hbar n}$$

(7)

Another limiting case takes place when the relative collision velocity of atoms is large compared to the electron velocity and we have

$$\Delta E = \vec{v}_a \Delta \vec{P} = m\vec{v}_a (\vec{v}_a - \vec{v}') = mv_a^2(1-\cos\theta)$$

where $v'$ is the relative velocity of the electron and the atom after the collision. Using Eq. 3 we find the ionization cross section

$$\sigma_{ion} = \int_{\Delta E \geqslant J} d\sigma = \sigma_o(1 - \frac{J}{2mv_a^2})$$

(8)

Summarizing the above results for two limiting cases we represent the ionization cross section in the following form:

$$\sigma = \sigma_o \phi(\chi), \quad \chi = (\frac{mv_a^2}{2J})^{1/2} = \frac{\hbar v_a n}{e^2}$$

$$\phi(\chi) = \begin{cases} 2.5\chi^3, & \chi << 1 \\ 1-\frac{1}{4\chi^2}, & \chi >> 1 \end{cases}$$

(9)

The first limiting case takes place in the range of thermal collision energies when $n >> 10^3 \div 10^4$.

The ionization cross section (9) of a highly excited atom corresponds to the process of elastic scattering of a weakly bound electron on the atom. In this case the value of the ionization cross section is limited by the elastic cross section of a free, slow electron on the atom. Let us consider another mechanism of ionization which is conditioned by the scattering of a nucleus. In the frame of this mechanism, the momentum of the incident atom is transferred to the nucleus of the excited atom, so there is an electron momentum change in the coordinate system which is related to a nucleon. That leads to the ionization.

Let the nuclear velocity change caused by the elastic scattering of atoms be equal to $\Delta\vec{v}$. For the electron energy change it gives

$$\Delta\varepsilon = mv\Delta v\cos\theta' + \frac{m(\Delta v)^2}{2}$$

Here m is the electron mass, $\theta'$ is the angle between the direction of electron velocity $\vec{v}$ and the vector of nuclear velocity change $\Delta\vec{v}$. Averaging the result over the direction of electron velocity, we finally receive the expression for the ionization probability $W_i$ for electron with initial velocity v:

$$W_i = \begin{cases} 1, & \Delta v \geqslant \sqrt{v^2 + \dfrac{2J}{m}} + v \\[2ex] \dfrac{1}{2} + \dfrac{\Delta v}{4v} - \dfrac{J}{2mv\Delta v}, & \sqrt{v^2 + \dfrac{2J}{m}} + v \geqslant \Delta v \geqslant \sqrt{v^2 + \dfrac{2J}{m}} - v \\[2ex] 0, & \sqrt{v^2 + \dfrac{2J}{m}} - v \geqslant \Delta v \end{cases}$$

To keep the following algebra simple we reduce these expressions to the other ones which are closest to them

$$W_i = \begin{cases} 1, & \Delta v \geqslant \sqrt{v + \dfrac{2J}{m}} \\[2ex] 0, & \Delta v \leqslant \sqrt{v + \dfrac{2J}{m}} \end{cases}$$

We assume that the electron does not screen the ion field. Then the elastic scattering of the nucleus is determined by the polarization interaction and small angle scattering takes place. We have for the ion momentum change

$$\Delta P = M\Delta v = \frac{8\beta e^2}{3v_a\rho^4} .$$

Here M is the ion mass, $\beta$ is a polarizability of the atom, $v_a$ is the relative nuclear velocity and $\rho$ is the impact parameter. Hence it follows for the ionization cross section

$$\sigma_{ion} = <\int 2\pi\rho d\rho W_i> = \pi(\frac{\beta e^2}{3Mv_a})^{1/2} \ <(v^2+\frac{2J}{m})^{-1/4}>$$

where brackets mark averaging over the electron velocities. With the aid of the velocity distribution function of electrons[2]

$$f(z)dz = \frac{32z^2 dz}{\pi(z^2+1)^4}, \quad z = \frac{\hbar vn}{e^2} \tag{10}$$

we receive the ionization cross section in the form

$$\sigma_{ion} = 4.8 \ (\frac{\beta n\hbar}{Mv_a})^{1/2} \ . \tag{11a}$$

The expression (10) is valid, if the scattering takes place in the range of small angles, that is $v_a >> e^2/\hbar n$. In the other limiting case, when $v_a << e^2/\hbar n$, the ionization cross section approaches zero and equals

$$\sigma_{ion} = \frac{128\sqrt{2}}{30\pi} \ (\frac{\hbar n\Delta v}{e^2})^5 \sigma_{capt} \tag{11b}$$

where $\Delta v$ is the velocity change of nuclei and $\sigma_{capt}$ is the cross section for the ion capture by atoms. The values of the cross sections in Eqs. 9 and 11, which are defined by the above two mechanisms, may be found in any relation.

The ionization processes of highly excited atoms by incoming molecules mainly correspond to inelastic scattering of electrons on the molecule. At ordinary temperatures the molecules are found in the excited rotational states. The molecular transitions to the lower rotational state, initiated by the collision with highly excited atoms, leads to the ionization of the atom.

Let us derive the ionization cross section of the highly excited atom by an incident molecule. We assume that the energy of rotational transition in the molecule is large compared with the ionization potential of the excited atom. Then the ionization cross section according to Eq. 3 equals

$$\sigma_{ion} = \frac{<|\vec{v}-\vec{v}_a|\sigma_{rot}>}{v_a} \tag{12}$$

where $v$ is the electron velocity, $v_a$ is the relative velocity of nuclei, and $\sigma_{rot}$ is the cross section for molecular rotational transitions caused by a collision with the free electrons. The brackets mean averaging over electron velocities and rotational states of the molecule.

The transition between rotational states of molecules initiated by collisions with slow electrons is related to the long range components of the interaction between an electron and a molecule. The cross section for the rotational transition for a dipole molecule is equal[3] to

$$\sigma_{rot}(j \to j-1, \ E) = \frac{4\pi D^2}{3Ea_0} \frac{j}{(2j+1)} \ \ln \frac{(\sqrt{1+ \frac{2Bj}{E}} +1)}{(\sqrt{1+ \frac{2Bj}{E}} -1)} \quad . \quad (13)$$

Here j is the rotational quantum number of the molecule before the collision, B is the rotational constant, E is the energy of incoming electron, D is the dipole moment of the molecule, and $a_0$ is the Bohr radius. In this case the rotational quantum number change is equal to one and the electron obtains the energy that is equal to 2Bj. The cross section of the rotational transition for a quadrupole molecule is given by [4]

$$\sigma_{rot}(j \to j-2) = \frac{8\pi Q^2}{15e^2 a_0^2} \frac{j(j-1)}{(2j-1)(2j+1)} \sqrt{1+ \frac{2B(2j-1)}{E}} \quad (14)$$

where Q is the molecular quadrupole moment, e is the electronic charge and the electron gains the energy 2B(2j-1).

Let us consider the ionization of highly excited atoms caused by collision with molecules at thermal energies. To simplify the results we consider the following relation between the principal quantum number n of an excited electron and the relative nuclei velocity $v_a$, which practically is fulfilled:

$$(\frac{e^2}{\hbar v_a})^{1/2} << n << \frac{e^2}{\hbar v_a} \quad (15)$$

According to the right inequality (practically it gives $n << 10^3$) Eq. 12 for the ionization cross section becomes

$$\sigma_{ion} = \frac{<v\sigma_{rot}>}{v_a} \quad (16)$$

The left inequality is analogous to the following inequality:

$$J << \sqrt{BT}$$

It allows us to simplify Eqs. 13 and 14. Using Eqs. 10, 16, 13 and 14 and these inequalities we find for the ionization cross section for the case of dipole molecule:

$$\sigma_{ion} = \frac{8\pi D^2}{3a_o(2j+1)} \sqrt{\frac{j}{Bm}} \frac{1}{v_a} \tag{17}$$

and for the case of a quadrupole molecule

$$\sigma_{ion} = \frac{16\pi Q^2}{15e^2 a_o^2} \frac{j(j-1)}{(2j+1)(2j-1)} \frac{\sqrt{B}}{mv_a}. \tag{18}$$

At thermal collision energies for most of the molecules the relation $T \gg B$ is fulfilled. Using this relation and averaging the ionization cross sections over the initial rotational quantum numbers, we obtain for the collisions of the dipole molecules

$$\sigma_{ion} = C_1 \sqrt{\frac{\mu}{m}} \frac{D^2}{a_o B^{1/4} T^{3/4}} \tag{19}$$

and for the collisions of quadrupole molecules

$$\sigma_{ion} = C_2 \sqrt{\frac{\mu}{m}} \frac{Q^2}{e^2 a_o^2} \left(\frac{B}{T}\right)^{1/4}. \tag{20}$$

Here the constants are $C_1 = \frac{\pi\sqrt{2\pi}}{3} \Gamma(\frac{3}{4}) = 3.2$ and $C_2 = \frac{2\pi^{3/2}}{15} \Gamma(\frac{5}{4}) = 0.74$; $\mu$ is the reduced mass of nuclei.

In another limiting case compared with Eq. 15 $n \gg e^2/\hbar v_a$ Eq. 12 for the ionization cross section has the form

$$\sigma_{ion} = \langle\sigma_{rot}\rangle.$$

Then the cross section for ionization defined by Eqs. 19 and 20 ought to be multiplied by the value $\frac{8}{\pi} \frac{e^2 n}{\hbar v_a}$.

Equations 19 and 20 give the limiting relations for the cross sections for ionization of highly excited atoms initiated by collisions with molecules. Let us estimate the value for the ionization cross section at these extremes. From Eq. 19 one sees that the ionization cross section is on the order of $10^{-11} cm^2$ for collisions between an highly excited atom and a water molecule and is of the order of $10^{-13} cm^2$ for a CO molecule at thermal energies. For collisions with a quadrupole nitrogen molecule the ionization cross section is on the order of $10^{-15} cm^2$ and for collision with a hydrogen molecule it is on the order of $10^{-16} cm^2$.

In table I the experimental results for the cross sections for ionization of a highly excited atom by molecular impact are given. Comparing with the experimental data we find that the results of theory and experiment disagree for the case of a hydrogen molecule. We note that the measurements have been made in rather difficult conditions, so that a small water admixture (0,001%) may cause a change in the result.

Table I.

The experimental cross sections for ionization by collisions between highly excited atoms and molecules, $10^{-13}cm^2$.

| Molecule | Highly excited atom | | | | |
|----------|----------|----------|----------|----------|----------|
|          | He** | Ne** | Ar** | O** | C** |
| $H_2O$ | 2.9 [5] | 7.8 [5] | 12  [5] | | |
| $SO_2$ | 1.1 [5] | 2.7 [5] | 4   [5] | | |
| $NH_3$ | 1.5 [5] | | 7.4 [5] | | |
| CO | | | -- | 2 [8] | 1.5 [8] |
| $H_2$ | | | 0.4 [6] | | |
| | | | 0.1 [7] | | |

Finally, we consider the ionization of a highly excited atom by a collision with a polyatomic molecule. Collisions of the free electron with a polyatomic molecule may lead to the formation of a negative ion through an autoionizing state with a large lifetime (on the order of $10^{-5}$-$10^{-4}sec$[9-12]). This process is effective at small electron energies. Then the ionization process of a highly excited atom by a collision with a polyatomic molecule consists as follows. The weakly bound electron attaches to the polyatomic molecule and the negative ion in the autoionization state is formed. If the kinetic energy of the nuclei exceeds the energy of the Coulombic attraction of the nuclei, the process proceeds according to scheme:

$$A^{**} + B \rightarrow A^{+} + B^{-*} \tag{21}$$

The ionization cross section of process 21 can be obtained with using Eq. 3.

$$\sigma_{ion} = \frac{<|\vec{v}-\vec{v}_a|\sigma_{at}>}{v_a} \tag{22}$$

Here $\sigma_{at}$ is the cross section of electron attachment to the molecule, the brackets mean averaging over electron velocities. In the case $v_a >> \frac{e^2}{hn}$ this formula gives

$$\sigma_{ion} = <\sigma_{at}> . \qquad (23)$$

At small electron energies $\varepsilon$ the attachment cross section has the form [16-18] $\sigma_{at} = \frac{A}{\varepsilon \gamma}$, where the value $\gamma \sim I$ (for $SF_6$ $\gamma = 1.12$). Then using the distribution function Eq. 10 we receive from Eq. 23

$$\sigma_{ion} = A(\frac{\hbar n}{e^2})^{\gamma} \frac{16}{\pi} \frac{\Gamma(\frac{5}{2}-\gamma)\Gamma(\frac{3}{2}-\gamma)}{\Gamma(4-2\gamma)} \qquad (24)$$

For example, in the case $SF_6$ the Eq. 24 gives on the bases of the experimental attachment cross section[16]

$$\sigma_{ion} = 2n^{2 \cdot 24} \overset{\circ}{A}{}^2$$

Comparing it with the experimental data[5,13-15] we find that the highly excited atoms are in states with principal quantum number n = 40-80.

We have used the simple semiclassical method to a number of problems. In the paper of M. Matsuzawa[19-24] the Born approach to these problems is developed. It is correct for $v_a >> e^2/hn$. In this case, and in a similar one, both the semiclassical method and the Born approach must give the same result. The similar situation takes place in the case of ionization by collision between an electron and an atom. It is shown[22,23] that the semiclassical method and the Born approach give the same results for ionization of a highly excited atom. The semiclassical method is correct in a wider range of the collision velocities and it is simpler than the Born approach. The simplicity of the semiclassical method allows one to understand the physical meaning of the mechanisms for the processes considered.

The main problem for experiments with highly excited atoms is connected with defining the ionization potential of an excited atom. The results presented may help to form the basis for new experimental methods which permit one to find the ionization potential of highly excited atoms.

REFERENCES

1. B. M. Smirnov. Ions and excited atoms in plasma.
   Atomisdat, Moscow 1974
2. H. A. Bethe, E. E. Salpeter.  Quantum mechanics of one and
   two-electron atom.  Springer Verlag, Berlin 1957.
3. O. H. Crawford, A. Dalgarno, P. B. Hays: Molec. Phys 13,
   181 (1967)
4. E. Gerjoy, S. Stein. Phys. Rev. 97, 1671 (1955)
5. H. Hotop, A. Niehaus. J. Chem. Phys. 47, 2506 (1967)
6. S. E. Kuprianov. J. Exp. and Theor. Phys 51, 1011 (1966)
7. H. Hotop, A. Niehaus. Zs. Phys 215, 395 (1968)
8. S. E. Kuprianov. J. Exp. and Theor. Phys 55, 460 (1968)
9. R. N. Compton et al. J. Chem. Phys 45, 4634 (1966)
10. W. T. Naff, C. D. Cooper, R. N. Compton. J. Chem. Phys 49,
    2784 (1968)
11. R. N. Compton, R. H. Huebner, J. Chem. Phys 51, 3132 (1969)
12. W. T. Naff, R. N. Compton, C. D. Cooper, J. Chem. Phys 54,
    212 (1971)
13. S. E. Kuprianov. V  Sovjet Conference on Physics Electron
    and Atomic Collisions, Ujgorod 1972, p151
14. T. Sigiura, K. Arakawa. Recent developments in mass
    spectroscopy.  Editor K. Agata, T. Hayakawa.  Univ. Tokyo
    Press, Tokyo 1970, p848
15. J. A. Stockdale et al. J. Chem. Phys. 60, 4279 (1973)
16. L. G. Christophorou, D. L. McCorlke, J. G. Carter. J. Chem.
    Phys 54, 253 (1971)
17. A. A. Christodoulides, L. G. Christophorou.  J. Chem. Phys
    54, 469 (1971)
18. L. G. Christophorou, D. L. McCorkle, V. E. Anderson. J. Phys
    4B, 1163 (1971)
19. M. Matsuzawa. J. Chem. Phys 55, 2685 (1971)
20. M. Matsuzasa. J. Phys. Soc. Jap 32, 1088 (1972); 33, 1108
    (1972)
21. M. Matsuzawa. Phys. Rev. A9, 241 (1974)
22. J. G. Garcia, E. Gerjoy. Proc. of 5ICPEAC, Leningrad Nauka
    1967, p655
23. A. E. Kingston. J. Phys. 1B, 559 (1968)

# HYPERFINE STRUCTURE OF THE HIGHLY EXCITED STATES OF ALKALI ATOMS

R. Gupta

Columbia Radiation Laboratory, Columbia University

New York, New York  10027

The work that I would like to describe here has been done at the Columbia Radiation Laboratory in collaboration with Professor William Happer and a number of former and present graduate students.  They are S. P. Chang , C. Y. Tai, K. H. Liao, L. K. Lam, P. Tsekeris and J. Farley.

Alkali-metal atoms are perhaps most interesting theoretically due to their simple single-valence electron structure, and they are experimentally easy to handle.  When we started our work a few years ago, extensive work on the hyperfine structure (hfs) measurements in the P states of alkali atoms using optical double resonance and level-crossing spectroscopy had been done by our European colleagues.  However, almost no information on the hfs in the S, D, F, etc. states of these atoms was then available.  Using the techniques of cascade fluorescence spectroscopy and stepwise laser spectroscopy, we have been able to investigate these states and some of the results that we have obtained are very surprising.  Therefore, I will concentrate mainly on the non-P states of the alkali atoms.

The hfs in almost all of the excited states of alkali atoms are smaller than the doppler width of the spectral lines. Therefore, only those techniques which are primarily limited by the natural linewidth (e.g. rf spectroscopy and the level-crossing spectroscopy) are suitable for an investigation of the hfs of these states.  The problem with the S and D states is that they cannot be populated by direct optical excitation of the ground state atoms as these transitions are forbidden by the electric dipole selection rules.  Archambault et al (1)

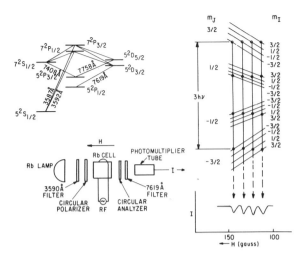

Fig. 1:   Schematic illustration of the cascade radiofrequency
          spectroscopy experiment.

had tried electron excitation but they were only able to obtain rough limits on the hfs of $5^2D_{5/2}$ state of Na and $9^2D_{5/2}$
state of Cs.  Because of the problems associated with the
affect of DC magnetic fields and rf fields (which are required
in these experiments) on the electron beams and electrical discharges, we have used an optical excitation scheme to get
around the selection rules.  Our method, for the particular
case of $7^2S_{1/2}$, $5^2D_{3/2}$ and $5^2D_{5/2}$ states of Rb, is shown in
Fig. 1.  Ground state Rb atoms are excited to the $7^2P$ states
by third resonance lines from a  Rb lamp.  Atoms in the $7^2P$
states have about 25% probability of decaying to the $7^2S$ state.
$7^2P_{3/2}$ and $7^2P_{1/2}$ states have about 25% and 30% branching probabilities to the $5^2D_{5/2}$ and the $5^2D_{3/2}$ states, respectively.
If circularly polarized exciting light is used, part of the
photon's angular momentum is transferred into the electronic
polarization of the $7^2P$ states.  When the atoms decay to the
$7^2S$ or the $5^2D$ states, this polarization is partially carried
over to these states.  The electronic polarization of the $7^2S_{1/2}$,
$5^2D_{3/2}$ and $5^2D_{5/2}$ states can be monitored by observing the
circular polarization of the 7408Å , 7619Å , and 7758Å  fluorescent light, respectively(2,3).  The Rb atoms are placed in
a longitudinal DC magnetic field as shown in Fig. 1.  The
energy levels of the atoms are then split into a number of
magnetic sublevels.  Fig. 1 shows these sublevels for the
$D_{3/2}$ state of Rb87 (nuclear spin I = 3/2), grouped according
to their high field quantum numbers.  In some of our experiments, we apply an rf magnetic field to the Rb atoms perpendicular to the DC field.  The DC field is slowly varied
through the region of interest.  Magnetic resonance transition

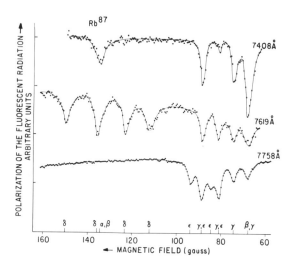

Fig. 2: Results of the cascade radiofrequency spectroscopy. $\alpha$, $\beta$, $\gamma$, $\delta$, and $\varepsilon$ represent resonances in the $7S_{1/2}$, $7P_{1/2}$, $7P_{3/2}$, $5D_{3/2}$ and $5D_{5/2}$ states, respectively.

between pairs of sublevels, whose magnetic quantum numbers differ by one, are detected as a decrease in the polarization of the fluorescent light (4,5). Some typical data are shown in Fig. 2. By using appropriate filters in front of the photomultiplier tube we observe resonances in the $7^2S_{1/2}$ (denoted by $\alpha$), $5^2D_{3/2}$ ($\delta$), or in the $5^2D_{5/2}$ ($\varepsilon$) states. Resonances are also observed in the $7P_{3/2}$ ($\gamma$) state with all three filters. $7P_{1/2}$ ($\beta$) resonances are too weak to be seen here. Most of these resonances are multiple quantum transitions, each one corresponding to a given value of $M_I$ as shown in Fig. 1.

I would like to point out that high field level crossing technique, which has yielded a wealth of information on the P states, is not a very useful technique in cascade experiments. High field cascade level crossing signals are extremely weak. This is because level crossing signals depend on transverse polarization which precesses to a zero average (at high fields) in the intervening stages of cascade (3). Zero field level crossing (Hanle effect) can however be easily seen in cascade experiments (6). Fig. 3 shows level crossing data in the $6^2D_{3/2}$ state of Cs. High field level crossings are indicated by the quantum numbers of the crossing sublevels. Even in this particularly favorable case, the high field signals are at least two orders of magnitude smaller than the zero field signal.

In addition to the rf and level crossing experiments, we

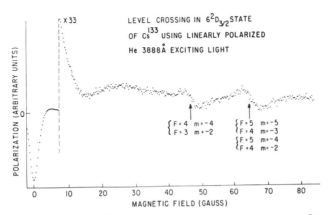

Fig. 3:   Cascase level crossing spectroscopy results on the $6^2D_{3/2}$ state of $Cs^{133}$.

have performed decoupling experiments.  The set up is exactly the same as in rf experiments except that we do not use the rf fields and simply observe the longitudinal polarization of the fluorescent light as a function of the DC magnetic field. Fig. 4 shows the results for the second excited S states of various alkali isotopes.  The polarization of the light is seen to increase with the increasing magnetic field.  The explanation is as follows (3):  At low magnetic fields the electronic and the nuclear angular momenta in the excited S state are coupled to each other due to hyperfine coupling. Therefore part of the electronic angular momentum is transferred into nuclear angular momentum and the electronic polarization is degraded.  On the other hand, at high magnetic fields, electronic and nuclear angular momenta are decoupled from each other and precess around the external magnetic field.  Since the polarization of the fluorescent light is proportional to the electronic polarization in the S state, we expect the polarization of the fluorescent light to increase with increasing external magnetic field.  The widths of these curves clearly depend on the strength of the hyperfine coupling. Hfs increases as we go from $K^{41}$ to $Cs^{133}$, as is shown in Fig. 4. As a matter of fact, we used these curves to deduce rough values of the hfs in these states even before we attempted rf experiments.

Decoupling experiments have the additional virtue that the shape of the curves are sensitive to the sign of the hfs.  When we looked for the sign of the hfs in the $5^2D$ states of Rb (at the suggestion of Dr. Allen Lurio), to our astonishment we found that the hfs of the $5^2D_{5/2}$ state was inverted while that of the $5^2D_{3/2}$ state was normal.  The decoupling curves are shown in Fig. 5.

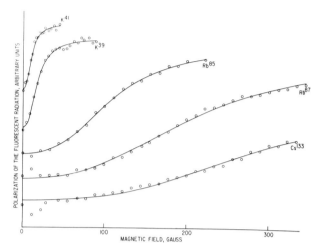

Fig. 4:  Signals due to decoupling of electronic and nuclear angular momenta in the second excited S states of alkali atoms.

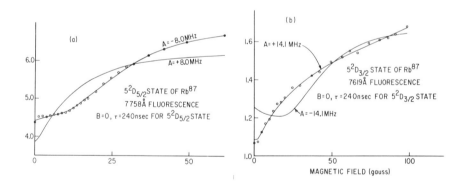

Fig. 5:  Decoupling signals in the 5D states of Rb.  Circular polarization of the fluorescent light is plotted against the applied magnetic field.  These curves show that the hfs in the $5D_{5/2}$ state is inverted while it is normal in the $5D_{3/2}$ state.

We then determined the hfs of the corresponding state in Cs ($6^2D$). Again, to our surprise, we found that the hfs in the $6^2D_{5/2}$ state was inverted while it was normal in the $6^2D_{3/2}$ state.

In order to make an independent check of these results, we have developed another technique for determining the signs of hfs which is based on creating nuclear polarization in the excited state by "nonwhite" excitation. In Rb, such excitation in produced by exciting Rb[85] by Rb[87] lamp. This method is discussed in detail by Tai et al (6). The results of these measurements confirm the decoupling results of inverted hfs.

At this point it was clearly desirable to do similar measurements in the first excited S and D states of Rb and Cs. Even the fine structure in the first excited D state ($4^2D$) of Rb was known to be inverted. Experimentally, however, these states presented a special problem as shown in Fig. 6. $4^2D_{5/2}$ state, for example, fluoresces at 15,290Å, well outside the range of conventional photomultiplier tubes. We solved the problem by making observations in the 7800Å fluorescent light, after an additional stage of cascading. It turns out that longitudinal polarization is carried from state to state with good efficiency in the cascading process and good signals are observable. Analysis problem, obviously, becomes very complicated due to many different cascading routes. Typical data for the case of $5^2D_{5/2}$ state of Cs is shown in Fig. 7. We find that the hfs of the $4^2D_{5/2}$ state of Rb and $5^2D_{5/2}$ state of Cs is inverted, while that of the $D_{3/2}$ states is normal (7).

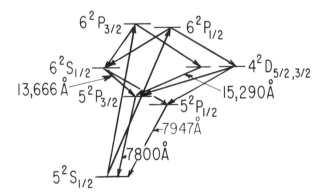

Fig. 6:  Hfs of the first excited S and D states are measured by observing the 7800Å or 7947Å fluorescent light, after an addition step in cascading.

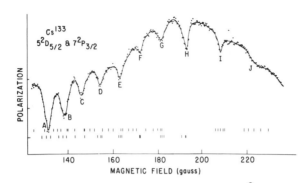

Fig. 7:  Rf spectroscopy results for the $5^2D_{5/2}$ state of Cs.
I and J are resolved $D_{5/2}$ state resonances.  Other
resonances overlap with the $7P_{3/2}$ (A-H) resonances.


In order to understand these inverted hfs further, we
have performed an experiment at such high magnetic fields
that even L and S begin to decouple from each other.  In this
way we have been able to determine the complete hfs Hamiltonian
for the $4^2D$ state of Rb.  The electric quadrupole and magnetic
dipole hfs Hamiltonian may be written as

$$\mathcal{H} = \mathcal{H}_q + \mathcal{H}_\ell + \mathcal{H}_d + \mathcal{H}_c$$

where q, ℓ, d and c refer to quadrupole, orbital, dipole-
dipole and the contact parts of the Hamiltonian.  We have
performed an anticrossing experiment, shown schematically in
Fig. 8, to determine the four parts separately.  In an external
field of about 4 kilogauss, two fine structure sublevels cross.
Due to hfs each fine structure sublevel is split into several
hfs sublevels.  Sublevels having same total magnetic quantum
numbers do not cross.  At these anticrossing points, due to
the hfs mixing of the sublevels, the fluorescent intensity
changes.  Typical signals are shown in Fig. 9.  Using this
data, we have found (8) that the electric quadrupole and orbital parts of the hfs have normal sign but the dipole-dipole
part has its sign reversed.  The contact term, which ideally
should be zero for a D state (ignoring the core), is very
large and negative.  These results are truly astonishing.
This is the first case we know in atomic physics where the
dipole-dipole term has been found to have reversed sign.

It was clearly very interesting to extend these hfs measurements to highly excited states of alkali atoms.  Although
cascade fluorescence spectroscopy is a very useful technique,

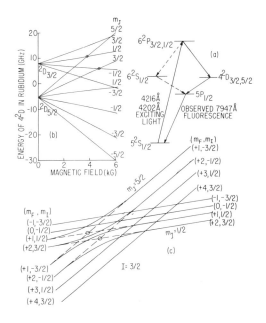

Fig. 8: Schematic illustration of the cascade anticrossing spectroscopy. Change in the 7947Å fluorescent light is observed at the anticrossing points (marked by circles in Fig. 8C).

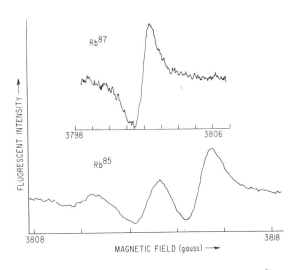

Fig. 9: Observed anticrossing signals in the $4^2D$ state of Rb.

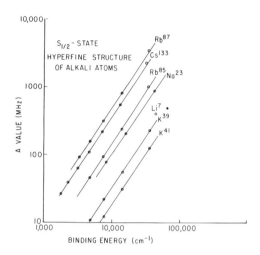

Fig. 10:   Hfs of the S states is found to be proportional to
the 3/2 power of the binding energy.

it is very difficult to get good signal-to-noise ratio for
highly excited states.  This is because the oscillator strengths
for ground state to high P state transitions are very low and
no cw dye lasers exist in the far uv region of spectrum.
Svanberg, Tsekeris and Happer (9) have developed a stepwise
excitation scheme using a cw dye laser to reach the highly
excited D states of alkali atoms.  In this method, the alkali
atoms are excited to their first excited P states by resonance
lines from an alkali discharge lamp.  These atoms are further
excited to the D states by a cw dye laser tuned to the appropriate wavelength.  Polarized laser light is used and polarization of the fluorescent light in the decay of D state atoms
is observed.  Level crossing and rf spectroscopy of used to
determine the hfs.  Although a stepwise excitation method with
conventional lamp was used earlier by Eck and Smith (10), use
of dye laser makes it possible to study large number of states
and do systematic study of the sequences of states.  Svanberg
and collaborators (11) have measured hfs in Cs up to $14^2D$
states and in Rb up to $8^2D$ states.  In most of these states
they have determined the sign of the hfs using Stark effect.
In all of the states investigated by them they find that $D_{3/2}$
hfs are normal while the $D_{5/2}$ states hfs are inverted.  Results
are shown in Table I and II.

TABLE I:  Hyperfine Structure in the $D_{3/2}$ States of Alakli Atoms (from Refs. 6,7 and 11).

A(MHz)

| n | $K^{39}$ | $Rb^{85}$ | $Rb^{87}$ | $Cs^{133}$ |
|---|---|---|---|---|
| 3 | <1.8 | | | |
| 4 | | +7.3(5) | +25.1(9) | |
| 5 | 0.44(10) | +4.18(20) | +14.43(23) | --- |
| 6 | 0.2(2) | +2.28(6) | +7.84(5) | +16.30(15) |
| 7 | | +1.34(1) | +4.53(3) | 7.4(2) |
| 8 | | +0.84(1) | +2.85(3) | +3.98(18) |
| 9 | | | | +2.37(3) |
| 10 | | | | +1.52(3) |
| 11 | | | | 1.055(15) |
| 12 | | | | 0.758(12) |
| 13 | | | | 0.566(8) |
| 14 | | | | 0.425(15) |

TABLE II: Hyperfine Structure in the $D_{5/2}$ States of Alkali Atoms (from Refs. 6, 7 and 11)

| n | $K^{39}$ | $Rb^{85}$ | $Rb^{87}$ | $Cs^{133}$ |
|---|---|---|---|---|
| | | A(MHz) | | |
| 3 | <2.2 | | | |
| 4 | | -5.2(3) | -16.9(6) | |
| 5 | ±0.24(7) | -2.12(20) | -7.44(10) | -22.1(5) |
| 6 | ±0.10(10) | -0.95(20) | -3.4(5) | -3.6(1.0) |
| 7 | | -0.55(10) | -2.0(3) | |
| 8 | | -0.35(7) | -1.2(2) | -0.85(20) |
| 9 | | | | -0.45(10) |
| 10 | | | | -0.35(10) |
| 11 | | | | ±0.24(6) |
| 12 | | | | ±0.19(5) |
| 13 | | | | ±0.14(4) |

TABLE III: Hyperfine Structure (A in MHz) in the S States of Alkali Atoms. All Unsuperscripted Results are from References 5 and 12.

| State | Na²³ | K³⁹ | K⁴¹ | Rb⁸⁵ | Rb⁸⁷ | Cs¹³³ |
|---|---|---|---|---|---|---|
| 3s | 885.8ᵃ | | | | | |
| 4s | 202(3) | 230.85ᵃ | 127.00ᵃ | | | |
| 5s | 79.5(3.0)ᵇ | 55.50(60) | 30.75(75) | 1011.9ᵃ | 3417.3ᵃ | |
| 6s | 39(3)ᶜ | 21.81(18) | 12.03(40) | 239.3(1.2) | 809.1(5.0) | 2298.1ᵃ |
| 7s | | 10.85(15) | | 94.00(64) | 318.1(3.2) | 546.3(3.0) |
| 8s | | | | 45.5(2.0) | 159.2(1.5) | 218.9(1.6) |
| 9s | | | | | 90.9(8) | 109.5(2.0) |
| 10s | | | | | 56.3ᵈ | 63.2(3) |
| 11s | | | | | | 39.4(2) |
| 12s | | | | | | 26.31(10) |
| 13s | | | | | | 18.4ᵈ |
| 14s | | | | | | 13.4ᵈ |

a.  Mostly atomic beam measurements referenced by G. H. Fuller and V. W. Cohen, Nucl. Data, Sect. A 5, 433 (1969).

b.  H. T. Duong et al, Phys. Rev. Lett. 33, 339 (1974).

c.  M. D. Levenson et al, Abstracts of Contributed Papers, Fourth International Conference on Atomic Physics, Heidelberg, 1974.

d.  P. Tsekeris, preliminary results.

We have also used stepwise laser spectroscopy to measure the hfs in the S states of alkali atoms (12). We have been able to reach up to $14^2S_{1/2}$ in Cs and $10^2S_{1/2}$ state in Rb. The results are summarized in Table III. In contrast to the D state hfs, the S state hfs are very close to those predicted by simple Fermi-Segre' formula (5). This formula predicts that the hfs are proportional to the 3/2 power of the binding energy of the valence electron. In Fig. 10 we have plotted hfs (A value) against the binding energy on a log-log scale. All the experimental points are seen to lie on straight lines of slope approximately 3/2, as predicted. Similar conclusion can be drawn about the D state hfs, and these are discussed by W. Happer (13).

In conclusion, I would like to mention that hfs of the $5^2F$ state of Cs (9) and of several highly excited $P_{3/2}$ (14) and $P_{1/2}$ (15) states have been measured by a combination of the stepwise excitation and cascading techniques. Due to very small oscillator strengths for these highly excited P states, although in principle possible, the usual optical double resonance and level crossing techniques are impractical. So we see that tremendous new information on the excited states of alkali atoms has become available in the last few years.

This work was supported by the Joint Services Electronics Program under Contract No. DAAB07-74-C-0341.

## References

1.  Y. Archambault et al, J. Phys, Rad. 21, 677 (1960).

2.  S. Chang, R. Gupta, and W. Happer, Phys. Rev. Lett. 27, 1036 (1971).

3.  R. Gupta, S. Chang, and W. Happer, Phys. Rev. A, 6, 529 (1972).

4.  R. Gupta, S. Chang, C. Tai, and W. Happer, Phys. Rev. Lett., 29, 695 (1972).

5.  R. Gupta, W. Happer, L. K. Lam, and S. Svanberg, Phys. Rev. A 8, 2792 (1973).

6.  C. Tai, W. Happer, and R. Gupta, Phys. Rev. A (Sept. 1975).

7.  L. K. Lam and W. Happer, Phys. Rev. A (to be published).

8.  K. H. Liao, L. K. Lam, R. Gupta, and W. Happer, Phys. Rev. Lett. 32, 1340 (1974).

9.  S. Svanberg, P. Tsekeris, and W. Happer, Phys. Rev. Lett. 30, 817 (1973).

10.  R. L. Smith and T. G. Eck, Phys. Rev. A, 2, 2179 (1970).

11.  S. Svanberg and P. Tsekeris, Phys. Rev. A, 11, 1125 (1975);
     S. Svanberg and G. Belin, J. Phys. B, 7, L82 (1974);
     W. Hogervorst and S. Svanberg, Z. Phys. (to be published);
     G. Belin, L. Holngren, I. Lindgren, and S. Svanberg,
     Physica Scripta (to be published).

12.  P. Tsekeris, R. Gupta, W. Happer, G. Belin, and S. Svan-
     berg, Phys. Lett. 48A, 101 (1974); P. Tsekeris and R.
     Gupta, Phys. Rev. A, 11, 455 (1975).

13.  W. Happer, in Atomic Physics IV, Ed. G. zuPutlitz and
     E. W. Weber (Plenum, New York 1975).

14.  G. Belin and S. Svanberg, Phys. Lett. 47A, 5 (1974).

15.  P. Tsekeris, J. Farley, and R. Gupta, Phys. Rev. A, 11,
     2202 (1975).

# THE PRODUCTION AND DETECTION OF HIGHLY EXCITED STATES

James E. Bayfield

Physics Department, Yale University
New Haven, Connecticut

## I.   Introduction

The study of highly excited states of atoms and molecules continues to be stimulated by their importance for problems in astrophysics and plasma physics.  Much of the early experimental work was concerned with the production by electron transfer of fast beams of highly excited atoms for injection into and Lorentz ionization within magnetic mirror plasma devices. Recombining plasmas containing highly excited states are observed in H II regions by radioastromomers[1]; other cases of interest are the sun, as well as various terrestrial plasmas produced for instance by Ohmic heating[2] or by pulsed lasers.[3] The question of possible effects of highly excited states on plasma bulk properties has been raised.[4,5]  The possibility of this stems from their collision cross sections possibly being geometrical, proportional to $n^4$ where n is the principal quantum number; this can dominate over possible fractional populations varying typically as $n^{-3}$ for production processes depending primarily upon an oscillator strength for a transition from a ground state.[6]  This situation can be altered drastically by the presence of static and time varying fields in the plasma.

The development of experimental techniques for the production and detection of highly excited species has led to a number of new types of experiments on highly excited state properties and interactions.  Among these are measurements of atomic structure,[7] lifetimes,[8] probabilities for ionization by oscillating fields,[9] cross sections for electron

loss in eV energy ion collisions,[10] and cross sections for
ionization[11] and angular momentum mixing[12] in thermal energy
collisions with atoms. Experiments to test the theory of
highly excited atom-molecule reactions should be forthcoming.[13,14]

Because of their large radii, large polarizabilities and
small ionization potentials, highly excited atoms are used in
the study of the alteration of atomic structure and processes
by very strong external electromagnetic fields. Considerable
work has been done for the case of a static magnetic field[15]
and the case of a strong oscillating field is under investigation.[9,16]

Most of the methods for excited state production have been
extended by now to highly excited states: atom and molecule
excitation by electron impact, atom production in the dissociation of excited molecules, keV energy atom and molecule
production by electron transfer and excitation collisions with
gas and solid targets, and laser pumping of gases, thermal
atomic beams and fast atomic beams. Because the binding
energy of highly excited states is so much smaller than that
of normal atoms, the collisional methods are highly non-
resonant. Only the laser pumping approach is resonant. Nevertheless, since CW photon beam densities are not yet competitive
with the densities of collision targets, each of the different
techniques can be useful at present. Under some circumstances
a combination of collisional and laser transitions can yield
reasonable production efficiency combined with state selectivity.

Among the schemes used for detecting highly excited states
are the observation of optical decay[7,17,19] and a number of
ionization techniques permitting ion counting rather than the
less efficient photon counting. Ionization can be accomplished
by strong electric fields, interaction with a metal surface, low
energy collision with a molecule of high electron affinity,
and Penning ionization. Field ionization is particularly
useful in that it is fairly state selective, highly efficient,
and useful for both fast and slow atom detection. Microwave
fields are preferable to static fields in that field effects
on ion trajectories are greatly reduced. It may be possible
to use multiple microwave fields, one inducing a resonant
transition between lower and upper highly excited levels, and
the other field ionizing the upper level only. This would
yield both high detection efficiency and state selectivity.

The various types of highly excited state experiments
rely upon a knowledge of the atomic or molecular structure
involved. Experiments[7,17,18] and theory[20,21] both continue
to indicate that most aspects of the structure problem are

understood in the case of atoms. The situation for molecules is not so developed. A considerable amount of work has been done on the UV Rydberg absorption spectra of diatomic molecules[22,23], which has assisted the theoretical treatment of autoionization and predissociation of excited molecular states.[24-28]

## II.  Production of Fast Highly Excited Atoms and Molecules by keV Energy Electron Transfer Collisions

A number of developments in this area have extended the range of studies discussed in previous reviews.[29-31] Polyatomic molecular gas targets have been investigated, with highly excited hydrogen atom production cross sections found to be as large or larger than for Xe gas targets and incident protons with energies above 25 keV[32] (see Figure 1).

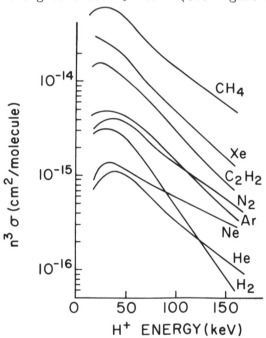

Figure 1. The $n^3$ weighted cross section for electron transfer into a highly excited state with principal quantum number n for protons incident on various gas targets.

At lower energies metal vapor targets produce large fractional amounts of highly excited atoms, with the alkalais and second column metals being useful (see Figure 2).[33] The use of solid film targets of C, Mg, Nb and Au has been studied using $D^+$ beams between 8 and 100 keV.[19] Thin film and optimal vapor targets of Mg are competitive, although the solid target is sensitive to surface contamination and produces atom beams having large fractional energy spreads. The $n^{-3}$ law for fractional population of the different states holds for these

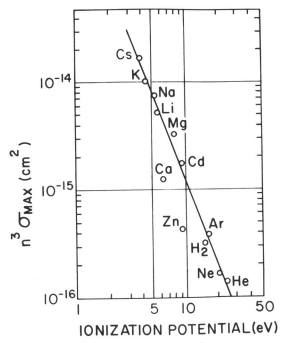

Figure 2. The maximum value of the $n^3$ weighted cross section for highly excited state production at the optimal collision energy, as a function of the ionization potential of various target gas atoms.[33]

cases for values of n in the region 6-18. For inert gas targets this law has been verified for values of n up to 70 (see Figure 3).[9,34] A study of the $H^+$–H case has underlined the importance of using the full Born approximation rather than the OBK approximation for estimating cross sections for highly excited state production in the 10-100 keV energy region.[34] Although work has concentrated on the production of highly excited hydrogen and helium atomic beams[31], extension to heavier atoms should be possible.

Production of highly excited $H_2$ and $H_3$ molecules by fast electron transfer onto $H_2^+$ and $H_3^+$ has also been studied.[35,36] Static electric field ionization was used for detection. Assuming a core ion model of these excited molecular states, the production cross section appears to vary as $n^{-4}$. The observed lifetimes $\geq 10^{-7}$ seconds for these molecules have been ascribed to sizeable values for the orbital angular momentum quantum number $\ell$.[28] This contrasts with OBK theoretical

predictions for the case of atom formation which favor s and
p states.[37,38] This latter prediction is experimentally
verified for $H^+$ collisions on gas targets and low values of
$n=3-5$,[39] but only scant evidence based upon field ionization

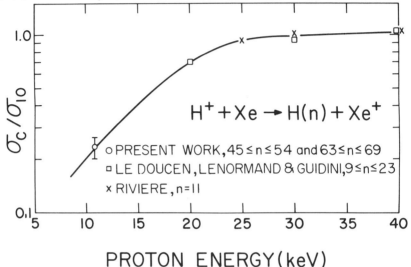

Figure 3. The $n^3$ weighted fractional electron transfer into
a given state n for $H^+$-Xe collisions and various n-values.

studies suggests it might be true at much higher n.  Much
work remains to be done to determine the nature of these
highly excited molecules formed by electron transfer.
Highly excited atoms are also formed by dissociative processes
when $H_2^+$ and $H_3^+$ collide with gas targets.[35,36,40]

III.  Production of Slow Highly Excited Atoms and Molecules
by Electron Impact

Following the observation of highly excited inert gas atom
production in electron collisions a decade ago,[41] a large
number of highly excited species have been produced this way,
both atoms,[42] inert gas ions[43] and metal ions.[44]  For electron
energies considerably above threshold an $n^{-3}$ rule for fractional
population is expected.[45]  An interesting recent prediction
is that for electron energies within 1 eV of threshold, states
with high angular momentum should be formed.[4]  Electron impact
was used in the highly excited state structure experiments in
He and Ne.[7]

Corresponding experiments on molecules have been performed,
with an emphasis on $H_2$,[46,47] $N_2$,[46,48-51] $O_2$,[46,52] and $CO$.[51,52]
Highly excited atomic fragments were seen in all cases, but only
in $H_2$ have highly excited molecules been observed.  Discrimina-

tion between atoms and molecules in these experiments is based on threshold behavior in the highly excited state signals as the electron energy is varied.  Primarily surface ionization possibly with static field ionization (see Figure 4) have been employed in the electron experiments, providing little information on the values of n detected.  Time of flight measurements of highly excited atom or molecule kinetic energy assists in their identification.[47]

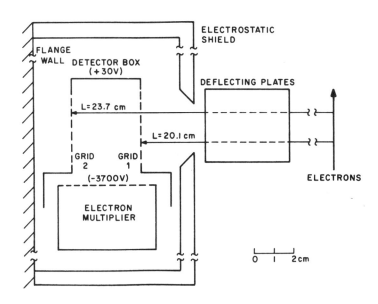

Figure 4.  The slow highly excited atom detector used at Bell Laboratories.[47,48,53]  The atoms produced by electron impact travel to the left and are detected in a gridded detector box.

The situation in $H_2$ is especially interesting, for the data on this simple system can be explained in terms of known molecular states and correlated with the Rydberg absorption spectra observed in UV photon impact.[22,23]  Although the observed production of highly excited states a little above the threshold of 18.1 eV can be understood in terms of an $H_2^+$ core ion model, further production in the 28 eV region and above involves doubly excited states such as $(2p\sigma_u)^2 {}^1\Sigma_g^+$, and the intersection of such states with the core ion states close to the $(1s\sigma_g)^2 \Sigma_g^+$ $H_2^+$ state limit (see Figure 5).

IV.  Detection of Highly Excited Atoms by Field Ionization

Since the latest review,[31] several developments have

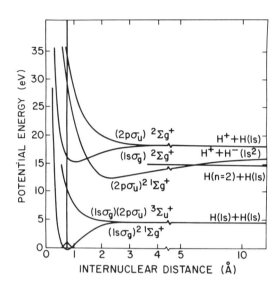

Figure 5.   Some states of $H_2$ involved in highly excited state production in e-$H_2$ collisions.

occurred.  For the case of static field ionization, the experiments using collisionally produced highly excited hydrogen atoms have indicated that higher threshold field strengths were required than the classical field value $1/(16n^4)$ a.u. where the field-free atomic level lies at the top of the barrier in the total potential (see Figure 6).[57] This can be understood if one assumes that the theoretically predicted s(and p) state production dominates and that the adiabatic correlation[54,19] with the Stark states depressed lowest in the well of the effective potential is followed. Then the classical field value is increased to $1/(9n^4)$ a.u., in accordance with the data.  Recent similar studies using laser excitation of the isolated s states of Na[18] and the probably mixed substates of H[55] also exhibit required field strengths for static field ionization larger than $1/(16n^4)$.

The field ionization technique has recently been extended to the use of microwaves.[9,55-57]  Less field strength is required (see Figure 7), ion deflection problems are absent, and ions formed within the microwave field structure can be selected from those produced elsewhere by means of a kinetic-energy ion labeling technique.[9,10,55,56]  This latter development involves a static potential on the field structure, with ions produced by ionization of a beam of atoms passing inside being accelerated as the beam leaves the structure.  Subsequent

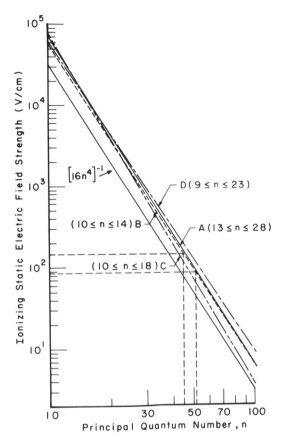

Figure 6.  A comparison of experimental[57] and classical static threshold electric fields required to ionize a hydrogen atom with principal quantum number n.

electrostatic kinetic energy analysis is then used to select a field ionization signal.

### V.  Laser Pumping of Highly Excited Atomic States

Two-step pulsed dye laser transitions have been used to pump ground state sodium atoms in a gas[8] or an atomic beam[18] via the $3p$ state into s and d excited states as high as n=37. The 3s-3p transition at 5890 Å is easily saturated, and the rate for the remaining transition near 4500 Å can also be made high.  The rare gases have also been pumped to highly excited states using a pulsed laser.[11]  The present limitations of this technique are the pulse duty factor $10^{-4}$ and the wavelength resolution of multi-mode pulsed dye lasers of about 1 cm$^{-1}$. The latter makes studies of individual substates with n≥50

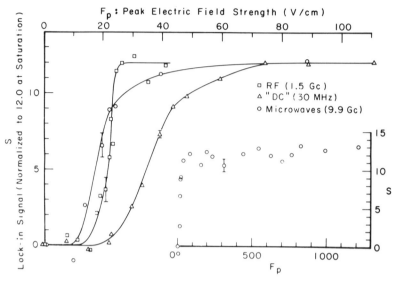

Figure 7. Ionization signals for hydrogen with n near 66, as a function of peak electric field strength and for three different frequencies. The corresponding transition rates are of order $10^7$ sec$^{-1}$.

difficult. Yet the signal count rates of $10^4$ sec$^{-1}$ are sufficiently high that many new experiments are possible.

In the case of fast atom beams a 50mW CW UV argon ion laser has been used to pump hydrogen atoms from the 2s metastable state into highly excited states with n as high as 55.[55] For the laser beam parallel to the keV atom beam, the 3638 Å line can be tuned over a wide range of n using the Doppler effect. A Doppler scan of n=44 and 45 obtained by varying the proton beam accelerator voltage before charge-exchange atom production is shown in Figure 8. The wavelength resolution limited by the 20 eV atom beam energy spread corresponds to 0.1 cm$^{-1}$. The overall laser pumping efficiency was only $10^{-6}$, but this can be increased using the watt level CW argon ion lasers now available. Signal count rates for the fast beam and thermal beam approaches are comparable. A more general fast beam technique under development is the production of n=10 states of H by charge exchange followed by high power CW $CO_2$ laser excitation of highly excited states. This should increase signal rates by a factor $10^3$, and permit fast beam studies on atoms other than hydrogen. Among the applications of this should be the extension of the merged beam atomic collision experiments of highly excited hydrogen atoms[10] to laser pumped states of a number of atomic species.

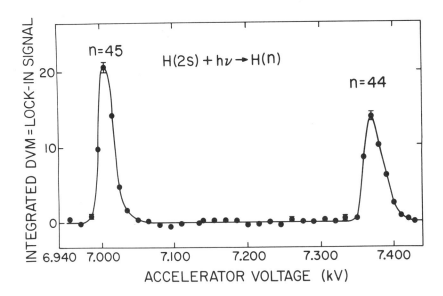

Figure 8.  The Doppler-tuned spectrum of UV laser pumped
fast highly excited hydrogen atoms.

### References

1.  A.K. Dupree and L. Goldberg, Ann. Rev. Astron. Astrophys.
    8, 231 (1970).
2.  L.C. Johnson. Phys. Rev. 155, 641 (1967).
3.  E. Ya. Kononov and K.N. Koshelev, Sov. J. Quant. Electron.
    4, 1340 (1975).
4.  U. Fano, J. Phys. B: Atom. Molec. Phys. 7, L401 (1974).
5.  M. Capitelli, C. Guidotti and U. Lamanna, J. Phys. B:
    Atom. Molec. Phys. 7, 1683 (1974).
6.  M. Matsuzawa, Phys. Rev. A 9, 241 (1974).
7.  W. Wing, R.R. Lea and W.E. Lamb Jr., in Atomic Physics 3,
    Proc. Third Int. Conf. on Atomic Physics, S.J. Smith and
    G.K. Walters, editors (Plenum Press, New York, 1973),
    pp. 309-326.
8.  T.F. Gallaher, S.A. Edelstein and R.M. Hill, Phys. Rev.
    A 11, 1504 (1975).
9.  J.E. Bayfield and P.M. Koch, Phys. Rev. Letters 33, 258
    (1974).
10. P.M. Koch and J.E. Bayfield, Phys. Rev. Letters 34, 448
    (1975).
11. T.B. Cook, W.P. West, F.B. Dunning and R.F. Stebbings,
    Bull. Am. Phys. Soc. 20, 253 (1975); see also contributed
    paper no. 847 given at this conference.

12. T.F. Gallagher, S.A. Edelstein and R.M. Hill, contributed paper no. 851 given at this conference.
13. M. Matsuzawa, J. Chem. Phys. 55, 2685 (1971).
14. M.R. Flannery, in Atomic Physics 3 (see ref. 7), pp. 143-153.
15. W.R.S. Garton and F.S. Tomkins, Astrophys. Jour. 158, 839 (1969).
16. J.E. Bayfield, Abstracts of the European Study Conference on Multiphoton Processes, Seillac, France, 14-17 April 1975; Bull. Am. Phys. Soc. 20, 826 (1975).
17. W. Happer, the previous invited paper in this volume.
18. T.W. Ducas, M.G. Littman, R.R. Freeman and D. Kleppner, Bull. Am. Phys. Soc. 20, 680 (1975); see also contributed paper no. 857 of this conference.
19. K.H. Berkner, I. Bornstein, R.V. Pyle and J.W. Stearns, Phys. Rev. A 6, 278 (1972).
20. T.N. Chang and R.T. Poe, Phys. Rev. A 10, 1981 (1974).
21. A. Lindgard and S.E. Nielsen, J. Phys. B: Atom. Molec. Phys. 8, 1183 (1975).
22. W.A. Chupka and J. Berkowitz, Jour. Chem. Phys. 51, 4244 (1969).
23. S. Takezawa, Jour. Chem. Phys. 52, 5793 (1970).
24. R.S. Mulliken, Jour. Am. Chem. Soc. 88, 1849 (1966).
25. J.N. Beardsley, Chem. Phys. Letters 1, 229 (1970).
26. R.S. Berry and S.E. Nielsen, Phys. Rev. A 1, 383 (1970); 395 (1970).
27. F.H.M. Faisal, Phys. Rev. A 4, 1396 (1971).
28. Y.B. Band, J. Phys. B: Atom. Molec. Phys. 7, 2072 (1974).
29. N.V. Fedorenko, V.W. Ankudinov and R.N. Il'in, Sov. Phys. Tech. Phys. 10, 461 (1965).
30. A.C. Riviere, in Methods of Experimental Physics, Vol. 7A ed. by B. Bederson and W.L. Fite (New York, Academic Press, 1968), p. 208.
31. R.N. Il'in, in Atomic Physics 3 (see ref. 7), pp. 309-326.
32. R. LeDoucen, J.M. Lenormand and J. Guidini, Le Jour. de Physique 31, 965 (1970).
33. V.W. Oparin, R.N. Il'in and E.S. Solov'ev, Sov. Phys. JETP 25, 240 (1967).
34. J.E. Bayfield, G.A. Khayrallah and P.M. Koch, Phys. Rev. A 9, 209 (1974).
35. E.S. Solov'ev, R.N. Il'in, V.A. Oparin and N.V. Fedorenko, Sov. Phys. - JETP 26, 1097 (1968).
36. C.F. Barnett, J.A. Ray and A. Russek, Phys. Rev. A 5, 2110 (1972); C.F. Barnett and J.A. Ray, A 5, 2120 (1972).
37. J.R. Hiskes, Phys. Rev. 137, A361 (1965).
38. A.V. Vinogradov, A.M. Urnov and V.P. Shevel'ko, Sov. Phys. - JETP 33, 1110 (1971).
39. See J.E. Bayfield, in Atomic Physics 4, G. zu Putlitz, E.W. Weber and A. Winnacker, editors, (New York, Plenum Press, 1975), pages 397-434.

40. K.H. Berkner, S.N. Kaplan, G.A. Paulikas and R.V. Pyle, Phys. Rev. 138, A410 (1965).
41. V. Cermak and Z. Herman, Collect. Czechosl. Chemic. Commun. 29, 953 (1964).
42. G.A. Surskii and S.E. Kupriyanov, Sov. Phys. - JETP 27, 61 (1968).
43. S.E. Kupriyanov and Z.Z. Latypov, Sov. Phys. - JETP 20, 361 (1965).
44. S.E. Kupriyanov, Sov. Phys. - JETP 21, 311 (1965).
45. D.W.O. Heddle, and R.G.W. Keesing, Adv. At. Molec. Phys. 4, 267 (1968).
46. S.E. Kupriyanov, Sov. Phys. - JETP 21, 311 (1965).
47. J.A. Schiavone, K.C. Smyth and R.S. Freund, Preprint, 1975.
48. K.C. Smyth, J.A. Schiavone, and R.S. Freund, Jour. Chem. Phys. 59, 5225 (1973).
49. C.E. Fairchild, H.P. Garg and C.E. Johnson, Phys. Rev. A 8, 796 (1973).
50. W.C. Wells, W.L. Borst and E.C. Zipf, J. Geophys. Res. 77, 69 (1972).
51. S.E. Kupriyanov, Sov. Phys. JETP 28, 240 (1969).
52. R.S. Freund, Jour. Chem. Phys. 54, 3125 (1971).
53. K.C. Smyth, J.A. Schiavone and R.S. Freund, Jour. Chem. Phys. 60, 1358 (1974).
54. J.S. Foster, Proc. Roy. Soc. (London) 117, 137 (1928).
55. P.M. Koch, L.D. Gardner and J.E. Bayfield, paper presented at the Fourth Int. Conf. on Beam-Foil Spectroscopy, Gatlinburg, Tennessee, 1975), to be published.
56. P.M. Koch, L.D. Gardner and J.E. Bayfield, contributed paper no. 473 to this conference.
57. P.M. Koch, and J.E. Bayfield, Abst. Fourth Int. Conf. on Atomic Physics, University of Heidelberg, Germany, 1974, pp. 336-337.

# SECONDARY ELECTRONS
# FROM ELECTRON AND ION IMPACT

# BASIC ASPECTS OF SECONDARY-ELECTRON DISTRIBUTIONS[*]

Yong-Ki Kim

Argonne National Laboratory

Argonne, Illinois 60439

## Abstract

Graphical methods proposed by Platzman and by Fano are applied to the analysis of the basic features observed in secondary electron spectra. These methods are useful not only in checking the consistency of experimental data, but also in extrapolating the spectra to the range of primary- and secondary-electron energies not covered by experiments. Illustrative examples are presented for He, Ne, and NO.

## Introduction

More than four decades after a pioneering experiment by Mohr and Nicoll,[1] there are now several experimental groups who have succeeded in producing extensive data on angular and energy distributions of secondary electrons ejected from gases by fast electrons and protons.[2-7] The secondary-electron data are essential in the study of energy deposition by energetic charged particles both in radiation and atmospheric research. (We use the customary definition of secondary electrons, i.e., when the incident particle is an electron, the slower of the two electrons that emerge after the collision is called secondary. The faster one is called primary.)

---

[*]Work performed under the auspices of the U.S. Energy Research and Development Administration.

In this report I shall review simple but powerful graphical methods that enable one to (a) check the consistency of
existing experimental data, (b) extrapolate to the primary- and
secondary-electron energies not covered by experiments, (c)
renormalize the data, if necessary, to be consistent with known
total ionization cross sections, (d) illustrate the similarities
and differences between the secondary electrons produced by
electron and proton impact, and finally (e) identify areas where
more experimental data are necessary.

For a slow incident particle, the colliding particles form
a transient compound state, and secondary-electron spectra from
such a state are strongly correlated to the corresponding distributions of the scattered particle and the resulting ionic state.
On the other hand, the action of a fast charged particle on an
atom can be treated as an impulse to the target atom, and then
the response of the target atom can be described as a function
of the impulse only. The first Born approximation is based on
such a model, and is valid for the description of the interaction
of a charged particle with an atom when the velocity of the incident particle is far greater than that of the atomic electron involved in the collision.[8,9]

The graphical method presented here originated from the
Born approximation, and hence it is best suited to analyze the
secondary-electron data produced by fast charged particles.
However, the same method can still provide useful guidelines
on the analysis of the data produced by intermediate to slow
incident particles when reliable measurements are available on
related quantities such as the total ionization cross section.

My discussions are limited to the angular and energy
distributions of secondary electrons produced by direct ionization. The energy loss and the scattering angle of the primary
particle are assumed to have been integrated. Also, the present
method is valid only when the incident particle can be taken as
structureless, e.g., electrons, protons, and α particles. After
a brief discussion of angular distribution, I shall illustrate the
application of the graphical method to the energy distribution
of He, Ne, and NO.

## Angular Distribution

The collisions of fast charged particles with an atom
can be qualitatively divided into two classes: "soft" collisions
with small momentum transfers, and "hard" collisions with large

momentum transfers. The former leads to an angular distribution similar in shape to that of photoelectrons ejected by unpolarized light,[10,11] whereas the latter produces a distribution
sharply peaked around a binary-collision direction determined
from the energy-momentum conservation laws. Actual angular
distribution $d^2\sigma/dWd\Omega$ (doubly differential cross section) with
secondary-electron energy W and its solid angle $d\Omega$ is a mixture of these two types of distribution. The binary peak dominates the angular distribution of fast secondaries, and for slow
secondaries, the photoelectron-like (dipole) angular distribution (see Eq. 1) is more significant than the hard-collision contribution. The dipole contribution to the angular distribution
can be separated from the rest[11] by using the Fano plot,[12]
i.e., a plot of the doubly differential cross section times the
incident energy $T_0$ versus the logarithm of the incident energy
in appropriate units. According to the Born approximation,[11]
such a plot should approach a straight line for high incident
energies with a slope $A(W,\theta)$ for a given secondary-electron
energy W and ejection angle $\theta$ (measured from the incident
beam direction). The slope is expressed in terms of the continuum dipole oscillator strength $df/dE$ and the asymmetry parameter $\beta(E)$:

$$A(W,\theta) = \sum_j \frac{1}{4\pi} \frac{df_j}{dE_j} [1 - \tfrac{1}{2}\beta(E_j) P_2(\cos\theta)] (R/E_j), \quad (1)$$

where $E_j$ is the photon energy required to eject a photoelectron
of kinetic energy W with the ion left in the jth state, and $P_2$
is the Legendre polynomial of the second order. According to
Eq. (1), the slope should be the same for a pair of supplementary angles (30° and 150°, 60° and 120°, etc.) because $P_2(\cos\theta)$
is symmetric with respect to $\theta = 90°$.

The electron-impact data by Opal et al.[6] in Fig. 1 for
He$(O, \triangle)$ show that the forward-angle ($\theta < 90°$) data do not exhibit a gradual approach to the Born limit as expected. Examination of the energy distribution by the graphical method to be
discussed later, indicates that the experimental values at
T = 200, 300, and 500 eV should be reduced by 26, 24, and 20%,
respectively. The corrections for the data at $T_0$ = 1 and 2 keV
were minor. These corrections are within the error limits ($\pm$ 25%).
The renormalized data ($\bullet$, $\blacktriangle$) clearly show a better approach toward the theoretical slope (solid line) which is based on the
$df/dE$ calculated by Jacobs from correlated wavefunctions.
Equation (1) is one of the examples where photoelectron data

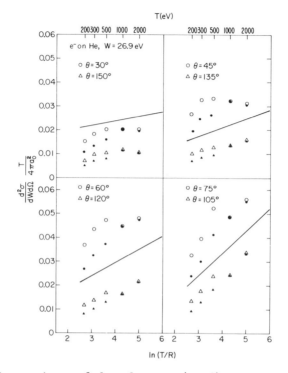

FIG. 1.--Comparison of the slopes A(W, θ) in the Fano plot for the angular distribution data by Opal et al.[6] on He. The open circles and triangles are the original experimental data, and they are expected to approach the theoretical slope (solid lines) by Jacobs[21] at high incident electron energies T. (The heights of the solid lines are arbitrary.) The filled circles and triangles are the same data renormalized to be consistent with the recommended energy distribution shown in Fig. 5. The kinetic energy W of the secondary electron is 26.9 eV.

can be directly related to secondary-electron data through the Born approximation.

Another feature of the angular distribution that is related to the energy distribution concerns positively charged, heavy incident particles. The positive particles literally drag out electrons in the forward direction through Coulomb attraction. This effect is most conspicuous for secondary electrons with similar velocity as that of the incident particle. For example, a 300 keV proton moves as fast as a 163 eV electron, and the angular distribution of ejected electrons with kinetic energies near 163 eV show a sharp increase in the forward direction.[13]

Theoretically, this process can be described as an electron capture by the proton forming a continuum state of the hydrogen atom,[14] but theoretical methods proposed so far[14-16] agree with experiment only qualitatively.

In recent years, many workers calculated angular distributions for He,[5b,17-22] mostly using the Born approximation. As expected, when good wavefunctions are used, these calculations agree well with experiment with fast incident particles.

## Energy Distribution

Two decades ago, Platzman used a graph (in a thesis by his student[23]) which turned out to be a powerful tool in analyzing the energy distribution $d\sigma/dW$ (singly differential cross section). The Platzman plot uses the ratio of measured $d\sigma/dW$ to the corresponding Rutherford cross section as the ordinate

$$Y(T,E) = \frac{d\sigma}{dW} \frac{TE^2}{4\pi a_0^2 z^2 R^2} \tag{2}$$

where $ze$ is the charge of the incident particle, $a_0 = 0.529$ Å, $R = 13.6$ eV, and $T = \frac{1}{2}mv^2$ with the electron mass $m$ and the incident-particle velocity $v$. Note that $T$ is the same as the incident energy $T_0$ for electrons and positrons, but $T = (m/M)T_0$ for heavy particles of mass $M$. The abscissa of the Platzman plot is the inverse of the energy transfer defined in terms of the lowest ionization potential $B_1$ of the target atom:

$$E = W + B_1 . \tag{3}$$

The last factor on the right-hand side of Eq. (2) is the inverse of the Rutherford formula slightly modified to account for the fact that the target electron is bound.[24]  Fast secondaries are ejected by the hard collisions which are well described by the Rutherford formula.  Hence, for fast secondaries, $Y(T,E)$ should approach the number of atomic electrons participating in the ionizing collision.  This number is of the order of valence orbital occupation numbers.

The abscissa $(E^{-1})$ was chosen so that the total area under the Platzman plot would be proportional to the total ionization cross section $\sigma_i$, i.e.,

$$\sigma_i(T) \frac{T/R}{4\pi a_0^2 z^2} = \int_{X_1}^{X_0} Y(T,E) \, d(R/E) , \tag{4}$$

where $x_1 = R/(W_{max} + B_1)$ and $x_0 = R/B_1$. For incident electrons, the upper limit for the secondary-electron energy $W_{max} = \frac{1}{2}(T - B_1)$, and for other particles, $W_{max} = T_0$. Thus, when reliable $\sigma_i$ is known, which is often the case, the Platzman plot is convenient for normalizing the energy distribution.

For the normalization of $d\sigma/dW$, however, its shape must be known for all values of W. As will be shown later, the lack of reliable experimental data for low W ($\lesssim 20$ eV) poses a greater problem than that for high W ($\gtrsim 200$ eV), because a large fraction of $\sigma_i$ comes from slow secondaries. Besides, $d\sigma/dW$ for fast secondaries can be estimated well by using one of the hard collision formulas[24] (Rutherford, Mott, or binary encounter theory). To supplement this deficiency, we once again turn to the Born approximation:[24]

$$\left(\frac{d\sigma}{dE}\right)_{Born} = \frac{4\pi a_0^2 z^2}{T} \left\{ A(E) \ln[4TRC(E)/E^2] + B(T,E) \right\},$$

(5)

where the first term in the braces represents the effects of soft collisions, and $B(T,E)$ those of hard collisions. The functions $A(E)$ and $C(E)$ depend only on the properties of the target atom, and $A(E)$ is defined in terms of the dipole oscillator strengths for ionization $df/dE$:

$$A(E) = (df/dE)(R^2/E).$$

(6)

For slow secondary electrons, the soft-collision contribution dominates. Hence, for fast incident particles, the shape of the Platzman plot for slow secondaries is expected to resemble that of the corresponding optical function

$$F(E) = A(E)(E/R)^2 = E(df/dE).$$

(7)

When necessary oscillator strengths are known, therefore, the shape of $F(E)$ should be a good guide for extrapolating $d\sigma/dW$ for slow secondaries. This is one of the most useful advantages of the Platzman plot.

Formulas introduced so far did not refer to inner orbitals. For instance, the energy transfer defined by Eq. (3) is not adequate for the electrons ejected from inner orbitals. For brevity, however, we shall still use E defined by Eq. (3) for atoms and molecules more complex than He and $H_2$. When ionization potentials of the inner orbitals are well separated from that of the outermost one and the dipole oscillator strengths for the production of slow secondaries from the inner orbitals are small (e.g., Ne and Ar), then most features of the Platzman plot described above should remain unaffected. On the other hand, we will see that some features of the Platzman plot are seriously affected when there are several alternative ionic states which are close in ionization potentials and competitive in partial cross sections (e.g., $N_2$ and NO).

In Fig. 2a, we present the Platzman plot of the electron-impact data on Ne by Opal et al.[6] at T = 500 eV, and in Fig. 2b the corresponding optical function F(E) based on the photoabsorption data compiled by Berkowitz.[25] [For most atoms, ionization efficiency is unity and photoabsorption data may be used instead of photoionization data. In molecules, however, ionization efficiency is often substantially lower than unity, and F(E) must be evaluated from photoionization data.]

Comparison of the Opal data[6] and F(E) confirms that the shape of the electron-impact data is basically correct. The dip near R/E = 0.07 in the electron-impact data is a consequence of the exchange effect (indistinguishability of the scattered and ejected electrons). In Fig. 2a, the same effect is clearly demonstrated in a Mott cross section[24] curve marked M, modified to account for the ejection of bound electrons. The curve marked R is the Rutherford cross section.[24] Grissom, Compton, and Garrett[26] measured the cross section for the ejection of secondary electrons with $W \approx 0$ by T = 500 eV electron. The same cross section has been calculated by extrapolating Born cross sections for discrete excitations to the ionization limit.[27] Although this cross section provides only one point in the Platzman plot, it is an important one because it reduces the uncertainty in the normalization of $d\sigma/dW$.

The chained curve marked $e^-$ in Fig. 2a represents the expected cross section for electron-impact data renormalized to agree with $\sigma_i = 0.667\ \pi a_0^2$ at T = 500 eV measured by Rapp and Englander-Golden.[28] Although the difference between the renormalized data and the original experimental data is $\sim 16\%$, well within the experimental error limits of $\pm\ 25\%$, the renormalized data are not only consistent with well-known values of $\sigma_i$, but also exhibit a shape consistent with other related experimental and theoretical data for all values of W. This is a necessary step before we can extrapolate the energy distribution to other incident energies. The contribution of fast Auger electrons to $\sigma_i$ is insignificant as is evident from Fig. 2a.

As is well known,[8,9] the Born cross sections for total ionization by fast electrons and protons of the same velocity are almost the same. The difference comes from different kinematic

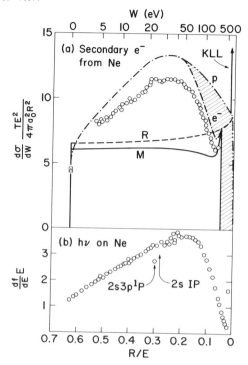

FIG. 2.--(a) The Platzman plot of secondary electrons from Ne. Circles are experimental data by Opal et al.[6] with 500 eV incident electrons. Chained curve marked e⁻ is renormalized energy distribution for 500 eV incident electrons, and that marked p is for a 0.92 MeV proton whose velocity is the same as that of a 500 eV electron. The dashed curve marked R represents the Rutherford cross section, and the solid curve marked M the Mott cross section.[24] The square and triangle at ejected electron energy W = 0 correspond to the values from the experiment by Grissom et al.[26] and the theory by Kim et al[27]. The shaded area indicates the difference in $\sigma_i$ for a 500 eV incident electron and a 0.92 MeV proton. The KLL Auger electrons should appear as a sharp peak marked KLL. (b) Optical function F(E) defined by Eq. (7) for Ne, derived from photoabsorption data compiled by Berkowitz.[25] A small fraction of F(E) to the right of the 2s ionization potential (IP) should be shifted to the region of low secondary-electron energy to construct shifted F(W) defined by Eq. (8).

limits in energy and momentum transfers, as marked by the
hatched area in Fig. 2a. With the aid of the Rutherford cross
section for fast secondaries and ignoring the contribution of KLL
Auger electrons, we can easily construct an approximate energy
distribution for $T = 500$ eV protons ($T_0 = 0.92$ MeV). The recommended curve for 0.92 MeV proton is marked p in Fig. 2a. The
area under the chained curve for the proton leads to $\sigma_i = 0.74\,\pi a_0^2$
compared to $\sigma_i = 0.72\,\pi a_0^2$ obtained from the experimental data
by Hooper et al.[29]

An important implication of the Ne data in Fig. 2 is that
the energy distribution seems to be roughly the sum of the hard-
collision contribution (Rutherford or Mott) and $F(E)$ scaled with
some slowly varying function of E [as expected from the Born
formula, Eq. (5)]. The fact that $d\sigma/dW$ at $W = 0$ goes below
the Mott and Rutherford cross sections implies that the hard-
collision contribution must also be scaled down for slow secondaries. This approach has been used extensively in constructing $d\sigma/dW$ by Miller[23] (without scaling the hard collision
contribution), Khare and co-workers,[30] and recently by Eggarter[31]
and Gerhart.[32]

When there are several ionic states with ionization potentials close to that of the lowest ionic state, the shape of $Y(T,E)$
for slow secondaries ($W \lesssim 100$ eV) is distorted from that of $F(E)$
defined by Eq. (7) as a function of the absorbed photon energy
$E = h\nu$. To compare the optical data with $Y(T,E)$, which is defined as a function of the ejected electron energy W, we must
partition $F(E)$ according to a given W. This is done by defining
a "shifted" optical function in terms of partial oscillator
strengths $df_j/dE_j$:

$$F(W) = (E_1/R)^2 \sum_j \frac{R^2}{E_j} \frac{df_j}{dE_j} , \qquad (8)$$

where the summation is over all energy transfers $E_j = W + B_j$
corresponding to alternative ionic states with ionization
potentials $B_j$. From available experimental partial cross
sections[33-35] and the photoionization cross sections compiled by Berkowitz,[25] we constructed a shifted optical function[36] $F(W)$ for NO. When we combine the shape of $F(W)$ in
Fig. 3b with the Mott cross section in Fig. 3a, the shape of
the electron-impact data by Opal et al.[6] is well reproduced,
in contrast to the "unshifted" $F(E)$.

For multiple ionization, the corresponding partial cross
section has to be multiplied by the ionicity because the integral
of $d\sigma/dW$ would lead to a gross ionization cross section

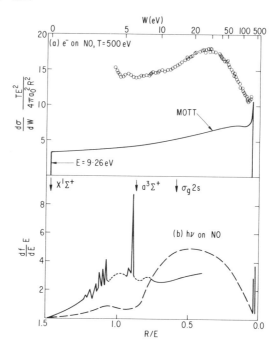

FIG. 3.--(a) The Platzman plot of secondary electrons from NO. Circles are experimental data by Opal et al.[6] with 500 eV incident electrons. The solid curve is the Mott cross section.[24] The lowest energy loss E is 9.26 eV. (b) Shifted optical function $F(W)$ defined by Eq. (8) (solid curve) and unshifted $F(E)$ defined by Eq. (7) (long dashed). The curves are based on the photoionization cross section compiled by Berkowitz[25] and experimental partial cross sections in the literature.[33-35] For clarity, auto-ionization peaks have been omitted in the dotted parts of $F(W)$ and also in $F(E)$. Some ionization potentials for states of $NO^+$ are indicated by arrows.

(= number of electrons produced). We have neglected multiple ionization in the shifted $F(W)$ for NO in Fig. 3b. Note that the areas under the optical functions $F(E)$ and $F(W)$, when multiple ionization is neglected, should be the same. The area, in fact, is the slope of the Fano plot for so-called counting ionization cross section (= number of ionizing events). The integral of $F(E)$ based on the same optical data have been reported earlier.[37]

The electron-capture into continuum states by protons, which was mentioned earlier, can be identified easily by using the Platzman plot.[38] The proton-impact data on $N_2$ by Crooks and Rudd[3] are presented in Fig. 4. We see clearly there the peaks corresponding to continuum-captured electrons with velocities similar to those of the incident protons. Furthermore, the importance of the continuum-capture contribution to $\sigma_i$ decreases as the proton energy is decreased. The continuum-capture peaks are hardly discernible in the Platzman plots of the fast proton ($T_0 \gtrsim 1$ MeV) data.[4,5b]

## Extrapolation to Higher Incident Energies

From Eq. (5), we see that the Fano plot $[(d\sigma/dE)(T/4\pi a_0^2 z^2)$ vs. $\ln T]$ for a given E will approach for high T a straight line with slope A(E). If we make the Fano plot of Y(T,E) defined by Eq. (2), at a fixed W, the asymptotic slope will be F(W) defined by Eq. (8). Thus, when some experimental data on $d\sigma/dW$ are available at various values of T, we can extrapolate $d\sigma/dW$ at a given W as a function of T. With a three-dimensional plot of Y(T,E) as a function of $E^{-1}$ in one direction (the Platzman plot), and as a function of $\ln T$ in another direction (the Fano plot), we can determine a set of $d\sigma/dW$ that satisfy consistency requirements such as (a) correct $\sigma_i$,

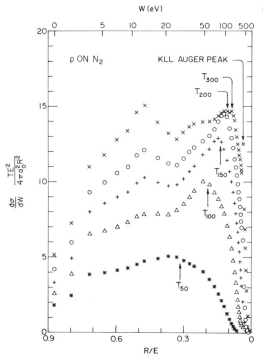

( Reproduced from Ref. 38 with permission of the publisher)

FIG. 4.--Continuum-capture peaks in the Platzman plot of the proton-impact data by Crooks and Rudd.[3] The arrow marked $T_{300}$ indicates the kinetic energy of an ejected electron which is moving with the same speed as that of a 300 keV proton. The incident proton energies are 50 keV (*), 100 keV ($\triangle$), 150 keV (+), 200 keV (O), 300 keV (X).

(b) correct slope for high T, and (c) smooth transition to the low T region where the resemblance to the shape of the optical function F(W) disappears. In addition to the optical data for F(W), the availability of trustworthy $\sigma_i$ is critical for the extrapolation along the values of T. For this purpose, the Fano plot again proves to be a powerful tool[9,37,38] for selecting trustworthy $\sigma_i$.

In Fig. 5 we show an example of the recommended electron-impact spectra for He constructed by the three-dimensional graphical method.[38] The experimental data by Grissom et al.[26] and theoretical data[39] at W = 0 were essential in the determination of the shape for low W < 10 eV. The shaded area in Fig. 5 shows the proportion of secondary electrons too slow to ionize again. In comparison with Fig. 5, we found that the Opal data[6] on angular distribution at W = 26.9 eV and T = 200, 300, and 500 eV should be reduced as indicated in Fig. 1.

A similar extrapolation by graphical method[38] has been done for $N_2$, but in view of the previous discussion on shifted F(W), we must modify the shape of Y(T,E) for slow secondaries (W $\lesssim$ 5 eV). A preliminary result[36] indicates that there should be a prominent peak in $d\sigma/dW$ near W = 1.1 eV, in addition to many autoionization peaks.

## Conclusion

I have demonstrated that simple graphical methods can be powerful in exposing various features in the angular and energy distributions of secondary electrons. The secondary electron spectra can be corrected, renormalized, and extrapolated to exhibit certain characteristics expected from theory. For this purpose, we need experimental data on (a) $\sigma_i$ on absolute scale, (b) df/dE for ionization on absolute scale, (c) partial $df_j/dE_j$ on relative scale, particularly for $h\nu \lesssim$ 100 eV, and (d) $d\sigma/dW$ on relative scale at various incident energies (T $\lesssim$ 5 keV). More specific description of desirable experiments are given elsewhere.[24,38]

On the theoretical side, it is desirable to find a procedure better than the empirical approach used so far[30-32] for the evaluation of $d\sigma/dW$ as scaled sums of the soft- and hard-collision contributions.

The large volume of data on secondary electrons[3-7] can now be scrutinized systematically, and reliable cross sections can be filtered out by the graphical method. Such

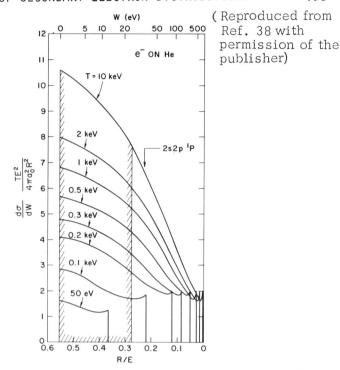

(Reproduced from Ref. 38 with permission of the publisher)

FIG. 5.--The Platzman plot of the recommended energy distributions of secondary electrons from He by electron impact. The shaded area corresponds to secondary electrons too slow to ionize further.

data would remove a major obstacle in the study of energy degradation in gases.

## Acknowledgements

I am very grateful to Dr. H. C. Tuckwell for his assistance in the preparation of materials presented here, particularly on NO.

## References

1.  C. B. O. Mohr and F. H. Nicoll, Proc. Roy. Soc. (London) A144, 596 (1934).
2.  H. Ehrhardt, K. H. Hesselbacher, K. Jung, E. Schubert, and K. Willmann, J. Phys. B: Atom. Molec. Phys. 7, 69 (1974), and references therein. See also, D. A. L. Paul, this symposium.

3.  J. B. Crooks and M. E. Rudd, Phys. Rev. A $\underline{3}$, 1628 (1971) and references therein.

4.  L. H. Toburen, Phys. Rev. A $\underline{9}$, 2505 (1974) and references therein.

5.  (a) N. Stolterfoht, Z. Physik $\underline{248}$, 81 (1971); $\underline{248}$, 92 (1971). (b) S. Manson, L. H. Toburen, D. H. Madison, and N. Stolterfoht, Phys. Rev. A $\underline{12}$, in press.

6.  C. B. Opal, E. C. Beaty, and W. K. Peterson, Atom. Data $\underline{4}$, 209 (1972), and references therein.

7.  N. Oda, in Invited Lectures and Progress Reports, VIII ICPEAC, Beograd. Eds. B. C. Čobić and M. V. Kurepa (Institute of Physics, Beograd, 1973), p. 443 and references therein.

8.  H. Bethe, Ann. Physik $\underline{5}$, 325 (1930).

9.  M. Inokuti, Rev. Mod. Phys. $\underline{43}$, 297 (1971).

10.  H. Bethe, in Handbuch der Physik, Vol. 24 (Springer, Berlin, 1933), p. 273. See Sec. 55 in particular.

11.  Y.-K. Kim, Phys. Rev. A $\underline{6}$, 666 (1972).

12.  U. Fano, Phys. Rev. $\underline{95}$, 1198 (1954).

13.  G. B. Crooks and M. E. Rudd, Phys. Rev. Letters $\underline{25}$, 1599 (1970).

14.  J. Macek, Phys. Rev. A $\underline{1}$, 235 (1970).

15.  A. Salin, J. Phys. B: Atom. Molec. Phys. $\underline{5}$, 979 (1972).

16.  Y. B. Band, J. Phys. B: Atom. Molec. Phys. $\underline{7}$, 2557 (1974).

17.  C. E. Kuyatt and T. Jorgensen, Jr., Phys. Rev. $\underline{131}$, 666 (1963).

18.  T. F. M. Bonsen and L. Vriens, Physica $\underline{47}$, 307 (1970).

19.  W. J. B. Oldham, Jr. and B. P. Miller, Phys. Rev. A $\underline{3}$, 942 (1971), and references therein.

20.  D. H. Madison, Phys. Rev. A $\underline{8}$, 2449 (1973), and references therein.

21.  V. Jacobs, Phys. Rev. A $\underline{10}$, 499 (1974).

22.  W. D. Robb, S. P. Rountree, and T. Burnett, Phys. Rev. A $\underline{11}$, 1193 (1975). This and Ref. 21 discuss triply differential cross section, but it can be converted to the singly differential cross section after appropriate integrations.

23.  W. F. Miller, Ph.D. Thesis, Purdue University, 1956.

24.  Y.-K. Kim, Radiat. Res. $\underline{61}$, 21 (1975).

25.  J. Berkowitz, private communication. The oscillator strengths are mostly based on available experimental data but have been readjusted to satisfy known sum rules to 10%.

26. J. T. Grissom, R. N. Compton, and W. R. Garrett, Phys. Rev. A 6, 977 (1972).
27. Y.-K. Kim, M. Inokuti, and R. P. Saxon, Abstracts of Papers, VIII ICPEAC, Beograd, Eds. B. C. Čobić and M. V. Kurepa (Institute of Physics, Beograd, 1973), p. 688.
28. D. Rapp and P. Englander-Golden, J. Chem. Phys. 43, 1464 (1965).
29. J. W. Hooper, D. S. Harmer, D. W. Martin, and E. W. McDaniel, Phys. Rev. 125, 2000 (1962).
30. R. Shingal, B. B. Srivastava, and S. P. Khare, J. Chem. Phys. 61, 4656 (1974) and references therein.
31. E. Eggarter, J. Chem. Phys. 62, 833 (1975).
32. D. E. Gerhart, J. Chem. Phys. 62, 821 (1975).
33. J. L. Bahr, A. J. Blake, J. H. Carver, J. L. Gardner, and V. Kumar, J. Quant. Spectrosc. Radiat. Transfer 12, 59 (1972).
34. J. A. R. Samson, Phys. Letters 28A, 391 (1968).
35. O. Edqvist, E. Lindholm, L. E. Selin, H. Sjögren, and L. Åsbrink, Ark. Fysik 40, 439 (1970).
36. H. C. Tuckwell and Y.-K. Kim, to be published.
37. J. Berkowitz, M. Inokuti, and J. C. Person, Abstracts of Papers, VIII ICPEAC, Beograd, Eds. B. C. Čobić and M. V. Kurepa (Institute of Physics, Beograd, 1973), p. 561.
38. Y.-K. Kim, Radiat. Res., in press.
39. Y.-K. Kim and M. Inokuti, Phys. Rev. A 7, 1257 (1973).

# RESONANCES SEEN IN SECONDARY ELECTRON SPECTRA

W. Mehlhorn

Fakultät für Physik, Universität Freiburg

78 Freiburg, W.-Germany

## INTRODUCTION

Resonances which can be studied in secondary electron spectra are quasi-discrete states of a system embedded in the continuum of the next higher charge state of the system. They decay predominantly or exclusively through electron emission. Depending on whether the resonance is due to inner shell ionization or excitation of the neutral or negative system we speak of Auger, autoionizing or autodetaching states. The experimental procedure and results of autodetaching states will be discussed by Edwards[1]. Also the negative resonances, suggested by Taylor and Yaris[2], to explain the near threshold excitation phenomena of autoionizing states by electron impact[3] are discussed elsewhere[4]. Experimentally the resonances are excited by collision of charged beam particles B (e, $H^+$, heavy ions) with target gas atoms T. In the case of ion-atom collisions resonances of the target as well as of the beam are excited, the ejected electrons of the latter are Doppler-shifted in energy.

Examples of Auger transitions are:

1. Single inner shell ionization of the target atom T

$$B + T \rightarrow T^+ + e + B$$

$$\begin{array}{l} \rightarrow T^{++} + e \qquad \text{diagram Auger transition} \\ \rightarrow T^{++*} + e \\ \rightarrow T^{+++} + 2e \end{array} \Big\} \quad \text{double Auger transitions}$$

2. Multiple initial ionization of the target atom T

$$B + T \rightarrow T^{n+} + n \cdot e + B$$

$$\rightarrow T^{(n+1)+} + e \quad \text{satellite Auger transition}$$

    3. Single and multiple ionization of the beam atom B
    leading to corresponding Auger transitions of B.

Examples of excitations leading to autoionizing transitions are:
    1. Inner shell excitation of the target atom T,
    2. Double outer shell excitation of the target atom T,
    3. Excitation of the beam atom B.

From the energies, intensities and line widths of Auger
and autoionizing spectra one gets information on the energies
of resonance states, the cross sections of the various excitation processes leading to the resonance states and the relative and total decay probabilities into final states.

Instead of discussing the physics of resonances in
general (see e.g. the reports by Moore[5] on Auger transitions
and by Morgenstern[6] and Ogurtsov[7] on autoionizing transitions
resulting in heavy ion-atom collisions) I will concentrate
in the following on two selected topics where I feel important
progress has been made.

## CORRELATION EFFECTS IN THE DECAY PROCESS OF AUGER STATES

    Correlation effects manifest themselves through deviations
between experimental and theoretical decay probabilities
where the latter have been calculated with independent particle
HF wave functions. In earlier reports[8] it has been shown that
the relative decay probabilities of Auger states into the
various final states depend rather sensitively on electron
correlation when the outermost or the next inner shell is
involved. This can be best seen using the KLL transitions as
example. In the meantime decisive progress has been made. On
the experimental side: In the region of atomic number Z where
correlation effects are expected to be large, the KLL spectra
of free Na[9], Mg[10] and Ar[11] atoms have been measured. On the
theoretical side: New transition probabilities using HFS wave
functions have been calculated by Bhalla et al.[12] and Crasemann
et al.[13] for $Z \leq 54$ and many body perturbation theory was
introduced by Kelly[14] to the KLL transitions of neon. In Fig. 1
the experimental relative group intensities $I(KL_1L_1)/I(KLL)$,
$I(KL_1L_{2,3})/I(KLL)$ and $I(KL_{2,3}L_{2,3})/I(KLL)$ are compared to
theoretical HFS values of Bhalla et al.[12] (dashed curves),
the values of Crasemann et al.[13] agree with those of Bhalla
et al.[12] within the widths of the curves. The HFS values of
McGuire[15] have not been included for comparison because of
their unrealistic oscillations for smaller Z. Introduction of
CI between final configurations $2s^{-2}$ and $2p^{-2}$ improves the
agreement considerably (solid curves), but only the results

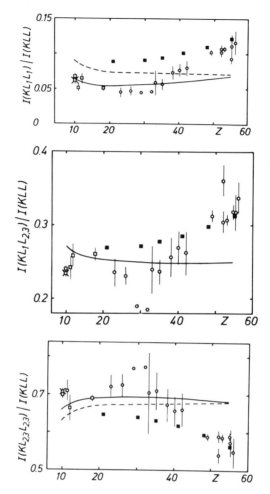

Fig. 1. Comparison of experimental intensities $I(KL_1L_1)/I(KLL)$, $I(KL_1L_{2,3})/I(KLL)$ and $I(KL_{2,3}L_{2,3})/I(KLL)$ with theory. Experiment: ⌀ = solid state target, ⌀ gaseous atomic target. Nonrelativistic theory: ----- = Bhalla et al.[12] without CI, ——— = Bhalla et al.[12] with CI, x = Kelly[14] (Z = 10). Relativistic theory: ■ = Bhalla and Ramsdale[16] without CI.

of Kelly's many body calculation[14] are in perfect agreement with the experiment. For $Z \leq 18$ only the experimental results obtained with gaseous atomic targets are shown in Fig. 1. Systematic deviations at higher Z are clearly due to relativistic effects[16].

The occurence of double Auger transitions is due either

to electron correlation or to shake transitions. From the results of discrete double Auger transitions in noble gases[17-19] and free Na[9] and Mg[10] atoms the following rule seems to be valid (see table 1): In case the electrons, involved in the double Auger process, are from the same outer shell, electron correlation plays a dominant role. On the other hand, if the excited electron in the double Auger transition stems from a different shell, then shake theory accounts for the larger part of the measured intensity. This rule is also valid for double Auger transitions with two ejected electrons[20].

Table 1. Intensities of discrete double Auger transitions relative to parent transition (P) in per cent and comparison to shake theory.

| Element | Transition | Exp.Int. | Shake Theory | Ref. |
|---------|-----------|----------|--------------|------|
| Kr | $3d^{-1}-4s^2 4p^3 \begin{smallmatrix} ns \\ nd \end{smallmatrix}(^1P)$ | 40-50 | 0 | 17-19 |
|    | P: $3d^{-1}-4s4p^5(^1P)$ | | | |
| Mg | $1s^{-1}-2s^2 2p^4(^1D)3s4s$ | 7.1 (8) | 4.3 | 10 |
|    | P: $1s^{-1}-2s^2 2p^4(^1D)3s^2$ | | | |
| Na | $1s^{-1}-2s^2 2p^4(^1D)4s$ | 7.8 (8) | 4.7 | 9 |
|    | P: $1s^{-1}-2s^2 2p^4(^1D)3s$ | | | |

## SHAPE AND INTENSITY OF RESONANCES

Autoionizing states in ejected electron spectra have in most cases asymmetric line shapes. Shape parameters and intensities as function of ejection angle $\theta$ and impact energy $E_o$ have been measured for the $2s^2$ $^1S$, $2s2p$ $^1P$ and $2p^2$ $^1D$ states of He for electron impact[21-25], proton impact[26-30], He$^+$ and H$_2$ impact[27]. Balashov et al.[31] and Lipovetsky and Senashenko[32,33] applied Fano's theory[34] of interference between discrete and continuum states to the case where the resonance is observed in the ejected electron spectrum. They obtained the result[32]

$$\frac{d^2\sigma}{dEd\Omega} = f(E_o,\theta) + \frac{a(E_o,\theta)\epsilon + b(E_o,\theta)}{\epsilon^2 + 1} , \qquad (1)$$

with $\epsilon = (E - E_r)/\tfrac{1}{2}\Gamma$.

The first term $f(E_0,\theta)$ describes the angular and energy distribution of ejected electrons from the direct ionization, $a(E_0,\theta)$ characterizes the resonance asymmetry, and $b(E_0,\theta)$ is the resonance yield describing the angular distribution of electrons from autoionization. The parameters a and b are given in first Born approximation by

$$a(E_0,\theta) = 8a_0^2 \frac{k}{k_0} \int \frac{d\Omega_k}{Q^4} \text{ Re } [t^*(\vec{k}',\vec{Q})t^L(\vec{k}',\vec{Q})(q(Q)-1)]$$

$$b(E_0,\theta) = 4a_0^2 \frac{k}{k_0} \int \frac{d\Omega_k}{Q^4} \{(q^2(Q)+1)|t^2(\vec{k}',\vec{Q})|^2 \qquad (2)$$

$$+ 2 \text{ Im } [t^*(\vec{k}',\vec{Q})t^L(\vec{k}',\vec{Q})(q(Q)-1)]\}.$$

They depend on the interference of the amplitude $t^L(\vec{k}',\vec{Q})(q(Q)-i)$, given by the interaction of the Lth component of the continuum state with the discrete state, with all amplitudes $t = \Sigma\, t^\ell$ for direct ionization. Although eq. (1) is the natural result for the theory of ejected electrons, the parametrization of the experimental spectra of ejected electrons is generally performed using equation

$$\frac{d^2\sigma}{dEd\Omega} = f_{backg}(E_0,\theta) + f_{res}(E_0,\theta) \frac{(\tilde{q}(E_0,\theta) + \epsilon)^2}{1 + \epsilon^2}, \quad (3)$$

which is formally equivalent to Fano's formula[34] for the resonance structure of scattered electrons. The Fano parameters $f_{backg}$, $f_{res}$ and $\tilde{q}$ are functions of f, a and b of eq. (1). From eq. (1) and (2) it can be seen that the general resonance structure in the ejected electron spectra is much more complicated than in the scattered particle spectra, where interference occurs only between the discrete state amplitude and that part of the direct amplitude having the same symmetry. Only for the case of direct amplitude to be $t \equiv 0$ eq. (1) breaks down to yield a simple angular distribution

$$\frac{d^2\sigma}{dEd\Omega} = I(E_0,\theta)/\frac{\Gamma}{2\pi}(1 + \epsilon^2). \quad (4)$$

This case we will meet for electrons ejected from Auger states.

Examples of different resonance shapes in the ejected electron spectra from He resonances excited by 100 keV protons[30] and 400 eV electrons[25] are shown in Fig. 2. The deduced parameters a and $\tilde{q}$ and the yield b of the $2s^2\ ^1S$ resonance for 100 keV protons are given in Fig. 3.

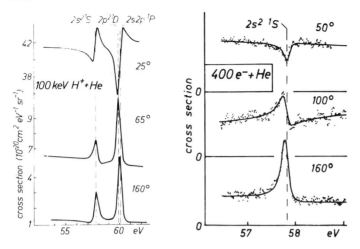

Fig. 2. Examples of ejected electron spectra from He resonances excited by 100 keV protons (left side) and 400 eV electrons (right side).

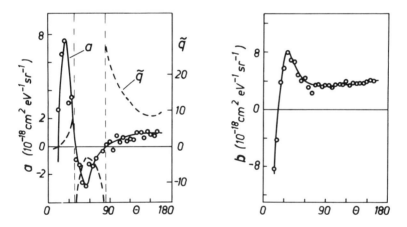

Fig. 3. Parameters a, $\tilde{q}$ and b of the $2s^2$ $^1S$ resonance of He for 100 keV protons.

The most prominent feature of the shape parameter $\tilde{q}$ are the singularities $\tilde{q} = \infty$. These correspond theoretically to $a = 0$, $b > 0$, and experimentally to symmetric line shapes with positive yields. Symmetric line shapes with negative yields (window shape) have $a = 0$, $b < 0$ and correspond to $\tilde{q} = 0$. Vanishing of $a(E_0, \theta)$ is purely an interference effect and may occur for certain values of $E_0, \theta$. The lines through these distinguished values in the $(E_0, \theta)$ plane have been called annulment lines[29]. Whether the annulment diagram of a resonance has certain characteristic features common to a whole class of resonances is not yet clear. At least the annulment diagram offers a convenient method to check the overall consistency between the different experimental results and theoretical calculations[29]. The close agreement of theoretical values for $\tilde{q}$ for proton impact ($E_0$ = 300 and 500 keV) and electron impact ($E_0$ = 400 and 1000 eV) for the $2s^2\ ^1S$ resonance[32] as function of $\theta$ with experimental values[28,25] is demonstrated in Fig. 4. At smaller proton energies (100 keV) theory[33] fails to calculate the observed variation of $\tilde{q}$ and b. This can be clearly seen in Fig. 5 for the $2s2p\ ^1P$ resonance and is also true for the $2s^2\ ^1S$ resonance (see ref. 32). Here first Born approximation seems to be inapplicable and a more rigorous treatment should be employed.

Line shapes of Auger electrons have always been found to be symmetrical. The reason for this is that the amplitude for direct double ionization in the vicinity of the Auger state is practically zero. Then eq. (1) collapses to eq. (4). Here $I(E_0,\theta)$ is given for impact ionization by unpolarized beams by[35]

$$I(E_0,\theta) = \frac{I_0(E_0)}{4\pi}\left[1 + \sum_{n=1}^{k} A_{2n}(E_0)P_{2n}(\cos\theta)\right] \qquad (5)$$

with $k \le L$. In general the anisotropy coefficients $A_{2n}$ are functions of population probabilities $P(SLJ|M|)$ of magnetic substates M of the Auger state SLJ, Auger transition ampli-

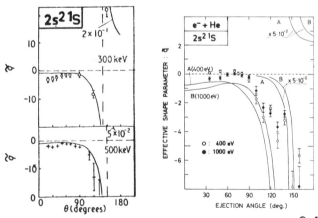

Fig. 4. Experimental shape parameter $\tilde{q}$ of the $2s^2\ ^1S$ resonance of He and comparison with theory. Left side: Proton excitation. o,+ = experimental values[28]; ——— = theory[32]. Right side: Electron excitation. o,● = experimental values[25]; ——— = theory[32].

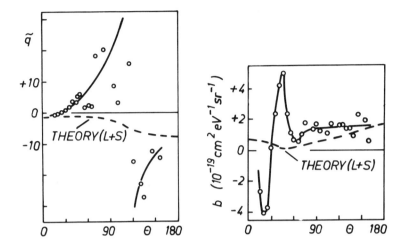

Fig. 5. Experimental shape parameter $\tilde{q}$ and yield $b$ of the 2s2p $^1$P resonance of He excited by 100 keV protons[30] and comparison with theory[33].

tudes and phases $\delta^\ell$ of ejected Auger partial waves. In case only one Auger partial wave is ejected, the coefficients $A_{2n}$ are only functions of $P(SLJ|M|)$, and from the anisotropy of $I(E_0,\theta)$ the alignment of inner shell ionized atoms can be determined. For inner shell ionization in the $2p_{3/2}$ shell the expected distribution of Auger electrons leaving the atom in a $^1S_0$ state is given by[35]

$$I(E_0,\theta) = \frac{I_0(E_0)}{4\pi}\left[1 + A_2(E_0)P_2(\cos\theta)\right],$$

$$\text{with } A_2(E_0) = \frac{P\left(\frac{3}{2}|\frac{1}{2}|,E_0\right) - P\left(\frac{3}{2}|\frac{3}{2}|,E_0\right)}{P\left(\frac{3}{2}|\frac{1}{2}|,E_0\right) + P\left(\frac{3}{2}|\frac{3}{2}|,E_0\right)} \qquad (6)$$

where the $P(SLJ|M|,E_0)$ are given in the notation $P(J|M|,E_0)$.

The coefficients $A_2(E_0)$ have been measured for electron impact ionization in the $2p_{3/2}$ shell of Ar[36] and Mg[37] via the angular distribution of the $L_3M_{2,3}M_{2,3}(^1S_0)$ transition of Ar and the $L_3M_1M_1$ transition of Mg. In Fig. 6 the experimental results are compared to theoretical values, calculated by McFarlane[38] using Born approximation and screened hydrogenic wave functions. There is good agreement for Ar but not for Mg, most probably due to the approximation of wave functions. In general the experimental alignment in inner shell ionization is rather small. This has been shown also for the 3d ionization in Kr[39,40], where theory[41] predicts a much larger alignment.

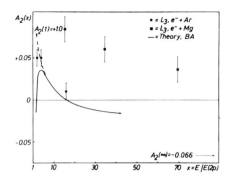

**Fig. 6.** Experimental anisotropy coefficients $A_2(x)$ for electron impact ionization in $L_3$ shell of Ar and Mg as function of reduced electron energy in units of the binding energy $E(2p)$ and comparison with theory[38].

# REFERENCES

1. A.K. Edwards, Invited papers of the IX ICPEAC (University of Washington Press, Seattle, 1975) p.     •
2. H.S. Taylor and R. Yaris, J. Phys. B: Atom. Molec. Phys. <u>8</u>, L 109 (1975).
3. A.J. Smith, P.J. Hicks, F.H. Read, S. Cvejanovic, G.C.M. King, J. Comer and J.M. Sharp, J. Phys. B: Atom. Molec. Phys. <u>7</u>, L 496 (1974).
4. F.H. Read, Invited Papers of the IX ICPEAC (University of Washington Press, Seattle, 1975) p.     •
5. C.F. Moore, Invited Papers of the IX ICPEAC (University of Washington Press, Seattle, 1975) p.     •
6. R. Morgenstern, Invited Papers of the IX ICPEAC (University of Washington Press, Seattle, 1975) p.     •
7. G.N. Ogurtsov, Invited Papers of the IX ICPEAC (University of Washington Press, Seattle, 1975) p.     •
8. W. Mehlhorn, Invited Papers of the VII ICPEAC (North-Holland, Amsterdam, 1971) p. 169; Invited Paper of the International Conference on X-Ray Processes in Matter, Phys. Fenn. <u>9</u>, Suppl. S1, 223 (1974).
9. H. Hillig, B. Cleff, W. Mehlhorn and W. Schmitz, Z. Physik <u>268</u>, 225 (1974).
10. B. Breuckmann and V. Schmidt, Z. Physik <u>268</u>, 235 (1974).
11. M.O. Krause, Phys. Rev. Lett. <u>34</u>, 633 (1975).
12. D.L. Walters and C.P. Bhalla, Atomic Data <u>3</u>, 301 (1971).
13. M.H. Chen and B. Crasemann, Phys. Rev. A <u>8</u>, 7 (1973).
14. H.P. Kelly, Phys. Rev. A <u>11</u>, 556 (1975).
15. E.J. McGuire, Phys. Rev. <u>185</u>, 1 (1969); A<u>2</u>, 273 (1970).
16. C.P. Bhalla and D.J. Ramsdale, Z. Phys. <u>239</u>, 95 (1970).
17. W. Mehlhorn, D. Stalherm and W. Schmitz, Z. Physik <u>252</u>,

399 (1972).

18. L.O. Werme, T. Bergmark and K. Siegbahn, Physica Scripta 6, 141 (1972).

19. E.J. McGuire, Phys. Rev. A 11, 17 (1975).

20. T. Åberg in Atomic Inner-Shell Processes, vol. I, p. 353, (ed. B. Crasemann), Acad. Press, 1975.

21. W. Mehlhorn, Phys. Letters 21, 155 (1966).

22. N. Oda, S. Tahira, F. Nishimura, J. Phys. B: Atom. Molec. Phys. 6, L 309 (1973).

23. H. Suzuki and K. Wakiya, Abstracts of IV International Conference on Atomic Physics, Heidelberg 1974, p. 473.

24. G.B. Crooks and M.E. Rudd, Abstracts of VII ICPEAC (Amsterdam, 1971), Vol 2, 1035.

25. H. Suzuki, Y. Jimbo, T. Takayanagi and K. Wakiya, Abstracts of IX ICPEAC (Seattle, 1975), p.       .

26. F.D. Schowengerdt and M.E. Rudd, Phys. Rev. Lett. 28, 127 (1972).

27. F.D. Schowengerdt, S.R. Smart and M.E. Rudd, Phys. Rev. A7, 560 (1973).

28. N. Stolterfoht, D. Ridder and P. Ziem, Phys. Letters 42A, 240 (1972); 6th Nat. Conf. on Electronic and Atomic Collisions, Liège, 1972; Abstracts of VIII ICPEAC (Belgrade, 1973) vol. 2, p. 521.

29. A. Bordenave-Montesquieu, P. Benoit-Cattin, M. Rodière, A. Gleizes and H. Merchez, J. Phys. B: Atom. Molec. Phys. 6, 1997 (1973); 7, L254 (1974); 8, 874 (1975).

30. A. Bordenave-Montesquieu, private communication, 1975, and Ph.D. Thesis, Université Toulouse, 1973.

31. V.V. Balashov, S.S. Lipovetsky and V.S. Senashenko, Sov. Phys. - JETP 36, 858 (1973).

32. S.S. Lipovetsky and V.S. Senashenko, J. Phys. B: Atom. Molec. Phys. 7, 693 (1974).

33. S.S. Lipovetsky and V.S. Senashenko, J. Phys. B: Atom. Molec. Phys. 5, L183 (1972).

34. U. Fano, Phys. Rev. 124, 1866 (1961). U. Fano, J.W. Cooper, Phys. Rev. 137, A1364 (1965).

35. B. Cleff and W. Mehlhorn, J. Phys. B: Atom.Molec. Phys. 7, 593 (1974).

36. B. Cleff and W. Mehlhorn, J. Phys. B: Atom. Molec. Phys. 7, 605 (1974).

37. B. Breuckmann and V. Schmidt, Verhandl. DPG (VI) 9, 411 (1974).

38. S.C. McFarlane, J. Phys. B: Atom. Molec. Phys. 5, 1906 (1972).

39. E. Döbelin, W. Sandner and W. Mehlhorn, Phys. Letters 49A, 7 (1974).

40. F. Nishimura, S. Tahira and N. Oda, J. Phys. Soc. Japan 35, 861 (1973).

41. K. Omidvar, private communication (1972).

# THE (e,2e) REACTION

I.E. McCarthy

School of Physical Sciences, The Flinders
University of South Australia, Bedford Park,
S.A. 5042, Australia.

## INTRODUCTION

In the (e,2e) reaction the momenta of two electrons in coincidence after an ionizing collision are measured, thus completely defining the momentum transfer $\underset{\sim}{q}$ to the ion.

$$\underset{\sim}{q} = \underset{\sim}{k}_0 - \underset{\sim}{k}_A - \underset{\sim}{k}_B. \tag{1}$$

The momentum variables and geometry are defined in fig. 1.

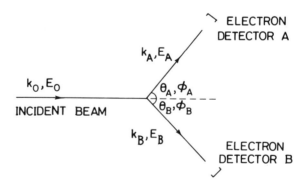

Fig. 1.   Definition of kinematic variables.

If the energies observed by the detectors A and B are about equal, the two electrons leave the interaction region quickly so that, at least to a first approximation, the momentum transfer $\underset{\sim}{q}$ is the momentum that the electron had in the target system (atom or molecule) before it was removed.

The (e,2e) experiment can therefore be considered as a measurement of the momentum distribution of electrons in matter. The first experiment with this point of view was performed by the Frascati group[1]. It observed tightly bound electrons in solid carbon.

I will be concerned with experiments designed to observe the properties of valence electrons in gas targets. The first such experiment was performed in 1973 at Flinders University by Weigold, Hood and Teubner[2].

In the Flinders experiments the kinematics has been of two kinds.

Coplanar symmetric[3]: $E_A = E_B$, $\theta_A = \theta_B = \theta$, $\phi_A = \phi_B = 0$;

$\theta$ varied.

Noncoplanar symmetric[4]: $E_A = E_B$, $\theta_A = \theta_B = 42.3^\circ$, $\phi_B = 0$,

$\phi_A = \phi$; $\phi$ varied.

I will discuss noncoplanar symmetric experiments.

## INFORMATION OBTAINED

The experiments obtain three important types of information about quantum states of the target and the ion.

1. Energy eigenvalues of the ion. These are obtained from the spectrum of the separation energy $\varepsilon$,

$$\varepsilon = E_0 - E_A - E_B, \tag{2}$$

measured for a fixed angle and a fixed total energy E.

$$E = E_A + E_B. \tag{3}$$

This aspect of the information is the same as that obtainable from photoelectron spectroscopy (ESCA). In general there are more ion eigenvalues than single-particle states of the electron in the target. We say that interelectron correlations split the independent-particle states.

2.    The momentum distribution of electrons, and hence the
orbital wave function, for resolved single-particle states of
the target. This is obtained by fixing $\varepsilon$ and varying $\theta$ or $\phi$.
The angular correlation shape is characteristic of the single
particle orbital and very sensitive to its details.

3.    Information relating the orbitals to the actual eigenstates of the target and the ion. This information is
obtained from the ratios of cross sections for states with
different separation energies. It is of three kinds:
a)    The normalized ratios of cross sections for ion eigen-
      states with similar angular correlation shapes give the
      probability of the single-particle configuration (hole
      plus target ground state) occuring as one of the inter-
      acting configurations in the many-body wave function of
      the ion. This probability is the <u>spectroscopic factor</u>.
      It may be understood as the ratio in which the electrons
      of the corresponding orbital are shared among the eigen-
      states among which the orbital is split.
b)    The centroid of energy eigenvalues for ion eigenstates
      with similar angular correlation shapes, weighted by the
      spectroscopic factors, is the corresponding single
      particle eigenvalue of the target.
c)    In cases where the ion wave function is accurately known,
      excitation of states unoccupied in the Hartree-Fock model
      gives sensitive information about target ground state
      correlations.


REACTION THEORY

The basic approximation used to analyse the reaction is
the distorted-wave off-shell impulse approximation.

$$T = A<\chi_A^{(-)}(\underset{\sim}{k}_A)\chi_B^{(-)}(\underset{\sim}{k}_B) \ (f|t_C(\varepsilon)|g) \ \chi_0^{(+)}(\underset{\sim}{k}_0)>. \qquad (4)$$

The distorted waves $\chi^{(\pm)}(\underset{\sim}{k},\underset{\sim}{r})$ are elastic scattering wave
functions for each electron-ion subsystem. $t_C(\varepsilon)$ is the
Coulomb t-matrix. The target ground state is $|g)$. The ion is
observed in one of a set of eigenstates $|f)$. The operator A
antisymmetrizes with respect to electron coordinates and spins.

If we make the quasi-three-body (electron-electron-ion)
approximation we can derive (4) from three-body theory by
neglecting only second and higher derivatives of the electron-
ion potential. It is an excellent approximation to the three-
body amplitude.

For very high electron energies the distorted waves may
be approximated by plane waves. In this case the amplitude

(4) factorizes into a Mott scattering t-matrix element (half off shell) and the Fourier transform of the target-ion overlap $(f|g)$.

$$T = \langle \tfrac{1}{2}(\mathbf{k}_A - \mathbf{k}_B) | t_{Mott}(\tfrac{1}{4}|\mathbf{k}_A - \mathbf{k}_B|^2) | \tfrac{1}{2}(\mathbf{k}_0 + \mathbf{q}) \rangle$$

$$\times \int d^3r \, \exp(i\mathbf{q}.\mathbf{r})(f|g). \tag{5}$$

For lower energies we take into account the degrees of freedom of the ion by using optical model wave functions[5] for elastic scattering at the appropriate energy. The experiments have been performed at energies where these functions may be approximated by an averaged eikonal approximation

$$\chi^{(\pm)}(\mathbf{k},\mathbf{r}) \cong \exp(-\gamma kR) \exp\left[i(1+\beta\pm i\gamma)\mathbf{k}.\mathbf{r}\right]. \tag{6}$$

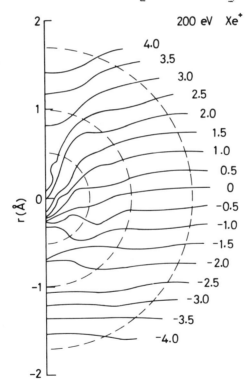

Fig. 2. The surfaces of equal phase for the optical model wave function for 200eV electron elastic scattering on the xenon ion. The surfaces are axially symmetric. The plot is a section through the scattering axis. The dashed circles are the radii for the outermost node in the 5p orbital, the outermost peak in the 5p orbital, and the region where the electron density is 1/10 of its maximum value.

The parameter β modifies the wave number, γ describes attenuation due to excitation of reaction channels other than (e,2e), R normalizes the function to unity just before the electron enters the interaction region.  The approximation (6) still permits the factorization (5), with small complex modifications to the wave numbers k̰.

With approximation (6) the surfaces of constant phase in the distorted wave are flat and equally spaced as for a plane wave.  Fig. 2 shows the optical model surfaces of equal phase in a section through the scattering axis for the xenon ion at 200eV.  The averaged eikonal approximation is clearly justified for the outer part of the atom.

The validity of the reaction theory is confirmed by testing it on a simple case, helium, whose structure wave functions are well-known.  The simple theory predicts that for non-coplanar symmetric geometry the t-matrix element is independent of q and the cross section shape depends only on q, not on the total energy E.  Fig. 3 shows the verification for E = 200eV, 400eV, 800eV.

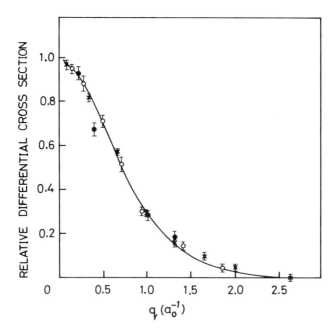

Fig. 3.  Distribution of the momentum transfer q for helium at 200eV (filled circles), 400eV (open circles) and 800eV (crosses).  The theory uses the wave function of Froese-Fischer[6].

## STRUCTURE WAVE FUNCTIONS

The many-body wave functions of the target and ion are expanded in independent-particle configurations $|\alpha)$ for the target system and single-hole orbitals $\psi_j$ (with angular momentum coupling understood).

$$|g) = \Sigma_\alpha \, a_\alpha |\alpha),$$
$$|f) = \Sigma_{\beta j} \, t_{\beta j}^{(f)} \, \psi_j |\beta). \tag{7}$$

For many closed-shell systems we have found that configuration interaction in the target is not observable. We may make the independent-particle approximation

$$a_0 = 1, \quad |\alpha) = |0), \tag{8}$$

where the Hartree-Fock wave function of the target is denoted by $|0)$.

In this approximation the overlap function $(f|g)$ is simply proportional to the target orbital $\psi_i$. This orbital is called the characteristic orbital of the ion eigenstate f.

$$(f|g) = t_{0i}^{(f)} \, \psi_i. \tag{9}$$

The (e,2e) cross section is proportional to the spectroscopic factor

$$S_{0i}^{(f)} = \left| t_{0i}^{(f)} \right|^2. \tag{10}$$

The normalization and closure relations give two important facts. The first is the sum rule for spectroscopic factors corresponding to an orbital $\psi_i$.

$$\Sigma_f \, S_{0i}^{(f)} = 1. \tag{11}$$

The second is the definition of the single-particle eigenvalue as the expectation value of the ion Hamiltonian in the state $\psi_i |0)$.

$$E_i = (0|\psi_i^* \, H_{ion} \, \psi_i |0) = \Sigma_f \, S_{0i}^{(f)} \, E_f, \tag{12}$$

where

$$(E_f - H_{ion})|f) = 0. \tag{13}$$

The single particle eigenvalue is the centroid of the ion eigenvalues for the states with the same characteristic orbital, weighted by the spectroscopic factors.

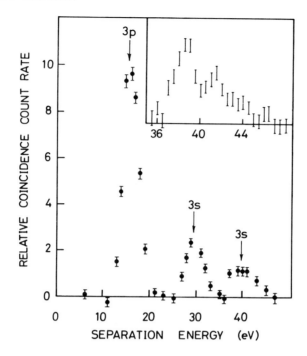

Fig. 4. The separation energy spectrum for argon at $\phi = 10^o$. Inset: detail near 40eV.

EXAMPLE: THE ARGON ION

The argon atom is a simple non-trivial example. The peaks in the separation energy spectrum, fig. 4, correspond to ion eigenstates, which are first assigned to a representation of the symmetry group by their angular correlations. 3s and 3p angular correlations for argon are compared in fig. 5.

Only the 15.76 eV peak belongs to the 3p representation, so for this eigenstate the spectroscopic factor is unity. This has been found to be true for the least-bound eigenstate of all systems observed so far. Spectroscopic factors for correlated states are normalized with reference to the least-bound state. This removes the one arbitrary multiplicative factor in the experiment.

The first aspect of the structure wave function to be tested is the single particle orbital. With approximations (5) and (8) the cross section is proportional to the square of the orbital wave function in momentum space. Not only may the orbital be identified, but the cross section is very

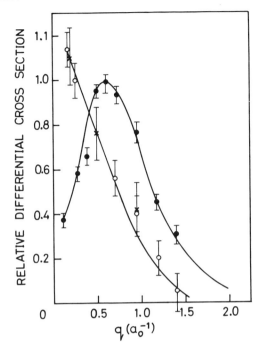

Fig. 5. Distributions of the momentum transfer q for 3p
(filled circles) and 3s (29.3eV open circles; states near 40eV,
crosses) states of argon at 400eV.  The theory uses the wave
functions of Froese-Fischer[6].  Finite acceptance angles (7°)
are folded into the theory.

sensitive to detail as shown in fig. 6 for three approximations
to the argon 3p orbital, namely the Hartree-Fock functions of
Froese-Fischer[6], and Lu *et al*[7], and the best variational
Slater-type orbital.  The (e,2e) reaction is a very sensitive
probe for the single-particle orbitals.

   It is evident from equations (7) and (10) that the
spectroscopic factors depend on the particular forms of the
orbitals $\psi_j$ used in the basis for expanding the ion wave
function.  It is significant that the Hartree-Fock and
experimental definitions of $\psi_i$ coincide.  Experimental and
theoretical definitions of the spectroscopic factors are
therefore identical.

   If one has accurate enough orbitals to fit the angular
correlation shape, one may then compare ratios for states
belonging to the same representation, i.e. states that have
the same characteristic orbital.

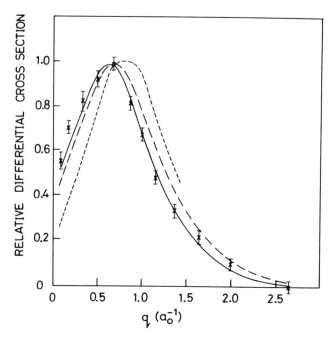

Fig. 6.  Comparison of 3p orbital models for the 15.76eV state
of argon.  The total energy is 800eV.  The curves are
described in the text.

In argon four states have been identified belonging to the
3s representation.  Their spectroscopic factors relative to

$$S_{3p}^{(15.76)} = 1$$

are

$$S_{3s}^{(29.3)} = 0.50 \pm 0.05,$$

$$S_{3s}^{(38.4)} = 0.24 \pm 0.01,$$

$$S_{3s}^{(41.3)} = 0.13 \pm 0.02,$$

$$S_{3s}^{(43.6)} = 0.06 \pm 0.01.$$

These factors sum to unity within experimental error,
thus verifying the whole analysis.  Similar verification
occurs for all inert gases.

The 3s single-particle eigenvalue is 34.0 ± 1.0 eV,
compared with the Hartree-Fock[6] value of 34.8 eV.

EXAMPLE:  THE METHANE MOLECULE

For small molecules, e.g. methane and nitrogen, we have been able to identify correlated ion states by their angular correlations and to check the analysis by the sum rule for spectroscopic factors.  The case of methane is illustrated in figs. 7 and 8.

Here we do not have accurate orbital wave functions. The functions of Snyder and Basch[8] are accurate enough to enable identification of the representation.  For methane the 2a orbital is split into eigenstates at 23 eV and 32 eV, with spectroscopic factors in the ratio 9:1.  Since the theoretical orbitals do not give the correct shape for the angular correlation, but are of course normalized, we can

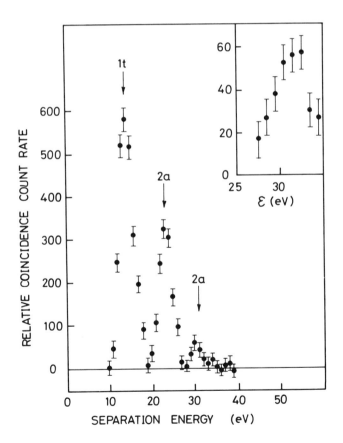

Fig. 7.  The separation energy spectrum for methane at $\phi = 10^\circ$.  Inset:  detail near 30eV.

check the analysis by comparing the integrals of $\sigma(q)q^2dq$. This has been done for the 1t states in fig. 8.

The analysis for molecules is complicated by the existence of vibrational and rotational degrees of freedom. The experiment does not resolve such states and closure is assumed in the analysis. The cross section is proportional to

$$\sigma = |t_{Mott}|^2 \Sigma_{nm} \phi_n^*(\underset{\sim}{q}, R_e) \phi_m(\underset{\sim}{q}, R_e) \, j_0(qR_{nm}), \qquad (14)$$

where $\phi_n$ is the Fourier transform of the whole term in the molecular orbital (expressed as a linear combination of atom-

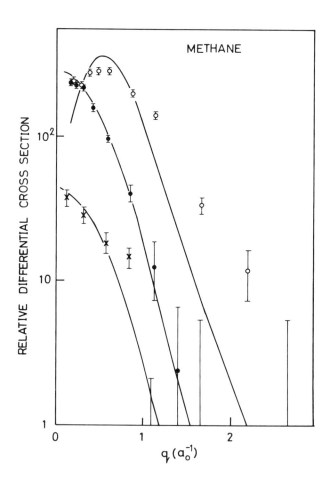

Fig. 8. Logarithmic plot of the angular correlations for the observed states of the methane ion. The orbital wave functions of Snyder and Basch[8] are used in the theory. Normalization is described in the text.

centered orbitals in spherical polar coordinates) representing
functions centered on the nucleus n.  The cross section depends
on the set $R_e$ of equilibrium values of the nuclear positions.

The validity of (14) has been verified in detail for $H_2$
and $D_2$ using a configuration-interaction wave function of
McLean *et al*[9] for the target and the accurate variational
function of Guillemin and Zener[10] for the ion.  Angular
correlations are identical for $H_2$ and $D_2$, confirming the Born-
Oppenheimer approximation.  In fig. 9 the data are compared
with the accurate calculation and with one making the extreme
single-particle approximation and using the wave function of
Snyder and Basch[8].  The latter functions, using a gaussian
basis, seem to fail in the same way for all states observed.
Their momentum variation is too sharp, underweighting high
and low momentum components.

## TARGET CORRELATIONS

Finally we turn to the most sensitive test of the theory
that the reaction has yet been subjected to.  This is the
comparison of the cross sections for exciting the n=2
(forbidden in the extreme independent-particle model) and n=1

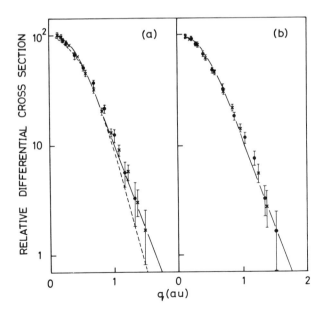

Fig. 9. The distribution of momentum transfer q for (a) $H_2$
and (b) $D_2$ at 600 eV (filled circles) and 400 eV (crosses).
The solid curve is the present theory.  The dashed curve is
computed with the wave function of Snyder and Basch[8].

| $\theta$ | Hartree-Fock | Configuration sum | Experiment |
|-----|-----|-----|-----|
| $45^\circ$ | 2.75 | 0.67 | $0.72\pm0.04$ |
| $49^\circ$ | 2.07 | 0.88 | $0.97\pm0.10$ |
| $53^\circ$ | 1.94 | 1.74 | $1.54\pm0.30$ |

Table I.  Differential cross section ratios for the excitation of $n=2$ and $n=1$ states of the helium ion, expressed as a percentage.

states of helium[11].  We know the ion wave function exactly, so we are able to use different approximations for the target wave function.  Table I shows the ratios at several angles for the Hartree-Fock approximation[6] and for a 32-configuration super-position computed by Joachain and Vanderpoorten[12].  The latter are within experimental error.  The former are not only incorrect in magnitude but have the opposite variation with angle.

## REFERENCES

1. R. Camilloni, A. Giardini-Guidoni, R. Tiribelli and G. Stefani, Phys. Rev. Letters 29, 618 (1972) and references therein.
2. E. Weigold, S.T. Hood and P.J.O. Teubner, Phys. Rev. Letters 30, 475 (1973).
3. A. Ugbabe, E. Weigold and I.E. McCarthy, Phys. Rev.A 11, 576 (1975).
4. E. Weigold, S.T. Hood and I.E. McCarthy, Phys. Rev.A 11, 566 (1975), and references therein.
5. J.B. Furness and I.E. McCarthy, J. Phys. B6, 2280 (1973).
6. C. Froese-Fischer, Atomic Data 4, 302 (1972).
7. C.C. Lu, T.A. Carlson, F.B. Malik, T.C. Tucker and C.W. Nestor Jr., Atomic Data 3, 1 (1971).
8. L.C. Snyder and H. Basch, Molecular Wave Functions and Properties, John Wiley, New York, 1972.
9. A.D. McLean, A. Weiss and M. Yoshimine, Rev. Mod. Phys. 32, 211 (1960).
10. V. Guillemin, Jr. and C. Zener, Proc. N.A.S. 15, 314 (1929).
11. I.E. McCarthy, A. Ugbabe, E. Weigold and P.J.O. Teubner, Phys. Rev. Letters 33, 459 (1974).
12. C.J. Joachain and R. Vanderpoorten, Physica 46, 211 (1960).

# DECAY OF AUTOIONIZATION STATES

# IN COLLISIONS OF HEAVY ATOMIC PARTICLES

G.N. OGURTSOV

A.F.Ioffe Physico-Technical Institute

Leningrad, K-21 , USSR

The study of structure in energy spectra of
electrons ejected in atomic collisions can give an
important information on autoionization transition
probabilities and cross sections for inner shell
excitation. In collisions of heavy atomic particles,
energy spectra of ejected electrons have some specific features connected with collision kinematics.
Due to considerable momentum transfer, the lines in
energy spectra experience fairly large Doppler shift
and Doppler broadening /1/. These effects as well as
the results on energy spectra of electrons ejected
in ion-atom collisions obtained during last decade
have been discussed in a few review papers /2-4/.

In this paper another effect intrinsic of atomic collisions is discussed connected with possibility of autoionization transitions in quasimolecule, at small and medium internuclear distances.
During long enough period, this effect was considered as negligible because of very small collision
time as compared to the life time of autoionization
transitions known from spectroscopy. In 1967 Kishinevsky and Parilis /5/ showed that this was not the
case indeed. The authors calculated the probability
of an autoionization transition as a function of
internuclear distance, for a particular case of two
electrons in the field of two Coulomb centers (Fig.1).
Results of the calculation /5/ indicate a considerable increase in probability of autoionization transitions in quasimolecule as compared to the probability
of transitions both in separated atoms and in united

atom. This increase reaches the values of about two
orders of magnitude at internuclear distances corresponding to the minimum energy of ejected electrons. The result obtained induces us to revise our
concepts concerning autoionization transitions in
quasimolecule and their influence on the character
of energy spectra of electrons ejected in autoionization processes in atomic collisions.

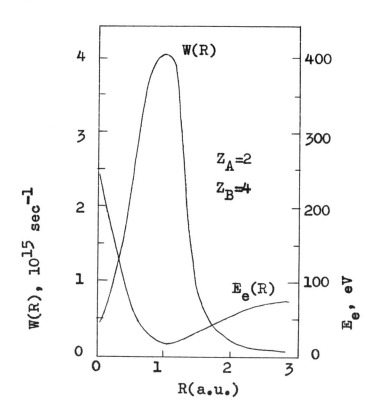

Fig.1.   Autoionization transition probability
and ejected electron energy as functions of internuclear distance, for a system of two electrons in
the field of two Coulomb centers $Z_A$ and $Z_B$.

( From Kishinevsky and Parilis /5/ )

Consider the energy distribution of electrons
ejected in autoionization transitions in quasimolecule. These transitions can occur at different internuclear distances. Since the transition energy is a
function of internuclear distance, this gives rise

to a very broad energy distribution of ejected electrons (Fig.2).

U(R)

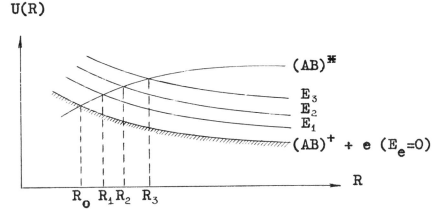

Fig.2.    Schematic diagram illustrating the
autoionization transitions in quasimolecule. A potential curve corresponding to the initial autoionization state can cross an infinite number of curves
corresponding to the final continuum states with
different electron energies.


Starting from the decay equation:

$$N = N_0 \left\{ 1 - \exp\left[ -\int^t W(R)dt \right] \right\} =$$

$$= N_0 \left\{ 1 - \exp\left[ -\int^{E_e} W(E_e)(dR/dE_e)v_R^{-1}\, dE_e \right] \right\} ,$$

where $W$ is the autoionization transition probability, $v_R$ is the radial velocity, the energy distribution of electrons can be written as:

$$f(E_e) = N_0^{-1}(dN/dE_e) = WR'v_R^{-1} \exp\left[ -\int^{E_e} WR'v_R^{-1}dE_e \right] ,$$

where $R' = dR/dE_e$. Thus, the shape of energy distribution depends on the function $X(E_e) = WR'/v_R$. This
distribution can have maxima. According to /5/, one
of these maxima should occur at the minimum energy
of ejected electrons corresponding to the maximum
transition probability. This energy is often equal
to zero ( the autoionization level crosses the boundary of continuum ). The energy distribution of

electrons usually ends by a sharp peak corresponding to an autoionization transition in isolated atom. If decay of the autoionization state occur in the Coulomb field of the receding ion then this peak experiences a shift toward lower energies and broadening inversely proportional to the incident ion velocity /6/:

$$E_m = E_\infty - W_\infty/2v \quad (a.u.)$$

$$\gamma = 1.07 \, W_\infty/v \quad (a.u.)$$

where $E_m$ is the energy corresponding to the maximum

of the peak, $E_\infty$ is the energy of transition in an isolated atom, $\gamma$ is width of the distribution, $W_\infty$ is the transition probability in an isolated atom.
    There is one particular case when an additional maximum appears in the energy distribution of electrons ejected in autoionization transitions in quasi-molecule, at high enough electron energies. This situation takes place when the orbital corresponding to the initial autoionization state is sharply promoted in a certain range of internuclear distances, so that $dR/dE_e \to 0$. In this range the function $f(E_e)$

vanishes, and a maximum of distribution appears at electron energies corresponding to larger internuclear distances where the function $E_e(R)$ becomes more

smooth. In general case, the maximum position is defined by the equation

$$dX/dE_e(E_m) = X^2(E_m)$$

    A typical example of sharp dependence of the orbital energy on the internuclear distance is the promotion of 4fб orbital in Ar-Ar system which is responsible for production of 2p-vacancies after collision. For a qualitative study of energy distribution of electrons ejected in Auger transitions to a vacancy in 4fб orbital, we considered a simplified case when 4sб and 3pπ electrons are involved in the transition. The functions $E_e(R)$ and $W(R)$ were obtained from the correlation diagram and from the calculations of Kishinevsky and Parilis /5/, respectively. Potential energy which appears in the expression for the radial velocity was estimated using Csavinszky formula /7/.
    The calculated energy distribution integrated over the impact parameter is shown in Fig.3.

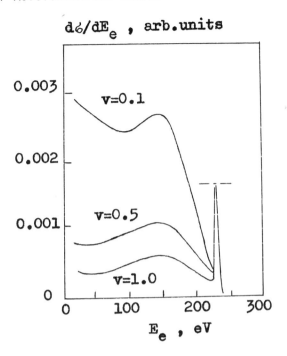

Fig.3. Energy distribution of electrons ejected in the transition 4fб - (3pπ)(4sб) in Ar-Ar system, at various velocities v ( v in atomic units ).

The distribution given in Fig.3 demonstrates the main features of a real distribution of electrons ejected in autoionization transitions in quasimolecule. Now we pass to the discussion of some experimental data in which the influence of autoionization transitions in quasimolecule can manifest itself.

1. Gerber, Morgenstern and Niehaus /8/ observed spectral lines in the ejected electron energy range 14 - 20 eV in He-He collisions. The authors ascribed broad lines at 15 and 17.5 eV to the Penning ionization in interaction between two receding helium atoms each of them being excited after collision. Generally speaking, this process can be considered as autoionization transition in quasimolecule. In contrast to the previous case, here the transition probability approaches zero at $R \rightarrow \infty$ rather than a finite value of transition probability in an isolated atom. This is caused by the fact that the electrons involved in the transition belong to different nuclei. The broad peaks at 15 and 17.5 eV can be associated with the transitions 2pб - 2sб2sб and 2pб - 2pπ2pπ, respect-

ively. At the incident atom energy 200 eV, only the
first line is excited, so a fit of experimental data
and calculations can be made without serious ambiguity. Known the probability of transition between the
orbitals relevant to the problem, and the absolute
values of cross sections for electron ejection, this
fit can give information on the probability of $2p\acute{o}$ -
$2s\acute{o}2s\acute{o}$ transition in He-He system. Since the data
needed are absent, we have attempted to find relative
probability of the above transition as a function of
internuclear distance using the experimental data /8/
and the correlation diagram for He-He system and calculating the function $X(E_e)$ which define the profile
of spectral line. The result obtained is given in
Fig.4.

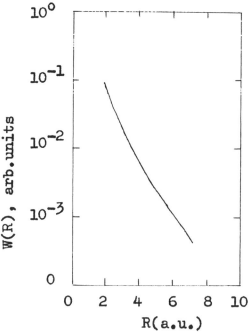

Fig.4.   Relative probability of the transition
$2p\acute{o}$ - $2s\acute{o}2s\acute{o}$ in He-He system as a function of internuclear distance.


2. In connection with autoionization processes
in quasimolecule, of interest is the fact of different energy dependence of the cross sections
for Auger electron ejection $\acute{o}_A$ , and for X-ray emission $\acute{o}_X$ in excitation of $L_{2,3}$ subshell of argon by
$Ar^+$ ions ( Fig.5 ).

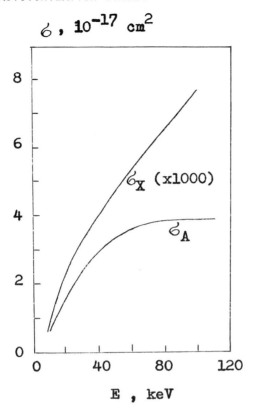

$\sigma$ , $10^{-17}$ cm$^2$

Fig.5.    Cross sections for X-ray emission /12/ and for Auger electron ejection /13,14/ in excitation of $L_{2,3}$ subshell of argon in Ar$^+$-Ar collisions.

The function $\sigma_A(E)$ have a plateau at $E \gtrsim 50$ keV whereas the function $\sigma_X(E)$ increases continuously. The fluorescence yield $\omega(E)$ obtained from the above data is represented by a curve with maximum ( Fig.6 ). The fluorescence yield can be calculated using the theoretical results /9/ on the fluorescence yields for different outer shell configurations and experimental data on relative probabilities for formation of these configurations in collisions. The latter data were obtained by Fortner and Garcia /10/ from the relative intensities of lines in X-ray spectrum. In this work we have used the data /11/ on charge distribution of colliding particles in excitation of the second and third lines of energy loss spectrum. The calculated values of fluorescence yield are essentially lower than experimental ones ( point in

Fig.6 ).

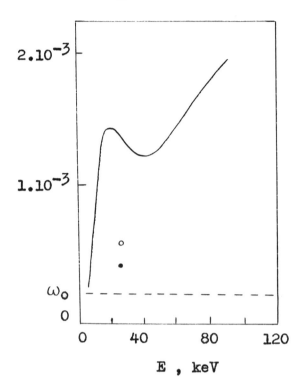

Fig.6.   Fluorescence yield $\omega_{L_{2,3}}$ as a function of ion energy ( $Ar^+$– Ar collisions ). Point – the value calculated from the data /9,11/. Circle – the value corrected for dependence of X-ray transition probability on the effective nuclear charge. $\omega_0$ – the fluorescence yield in isolated argon atom /15/.

The reason for such discrepancy may be connected with uncertainty of the calculation /9/ in which probabilities of individual X-ray transitions were supposed to be independent of the number of outer shell electrons, i.e. of a change in effective nuclear charge in ejection of the outer shell electrons. We have made an attempt to improve the data obtained in /9/ using somewhat more realistic approximation for X-ray transition probability: $W_X \sim Z_{eff}^4$ .

It has been assumed that the effective charge increases by $\Delta Z \simeq 0.3$ in ejection of each outer shell electron ( as follows from hydrogenic approximation ). However, the "improved" data still remain considerably lower than the experimental ones ( circle in Fig.6 ).

Fortner and Garcia /10/ suggested that the disagreement between experimental data and calculations resulted from systematic errors in measuring cross sections for X-ray emission. However, Saris and Bhalla /16/ have disproved such a suggestion. Here we give another explanation of the observed disagreement.

The cross sections $\sigma_A$ and $\sigma_X$ were estimated by integrating the structure in the spectra over energy ( or wave length ) and solid angle. The inner shell vacancy can be filled not only after departure of colliding particles but also in quasimolecule. Thus the spectra of X-ray emission and Auger electron emission will contain both discrete structure and continuous background ( see Fig.3 ). The latter part of distribution can not be taken into account in determination of $\sigma_A$ and $\sigma_X$. In the case of X-ray transitions, the contribution of transitions in quasimolecule is negligible because of very large life time. On the contrary, the contribution of Auger-transitions in quasimolecule is considerable, and it can even exceed the contribution of transitions in isolated atoms at low collision energies. So a considerable fraction of transitions produces continuous background which can not be taken into account in determination of cross sections for ejection of Auger electrons. As a result, the values $\sigma_A$ are underestimated. The experimental data on the fluorescence yield given in Fig.6 represent actually the values
$$\omega_{exp} = \omega_{true}(E) k(E)$$ where $\omega_{true}(E)$ is the true fluorescence yield which increases with increasing energy due to increase in the mean charge of colliding particles /9/, $k(E)$ is a function inverse to the fraction of Auger transitions in isolated atoms which decreases with increasing energy. In principle, the product $\omega_{exp}(E)$ can have maximum that is not inconsistent with the data given in Fig.6.

Comparison between experimental data and calculations can give information on the contribution of Auger transitions in quasimolecule. This contribution is about 60% at an ion energy 25 keV which is consistent with the data obtained by Afrosimov and co-workers /17/ from the correlation of charge states

of colliding particles. One can suppose that different energy dependence of the cross sections $\sigma_A$ and
$\sigma_X$ is also connected with the influence of Auger
transitions in quasimolecule. For explanation of this
difference, it is necessary to accept that the contribution of Auger transitions in quasimolecule is
about 30% even at 100 keV.

The examples considered in this paper show that
autoionization transitions in quasimolecule can play
an important role in ionization processes at low
and medium energies of colliding atomic particles.
The study of energy spectra of electrons and spectra
of X-ray emission makes it possible to obtain information on the probability of autoionization transitions in quasimolecule .

References

1.  M.E.Rudd, T.Jorgensen, Jr., D.J.Volz,
    Phys.Rev.Lett., 16, 929, (1966)
    Yu.S.Gordeev, G.N.Ogurtsov, Zh.Eksp.Teor.Fiz.,
    60, 2051, (1971)

2.  G.N.Ogurtsov, Rev.Mod.Phys., 44, 1, (1972)

3.  M.E.Rudd, J.H.Macek, "Case Studies in Atomic
    Physics", v.3, 49, (1972)

4.  J.D.Garcia, R.J.Fortner, T.M.Kavanagh,
    Rev.Mod.Phys., 45, 111, (1973)

5.  L.M.Kishinevsky, E.S.Parilis, V ICPEAC,
    Abstracts of Papers, p.100, Leningrad, (1967)

6.  H.W.Berry, Phys.Rev., 121, 1714, (1961)

7.  P.Csavinszky, Phys.Rev., 166, 53, (1968)

8.  G.Gerber, R.Morgenstern, A.Niehaus, J.Phys.,
    B6, 493, (1973)

9.  F.P.Larkins, J.Phys., B4, L29, (1971)

10. R.J.Fortner, J.D.Garcia, J.Phys., B6, L346, (1973)

11. Q.C.Kessel, E.Everhart, Phys.Rev.,146, 16, (1966)

12. F.W.Saris, D.Onderdelinden, Physica, 49, 441,
    (1970)

13. R.K.Cacak, Q.C.Kessel, M.E.Rudd, Phys.Rev., A2,
    1327, (1970)

14. G.N.Ogurtsov, VII ICPEAC, Abstracts of Papers,
    p.400, Amsterdam, (1971)

15. E.J.McGuire, Phys.Rev., A3, 1801, (1971)
16. F.W.Saris, C.P.Bhalla, J.Phys., B7, L115, (1974)
17. V.V.Afrosimov, Yu.S,Gordeev, A.M.Polyanski, A.P.Shergin, Zh.Eksp.Teor.Fiz., 63, 799, (1972)

# COLLISIONALLY PRODUCED AUTOIONIZING AND AUTODETACHING

# STATES OF NEUTRAL ATOMS AND THEIR NEGATIVE IONS

A. K. Edwards

Dept. of Physics and Astronomy, Univ. of Georgia

Athens, Georgia  30602  U.S.A.

## INTRODUCTION

In this paper experimental procedures and results are discussed of collisionally produced autodetaching states of negative ions and autoionizing states of atoms.  Most of the ions and atoms reviewed here are those that exist naturally in molecular form and, as such, cannot be readily investigated by electron scattering experiments[1] or by positive ion bombardment[2,3].  To study these systems a negative ion beam of an atomic species is accelerated, stripped to form a neutral beam or left as an ion, then excited to the states of interest by collisions with a target gas.  Electrons ejected from the collision region are energy analyzed and a spectrum recorded.  From the electron spectrum the autoionizing and autodetaching states can be identified.

Illustrative examples of the type reactions that produce electron spectra from negative ion beams are:

1.  stripping
    $O^- + He \rightarrow O + He + e$

2.  ionization
    $H^- + Ar \rightarrow H^- + Ar^+ + e$

3.  excitation of the beam
    $O^- + He \rightarrow O^-* + He \rightarrow O + He + e$

4.  excitation of the target
    $H^- + Ar \rightarrow H^- + Ar^* \rightarrow H^- + Ar^+ + e$

5. stripping and excitation of the beam
$$C^- + He \rightarrow C^* + He + e \rightarrow C^+ + He + 2 e$$

6. charge exchange
$$O^- + Ar \rightarrow O + Ar^-{}^* \rightarrow O + Ar + e$$

7. exchange and dissociation
$$O^- + H_2 \rightarrow O + H + H^-{}^* \rightarrow O + H + H + e$$

The first two processes are mainly responsible for the continuum of the electron spectra while the others produce peaks in the spectra.

Illustrative examples for neutral beams are:

1. ionization of the beam
$$O + He \rightarrow O^+ + He + e$$

2. ionization of the target
$$O + He \rightarrow O + He^+ + e$$

3. excitation of the beam
$$O + He \rightarrow O^* + He \rightarrow O^+ + He + e$$

4. excitation of the target
$$H + Ar \rightarrow H + Ar^* \rightarrow H + Ar^+ + e$$

5. charge exchange
$$He + He \rightarrow He^+ + He^-{}^* \rightarrow He^+ + He + e$$

Again, the first two reactions produce the background of the spectra and the others, the peaks of the spectra. These lists are not all inclusive but do show many of the processes of interest. Other investigations that can be made using the collisional excitation method are studies of the excitation cross sections[4] and of the beam-target interactions[5].

## EXPERIMENTAL PROCEDURE

The negative ions constituting the accelerated beam are formed in an arc discharge of a duoplasmatron ion source. The elements to be extracted can be introduced to the discharge by several methods; in a molecular form as the discharge gas itself, as a doping substance through gas mixing or by evaporation from a solid compound, or by sputtering from the anode. As examples, $Cl^-$ ions[6] were formed in a hydrogen discharge by doping with $CCl_4$ vapors and $I^-$ ions[7] were formed by the evaporation of KI (from an extra filament added to the source). Carbon negative ions were formed by sputtering from a carbon anode. The beam particles are

normally ground state negative ions when they reach the
interaction region since any excited states will decay to
neutral particles before reaching an analyzing magnet placed
along the beam path.  It is possible for long-lived metastables to exist in the beam.  The beam energies have usually been in the range from several hundred eV to 10 keV.  The
energy in the center of mass frame of the collision depends
on the colliding pairs.  For example, for Cl⁻ on He collisions the center of mass energy is about one-tenth the beam
energy.

To form neutral beams[8] the negative ion beams are
stripped by a gas cell placed in front of the collision
region.  At the end of the stripping cell are a set of
deflection plates to remove the charged particles from the
beam.  The amount of deflected beam is measured to approximate the amount of neutral beam entering the collision
region.  The neutral beam may have a metastable component,
some of which may be reduced by the electric field of the
deflection plates.

One method of recording the electron spectra is shown in
Fig. 1.  The electrons leaving the collision region are
energy analyzed and detected.  The detector pulses are amplified, discriminated, then stored in a multichannel scaler
(MCS).  A ramp voltage that is generated by the MCS and is
related to the channel address is amplified and added to the
starting-point voltage of the electron energy analyzer.  This
process correlates the electron energy to the channel address
of the MCS.  This channel address can be stepped by an internal clock or by an external address advance driven by a beam
integrator.  This last method yields the extra information
of the number of counts per unit beam charge passing through
the collision region.

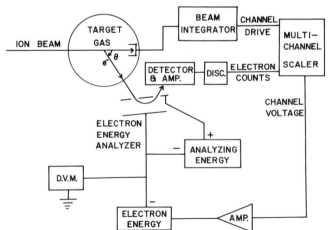

Fig. 1.   Schematic of the apparatus used to record the
electron spectra.

The energy analyzer and detector system can be constructed so that the angular distribution of electrons can be measured. These measurements may yield additional information about such things as the collision process or angular momentum of the electrons[5,9]. For the electrons ejected by the moving beam particles, their observed energies must be corrected for this motion. This is easily done by considering the vector addition of the beam velocity and the ejected electron velocity and has been discussed in detail elsewhere[10].

## AUTODETACHING STATES

For the direct excitation of the negative-ion beam particle, helium usually serves as the best target. It requires the most energy of any other target to be excited itself and has the highest ionization potential of other gases which means it does not contribute many electrons to the background. For low Z elements the multiplicity of the excited states is found to be the same as the multiplicity of the ground state when helium is the target. The helium would have to be excited to its first triplet state in order to change the spin of the negative ion. When $H_2$ was used as a target for a $Cl^-$ beam a triplet $Cl^-$ state was strongly excited through spin exchange. For high Z elements such as $Br^-$ and $I_2^-$, triplet states are readily excited with a helium target[11,7].

In general, the autodetaching states are doubly-excited states with dominant configurations of (positive core)$n\ell^2$ or (positive core)$n\ell n\ell'$. The strongest excited state is usually the first available (positive core)$ns^2$ configuration; multiplicity being considered for low Z elements. One identifiable exception[8] to this is the $2s2p^6$ level of $O^-$ whose identification has been confirmed theoretically by Chase and Kelly[12] and by the electron scattering work of Spence and Chupka[13]. Many of the doubly-excited states have been calculated by Matese, et. al.,[14-17] using configuration interaction techniques. Excellent agreement with experiment has occurred for their calculations of $C^-$[17], $O^-$[15], $F^-$[18], and $Cl^-$[16] ion states.

Autodetaching states of the target gas can be studied through the charge exchange reactions given in the introductory listing. The states of $Ar^-$ formed by exchange have served as calibrators for other spectra and the angular distribution of its detached electrons are being investigated[5]. Molecular states of $N_2^-$ have also been produced by charge exchange of negative ions with $N_2$[19,20] and a $He^-$ state was formed[21] in He on He collisions. However, for measurements of the energies of autodetaching states of target gases, the electron scattering experiments are superior in sensitivity and number of states observed.

A typical electron spectrum is shown in Fig. 2. This one was produced by 2 keV O[8] on He collisions and the electrons ejected at 15° for the beam direction. Table I is a listing of the negative ions studied, the measured energies of the identified autodetaching states, and the calculated values. Also listed is the measured binding energy of the excited pair of electrons to the positive core (grandparent) for the doubly-excited states. This is found by subtracting the measured energy for the doubly-excited state from the ionization potential of the neutral atom[22]. For the $ns$ configurations the binding energy ranges from 4 to 5 eV, for the $nsnp$ it is about 3.8 eV, and about 2.8 eV for $np^2$. The binding energy of the last electron to the excited parent ranges from very small (the order of meV) to about 0.6 eV.

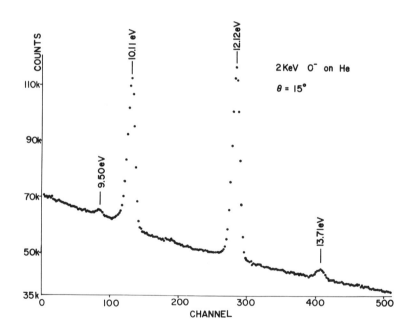

Fig. 2. Electron spectrum produced by the collision of O[-] ions with a helium target.

Table I.  Autodetaching states of negative ions measured
using the collisional excitation method.

| Ion | Configuration | Energy Exp. (eV) | Energy Calc. (eV) | Binding Energy | | |
|---|---|---|---|---|---|---|
| | | | | $ns^2$ (eV) | $nsnp$ (eV) | $np^2$ (eV) |
| H⁻ | $2s^2$ | 9.59±0.03 | 9.56[a] | 4.01 | | |
| | $2s2p$ | 9.76±0.03 | 9.75[a] | | 3.84 | |
| C⁻ | $(^2P)3s3p$ | 7.44±0.07[b] | 7.45 | | 3.83 | |
| O⁻ | $(^4S)3s3p$ | 9.50±0.02 | 9.50 | | 4.12 | |
| | $2s2p^6$ | 10.11±0.02 | | | | |
| | $(^2S)3p^2$ | 10.87±0.02 | 10.88 | | | 2.75 |
| | $(^2D)3s^2$ | 12.12±0.02 | 12.05 | 4.82 | | |
| | $(^2P)3s^2$ | 13.71±0.02 | 13.65 | 4.93 | | |
| F⁻ | $(^1D)3s^2$ | 14.85±0.04[c] | | 5.16 | | |
| Cl⁻ | $(^3P)4s^2$ | 8.53±0.05 | 8.53 | 4.49 | | |
| | $(^3P)4s4p$ | 9.15±0.05 | 9.16 | | 3.87 | |
| | $(^1D)4s^2$ | 9.97±0.04 | 9.98 | 4.49 | | |
| | $(^3S)4s^2$ | 12.09±0.06 | 12.14 | 4.39 | | |
| Br⁻ | $(^3P_2)5s^2$ | 7.39±0.06 | | 4.42 | | |
| | $(^1P_{1\,or\,0})5s^2$ | 7.84±0.06 | | (4.36-4.45) | | |
| | $(^1D)5s^2$ | 8.85±0.06 | | 4.46 | | |
| I⁻ | $(^3P_2)6s^2$ | 6.41±0.06 | | 4.04 | | |
| | $(^3P_2)6s6p$ | 6.75±0.06 | | | 3.70 | |
| | $(^3P\,or\,0)6s^2$ | 7.15±0.06 | | (4.10-4.18) | | |
| | $(^1D)6s^2$ | 8.06±0.06 | | 4.09 | | |

a.  See ref. 10 for a listing of calculated values of H⁻.
b.  Reference 23
c.  Reference 24

## AUTOIONIZING STATES

By observing the spectra produced by both the negative
ion and neutral atom of an element, the peaks due to autodetaching states are separated from peaks due to autoionizing
states.  With the ion beam it is possible for both to appear
in a spectrum whereas with the neutral beam only the autoionizing states are found.  Since these states are collisionally excited both optically allowed and forbidden excitations are formed.  Again, helium is found to be the most
efficient target for exciting these states.

The autoionizing states observed have been of three
types:
1.  those immediately above the first ionization potential,
such as the $O^+(^2D)n\ell$ and $O^+(^2P)n\ell'$ levels which lie in
the $O^+(^4S)$ continuum,

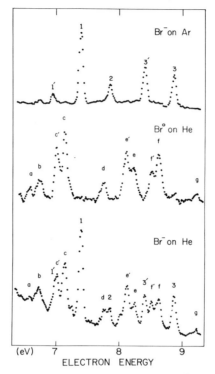

Fig. 3. Electron spectra produced by Br° and Br⁻ beams with helium and argon targets. The numbered peaks are due to Br⁻ states and the lettered peaks are due to Br° states.

2.  inner shell excitations, such as the $1s^2 2s 2p^2 n\ell$ configurations of carbon, and

3.  doubly-excited states, such as the $3p^4 n\ell n\ell'$ configurations of argon.

Identification of these states is made by comparison with photoionization work, whenever possible, and by fitting transition energies into Rydberg series. In making Rydberg series the quantum defects of excited states of the element (Z+1) adjacent to the atom of interest in the periodic table are used as guides for the quantum defects of the autoionizing states. Also, the separation of the parents and grandparents are found to be nearly the same as those listed for optical spectra[22].

Figure 3 shows the electron spectra produced by Br⁻ on He, Br° on He, and Br⁻ on Ar. For a Br⁻ beam both autodetaching and autoionizing transitions are observed with a He target, but for Br° on He only the autoionizing states

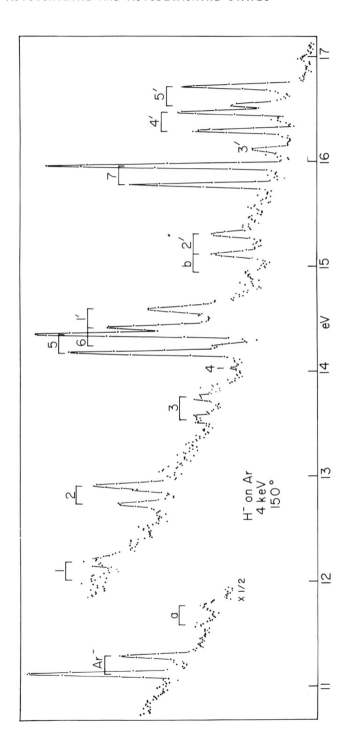

Fig. 4. Electron spectrum produced by $H^-$ on Ar collisions. The two strong transitions near 11 eV are from $Ar^-$ and the others are autoionizing transitions of argon.

appear. With an argon target only negative ion states are excited with the Br⁻ beam.

Figure 4 shows the electron spectrum produced by H⁻ on Ar collisions and the identifications of the transitions are given in Table II. The pairing of the autoionizing transitions in Fig. 4 indicates decay from a single level to the Ar⁺ ²P ground state. The two strongest transitions are those of Ar⁻ which is formed through charge exchange. The many autoionizing transitions which appear are tentatively identified as having as parents the configuration (³P)3d or (³P)4s and attaching another 3d or 4s electron to either parent (only two exceptions have been identified). The separations of the autoionization states formed by adding a 3d (or 4s) is nearly the same as the separation of the parents. Also, the energy difference between (parent)4s and (parent)3d is about 2.4 eV for the doubly excited states compared to 2.3 eV for the singly excited optical states. The spectra produced by H⁻ and H° on rare gases is now being studied.

Table II. Autoionizing States of Argon excited by 4 keV H⁻.

| Peak | Configuration | Trans. Energy (±0.05eV) [c] | Separ. (eV) | Separ. from peaks 1 & 1' | Parent Separ. [d] |
|------|---------------|------------------|-------------|--------------------------|-----------------|
| 1  | $(^3P)3d(^4D)4s$ | 12.19 | 2.23 | | |
| 1' | $(^3P)3d(^4D)3d$ | 14.42 | | | |
| 2  | $(^3P)4s(^2P)4s$ | 12.91 | 2.40 | 0.72 | 0.73 |
| 2' | $(^3P)4s(^2P)3d$ | 15.31 | | 0.89 | |
| 3  | $(^3P)3d(^2P)4s$ | 13.75 | 2.37 | 1.56 | 1.54 |
| 3' | $(^3P)3d(^2P)3d$ | 16.12 | | 1.70 | |
| 4  | $(^3P)3d(^4P)4s$ | 14.03 | 2.44 | 1.84 | 1.85 |
| 4' | $(^3P)3d(^4P)3d$ | 16.47 | | 2.05 | |
| 5  | $(^3P)3d(^2F)4s$ | 14.36 | 2.37 | 2.17 | 2.09 |
| 5' | $(^3P)3d(^2F)3d$ | 16.73 | | 2.31 | |
| 6  | $(^3P)3d(^2D)4s$ | 14.42 | | 2.23 | 2.25 |
| 7  | $(^3P)3d(^4F)3d$ | 15.96 | | 1.54 | 1.22 |
| a  | $3s3p^63d_4$ | 11.77 | | | |
| b  | $(^3P)3d(^4P)4p$ | 15.12 | | | |

c. The energy of the state is found by adding 15.76 eV.
d. Reference 22

BIBLIOGRAPHY

1.  G. J. Schulz, Rev. Mod. Phys. 45, 378 (1973).
2.  Gennadi N. Ogurtsov, Rev. Mod. Phys. 44, 1 (1972).
3.  M. E. Rudd, Invited Papers of the Seventh Internationl
    Conference on the Physics of Electronic and Atomic
    Collisions (North-Holland, Amsterdam, 1971) p. 107;
    M. E. Rudd and J. H. Macek, Case Studies in Atomic
    Physics 3, 47 (1972).
4.  J. S. Risley, Ph.D. dissertation (University of
    Washington, 1973) (unpublished).
5.  A. K. Edwards, Phys. Rev. A (to be published).
6.  D. L. Cunningham and A. K. Edwards, Phys. Rev. A 8,
    2960 (1973).
7.  D. L. Cunningham and A. K. Edwards, Phys. Rev. Lett. 32,
    915 (1974).
8.  A. K. Edwards and D. L. Cunningham, Phys. Rev. A 8,
    168 (1973).
9.  J. S. Risley and R. Geballe, Phys. Rev. A 10, 2206 (1974).
10. J. S. Risley, A. K. Edwards, and R. Geballe, Phys. Rev A
    1115 (1974).
11. A. K. Edwards and D. L. Cunningham, Phys. Rev. A 10, 448
    (1974).
12. Robert L. Chase and Hugh P. Kelly, Phys. Rev. A 6, 2150
    (1972).
13. D. Spence and W. A. Chupka, Phys. Rev. A 10, 71 (1974).
14. A. C. Fung and J. J. Matese, Phys. Rev. A 5, 22 (1972).
15. J. J. Matese, S. P. Rountree, and R. J. W. Henry, Phys.
    Rev. A 7, 846 (1973).
16. J. J. Matese, S. P. Rountree, and R. J. W. Henry, Phys.
    Rev. A 8, 2965 (1973).
17. John J. Matese, Phys. Rev. A 10, 454 (1974).
18. J. J. Matese, S. P. Rountree, and R. J. W. Henry,(unpub-
    lished results).
19. J. S. Risley, Bull. Am. Phys. Soc. 19, 1195 (1974).
20. D. L. Quarterman and A. K. Edwards (to be published).
21. F. D. Schowengerdt, S. R. Smart, and M. E. Rudd, Phys.
    Rev. A 7, 560 (1973).
22. C. Moore, Atomic Energy Levels, Natl. Bur. Std., Circ.
    No. 467 (U.S. GPO, Washington, D. C., 1949), Vol. I, II,
    and III.
23. Naii Lee and A. K. Edwards, Phys. Rev. A 11, 1768 (1975).
24. A. K. Edwards and D. L. Cunningham, Phys. Rev. A 9, 1011
    (1974).

# COLLISION PROCESSES
# IN GAS LASERS AND PLASMAS

# DETERMINATION OF EFFECTIVE CROSS-SECTIONS OF VARIOUS ELEMENTARY PROCESSES FROM LOW TEMPERATURE PLASMA DATA

N.P. Penkin, State University of Leningrad,

Leningrad 199164, USSR

Studies of gas-discharge low-temperature plasma of low-pressure by the combination of different optical and electric methods allows the determination of effective cross-sections of different elementary processes. For a number of some processes, the precision of the effective cross-sections of different elementary processes. For a number of some processes, the precision of the effective cross-section determinations from the data on plasma studies is not worse, but sometimes even better, compared to the precision of measurements performed by the beam-methods.

"Plasma" measurement and experiments with beams should not be opposed to one another, since complementing and supplementing each other, they permit to obtain more complete data on effective cross-sections of different collision processes.

To-day I shall report some of the results of determination of effective cross-sections of electron-atom and atom-atom collisions, as obtained from plasma studies, performed at the Optics Department of the Leningrad University.

For the purpose of determining the effective cross-sections of collisions, one may use either stationary or decaying gas-discharge low-temperature plasma of low pressure. In the decaying plasma, the damping of different processes does not occur simul-

taneously, which allows to separate them from each
other. That is why the interpretation of the experimental data in this case is easier than in the case
of stationary plasma. In a general case, stationary
equation has so many terms that it does not allow to
separate the term of interest, and hence, to determine the cross-section of the process. Nevertheless
some preliminary investigations permit to choose such
discharge conditions, that the equation has only a
small number of terms, the term corresponding to the
process under investigation being the main, the determining one. That is why, in all our studies, the
determination of effective cross-sections was always
preceeded by the study of gas-discharge plasma using
different optical and electrical methods /1/.

In a series of experimental and theoretical
studies made at our department under Prof. Ju.M.Kagan
/2/, it has been shown that low and medium pressure plasm
may show strong deviations of electron energy distribution function F(V) from the Maxwellian one. The
characteristic feature of these deviations is the
lack of high-speed electrons in the real distribution
function.

To obtain trustworthy data on effective cross-
-sections of electron-atom collisions, it is necessary to know the real electron energy distribution
function. That is why, in all our studies of that
kind, measurements were made of the electron energy
distribution function. On the other hand, as will be
further discussed in more detail, the distribution
function itself, being influenced by atom-electron
and atom-atom collisions, may be used for the determination of effective cross-sections of such kind
collisions.

1. EFFECTIVE CROSS-SECTIONS OF EXCITATION
   AND MIXING OF THE $n^3P_{0,1,2}$ LEVELS OF
   Zn I, Cd I AND Hg I IN ELECTRON AND
   ATOM COLLISIONS

In my papers with T.P.Redko /3/ effective cross-
-sections of the following processes are determined

$$A(n\,{}^1S_0) + e \rightarrow A(n^3P_j) + e; \quad j = 0,1,2 \qquad (1)$$

$$A(n^3P_j) + e \rightarrow A(n^3P_{j'}) + e; \quad j' = 0,1,2; \; j' \neq j \qquad (2)$$

$$A(n\,^3P_j) + A(n'S_o) \rightarrow A(n\,^3P_{j'}) + A(n'S_o); \quad j \neq j' \tag{3}$$

Which this purpose stationary and decaying zinc, cadmium and mercury plasmas of low pressure have been investigated using optical and probe methods. It is possible to choose such discharge conditions for which a stationarity equation for some level j has the following form:

$$\Delta(N_o\,N_e) + \Delta(N_o\,N_{3P_{j'}}) + \Delta(N_e N_{3P_{j'}}) + \Delta(N_k\,A_{kj}) = \frac{N_{3P_j}}{\mathcal{T}_{eff}\,^3P_j} \tag{I}$$

The left-hand terms of the equation represent the number of excitations of the level j resulted from processes (1), (2) and (3) and cascade transitions. The influence of the cascade can be taken into account through the measurement of the intensity of spectral lines terminating at the level $^3P_j$. The effective lifetime at the level $j - \mathcal{T}_{eff}\,^3P_j$ is due to all the processes destroying this level.

For the measurement of $\mathcal{T}_{eff}\,^3P_j$ the decaying plasma was investigated by absorption or emission methods. The pressure of metal vapours and the initial strength of the discharge current varied within the wide limits (p = $10^{-3}$–$10^{-1}$mm Hg; i = 50ma–2a).

The experimental installation is described in /3/. Processing the curves of the fall decrease of absorption it is possible to determine the value

$$A_\alpha = \frac{\int E_v\,(1 - e^{-K_v\cdot\ell})\,d\nu}{\int E_v\,d\nu} \tag{II}$$

which is the function of the optical density $K_o\ell \sim Nf\ell$ As an example, Fig.1 presents the dependences of the value $\ell n\,(K_o\ell)$ on t for the $5\,^3P_o$; $5\,^3P_1$ and $5\,^3P_2$ levels of $CdI$ In the late afterglow (t < 100 $\mu$sec) the decaying curves of the $5\,^3P_o$ and $5\,^3P_1$ levels, are parallel. This indicates that the decay of the emitting level $5\,^3P_1$ at such times, with small concentrations of electrons is determined by the following process:

$$Cd(5\,^3P_o) + Cd(5'S_o) \rightarrow Cd(5\,^3P_1) + Cd(5'S_o) \tag{4}$$

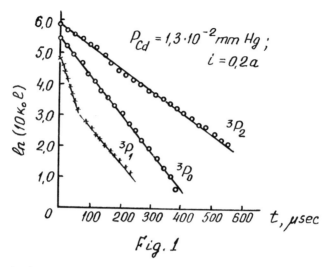

$$P_{Cd} = 1,3 \cdot 10^{-2} mm \, Hg \; ;$$
$$i = 0,2 a$$

Fig. 1

From such kind of curves it is possible to determine cross-sections $Q_{3P_j \rightarrow {}^3P_{j'}}$ of similar processes:

$$N_{3P_o} \cdot N_{1S_o} \cdot Q_{3P_o \rightarrow {}^3P_1} \cdot \overline{v} = \frac{N_{3P_1}}{\tau_{eff\,{}^3P_1}} \qquad (III)$$

The cross-sections of zinc and cadmium atoms thus determined are given below.

The experiments made with different currents and pressures permit the determination of the cross-sections of mixing of the $n^3P_{0,1,2}$ levels by the atom and electron collisions. The results of such a determination are given in Table 1. One should pay attention to large cross-sections ($10^{-14}$ cm$^2$) for mixing of the levels of a zinc atom in the collisions with electrons.

In the case of a low-pressure discharge and small densities of the current for cadmium and mercury in the left part of equation (1) only the first term remains, and it is possible to write:

$$N_o N_e \langle \sigma_{oj} \, v_e \rangle = \frac{N_{3P_j}}{\tau_{eff}} \qquad (IV)$$

In the case of zinc:

$$N_o N_e \langle \sigma_{oj} \, v_e \rangle + \sum_{j'} N_o N_{3P_{j'}} \langle Q_{j'j} \, v \rangle = \frac{N_{3P_j}}{\tau_{eff\,{}^3P_j}} \qquad (V)$$

Table 1

| Transi-tion | Mixing by the atomic impact | | Mixing by the electronic impact | |
|---|---|---|---|---|
| | Zn | Cd | Zn | Hg |
| | $Q \cdot 10^{16} cm^2$ $T_{at} = 500°C$ | $Q \cdot 10^{16} cm^2$ $T_{at} = 350°C$ | $\bar{\sigma} \, 10^{16} cm^2$ $T_e = (3-4) \cdot 10^3 K$ | $\sigma(v_{max}) \cdot 10^{16} cm^2$ |
| $3P_{0-1}$ | >10 | 8 | | 25 |
| $3P_{0-2}$ | 1.0 | | 160 | |
| $3P_{1-0}$ | >10 | 3 | | 10 |
| $3P_{1-2}$ | 0.6 | | 120 | 14 |
| $3P_{2-0}$ | 0.4 | | 40 | |
| $3P_{2-1}$ | 0.6 | | 80 | 26 |

The second term in the left can be calculated because the cross-sections $Q_{jj'}$ are known from the decay curves constructed with small concentrations of electrons (Ne < $10^{10}$cm$^{-3}$).

From the results of our measurements for electron-atom collisions the value of

$$\langle \sigma_{oj} \, v_e \rangle = \int_{v_a}^{\infty} \sigma_{oj} (v) \cdot F(v) \sqrt{v} \cdot dV$$

is determined. Hence, measuring by the probe method the function of electron energy distribution and knowing the shape of the excitation function it is possible to obtain the cross-section in the maximum of the excitation function $\sigma_{oj} (V_{max})$. If the shape of the excitation function is unknown, one can obtain only $\bar{\sigma}_{oj}$ - the effective cross-section averaged over the Maxwell distribution.

Table 2 demonstrates the values for $\sigma_{oj} (V_{max})$ obtained by ourselves. It also presents the values of $\sigma_{oj} (V)_{max}$ for the mercury atom obtained by other authors and by other methods. A satisfactory agreement between our data and those of other authors is seen.

Table 2

$$A(n'S_0) + e \rightarrow A^*(n^3P_{0,1,2}) + e$$

| Level | Zn | Cd | $H_g$ | | | | |
|---|---|---|---|---|---|---|---|
| | | | This work | [4] Borst | [5] Iongerius | [6] Zapesochny | [7] Korotkov |
| $^3P_2$ | 1.5 | 1.8 | 2.2 | 3.0 | – | – | 2.9 |
| $^3P_1$ | 1.0 | 1.6 | 1.3 | – | 0.9 | 1.3 | – |
| $^3P_0$ | 0.5 | 1.2 | 0.6 | – | – | – | 0.5 |

## 2.   DIFFUSION OF EXCITED ATOMS IN THE PROPER GAS AND GASES ADMIXTURES

As known in the cases of no quenching of metastable atoms by the collisions with electrons and atoms one can easily determine diffusion coefficients from the decay curves.

For this purpose one can use the installation described in our paper /3/. In each particular case it is necessary to make an analysis of volume losses and to choose such conditions for the experiment that $\frac{1}{\tau} = \frac{D}{\Lambda^2}$ . For the cylindrical tube:

$$\frac{1}{\Lambda^2} = \frac{1}{D\tau} = \left(\frac{2,405}{\tau}\right)^2 + \left(\frac{\pi}{\ell}\right)^2$$

The values of D of excited Hg I and Cd I in the proper gas and in some inert gases have been measured by myself, T.P.Redko and N.A.Krykov /8/. We also collected and analysed data on diffusion coefficients of excited atoms. The analysis of the data shows: 1) that systematic measurements in mixtures are scarce, 2) in the case of the diffusion of the excited atoms in the proper gas, the diffusion coefficient is approximately twice smaller than for the atoms in normal state. In the diffusion in the gaseous mixture the diffusion coefficients of excited and normal atoms are close in value.

Temperature dependence of diffusion coefficients of the metastable atoms of mercury $6^3P_0$ and $6^3P_2$ in the mercury vapour and in inert gases (neon and argon) is studied in our works.

The results of measurements in the mixture of mercury and neon and argon are given in Fig.2.  A more rapid growth of diffusion coefficients with temperature than that expected for the Lennard-Jones potential is observed.

It should be noted that theoretical and experimental data on diffusion coefficients of excited atoms are too scarce to satisfy the requirements of physics and chemistry of plasma.

3.  EFFECTIVE CROSS-SECTIONS FOR STEP-BY-STEP EXCITATION OF HELIUM AND NEON ATOMS IN COLLISIONS WITH ELECTRONS

To measure the effective cross-sections for the reactions:

$$A_m^* + \bar{e} \longrightarrow A^{**} + e \qquad (5)$$

$$A_m^* + \bar{e} \longrightarrow A^{+*} + 2e \qquad (6)$$

A.A.Mitureva and I used the experimental arrangement schematically presented in Fig.3.  Metastable helium or neon atoms diffused from the discharge tube into the excitation tube, whose collision space was penetrated by a modulated electron beam.  The monokineticity of the electron beam was equal to 1.5 eV, the velocity of electrons was changed from the threshold of the process under investigation up to

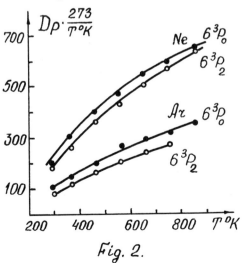

Fig. 2.

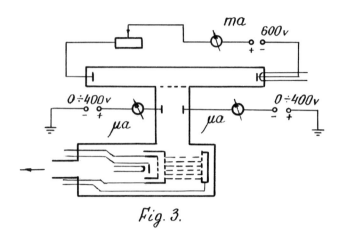

*Fig. 3.*

110 V. The electric field between the electrodes,
located in the branch pipe connecting the discharge
tube with the excitation tube, did not allow the
charged particles to penetrate into the collision
space from the plasma. The pressure of the gas under
study was equal to 0.1 mm Hg. Intensities of spectral lines originating with the decay of the excited
atoms ($A^{**} \rightarrow A^* + h\nu$) and also the concentration of the
metastable atoms $A_m^*$ were measured using a spectral
arrangement. The concentration of the metastable
atoms was measured by the absorption method: for this
purpose, the collision space was illuminated by a
line light source. Using this arrangement, study has
been made of spectra originating due to the direct
and step-by-step excitation of helium and neon atoms
by the electron impact. As an example, Figs. 4, 5
and 6 show the step-by-step and direct excitation
functions for the lines He I ( $\lambda$ 501.6 nm), Ne I ( $\lambda$
640.2 nm) and Ne II ( $\lambda$ 332.4 nm). Step-by-step excitation functions are characterized by a narrow maximum near the threshold. The absolute cross-section
values were obtained by comparison of direct and
step-by-step excitation of spectral lines and from
measurements of metastable atom concentration in the
electron beam zone by the absorption method.

It is seen that the effective cross-sections for
the step-by-step excitation processes are by some orders of magnitude greater than the effective cross-
-sections for the direct processes.

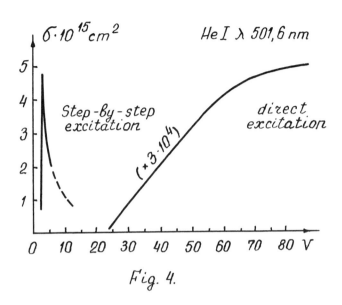

Fig. 4.

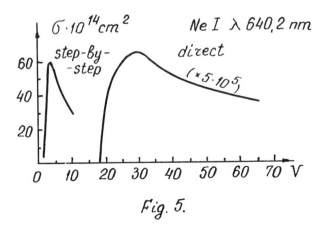

Fig. 5.

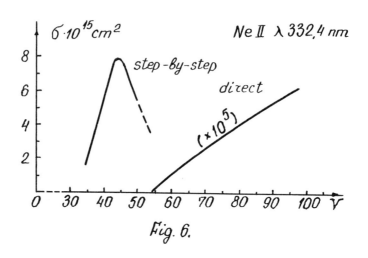

Fig. 6.

## 4. DETERMINATION OF ATOM-ATOM COLLISIONS FROM PLASMA STUDIES USING OPTICAL PUMPING

Effective cross-sections for different atom-atom collisions at thermal energies may be determined with high enough precision from the investigation of plasma, if use is made of optical pumping. The latter changes concentration of excited atoms, $A^*$, in reactions of the following type:

$$A^* + B \longrightarrow A + B^* \pm \Delta E \qquad (7)$$

$$A^* + A \longrightarrow A + A'^* \pm \Delta E \qquad (8)$$

$$A^* + B \longrightarrow A'^* + B \pm \Delta E \qquad (9)$$

Measuring concentration changes in $A^*$, $B^*$ and $A^*$ atoms by the absorption of emission methods and knowing concentration of A and B atoms ($p = N \kappa T$), one can easily obtain effective cross-sections for these reactions. It is possible to choose such experimental conditions, under which optical pumping will cause no change in the electron kinetic characteristics of plasma. The above process can, therefore, be separated in their pure form. That is why, recent years have produced a much greater number of works, using optical pumping.

As an example, I shall give a brief account of works which were performed at the Optics Department of Leningrad University.

In the work carried out jointly by J.Z.Jonih, A.L.Kuranov and myself /9/, study has been made of the following reactions:

$$He\,(2\,{}^1S_0) + Ne_o \rightarrow He_o + Ne\,(3s) \qquad (10)$$

$$He\,(2\,{}^1S_0) + Ne_o \rightarrow He_o + Ne\,(4d) \qquad (11)$$

These reactions, as is well-known, are of great importance for the formation of the inversion population in the neon-helium laser.

The change of population of $2\,{}^1S_0\,He\,I$ level was done by optical pumping ( $\lambda$ 2058 nm He I, $2\,{}^1S_0 - 2\,{}^1P_1$ ). Electron temperature and electron concentration ($10^9 - 10^{10}$ cm$^{-3}$) did not change, by the optical pumping.

As a result of the absorption of the line 203.8 He I, the population of the level 2'So was decreasing, which led, according to the two latter reactions, to the decrease of population of the 3 s and 4 d levels Ne I and, hence, to the attenuation of spectral lines going from these levels to the 2 p levels. The change in the 2'So level population ( $\delta$ Nm) was measured by the reabsorption method, and the change of the transitions from the 3 s and 4 d levels ( $\delta$ Nik) was determined from the decrease of brightness Ne I spectral lines.

Effective cross-sections for reactions (10) and (11) $Q_i(v)$ and $\bar{Q}_i$ , can be found from the stationary equation:

$$N_o \langle Q(v)\,v \rangle\, \delta N_m = \sum_K \delta N_{iK} \qquad (VI)$$

$$\bar{Q}_i = \frac{\langle Q_i(v)\,v \rangle}{\int\limits_{v_i}^{\infty} v f(v)\,dv} \qquad (VII)$$

Results of $Q_i$ measurements are presented in the Table.3. From the Table it is seen, that the value of the effective cross-sections for reactions (10) and (11) is of the order of $10^{-16}$ cm$^2$. For the

Table 3

| i | $\bar{Q}_i \cdot 10^{17}$ cm$^2$ | | i | $\bar{Q}_i \cdot 10^{17}$ cm$^2$ |
| --- | --- | --- | --- | --- |
| | this work | other data | | this work |
| $3s_2$ | 57.8 | 22 30 7 | $4d_1' + 4d_1''$ | 9.3 |
| $3s_3$ | 0.13 | 0.29 | $4d_2$ | 3.3 |
| $3s_4$ | 3 | < 0.1 50 0.8 | $4d_3$ | 5.0 |
| | | | $4d_4$ | 11.0 |
| | | | $4d_4'$ | 9.7 |
| $3s_5$ | 3 | 1.3 | $4d_5 + 4d_6$ | 6.7 |

second reaction the values are of the same order in spite of the fact that 4 d levels are by 0.1 eV higher than 2'So level of He I.  Our results permit determination of total effective cross-section for the destruction of the level 2'So He I by neon. Averaged by rates, it is equal to 5.2 10$^{-16}$cm$^2$ ; destruction of this value 1.1 10$^{-16}$cm$^2$ falls to the de-excitation with the energy transfer to the 4 d levels.

A.M.Shuhtin, M.K.Shevtsov and A.S.Tibilov /10/ have made an experimental study of the process of excitation transfer from 6'D$_2$ and 4 $^3$F$_{2,3,4}$ levels of cadmium atoms to its neighbouring levels ( E < 0.1 eV) by collisions with helium, neon, argon and cadmium atoms.

$$Cd(6'D_2) + A \rightarrow Cd(7^3P_{0,1,2}; 6^3D_{1,2,3}) + A \pm \Delta E \qquad (12)$$

$$Cd(4^3F_{2,3,4}) + A \rightarrow Cd(7^3P_{0,1,2}; 6^3D_{1,2,3}; 6'D_2; 7'P_1) \pm \Delta E \qquad (13)$$

$$Cd(6\,{}^1D_2; 4\,{}^3F_{2,3,4}) + Cd(5\,{}^1S_0) + Cd(7\,{}^3P_2; 7\,{}^3P_{2,1,0}; 7\,{}^1P_1; 6\,{}^1D_2) + Cd(5\,{}^1S_0) \pm \Delta E \quad (14)$$

Here A is the inert gas atom.
To study these reactions they employed the method of
optical pumping of cadmium plasma by means of impulse
cadmium laser operating by atomic transitions.  Let
me point out here that the above investigators also
created impulse zink and magnesium vapours lasers,
generating a large number of atom lines /11/.

    Their experimental device is schematically pre-
sented in Fig.7.  The discharge tube was placed with
the resonator, coaxially with the laser tube (7).
The discharge tube was fed from the stabilized source
of current (2; i=30 ma) and was filled with the mixture of cadmium vapour ($P_{Cd}$ = 2·10⁻² mm Hg) with helium, neon or argon.  The pressure of the inert gas
was changed from 0.3 to 3 mm Hg.  The prism (on the
figure not indicated) placed into the resonator, separated from the laser emission, pumping lines
    $\lambda$ 1912 nm ($6{}^1D_2-6{}^1P_1$) or   $\lambda$ 1648 nm ($4\,{}^3F_{2,3,4}$ -
- $5\,{}^3D_3$).

    The corresponding double-channel scheme (3-5
and 4-6) permitted determination of intensities for
lines, going from $6{}^1D_2$ and $4\,{}^3F_{2,3,4}$ levels and the
levels to which the excitation was transferred by
collisions.

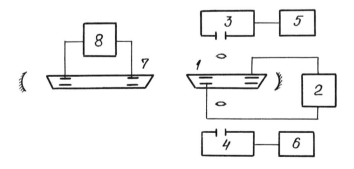

*Fig. 7.*

Table 4

| Transition | $\Delta E$ eV | $Q\ (\text{Å}^2)$ | | | |
|---|---|---|---|---|---|
| | | He | Ne | Ar | Cd |
| $6^1D_2 - 7^3P_2$ | +0.005 | 2.6 | 2.8 | 15 | 120 |
| $6^1D_2 - 7^3P_1$ | +0.01 | 1.7 | 2.2 | 5 | |
| $6^1D_2 - 7^3P_0$ | +0.02 | <0.5 | <0.5 | <2 | |
| $6^1D_2 - 6^3D_3$ | −0.03 | 0.3 | 0.6 | 3.0 | |
| $6^1D_2 - 6^3D_2$ | −0.03 | 0.9 | 0.9 | 6.0 | |
| $6^1D_2 - 6^3D_1$ | −0.03 | 0.2 | 0.9 | 1.0 | |
| $4^3F_{2,3,4} - 7^3P_2$ | +0.05 | 0.32 | | | 100 |
| $4^3F_{2,3,4} - 7^3P_1$ | +0.06 | 0.21 | | | 90 |
| $4^3F_{2,3,4} - 7^3P_0$ | +0.07 | 0.20 | | | 90 |
| $4^3F_{2,3,4} - 7^1P_1$ | +0.01 | 0.003 | | | 30 |
| $4^3F_{2,3,4} - 6^1D_2$ | +0.05 | 0.04 | | | 20 |
| $4^3F_{2,3,4} - 6^3D_{1,2,3}$ | +0.02 | 0.34 | | | |

Results of cross-section measurements for all the above reactions are presented in the Table 4. From these data it follows that:

1. The values of effective cross-sections for energy transfer from $6'D_2$ and $4^3F_{2,3,4}$ levels of the cadmium atom to its neighbouring levels ($\Delta E < 0.1$eV) by collisions with atoms of inert gases are in the range of $10^{-19}$ to $10^{-15}$cm$^2$.

2. For all cadmium levels, the cross-sections for the energy transfer increase from helium to argon.

3. Effective cross-sections for cadmium-cadmium collisions are by 2-4 orders of magnitude higher than effective cross-sections for collisions of cadmium atoms with atoms of inert gases and have the values of $10^{-15} - 10^{-14}$ cm$^2$.

5.  DETERMINATION OF EFFECTIVE CROSS-SECTIONS OF
    COLLISIONS FROM THE STUDY OF THE ELECTRON
    ENERGY DISTRIBUTION FUNCTION IN THE DECAYING
    PLASMA

Collisions of metastable atoms with electrons and between each other result in the production of high-speed electrons. For helium reactions of such collisions may be written in the following way:

$$He\,(2\,^3S_1)\ +\ e\ \rightarrow\ He\,(1\,^1S_0)\ +\ \overline{e} \tag{15}$$

$$He\,(2\,^3S_1)\ +He\,(2\,^3S_1)\ \rightarrow\ He_2^+\ +e \tag{16}$$

$$He\,(2\,^3S_1)\ +He\,(2\,^3S_1)\ \rightarrow\ He^+ + He + e \tag{17}$$

These reactions occur both in the stationary and decaying plasma. They must tell upon the form of the electron energy distribution function. In the decaying plasma, in which the field is absent, reactions (15)-(17) must result in the formation of maxima on the electron energy distribution function.

In fact, investigations of the decaying helium plasma, performed by Prof. Y.M.Kagan's and his collaborators /12/ have shown that the electron energy distribution function in the range from 4 to 20 eV is of a complex nonmonotonous character. As an example, Fig.8 gives electron distribution functions at different late afterglow time (from 50 to 450 μsec). Helium pressure in the discharge tube (d = = 2.7 cm; L = 35 cm) was 1 mm Hg; initial discharge currents were 0.45 and 1.8a.

The figure shows that the distribution functions have two maxima: one at about 20, and the other at about 15 eV. The first maxima is due to electrons, resulting from the reaction (15), the second one results from reactions (16) and (17). The latter assertion is supported by calculation data of Ingraham and Brown /13/, who found that reaction (16) produces monokinetic electrons with the energy of 15 eV, while reaction (17) gives electrons with a continuous energy spectrum in the range from 0 to 14.5 eV.

Measurements performed by the afterglow absorption method, have shown that $N_{2^1S_0} \ll N_{2^3S_1}$ . That is why, reactions analogous to those of (15)-(17) with the participation of 2'So HeI do not affect the form

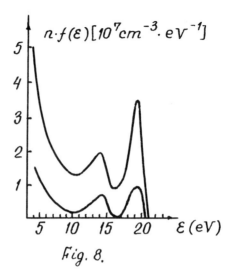

*Fig. 8.*

of the electron distribution function in the afterglow.

On measuring absolute concentrations of electrons by the probe method those of the metastable
atoms by the absorption method, one can find reaction rate constants for (15)-(17).

Rate constant, $\beta_1$ , for the reaction
$$He(2^3S_1) + e \rightarrow He(1^1S_0)' + e$$
may be found from the ratio
$$\beta \cdot N_m \, Ne = \frac{Ne'}{\tau} \qquad\qquad (VIII)$$

Here Ne' is the electron concentration, corresponding to the maximum, lying at about 20 eV, $\tau$ - is
the life-time of these electrons, Nm and Ne are the
concentrations of the metastables and electrons,
respectively (Ne $\approx 10^3$Ne'). Thus defined, $\beta_1$ -value,
depending on the experimental conditions changed
from 2.7 $10^{-9}$ to 0.8 $10^{-9}$cm$^3$ sec$^{-1}$, which gives the
effective cross-section value for this reaction at
0.8 $10^{-16}$cm$^2$.

The above curves do not allow to determine (16)
and (17) reaction rates, $\beta_2$ and $\beta_3$ , separately.
Nevertheless, they permit to find $\beta_m = \frac{\beta_2 + \beta_3}{3}$ . The
value of this constant, $\beta_m = 4.2 \ 10^{-9}$cm$^3$sec$^{-1}$, is
in fairly good agreement with values cited in literature (1.3-4.6) $10^{-9}$cm$^3$sec$^{-1}$).

There are reasons for believing that (choosing the discharge conditions) through a special arrangement of the discharge conditions and using, in the probe method, differentiating signal of a small amplitude, it will be possible to determine values of coefficients $\beta_2$ and $\beta_3$.

1. N.P.Penkin, Spektroskopiia gazorazriadnoi plazmy, 274; "Nauka", 1970. 2. Yu.M.Kagan, Spektroskopiia gazorazriadnoi plazmy, 201; "Nauka", 1970. 3. N.P. Penkin, T.P.Redko, Opt. i spektr. 22, 699, 1967; 23, 475, 1967; 30, 3, 1971; 30, 359, 1971; 36, 447, 1974. 4. W.L.Borst, Phys.Rev. 181, 257, 1969. 5. H.M.Jongerius, Proefchrift Utrecht, 1961. 6. O.B.Shpenik, I.P.Zapesochnyi, Opt. i spektr. 23, 15, 1967. 7. A.I.Korotkov, N.A.Prilezhaeva, Izv. vuzov, fizika 12, 85, 1970. 8. N.P.Penkin, T.P.Redko, Opt. i spek spektr. 36, 226, 1974. 9. Yu.Z.Ionich, N.P.Penkin, Opt. i spektr. 31, 837, 1971; Yu.Z.Ionich, N.P.Penkin, A.L.Kuranov, Opt. i spektr. 34, 814, 1973. 10. A.C. Tibilov, M.K.Shevtsov, A.M.Shuhtin, Opt. i spektr. 35, 626, 1973. 11. A.N.Dubrovin, A.S.Tibilov, M.K. Shevtsov, Opt. i spektr. 32, 1252, 1972. 12. A.B. Blagoev, Yu.M.Kagan, N.B.Kolokolov, R.I.Liaguschenko J.Tech.Phys. 44, 333, 1974; 44, 339, 1974. 13. J.C. Ingraham, S.C.Brown, Phys.Rev. 138, 1015, 1965.

STUDY OF MOMENTUM TRANSFER DISTRIBUTIONS IN ROTATIONALLY
INELASTIC COLLISIONS[*]

William K. Bischel, University of California, Dept.
of Applied Science, Davis--Livermore, Calif. 94550

Charles K. Rhodes,[†] Lawrence Livermore Laboratory,
Livermore, Calif. 94550

ABSTRACT

We present an experimental technique utilizing the methods
of laser saturation spectroscopy that enables a direct observation of the momentum transfer distribution generated by rotationally inelastic molecular collisions. This method has been applied
to $CO_2$ in collision with a wide variety of partners including He,
$H_2$, $CO_2$, $CH_3F$, and Kr. The observations indicate that in a large
number of cases the rotational transition results primarily from
peripheral collisions which are effective in transferring angular
momentum, but which communicate a relatively small linear momentum
transfer.

I   INTRODUCTION

Fundamental knowledge concerning the detailed aspects of
atomic and molecular collisions is of considerable scientific and
practical importance. Although studies involving molecular
beams[1] provide us with the most refined data, many other methods
including coherent spectroscopy[2] and the consideration of collisional effects on spectral line profiles[3,4] have been used.
These latter techniques, although relatively simple to implement
experimentally, are not sufficiently precise in the state selection of the colliding particles, and therefore, are often insensitive to many details of the interaction potential. For
example, total cross-section data do not exhibit interferences
between partial waves, an aspect which is both an informative

and striking characteristic of the angular distributions available from molecular-beam experiments.

Molecular interactions are complicated phenomena. One of the essential complexities arises from their normally non-central nature, a fact which significantly multiplies both experimental and theoretical considerations.[5] The central and noncentral components of the potential differ, however, in an important way. Central forces are incapable of developing a torque and are, consequently, ineffective in producing transitions involving a change in the angular momentum of the colliding systems. Such transitions are generated entirely by the anisotropic component of the force field. A common example of this phenomenon is rotational relaxation of molecules in the gas phase. An understanding of these processes is important for a broad range of problems ranging from astrophysical[6] questions to coherent pulse propagation.[7]

We report below the characteristics of an experimental approach[8,9] which isolates some of the properties of the non-central interaction of molecular systems. In this work we observe the linear momentum distribution generated by rotationally inelastic encounters of $CO_2$ with a variety of collision partners. From the previous discussion, the rotational inelasticity signifies the participation of the noncentral forces. We note that the extension of this experimental technique, which is one dimensional in nature, to three spatial dimensions will enable studies equivalent to molecular-beam investigations in the examination of differential scattering cross sections.

The general process under examination is

$$CO_2(j,\vec{v}) + M \rightleftarrows CO_2(j',\vec{v}') + M \tag{1}$$

where the pair $(j,\vec{v})$ labels, respectively, the rotational quantum number and velocity of the $CO_2$ molecule, and M represents the appropriate collision partner. In the experimental technique described below, the methods of saturation spectroscopy are utilized to detect the velocity distribution arising from rotationally inelastic collisional processes and therefore provide an evaluation of the mean change of one component (z-component) of the molecular velocity, viz., $(\vec{v} - \vec{v}')_z\big|_{ave} \equiv \langle \Delta v_z \rangle$, in these collisions. These data provide direct information on the characteristics of the intermolecular forces[5] and supplement the findings of alternative methods such as pressure broadening and kinetic studies.[10]

## II EXPERIMENTAL METHOD

Basically, the experimental technique involved the generation of a perturbed velocity distribution of a particular rotational level by velocity selective laser-induced saturation of
specific molecular transition and the detection of the collisional transfer of this perturbation to a nearby rotational
state using the apparatus illustrated in Figure 1.  Two pairs
of stable single $TEM_{oo}$-mode $CO_2$ sources[11] (1,2) and (3,4) were
used; the two labeled 1 and 4 served as local oscillators
stabilized[12] to the centers of their respective P(j) and P(j′)
operating transitions.  The beams from oscillators 2 and 3, with
intensities of 20 and 2 $W/cm^2$, respectively, were passed co-
linearly through an absorption cell which contained a gaseous
mixture of $CO_2$ and the collision partner M.  Part of the beams
$\omega_2$ and $\omega_3$ were heterodyned with their respective local oscillators,
establishing the frequency measurements $|\omega_1 - \omega_2|$ and $|\omega_3 - \omega_4|$.

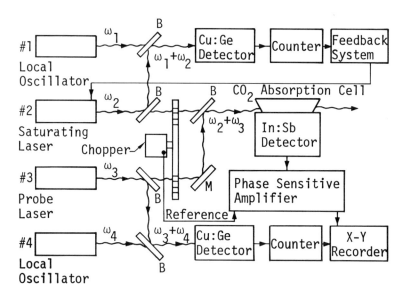

Figure 1.  Schematic diagram of the experimental apparatus
illustrating the two pairs (1,2) and (3,4) of stable probes
and local $CO_2$ oscillators operating on P(j) and P(j′) transitions, respectively.  The probe beams $\omega_2$ and $\omega_3$ are combined by
a beam splitter (B) and pass in spatial coincidence through
the absorption cell.  Modulation of the two probe beams is provided by the chopper.  The oscillator frequencies are established
by heterodyne techniques.

The 4.3-μm fluorescence arising from the $00^O1 - 00^O0$ transition in the $CO_2$ contained in the cell falls on a liquid-nitrogen-cooled InSb detector and is observed by mechanically chopping the $\omega_2$ beam at 780 Hz and the $\omega_3$ beam at 540 Hz, and detecting synchronously at the sum frequency of 1320 Hz as oscillator 3 was scanned in frequency. Detection of the component at the sum frequency greatly reduces the background of 4.3-μm fluorescence which is of no interest in this experiment. The 4.3-μm intensity and the frequency of oscillator 3 were simultaneously recorded on magnetic tape for subsequent numerical data reduction.

## III    RESULTS

Figure 2 illustrates typical data obtained with this experimental method for a mixture of $CO_2$ and $CH_3F$. In this example, the total pressure was 30 mTorr in a mixture with the ratio $CO_2:CH_3F = 1:3$ and the oscillators 2 and 3 were operating on the 10.6-μm P(20) and P(18) transitions, respectively. We are, therefore, examining a case in which $\Delta j \equiv |j - j'| = 2$. Furthermore, oscillator 2 was locked 12 MHz above line center causing the perturbed velocity distribution to be displaced in the manner shown. We note that the observed width of 4.7 MHz is significantly greater than that due to the radiatively produced perturbation ($\sim$1.7 MHz for these conditions[8,13]) as a result of collisionally induced changes in the velocity. From a study of the pressure and concentration dependence of these collisionally induced profiles it is possible to deduce the distribution of linear momentum transfer in rotationally inelastic collisions, and therefore obtain direct information on anisotropic molecular interactions.

This method is capable of isolating the contribution from a single rotational level. Data demonstrating this effect are illustrated in Figure 3. For the experimental configuration illustrated in Figure 1 the influence of the $00^O1$ level is isolated when oscillator 2 is operated on the 10.6-μm P(20) transition and oscillator 3 is operated on the 9.6-μm P(18) transition. No contribution from the lower $10^O0$ and $02^O0$ levels is expected under these conditions, since the $10^O0 - 02^O0$ collisional rate[14] is much less than that associated with rotational transfer and essentially complete velocity redistribution will be generated in collisionally induced transitions between these two states. The data of Figure 3 correspond to

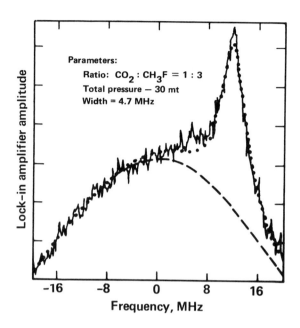

Figure 2.   Saturated resonance observed in the 4.3-μm $CO_2$
fluorescence for a mixture of $CO_2$ and $CH_3F$ using the apparatus
of Figure 1.   The abscissa is the frequency $\omega_3 - \omega_4$.   Oscillators 2 and 3 are operating, respectively, on the 10.6-μm P(20) and
P(18) transitions with oscillator 2 locked 12 MHz above line
center.   Dotted line, Gaussian plus Lorentzian (centered at
+12 MHz) fit to the total signal.   Dashed Line, computer determined Gaussian background contribution to the total signal.

a gas mixture composed of 10 mTorr of $CO_2$ and 20 mTorr of $H_2$.
For these data oscillator 2 was locked 12 MHz above the 10.6-μm
P(20) line center while oscillator 3 was scanned slowly in
frequency across the gain profile of the 9.6-μm P(18) transition.
The resonant feature centered at 12 MHz is clearly visible and
has a measured width of ~11.5 MHz.   This result is attributed
entirely to rotationally inelastic processes occurring in the
$00^01$ manifold.

   A brief investigation has also been made of the dependence
of the four-level, collision-induced resonance on the value of
$|j - j'|$.   For the case of $CO_2$ in collision with $H_2$, the velocity-
selective resonance has been observed for[15] $|j - j'| = 2$ and 4,

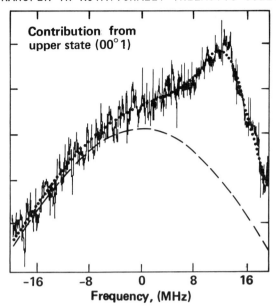

Figure 3.  Saturated resonance observed in the 4.3-µm fluorescence (solid line) using the apparatus described in Ref. 9.
The abscissa is the beat frequency between the probe and reference oscillators [9.6-µm P(18) line].  The saturating
oscillator was locked at +12 MHz off line center [10.6-µm
P(20) line].  Dotted line, Gaussian plus Lorentzian (centered
at +12 MHz) fit to the total signal.  Dashed line, computer-
determined Gaussian background contribution to the total signal.


but was not detected for $|j - j'| > 4$.  This observation is
related to data obtained from line-broadening studies[16] which
can be used to estimate the maximum orbital angular momentum
$l_{max}$ operative in the collision.  For $CO_2$-$H_2$ collisions at $300°K$,
$l_{max} \cong 4$, a value which is consistent with our observations.

     In an attempt to describe these four-level, collision-
induced resonances, we are employing an analysis similar to that
of Freed and Haus,[17] but generalized to include rotationally
inelastic collisions.  A similar treatment has been given by
Johns et al.[18]  Ignoring the roles of particle diffusion and
polarization effects,[19] we may write the equations of motion
for the velocity-distribution functions of $f_j(\vec{v})$ and $f_{j'}(\vec{v})$ of
the lower two levels, including the influence of the saturating
radiative fields $I_j$ and $I_{j'}$, as

$$\partial f_j(\vec{v})/\partial t = \sum_{i=j,j'} \left\{ \int d^3 v' \left[ K_{ij}(\vec{v}',\vec{v}) f_i(\vec{v}') - K_{ji}(\vec{v},\vec{v}') f_j(\vec{v}) \right] \right.$$

$$\left. + g_j(\vec{v}, I_j, \omega_j) + R_j(\vec{v}) - \gamma_j(\vec{v}) f_j(\vec{v}) \right. \quad . \tag{2}$$

A similar equation for $\partial f_{j'}(\vec{v})/\partial t$ is obtained with the interchange $j \to j'$. The quantity $K_{rs}(\vec{v},\vec{v}')$ is the collisional matrix of which the off-diagnoal elements ($r \neq s$) describe rotationally inelastic scattering and which are directly related to
the angle-dependent forces occurring during a collision.  The
functions $g_j(\vec{v}, I_j, \omega_j)$ describe the perturbation generated by
the radiation field of the P(j) transition at frequency $\omega_j$ with
intensity $I_j$.  Since the transitions are Doppler broadened,
the radiative excitation is highly selective in velocity space.
The quantities $R_j(\vec{v})$ and $\gamma_j(\vec{v})$ describe, respectively, the
excitation and decay processes coupling the population of the
$j^{th}$ level to all other molecular states exclusive of $j'$.  The
resulting steady-state distribution derived by equating Eq. (2)
to zero is established by the competition between the strong
velocity-selective perturbing influence of the radiation fields
$I_j$ and $I_{j'}$ and both elastic and inelastic collisions which tend
to restore the equilibrium Boltzmann distribution.  For all
other conditions fixed, an increase in the perturber gas pressure enables the collisional redistribution to compete more
effectively with the selective radiative processes, causing
the width $\Delta\nu$ of the observed resonance to exhibit a dependence
on the partial pressures of the constituent gases.

It is possible to formulate a closed form solution[13] for
the width $\Delta\nu$ under conditions that are valid for our experimental situation.  The result is given by expression (3) below
for $CO_2/H_2$ mixtures as

$$\Delta\nu = 2 \left\{ (\Gamma_1 + \Gamma_2)^2 + \frac{\omega^2}{2} \left[ \Delta\beta_{CO_2}^2 (1 + C) + \Delta\beta_{H_2}^2 (1 - C) \right] \right\}^{1/2} \tag{3}$$

where

$$C = \frac{1 - Rk}{1 + Rk} \tag{4}$$

In Eqs. (3) and (4) R is equal to the ratio of the partial pres-
sures $(P_{H_2}/P_{CO_2})$, $\Delta\beta_{H_2}$ and $\Delta\beta_{CO_2}$ are the velocity changes
associated with the collision partners $H_2$ and $CO_2$, respectively,
and k is a constant which depends on the rotationally inelastic

collision rates.  The parameters $\Gamma_i$ (i = 1,2) are the power broadened widths and $\omega$ is the laser frequency.  This functional form is found to be a good fit to the data over a wide range of experimental conditions.  As evaluated from our data, we obtain $\Delta\beta_{CO_2}$ = 2.6 $\times$ $10^3$ cm/s and $\Delta\beta_{H_2}$ = 6.8 $\times$ $10^3$ cm/s, a finding which indicates that a rather close encounter is necessary to produce a rotational transition with $H_2$.

Considerable velocity selectivity was observed for $CO_2$ in collision with reasonably massive perturbers such as $CO_2$, Kr, and $CH_3F$.  For the latter system see the data in Figure 2. This strongly suggests that the rotationally inelastic colli- sions are mainly peripheral encounters occurring at a large interaction radius, favoring the transfer of angular momentum with a relatively small linear momentum transfer.  We note that at sufficiently large intermolecular separations, allowing the neglect of exchange forces arising from electron overlap, the main contributions to the angle-dependent interaction will arise from anisotropic dispersion forces in $CO_2$-He and $CO_2$-Kr scattering and the quadrupole-quadrupole interaction in addi- tion to the anisotropic dispersion forces for $CO_2$-$H_2$ and $CO_2$-$CO_2$ scattering.[20]  The dipole-quadrupole interaction is presumed to dominate in the $CO_2$-$CH_3F$ scattering.  Finally, the $CO_2$-$CO_2$ case has the additional possibility of exchange processes.

## IV  CONCLUSIONS

In summary a new experimental technique has been developed enabling a direct observation of the momentum-transfer distri- bution in rotationally inelastic molecular collisions.  With this method, velocity-selective rotational transfer has been detected for $CO_2$ in collision with $H_2$, He, $CO_2$, $CH_3F$, and Kr. Specifically, for $H_2$ and $CO_2$ the data indicate values of $\langle\Delta v_z\rangle$ for j = 20 $\rightleftarrows$ j$'$ = 18 transitions at $300^\circ$K which are far less than the mean thermal speed.  These data provide direct infor- mation on the anisotropic component of the intermolecular inter- action and also demonstrate that collision-induced transitions between closely spaced levels can occur with a small effect on the molecular velocity for states that have no transition di- pole moment.  This is an extension of the conclusion reached by Anderson[21] for cases involving a transition dipole moment.

## V ACKNOWLEDGMENTS

The authors gratefully acknowledge helpful conversations with J. E. Bayfield and the expert technical assistance of B. R. Schleicher.

\* Work partially performed under the auspices of the U.S. Energy Research and Development Administration.

† Present address: Molecular Physics Center, Stanford Research Institute, Menlo Park, California 94025.

[1] Faraday Discussions of the Chemical Society, No. 55, Molecular Beam Scattering (Faraday Division of the Chemical Society, London, 1973).

[2] J. Schmidt, P. R. Berman, and R. G. Brewer, Phys. Rev. Lett. 31, 1103 (1973).

[3] P. R. Berman, Phys. Rev. A6, 2157 (1972).

[4] A. P. Kolchenko, A. A. Pukhov, S. G. Rautian, and A. M. Shalagin, Zh. Eksp. Teor. Fiz. 63, 1173 (1973) [Sov. Phys.-JETP 36, 619 (1973)].

[5] M. L. Goldberger and K. M. Watson, Collision Theory (Wiley, New York, 1964); M. A. D. Fluendy, I. H. Kerr, and K. P. Lawley, Molec. Phys. 28, 69 (1974).

[6] C. H. Townes in Fundamental and Applied Laser Physics, edited M. S. Feld, A. Javan, and N. A. Kurnit (Wiley, New York, 1973) p. 739.

[7] F. H. Hopf and C. K. Rhodes, Phys. Rev. A8, 912 (1973).

[8] Thomas W. Meyer and Charles K. Rhodes, Phys. Rev. Lett. 32, 637 (1974).

[9] Thomas W. Meyer, William K. Bischel, and Charles K. Rhodes, Phys. Rev. A10, 1433 (1974).

[10] J. S. Murphy and J. E. Boggs, J. Chem. Phys. 47, 691 (1967); R. G. Gordon, J. Chem. Phys. 44, 3083 (1966).

[11] C. Freed, IEEE J. Quant. Electron. 4, 404 (1968).

[12] C. Freed and A. Javan, Appl. Phys. Lett. 17, 53 (1970).

[13] William K. Bischel, Ph.D. thesis, Application of Nonlinear Optical Techniques for the Investigation of Molecular Properties and Collisional Processes, University of California, 1975 (unpublished).

[14] For a discussion of the appropriate vibrational transfer rates, see P. K. Cheo, in Lasers, Vol. 3, edited by A. K. Levine and A. J. DeMaria (Marcel Dekker, New York, 1971), p. 111.

[15] For the cases examined, the oscillators were operated on 10.6-$\mu$m P-branch transitions with $j = 20$ and $j' = 18$, 16, 14, and 12.

[16] T. W. Meyer, Ph.D. thesis, Line Broadening and Collisional Studies of $CO_2$ Using the Techniques of Saturation Spectroscopy, University of California, 1974 (unpublished).

[17] C. Freed and H. A. Haus, IEEE J. Quant. Electron. 9, 219 (1973).

[18] J. W. C. Johns, A. R. W. McKellar, T. Oka, and M. Römheld, J. Chem. Phys. 62, 1488 (1975).

[19] S. G. Rautian, G. I. Smirnov, and A. M. Shalagin, Zh. Eksp. Teor. Fiz. 62, 2097 (1972) [Sov. Phys. JETP-35, 1095 (1972)].

[20] A. D. Buckingham, in Advances in Chemical Physics, Vol. 12, edited by J. O. Hirschfelder (Interscience, New York, 1967), p. 107; H. Margenau and N. R. Kester, Theory of Intermolecular Forces (Pergamon, New York, 1971), 2nd ed.

[21] P. W. Anderson, Phys. Rev. 76, 647 (1949).

ATOMIC COLLISION ASPECTS
OF ENERGY RELATED RESEARCH

# ATOMIC COLLISIONS AND FISSION TECHNOLOGY

Sheldon Datz

Oak Ridge National Laboratory

Oak Ridge, Tennessee 37830, U.S.A.

## INTRODUCTION

Of all the technologies presently used in the generation of significant amounts of electrical energy, fission is the only one to emerge as a result of 20th century science applied on a massive scale. Almost all of the steps involved in the technology from the winning of a useful fuel to the disposal of its waste products have required the generation and exploitation of concepts which did not exist 50 years ago. Moreover, because the birth of the industry occurred as a result of the development of nuclear weapons, and because of the lack of understanding of the potential dangers of radioactive materials, the development of nuclear power technology in all its aspects has been closely controlled and managed by a single government agency in the U. S. and elsewhere in the world. This is unique in the history of technology. For example, environmental impact responsibilities have had to be faced prior to rather than after the de facto existence of a massive industry and the agencies involved have had as an important part of their mission the responsibility for determining the effects of radiation not only on materials but upon living systems.

The purpose of the above paragraph is merely to emphasize that there is a great deal more involved in nuclear reactors than nuclear reactions. With this in mind (together with some standard cartoons on, e.g., a map of the U. S. as seen by a Bostonian, a Texan, a New Yorker, etc.) let us proceed to briefly view fission technology as seen by an atomic collisions physicist. If we dispense with prospecting, mining, and

metallurgy we come first upon the problem of isotope enrichment, an area almost exclusively in the domain of collision
physics.  Given the proper isotopic ratio we proceed to fabri-
cate the fuel, reactors, etc., and arrange things so that the
requisite number of neutrons cause the requisite number of
fission events.  Following the nuclear fission event we are
dealing with energetic heavy ions (fission fragments), neutrons,
helium ions, electrons, and photons which undergo a series of
complex collision processes giving rise ultimately to the heat
used in power generation and to radiation damage.  Similarly
the underlying science of the interaction of radiation from
radioactive wastes with living systems and materials is properly
in the area of atomic collision physics.


## ISOTOPE SEPARATION

    The primary market for isotopic enrichment processes has,
of course, been for the preparation of uranium fuels.  Although
∿3% enriched material is generally used in most present reactors,
higher enrichments may be desirable in high temperature gas
cooled reactors and, if efficient methods can be developed,
isotopic clean up may be desirable in future reactors fueled
from plutonium breeders.  Other possible applications exist in
the treatment of radioactive wastes to reduce the storage
capacity required.

    Historically, the first method used for separation of
uranium isotopes on a macroscopic scale was the electromagnetic
process which relied heavily on atomic physics and gaseous
electronics.  This was replaced by the more economic gaseous
diffusion process.  In this regard it should be realized that
the efficiency of the gaseous diffusion process in terms of
actual energy consumption to the thermodynamic value necessary
for obtaining 3% enriched material from the 0.7% naturally
abundant material is only $10^{-8}$.  The energy used in isotope
separation corresponds to ∿3% of the energy, but in nuclear
power based economy this constitutes considerable consumption.
The gas centrifuge method improves this power utilization figure by a factor of >10 but there is clearly still much room
for improvement.  The latter two methods and the aerodynamic
jet approach utilize information on gas transport properties
supplied by collision physics.

    New techniques utilizing laser excitation are presently
under study and require considerable and more detailed input
from collision physics.  These methods generally rely upon
selective excitation of either an atom of the desired element
or a molecule containing it followed by either photoionization
or reaction of the excited particle.

Using atomic species, the first step involves isotopically selective electronic excitation

$$X + h\nu \rightarrow X^*$$ (1a)

followed either by single or multiple photon excitation to ionization

$$X^* + h\nu' \rightarrow X^+ + e \ .$$ (1b)

Alternatively associative ionization of the excited state atoms might be possible;

$$X^* + R \rightarrow XR^+ + e$$ (1c)

or by chemical reaction

$$X^* + RY \rightarrow XR + Y \ .$$ (1d)

For processes (1b) and (1c) the separation is simply accomplished by deflection in an electric field and collection on a surface. The photoionization process, (1b), has been investigated by a group at the Lawrence Livermore Laboratory[1] which has demonstrated macro separation of uranium isotopes.[2] In their scheme an atomic beam of uranium is bombarded simultaneously by tuneable dye laser beam (1a) and high intensity ultra-violet source (1b). When the beam is injected into a quadrupole mass filter, the spectrum of photoionized species is shown in Fig. 1. In the uppermost trace the laser is tuned to excite the $U^{235}$ isotope and in the next trace down, $U^{238}$ is the excited species.

Process (1a), (1d) requires that the excited state can undergo chemically reactive collisions while the ground state atom is non-reactive. This type of selective reactivity has been amply demonstrated and the subject is discussed in some papers at this conference. An isotopic separation process based on this principle has been demonstrated in the case of mercury,

$$Hg(6\,^3P_1) + H_2O \rightarrow HgO + H_2 \ .$$

Selective laser excitation of molecular species offers other possibilities for isotope separation schemes. Here the first step would be excitation either to higher vibrational or electronic levels

$$XY_n + h\nu \rightarrow XY_n^*$$ (2a)

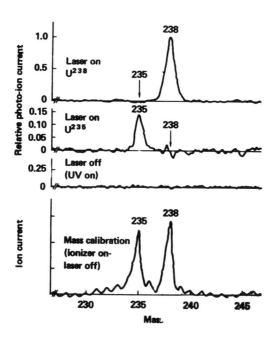

Fig. 1.  Laser separation of uranium isotopes by laser excitation of atomic $U^{235}$ and $U^{238}$ followed by photoionization, see
Refs. 1 and 2.

followed by photodissociation

$$XY_n^* + h\nu' \rightarrow XY_{(n-m)} + Y_m \tag{2b}$$

or by chemical reaction

$$XY_n^* + Z \rightarrow XY_{(n-m)} + ZY_m \tag{2c}$$

or by chemi-ionization

$$XY_n^* + Z \rightarrow XY_n^- + Z^+ . \tag{2d}$$

An interesting example of process (2a) - (2b) has recently been demonstrated in the Soviet Union[3] and at Los Alamos[4] in the U. S. in which $SF_6$ is selectively photodissociated into $SF_5$ + F under irradiation by the P-20 line of a $CO_2$ laser which vibrationally excites the $S^{32}F_6$ molecule (the resultant fluorine is scavenged by $H_2$ which is added to the system). A mass spectrometer trace of the enrichment process is shown in Fig. 2. An interesting point here is that no other radiation source is used and multiphoton process must be involved in the dissociation process. Although the exact mechanism has not been determined, it is known that the separation is sensitive to laser pulse intensity and duration.

Clearly, numerous other schemes of the type described above could be devised, but all of these approaches need detailed information not only on the primary collision processes involved but also on those collision processes which could interfere with the separation. Excitation transfer from selectively excited states (1a) and (2a) to other isotopic species could be troublesome. Charge transfer prior to product collection in step (1b) or (2d) could decrease product selectivity. Chemical exchange reactions can reduce yields for processes such as (1d), (2b) and (2c), etc.

The economic feasibility of any of these methods has yet to be demonstrated but none of these possibilities would exist

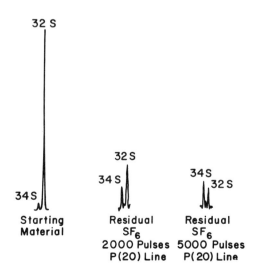

Fig. 2. Laser isotope separation of sulfur isotopes by photodissociation with a $CO_2$ laser.

without previous broad support for collision physics and chemistry. The opportunities for collision physics to contribute in these areas are very broad and the possible payoffs are very large.

## ATOMIC COLLISIONS IN SOLIDS:  STOPPING POWER AND RADIATION DAMAGE

Radiation damage following a nuclear fission event is caused principally by the collisions of fast heavy ions in solids. The heavy ions in question are either fission fragments which range in energy up to $\sim$0.5 MeV/nucleon and are localized in the fuel region, or they arise from knock-on collisions of fast neutrons which leave the fuel and collide in the containment material. These knock-ons can range in energy from 10-100 keV. In this latter regard the problem has the same elements in nuclear fusion reactors where large fast neutron fluxes will exit. In addition, the 14 MeV D-T neutrons can create knock-ons with energies up to $\sim$500 keV.

Numerous experiments designed to simulate the effects of the enormous neutron fluences which will obtain in the use of fast breeder reactors (and CTR reactors) are being carried out using accelerated heavy ion beams. For example, 6 MeV Ni ions are injected into stainless steel, and in the region below the surface where they have slowed to $\sim$0.5 MeV they simulate the effects of 14 MeV neutron knock-ons. The initial high energy is necessary to penetrate below the surface region where other effects may interfere. If accurate simulation can be achieved, the time scale for materials testing can be shrunk by orders of magnitude. This, however, requires an understanding of the stopping over the entire range.

The processes attending the slowing of these ions in all cases is properly the domain of atomic collision physics. The ions lose energy by 1) inelastic loss due to ionization and excitation of the electrons on the target atom and projectile ion (electronic stopping) and 2) by "nuclear stopping," i.e., elastic loss to the target atoms. At low velocities elastic scattering causes collision cascades in the host and ends in an aggregate of point defects ("nascent" damage state). For the entire slowing down process the time scale is $\sim 10^{-13}$ – $10^{-12}$ seconds. Following this, the processes of solid state physics are dominant (diffusion, nucleation, precipitation, void formation, etc.). The time scale here can be quite large. In metals the nascent damage state is produced entirely by the atomic displacement, but in insulators and possibly in semiconductors, electron excitation can also lead to damage. For a detailed understanding of the elastic scattering cascade a

knowledge of the relevant interatomic scattering potentials is
needed and for the electronic stopping, a knowledge of inelastic
events suffered by fast heavy ions of the type supplied by
scores of papers at this conference is vital to a better understanding.  As most of you know, the stopping power (energy loss
per atomic collision) initially rises with velocity, tops out
and then decreases.  Most of the ions we are dealing with here
have initial velocities in the lower region where the stopping
power curve is rising with velocity.  This region has been
handled theoretically by Lindhard, Scharff, and Schiøtt[5] in
terms of electron gas interactions and by Firsov[6] in terms of
charge exchange.  Both lead to a linear dependence on velocity
(S=kv) and to a monotonic dependence on the atomic number of
the projectile in a given target.  Both theories have been
extremely valuable but they were formulated before the recent
burst of activity in heavy ion collision physics and before
the advent of "channeling" as a tool for understanding the
motion of ions in solids.[7]

     The study of atomic collisions in solids has been greatly
aided by discovery and application of the channeling phenomenon.
Channeling occurs when an ion enters a crystalline solid at a
small angle with respect to atomic rows or planes.  It undergoes a set of correlated small angle scattering events with the
closely spaced atoms which tends to steer the ion's trajectory
away from close collisions with a lattice atom.  The potential
which controls the motion of the ion can be viewed as a continuum potential made up of an orderly sum of the individual
ion-atom potentials.  Under these conditions the ion is said
to be "channeled" and, under easily achieved experimental
conditions, more than 90% of ions entering a crystal can be
channeled.

     There are many important consequences of this effect.
Among them is the elimination of collision phenomena involving
small impact parameters, i.e., nuclear stopping is virtually
eliminated as well as Rutherford scattering, inner-shell
ionization, and nuclear reactions.  Second, the ion's trajectory is closely circumscribed, e.g., for channeling between
two planes (sheets) of atoms in a crystal the simple two dimmen-
sional potential leads to discrete oscillatory trajectories
for the channeled ions which may be identified by a characteristic energy loss due to the regions of electron density
encountered along the ion's path.[8]  Third, particles injected
with transverse energy less than the potential barrier between
atomic rows (hyperchanneling) may be used to map interatomic
potentials out to valence electron distances.  An example of
this effect is given in Fig. 3 which shows the result of different entrance points for ions directed at the same angle with
respect to a set of atomic planes.  The energy loss for a<b<c

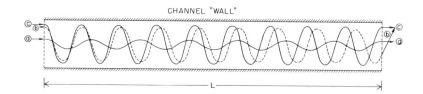

CHANNEL "WALL"

Fig. 3. Oscillatory motion of ions in a potential set up
by two planes of atoms in a crystal. For trajectories a and
c the ions exit the crystal in the same direction as they
enter.

in accord with the higher electron densities penetrated at
larger amplitude and the number of times the particle penetrates
these regions, i.e., shorther wavelength. In the particular
configuration shown, trajectories a and c exit at the same
angle as the entrance angle and may be measured by an appropriately placed detector.[8]

A typical set of energy loss spectra is shown in Fig. 4
for 60 MeV I ions channeled in Au. The peaks are cuased by
particles with discrete trajectories such that when the length,
L, is changed, the wavelength $\lambda$ of the transmitted particle is
changed in accord with $n\lambda = L$ where n is the number of oscillations. Measurements of this sort lead to determination of
the quasi-elastic interatomic scattering potential. For
numerous ions it was found that the proper description is the
free atom potential of the lattice atom interacting with a
Coulomb probe (the penetrating ion) of charge appropriate to
its velocity.[8,9] Impact parameter dependence of stopping
power (inelastic loss) is also obtained from these measurements.

Figure 5 shows some recent measurements of stopping
power for U ions in C,[10] with a comparison to the Lindhard et
al.[5] and Firsov[6] theories. The linear dependence on velocity
is evident but the intercept is clearly not at zero. These
measurements have been corrected for the anticipated nuclear
stopping. However, to extend the measurements to lower velocities, where the nuclear stopping correction becomes more important, the channeling technique has been used. Bøttiger and
Basson[11] showed that for a number of elements (17), low energy
electronic stopping powers were not proportional to v but instead to $E^p$ where p varied from 0.2 to 0.9. The results of
Moak et al.[12] are shown in Fig. 6 for I ions from 0.6 to
60 MeV. The behavior is clearly more complex than simple

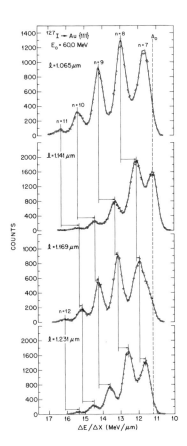

Fig. 4.  Energy loss (dE/dx) spectra of 60 MeV I ions transmitted through Au channels (111) with exit angle equal to
the entrance angle.  The peaks correspond to trajectories
with n integral oscillations in a given crystal length L.

linearity with velocity, and reflects the velocity dependence
of the ion's charge and excitation state in the medium.

At low velocities ($v \sim v_0$) the dependence of electronic
stopping on projectile atomic number $Z_1$ is far from the monotonic behavior predicted by earlier theories as shown in
Fig. 7.  Here again the channeling effect is used to eliminate
nuclear stopping and inner shell electron interactions.  (The
oscillations shown in Fig. 7 are also observed in non-channeled
experiments but are not as pronounced.)  Several explanations

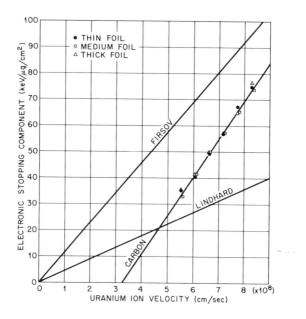

Fig. 5.   Comparison of U electronic stopping component in C
compared with Firsov[5] and Lindhard et al.[4] predictions.

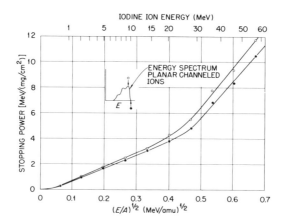

Fig. 6.   Electronic stopping power for 0.6 to 60 MeV I ions
in Ag (111) channel.

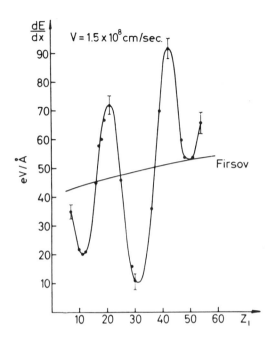

Fig. 7. Oscillation of electronic stopping power for ions
with atomic number Z at a velocity 1.5 x $10^8$ cm/sec channeled
in Au compared with Firsov theory (Ref. 11).


of the observed oscillations have been forwarded. In some
the Firsov model is modified using more realistic atomic
models[13] and agreement is qualitatively good but the peak to
valley ratio predicted is not large enough. Another approach
proposed by Lindhard and Finneman attributes the oscillations
to a variation in the scattering cross section of target
electrons near the Fermi level from the field of the incident
ion. This corresponds to a variation in the phase shift induced
in the electron wave function by the presence of the field of
the moving ion. The stopping power in this description is
proportional to electron momentum transfer cross section which
can display transparencies (deep minima) similar to those
observed in the Ramsauer-Townsend effect. Calculations on
this basis[14] predict the observed effect rather well.

From the above and other experiments it is clear that much
remains to be discovered about simple stopping processes in
condensed media. Numerous experiments relating to the charge
and excitation states of ions moving in solids have been

performed in the past few years, and more are presently in progress.  The application of these findings to the effects of radiation on materials remains to be completed.

Because of the limitation in time, I have limited the scope of these remarks to the primary ion.  In insulators, semiconductors, and biological material, much of the damage is caused by interactions of the photons and electrons generated along the ion track.  In biological materials especially, the "energy pathways," i.e., the detailed paths by which the energy is finally degraded, is of great importance.  A complete knowledge here demands information on whole hosts of atomic collision phenomena ranging from ionization and excitation cross sections to electron attachment cross sections.

In conclusion, fission is operating technology but it is one in which the contributions from atomic collision science can aid greatly in lowering the operating and capital costs, in increasing system reliability, and in understanding the environmental effects of the attendant ionizing radiation.

## ACKNOWLEDGMENTS

I should like to acknowledge the aid of Drs. R. Stern and J. Dubrin of Lawrence Livermore Laboratory and Drs. N. Blaise and G. Kwei of the Los Alamos Scientific Laboratory in assembling material on laser isotope separation, and of Dr. M. T. Robinson of Oak Ridge National Laboratory for discussion concerning the collisional basis of radiation damage.

## REFERENCES

1.  B. B. Snavely, "Isotope Separation Using Tunable Lasers," Proc. Int. Conf. on Lasers (LASER 75), Munich, W. Germany, June 1975 (in press).
2.  S. A. Tuccio, R. J. Foley, J. W. Dubrin and O. Krikorian, in Abstr. 1975 IEEE/OSA Conf. on Laser Engineering and Applications, Washington, D. C., May 28-30, 1975.
3.  Laser Focus, June, 1975, p. 10.
4.  C. Paul Robinson, "Laser Isotope Separation," Proc. New York Acad. of Sci. Third Conf. on the Laser, 1975 (in press).
5.  J. Lindhard, M. Scharff, and H. E. Schiøtt, Kgl. Danske Videnskab. Selskab., Mat.-Fys. Medd. 33, 10 (1963).
6.  O. B. Firsov, Zh. Eksp. Teor. Fiz. 36, 1517 (1959); Sov. Phys. JETP 36, 1076 (1959).
7.  See, e.g., D. Gemmel, Revs. Mod. Phys. 46, 129 (1974).

8.  S. Datz, B. R. Appleton and C. D. Moak, "Detailed Studies of Channeled Ion Trajectories and Associated Channeling Potentials and Stopping Powers," in Channeling, D. V. Morgan, Ed., J. Wiley & Sons, London (1973), pp. 153-181.
9.  M. T. Robinson, Phys. Rev. 179, 327 (1969).
10. M. D. Brown and C. D. Moak, Phys. Rev. B6, 90 (1972).
11. J. Bøttiger and F. Basson, Radiation Effects 2, 105 (1969).
12. C. D. Moak, S. Datz, B. R. Appleton, J. A. Biggerstaff, T. S. Noggle and H. Verbeek (manuscript in preparation).
13. See, e.g., I. M. Cheshire, G. Dearnally and J. M. Poate, Proc. Roy. Soc. A311, 47 (1969).
14. J. S. Briggs and A. P. Pathak, Solid State Physics 7, 1929 (1974).

ATOMIC PHYSICS IN THE CONTROLLED THERMONUCLEAR RESEARCH
PROGRAM

C. F. Barnett

Oak Ridge National Laboratory[*]

Oak Ridge, Tennessee 37830

## A. INTRODUCTION

Since the beginning of thermonuclear research in the USA in the early 1950s, high temperature plasma parameters have been dominated by atomic processes. For the first twenty years the plasma physicists were preoccupied with plasma instabilities and diverted little of their attention to atomic processes occurring in the plasma. Only in the past few years with the success of toroidal plasmas of the tokamak type has it become evident that atomic physics is of vital importance to the understanding of plasma heating, cooling, and diagnostic. More recently, as plasma engineers have started conceptual designs of proto-type fusion reactors, the need for the solution of atomic physics problems occurring in high temperature plasmas has become urgent.

The importance of atomic physic problems becomes very real if we compare the atomic and nuclear cross sections for particles in a thermonuclear reactor. For a 10 keV ion temperature D-T plasma the nuclear cross section to produce a neutron with 14 MeV energy is $10^{-29}$ cm$^2$. The charge exchange cross section for a resonance collision is approximately $10^{-15}$ cm$^2$. Thus, if we equate the reaction rate for production of neutrons to the loss of a deuteron we have

---

[*]Operated by Union Carbide Corporation for the Energy Research and Development Administration.

$$n_D n_T \sigma_N v_D = n_0 n_D \sigma_A v_D$$

where $n_T$ and $n_D$ are total tritium and deuteron densities and $n_0$ is the density of neutral particles. The velocities are the same so that $n_T \sigma_N = n_0 \sigma_A$. For a plasma with a deuteron density of $10^{14}$ cm$^{-3}$ we are left with the requirement that $n_0 = 1$ cm$^{-3}$. If we only require an energy breakeven, we can relax the neutral density requirement by the ratio of energy gained to energy lost, which is $10^3$. Thus, the neutral density for energy breakeven is $10^3$ cm$^{-3}$. These conditions are extremely difficult to achieve in present-day plasmas.

In this report the atomic processes involving plasma heating, cooling and diagnostics will be discussed. Also, of interest are those atomic processes involved in the interaction of a high temperature plasma with the vacuum wall. Due to time and space limitations these processes will not be discussed. In the interest of brevity remarks will be confined to tokamak type plasmas, whose present parameters are

| | | |
|---|---|---|
| $T_e$ (electron temperature) | - 1 - 2 keV | |
| $T_i$ (ion temperature) | - 0.5 - 1.0 keV | |
| $n_e$ (electron density) | - 2.5 x $10^{13}$ electrons/cc | |
| $\tau_E$ (energy containment time) | - 5 - 10 x $10^{-3}$ sec | |
| T (pulse time) | - 50 - 300 x $10^{-3}$ sec | |
| r (minor torus radius) | - 10 - 30 cm | |
| R (major torus radius) | - 0.5 - 1.0 m | |
| $I_p$ (ohmic heating current) | - 100 - 300 kA | |

The corresponding parameters of planned proto-type reactors are

| | |
|---|---|
| $T_e$ | - 5 - 10 keV |
| $T_i$ | - 10 keV |
| $n_e$ | - 5 x $10^{13}$ - 5 x $10^{14}$ cm$^{-3}$ |
| $\tau_E$ | - 1-4 sec |
| r | - 1 m |
| R | - 4 m |
| $I_p$ | - 2 - 3 MA |

### B.  PLASMA HEATING

In the past, heating of toroidal plasmas has been accomplished by applying a current pulse to a set of toroidally-wound coils. The plasma acts as the secondary winding of this set of coils and a voltage of 2 - 5 volts is induced around the torus. This induced voltage supplies the driving force to heat the electrons which in turn transfers energy

to the plasma ions.  As the electrons approach energies of
1-2 keV the effective energy transfer through coulomb collisions from electrons to ions approaches zero such that the
ions are no longer heated.  Thus, supplementary ways must be
found to heat the ions.

One approach to supplementary heating is to inject into
the plasma large currents of neutral atoms and through resonant charge exchange collisions with plasma ions, the
resulting energetic protons are captured by the confining
magnetic field.  These protons thermalize by energy transfer
to both ions and electrons.  In injection heating, it is
necessary to have knowledge of the collisional processes in
the ion source and the conversion cell where neutrals are
formed.

For present day injection heating experiments pulsed ions
sources have been developed for ion currents up to 50 amps
and energies of 10-25 keV with a pulse length of 20 millisec.
In this energy range the atomic and molecular ions are converted in gas conversion cells through charge exchange and
dissociative collisions.  Of importance to the injection process is the rate at which impurity atoms from the ion source
enter the plasma confinement region.  Due to space charge
blowup it is not practical to use magnetic analysis to pre-
select the beam current before neutralization.  It is
imperative that we know the electron capture cross sections
of $C^+$, $O^+$, $N^+$, $Pt^+$, $Fe^+$ and $Ta^+$ in gases of $H_2$, He and $N_2$.
Also, needed in order to assess the rate at which these
impurity atoms enter the plasma are the electron stripping
cross sections for atomic species of these ions in the energy
range 10-150 keV.

As experiments proceed to the proto-type fusion reactor
it is necessary to increase the injected particle energy to
the region of 150-200 keV.  Two methods have been proposed
and are being developed to obtain neutral hydrogen beams at
these high energies.  (1) Negative $D^-$ ions are obtained directly from an ion source, accelerated to the desired energy,
and stripped in a suitable gas target.  (2) Conventional
positive ion sources are used, hydrogen beams are extracted
at 1-2 keV, and passed through an alkaline vapor cell where
10-20% of the incident ions are converted to $D^-$.  Our knowledge of the mechanisms producing $D^-$ in the ion source is very
limited.  Collision cross sections involving electrons on
$D_2$ or $D_2^+$ in various states of vibrational excitation need to
be known over the energy range from threshold to 500 eV.
Current experiments indicate that tremendous gains in negative
ion source output can be obtained by contaminating the walls
and cathode of the ion source with alkaline vapors.  The

reaction kinetics must be known for deuterium atomic and
molecular neutrals and ions interacting with surfaces contaminated with Cs, Mg, Na and Li in the energy range 5-500 eV.

Conversion of atomic and molecular ions to $D^-$ ions in a
alkaline vapor cell demands a multiplicity of cross sections
that are not presently available.  Considerable work has been
done for $D^+$ ions interacting with Cs, Na, and Li vapor.  Of
more interest than cross sections are the $D^-$ equilibrium fractions formed in these vapor cells.  Discrepancies of a factor
of three exist in the $D^-$ equilibrium fraction from Cs cells.
Very little information is available for the formation of $D^-$
from collisions of $D_2^+$ and $D_3^+$.  Since the low energy $D^-$ ions
formed must be accelerated, it is imperative that we know the
differential scattering cross sections for $D^-$ production from
both atomic and molecular species.

Cross sections are available for the stripping of $D^-$ in
many gases in the energy region 3-50 keV and above 300 keV.
This intermediate region needs to be filled in.  Of more
importance the peak fraction of $D^0$ resulting from $D^-$ stripping
needs to be measured.  As the stripping gas density is increased from $10^{13}$ $cm^{-3}$ to $10^{16}$ $cm^{-3}$ the $D^0$ fraction goes
through a maximum and then decreases.  These maxima need to
be identified for gases such as $H_2$, He, $N_2$, $O_2$, Ne, Ar, and
$H_2O$ in the energy region 0.1-1 MeV.

In attempts to design and predict the behavior of the
next generation plasma experiments, physicists must understand the trapping processes of the injected beam.  If only
hydrogen ions occur in the plasma, one can show that the
injected neutrals will be trapped approximately uniformily
across the magnetic field.  However, with impurity ions present with moderate densities at the plasma boundary, the
neutrals particles may be ionized locally, setting up an instability.  This instability may occur in two ways:  (1) large
pressure and electric field gradients at the plasma boundary;
(2) trapping at plasma boundary may lead to increased charge
exchange loss.  These particles will bombard the vacuum wall,
giving more impurities, and resulting in a cascade process
that will prevent any heating from the injected beam.  We
must know charge exchange cross sections for multiply-charged
atoms of O, C, Fe, Au, and W with H and $H_2$ in the energy range
5-200 keV.

## C.  PLASMA COOLING

Atomic physic processes in a hot fully ionized plasma
contribute an important cooling or loss mechanism which may
limit the obtainable temperature in an operating thermonuclear
plasma.

Ideally, a thermonuclear plasma would consist of deuterons and tritons. However, in existing plasmas, impurity ions and atoms are present in concentrations up to 5%. In some of the experiments it has been shown that up to 40% of the power loss of today's plasma was through line radiation from these impurities which have been estimated to be up to 40 times ionized. These impurities arise through particle and photon bombardment of the surrounding surfaces and from charged particle bombardment of the limiter. The limiter is an annular W or $M_O$ ring placed at one point at the periphery of the plasma to prevent circulating particles from burning a hole in the plasma vacuum wall. Under fusion conditions the temperature will be greater and these heavy impurities may be up to 70 times ionized. Thus, for heavy ions such as W the power loss may be so great that the temperature necessary for fusion may not be obtainable.

Transitions between atomic energy levels in fusion type plasmas involve excitation or ionization by electron collisions and subsequent de-excitation through radiative transitions. For energies several times greater than threshold the Born approximation may be used to compute the desired information. In the range of threshold the resonant transitions are the highest and are not amendable to Born calculations although some success has been achieved by using close coupling computations for predicting excitation cross sections. For contemporary plasmas a need exists for excitation and ionization cross sections for multiply-charged O and C particles plus the heavier impurity ions that may be present.

Two recombination processes are of critical importance in plasma cooling. These are (1) radiative recombination and (2) dielectronic recombination. The importance of radiative recombination lies in the $Z^4$ dependence of the recombination radiation. For impurity ions in the plasma the effective "Z" of partially stripped atoms will be the net charge of the ion before recombination. Some of the impurity ions (W, Nb, Mo, Au, Fe) will be only partially stripped for 10 keV plasmas. Thus, it is important to obtain recombination rates for these impurities in all states of ionization for plasma electron temperatures up to 20 keV.

Dielectronic recombination has been known to the astrophysicists for several years. In this process an electron colliding with an ion with bound electrons may be captured into a doubly excited state, which may decay into a bound state by emission of two or more photons and can be represented as follows

$$e + x^{n+} \longrightarrow [x^{(n-1)^+}]^{**} \longrightarrow x^{n+} + e$$

$$\longrightarrow x^{(n-1)^+} + hv + hv'.$$

The upper reaction is an excitation collision while the
lower branch is a dielectronic recombination process and can
be much faster in high temperature plasmas than radiative
recombination collisions.  In solar corona it is well known
that dielectronic recombination is $10^2$ times faster than radiative recombination.  The damaging aspect of dielectronic recombination is the reduction of the equilibrium charge state
and increased power losses from the impurities.  Thus, this
process may demand that impurities be reduced to an impracti-
cable level.

When a free electron is accelerated and deflected as it
passes a positively charge particle a photon is emitted.  This
process is referred to as free-free Bremsstrahlung and is an
important process in plasma cooling.  Bremsstrahlung losses
due to heavy charged impurities will be very harmful or if
a plasma is heated by wave interactions such that the applied
power is to the electrons the Bremsstrahlung loss will be
damaging.  Computations are needed to predict Bremsstrahlung
losses from partially stripped impurities found in 10 keV
plasmas.


D.  DIAGNOSTICS

The two previous sections were directed toward understanding the heating and cooling of a plasma.  Plasma diagnostics attempt to quantitatively determine plasma parameters (i.e. temperature, density, fields, etc.) as well as
the properties of impurities present.  A high degree of
sophistication has been achieved in the measurement of
spatial electron temperatures and densities through Thomson
scattering of laser beams.  However, many plasma parameters
remain to be measured with this degree of precision.

For measurements of concentrations of heavy elements in
CTR plasmas, it is essential to know the energy levels and
wavelengths of the strong resonant lines, especially those
with $\Delta n=0$.  Two problem areas need to be investigated:  (1)
resonance lines of the copper and zinc isoelectronic sequence
with particular attention to tungsten and gold need to be
investigated to the upper end of the periodic table; (2) atomic
structure of the first 40 states of ionization of W and Au for
simple electronic configurations that lead to resonance lines

need to be investigated.   These tasks are both theoretical and experimental.

For quantitative measurements of impurity concentrations to predict power loss, it is necessary to know the transition probabilities of light elements ($C^{n+}$, $O^{n+}$) and the heavier impurities.   Not only are transitions with zero change in principle quantum numbers important, but also the strong transitions with $\Delta n \neq 0$.

Heavy ion beam probes have been developed in the past few years to measure spatial plasma potential and densities.   In this method a heavy ion beam with sufficient momentum is projected across the plasma.   At some point in the plasma a fraction of the ions are converted to doubly charged ions by electron or other type ionization collisions.   If the ion beam is deflected across the plasma, a detector line can be found such that doubly charged ions formed on this line will cross over at some point exterior to the plasma.   Placing an electrostatic analyzer at this cross-over measures the change in particle energy which is directly proportional to the space potential at the point of ionization.   The analyzer detector currents are proportional to the plasma density.   This determination can be made quantitative if the electron and deuteron ionization cross sections are known for ions of Tl, Rb, K, and Ba.   This information is needed in the energy range 0.05 to 1 MeV.

One technique useful for ion temperature measurements by Doppler broadening is to inject an impurity atom such as He, Ne or Ar.   The excitation of triplet states in He-like configurations and subsequent radiative decay is being used for high temperature plasmas.

Two additional techniques being developed using atomic physics to determine plasma parameters deserve to be mentioned.   In one a lithium beam is projected across a toroidal plasma.   An electron collision excites the atom which radiates.   By measuring the direction of polarization of this radiation the magnetic field direction at the point of ionization can be determined.   By unfolding the magnetic field distribution the plasma heating current distribution can be determined.   Theoretically, current distributions are important in tokamak stability and studies indicate that with fusion type reactors the large heating currents will flow along the plasma skin leading to an unstable configuration.

Present laser scattering involves short-wave length (8,000 Å) lasers in which the scattered spectrum provides information about the electron temperature and density.   For

long-wave (300-500 μm) lengths the scattered spectrum arises
principally from ions and ion temperatures can be determined
in this manner.  Several laboratories are pumping $CH_3F$ gas
with $CO_2$ lasers to obtain coherent radiation at 496 μm.
Radiation at this wave length is also useful as a μ-wave
interferometer and has possible use through photon echo
techniques to study trapped electron lifetimes.

In summary, one can conclude that indeed atomic physics
plays a dominant role in plasma physicist's and engineer's
quest for thermonuclear power.  Only through solution and
understanding of some of the problems listed in the preceeding
pages will the scientific community meet the temperature and
confinement criteria demanded of fusion plasmas.

METHODS IN LOW ENERGY COLLISIONS

# THE ION-STORAGE COLLISION TECHNIQUE

Hans G. Dehmelt

University of Washington

Seattle, Washington 98195, U.S.A.

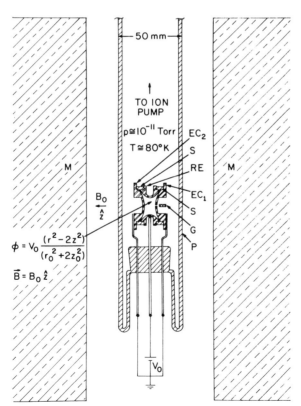

Fig. 1.  Penning trap, EC = end caps, S = glass spacer, G = electron gun, P = Pyrex envelope, M = magnet. from (Wineland and Dehmelt, 1975b).

Orientation dependent collision processes such as
e↑ + Na↓ → e↓ + Na↑, (Dehmelt, 1958; 1961, & 1962; Graff et al.,
1968, 1969 & 1972; Church & Mokri, 1971), e↑ + Na↓($^2$S) →e
+ Na($^2$P) − ΔE, (Graff et al., 1968, 1969, Church & Mokri, 1971),
see Fig. 5., He↑ + Cs↓ → He$^+$↓ + Cs↑, He$^+$↑(1s$^2$S) + Cs↓ → He(2s$^1$S)
+ Cs$^+$ + ΔE (Dehmelt & Major, 1962; Major & Dehmelt, 1968;
Schuessler et al., 1969) see Fig. 2, and hν(↔) + (H-H)$^+$ → H
+ H$^+$ + ΔE, (Dehmelt & Jefferts, 1962; Richardson et al., 1967;
Jefferts, 1968 & 1969), see Fig. 3 & 4, have been used in past
spin- and hfs- resonance studies on stored ion. The special
forte of the

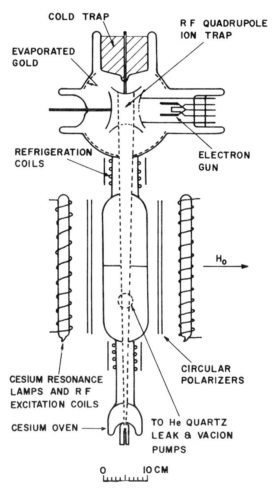

Fig. 2. Apparatus in which the first collision experiments with
(polarized) stored ions (He$^+$) and a polarized atomic beam (Cs)
were carried out, from (Dehmelt & Major, 1962).

technique here was orientation of the stored ions by collisions
with an incoming beam of oriented projectiles.  The early
experiments of the author (1961) on electrons stored in a low-
magnetic field Penning trap interacting with an optically
polarized Na-Beam were undertaken with an apparatus resembling
that shown in Fig. 2.  Drastic reduction of the electron storage time in the presence of the Na-Beam or a variable pressure
He gas background was observed, especially when a forced non-
resonant oscillation of the electrons was excited.  However,
no spin dependent effects were seen at this time.  As valuable
by-products of all these studies some information on the relevant cross sections often was obtained also.  Paul traps (rf
quadrupole) (Fischer, 1959), were used for the atomic or molecular ions and Penning traps, cf. (Dehmelt, 1967), Fig. 1., for
the electrons.  The traps were filled by creating the ions,
e.g. $He^+$ , e, inside them.  In Paul traps the rf heating associated with ion-atom and ion-ion collisions accelerates evaporation,
cf. (Dehmelt, 1967), of the ions out of the trap.  A single ion
in a perfect vacuum will presumably be stable.  In the Penning
trap ion-atom collisions also cause a biased random walk (Walls,
1970) of the cyclotron motion guiding centers towards the ring.
The ions were "counted" primarily by interaction with an LC
circuit tuned to their axial oscillation frequency and excited
externally or merely thermally, Figs. 6, 7 & 8.  This interac-

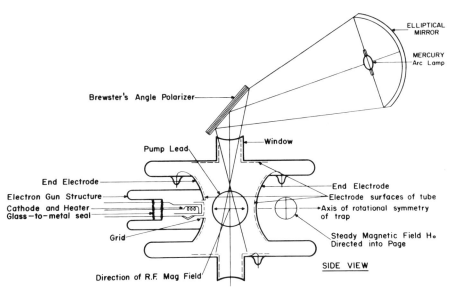

Fig. 3.  Apparatus in which photodissociation experiments on
stored (aligned) $H_2^+$ ions bombarded by polarized photons were
carried out, from (Jefferts & Dehmelt, 1962).

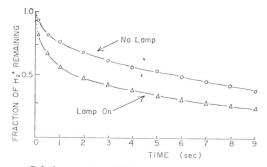

Relative number of $H_2^+$ ions trapped at $t$ seconds after end of ionization pulse, for lamp on and off.

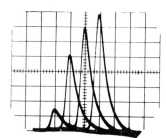

Oscilloscope presentation of the detection of ions by the resonance method. Peaks represent number of $H^+$ ions in the trap. The production is by photodissociation of the simultaneously trapped $H_2^+$. The four peaks are for $H^+$ collection times of 50, 250, 450, and 650 msec. The largest represents 10% damping. Time base is 1 msec/division.

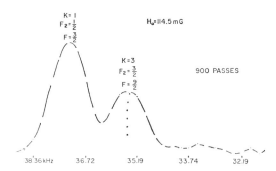

Magnetic resonance in $H_2^+$ in a magnetic field $H_0 = 114.5$ mG. Peaks correspond to transitions among the Zeeman sublevels of the states shown. Integration time was 3.5 h. The locus of points under the low-frequency peak represents the centroid of a third peak, not shown, which has been replotted after all frequencies have been scaled 5/9, the theoretical value of the ratio of magnetic moments of the states $K=3$, $F=\frac{9}{2}$, and $K=1$, $F=\frac{3}{2}$.

Fig. 4.   Some experimental results obtained with an apparatus similar to that of Fig. 3, from (Richardson et al., 1968).

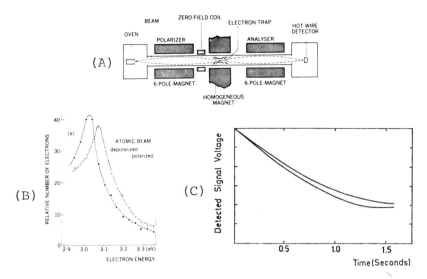

Fig. 5. Stored electron cloud/Na-Beam experiment. (A) Shows apparatus in which collisions between electrons stored in a Penning trap and a beam of polarized Na-Atoms were studied. (B) Shows spin-dependent change of energy distribution in stored electron cloud due excitation of Na D-lines. (C) Shows analogous temperature observed by noise thermometry. (A,B) from (Graff et al., 1968), (C) from (Church & Mokri, 1971).

## SINGLE HOT ION INTERACTING WITH TUNED CIRCUIT

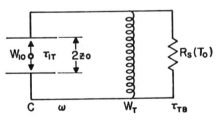

**THERMALIZATION OF ION**

$$W_I = kT_0 + (W_{IO} - kT_0)\exp(-t/\tau_{IT})$$

$$\boxed{\tau_{IT} = (4M\,z_0^2)/(e^2 R_s)}$$

**OPTIMUM SIGNAL TO NOISE RATIO**

INITIAL ENERGY OF ION, $W_{IO}$, FLOWS SLOWLY INTO TANK, FAST INTO BATH, $\tau_{IT} \gg \tau_{TB}$. RETAINED IN TANK FOR INTERVAL $\approx \tau_{TB}$, $W_T \approx (\tau_{TB}/\tau_{IT})\,W_{IO}$. THERMAL FLUCTUATIONS OF TANK ENERGY FOR OBSERVATION TIME $\approx \tau_{IT}$ AVERAGE OUT TO $\Delta W_T \approx (\tau_{TB}/\tau_{IT})\,kT_0$, S/N $= W_T/\Delta W_T$ ;

$$\boxed{S/N \approx W_{IO}/kT_0}$$

### NUMERICAL EXAMPLE

$M = 100\,M_H$ ; $2z_0 = 0.5$ cm

$C \approx 10^{-11}$ F ; $Q = 100$

$\omega \approx 5 \times 10^5$ CPS; $R_s \approx 2 \times 10^7\ \Omega$

$\tau_{IT} \approx 13$ sec ; $W_{IO} \approx 3$ eV
S/N $\approx 100$, $kT_0 \approx 0.03$ eV

Fig. 6. Brief analysis of hot oscillating ion interacting with resonant tuned circuit, from (Dehmelt, 1962).

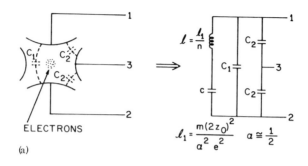

ELECTRONS

(a)

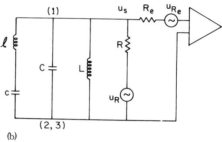

(b)

Fig. 7. Electrical equivalent representations of electrons in the Penning trap structure. (a) Pictorial representation of electrons in the Penning trap and electrical equivalent representation. (b) Electrical equivalent representation of electrons ($\ell$, c), Penning trap, and external circuit (L, C) showing noise voltages associated with real resistance R and fictitious transistor in-put noise resistor $R_e$, from (Wineland & Dehmelt, 1975b)

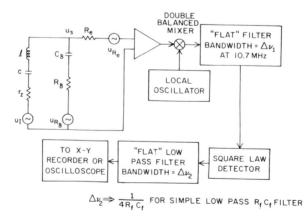

$$\Delta \nu_2 \Rightarrow \frac{1}{4 R_f C_f} \text{ FOR SIMPLE LOW PASS } R_f C_f \text{ FILTER}$$

Fig. 8. Narrow frequency band model for electrons/ions ($\ell$, c, $r_z$) interacting with LC circuit ($C_\delta$, $R_\delta$) showing block diagram of detection electronics, from (Wineland & Dehmelt, 1975b).

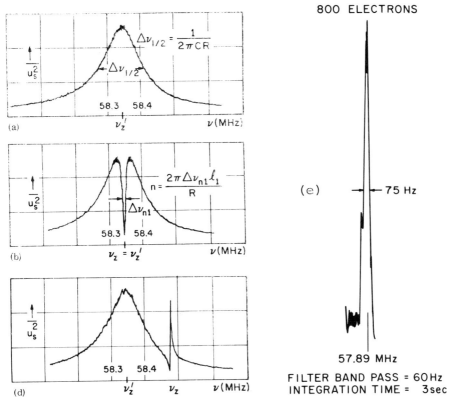

Fig. 9. Noise spectra associated with electrons and Penning trap external LC circuit. (a) Spectrum with electrons absent from trap. (b) Spectrum of ^n ≃5000 electrons when electron axial resonant frequency $\nu_z = \nu_z'$ , the LC circuit frequency. (d) Same for $\nu_z = \nu_z' + ~130$ kHz. Observation filter width $\Delta\nu_1$ ~2 kHz. From (Wineland & Dehmelt, 1975b). (e) Electron parallel resonance ($\nu_z > \nu_z'$) signal obtained with anharmonicity-compensated trap; Fig. 10., (Van Dyck et al., 1975).

tion is also being used to cool the ions. Cloud temperatures of ~80°K for e's in a Penning trap (Dehmelt & Walls, 1968), and ~800°K for protons in a Paul trap (Church & Dehmelt, 1969), have been found by measurements on the LC ciruit, whose noise spectrum is grossly modified by the ions, Fig. 9. All the processes involving collisions listed above may be used to infer cross sections. Further, the forced center of mass oscillation of an ion cloud is not broadened by like ion interactions, (Wineland & Dehmelt, 1975a). The equations of motion of a single particle in a Penning trap under forced cyclotron/axial excitation $f_x(t)/f_z(t)$ may be written

$$m\ddot{x} - m\omega_z^2 x/2 + m\omega_c \dot{y} = f_x(t), \quad (-m\omega_z^2/2 = e\phi_{xx})$$
$$m\ddot{y} - m\omega_z^2 y/2 - m\omega_c \dot{x} = 0, \quad (-m\omega_z^2/2 = e\phi_{yy})$$
$$m\ddot{z} + m\omega_z^2 z = f_z(t), \quad (m\omega_z^2 = e\phi_{zz})$$

Electrostatic interactions between like particles in a cloud very small compared to the wavelength of the exciting r.f. field do not shift or broaden the cyclotron resonance at $\omega_c - \omega_m$ or the axial resonance at $\omega_z$,

$$\omega_c = eH_o/mc, \quad \omega_c\omega_m - \omega_m^2 = \omega_z^2/2 = -e\phi_{xx}/m$$

Rather, from the equations of the z-motion of two interacting particles

$$m\ddot{z}_1 + m\omega_z^2 z_1 = F_{z12} + f_z(t)$$

$$m\ddot{z}_2 + m\omega_z^2 z_2 = F_{z21} + f_z(t)$$

it follows by addition that the center of mass coordinate $Z = (z_1 + z_2)/2$ obeys the same equation as a single particle,

$$m\ddot{Z} + m\omega_z^2 Z = f_z(t)$$

The same argument may be extended to the x and y co-ordinates and to an arbitrary number of identical particles.[†] Experimentally for e-clouds in a Penning trap with compensated anharmonicity, Fig. 10, widths of 20 Hz have been realized, Fig. 9(e), (Van Dyck et al., 1975 ), making broadening due to e-atomic beam collisions detectable. Earlier a bolometric technique was proposed for the detection of energy transfer from other degrees of freedom to the axial motion, and such transfer from the microwave excited cyclotron motion via e-e or e-atom collisions was demonstrated (Dehmelt & Walls, 1968; Wineland & Dehmelt, 1975 b). Estimates indicate that the sensitivity of the electron calorimeter realized should be sufficient to study such exothermic reactions as $e + H(F=1) \rightarrow e(\Delta m_s = \pm 1) + H(F=o) +$ 5 μeV, in a stored electron cloud/H-beam apparatus. Most recently D. Wineland et al. have observed cyclotron resonance in the monoelectron oscillator in a similar fashion after the collision sensitive forced axial oscillation at ∼60MHz of a single e ,had been observed continuously, Fig. 11, (Wineland et al., 1973). The detection of the cyclotron resonance in the slightly anharmonic monoelectron oscillator was based on a trigger technique relying on off-resonance parametric excitation near $2\nu_z$. Energy transfer from the excited cyclotron motion to the axial motion via an electron/background-gas-atom collision moved the axial frequency $\nu_z$ within the regeneration range building up a large detectable forced oscillation, Fig. 12. Here operation with sharp cyclotron energies in the range .001-1 eV and axial energies <1meV seems feasible eventually

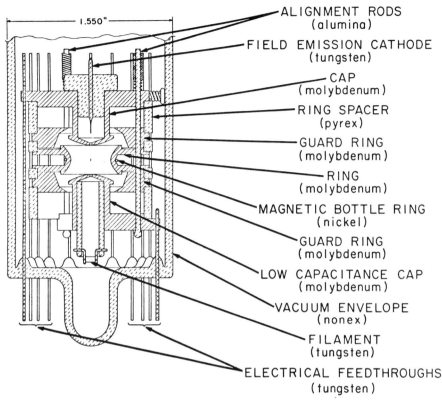

Fig. 10. Anharmonicity compensated Penning trap, from (Van Dyck et al., 1975)

for electrons or positrons (Dehmelt, 1974). In other developments Walls & Dunn (1974) Fig. 13; and Walls (1974) have carried
out quantitative measurements of recombination cross sections
for $O_2^+$, $NO^+$, $H_3O^+$ and $NH_4^+$ ions practically at rest at the
bottom of a Penning trap and in their vibrational ground states
bombarded with an electron beam in the $\sim.1$ to 8 eV range. The
ions were stored in an apparatus similar to that shown in Fig. 1
and a detection circuit as shown in Fig. 8 was used. The decay
of the ion number in the same sample was followed by repeatedly
observing parallel resonance signals as shown in Fig. 9(e). In
an apparatus similar to that shown in Fig. 5, McQuire and
Fortson (1974), have been able to demonstrate the spindependence
of the elastic collision cross section for thermal electrons and
K-atoms Fig. 14. Their method is based on the shift of the
frequency $\nu_z$ of the series resonant notch noise signal Fig. 9(b),
occurring away from the observation window at $\nu_z'$ when electrons
stored and thermalized in a slightly anharmonic Penning trap

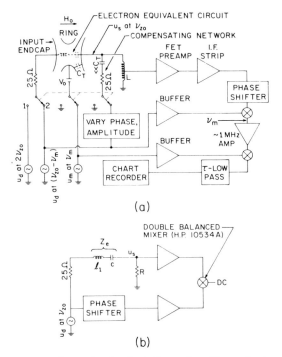

(a)

(b)

Apparatus for isolating and continuously observing the forced oscillation of a single elastically
bound electron. A block diagram is given in (a). Switch
position 2 is for direct excitation, position 1 for parametric excitation of the forced oscillation at $\nu_{z0} \approx 55.7$
MHz. In (b) an equivalent circuit is shown for switch
position 2, $Z_e$ representing the electron.

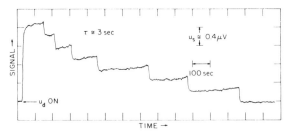

Recorder trace of forced-oscillation signal
versus time. The signal at $\nu_{z0} \approx 55.7$ MHz for an initially injected bunch of electrons decreases discontinuous-
ly as the electrons are successively boiled out of the
trap by the drive at $\nu_z' \approx 54.7$ MHz. The last plateau
corresponds to a single electron.

Fig. 11.   Apparatus for observing forced oscillation of single
electron stored in Penning trap and signal obtained with it,
from (Wineland et al., 1973).

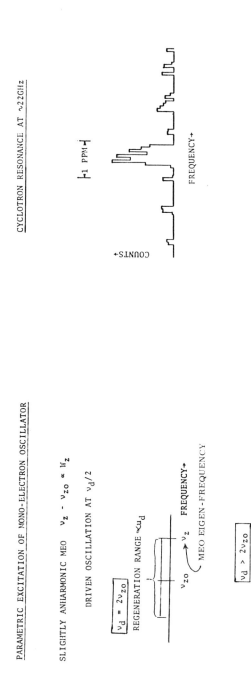

Fig. 12. Parametric excitation of monoelectron oscillator and application for detection of single 1 - 1000 meV collisions, from (Dehmelt, 1974).

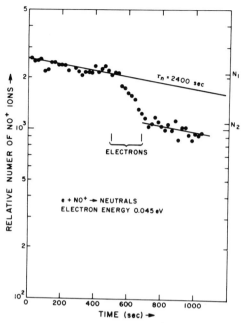

Fig. 13. Recombination data for $NO^+$ at an electron energy of
0.045 eV. The data from 500 to 700 s show the decay of ion
signal in the presence of electrons. Measurements at other
times show          residual decay mechanisms, from (Walls & Dunn,
1974).

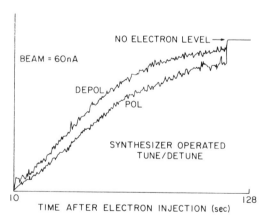

Fig. 14. Difference in electron out-of-observation-channel-
diffusion signals due to collisions with potassium atoms when
the potassium spin polarization is changed, (McGuire &
Fortson, 1974).

diffuse radially due to e-atom collisions for a distance of
$\sim.01$ cm. These authors have also observed the cooling of the
electron cloud at $kT_e \ll 1$ eV due to inelastic collisions with
molecules, cf. (Church & Mokri, 1971), Fig. 5(c). In the way of
new schemes of some interest to collision physics it has been
proposed to use laser excited resonance fluorescence at $\nu_1$ to
(A) make a single ion stored in a miniature Paul trap visible to
the naked eye (Dehmelt & Toschek, 1975), (Dehmelt & Walther,
1975) and (B) freeze out the secular oscillation at $\nu_v$ of
the ion in the trap completely. The latter trick is accomplished by making the ion assorb photons at the Doppler effect
induced sideband frequency $\nu_1 - \nu_v$. Since reemission will
occur symmetrically at $\nu_1 \pm n\nu_v$ energy is extracted from the
vibrational motion on the average, (Wineland & Dehmelt, 1975c).
Hereby a solution of the previously posed problem of how to make
an isolated (charged) atom float at rest in free space (Dehmelt,
1967), appears to have been brought within reach. In a related
scheme (Wineland & Dehmelt,1975a,Errata & Addenda), it has been
proposed to pump energy into and extract angular momentum from
the magnetron motion of an electron cloud carrying out a <u>damped</u>
oscillation at $\omega_z$ in a Penning trap by irradiating it with an
<u>inhomogeneous</u> rf field at $\omega_z + \omega_m$, and thereby make it contract
radially. There is the possibility that the underlying princi-
ple is of broader applicability and suitable for the containment of fusion plasmas.

Ion storage techniques as well as collision experiments
based on them previously have been reviewed in (Dehmelt, 1967
& 1969), (Dawson & Whetten, 1969), (Walls & Dunn, 1974),
(Dehmelt, 1975), and (Dunn, 1974).

The author thanks his coworkers, Drs. David Wineland and
Robert Van Dyck and Messrs. Paul Schwinberg and Frank Gorecki
for reading the manuscript and Ms. Lyn Maddox for typing it.

---

†) An important consequence of this is that in a perfectly
   harmonic trap excitation of the center-of-mass axial motion
   of an electron cloud will not lead via e-e collisions to
   excitation of the center-of-mass cyclotron motion. The
   center-of-mass of the cloud behaves like a single particle!
   By contrast energy transfer via e-atom collisions will of
   course take place.

CHURCH, D. A. & DEHMELT, H. G. (1969). Journ. Appl. Phys. 40, 3421.
CHURCH, D. A. & MOKRI, B. (1971). Z. Physik 244, 6.
DEHMELT, H. G. (1958). Phys. Rev. 109, 381.
DEHMELT, H. (1961). Progress Report NSF-G5955, May 1961, "Spin Resonance of Free Electrons".
DEHMELT, H. (1962), Bull. A.P.S. 7, 470.
DEHMELT, H. (1967 & 1969). Adv. Atom. & Mol. Physics, 3 & 5, (Academic Press).
DEHMELT, H. G. (1974). Bull. A.P.S. 19, 14.
DEHMELT, H. G. (1975), Proceedings, Fifth Int. Conf. on Atomic Masses and Fund. Constants, Paris (Plenum Press)
DEHMELT, H. G. & JEFFERTS, K. B. (1962). Phys. Rev. 125, 1318.
DEHMELT, H. & MAJOR, F. G. (1962). Phys. Rev. Letters 8, 213.
DEHMELT, H. & WALLS, F. (1968). Phys. Rev. Letters 21, 127.
DEHMELT, H. & TOSCHEK, P. (1975). Bull. A.P.S. 20, 61.
DEHMELT, H. & WALTHER, H. (1975). Bull. A.P.S. 20, 61.
DUNN, G. H. (1974). Atomic Physics 4, p. 575 (Plenum Press)
FISCHER, E. (1959). Z. Physik 156, 1.
GRAEFF, G., KLEMPT, E., & WERTH, G. (1969). Z. Physik 222, 201.
GRAEFF, G., MAJOR, F. G., ROEDER, R. W., & WERTH, G. (1968). Phys. Rev. Letters 21, 340.
GRAEFF, G., HUBER, K., KALINOWSKY, H. & WOLF, H., (1972). Phys. Lett. A41, 277.
JEFFERTS, K. B. (1968). Phys. Rev. Letters 20, 39.
JEFFERTS, K. B. (1969). Phys. Rev. Letters 23, 1476.

JEFFERTS, K. B. & DEHMELT, H. G. (1962). Bull. A.P.S. 7, 432.
MCGUIRE, M. D. & FORTSON, E. N. (1974). Phys. Rev. Letters 33, 737.
RICHARDSON, C. B., JEFFERTS, K. B., & DEHMELT, H. G. (1968). Phys. Rev. 165, 80.
SCHUESSLER, H. FORTSON, N. & DEHMELT, H. (1969). Phys. Rev. 187, 5.

VAN DYCK, R., EKSTROM, P. & DEHMELT, H. (1975). Bull. A.P.S. 20, 492.
WALLS, F. (1970). Thesis, University of Washington.
WALLS, F. & DUNN, G. (1974). Physics Today 27, No 8, 30.
WINELAND, D., EKSTROM, P., & DEHMELT, H. (1973). Phys. Rev. Lett. 31, 1279.
WINELAND, D., & DEHMELT, H. (1975a). Int. Journ. Mass Spectroscopy & Ion Phys. 16, 338.
WINELAND, D., & DEHMELT, H. (1975b). Journal Appl. Phys. 46, 919.
WINELAND, D. & DEHMELT, H. (1975c). Bull. A.P.S. 20, 637.

# APPLICATION OF ION CYCLOTRON RESONANCE

# SPECTROSCOPY TO STUDIES OF COLLISION PROCESSES

J. L. Beauchamp

California Institute of Technology

Pasadena, California 91125

Many diverse applications of ion cyclotron resonance spectroscopy (ICR) are evident in the numerous available review articles which chronicle the development of this instrumental method and the contributions which it has made to the solution of problems of general interest to physical scientists [1-5]. Although many of the principles of Lawrence's cyclotron are conceptually evident, the precursor to the modern ICR spectrometer is the omegatron [6]. Originally developed as part of an experimental program to measure the magnetic moment of the proton, the omegatron as a simple mass spectrometer has been widely used as a partial pressure gauge. Studies initiated at the Cornell Aeronautical Laboratory in the early 1960's directed attention to many of the applications of ICR to the study of atomic and molecular processes in partially ionized gases [7]. An expansion of the field which has continued until present commenced with the introduction of commercial ICR instrumentation by Varian in 1965, and even today most "home built" spectrometers with specialized capabilities retain design innovations introduced by Varian.

A cutaway drawing of an ICR probe assembly recently constructed in our laboratory is shown in Figure 1. A directly heated filament produces an electron beam which crosses the source region of the ICR cell in the direction of the primary magnetic field H. Ions formed by electron impact are constrained to move in circular orbits in the plane perpendicular to H, at the cyclotron frequency $\omega_c = qH/mc$ where $q/m$ is the ion charge to mass ratio and c is the speed of light. An appropriate bias on the trapping electrodes restricts the motion of either positive or negative ions in a direction

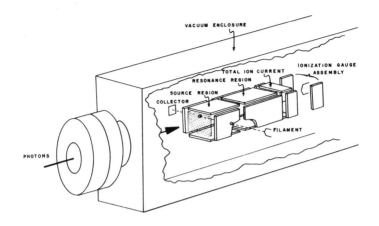

**Figure 1.** Cutaway view of ICR probe assembly.

parallel to the magnetic field. Ions move from the source into the resonance region under the influence of a constant electric field $E_D$ applied perpendicular to H. In this crossed field geometry ions move with a drift velocity $v_D = cE_D/H$ in a direction perpendicular to $E_D$ and H, with the center of the ion orbit remaining on a equipotential of the electrostatic field. In the resonance region ions absorb energy from a radio frequency field $E_1 \sin\omega_1 t$ applied perpendicular to H. In the absence of collisions the power absorption is given by equation (1) where P(0) is the rate of ion formation at the

$$A(\omega_c) = \frac{P(0)q^2E_1^2}{4m(\omega_1-\omega_c)^2} \left[1 - \cos(\omega_1-\omega_c)\tau\right] \tag{1}$$

electron beam and $\tau$ is the time which an ion spends in the resonance region of the cell [1]. If the power absorption is interrupted by collisions the calculated lineshape adopts the Lorentzian form given by equation (2) where $\xi$ is the collision

$$A(\omega_c) = \frac{P(0)\tau q^2E_1^2}{4m} \frac{\xi}{(\omega_1-\omega_c)^2 + \xi^2} \tag{2}$$

frequency for momentum transfer. Assuming only elastic collisions, the collision frequency $\xi$ is given by equation (3)

$$\xi = \frac{nM}{m + M} \langle v\sigma_d(v)\rangle \tag{3}$$

where n and M are the number density and mass of the neutral, and $\sigma_d(v)$ and v are the diffusion cross section and relative velocity of the ion-neutral pair. The brackets indicate an average over the distribution of relative velocities. Information concerning ion-neutral collision processes can thus be obtained from experimentally determined linewidths [1, 3, 7, 8].

Recent developments in ICR instrumentation include most importantly trapped ion methods [9, 10]. Trapping is effected by having the equipotentials of the electrostatic field in the plane perpendicular to H close on themselves. This is accomplished in the cell shown in Figure 1 by electrode biasing. After a suitable delay, ion concentrations are sampled by switching on the drift field and integrating the transient power absorption. Other relatively new developments include Fourier transform techniques which provide fast response and high resolution [11],variable temperature capabilities [12], the adaptation to ionization sources other than electron impact [13], the intracavity incorporation of an ICR cell in a tunable dye laser [14], and tandem ICR spectrometers which make it possible to study the reactions of mass selected ions [15].

In this brief review many of the newer experimental techniques are exemplified in their application to a range of problems under investigation in our laboratory and elsewhere. Although the main focus of our research involves problems in physical organic and inorganic chemistry, the many applications which ICR has in the field of atomic and molecular physics will be evident from the examples given.

## ELECTRON MOLECULE COLLISIONS

Elastic Encounters. Collision broadened electron cyclotron resonance spectra have been analyzed to determine collision frequencies for momentum transfer with a range of polar and non-polar gases [16]. Although the cell design shown in Figure 1 is not suitable for operation at the very high electron cyclotron frequencies, it is of interest to note that electrons can be detected by observing resonant power absorption resulting from excitation of their harmonic motion in the trapping well [17]. The analysis of spectral lineshapes obtained in this manner can also yield electron neutral collision frequencies at near thermal energies and remains to be explored as an experimental method.

Inelastic Encounters. No modification of the standard ICR apparatus is required to obtain trapped electron (TE) spectra of molecules [18]. Threshold excitation of an excited electronic state leaves electrons with insufficient translational energy to escape the trapping well. The yield of trapped electrons is measured using $SF_6$ as a scavenger. TE spectra are shown for formaldehyde and two of its derivatives in Figure 2 [19]. The two peaks at 3.7 and 6.2 eV correspond to the lowest $^3(n \rightarrow \pi^*)$ and $^3(\pi \rightarrow \pi^*)$ excitations.

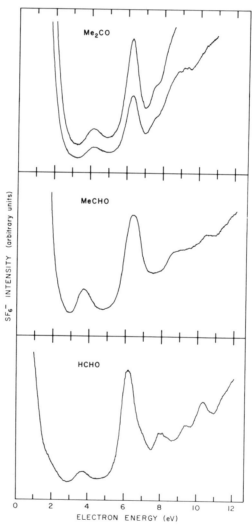

Figure 2.  Trapped electron spectra of acetone ($Me_2CO$), acetaldehyde (MeCHO) and formaldehyde (HCHO).

The shoulder at 7.0 eV is due to the $^3(n \rightarrow 3s)$ excitation. Hence the three lowest triplet states are observed to the apparent exclusion of the lowest $^1(n \rightarrow \pi^*)$ excitation observed optically at 4.1 eV. A general feature of triplet states is that their excitation cross sections peak at low electron energies relative to singlet states. As a result it can be expected that the derivative of the cross section with respect to electron energy evaluated at threshold, $(d\sigma/dE)_0$, will be larger for triplet states than for singlet states. A necessary but not sufficient condition for an intense peak in the TE spectrum is that $(d\sigma/dE)_0$ be large. Thus it is not surprising that triplet states are often dominant features of such spectra. The TE technique for obtaining electronic spectra is made attractive by the simplicity of the method, provides coverage of the vacuum ultraviolet, and in conjunction with optical spectra and theoretical calculations affords information relating to triplet states.

Electron Attachment. Trapped ion capabilities facilitate the investigation of electron attachment processes [20]. For example, the variation of ion abundance with time in $UF_6$ ionized with a 10 msec electron beam pulse at 11 eV and at $3.5 \times 10^{-6}$ torr is shown in Figure 3 [21]. $UF_5^-$ is formed directly during the electron beam pulse and reacts rapidly with $UF_6$ to form $UF_6^-$ (reaction 4). The majority of the $UF_6^-$ is formed by the attachment of inelastically scattered electrons (process 5) which are trapped in the source region during the electron beam pulse. The minor ion $UF_7^-$ results from

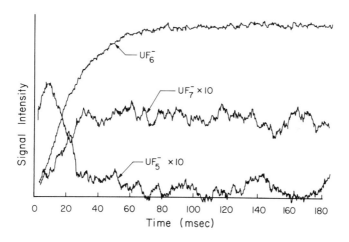

Figure 3. Temporal variation of negative ion concentrations in $UF_6$ at $3.5 \times 10^{-6}$ torr. The electron beam is gated on for 10 msec at 11 eV.

$$UF_5^- + UF_6 \rightarrow UF_6^- + UF \tag{4}$$

$$e^- + UF_6 \rightarrow UF_6^- \tag{5}$$

reaction 6 which involves a portion of the $UF_6$ which is formed with excess internal excitation.  The buildup of $UF_6^-$ in

$$(UF_6^-)^* + UF_6 \rightarrow UF_7^- + UF_5 \tag{6}$$

Figure 3 yields an electron attachment rate of $5 \times 10^{-10}$ $cm^3molecule^{-1}sec^{-1}$, which is exceptionally slow.  In a 1:1 mixture of $CCl_4$ and $UF_6$, all of the electrons are scavenged by $CCl_4$ (process 7, $k = 3 \times 10^{-7}$ $cm^3molecule^{-1}sec^{-1}$) to form $Cl^-$ which undergoes the electron transfer reaction 8 to form

$$CCl_4 + e \rightarrow CCl_3 + Cl^- \tag{7}$$

$$Cl^- + UF_6 \rightarrow UF_6^- + Cl \tag{8}$$

$UF_6^-$.  Reaction 8 is consistent with the remarkably high electron affinity of $UF_6$ ($\sim 5.7$ eV).

It has been shown that species such as $SF_6$ have a limiting low pressure rate for non-dissociative electron attachment which results from radiative stabilization of the initially formed negative ions [20].  Rates for dissociative electron capture (e.g., process 7) depend markedly on the electron energy, with the rate limiting step often being the relaxation of an epithermal electron energy distribution.  The efficiencies of various neutrals in thermalizing electrons can be investigated in this fashion.

## ION MOLECULE COLLISIONS

Ion-Neutral Interactions.  The prevalence and rapid rates of ion-molecule reactions can be attributed to the very strong and long range attractive interactions between ions and neutrals.  Potential energy curves for several systems are indicated in Figure 4 [4].  The interaction of $CH_5^+$ with $CH_4$ is typical of systems involving non-polar neutrals with no specific chemical bonding between the collision partners. The interaction of $Li^+$ and $H_3O^+$ with $H_2O$ are comparable (Figure 4), and electrostatic interactions play an important role in providing a relatively deep potential well at the equilibrium heavy atom separation.  Chemical bonding leads to even stronger interactions and collision of $CH_3^+$ with $NH_3$ leads to the formation of a chemically activated intermediate with an internal energy of 106 kcal/mole (Figure 4).

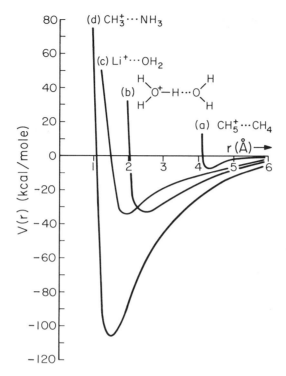

Figure 4. Potential energy curves for ion-neutral interactions. The separation is taken to be the spacing of the
two heavy atoms.

Elastic Encounters. Long range interactions between
ions and neutrals can be probed by examining the collision
broadening of the ICR absorption of any ion interacting with
any neutral with which no rapid reaction occurs. Lineshape
analysis yields, in accordance with equation 2, collision
frequencies for momentum transfer. These are inversely
proportional to ion mobilities and the determination of an
ICR lineshape may be regarded as a resonance mobility
experiment [1, 3, 7, 8 ].

Low energy collisions of ions with non-polar neutrals
which result in momentum transfer are almost entirely
dominated by the long range polarization interaction. The
more interesting case of non-reactive encounters of ions
with polar molecules requires the addition of a long range
ion-dipole interaction as indicated in equation 9, where
$\alpha$ and $\mu_D$ are the angle averaged polarizability and dipole

$$V(r) = - \frac{\alpha q^2}{2r^4} - \frac{q\mu_D \cos\theta}{r^2} \tag{9}$$

moment of the neutral and $\theta$ is the angle between the dipolar axis and the radius vector between the collision partners. Rather straightforward arguments suggest that the collision frequency for momentum transfer will be given by equation 10 where $\mu$ is the reduced mass and $\beta$ is a constant between 0

$$\xi/n = (2\pi e \mu^{\frac{1}{2}}/m)[\alpha^{\frac{1}{2}} + \beta\mu_D(2/\pi kT)^{\frac{1}{2}}] \tag{10}$$

and 1 which describes the importance of the ion-dipole interaction. We have examined the validity of this analysis by measuring $\xi/n$ for $Na^+$ ions interacting with a series of $C_3H_6O$ isomers which have nearly the same polarizability but dipole moments which vary considerably [22]. As shown in Figure 5, $\xi/n$ exhibits a linear variation with dipole moment, and equation 10 reproduces the observed slope with $\beta = 1.05$. The intercept is somewhat below the polarization limit of $1.02 \times 10^{-9}$ cm$^3$molecule$^{-1}$sec$^{-1}$.

Reactive Encounters. Investigations of ion-molecule reactions comprise the majority of the published ICR studies [1-5]. ICR is ideally suited for investigating ion-molecule reactions at near thermal ion-energies. Double resonance experiments provide positive identification of reaction pathways in complex systems. Selective acceleration of reactant ions in a time short compared to the time between collisions makes it possible to examine the variation of reaction rate constants and product distributions with ion energy in the important range between thermal and 20 eV. Trapped ion experiments such as illustrated in Figure 3 allow for many collisions and the observation of extremely slow reactions.

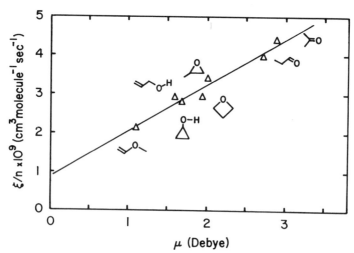

Figure 5. Variation of $\xi/n$ with dipole moment for $Na^+$ interacting with $C_3H_6O$ isomeric neutrals. The line is a least squares fit to the data.

Relaxation of the system to thermal equilibrium facilitates the determination of equilibrium constants and related ion thermochemical properties [23, 24].

Exemplifying these studies are reactions of metal ions such as $Li^+$ with organic molecules [25]. Figure 6a illustrates the ICR spectrum of ions emitted from a thermionic source doped with a lithium salt and mounted inside the source region of the ICR cell. With the addition of propionyl bromide (Figure 6b), the $Br^-$ transfer reaction 11 is

$$Li^+ + CH_3CH_2C \underset{Br}{\overset{O}{\diagup}} \longrightarrow \begin{cases} CH_3CH_2C=O^+ + LiBr & (11) \\ \\ [CH_3CHC=O]Li^+ + HBr & (12) \end{cases}$$

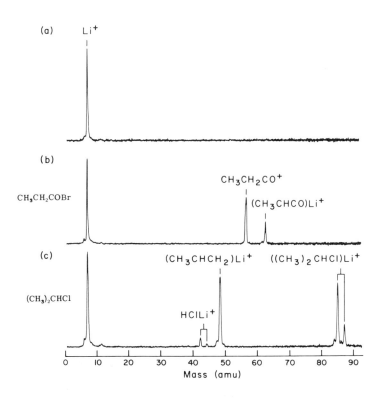

Figure 6. (a) ICR spectrum of ions emitted from thermionic source doped with LiOH. (b) Product ions observed with addition of $10^{-5}$ torr of propionyl bromide. (c) Product ions observed with addition of $10^{-5}$ torr of isopropyl chloride.

observed in addition to the HBr elimination process 12. With
isopropyl chloride, the endothermic halide transfer reaction 13
is not observed. Instead the elimination of HCl occurs with

$$X \rightarrow (CH_3)_2CH^+ + LiCl \qquad (13)$$

$$Li^+ + (CH_3)_2CHCl \longrightarrow HClLi^+ + CH_3CH{=}CH_2 \qquad (14)$$

$$\longrightarrow [CH_3CH{=}CH_2]Li^+ + HCl \qquad (15)$$

$Li^+$ remaining bound to either HCl or propylene (reactions
14 and 15, respectively). The products of reactions 14 and 15
both transfer $Li^+$ to isopropyl chloride, forming the cluster
ion $[(CH_3)_2CHC\ell]Li^+$. No further reaction occurs in this
system at low pressures. Reactions such as these provide
bimolecular pathways to form complexes of $Li^+$ with various
molecules. Trapped ion studies of equilibrium $Li^+$ transfer
reactions provide the $Li^+$ binding energies shown in Figure 7
[26]. This experimental methodology complements ion beam
scattering experiments for studying the interaction of metal
ions with small molecules [27].

## PHOTOEXCITATION OF IONS

Photodetachment of Electrons. ICR ion storage capabilities facilitate the investigation of photophysical and photo-
chemical processes involving ions [3]. Interactions of
photons with ions are inferred from observed changes in ion
abundances. By monitoring the decrease in negative ion
abundance at different wavelengths, photodetachment cross
sections can be measured. In favorable cases the threshold
photodetachment energy yields the electron affinity of the
nautral species. Results for $Cl^-$ obtained using the
apparatus shown in Figure 1 are presented in Figure 8 [28].
The threshold for photodetachment at 3400 Å corresponds
to an electron affinity of 3.65 ± 0.1 eV, in good agreement
with the absorption value of 3.61 eV. Refinements of these
measurements should permit observation of the spin orbit
coupling in Cl.

ICR photodetachment studies have provided a large number of molecular electron affinities and related thermodynamic
parameters. For example, studies of $HSe^-$ at moderate
resolution give EA(HSe) = 2.21 ± 0.03 eV and analysis of fine
structure in the photodetachment cross section yields a spin
orbit coupling constant A(HSe) = -1815 ± 100 cm$^{-1}$ [29]. Laser
photodetachment of $HS^-$ reveals an interesting and thus far
unexplained structure in the photodetachment cross section
near threshold [30]. Incorporation of the ICR cell

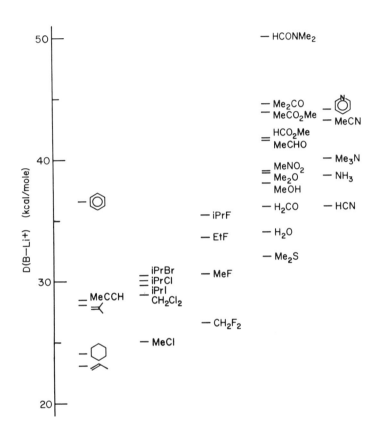

Figure 7.  Binding energies $D(B-Li^+)$ of $Li^+$ to a variety of bases B in the gas phase.

intravacity in a dye laser system promises even higher sensitivity than has been possible thus far in these experiments [14].

Photochemistry of Ions.  Information relating to electronic excitation energies of ions can be determined by monitoring the wavelength dependence of processes such as generalized in equation 16.  The measured quantity is the

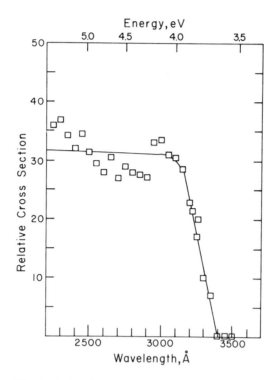

Figure 8.   Photodetachment of electrons from Cl⁻. The light source employed is a xenon-mercury arc dispersed with a 0.25 m monochrometer set for a bandpass of 100 Å.

photodissociation cross section $\sigma(\lambda)$ which is proportional to

$$A^+ + h\nu \rightarrow B^+ + C \tag{16}$$

the product of the gas phase extinction coefficient, $\epsilon_g(\lambda)$, and the photodissociation quantum yield, $\varphi_d(\lambda)$.

The photodissociation of benzene radical ions [31] is illustrated in Figure 9, which displays the temporal variation of $C_6H_6^+$ and $C_6H_5^+$ in the presence and absence of 1 watt

$$C_6H_6^+ + h\nu \rightarrow C_6H_5^+ + H \tag{17}$$

continuous 4579 Å irradiation from an argon ion laser. The thermodynamic threshold of 3.9 eV for process 17 is considerably above the photon energy of 2.71 eV. Studies varying irradiation time, neutral gas pressure, and light intensity indicate that the sequential absorption of two photons

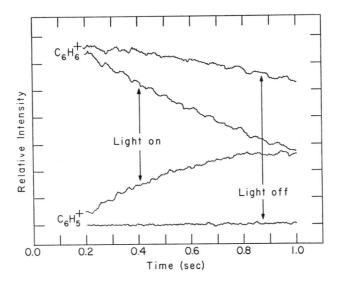

Figure 9. Photodissociation of benzene radical cations, illustrated by the temporal variation of $C_6H_6^+$ and $C_6H_5^+$ in the presence and absence of 1 watt continuous 4579 Å irradiation from an argon ion laser.

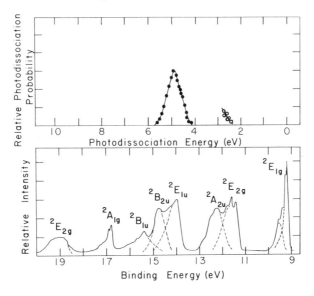

Figure 10. Comparison of the photoelectron spectrum of benzene to the partial photoexcitation function for the two photon dissociation of $C_6H_6^+$ (open circles) and the complete excitation function for the direct process (filled circles). The latter was recorded at 50 Å resolution.

leads to dissociation [31]. Figure 10 compares the deconvoluted photoelectron spectra [32] of benzene to the partial photodissociation cross section (open circles) for the two photon process, obtained using eight argon ion laser lines between 4545 Å and 5145 Å (2.4 - 2.7 eV). Included in the figure is the complete photodissociation cross section (filled circles) for a second process which leads to dissociation of $C_6H_6^+$ in accordance with equation 17 above thermodynamic threshold. The energy axis of the photoelectron spectrum is adjusted such that the first adiabatic ionization potential of benzene is zero on the photodissociation energy scale. The open circles identify with the optically allowed $^2E_{1g} \rightarrow {}^2A_{2u}$ transition. The $^2A_{2u}$ state reverts to a vibrationally excited ground state which absorbs a second photon, thereby pooling sufficient internal energy to exceed the dissociation threshold. Although the direct process at higher energy closely matches the $^2E_{1u}$ state observed in the photoelectron spectrum, it is likely that the observed excitation is the $^2E_{1g} \rightarrow {}^2E_{2u}{}^2(\pi \rightarrow \pi^*)$ excitation calculated to have a vertical excitation energy of 5.2 eV [33]. The same transition is apparent in many spectra of substituted benzene radical ions where it has no counterpart in the corresponding photoelectron spectra. Photodissociation spectra thus provide information relating to excited states of radical ions which are not observed by photoelectron spectroscopy.

Another example of interest is the benzoyl cation, $C_6H_5CO^+$, which undergoes the photoreaction 18 [34]. In Figure 11 the gas phase photodissociation spectrum is

$$C_6H_5CO^+ \rightarrow C_6H_5^+ + CO \tag{18}$$

compared to the solution absorption spectrum. The observed onset for reaction 18 at 340 ± 10 nm (3.7 eV) is well above the calculated thermodynamic threshold of 2.3 eV. From a comparison of the gas phase and solution spectra in Figure 11 it is evident that the quantum yield for dissociation $\varphi_d(\lambda)$, is relatively constant over the wavelength region considered. Also significant and somewhat surprising is the absence of a solvent shift which would indicate differential interaction of the ion with the solvent in its ground and excited electronic states.

## PROGNOSIS

ICR techniques are generally suited for studying processes which involve the formation of ions over a large volume, particularly where their subsequent interactions with neutral molecules are of interest. This suggests many

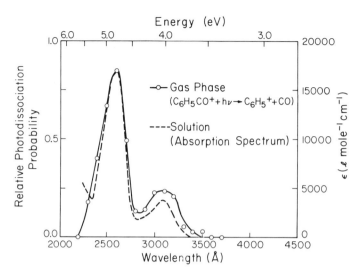

Figure 11.  Comparison of the gas phase photodissociation spectrum of benzoyl cation to the uv absorption spectrum obtained in solution.

applications as a detector for charged particles resulting from photon and particle impact processes.  Spectroscopic studies of ions using both photon and electron impact excitation will continue as a major application in the area of atomic and molecular physics.  It is likely in the near future that techniques will be developed for obtaining high resolution infrared and microwave spectra of ions and hence precise structural information.  Optical pumping experiments will allow for the generation of reactant ions in well defined internal energy states.  Barium ions, for example, can be produced by surface ionization and pumped into the metastable $^2D_{5/2}$ and $^2D_{3/2}$ states.  The simultaneous production and control of positive and negative ions remains an intriguing experimental problem which when solved will permit study of ion recombination processes.  Trapped ion experiments at low pressures present the interesting situation where the time between collisions greatly exceeds infrared radiative lifetimes.  Molecular aggregation by radiative mechanisms under these conditions may provide significant insights into the processes which lead to the formation of complex molecules in interstellar space.  Finally, taking another direction in changing the parameters which appear to confine present experimentation, ion cyclotron resonance will likely be developed for diagnostic and control purposes in the area of plasma physics.

Acknowledgment.  This work was supported by the United States Energy Research and Development Administration.

# REFERENCES

1)  J. L. Beauchamp, Ann. Rev. Phys. Chem., 22, 527 (1971).

2)  M. T. Bowers and T. Su, Adv. in Electronics and Electron Phys., 34, 223 (1973).

3)  (a) R.C. Dunbar, "Photodissociation of Gas Phase Ions," in "Interactions Between Ions and Molecules," P. Ausloos, Ed., Plenum Press, New York, N.Y., 1975, p 579; (b) "Ionic Reactions in the Gas Phase; Study by Ion Cyclotron Resonance," in "Chemical Reactivity and Reaction Paths," G. Klopman, Ed., Wiley, New York, N.Y., 1974, p 339.

4)  J. L. Beauchamp, "Reaction Mechanisms of Organic and Inorganic Ions in the Gas Phase," ref 3a, p 413.

5)  J. D. Baldeschwieler, Science, 159, 263 (1968).

6)  H. Sommer, H. A. Thomas, and J. A. Hipple, Phys. Rev., 82, 697 (1951).

7)  D. Wobschall, J. R. Graham, and D. P. Malone, Phys. Rev., 131, 1565 (1963).

8)  J. L. Beauchamp, J. Chem. Phys., 46, 1231 (1967).

9)  R. T. McIver, Rev. Sci. Instrum., 41, 555 (1970).

10) T. B. McMahon and J. L. Beauchamp, Rev. Sci. Instrum., 43, 509 (1972).

11) M. B. Comisarow and A. G. Marshall, Chem. Phys. Lett., 25, 282 (1974).

12) S. E. Buttrill, J. Chem. Phys., 58, 656 (1973).

13) Surface ionization provides a variety of atomic and molecular positive and negative ions which are otherwise difficult to produce [J. L. Beauchamp, unpublished results; see also ref 25].

14) J. R. Eyler, Rev. Sci. Instrum., 45, 1154 (1974).

15) D. L. Smith and J. H. Futrell, Int. J. Mass Spectrom. and Ion Phys., 14, 171 (1974).

16) K. D. Bayes, D. Kivelson, and S. C. Wong, J. Chem. Phys., 37, 1217 (1962).

17) W. T. Huntress, Jr., and W. T. Simms, Rev. Sci. Instrum., 44, 1274 (1973).

18) D. P. Ridge and J. L. Beauchamp, J. Chem. Phys., 51, 470 (1969).

19) R. H. Staley, L. B. Harding, W. A. Goddard III, and J. L. Beauchamp, Chem. Phys. Lett., submitted for publication.

20) M. S. Foster and J. L. Beauchamp, Chem. Phys. Lett., 31, 482 (1975).

21) J. L. Beauchamp, J. Chem. Phys., submitted for publication.

22) D. P. Ridge, Ph.D. Thesis, California Institute of Technology (1973).

23) M. T. Bowers, D. H. Aue, H. M. Webb, and R. T. McIver Jr., J. Am. Chem. Soc., 93, 4314 (1971).

24) R. H. Staley and J. L. Beauchamp, J. Chem. Phys., 62, 1998 (1975).

25) R. D. Wieting, R. H. Staley, and J. L. Beauchamp, J. Am. Chem. Soc., 97, 924 (1975).

26) R. H. Staley and J. L. Beauchamp, J. Am. Chem. Soc., in press.

27) E. Bottner, W. L. Dimpfl, U. Ross, and J. P. Toennies, Chem. Phys. Lett., 32, 197 (1975).

28) B. S. Freiser and J. L. Beauchamp, unpublished results.

29) K. C. Smyth and J. I. Brauman, J. Chem. Phys., 56, 5993 (1972).

30) J. R. Eyler and G. H. Atkinson, Chem. Phys. Lett., 28, 217 (1974).

31) B. S. Freiser and J. L. Beauchamp, Chem. Phys. Lett., in press.

32)   T. A. Carlson and C. P. Anderson, Chem. Phys. Lett., 10, 561 (1971).

33)   P. J. Hay and I. Shavitt, J. Chem. Phys., 60, 2865 (1974).

34)   B. S. Freiser and J. L. Beauchamp, J. Am. Chem. Soc., submitted for publication.

# APPLICATION OF THE VARIABLE-TEMPERATURE FLOWING AFTERGLOW AND FLOW-DRIFT TUBE TECHNIQUES TO STUDIES OF THE ENERGY DEPENDENCE OF ION-MOLECULE REACTIONS

F. C. Fehsenfeld and D. L. Albritton

Aeronomy Laboratory, NOAA Environmental Research
Laboratories
Boulder, Colorado 80302, U.S.A.

## INTRODUCTION

The past two decades have seen a considerable amount of research on ion-molecule reactions. The information gained from these studies is concentrated in energy either near room temperature or at energies above 1 eV. The room-temperature studies have employed experimental techniques such as high-pressure mass-spectrometer ion sources, ion-cyclotron resonance methods, and flowing afterglows. Studies at the higher energies typically have employed conventional beam methods. In recent years, experimental methods, such as the drift tube, have been applied to the study of ion-molecule reactions and offer the possibility of bridging the important energy gap between the room-temperature and higher-energy data. Consequently, it is important to understand what drift tubes measure. Being a non-thermal swarm technique, characterizing the relative speed distribution of the ions and the reactant molecules is not straight-forward and hence the relationship between drift-tube data and those taken using other experimental techniques is nontrivial. Furthermore, the excitation of internal energy states of molecular ions in energetic swarms is a possibility and its occurrence would further complicate the interpretation of drift-tube results and their comparison with those from other techniques.

Recently in our laboratory, a flow-drift tube has been developed, which, being a combination of a flowing afterglow and a drift tube, combines the chemical versatility of the former with the energy variability of the latter. This new apparatus,

in some instances used in concert with a variable-temperature
flowing afterglow, has been used to study the effects of drift-
tube speed distributions and ion internal energy states on
ion-molecule reactions.  These results are summarized here.

## EXPERIMENTAL

### Variable-Temperature Flowing Afterglow

The flowing afterglow technique[1,2] has been used to investigate a wide variety of ion-molecule reactions.  Many of these
studies have been made over a range of temperatures.[3,4] A simple
schematic diagram of the flowing afterglow apparatus is given in
Fig. 1.  A buffer gas is introduced into the 8 cm diam., 1 m
long flow tube.  The flow rate and the speed of the pumps can be
adjusted to yield gas pressures in the range from 0.1 to 2 Torr
and gas flow velocities in the range from $10^3$ to $10^4$ cm/sec.
The buffer gas flows past an electron gun where it is excited and
ionized.  Selected source gases may be added to produce a wide
variety of ions of interest.  The flowing buffer gas carries
these reactant ions downstream past an inlet where the selected
neutral reactant gas can be added.  Prior to reaching this inlet,
gas phase collisions have equilibrated the ions to the temperature of the flow tube wall.  After its introduction, the neutral
reactant gas mixes into the buffer gas and reacts with the ions
as the flow carries the mixture through the reaction region.  A
mass spectrometer sampling orifice marks the termination of the
reaction region.  The variation of the ion signals as a function
of the reactant neutral addition rate yields the rate constant.

The temperature dependence of rate constants is studied by
enclosing the flow tube in a radiation-shielded vacuum jacket
and by placing cooling coils and heaters in close contact with
the tube.  The various gases are introduced such that they enter

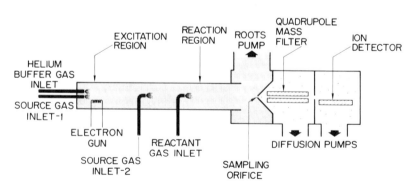

Figure 1.  Schematic diagram of the flowing afterglow.

in equilibrium with the cooled or heated tube walls. The present apparatus has a temperature range from 80 to 900°K with a stability and uniformity at the wall of 1% near room temperature and within 5% above and below room temperature. Thus, the mean energy $KE_{cm}$ is well-characterized by $3kT/2$, where k is Boltzmann's constant and T is the gas temperature. The ion-molecule joint speed distribution is Maxwellian and the rotational and vibration populations are in thermal equilibrium.

### Flow-Drift Tube

Figure 2 gives a schematic diagram of the flow-drift tube,[5] which in most ways operates like the flowing afterglow. The reactant ions are created as described above in the upstream unipotential ion production section. However, the buffer gas flow carries these ions into the separation section, where a potential gradient separates the positive ions from the negative ions and electrons. Thus, ions of only one charge sign reach the electric shutter, a pair of closely spaced, high-transparency grids that mark the beginning of the drift-reaction region, in which a uniform dc electric field is established.

When measuring a rate constant, this shutter is held open so that a continuous stream of ions enters the drift-reaction region. The acceleration of the ions during the time between collisions with the buffer gas neutrals and the deceleration occurring during these collisions rapidly equilibrate and the ion swarm maintains a mean drift velocity $v_d$ through the buffer gas and hence a mean suprathermal kinetic energy $KE_{lab}$ in the laboratory frame, both being characterized by the ratio of the electric field strength E to the buffer gas density N. A reactant neutral is then mixed with these suprathermal ions and the reaction rate is determined in the same manner as described above, for different values of E/N.

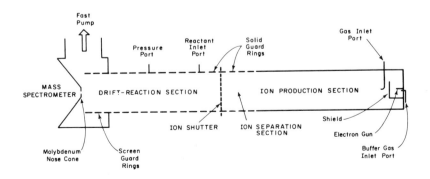

Figure 2. Schematic diagram of the flow-drift tube.

While E/N is a proper characterization of the motion of the suprathermal ions, the mean energy is clearly a more useful one. However, the speed distribution of drifting ions is not Maxwellian at large E/N and is exceedingly difficult to measure accurately. Hence, theory has been the sole guide for determining the mean ion energy. The expression commonly used is the one derived by Wannier[7] for the point-charge, induced-dipole model

$$KE_{lab} = \frac{1}{2} mv_d^2 + \frac{1}{2} M_b v_d^2 + \frac{3}{2} kT, \qquad (1)$$

where m and $M_b$ are the masses of the ion and buffer gas neutral, respectively. This mean ion energy in the laboratory frame is converted to the more meaningful mean energy in the center-of-mass frame of the ion and the reactant neutral, $KE_{cm}$, with the usual tranformation.[5]

Equation 1 uses the experimental drift velocity as input, which can be obtained from experimental ion mobility data. While a substantial quantity of such data are available (ref.6, Chapt. 7), most of it pertains to ions in their parent gases; hence, mobilities of a wide variety of atomic and molecular ions in the commonly used buffer gases helium and argon are not generally available over a large E/N range. Consequently, we have used[5,8] the flow-drift tube to measure the required $v_d$ values.

## DISCUSSION

### Atomic Ions

The question of internal excitation does not arise for drifting atomic ions; hence, this class of ion is well-suited for studies of the effects of the speed distribution alone. The reliability of Eq. 1 has long been open to question, since the polarization interaction on which it is based is not always the dominant interaction between the ion and the buffer gas neutral. Nevertheless, there is now recent experimental and theoretical evidence that Eq. 1 gives a remarkably accurate value.

Figure 3 shows experimental rate constants for the reaction

$$O^+ + N_2 \rightarrow NO^+ + N \qquad (2)$$

plotted as a function of the mean relative $O^+$ - $N_2$ energy. Data from two very different experimental techniques are represented. First, the open triangles represent the static drift tube data of Johnsen and Biondi[9] and the closed figures represent the flow-drift tube data from this laboratory.[5] While the agreement here is indeed gratifying, it makes no statement about the validity of the $KE_{cm}$ scale, since Eq. 1 was used to represent the mean ion energy in both drift tube experiments. The second type of

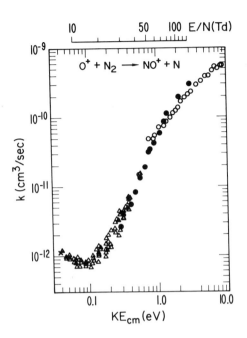

Figure 3.   Rate constants for the reaction $O^+ + N_2 \rightarrow NO^+ + N$ as a function of $KE_{cm}$ and E/N.   (1 Td = $10^{-17}$ $V \cdot cm^2$).

data in Fig. 3 are the crossed-beam results of Rutherford and Vroom,[10] denoted by the open circles. The agreement here is very significant since it indicates that Eq. 1 does not give a grossly inaccurate estimate of the ion mean energy. For other $O^+$ reactions, similar comparisons[5] also support the drift-tube mean energy scale given by Eq. 1.

Perhaps the best tests of Eq. 1 to date have been theoretical Monte Carlo calculations of ion speed distributions. Skullerud[11] calculated the speed distributions associated with several different models of the ion-neutral interaction and found that Eq. 1 predicts the mean energy within a few percent. Recently, Lin and Bardsley[12] have obtained similar results for $O^+$ ions in helium and argon. Thus, despite the fact that Eq. 1 was derived for the polarization interaction, it yields a very accurate value for the mean ion energy, most likely due to the fact that it employs as input the underlined experimental drift velocity (as opposed to the theoretical drift velocity appropriate to the polarization interaction). Furthermore, Viehland, Mason and Whealton[13] have shown that even the small inaccuracy of Eq. 1 can now be corrected based on the E/N dependence of experimental mobility data.

However, despite the fact that the mean ion energy can be determined very accurately, it corresponds to only one moment of the squared velocity distribution. Clearly for a reaction with an energy dependence as strong as that in Fig. 3, the relation between the rate constant and the cross section will be a sensitive function of the joint speed distribution itself. Hence, the rate constants determined in drift tubes even at accurately known mean energies do not straightforwardly define the reaction cross section.

The ion speed distribution depends on the scattering potential of the ion and the buffer gas neutral. Generally, the mean velocity component directed along the field is larger than the scattered component perpendicular to the field when the ratio of the ion mass to the neutral mass is large. As this ratio decreases, the scattered component increases and the width of the speed distribution increases. Thus, the character of the speed distribution of an ion may be significantly different in buffer gases of sizably different masses. Consequently, the rate constants of extremely energy-dependent reactions between atomic ions and neutral molecules measured in very different buffer gases should show the effects of different speed distributions.

Figure 4 presents such measurements made recently in our laboratory. The solid figures are the same data given earlier in Fig. 3, namely, flow-drift tube data for $O^+$ reacting with $N_2$ while drifting in a helium buffer. The open figures are flow-drift tube data for the same reaction in an argon buffer. Both sets of data were plotted as a function of $KE_{cm}$, as given by Eq. 1. The argon-buffered data lie significantly above the helium-buffered data at the lower energies and the two sets merge together at the higher energies. This strongly suggests that the $O^+$ speed distribution in argon is broader than that in helium, corresponding to the increased random scattering of $O^+$ in argon compared to $O^+$ in helium, and that the broader distribution yields a larger rate constant as it penetrates the steeply rising cross section.

Fortunately, it is possible to go beyond these qualitative statements. Lin and Bardsley[12] have made Monte Carlo calculations of the speed distributions of $O^+$ ions in helium and argon as a function of $E/N$, based on scattering potentials determined by fitting the $E/N$ dependence of experimental mobility data. At large values of $E/N$, the distribution in argon is distinctly broader than that in helium. Using the $O^+$ speed distribution in helium, the reaction cross section was determined from the helium-buffered rate constant data in Fig. 3. The + symbols denote the rate constants "back-calculated" using the $O^+$ speed distributions in helium and the cross section as a check on the latter. Then this cross section was folded into the $O^+$ speed

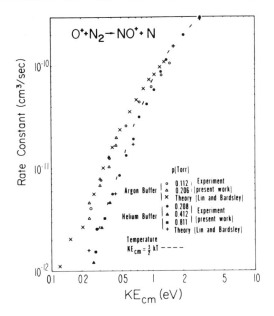

Figure 4.  Comparison of experimental and theoretical rate
        constants.

distribution in <u>argon</u> to obtain the rate constants appropriate
to an argon-buffered drift tube.  These results are given by the
x symbols and are in remarkable agreement with the experimental
data.  Similar agreement was found for other $O^+$ reactions that
have relatively large energy dependences.  The dashes are the
predicted rate constants corresponding to a Maxwellian speed
distribution.  Their close proximity to the helium-buffered data
is indicative of the fact that the $O^+$ speed distribution in
helium has much more of a Maxwellian form than the distribution
in argon.


## Molecular Ions

    In drift-tube studies of reactions involving molecular ions,
there is the added complexity of the possible excitation of internal energy modes in the ion drifting  through the buffer gas
at high E/N.  Little is known about the effect of rotational or
vibrational excitation of the ion on ion-molecule reaction rates.
We have found recently several cases where it apparently has a
profound effect.

    Figure 5 shows the rate constants for the charge-transfer
reaction

$$CO_2^+ + O_2 \rightarrow O_2^+ + CO_2 \qquad (3)$$

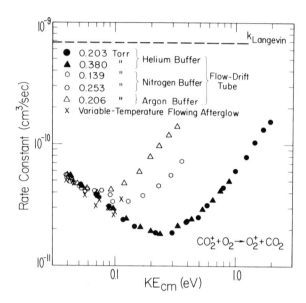

Figure 5.   Rate constants for the reaction $CO_2^+ + O_2 \rightarrow O_2^+ + CO_2$ as a function of $KE_{cm}$.

measured in the flow-drift tube with helium, nitrogen, and argon buffer gases and over a small range in the variable-temperature flowing afterglow, all plotted as a function of the mean kinetic energy in the $CO_2^+$-$O_2$ center-of-mass system. In all four cases, the rate constant shows a pronounced minimum as a function of energy. However, the variation with energy is very different for the three buffer gases, apparently much more so than could possibly be expected on the basis of different speed distributions alone. We interpret these differences as being largely due to the enhancement of the rate constant by vibrational excitation of $CO_2^+$ in the heavier buffer gases.

First, it should be noted that the energy available for vibrational excitation of $CO_2^+$ is that in the center-of-mass system of $CO_2^+$ and the buffer gas neutral. Consequently, if vibrational excitation is playing a dominant role in reaction 3, then one would indeed expect different behavior in different buffer gases. Furthermore, one also expects that the energy dependences would be much more similar if the rate constant data were plotted as a function of the mean kinetic energy in the ion-buffer center-of-mass rather than the ion-reactant center-of-mass. While the data do not exactly coincide in this frame, perhaps due to different velocity distributions, strong similarities do indeed then exist. For example, the rate constant minima then all occur at approximately 0.09 eV in the ion-buffer center-of-mass frame. Since the energies[14] of the first three fundamental vibrational modes of $CO_2^+$ are $\nu_1$ (symmetric stretch) = 0.16 eV, $\nu_2$ (bending) = 0.07 eV (estimated from $\nu_2$ of $CO_2$), and $\nu_3$ (asymmetric stretch) = 0.18 eV, there is ample energy at high E/N to excite these modes. Therefore, it appears inescapable that there is some vibrational excitation of the drifting $CO_2^+$ ion that is influencing the rate constant. However, we remain at a loss to answer the more difficult question as to why $CO_2^+$ vibrational energy should have such a strong effect on the rate constant of this charge-transfer reaction.

Further evidence of the vibrational excitation of drifting molecular ions is provided in the recent variable-temperature flowing afterglow and flow-drift tube studies of the proton-transfer reaction

$$N_2OH^+ + CO \rightleftarrows COH^+ + N_2O \qquad (4)$$

in our laboratory. The data presented below on the temperature variation of the forward and reverse rate constants, $k_f$ and $k_r$ respectively, establishes that the forward direction is exothermic by 0.19 eV for the transfer of a proton from $N_2O$ to CO. Since this exothermicity is small, the reaction can be measured readily in the forward and the reverse directions separately. Figure 6 shows the rate constant data obtained in the variable-temperature flowing afterglow and in the flow-drift tube, using

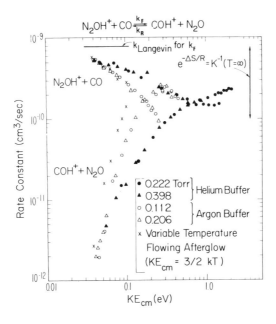

Figure 6. Rate constants for the reaction $N_2OH^+ + CO \rightleftarrows COH^+ + N_2O$ as a function of $KE_{cm}$.

helium and argon buffer gases in the latter. At room temperature, $k_f$ is over 200 times larger than $k_r$. With increasing
energy, all of the experimental situations show that $k_f$ decreases and that $k_r$ increases dramatically. However, these
variations are again different for the two buffer gases and
both of these are also different from the variation of the
thermal data. We again attribute this to different vibrational
excitation in the drifting molecular ions at high E/N.

For ions in thermal equilibrium with the buffer gas at
temperature T, the equilibrium constant $K \equiv k_f/k_r$ is related to
the changes in enthalpy $\Delta H^O$ and entropy $\Delta S^O$ by[15]

$$RT \ln K = - \Delta H^O + T\Delta S^O, \qquad (5)$$

where R is the gas constant. $\Delta H^O$ is the exothermicity of the
reaction and $\Delta S^O$ is related to the change in the density of
states in phase space. Equation 5 shows that $\ln K$ should be a
linear function of $1/T$ over a limited temperature range. Figure
7 shows such a van't Hoff plot for some of the data of Fig. 6.
The variable-temperature flowing afterglow results are indeed
linear, but the helium-buffered drift-tube data lie considerably
higher and do not exhibit the linear relation. This behavior
indicates that the ions $N_2OH^+$ and $COH^+$ are not in thermal equilibrium at effective "temperatures" corresponding to their mean
kinetic energies. It seems likely to be due to non-equilibrium

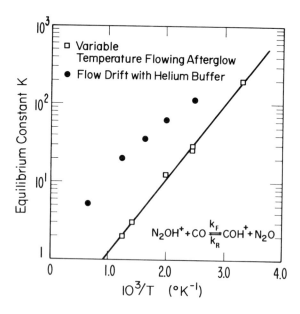

Figure 7.  van't Hoff plot for the reaction $N_2OH^+ + CO \rightleftarrows$
COH$^+$ + N$_2$O.

vibrational effects and that the drifting ions are not as likely to be vibrationally excited by collisions with the lighter helium atoms as they are with the heavier argon atoms. Thus, the argon-buffered data lie closer to the temperature data.

The argon-buffered data in Fig. 6 for the reverse reaction increase above the same data for the forward direction and then show an interesting maximum. This is probably due to the fact that the van't Hoff plot in Fig. 7 yields $\Delta S^o = -4.0$ Gibbs/mole, which means that the entropy decreases in the exothermic direction; i.e., entropy favors the endothermic direction. Thus, at very high temperatures, Eq. 5 shows that $k_f/k_r$ must be $\exp(\Delta S/R)$ = 1/7.5, or $k_r = 7.5 \ k_f$, a ratio that is indicated by the length of the arrow in Fig. 6. Thus, while $k_r > k_f$ at high temperatures, the limit on this ratio implies that $k_r$ cannot continue to increase with energy as $k_f$ decreases. While one certainly does not expect the nonequilibrium flow-drift tube data to adhere rigorously to these thermodynamic limits, this seems to be the reason for the general behavior of the argon-buffered data.

## CONCLUSIONS

The results presented here indicate that ion-molecule reaction rate constants measured in drift tubes can show the effects of different ion speed distributions and ion vibrational excitation as a result of different buffer gases. For atomic ions in rare gas buffers, it is now becoming possible to calculate the ion speed distributions, thereby providing the means to determine the reaction cross section from drift-tube reaction rate data. In the case of possible vibrational excitation of drifting ions, the understanding is far less quantitative. However, the systematic use of different buffer gases in drift tubes does provide a qualitative way for testing the sensitivity of many ion-molecule reactions to ion vibrational excitation, an important effect for which there is presently a dearth of information.

## REFERENCES

1.  E. E. Ferguson, F. C. Fehsenfeld, and A. L. Schmeltekopf, Adv. Atomic Molecular Phys. 5 1 (1969).
2.  F. C. Fehsenfeld, Int. J. Mass Spectrom. Ion Phys. 16 151 (1975).
3.  D B. Dunkin, F. C. Fehsenfeld, A. L. Schmeltekopf, and E. E. Ferguson, J. Chem Phys. 49 1365 (1968).
4.  W. Lindinger, F. C. Fehsenfeld, A. L. Schmeltekopf, and E. E. Ferguson, J. Geophys. Res. 79 4753 (1974).

5.  M. McFarland, D. L. Albritton, F. C. Fehsenfeld, E. E. Ferguson, and A. L. Schmeltekopf, J. Chem. Phys. $\underline{59}$ 6610, 6620, 6629 (1973).

6.  E. W. McDaniel and E. A. Mason, "The Mobility and Diffusion of Ions in Gases," John Wiley and Sons, New York, 1973, Sec. 5.1B.

7.  G. H. Wannier, Bell Syst. Tech. J. $\underline{32}$ 170 (1953).

8.  W. Lindinger and D. L. Albritton, J. Chem. Phys. $\underline{62}$ 3517 (1975).

9.  R. Johnsen and M. A. Biondi, J. Chem. Phys. $\underline{59}$, 3504 (1973).

10. J. A. Rutherford and D. A. Vroom, J. Chem. Phys. $\underline{55}$ 5622 (1971).

11. H. R. Skullerud, J. Phys. B. $\underline{6}$ 728 (1973).

12. S. Lin and J. N. Bardsley, J. Phys. B. to be submitted.

13. L. A. Viehland, E. A. Mason, and J. H. Whealton, J. Phys. B. $\underline{7}$ 2433 (1974).

14. G. Herzberg, "Electronic Spectra and Electronic Structure of Polyatomic Molecules," D. Van Nostrand, Princeton, 1966, p. 594.

15. H. S. Johnston "Gas Phase Reaction Rate Theory" Ronald Press, New York, 1966, p. 128.

| | | | |
|---|---|---|---|
| Albritton, D.L. | 889 | Kleinpoppen, H. | 641 |
| Armbruster, P. | 501 | Klemperer, William | 62 |
| | | Kraft, G. | 501 |
| Bardsley, J.N. | 151 | Kruglova, I.M. | 419 |
| Barnett, C.F. | 846 | Kuppermann, Aron | 259 |
| Bayfield, James E. | 726 | | |
| Beauchamp, J.L. | 871 | Lambropoulos, P. | 593 |
| Bell, F. | 520 | Lineberger, W.C. | 584 |
| Beloshitsky, V.V. | 419 | Lutz, H.O. | 432 |
| Betz, H.-D. | 520 | | |
| Betz, W. | 531 | McCarthy, I.E. | 766 |
| Bischel, William K. | 820 | McDowell, M.R.C. | 126 |
| Blum, K. | 641 | Macdonald, James R. | 408 |
| Briggs, J.S. | 384 | Macek, Joseph | 627 |
| Bromberg, J. Philip | 98 | Madden, R.P. | 563 |
| Byron, F.W., Jr. | 675 | Massey, Prof. Sir Harrie | 3 |
| | | Mehlhorn, W. | 756 |
| Church, D.A. | 660 | Meyerhof, W.E. | 470 |
| | | Mokler, P.H. | 501 |
| Datz, Sheldon | 833 | Moore, C. Fred | 447 |
| de Heer, F.J. | 79 | Morgenstern, R. | 345 |
| Dehmelt, Hans G. | 857 | Müller, B. | 481 |
| Demkov, Yu. N. | 313 | | |
| Drukarev, G. | 231 | Nikitin, E.E. | 275 |
| Durup, J. | 609 | Nikolaev, V.S. | 419 |
| Edwards, A.K. | 790 | Ogurtsov, G.N. | 779 |
| Ehrhardt, H. | 194 | | |
| | | Panke, H. | 520 |
| Fano, U. | 27 | Paul, Derek | 194 |
| Fastrup, Bent | 361 | Peierls, Prof. Sir Rudolf | 40 |
| Fehsenfeld, F.C. | 889 | Penkin, N.P. | 803 |
| Folkmann, F. | 501 | Petukhov, V.P. | 419 |
| Greiner, Walter | 531 | Read, Frank H. | 176 |
| Gupta, R. | 712 | Reiland, W. | 158 |
| | | Reinhardt, J. | 531 |
| Hagmann, S. | 501 | Rhodes, Charles K. | 820 |
| Heiligenthal, G. | 531 | Romanovsky, E.A. | 419 |
| Henley, Ernest M. | 51 | | |
| Herbst, Eric | 62 | Schubert, E. | 194 |
| Hermann, H.W. | 158 | Sergeev, V.A. | 419 |
| Hertel, I.V. | 158 | Sidis, V. | 295 |
| | | Smirnov, B.M. | 701 |
| Jung, K. | 194 | Smith, R.K. | 531 |
| | | Spence, David | 241 |
| Kempter, V. | 327 | Spindler, E. | 520 |
| Kessler, Joachim | 112 | Spruch, Larry | 685 |
| Kim, Yong-Ki | 741 | Stamatović, A. | 158 |
| Kleber, M. | 520 | Standage, M.C. | 641 |

Stehling, W.                    520
Stein, H.J.                     501
Stoll, W.                       158

Takayanagi, Kazuo               219

Williams, J.F.                  139